INSTRUCTOR'S GUIDE
FOR CAMPBELL'S

BIOLOGY

FOURTH EDITION

NINA CARIS, Ph.D.
HAROLD UNDERWOOD, Ph.D.

Texas A&M University

The Benjamin/Cummings Publishing Company, Inc.

Menlo Park, California • Reading, Massachusetts
New York • Don Mills, Ontario • Wokingham, U.K. • Amsterdam
Bonn • Paris • Milan • Madrid • Sydney • Singapore • Tokyo
Seoul • Taipei • Mexico City • San Juan, Puerto Rico

About the Authors

Nina Caris is Director of the Introductory Biology Program at Texas A&M University, a position she has held since 1984. From 1981 to 1995, Harold Underwood taught in the Department of Biology at Texas A&M, where he served as the Coordinator of Undergraduate Advising for nine years. Dr. Caris and Dr. Underwood have taught thousands of introductory biology students over the past decade. Committed to improving undergraduate education, they have both been honored with Distinguished Teaching Awards by the Association of Former Students at Texas A&M University and have received joint funding from the National Science Foundation to enhance the introductory level laboratory courses.

ISBN 0-8053-1943-3

1 2 3 4 5 6 7 8 9 10—VG—00 99 98 97 96

The Benjamin/Cummings Publishing Company, Inc.
2725 Sand Hill Road
Menlo Park, California 94025

CONTENTS

TO THE INSTRUCTOR

The idea for a comprehensive lecture outline came from the observation that the most arduous and time-consuming task for a new instructor is reading and outlining the textbook. We thought it would be helpful to provide extensive lecture notes on material covered in the text, so that more time could be devoted to lecture enhancement – a luxury usually afforded only to experienced teachers. This set of lecture notes is only intended to be a beginning. It is *not* recommended that you lecture on *all* material presented here or lecture *only* from material presented in the text. For experienced teachers who are first-time users of Neil Campbell's *Biology*, we hope these notes will ease the transition to a new book.

Each chapter in the instructor's guide has five major components. The lecture *Outline* is a complete outline of the textbook chapter, usually down to level two headings. *Objectives* are provided for all major topics covered in the text. These objectives are primarily at the knowledge and comprehension level, but whenever possible, higher-order categories are included. *Key Terms* for students to define or identify are listed at the end of the objectives. In most cases they are in order of their appearance in the chapter. The *Lecture Notes* are a summary of chapter content in outline form. New terms are defined and relevant points listed below in bulleted format for ease in lecturing. *References* listed at the end of each chapter are by no means an inclusive list of readings, but rather the references we have actually used to prepare lecture notes. For practical reasons, we relied heavily on references already at our disposal; there are many excellent texts not cited here that would serve just as well.

How To Use These Notes

Read the Textbook Read the textbook chapter to become familiar with material your students will read and to note pedagogical aids that can be incorporated into a lecture. The artwork in Campbell's *Biology* is skillfully integrated with the text, and almost an entire lecture can be centered around one or two effective diagrams, charts or tables. Additionally, the end-of-chapter challenge questions can be used to generate class discussion or to help students synthesize newly presented material.

Select Course Objectives Select the course objectives that suit your specific goals for the course and that are appropriate in the context of your department, college or university. Once the objectives for your course are mapped out, they can be used as a guide for:

- *Lecture Preparation*. Preparing lectures with specific objectives in mind will help focus lecture material on important concepts and avoid overemphasis on the trivial – a common criticism of introductory science courses. Try to strike a balance between knowledge level objectives and those requiring higher-order thought processes.

- *Advanced Organizers and Transitions for Lecturing*. An effective way to provide organizational cues is to change objectives from declarative statements to questions and use them in lecture as you make a transition from one topic to another. Prefacing a topic with a question creates anticipation. As students wait to hear the answer they attend better to the material.

- *Reading Assignments.* If students are not held responsible for all material in the text, you can use the course objectives to determine selected reading assignments.

- *Study Questions.* Transforming course objectives into study questions can inform students of your expectations and enhance learning in the process. As students answer questions, they not only take a more active role in learning, but their attention is focused on the concepts you believe are important.

- *Exam Preparation.* The course objectives can be used as a guide for both the content and cognitive level of exam questions. We highly recommend writing or selecting test questions that address specific course objectives and that are at or below the cognitive level of the objective being tested. Our exams generally contain 65-70% knowledge level questions and 30-35% comprehension and application level questions. For our students, we have found that an exam with 100% simple recall questions is not challenging enough, encourages rote memorization and does not test for understanding or complex learning outcomes such as the ability to solve problems or to apply knowledge in new situations. An exam with more than 35-40% application type questions is perceived by the students as too difficult, and some slower students have difficulty completing the exam within the allotted time.

  Test questions from the new *Testbank for Campbell's Biology* are classified into three groups corresponding to cognitive levels in Bloom's Taxonomy of Educational Objectives: 1) *factual recall* (knowledge level); 2) *conceptual understanding* (comprehension level); and 3) *application* (application level). This added feature is extremely useful for constructing an exam. For example, it can be used to produce a table of teaching objectives versus cognitive levels – a practical tool to readily check the test for an even coverage of material, as well as for an appropriate number of higher-order questions. A sample table is provided for your use.

Modify Lecture Outline and Lecture Notes Once you have selected course objectives, modify the lecture outline and lecture notes to include only the material you wish to cover in the course. One measure of success for these lecture notes is how well they are used. It would delight us to see dog-eared versions, rearranged with cut and paste, three-hole punched in binders, highlights and scribbles in the margins. Customize the notes into your own.

Integrate Visual Ancillaries After the basic lecture notes are assembled, integrate the color acetates or slides you will use during the lecture presentation. The new set of beautiful color transparencies for *BIOLOGY*, fourth edition, is well done and inclusive. The type has been enlarged for large lecture rooms and we have found them invaluable for making effective presentations. For those who prefer to use computer-based presentation programs, there is a new ancillary – *The Art CD-ROM*. You can now easily incorporate text art into a customized presentation. An added dimension to visual ancillaries is *BioShow II: The Videodisc*, which allows you to show motion or animation sequences of dynamic processes that are difficult for students to visualize, and thus difficult to teach.

LECTURE ENHANCEMENT

We hope that by providing comprehensive lecture notes over the material covered in the textbook, instructors will have time to concentrate on the more enjoyable task of lecture enhancement. One simple way to generate student enthusiasm and interest is to humanize the course. Neil Campbell's interviews with contemporary biologists can be the basis for a reading assignment or class discussion to illustrate that science is a human endeavor and an ongoing process that is a "self-correcting way of knowing". Comb the newspapers and periodicals for current events that can illustrate to students *why* biological principles are relevant to them. Actively involve students in the search as well. Finally, do not overlook the valuable resource of experienced teachers who can share some of their own successful stories, examples and analogies.

USEFUL REFERENCES

1. Borror, Donald J. *Dictionary of Word Roots and Combining Forms*. Palo Alto, California: Mayfield Publishing Company, 1960.

2. Marchuk, William N. *A Life Science Lexicon*. Dubuque, Iowa: Wm. C. Brown Publishers, 1992. An unavoidable problem in teaching introductory biology is providing students with a working vocabulary of technical terms without overwhelming them with unnecessary memorization. To help retention and understanding, we recommend that students learn basic word roots commonly found in biological terminology.

3. King, Robert C. and William D. Stansfield. *A Dictionary of Genetics*. New York: Oxford University Press, 1990. More than just a dictionary, this is a really a handbook, which we have used not only for the lectures on genetics, but also for the topics of evolution, ecology, systematics, biodiversity, microbiology and immunology.

4. Mayr, Ernst. *The Growth of Biological Thought; Diversity, Evolution, and Inheritance*. Cambridge, Massachusetts: Belknap Press of Harvard University Press, 1982. This book provides the historical context for many of the topics that are covered in an introductory course, such as genetics, evolution, systematics and taxonomy.

ACKNOWLEDGMENTS

The authors wish to thank faculty colleagues from Texas A&M who offered many helpful suggestions for improving the fourth edition. Each time we write the Instructor's Guide, we are amazed at how labor intensive it is to produce notes that faithfully reflect the text. We have been heartened by feedback from professors across the country who tell us that the objectives and lecture notes are helpful. This makes the effort worthwhile. It is a pleasure to acknowledge the staff from Benjamin/Cummings Publishing Company, especially Tabinda Khan, Editorial Assistant and Kim Johnson, Project Editor. Their patience and amiable support cheered us on to the finish line, and we thank them for their help through all phases of this project. We also thank Larry Olsen, Production Editor, who gave us "non-techies" technical advice; and the Sponsoring Editors, Don O'Neal and Johanna Schmid who endorsed our vision of the Instructor's Guide. Finally, we thank Neil Campbell for writing such a wonderful book and for inspiring us to be better teachers.

QUESTION CLASSIFICATION TABLE

Objectives to Be Tested	Questions That Measure Factual Recall	Questions That Measure Conceptual Understanding	Questions That Measure Application	Number of Questions	% of Test Devoted to Each Content Area
Number of Questions					
% of Test Devoted to Each Cognitive Level					

CHAPTER 1

INTRODUCTION: THEMES IN THE STUDY OF LIFE

OUTLINE

OBJECTIVES

After reading this chapter and attending lecture, the student should be able to:
1. Briefly describe unifying themes that pervade the science of biology.
2. Diagram the hierarchy of structural levels in biology.
3. Explain how the properties of life emerge from complex organization.
4. Describe seven emergent properties associated with life.
5. Distinguish between holism and reductionism.
6. Explain how technological breakthroughs contributed to the formulation of the cell theory and our current knowledge of the cell.
7. Distinguish between prokaryotic and eukaryotic cells.
8. Explain, in their own words, what is meant by "form fits function".
9. List and distinguish among the five kingdoms of life.
10. Briefly describe how Charles Darwin's ideas contributed to the conceptual framework of biology.
11. Outline the scientific method.
12. Distinguish between inductive and deductive reasoning.
13. Explain how science and technology are interdependent.

KEY TERMS

emergent property	holism	evolution	control group
population	reductionism	natural selection	variable
community	prokaryotic	scientific method	experimental group
ecosystem	eukaryotic	hypothesis	deductive reasoning
biome	taxonomy	inductive reasoning	scientific theory
biogenesis			

LECTURE NOTES

Biology, the study of life, is a human endeavor resulting from an innate attraction to life in its diverse forms (E.O. Wilson's *biophilia*).

The science of biology is enormous in scope.
*   It reaches across size scales from submicroscopic molecules to the global distribution of biological communities.
*   It encompasses life over huge spans of time from contemporary organisms to ancestral life forms stretching back to nearly four billion years.

As a science, biology is an ongoing process.
*   As a result of new research methods developed over the past few decades, there has been an information explosion.
*   Technological advances yield new information that may change the conceptual framework accepted by the majority of biologists.

With rapid information flow and new discoveries, biology is in a continuous state of flux. There are, however, enduring unifying themes that pervade the science of biology:

- A hierarchy of organization.
- Emergent properties.
- The cellular basis of life.
- Heritable information.
- A feeling for organisms.
- The correlation between structure and function.
- The interaction of organisms with their environment.
- Unity in diversity.
- Evolution: the core theme.
- Scientific process: the hypothetico-deductive method.

I. Life is organized on many structural levels

A characteristic of life is a high degree of order. Biological organization is based on a hierarchy of structural levels, with each level building on the levels below it.

Atoms

↓

Complex biological molecules.

↓

Subcellular Organelles

| are ordered into
↓

Cells

| in multicellular organisms similar cells are organized into
↓

Tissues

↓

Organs

↓

Organ systems

↓

Complex organism

There are levels of organization beyond the individual organism:

Population	=	Localized group of organisms belonging to the same species.
Community	=	Populations of species living in the same area.
Ecosystem	=	An energy-processing system of community interactions that include abiotic environmental factors such as soil and water.
Biomes	=	Large scale communities classified by predominant vegetation type and distinctive combinations of plants and animals.
Biosphere	=	The sum of all the planet's ecosystems.

II.   Each level of biological organization has emergent properties

Emergent property = Property that emerges as a result of interactions between components.
- With each step upward in the biological hierarchy, new properties emerge that were not present at the simpler organizational levels.
- Life is difficult to define because it is associated with numerous emergent properties that reflect a hierarchy of structural organization.

Some of the emergent properties and processes associated with life are:
1. *Order*. Organisms are highly ordered, and other characteristics of life emerge from this complex organization.
2. *Reproduction*. Organisms reproduce; life comes only from life (*biogenesis*).
3. *Growth and Development*. Heritable programs stored in DNA direct the species-specific pattern of growth and development.
4. *Energy Utilization*. Organisms take in and transform energy to do work, including the maintenance of their ordered state.
5. *Response to Environment*. Organisms respond to stimuli from their environment.
6. *Homeostasis*. Organisms regulate their internal environment to maintain a steady-state, even in the face of a fluctuating external environment.
7. *Evolutionary Adaptation*. Life evolves in response to interactions between organisms and their environment.

Because properties of life emerge from complex organization, it is impossible to fully explain a higher level of order by breaking it into its parts.

Holism = The principle that a higher level of order cannot be meaningfully explained by examining component parts in isolation.
- An organism is a living whole greater than the sum of its parts.
- For example, a cell dismantled to its chemical ingredients is no longer a cell.

It is also difficult to analyze a complex process without taking it apart.

Reductionism = The principle that a complex system can be understood by studying its component parts.
- Has been a powerful strategy in biology.
- For example, Watson and Crick deduced the role of DNA in inheritance by studying its molecular structure.

The study of biology balances the reductionist strategy with the goal of understanding how the parts of cells, organisms and populations are functionally integrated.

III.   Cell's are an organism's basic units of structure and function

The cell is an organism's basic unit of structure and function.
- Lowest level of structure capable of performing all activities of life.
- All organisms are composed of cells.
- May exist singly as unicellular organisms or as subunits of multicellular organisms.

The invention of the microscope led to the discovery of the cell and the formulation of the cell theory.
- Robert Hooke (1665) reported a description of his microscopic examination of cork. Hooke described tiny boxes which he called "cells" (really cell walls). The significance of this discovery was not recognized until 150 years later.
- Antonie van Leeuwenhok (1600's) used the microscope to observe living organisms such as microorganisms in pond water, blood cells and animal sperm cells.
- Matthias Schleiden and Theodor Schwann (1839) reasoned from their own microscopic studies and those of others, that all living things are made of cells. This formed the basis for the *cell theory*.
- The cell theory has since been modified to include the idea that all cells come from preexisting cells.

Over the past 40 years, use of the electron microscope has revealed the complex ultrastructure of cells:
- Cells are bounded by *plasma membranes* that regulate passage of materials between the cell and its surroundings.
- All cells, at some stage, contain DNA.

Based on structural organization, there are two major kinds of cells: *prokaryotic* and *eukaryotic*.

<u>Prokaryotic cell</u> = Cell lacking membrane-bound organelles and a membrane-enclosed nucleus.
- Found only in the Kingdom Monera, the bacteria and blue-green algae.
- Generally much smaller than eukaryotic cells.
- Contains DNA that is <u>not</u> separated from the rest of the cell, as there is no membrane-bound nucleus.
- Lacks membrane-bound organelles.
- Almost all have tough external walls.

<u>Eukaryotic cell</u> = Cell with a membrane-enclosed nucleus and membrane-enclosed organelles.
- Found in protists, plants, fungi and animals.
- Subdivided by internal membranes into different functional compartments called *organelles*.
- Contains DNA that is segregated from the rest of the cell. DNA is organized with proteins into *chromosomes* that are located within the *nucleus*, the largest organelle of most cells.
- *Cytoplasm* surrounds the nucleus and contains various *organelles* of different functions.
- Some cells have a tough *cell wall* outside the plasma membrane (e.g. plant cells). Animal cells lack cell walls.

Though structurally different, eukaryotic and prokaryotic cells have many similarities, especially in their chemical processes.

IV. The continuity of life is based on heritable information in the form of DNA

Biological instructions for an organism's complex structure and function are encoded in DNA.
- Each DNA molecule is made of four types of chemical building blocks called *nucleotides*.
- The linear sequence of these four *nucleotides* encode the precise information in a *gene*, the unit of inheritance from parent to offspring.
- An organism's complex structural organization is specified by an enormous amount of coded information.

Inheritance itself is based on:
- A complex mechanism for copying DNA.
- Passing the information encoded in DNA from parent to offspring.

All forms of life use essentially the same genetic code.
- A particular nucleotide sequence provides the same information to one organism as it does to another.
- Differences among organisms reflect differences in nucleotide sequence.

V.   A feeling for organisms enriches the study of life

We may study life at structural levels below or above the organism, but it is organisms that first attract us. A feeling for the organisms we seek to understand enriches our study of life at all levels.

Barbara McClintock is a good example of a biologist whose feeling for the organism contributed to an important scientific discovery.

- McClintock investigated inheritance of kernel color in Indian corn and uncovered a type of genetic change not yet recognized by the scientific community in the 1940's and 50's.
- In McClintock's biography, *A Feeling for the Organism* by Evelyn Keller, the author highlights that McClintock's success was due to her familiarity with corn. McClintock knew the individual corn plants in her experiments and enjoyed these attachments.
- Because she was familiar with the corn plants in her experiments, McClintock was able to recognize unexpected results from corn breeding experiments and deduce the existence of transposable genetic elements.
- The significance of McClintock's discovery was not recognized until researchers in the 1970's found a similar phenomenon in bacteria. She later received due recognition with the Nobel Prize in 1983.

VI.   Structure and function are correlated at all levels of biological organization

There is a relationship between an organism's structure and how it works. Form fits function.

- Biological structure gives clues about what it does and how it works.
- Knowing a structure's function gives insights about its construction.
- This correlation is apparent at many levels of biological organization. (See Campbell, Figure 1.7)

VII.   Organisms are open systems that interact continuously with their environments

Organisms interact with their environment, which includes other organisms as well as abiotic factors.

- Both organism and environment are affected by the interaction between them.
- Ecosystem dynamics include two major processes:
    1. Nutrient cycling.
    2. Energy flow.

VIII. Diversity and unity are the dual faces of life on Earth

Biological diversity is enormous.

- Estimates of total diversity range from 5 million to over 30 million species.
- About 1.5 million species have been identified and named, including approximately 260,000 plants, 50,000 vertebrates and 750,000 insects.

To make this diversity more comprehensible, biologists classify species into categories.

Taxonomy = Branch of biology concerned with naming and classifying organisms.

- Taxonomic groups are ranked into a hierarchy from the most to least inclusive category: *kingdom, phylum, class, order, family, genus, species.*
- The kingdoms of life recognized in the five-kingdom system are: Monera, Protista, Plantae, Fungi and Animalia. (See Campbell, Figure 1.10)
- A six-kingdom system recognizes two prokaryotic groups and divides the Monera into the Archaebacteria and Eubacteria.

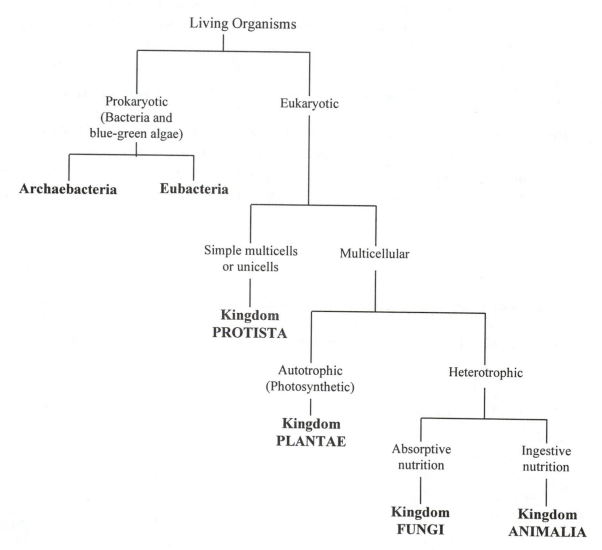

There is unity in the diversity of life forms at the lower levels of organization. Unity of life forms is evident in:

- A universal genetic code.
- Similar metabolic pathways (e.g. glycolysis).
- Similarities of cell structure (e.g. flagella of protozoans and mammalian sperm cells).

IX.   Evolution is the core theme of biology

Evolution is the one unifying biological theme.

- Life evolves. Species change over time and their history can be described as a branching tree of life.
- Species that are very similar share a common ancestor at a recent branch point on the phylogenetic tree.
- Less closely related organisms share a more ancient common ancestor.
- All life is connected and can be traced back to primeval prokaryotes that existed more than three billion years ago.

In 1859, Charles Darwin published *On the Origin of Species* in which he made two major points:

1. Species change, and contemporary species arose from a succession of ancestors through a process of "descent with modification."
2. A mechanism of evolutionary change is *natural selection*.

Darwin synthesized the concept of natural selection based upon the following observations:

- Individuals in a population of any species vary in many inheritable traits.
- Populations have the potential to produce more offspring than will survive or than the environment can support.
- Individuals with traits best suited to the environment leave a larger number of offspring, which increases the proportion of inheritable variations in the next generation. This differential reproductive success is what Darwin called *natural selection*.

Organisms' adaptations to their environments are the products of natural selection.

- Natural selection does <u>not create</u> adaptations; it merely increases the frequency of inherited variants that arise by chance.
- Adaptations are the result of the editing process of natural selection. When exposed to specific environmental pressures, certain inheritable variations favor the reproductive success of some individuals over others.

Darwin proposed that cumulative changes in a population over long time spans could produce a new species from an ancestral one.

Descent with modification accounts for both the unity and diversity of life:

- Similarities between two species may be a reflection of their descent from a common ancestor.
- Differences between species may be the result of natural selection modifying the ancestral equipment in different environmental contexts.

X. Science as a process of inquiry often involves hypothetico-deductive thinking

As the science of life, biology has the characteristics associated with science in general.

Science is a way of knowing. It is a human endeavor that emerges from our curiosity about ourselves, the world and the universe. Good scientists are people who:

- Ask questions about nature and believe those questions are answerable.
- Are curious, observant and passionate in their quest for discovery.
- Are creative, imaginative and intuitive.
- Are generally skeptics.

Scientific method = Process which outlines a series of steps used to answer questions.

- Is not a rigid procedure.
- Based on the conviction that natural phenomena have natural causes.
- Requires evidence to logically solve problems.

The key ingredient of the scientific process is the *hypothetico-deductive method*, which is an approach to problem-solving that involves:

1. asking a question and formulating a tentative answer or *hypothesis* by *inductive reasoning*.

2. using *deductive reasoning* to make predictions from the hypothesis and then testing the validity of those predictions.

Hypothesis = Educated guess proposed as a tentative answer to a specific question or problem.

Inductive reasoning = Making an inference from a set of specific observations to reach a general conclusion.

Deductive reasoning = Making an inference from general premises to specific consequences, which logically follow if the premises are true.

- Usually takes the form of *If...then* logic.
- In science, deductive reasoning usually involves predicting experimental results that are expected <u>if</u> the hypothesis is true.

NOTE: Some students cannot make the distinction between inductive and deductive reasoning. An effective teaching strategy is to let them actually experience both processes. To illustrate inductive reasoning, provide an everyday scenario with enough pieces of information for student to hypothesize a <u>plausible</u> explanation for some event. Demonstrate deductive reasoning by asking students to solve a simple problem, based upon given assumptions.

Useful hypotheses have the following characteristics:

- *Hypotheses are possible causes.* Generalizations formed by induction are not necessarily hypotheses. Hypotheses should also be tentative explanations for observations or solutions to problems.

- *Hypotheses reflect past experience with similar questions.* Hypotheses are not just blind propositions, but are <u>educated</u> guesses based upon available evidence.

- *Multiple hypotheses should be proposed whenever possible.*   The disadvantage of operating under only one hypothesis is that it might restrict the search for evidence in support of this hypothesis; scientists might bias their search, as well as neglect to consider other possible solutions.

- *Hypotheses must be testable via the hypothetico-deductive method.*   Predictions made from hypotheses must be testable by making observations or performing experiments.   This limits the scope of questions that science can answer.

- *Hypotheses can be eliminated, but not confirmed with absolute certainty.*   If repeated experiments consistently disprove the predictions, then we can assume that the hypothesis is false.   However, if repeated experimentation supports the deductions, we can only assume that the hypothesis <u>may</u> be true; accurate predictions can be made from false hypotheses.   The more deductions that are tested and supported, the more confident we can be that the hypothesis is true.

Another feature of the scientific process is the *controlled experiment* which includes *control* and *experimental groups*.

<u>Control group</u> = In a controlled experiment, the group in which all *variables* are held constant.

- Controls are a necessary basis for comparison with the experimental group, which has been exposed to a single treatment variable.
- Allows conclusions to be made about the effect of experimental manipulation.
- Setting up the best controls is a key element of good experimental design.

<u>Variable</u> = Condition of an experiment that is subject to change and that may influence an experiment's outcome.

<u>Experimental group</u> = In a controlled experiment, the group in which one factor or treatment is varied.

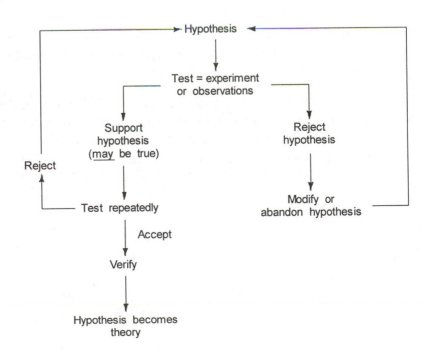

Science is an ongoing process that is a self-correcting way of knowing. Scientists:

- Build on prior scientific knowledge.
- Try to replicate the observations and experiments of others to check on their conclusions.
- Share information through publications, seminars, meetings and personal communication.

What really advances science is not just an accumulation of facts, but a new concept that collectively explains observations that previously seemed to be unrelated.

- Newton, Darwin and Einstein stand out in the history of science because they synthesized ideas with great explanatory power.
- *Scientific theories* are comprehensive conceptual frameworks which are well supported by evidence and are widely accepted by the scientific community.

Biology is a multi-disciplinary science that requires knowledge of chemistry, physics and mathematics.

XI. Science and technology are functions of society

Science and technology are interdependent.

- Technology extends our ability to observe and measure, which enables scientists to work on new questions that were previously unapproachable.
- Science, in turn, generates new information that makes technological inventions possible.
- For example, Watson and Crick's scientific discovery of DNA structure led to further investigation that enhanced our understanding of DNA, the genetic code, and how to transplant foreign genes into microorganisms. The biotechnology industry has capitalized on this knowledge to produce valuable pharmaceutical products such as human insulin.

We have a love-hate relationship with technology.

- Technology has improved our standard of living.
- The consequence of using technology also includes the creation of new problems such as increased population growth, acid rain, deforestation, global warming, nuclear accidents, ozone holes, toxic wastes and endangered species.
- Solutions to these problems have as much to do with politics, economics, culture and values as with science and technology.

A better understanding of nature must remain the goal of science. Scientists should:

- Try to influence how technology is used.
- Help educate the public about the benefits and hazards of specific technologies.

REFERENCES

Campbell, N. *Biology*. 4th ed. Menlo Park, California: Benjamin/Cummings, 1996.

Moore, J.A. "Science as a Way of Knowing – Evolutionary Biology." *American Zoologist*, 24(2): 470-475, 1980.

OUTLINE

I.   Matter consists of chemical elements in pure form and in combinations called compounds

II. Life requires about 25 chemical elements

III. Atomic structure determines the behavior of an element
    A.   Subatomic Particles
    B.   Atomic Number and Atomic Weight
    C.   Isotopes
    D.   Energy Levels
    E.   Electron Orbitals
    F.   Electron Configuration and Chemical Properties

IV. Atoms combine by chemical bonding to form molecules
    A.   Covalent Bonds
    B.   Ionic Bonds

V.   Weak chemical bonds play important roles in the chemistry of life
    A.   Hydrogen Bonds

VI. A molecule's biological function is related to its shape

VII. Chemical reactions change the composition of matter

VIII. Chemical conditions on the early earth set the stage for the origin and evolution of life

OBJECTIVES

After reading this chapter and attending lecture, the student should be able to:
1. Define element and compound.
2. State four elements essential to life that make up 96% of living matter.
3. Describe the structure of an atom.
4. Define and distinguish among atomic number, mass number, atomic weight and valence.
5. Given the atomic number and mass number of an atom, determine the number of neutrons.
6. Explain why radioisotopes are important to biologists.
7. Explain how electron configuration influences the chemical behavior of an atom.
8. Explain the octet rule and predict how many bonds an atom might form.
9. Explain why the noble gases are so unreactive.
10. Define electronegativity and explain how it influences the formation of chemical bonds.
11. Distinguish among nonpolar covalent, polar covalent and ionic bonds.
12. Describe the formation of a hydrogen bond and explain how it differs from a covalent or ionic bond.
13. Explain why weak bonds are important to living organisms.
14. Describe how the relative concentrations of reactants and products affect a chemical reaction.
15. Describe the chemical conditions on early Earth and explain how they were different from today.

KEY TERMS

matter	electron	potential energy	covalent bond
mass	atomic number	electron shell	polar covalent bond
element	mass number	orbital	ionic bond
trace element	isotope	valence electrons	cation
compound	radioactive	chemical bond	anion
atom	half life	molecule	ion
proton	energy	electronegativity	hydrogen bond
neutron			

LECTURE NOTES

I. **Matter consists of chemical elements in pure form and in combinations called compounds**

Chemistry is fundamental to an understanding of life, because living organisms are made of *matter*.

Matter = Anything that takes up space and has mass.

Mass = A measure of the amount of matter an object contains.

NOTE: You might want to distinguish between mass and weight. *Mass* is the measure of the amount of matter an object contains, and it stays the same regardless of changes in the object's position. *Weight* is the measure of how strongly an object is pulled by earth's gravity, and it varies with distance from the earth's center.

The key point is that the mass of a body does not vary with its position, whereas weight does. So, for all practical purposes - as long as we are earthbound - weight can be used as a measure of mass.

II. Life requires about 25 chemical elements

Element = A substance that cannot be broken down into other substances by chemical reactions.

- All matter made of elements.
- 92 naturally occurring.
- Designated by a symbol of one or two letters.

About 25 of the 92 naturally occurring elements are essential to life. Biologically important elements include:

C	=	carbon	
O	=	oxygen	make up 96% of living matter
H	=	hydrogen	
N	=	nitrogen	

Ca	=	calcium	
P	=	phosphorus	
K	=	potassium	
S	=	sulfur	make up remaining 4% of an organism's weight
Na	=	sodium	
Cl	=	chlorine	
Mg	=	magnesium	
Trace elements			

Trace element = Element required by an organism in extremely minute quantities.

- Though in small quantity, are indispensable for life.
- For example: B, Cr, Co, Cu, F, I, Fe, Mn, Mo, Se, Si, Sn, V and Zn.

Elements can exist in combinations called *compounds*.

Compound = A pure substance composed of two or more elements combined in a fixed ratio.

- For example: NaCl (sodium chloride).
- Has unique emergent properties beyond those of its combined elements. (Na and Cl have very different properties from NaCl.)

NOTE: Since a compound is the next structural level above element or atom, this is an excellent place to emphasize the concept of emergent properties, an integral theme found throughout the text and course.

III. Atomic structure determines the behavior of an element

Atom = Smallest possible unit of matter that retains the physical and chemical properties of its element.

- Atoms of the same element share similar chemical properties.
- Made up of *subatomic particles*.

A. Subatomic Particles

The three *most stable* subatomic particles are:

1. *Neutrons* [no charge (neutral)].
2. *Protons* [+1 electrostatic charge].
3. *Electrons* [-1 electrostatic charge].

NEUTRON	PROTON	ELECTRON
no charge	+1 charge	-1 charge
found together in a dense core called the *nucleus* (positively charged because of protons)		orbit around nucleus (held by electrostatic attraction to positively charged nucleus)
1.009 dalton	1.007 dalton	1/2000 dalton
masses of both are about the same (about 1 dalton)		mass is so small, usually not used to calculate atomic mass

NOTE: The *dalton* is a unit used to express mass at the atomic level. One dalton (d) is equal to 1.67×10^{-24} g.

If an atom is electrically neutral, the number of protons equals the number of electrons, which yields an electrostatically balanced charge.

B. Atomic Number and Atomic Weight

Atomic number = Number of protons in an atom of a particular element.

- All atoms of an element have the same atomic number.
- Written as a subscript to the left of the element's symbol (e.g. $_{11}Na$).
- In a neutral atom, # protons = # electrons.

Mass number = Number of protons and neutrons in an atom.

- Written as a superscript to left of an element's symbol (e.g. ^{23}Na).
- Is approximate mass of the whole atom, since the mass of a proton and the mass of a neutron are both about 1 dalton.
- Can deduce the number of neutrons by subtracting *atomic number* from *mass number*.

- Number of neutrons can vary in an element, but number of protons is constant.
- Is not the same as an element's *atomic weight*, which is the weighted mean of the masses of an element's constituent isotopes.

NOTE: In a large classroom with up to 300 people, it is difficult to interact with students. We put examples on an overhead transparency and solicit student input to complete the information. It is a quick way to check for understanding and to actively involve students at least in a limited way.

Examples:

(Mass #) ^{23}Na   # of electrons   _____
(Atomic #) $_{11}$

              # of protons   _____
              # of neutrons   _____

     $^{12}_{6}$C   # of electrons   _____

              # of protons   _____
              # of neutrons   _____

C.   Isotopes

<u>Isotopes</u> = Atoms of an element that have the same atomic number but different mass number.

- Have the same number of protons, but different number of neutrons.
- Under natural conditions, elements occur as mixtures of isotopes.
- Different isotopes of the same element react chemically in same way.
- Some isotopes are radioactive.

<u>Radioactive isotope</u> = Unstable isotope in which the nucleus spontaneously decays emitting sub-atomic particles and/or energy as radioactivity.

- Loss of nuclear particles may transform one element to another (e.g. $^{14}_{6}$C → $^{14}_{7}$N).
- Has a fixed *half life*.

<u>Half life</u> = Time for 50% of radioactive atoms in a sample to decay.

Biological applications of radioactive isotopes include:

1. Dating geological strata and fossils.
   - Radioactive decay is at a fixed rate.
   - By comparing the ratio of radioactive and stable isotopes in a fossil with the ratio of isotopes in living organisms, one can estimate the age of a fossil.
   - The ratio of ^{14}C to ^{12}C is frequently used to date fossils less than 50,000 years old.

2. Radioactive tracers

- Chemicals labelled with radioactive isotopes are used to trace the steps of a biochemical reaction or to determine the location of a particular substance within an organism.
- Radioactive isotopes are useful as biochemical tracers because they chemically react like the stable isotopes and are easily detected at low concentrations.
- Isotopes of P, N and H were used to determine DNA structure.
- Used to diagnose disease (e.g. PET scanner).
- Because radioactivity can damage cell molecules, radioactive isotopes can also be hazardous.

3. Treatment of cancer

- e.g. radioactive cobalt.

D. Energy Levels

Electrons = Light negatively charged particles that orbit around nucleus.

- Equal in mass and charge.
- Are the only stable subatomic particles directly involved in chemical reactions.
- Have *potential energy* because of their position relative to the positively charged nucleus.

Energy = Ability to do work.

Potential energy = Energy that matter stores because of its position or location.

- There is a natural tendency for matter to move to the lowest state of potential energy.
- Potential energy of electrons is not infinitely divisible, but exists only in discrete amounts called *quanta*.
- Different fixed potential energy states for electrons are called *energy levels* or *electron shells*.
- Electrons with lowest potential energy are in energy levels closest to the nucleus.
- Electrons with greater energy are in energy levels further from nucleus.

Electrons may move from one energy level to another. In the process, they gain or lose energy equal to the difference in potential energy between the old and new energy level.

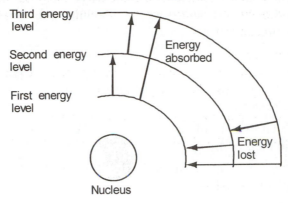

E.   **Electron Orbitals**

Orbital = Three-dimensional space where an electron will most likely be found 90% of the time.

- Viewed as a three-dimensional probability cloud - a statistical concept.
- No more than two electrons can occupy same orbital.

First energy level:

- Has one spherical *s* orbital (1*s* orbital).
- Holds a maximum of two electrons.

Second energy level:

- Holds a maximum of 8 electrons.
- One spherical *s* orbital (2*s* orbital).
- 3 dumbbell-shaped *p* orbitals each oriented at right angles to the other two ($2p_x$, $2p_y$, $2p_z$ orbitals).

Higher energy levels:

- Contain *s* and *p* orbitals.
- Contain additional orbitals with more complex shapes.

F.   **Electron Configuration and Chemical Properties**

An atom's *electron configuration* determines its chemical behavior.

Electron configuration = Distribution of electrons in an atom's electron shells.

The first 18 elements of a periodic chart are arranged sequentially by atomic number into 3 rows (periods). In reference to these *representative* elements, note the following:

- Outermost shell of these atoms never have more than 4 orbitals (one *s* and three *p*) or 8 electrons.
- Electrons must first occupy lower electron shells before the higher shells can be occupied. (A reflection of the natural tendency for matter to move to the lowest possible state of potential energy - the most stable state.)

- Electrons are added to each of the *p* orbitals singly, before they can be paired.

- If an atom does not have enough electrons to fill all shells, the outer shell will be the only one partially filled. For example: O_2 with a total of 8 electrons:

OXYGEN

$_8O$

Two electrons have the $1s$ orbital of the first electron shell.

First two electrons in the second shell are both in the $2s$ orbital.

Next three electrons each have a *p* orbital ($2p_x$, $2p_y$, $2p_z$).

Eighth electron is paired in the $2p_x$ orbital.

$1s$			
2			
$2s$	$2p_x$	$2p_y$	$2p_z$
2	2	1	1

Chemical properties of an atom depend upon the number of *valence electrons*.

<u>Valence electrons</u> = Electrons in the outermost energy shell (valence shell).

<u>Octet rule</u> = Rule that a valence shell is complete when it contains 8 electrons (except H and He).

- An atom with a complete valence shell is unreactive or *inert*.

- Noble elements (e.g. helium, argon and neon) have filled outer shells in their elemental state and are thus inert.

- An atom with an incomplete valence shell is chemically reactive (tends to form chemical bonds until it has 8 electrons to fill the valence shell).

- Atoms with the same number of valence electrons show similar chemical behavior.

NOTE: The consequence of this unifying chemical principle is that the valence electrons are responsible for the atom's bonding capacity. This rule applies to most of the representative elements, but <u>not all</u>.

IV. Atoms combine by chemical bonding to form molecules

Atoms with incomplete valence shells tend to fill those shells by interacting with other atoms. These interactions of electrons among atoms may allow atoms to form *chemical bonds*.

Chemical bonds = Attractions that hold molecules together.

Molecules = Two or more atoms held together by chemical bonds.

A. Covalent Bonds

Covalent bond = Chemical bond between atoms formed by sharing a pair of valence electrons.

- Strong chemical bond.
- For example: molecular hydrogen (H_2). When 2 hydrogen atoms come close enough for their $1s$ orbitals to overlap, they share electrons - thus completing the valence shell of each atom.

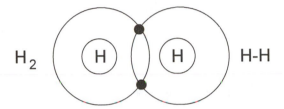

Structural formula = Formula which represents the atoms and bonding within a molecule (e.g. H-H). The line represents a shared pair of electrons.

Molecular formula = Formula which indicates the number and type of atoms (e.g. H_2).

Single covalent bond = Bond between atoms formed by sharing a single pair of valence electrons.

- Atoms may freely rotate around the axis of the bond.

Double covalent bond = Bond formed when atoms share *two* pairs of valence electrons (e.g. O_2).

Triple covalent bond = Bond formed when atoms share three pairs of valence electrons (e.g. N_2 or $N^{\circ}N$).

NOTE: Double and triple covalent bonds are rigid and do not allow rotation.

Valence = Bonding capacity of an atom which is the number of covalent bonds that must be formed to complete the outer electron shell.

- Valences of some common elements: hydrogen = 1, oxygen = 2, nitrogen = 3, carbon = 4, phosphorus = 3, sulfur = 2.

<u>Compound</u> = A pure substance composed of two or more elements combined in a fixed ratio.

- For example: water (H_2O), methane (CH_4).

- Note that two hydrogens are necessary to complete the valence shell of oxygen in water, and four hydrogens are necessary for carbon to complete the valence shell in methane.

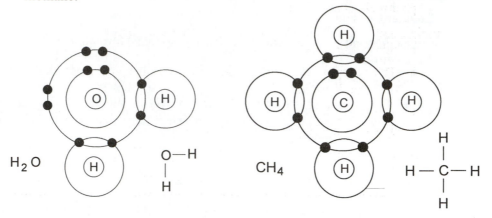

Nonpolar and Polar Covalent Bonds

<u>Electronegativity</u> = Atom's ability to attract and hold electrons.

- The more electronegative an atom, the more strongly it attracts shared electrons.

- Scale determined by Linus Pauling:

O	=	3.5
N	=	3.0
S and C	=	2.5
P and H	=	2.1

<u>Nonpolar covalent bond</u> = Covalent bond formed by an equal sharing of electrons between atoms.

- Occurs when electronegativity of both atoms is about the same (e.g. CH_4).

- Molecules made of one element usually have nonpolar covalent bonds (e.g. H_2, O_2, Cl_2, N_2).

<u>Polar covalent bond</u> = Covalent bond formed by an unequal sharing of electrons between atoms.

- Occurs when the atoms involved have different electronegativities.

- Shared electrons spend more time around the more electronegative atom.

- In H_2O, for example, the oxygen is strongly electronegative, so negatively charged electrons spend more time around the oxygen than the hydrogens. This causes the oxygen atom to have a slight negative charge and the hydrogens to have a slight positive charge.

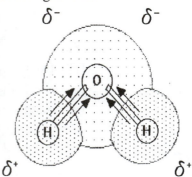

B.   Ionic Bonds

Ion = Charged atom or molecule.

Anion = An atom that has gained one or more electrons from another atom and has become negatively charged; a negatively charged ion.

Cation = An atom that has lost one or more electrons and has become positively charged; a positively charged ion.

Ionic bond = Bond formed by the electrostatic attraction after the complete transfer of an electron from a donor atom to an acceptor.

- The acceptor atom attracts the electrons because it is much more electronegative than the donor atom.
- Are strong bonds in crystals, but are fragile bonds in water; salt crystals will readily dissolve in water and dissociate into ions.
- Ionic compounds are called salts (e.g. NaCl or table salt).

NOTE: The *difference* in electronegativity between interacting atoms determines if electrons are shared equally (nonpolar covalent), shared unequally (polar covalent), gained or lost (ionic bond). Nonpolar covalent bonds and ionic bonds are two extremes of a continuum from interacting atoms with similar electronegativities to interacting atoms with very different electronegativities.

V.   Weak chemical bonds play important roles in the chemistry· of life

Biologically important weak bonds:

- Include hydrogen bonds, ionic bonds in aqueous solutions and other weak forces such as Van der Waals and hydrophobic interactions.
- Make chemical signaling possible in living organisms, because they are only temporary associations. Signal molecules can briefly and reversibly bind to receptor molecules on a cell, causing a short-lived response.
- Can form between molecules or between different parts of a single large molecule.
- Help stabilize the three-dimensional shape of large molecules (e.g. DNA and proteins).

A.   Hydrogen Bonds

Hydrogen bond = Bond formed by the charge attraction when a hydrogen atom covalently bonded to one electronegative atom is attracted to another electronegative atom.

- Weak attractive force that is about 20 times easier to break than a covalent bond.

- Is a charge attraction between oppositely charged portions of polar molecules.

- Can occur between a hydrogen that has a slight positive charge when covalently bonded to an atom with high electronegativity (usually O and N).

- For example: NH_3 in H_2O.

VI.   A molecule's biological function is related to its shape

The function of many molecules depends upon their shape.

- Molecules with only two atoms are linear.

- Molecules with more than two atoms have more complex shapes.

When an atom forms covalent bonds, orbitals in the valence shell rearrange into the most stable configuration. To illustrate, consider atoms with valence electrons in the *s* and three *p* orbitals:

- The *s* and three *p* orbitals *hybridize* into four new orbitals.

- The new orbitals are teardrop shaped, extend from the nucleus and spread out as far apart as possible.

- For example, if outer tips of orbitals in methane (CH_4) are connected by imaginary lines, the new molecule has a tetrahedral shape with C at the center. (See Campbell, Figure 2.15)

VII.   Chemical reactions change the composition of matter

Chemical reactions = process of making and breaking chemical bonds leading to changes in the composition of matter.

- Process where *reactants* undergo changes into *products*.

- Matter is conserved, so all reactant atoms are only rearranged to form products.

- Some reactions go to completion (all reactants converted to products), but most reactions are *reversible*. For example:

$$3H_2 + N_2 \rightleftharpoons 2NH_3$$

- The relative concentration of reactants and products affects reaction rate (the higher the concentration, the greater probability of reaction).

Chemical equilibrium = Equilibrium established when the rate of forward reaction equals the rate of the reverse reaction.

- Is a *dynamic* equilibrium with reactions continuing in both directions.

- Relative concentrations of reactants and products stay the same.

NOTE: Point out to students that *chemical equilibrium* does NOT mean that the concentrations of reactants and products are equal.

VIII. Chemical conditions on the early earth set the stage for the origin and evolution of life

Matter in our solar system formed from the debris of supernovae that exploded over 5 billion years ago.

- Most of the resulting swirling matter condensed in the center as the sun.
- The Earth formed about 4.6 billion years ago from peripheral material spinning around the early sun.

The early Earth was initially cold, but melted from heat generated by compaction, radioactive decay and meteorite impact. The resulting molten material sorted into layers of varying density.

- Most nickel and iron sank to the center.
- Less dense material concentrated into a mantle.
- Least dense material settled on the surface solidifying into a thin crust.
- Present continents are attached to plates of crust that float on the semisolid, flexible mantle.

Chemical conditions on the early earth were probably quite different from today. Evidence indicates that:

- The first early atmosphere was mostly hot hydrogen gas made of small molecules that escaped the earth's gravitational pull.
- The second early atmosphere consisted mostly of water vapor, carbon monoxide, carbon dioxide, nitrogen, methane and ammonia released by volcanoes and vents through the Earth's crust.
- As the earth cooled, water vapor condensed and rainfall formed the first seas.
- Lightning, volcanic activity and ultraviolet radiation were more intense than at present.

Under these conditions of early Earth, chemical evolution may have given rise to the first cells.

REFERENCES

Atkins, P.W. *Atoms, Electrons and Change*. W.H. Freeman and Company, 1991. An eloquent and very readable account of atomic structure and how electrons affect the chemical behavior of atoms.

Campbell, N. *Biology*. 4th ed. Menlo Park, California: Benjamin/Cummings, 1996.

Weinberg, S. *The Discovery of Subatomic Particles*. New York, San Francisco: W.H. Freeman and Company, 1983. A historical approach and excellent reference.

Whitten, K.W. and K.D. Gailey. *General Chemistry*. 4th ed. New York: Saunders, 1992. This is a good general book on the fundamentals of chemistry.

CHAPTER 3

WATER AND THE FITNESS OF THE ENVIRONMENT

OUTLINE

OBJECTIVES

After reading this chapter and attending lecture, the student should be able to:
1. Describe how water contributes to the fitness of the environment to support life.
2. Describe the structure and geometry of a water molecule, and explain what properties emerge as a result of this structure.

3. Explain the relationship between the polar nature of water and its ability to form hydrogen bonds.
4. List five characteristics of water that are emergent properties resulting from hydrogen bonding.
5. Describe the biological significance of the cohesiveness of water.
6. Distinguish between heat and temperature.
7. Explain how water's high specific heat, high heat of vaporization and expansion upon freezing affect both aquatic and terrestrial ecosystems.
8. Explain how the polarity of the water molecule makes it a versatile solvent.
9. Define molarity and list some advantages of measuring substances in moles.
10. Write the equation for the dissociation of water, and explain what is actually transferred from one molecule to another.
11. Explain the basis for the pH scale.
12. Explain how acids and bases directly or indirectly affect the hydrogen ion concentration of a solution.
13. Using the bicarbonate buffer system as an example, explain how buffers work.
14. Describe the causes of acid precipitation, and explain how it adversely affects the fitness of the environment.

KEY TERMS

cohesion	temperature	solvent	dissociation
surface tension	calorie	solute	acid
adhesion	kilocalorie	aqueous solution	base
hydrophilic	specific heat	mole	pH scale
hydrophobic	heat of vaporization	molecular weight	buffer
kinetic energy	evaporative cooling	hydronium ion	acid precipitation
heat	solution	hydroxide ion	

LECTURE NOTES

Water contributes to the fitness of the environment to support life.

- Life on earth probably evolved in water.
- Living cells are 70%-95% H_2O.
- Water covers about 3/4 of the earth.
- In nature, water naturally exists in all three physical states of matter - solid, liquid and gas.

Water's extraordinary properties are emergent properties resulting from water's structure and molecular interactions.

I. The polarity of water molecules results in hydrogen bonding

Water is a *polar* molecule. Its polar bonds and asymmetrical shape give water molecules opposite charges on opposite sides.

Unbonded electron pairs

- Four valence orbitals of O point to corners of a tetrahedron.
- 2 corners are orbitals with unshared pairs of electrons and weak negative charge.
- 2 corners are occupied by H atoms which are in polar covalent bonds with O. Oxygen is so electronegative, that shared electrons spend more time around the O causing a weak positive charge near H's.

Hydrogen bonding orders water into a higher level of structural organization.

- The polar molecules of water are held together by hydrogen bonds.
- Positively charged H of one molecule is attracted to the negatively charged O of another water molecule.
- Each water molecule can form a maximum of four hydrogen bonds with neighboring water molecules.

Water has extraordinary properties that emerge as a consequence of its polarity and hydrogen-bonding. Some of these properties are that water:
- has cohesive behavior
- resists changes in temperature
- has a high heat of vaporization and cools surfaces as it evaporates
- expands when it freezes
- is a versatile solvent

II. Organisms depend on the cohesion of water molecules

Cohesion = Phenomenon of a substance being held together by hydrogen bonds.
- Though hydrogen bonds are transient, enough water molecules are hydrogen bonded at any given time to give water more structure than other liquids.
- Contributes to upward water transport in plants by holding the water column together. *Adhesion* of water to vessel walls counteracts the downward pull of gravity.

Surface tension = Measure of how difficult it is to stretch or break the surface of a liquid.
- Water has a greater surface tension than most liquids; function of the fact that at the air/H_2O interface, surface water molecules are hydrogen bonded to each other and to the water molecules below.
- Causes H_2O to bead (shape with smallest area to volume ratio and allows maximum hydrogen bonding).

III. Water contributes to Earth's habitability by moderating temperatures

A. Heat and Temperature

<u>Kinetic energy</u> = The energy of motion.

<u>Heat</u> = Total kinetic energy due to molecular motion in a body of matter.

<u>Calorie (cal)</u> = Amount of heat it takes to raise the temperature of one gram of water by one degree Celsius.

<u>Kilocalorie (kcal or Cal)</u> = Amount of heat required to raise the temperature of one kilogram of water by one degree Celsius (1000 cal).

<u>Temperature</u> = Measure of heat intensity due to the *average* kinetic energy of molecules in a body of matter.

Celsius Scale at Sea Level	Scale Conversion
100°C (212°F) = water boils	$°C = \dfrac{5(°F - 32)}{9}$
37°C (98.6°F) = human body temperature	
23°C (72°F) = room temperature	$°F = \dfrac{9°C}{5} + 32$
0°C (32°F) = water freezes	
	$°K = °C + 273$

B. Water's High Specific Heat

Water has a high *specific heat*, which means that it resists temperature changes when it absorbs or releases heat.

<u>Specific heat</u> = Amount of heat that must be absorbed or lost for one gram of a substance to change its temperature by one degree Celsius.

<u>Specific heat of water</u> = One calorie per gram per degree Celsius (1 cal/g/°C).

- As a result of hydrogen bonding among water molecules, it takes a relatively large heat loss or gain for each 1°C change in temperature.
- Hydrogen bonds must absorb heat to break, and they release heat when they form.
- Much absorbed heat energy is used to disrupt hydrogen bonds before water molecules can move faster (increase temperature).

A large body of water can act as a heat sink - absorbing heat from sunlight during the day and summer and releasing heat during the night and winter as the water gradually cools. As a result:

- Water, which covers three-fourths of the planet, keeps temperature fluctuations within a range suitable for life.
- Coastal areas have milder climates than inland.
- The marine environment has a relatively stable temperature.

C.   Evaporative Cooling

Vaporization (evaporation) = Transformation from liquid to a gas.

- Molecules with enough kinetic energy to overcome the mutual attraction of molecules in a liquid, can escape into the air.

Heat of vaporization = Quantity of heat a liquid must absorb for 1 g to be converted to the gaseous state.

- For water molecules to evaporate, hydrogen bonds must be broken which requires heat energy.
- Water has a relatively high heat of vaporization at the boiling point (540 cal/g or 2260 J/g; Joule = 0.239 cal).

Evaporative cooling = Cooling of a liquid's surface when a liquid evaporates.

- The surface molecules with the highest kinetic energy are most likely to escape into gaseous form; the average kinetic energy of the remaining surface molecules is thus lower.

Water's high heat of vaporization:

- Moderates the earth's climate.
  - ⇒ Solar heat absorbed by tropical seas dissipates when surface water evaporates (*evaporative cooling*).
  - ⇒ As moist tropical air moves poleward, water vapor releases heat as it condenses into rain.
- Stabilizes temperature in aquatic ecosystems (*evaporative cooling*).
- Helps organisms from overheating by *evaporative cooling*.

IV.   Oceans and lakes don't freeze solid because ice floats

Because of hydrogen bonding, water is less dense as a solid than it is as a liquid. Consequently, ice floats.

- Water is densest at 4°C.
- Water contracts as it cools to 4°C.
- As water cools from 4°C to freezing (0°C), it expands and becomes *less dense* than liquid water (ice floats).
- When water begins to freeze, the molecules do not have enough kinetic energy to break hydrogen bonds.
- As the crystalline lattice forms, each water molecule forms a maximum of 4 hydrogen bonds, which keeps water molecules further apart than they would be in the liquid state.

Expansion of water contributes to the fitness of the environment for life:

- Prevents deep bodies of water from freezing solid from the bottom up.
- Since ice is less dense, it forms on the surface first. As water freezes it releases heat to the water below and insulates it.
- Makes the transitions between seasons less abrupt. As water freezes, hydrogen bonds form releasing heat. As ice melts, hydrogen bonds break absorbing heat.

V. Water is the solvent of life

Solution = A liquid that is a homogenous mixture of two or more substances.

Solvent = Dissolving agent of a solution.

Solute = Substance dissolved in a solution.

Aqueous solution = Solution in which water is the solvent.

Water is a versatile solvent owing to the *polarity* of the water molecule.

Hydrophilic

1. Ionic compounds dissolve in water.
   - Charged regions of polar water molecules have an electrical attraction to charged ions.
   - Water surrounds individual ions, separating and shielding them from one another.

2. Polar compounds in general, are water-soluble.
   - Charged regions of polar water molecules have an affinity for oppositely charged regions of other polar molecules.

Hydrophobic

3. Nonpolar compounds (which have symmetric distribution in charge) are NOT water-soluble.

A. Hydrophilic and Hydrophobic Substances

Ionic and polar substances are *hydrophilic*, but nonpolar compounds are *hydrophobic*.

Hydrophilic = (Hydro=water; philo=loving) Property of having an affinity for water.
- Some large hydrophilic molecules can absorb water without dissolving.

Hydrophobic = (Hydro=water; phobos=fearing) Property of not having an affinity for water, and thus not being water-soluble.

B. Solute Concentration in Aqueous Solutions

Most biochemical reactions involve solutes dissolved in water. There are two important quantitative properties of aqueous solutions: solute concentration and pH.

Molecular weight = Sum of the weight of all atoms in a molecule (expressed in daltons).

Mole = Amount of a substance that has a mass in grams numerically equivalent to its molecular weight in daltons.

For example, to determine a mole of sucrose ($C_{12}H_{22}O_{11}$):
- Calculate molecular weight:

C = 12 dal	12 dal × 12 =	144 dal
H = 1 dal	1 dal × 22 =	22 dal
O = 16 dal	16 dal × 11 =	176 dal
		342 dal

- Express it in grams (342 g).

<u>Molarity</u> = Number of moles of solute per liter of solution.

- For example, to make a 1M sucrose solution, weigh out 342 g of sucrose and add water up to 1L.

Advantage of measuring in moles:

- Rescales weighing of single molecules in daltons to grams, which is more practical for laboratory use.
- A mole of one substance has the *same* number of molecules as a mole of any other substance (6.02×10^{23}; Avogadro's number).
- Allows one to combine substances in fixed ratios of molecules.

VI. Organisms are sensitive to changes in pH

A. Dissociation of Water Molecules

Occasionally, the hydrogen atom that is shared in a hydrogen bond between two water molecules, shifts from the oxygen atom to which it is covalently bonded to the unshared orbitals of the oxygen atom to which it is hydrogen bonded.

- Only a *hydrogen ion* (proton with a +1 charge) is actually transferred.
- Transferred proton binds to an unshared orbital of the second water molecule creating a *hydronium ion* (H_3O^+).
- Water molecule that lost a proton has a net negative charge and is called a *hydroxide ion* (OH^-).

$$H_2O + H_2O \rightleftharpoons H_3O^+ + OH^-$$

- By convention, ionization of H_2O is expressed as the *dissociation* into H^+ and OH^-.

$$H_2O \rightleftharpoons H^+ + OH^-$$

- Reaction is reversible.
- At equilibrium, most of the H_2O is <u>not</u> ionized.

B. Acids and Bases

At equilibrium in pure water at 25°C:

- Number of H^+ ions = number of OH^- ions.
- $[H^+] = [OH^-] = \dfrac{1}{10,000,000}$ M = 10^{-7} M
- Note that brackets indicate molar concentration.

NOTE: This is a good place to point out how *few* water molecules are actually dissociated (only 1 out of 554,000,000 molecules).

ACID	BASE
Substance that *increases* the relative $[H^+]$ of a solution.	Substance that *reduces* the relative $[H^+]$ of a solution.
Also removes OH^- because it tends to combine with H^+ to form H_2O.	May alternately increase $[OH^-]$.
For example: (in water) $$HCl \rightarrow H^+ + Cl^-$$	For example: A base may reduce $[H^+]$ directly: $$NH_3 + H^+ \rightleftharpoons NH_4^+$$ A base may reduce $[H^+]$ indirectly: $$NaOH \rightarrow Na^+ + OH^-$$ $$OH^- + H^+ \rightarrow H_2O$$

A solution in which:

- $[H^+] = [OH^-]$ is a neutral solution.
- $[H^+] > [OH^-]$ is an acidic solution.
- $[H^+] < [OH^-]$ is a basic solution.

Strong acids and bases dissociate completely in water.

- For example, HCl and NaOH.
- Single arrows indicate complete dissociation.

$$NaOH \rightarrow Na^+ + OH^-$$

Weak acids and bases dissociate only partially and reversibly.

- For example, NH_3 (ammonia) and H_2CO_3 (carbonic acid)
- Double arrows indicate a reversible reaction; at equilibrium there will be a fixed ratio of reactants and products.

$$H_2CO_3 \rightleftharpoons HCO_3^- + H^+$$
Carbonic   Bicarbonate   Hydrogen
Acid   ion   ion

C. The pH Scale

In any aqueous solution:

$$[H^+][OH^-] = 1.0 \times 10^{-14}$$

For example:

- In a neutral solution, $[H^+] = 10^{-7}$ M and $[OH^-] = 10^{-7}$ M.
- In an acidic solution where the $[H^+] = 10^{-5}$ M, the $[OH^-] = 10^{-9}$ M.
- In a basic solution where the $[H^+] = 10^{-9}$ M, the $[OH^-] = 10^{-5}$ M.

pH scale = Scale used to measure degree of acidity. It ranges from 0 to 14.

pH = Negative $\log_{10}$ of the $[H^+]$ expressed in moles per liter.

- pH of 7 is a neutral solution.

  pH < 7 is an acidic solution.

  pH > 7 is a basic solution.

- Most biological fluids are within the pH range of 6 to 8. There are some exceptions such as stomach acid with pH = 1.5. (See Campbell, Figure 3.9)

- Each pH unit represents a *tenfold* difference (scale is logarithmic), so a slight change in pH represents a large change in actual $[H^+]$.

NOTE: To illustrate this point, we project the following on a transparency and cover the answer. The students will frequently give the wrong response (3×), and they are surprised when you unveil the solution.

How much greater is the $[H^+]$ in a solution with pH 2 than in a solution with pH 6?

ANS: pH 2 = $[H^+]$ of 1.0×10^{-2} = $\dfrac{1}{100}$ M

pH 6 = $[H^+]$ of 1.0×10^{-6} = $\dfrac{1}{1,000,000}$ M

10,000 times greater.

D.   Buffers

By minimizing wide fluctuations in pH, buffers help organisms maintain the pH of body fluids within the narrow range necessary for life (usually pH 6-8).

Buffer = Substance that prevents large sudden changes in pH.

- Are combinations of H^+-donor and H^+-acceptor forms of weak acids or bases.

- Work by accepting H^+ ions from solution when they are in excess, and by donating H^+ ions to the solution when they have been depleted.

For example: Bicarbonate buffer.

$$H_2CO_3 \underset{\substack{\text{response to a}\\\text{drop in pH}}}{\overset{\substack{\text{response to a}\\\text{rise in pH}}}{\rightleftharpoons}} HCO_3^- \quad + \quad H^+$$

| H^+ donor | | H^+ acceptor | Hydrogen |
| (weak acid) | | (weak base) | ion |

$$HCl + NaHCO_3 \longrightarrow H_2CO_3 + NaCl$$

strong   weak
acid   acid

$$NaOH + H_2CO_3 \longrightarrow NaHCO_3 + H_2O$$

strong   weak
base   base

VII.   Acid precipitation threatens the fitness of the environment

<u>Acid precipitation</u> = Rain, snow or fog more strongly acidic than pH 5.6.
- Has been recorded as low as pH 1.5 in West Virginia.
- Occurs when sulfur oxides and nitrogen oxides in the atmosphere react with water in the air to form acids which fall to Earth in precipitation.
- Major oxide source is the combustion of fossil fuels by industry and cars.

Acid rain affects the fitness of the environment to support life:
- Lowers soil pH which affects mineral solubility. May leach out necessary mineral nutrients and increase the concentration of minerals that are potentially toxic to vegetation in higher concentration (e.g. aluminum). This is contributing to the decline of some European and North American forests.
- Lowers the pH of lakes and ponds, and runoff carries leached out soil minerals into aquatic ecosystems. This adversely affects aquatic life. For example: In the Western Adirondack Mountains, there are lakes with a pH < 5 that have no fish.

What can be done to reduce the problem?
- Add industrial pollution controls.
- Develop and use antipollution devices.
- Increase involvement of voters, consumers, politicians and business leaders.

NOTE: The political issues surrounding acid rain can be used to enhance student awareness and make this entire topic more relevant and interesting to the students.

REFERENCES

Campbell, N. *Biology*. 4th ed. Menlo Park, California: Benjamin/Cummings, 1996.

Gould, R. *Going Sour: Science and Politics of Acid Rain*. Boston: Birkhauser, 1985.

Henderson, L.J. *The Fitness of the Environment*. Boston: Beacon Press, 1958.

Mohnen, V.A. "The Challenge of Acid Rain." *Scientific American*, August 1988.

This is an excellent resource and could be used for outside reading. We use a diagram similar to the one on atmospheric chemistry as an overhead transparency. The author clearly explains acid rain, the scope of the problem and potential solutions.

CHAPTER 4

CARBON AND THE
MOLECULAR DIVERSITY OF LIFE

OUTLINE

I.   Organic chemistry is the study of carbon compounds

II. Carbon atoms are the most versatile building blocks of molecules

III. Variation in carbon skeletons contributes to the diversity of organic molecules
     A.   Isomers

IV. Functional groups also contribute to the molecular diversity of life
     A.   The Hydroxyl Group
     B.   The Carbonyl Group
     C.   The Carboxyl Group
     D.   The Amino Group
     E.   The Sulfhydryl Group
     F.   The Phosphate Group

OBJECTIVES

After reading this chapter and attending lecture, the student should be able to:
1. Summarize the philosophies of *vitalism* and *mechanism*, and explain how they influenced the development of organic chemistry, as well as mainstream biological thought.
2. Explain how carbon's electron configuration determines the kinds and number of bonds carbon will form.
3. Describe how carbon skeletons may vary, and explain how this variation contributes to the diversity and complexity of organic molecules.
4. Distinguish among the three types of isomers: structural, geometric and enantiomers.
5. Recognize the major functional groups, and describe the chemical properties of organic molecules in which they occur.

KEY TERMS

organic chemistry
organic molecules
vitalism
mechanism
hydrocarbons
isomer

structural isomer
geometric isomer
enantiomer
asymmetric carbon
functional group
hydroxyl

alcohol
carbonyl
aldehyde
ketone
carboxyl
carboxylic acid

amino group
amine
sulfhydryl
thiol
phosphate

LECTURE NOTES

Aside from water, most biologically important molecules are carbon-based (*organic*).

The structural and functional diversity of organic molecules emerges from the ability of carbon to form large, complex and diverse molecules by bonding to itself and to other elements such as H, O, N, S and P.

I. **Organic chemistry is the study of carbon compounds**

Organic chemistry = The branch of chemistry that specializes in the study of carbon compounds.

Organic molecules = Molecules containing carbon.

Vitalism = Belief in a life force outside the jurisdiction of chemical/physical laws.
   - Early 19th century organic chemistry was built on a foundation of vitalism because organic chemists could not artificially synthesize organic compounds. It was believed that only living organisms could produce organic compounds.

Mechanism = Belief that all natural phenomena are governed by physical and chemical laws.
   - Pioneers of organic chemistry began to synthesize organic compounds from inorganic molecules. This helped shift mainstream biological thought from vitalism to mechanism.
   - For example, Friedrich Wohler synthesized urea in 1828; Hermann Kolbe synthesized acetic acid.
   - Stanley Miller (1953) demonstrated the possibility that organic compounds could have been produced under the chemical conditions of primordial Earth.

II. **Carbon atoms are the most versatile building blocks of molecules**

The carbon atom:
   - Has an atomic number of 6; therefore, it has 4 valence electrons.
   - Completes its outer energy shell by sharing valence electrons in four covalent bonds. (Not likely to form ionic bonds.)

Emergent properties, such as the kinds and number of bonds carbon will form, are determined by its *tetravalent* electron configuration.

- It makes large complex molecules possible. The carbon atom is a central point from which the molecule branches off into four directions.

- It gives carbon covalent compatibility with many different elements. The four major atomic components of organic molecules are:

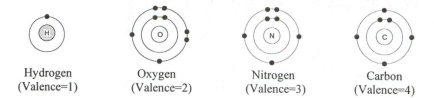

| Hydrogen | Oxygen | Nitrogen | Carbon |
| (Valence=1) | (Valence=2) | (Valence=3) | (Valence=4) |

- It determines an organic molecule's 3-dimensional shape, which may affect molecular function. For example, when carbon forms four single covalent bonds, the four valence orbitals hybridize into teardrop-shaped orbitals that angle from the carbon atom towards the corners of an imaginary tetrahedron.

NOTE: (See Campbell, Figure 4.3) Students have problems visualizing shapes of organic molecules in three dimensions. Specific examples can be enhanced by an overhead transparency of ball-and-stick or space-filling models. A large three-dimensional molecular model that can be held up in front of the class works best.

III. Variation in carbon skeletons contributes to the diversity of organic molecules

Covalent bonds link carbon atoms together in long chains that form the skeletal framework for organic molecules. These carbon skeletons may vary in:

- Length.
- Shape (straight chain, branched, ring).
- Number and location of double bonds.
- Other elements covalently bonded to available sites.

This variation in carbon skeletons contributes to the complexity and diversity of organic molecules. (See Campbell, Figure 4.4)

Hydrocarbons = Molecules containing only carbon and hydrogen

- Are major components of fossil fuels produced from the organic remains of organisms living millions of years ago, though they are not prevalent in living organisms.

- Have a diversity of carbon skeletons which produce molecules of various lengths and shapes.

- As in hydrocarbons, a carbon skeleton is the framework for the large diverse organic molecules found in living organisms. Also, some biologically important molecules may have regions consisting of hydrocarbon chains (e.g. fats).

- Hydrocarbon chains are hydrophobic because the C–C and C–H bonds are nonpolar.

A.   Isomers

<u>Isomers</u> = Compounds with the same molecular formula but with different structures and hence different properties.   Isomers are a source of variation among organic molecules.

There are three types of isomers (See Campbell, Figure 4.6):

<u>Structural isomers</u> = Isomers that differ in the covalent arrangement of their atoms.

- Number of possible isomers increases as the carbon skeleton size increases.
- May also differ in the location of double bonds.

<u>Geometric isomers</u> = Isomers which share the same covalent partnerships, but differ in their spatial arrangements.

$$
\begin{array}{ccc}
HO & & OH \\
\backslash & & / \\
C & = & C \\
/ & & \backslash \\
H & & H
\end{array}
\qquad
\begin{array}{ccc}
H & & OH \\
\backslash & & / \\
C & = & C \\
/ & & \backslash \\
HO & & H
\end{array}
$$

- Result from the fact that double bonds will not allow the atoms they join to rotate freely about the axis of the bonds.
- Subtle differences between isomers affects their biological activity.

<u>Enantiomers</u> = Isomers that are mirror images of each other.

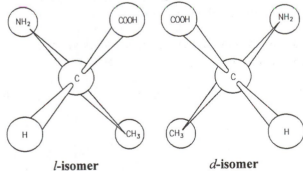

l-isomer   *d*-isomer

- Can occur when four different atoms or groups of atoms are bonded to the same carbon (*asymmetric carbon*).
- There are two different spatial arrangements of the four groups around the asymmetric carbon. These arrangements are mirror images.
- Usually one form is biologically active and its mirror image is not.

IV.   **Functional groups also contribute to the molecular diversity of life**

Small characteristic groups of atoms (functional groups) are frequently bonded to the carbon skeleton of organic molecules. These functional groups:

- Have specific chemical and physical properties.
- Are the regions of organic molecules which are commonly chemically reactive.
- Behave consistently from one organic molecule to another.
- Depending upon their number and arrangement, determine unique chemical properties of organic molecules in which they occur.

As with hydrocarbons, diverse organic molecules found in living organisms have carbon skeletons. In fact, these molecules can be viewed as hydrocarbon derivatives with functional groups in place of H, bonded to carbon at various sites along the molecule.

A.   **The Hydroxyl Group**

Hydroxyl group = A functional group that consists of a hydrogen atom bonded to an oxygen atom, which in turn is bonded to carbon (–OH).

- Is a *polar* group; the bond between the oxygen and hydrogen is a polar covalent bond.
- Makes the molecule to which it is attached *water soluble*. Polar water molecules are attracted to the polar hydroxyl group which can form hydrogen bonds.
- Organic compounds with hydroxyl groups are called *alcohols*.

B.   **The Carbonyl**

Carbonyl group = Functional group that consists of a carbon atom double-bonded to oxygen (–CO).

- Is a *polar* group. The oxygen can be involved in hydrogen bonding, and molecules with his functional group are *water soluble*.
- Is a functional group found in sugars.
- If the carbonyl is at the end off the carbon skeleton, the compound is an *aldehyde*.

$$
\begin{array}{ccc}
OH & OH & O \\
| & | & \mathbin{/\!/} \\
H\text{–}C & \!\!\!-\!\! C & \!\!\!-\!\! C \\
| & | & | \\
H & H & H
\end{array}
$$

Glyceraldehyde

- If the carbonyl is at the end of the carbon skeleton, the compound is a *ketone*.

$$
\begin{array}{ccc}
H & O & H \\
| & || & | \\
H\text{–}C & \!\!\!-\!\! C & \!\!\!-\!\! C\text{–}H \\
| & & | \\
H & & H
\end{array}
$$

Acetone

C. **The Carboxyl Group**

Carboxyl group = Functional group that consists of a carbon atom which is both double-bonded to an oxygen and single-bonded to the oxygen of a hydroxyl group (–COOH).

- Is a polar group and water soluble. The covalent bond between oxygen and hydrogen is so polar, that the hydrogen reversibly dissociates as H^+. This polarity results from the combined effect of the two electronegative oxygen atoms bonded to the same carbon.

$$
\begin{array}{ccccc}
& H \quad O & & H \quad O & \\
& | \quad // & & | \quad // & \\
H–C–C & & \rightleftharpoons & H–C–C & + \quad H^+ \\
& | \quad \backslash & & | \quad \backslash & \\
& H \quad OH & & H \quad O^- & \\
\text{Acetic} & & \text{Acetate} & & \text{Hydrogen} \\
\text{acid} & & \text{ion} & & \text{ion}
\end{array}
$$

- Since it donates protons, this group has acidic properties. Compounds with this functional group are called *carboxylic acids*.

D. **The Amino Group**

Animo group = Functional group that consists of a nitrogen atom bonded to two hydrogens and to the carbon skeleton (–NH₂).

- Is a polar group and soluble in water.
- Acts as a weak base. The unshared pair of electrons on the nitrogen can accept a proton, giving the amino group a +1 charge.

$$
\begin{array}{ccccc}
H & & & H & \\
/ & & & / & \\
R–N & + & H^+ \rightleftharpoons & R–{}^+N–H & \\
\backslash & & & \backslash & \\
H & & & H & \\
\text{Amine} & & & \text{Ammonium} & \\
& & & \text{ion} &
\end{array}
$$

- Organic compounds with this function group are called *amines*.

E. **The Sulfhydryl Group**

Sulfhydryl group = Functional group which consists of an atom of sulfur bonded to an atom of hydrogen (–SH).

- Help stabilize the structure of proteins. (Disulfide bridges will be discussed with tertiary structure of proteins in Chapter 5: Structure and Function of Macro-molecules.)
- Organic compounds with this functional group are called *thiols*.

F.   The Phosphate Group

<u>Phosphate group</u> = Functional group which is the dissociated form of phosphoric acid (H_3PO_4).

- Loss of two protons by dissociation leaves the phosphate group with a negative charge.

$$\begin{array}{ccc} & O & & & & O \\ & \parallel & & & & \parallel \\ R\text{–}O\text{–}P\text{–}OH & & \rightleftharpoons & & R\text{–}O\text{–}P\text{–}O^- & + \quad 2H^+ \\ & \mid & & & & \mid \\ & OH & & & & O^- \end{array}$$

- Has acid properties since it loses protons.
- Polar group and soluble in water.
- Organic phosphates are important in cellular energy storage and transfer. (ATP is discussed with energy for cellular work in Chapter 6: Introduction to Metabolism.)

NOTE: In our lecture course, we also include the methyl group ($-CH_3$) as an example of a nonpolar hydrophobic functional group. Later in the course, this is helpful in explaining how nonpolar amino acids contribute to the tertiary structure of proteins including integral membrane proteins.

To impress upon students how important functional groups are in determining chemical behavior of organic molecules, we use the following demonstration. Show a comparison of estradiol and testosterone and ask students to find the differences in functional groups. Ask one male and female student to stand up or show pictures of sexual dimorphism in other vertebrates. Point out that differences between males and females are due to slight variation in functional groups attached to sex hormones.

REFERENCES

Campbell, N. *Biology*. 4th ed. Menlo Park, California: Benjamin/Cummings, 1996.

Lehninger, A.L., D.L. Nelson and M.M. Cox. *Principles of Biochemistry*. 2nd ed. New York: Worth, 1993.

Whitten, K.W. and K.D. Gailey. *General Chemistry*. 4th ed. New York: Saunders, 1992.

CHAPTER 5

THE STRUCTURE AND FUNCTION
OF MACROMOLECULES

OUTLINE

OBJECTIVES

After reading this chapter and attending lecture, the student should be able to:

1. List the levels of biological hierarchy from subatomic particles to macromolecules.
2. Explain how organic polymers contribute to biological diversity.
3. Describe how covalent linkages are formed and broken in organic polymers.
4. Describe the distinguishing characteristics of carbohydrates, and explain how they are classified.
5. List four characteristics of a sugar.
6. Identify a glycosidic linkage and describe how it is formed.
7. Describe the important biological functions of polysaccharides.
8. Distinguish between the glycosidic linkages found in starch and cellulose, and explain why the difference is biologically important.
9. Explain what distinguishes lipids from other major classes of macromolecules.
10. Describe the unique properties, building block molecules and biological importance of the three important groups of lipids: fats, phospholipids and steroids.
11. Identify an ester linkage and describe how it is formed.
12. Distinguish between a saturated and unsaturated fat, and list some unique emergent properties that are a consequence of these structural differences.
13. Describe the characteristics that distinguish proteins from the other major classes of macromolecules, and explain the biologically important functions of this group.
14. List and recognize four major components of an amino acid, and explain how amino acids may be grouped according to the physical and chemical properties of the side chains.
15. Identify a peptide bond and explain how it is formed.
16. Explain what determines protein conformation and why it is important.
17. Define primary structure and describe how it may be deduced in the laboratory.
18. Describe the two types of secondary protein structure, and explain the role of hydrogen bonds in maintaining the structure.
19. Explain how weak interactions and disulfide bridges contribute to tertiary protein structure.
20. Using collagen and hemoglobin as examples, describe quaternary protein structure.
21. Define denaturation and explain how proteins may be denatured.
22. Describe the characteristics that distinguish nucleic acids from the other major groups of macromolecules.
23. Summarize the functions of nucleic acids.
24. List the major components of a nucleotide, and describe how these monomers are linked together to form a nucleic acid.
25. Distinguish between a pyrimidine and a purine.
26. List the functions of nucleotides.
27. Briefly describe the three-dimensional structure of DNA.

KEY TERMS

polymer
monomer
macromolecule
condensation reaction
hydrolysis
carbohydrate
monosaccharide
disaccharide
polysaccharide
glycosidic linkage
starch

glycogen
cellulose
chitin
α configuration
β configuration
lipid
fat
phospholipid
steroid
ester linkage
triacylglycerol

protein
amino acid
polypeptide
peptide bond
protein conformation
native conformation
primary structure
secondary structure
tertiary structure
quaternary structure
disulfide bridge

denaturation
nucleic acid
nucleotide
pyrimidine
purine
phosphodiester bond
DNA
RNA
pentose

LECTURE NOTES

The topic of macromolecules lends itself well to illustrate three integral themes woven throughout the text and course: (1) there is a natural hierarchy of structural level in biological organization; (2) as one moves up the hierarchy, new properties emerge because of interactions among subunits at the lower levels; and (3) form fits function.

I.   **Most macromolecules are polymers**

Polymer = (Poly = many; mer = part) Large molecule consisting of many identical or similar subunits connected together.

Monomer = Subunit or building block molecule of a polymer.

Macromolecule = (Macro = large) Large organic polymer.
   • Formation of macromolecules from smaller building block molecules represents another level in the hierarchy of biological organization.
   • There are four classes of macromolecules in living organisms:
      1. Carbohydrates.
      2. Lipids.
      3. Proteins.
      4. Nucleic acids.

Most *polymerization reactions* in living organisms are *condensation reactions*.

Polymerization reactions = Chemical reactions that link two or more small molecules to form larger molecules with repeating structural units.

<u>Condensation reactions</u> = Polymerization reactions during which monomers are covalently linked, producing net removal of a water molecule for each covalent linkage.

- One monomer loses a hydroxyl (–OH), and the other monomer loses a hydrogen (–H).
- Removal of water is actually indirect, involving the formation of "activated" monomers (to be discussed in Chapter 6: Introduction to Metabolism).
- Process requires energy.
- Process requires biological catalysts or enzymes.

<u>Hydrolysis</u> = (Hydro = water; lysis = break) A reaction process that breaks covalent bonds between monomers by the addition of water molecules.

- A hydrogen from the water bonds to one monomer, and the hydroxyl bonds to the adjacent monomer.
- For example: digestive enzymes catalyze hydrolytic reactions which break apart large food molecules into monomers that can be absorbed into the bloodstream.

II.   A limitless variety of polymers can be built from a small set of monomers

Structural variation of macromolecules is the basis for the enormous diversity of life.

- There is *unity* in life as there are only about 40 - 50 common monomers used to construct macromolecules.
- There is *diversity* in life as new properties emerge when these universal monomers are arranged in different ways.

III.   Organisms use carbohydrates for fuel and building material

<u>Carbohydrates</u> = Organic molecules made of sugars and their polymers.

- Monomers or building block molecules are simple sugars called *monosaccharides*.
- Polymers are formed by condensation reactions.
- Are classified based upon the number of simple sugars.

A.   Monosaccharides

<u>Monosaccharides</u> = (Mono = single; sacchar = sugar) Simple sugar in which C, H and O occur in the ratio of (CH_2O).

- Are major nutrients for cells. Glucose is the most common.
- Can be produced (glucose) by photosynthetic organisms from CO_2, H_2O and sunlight.
- Store energy in their chemical bonds which is harvested by cellular respiration.
- Their carbon skeletons are raw material for other organic molecules.
- Can be incorporated as monomers into disaccharides and polysaccharides.

Characteristics of a sugar:

1. An –OH group is attached to each carbon except one, which is double bonded to an oxygen (*carbonyl*).

<table>
<tr><td>Aldehyde
Terminal carbon forms a
double bond with oxygen.</td><td>Ketone
Carbonyl group is within
the carbon skeleton.</td></tr>
</table>

Glucose
(aldose)

Fructose
(ketose)

2. Size of the carbon skeleton varies from 3 to 7 carbons. The *most common* monosaccharides are:

Classification	Number of Carbons	Example
Triose	3	Glyceraldehyde
Pentose	5	Ribose
Hexose	6	Glucose

3. Spatial arrangement around asymmetric carbons may vary. For example, glucose and galactose are enantiomers.

Glucose

Galactose

The small difference between isomers affects molecular shape which gives these molecules distinctive biochemical properties.

4. In aqueous solutions, many monosaccharides form rings. Chemical equilibrium favors the ring structure.

Linear Form
of Glucose

Ring Form of Glucose

B. Disaccharides

<u>Disaccharides</u> = (Di = two; sacchar = sugar) A double sugar that consists of two monosaccharides joined by a *glycosidic linkage*.

<u>Glycosidic linkage</u> = Covalent bond formed by a condensation reaction between two sugar monomers. For example, maltose:

MALTOSE

Examples of Disaccharides:

Disaccharide	Monomers	General Comments
Maltose	Glucose + Glucose	Sugar important in brewing beer.
Lactose	Glucose + Galactose	Sugar present in milk.
Sucrose	Glucose + Fructose	Table sugar; most prevalent disaccharide; transport form in plants.

C. Polysaccharides

Polysaccharides = Macromolecules that are polymers of a few hundred or thousand monosaccharides.

- Are formed by linking monomers in enzyme-mediated condensation reactions.
- Have two important biological functions:
  1. Energy storage (starch and glycogen).
  2. Structural support (cellulose and chitin).

1. Storage Polysaccharides

Cells hydrolyze storage polysaccharides into sugars as needed. Two most common storage polysaccharides are *starch* and *glycogen*.

Starch = Glucose polymer that is a storage polysaccharide in plants.

- Helical glucose polymer with α 1-4 linkages. (See Campbell, Figure 5.6).
- Stored as granules within plant organelles called *plastids*.
- *Amylose*, the simplest form, is an unbranched polymer.
- *Amylopectin* is branched polymer.
- Most animals have digestive enzymes to hydrolyze starch.
- Major sources in the human diet are potato tubers and grains (e.g. wheat, corn, rice and fruits of other grasses).

Glycogen = Glucose polymer that is a storage polysaccharide in animals.

- Large glucose polymer that is more highly branched than amylopectin.
- Stored in the muscle and liver of humans and other vertebrates.

2. Structural Polysaccharides

Structural polysaccharides include *cellulose* and *chitin*.

Cellulose = Linear unbranched polymer of *D*-glucose in β 1-4 linkages.

- A major structural component of plant cell walls.
- Differs from starch (also a glucose polymer) it its glycosidic linkages. (See Campbell, Figure 5.7)

STARCH	CELLULOSE
Glucose monomers are in α *configuration* (–OH group on carbon one is *below* the ring's plane).	Glucose monomers are in β *configuration* (–OH group on carbon one is *above* the ring's plane).
Monomers are connected with α 1-4 linkage.	Monomers are connected with β 1-4 linkage.

- Cellulose and starch have different three-dimensional shapes and properties as a result of differences in glycosidic linkages.

- Cellulose reinforces plant cell walls. Hydrogen bonds hold together parallel cellulose molecules in bundles of *microfibrils*. (See Campbell, Figure 5.8)

- Cellulose cannot be digested by most organisms, including humans, because they lack an enzyme that can hydrolyze the β 1-4 linkage. (Exceptions are some symbiotic bacteria, other microorganisms and some fungi.)

Chitin = A structural polysaccharide that is a polymer of an amino sugar.

- Forms exoskeletons of arthropods.

- Found as a building material in the cell walls of some fungi.

- Monomer is an *amino sugar*, which is similar to *beta-glucose* with a nitrogen-containing group replacing the hydroxyl on carbon 2.

IV.   Lipids are mostly hydrophobic molecules with diverse functions

Lipids = Diverse group of organic compounds that are insoluble in water, but will dissolve in nonpolar solvents (e.g. ether, chloroform, benzene). Important groups are *fats*, *phospholipids* and *steroids*.

A.   Fats

Fats = Macromolecules constructed from:

1. Glycerol, a three-carbon alcohol.

2. Fatty acid (carboxylic acid).
   - Composed of a carboxyl group at one end and an attached *hydrocarbon chain* ("tail").
   - Carboxyl functional group ("head") has properties of an acid.
   - Hydrocarbon chain has a long carbon skeleton usually with an even number of carbon atoms (most have 16 – 18 carbons).
   - Nonpolar C–H bonds make the chain hydrophobic and not water soluble.

During the formation of a fat, enzyme-catalyzed condensation reactions link glycerol to fatty acids by an *ester linkage*.

Ester linkage = Bond formed between a hydroxyl group and a carboxyl group.

Each of glycerol's three hydroxyl groups can bond to a fatty acid by an ester linkage producing a fat.

Triacylglycerol = A fat composed of three fatty acids bonded to one glycerol by ester linkages (triglyceride).

Some characteristics of fat include:
- Fats are insoluble in water. The long fatty acid chains are hydrophobic because of the many nonpolar C–H bonds.
- The source of variation among fat molecules is the fatty acid composition.
- Fatty acids in a fat may all be the same, or some (or all) may differ.

- Fatty acids may vary in length.
- Fatty acids may vary in the number and location of carbon-to-carbon double bonds:

SATURATED FAT	UNSATURATED FAT
No double bonds between carbons in fatty acid tail.	One or more double bonds between carbons in fatty acid tail.
Carbon skeleton of fatty acid is bonded to maximum number of hydrogens (*saturated with hydrogens*).	Tail kinks at each C=C, so molecules do not pack closely enough to solidify at room temperature.
Usually a solid at room temperature.	Usually a liquid at room temperature.
Most animal fats.	Most plant fats.
e.g. bacon grease, lard and butter.	e.g. corn, peanut and olive oil.

- In many commercially prepared food products, unsaturated fats are artificially hydrogenated to prevent them from separating out as oil (e.g. peanut butter and margarine).

Fat serves many useful functions, such as:

- Energy storage. One gram of fat stores twice as much energy as a gram of polysaccharide. (Fat has a higher proportion of energy rich C–H bonds.)
- More compact fuel reservoir than carbohydrate. Animals store more energy with less weight than plants which use starch, a bulky form of energy storage.
- Cushions vital organs in mammals (e.g. kidney).
- Insulates against heat loss (e.g. mammals such as whales and seals).

B. Phospholipids

Phospholipids = Compounds with molecular building blocks of glycerol, two fatty acids, a phosphate group and usually an additional small chemical group attached to the phosphate. (See Campbell, Figure 5.12)

- Differ from fat in that the third carbon of glycerol is joined to a *negatively charged* phosphate group.
- Can have small variable molecules (usually charged or polar) attached to phosphate.
- Are diverse depending upon differences in fatty acids and in phosphate attachments.
- Show ambivalent behavior towards water. Hydrocarbon tails are hydrophobic, and the polar head (phosphate group with attachments) is hydrophilic.
- Cluster in water as their hydrophobic portions turn away from water. One such cluster, a *micelle*, assembles so the hydrophobic tails turn towards the water-free interior; and the hydrophilic phosphate heads arrange facing outward in contact with water.
- Are major constituents of cell membranes. At the cell surface, phospholipids form a bilayer held together by hydrophobic interactions among the hydrocarbon tails. Phospholipids in water will spontaneously form such a bilayer:

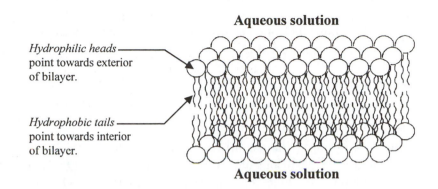

Aqueous solution

Hydrophilic heads point towards exterior of bilayer.

Hydrophobic tails point towards interior of bilayer.

Aqueous solution

C. Steroids

Steroids = Lipids which have four fused carbon rings with various functional groups attached.

Cholesterol, an important steroid:
- Is the precursor to many other steroids including vertebrate sex hormones and bile acids.
- Is a common component of animal cell membranes.
- Can contribute to atherosclerosis.

V. Proteins are the molecular tools for most cellular functions

Polypeptide chains = Polymers of amino acids that are arranged in a specific linear sequence and are linked by peptide bonds.

Protein = A macromolecule that consists of one or more *polypeptide chains* folded and coiled into specific conformations.
- Are abundant, making up 50% or more of cellular dry weight.
- Have important and varied functions in the cell:
    1. Structural support.
    2. Storage (of amino acids).
    3. Transport (e.g. hemoglobin).
    4. Signaling (chemical messengers).
    5. Cellular response to chemical stimuli (receptor proteins).
    6. Movement (contractile proteins).
    7. Defense against foreign substances and disease-causing organisms (antibodies).
    8. Catalysis of biochemical reactions (enzymes).
- Vary extensively in structure; each type has a unique three-dimensional shape (conformation).
- Though vary in structure and function, are commonly made of only 20 amino acid monomers.

VI. **A polypeptide is a polymer of amino acids connected in a specific sequence**

Amino acid = Building block molecule of a protein; most consist of an asymmetric carbon, termed the *alpha carbon*, which is covalently bonded to:

1. Hydrogen atom.
2. Carboxyl group.
3. Amino group.
4. Variable R group (side chain) specific to each amino acid. Physical and chemical properties of the side chain determine the uniqueness of each amino acid.

$$\begin{array}{c} H \quad O \\ | \quad // \\ H_3N^+-C-C \\ | \quad \backslash \\ R \quad O^- \end{array}$$

(At pH's normally found in the cell, both the carboxyl and amino groups are ionized.)

Amino acids contain both carboxyl and amino functional groups. Since one group acts as a weak acid and the other group acts as a weak base, an amino acid can exist in three ionic states. The pH of the solution determines which ionic state predominates.

$$\begin{array}{c} H \quad H \quad O \\ | \quad | \quad // \\ H-N^+-C-C \\ | \quad | \quad \backslash \\ H \quad H \quad OH \\ \text{Cation} \end{array} \quad \underset{\text{as pH decreases}}{\overset{\text{as pH increases}}{\rightleftharpoons}} \quad \begin{array}{c} H \quad H \quad O \\ | \quad | \quad // \\ H-N^+-C-C \\ | \quad | \quad \backslash \\ H \quad H \quad O^- \\ \text{Zwitterion} \\ \text{(dipolar ion)} \end{array} + H^+ \quad \underset{\text{as pH decreases}}{\overset{\text{as pH increases}}{\rightleftharpoons}} \quad \begin{array}{c} H \quad H \quad O \\ \backslash \quad | \quad // \\ N-C-C \\ / \quad | \quad \backslash \\ H \quad H \quad O^- \\ \text{Anion} \end{array} + H^+$$

The twenty common amino acids can be grouped by properties of side chains: (See Campbell, Figure 5.15)

1. Nonpolar side groups (hydrophobic). Amino acids with nonpolar groups are less soluble in water.
2. Polar side groups (hydrophilic). Amino acids with polar side groups are soluble in water. Polar amino acids can be grouped further into:
   a. Uncharged polar.
   b. Charged polar.
   * Acidic side groups. Dissociated carboxyl group gives these side groups a negative charge.
   * Basic side groups. An amino group with an extra proton gives these side groups a net positive charge.

Polypeptide chains are polymers that are formed when amino acid monomers are linked by *peptide bonds*.

Peptide bond = Covalent bond formed by a condensation reaction that links the carboxyl group of one amino acid to the amino group of another.

- Has polarity with an amino group on one end (*N-terminus*) and a carboxyl group on the other (*C-terminus*).

- Has a backbone of the repeating sequence –N–C–C–N–C–C–.

Polypeptide chains:
- Range in length from a few monomers to more than a thousand.
- Have unique linear sequences of amino acids.

VII. A protein's function depends on its specific conformation

A protein's function depends upon its unique *conformation*.

Protein conformation = Three-dimensional shape of a protein.

Native conformation = Functional conformation of a protein found under normal biological conditions.
- Enables a protein to recognize and bind specifically to another molecule (e.g. hormone/receptor, enzyme/substrate and antibody/antigen).
- Is a consequence of the specific linear sequence of amino acids in the polypeptide.
- Is produced when a newly formed polypeptide chain coils and folds spontaneously, mostly in response to hydrophobic interactions.
- Is stabilized by chemical bonds and weak interactions between neighboring regions of the folded protein.

A. Four Levels of Protein Structure

The correlation between form and function in proteins is an emergent property resulting from superimposed levels of protein structure:
1. Primary structure.
2. Secondary structure.
3. Tertiary structure.

When a protein has two or more polypeptide chains, it also has:

4. Quaternary structure.

1. Primary Structure

<u>Primary structure</u> = Unique sequence of amino acids in a protein.
- Determined by genes.
- Slight change can affect a protein's conformation and function (e.g. sickle-cell hemoglobin; See Campbell, Figure 5.19).
- Can be sequenced in the laboratory. A pioneer in this work was Frederick Sanger who determined the amino acid sequence in insulin (late 1940's and early 1950's). This laborious process involved:
  a. Determination of amino acid composition by complete acid hydrolysis of peptide bonds and separation of resulting amino acids by chromatography. Using these techniques, Sanger identified the amino acids and determined the relative proportions of each.
  b. Determination of amino acid sequence by partial hydrolysis with enzymes and other catalysts to break only specific peptide bonds. Sanger deductively reconstructed the primary structure from fragments with overlapping segments.
- Most of the sequencing process is now automated.

2. Secondary Structure

<u>Secondary structure</u> = Regular, repeated coiling and folding of a protein's polypeptide backbone.
- Contributes to a protein's overall conformation.
- Stabilized by hydrogen bonds between peptide linkages in the protein's backbone (carbonyl and amino groups).
- The major types of secondary structure are alpha (α) helix and beta (β) pleated sheet.

a. Alpha (α) Helix

<u>Alpha (α) helix</u> = Secondary structure of a polypeptide that is a helical coil stabilized by hydrogen bonding between every fourth peptide bond (3.6 amino acids per turn).
- Described by Linus Pauling and Robert Corey in 1951.
- Found in fibrous proteins (e.g. α-keratin and collagen) for most of their length and in some portions of globular proteins.

b. Beta (β) Pleated sheet

Beta (β) pleated sheet = Secondary protein structure which is a sheet of antiparallel chains folded into accordion pleats.
- Parallel regions are held together by either intrachain or interchain hydrogen bonds (between adjacent polypeptides).
- Make up the dense core of many globular proteins (e.g. lysozyme) and the major portion of some fibrous proteins (e.g. fibroin, the structural protein of silk).

3. Tertiary Structure

Tertiary structure = Irregular contortions of a protein due to bonding between side chains (R groups); third level of protein structure superimposed upon primary and secondary structure.

Types of bonds contributing to tertiary structure are weak interactions and covalent linkage (both may occur in the same protein).

a. Weak Interactions

Protein shape is stabilized by the cumulative effect of weak interactions. These weak interactions include:
- Hydrogen bonding between polar side chains.
- Ionic bonds between charged side chains.
- *Hydrophobic interactions* between nonpolar side chains in protein's interior.

Hydrophobic interactions = (Hydro = water; phobos = fear) The clustering of hydrophobic molecules as a result of their mutual exclusion from water.

b. Covalent Linkage

Disulfide bridges form between two cysteine monomers brought together by folding of the protein. This is a strong bond that reinforces conformation.

$$H_3N^+-\overset{\overset{\displaystyle H}{|}}{C}-\overset{\overset{\displaystyle O}{/\!/}}{C} \qquad -\overset{|}{C}-CH_2-S \longrightarrow S-CH_2-\overset{|}{C}-$$

Cysteine

Disulfide Bridge
(S of one cysteine sulfhydryl, bonds
to the S of a second cysteine.)

4. Quaternary Structure

Quaternary structure = Structure that results from the interaction among several polypeptides (subunits) in a single protein.

- For example: collagen, a fibrous protein with three helical polypeptides supercoiled into a triple helix. Found in animal connective tissue, collagen's supercoiled quaternary structure gives it strength.
- Some globular proteins have subunits that fit tightly together. For example: hemoglobin, a globular protein that has four subunits (two α chains and two β chains).

B. What Determines Protein Conformation?

A protein's three-dimensional shape is a consequence of the interactions responsible for secondary and tertiary structure.

- This conformation is influenced by physical and chemical environmental conditions.
- If a protein's environment is altered, it may become *denatured* and lose its native conformation.

Denaturation = A process that alters a protein's native conformation and biological activity. Proteins can be denatured by:

- Transfer to an organic solvent. Hydrophobic side chains, normally inside the protein's core, move towards the outside. Hydrophilic side chains turn away from the solvent towards the molecule's interior.
- Chemical agents that disrupt hydrogen bonds, ionic bonds and disulfide bridges.
- Excessive heat. Increased thermal agitation disrupts weak interactions.

The fact that some denatured proteins return to their native conformation when environmental conditions return to normal is evidence that a protein's amino acid sequence (primary structure) determines conformation. It influences where and which interactions will occur as the molecule arranges into secondary and tertiary structure.

C. The Protein-Folding Problem

Even though primary structure ultimately determines a protein's conformation, three-dimensional shape is difficult to predict solely on the basis of amino acid sequence. It is difficult to find the rules of protein folding because:

- Most proteins pass through several intermediate stages in the folding process; knowledge of the final conformation does not reveal the folding process required to create it.
- A protein's native conformation may be dynamic, alternating between several shapes.

Using recently developed techniques, researchers hope to gain new insights into protein folding:

- Biochemists can now track a protein as it passes through its intermediate stages during the folding process.
- *Chaperone proteins* have just been discovered that temporarily brace a folding protein.

Rules of protein folding are important to molecular biologists and the biotechnology industry. This knowledge should allow the design of proteins for specific purposes.

VIII. Nucleic acids store and transmit hereditary information

Protein conformation is determined by primary structure. Primary structure, in turn, is determined by *genes* — hereditary units that consist of DNA, a type of *nucleic acid*.

There are two types of nucleic acids.

1. Deoxyribonucleic Acid (DNA)
   - Contains coded information that programs all cell activity.
   - Contains directions for its own replication.
   - Is copied and passed from one generation of cells to another.
   - In eukaryotic cells, is found primarily in the nucleus.
   - Makes up *genes* that contain instructions for protein synthesis. Genes do not directly make proteins, but direct the synthesis of mRNA.
2. Ribonucleic Acid (RNA)
   - Functions in the actual synthesis of proteins coded for by DNA.
   - Sites of protein synthesis are on *ribosomes* in the cytoplasm.
   - Messenger RNA (mRNA) carries encoded genetic message from the nucleus to the cytoplasm.
   - The flow of genetic information goes from DNA → RNA → protein.

Nucleus	mRNA	*Cytoplasm*
Genetic message is *transcribed* from DNA onto mRNA	⟶ moves into cytoplasm	Genetic message *translated* into a protein.

IX. A DNA strand is a polymer with an information-rich sequence of nucleotides

<u>Nucleic acid</u> = Polymer of *nucleotides* linked together by condensation reactions.

<u>Nucleotide</u> = Building block molecule of a nucleic acid; made of (1) a five-carbon sugar covalently bonded to (2) a phosphate group and (3) a nitrogenous base.

1. **Pentose (5-Carbon Sugar)**

   There are two pentoses found in nucleic acids: ribose and deoxyribose.

   Ribose is the pentose in RNA.
   Deoxyribose is the pentose in DNA. (It lacks the –OH group at number two carbon.)

2. **Phosphate**

   The phosphate group is attached to the number 5 carbon of the sugar.

3. **Nitrogenous Base**

   There are two families of *nitrogenous bases*:

   Pyrimidine = Nitrogenous base characterized by a six-membered ring made up of carbon and nitrogen atoms. For example:
   - Cytosine (C)
   - Thymine (T) — found only in DNA
   - Uracil (U) — found only in RNA.

   Purine = Nitrogenous base characterized by a five-membered ring fused to a six-membered ring. For example:
   - Adenine (A)
   - Guanine (G)

Nucleotides have various functions:
- Are monomers for nucleic acids.
- Transfer chemical energy from one molecule to another (e.g. ATP).
- Are electron acceptors in enzyme-controlled redox reactions of the cell (e.g. NAD).

DNA is a polymer of nucleotides joined by *phosphodiester linkages* between the phosphate of one nucleotide and the sugar of the next.
- Results in a backbone with a repeating pattern of sugar-phosphate-sugar-phosphate.
- Variable nitrogenous bases are attached to the sugar-phosphate backbone.
- Each gene contains a unique linear sequence of nitrogenous bases which codes for a unique linear sequence of amino acids in a protein.

Phosphate
Base
Sugar

X.   Inheritance is based on precise replication of DNA

In 1953, James Watson and Francis Crick proposed the *double helix* as the three dimensional structure of DNA.

- Consists of two nucleotide chains wound in a double helix.
- Sugar-phosphate backbones are on the outside of the helix.
- Nitrogenous bases are paired in the interior of the helix and are held together by hydrogen bonds.
- Base-pairing rules are that adenine (A) always pairs with thymine (T); guanine (G) always pairs with cytosine (C).
- Two strands of DNA are complimentary and thus can serve as templates to make new complementary strands. It is this mechanism of precise copying that makes inheritance possible.
- Most DNA molecules are long — with thousands or millions of base pairs.

XI.   We can use DNA and proteins as tape measures of evolution

Closely related species have more similar sequences of DNA and amino acids, than more distantly related species. Using this type of molecular evidence, biologists can deduce evolutionary relationships among species.

NOTE: Chapters 15 and 16 are devoted to DNA and protein synthesis. Since any discussion of DNA function must include the details of DNA structure, it may be more practical and less time-consuming to cover nucleic acids later in the course.

REFERENCES

Atkins, P.W. *Molecules*. New York, Oxford: W.H. Freeman and Company, 1987. This delightful book on molecules is packed with useful information and "hooks" to add interest to your lectures.

Campbell, N. *Biology*. 4th ed. Menlo Park, California: Benjamin/Cummings, 1996.

Lehninger, A.L., D.L. Nelson and M.M. Cox. *Principles of Biochemistry*. 2nd ed. New York: Worth, 1993.

Whitten, K.W., K.D. Gailey and R.E. Davis. *General Chemistry*. 4th ed. New York: Saunders, 1992.

CHAPTER **6**

AN INTRODUCTION TO METABOLISM

OUTLINE

I. The chemistry of life is organized into metabolic pathways

II. Organisms transform energy

III. The energy transformations of life are subject to two laws of thermodynamics

IV. Organisms live at the expense of free energy
   A. Free Energy: A Criterion for Spontaneous Change
   B. Free Energy and Equilibrium
   C. Free Energy and Metabolism

V. ATP powers cellular work by coupling exergonic to endergonic reactions
   A. The Structure and Hydrolysis of ATP
   B. How ATP Performs Work.
   C. The Regeneration of ATP

VI. Enzymes speed up metabolic reactions by lowering energy barriers: *an overview*

VII. Enzymes are substrate-specific: *a closer look*

VIII. The active site is an enzyme's catalytic center: *a closer look*

IX. A cell's chemical and physical environment affects enzyme activity: *a closer look*
   A. Effects of Temperature and pH
   B. Cofactors
   C. Enzyme Inhibitors
   D. Allosteric Regulation
   E. Cooperativity

X. Metabolic order emerges from the cell's regulatory systems and structural organization
   A. Feedback Inhibition
   B. Structural Order and Metabolism

OBJECTIVES

After reading this chapter and attending lecture, the student should be able to:

1. Explain the role of catabolic and anabolic pathways in the energy exchanges of cellular metabolism.
2. Distinguish between kinetic and potential energy.
3. Distinguish between open and closed systems.
4. Explain, in their own words, the First and Second Laws of Thermodynamics.
5. Explain why highly ordered living organisms do not violate the Second Law of Thermodynamics.
6. Distinguish between entropy and enthalpy.
7. Write the Gibbs equation for free energy change.
8. Explain how changes in enthalpy, entropy and temperature influence the maximum amount of useable energy that can be harvested from a reaction.
9. Explain the usefulness of free energy.
10. List two major factors capable of driving spontaneous processes.
11. Distinguish between exergonic and endergonic reactions.
12. Describe the relationship between equilibrium and free energy change for a reaction.
13. Describe the function of ATP in the cell.
14. List the three components of ATP and identify the major class of macromolecules to which it belongs.
15. Explain how ATP performs cellular work.
16. Explain why chemical disequilibrium is essential for life.
17. Describe the energy profile of a chemical reaction including activation energy (E_A), free energy change (ΔG) and transition state.
18. Describe the function of enzymes in biological systems.
19. Explain the relationship between enzyme structure and enzyme specificity.
20. Explain the *induced fit* model of enzyme function and describe the catalytic cycle of an enzyme.
21. Describe several mechanisms by which enzymes lower activation energy.
22. Explain how substrate concentration affects the rate of an enzyme-controlled reaction.
23. Explain how enzyme activity can be regulated or controlled by environmental conditions, cofactors, enzyme inhibitors and allosteric regulators.
24. Distinguish between allosteric activation and cooperativity.
25. Explain how metabolic pathways are regulated.

KEY TERMS

metabolism	bond energy	equilibrium	cofactors
catabolic pathways	heat of reaction (ΔH)	catalyst	coenzymes
anabolic pathways	enthalpy	enzyme	competitive
kinetic energy	exothermic	activation energy	inhibitors
potential energy	endothermic	transition state	noncompetitive
thermodynamics	spontaneous reaction	substrate	inhibitors
closed system	free energy	active site	cooperativity
open system	exergonic	induced fit	feedback inhibition
entropy	endergonic	saturation	

LECTURE NOTES

I.   The chemistry of life is organized into metabolic pathways

Metabolism = Totality of an organism's chemical processes.
- Property emerging from specific molecular interactions within the cell.
- Concerned with managing cellular resources: material and energy.

Metabolic reactions are organized into pathways that are orderly series of enzymatically controlled reactions. *Metabolic pathways* are generally of two types:

Catabolic pathways = Metabolic pathways which *release energy* by breaking down complex molecules to simpler compounds. (e.g. Cellular respiration which degrades glucose to carbon dioxide and water; provides energy for cellular work.)

Anabolic pathways = Metabolic pathways which *consume energy* to build complicated molecules from simpler ones. (e.g. Photosynthesis which synthesizes glucose from CO_2 and H_2O; any synthesis of a macromolecule from its monomers.)

Metabolic reactions may be coupled, so that energy released from a catabolic reaction can be used to drive an anabolic one.

NOTE: Figure 9.1 is available as a transparency master in the Campbell set. We have found it useful to use it at this point to illustrate energy exchanges in metabolic reactions. When respiration is introduced, the transparency can be used again as a transition from the material in this chapter to that in chapters nine and ten.

II.   Organisms transform energy

Energy = Capacity to do work.

Kinetic energy = Energy in the process of doing work (energy of motion). For example:
- Heat (thermal energy) is kinetic energy expressed in random movement of molecules.
- Light energy from the sun is kinetic energy which powers photosynthesis.

Potential energy = Energy that matter possesses because of its location or arrangement (energy of position). For example:
- In the earth's gravitational field, an object on a hill or water behind a dam have potential energy.
- Chemical energy is potential energy stored in molecules because of the arrangement of nuclei and electrons in its atoms.

Energy can be transformed from one form to another. For example:
- Kinetic energy of sunlight can be transformed into the potential energy of chemical bonds during photosynthesis.
- Potential energy in the chemical bonds of gasoline can be transformed into kinetic mechanical energy which pushes the pistons of an engine.

III. The energy transformations of life are subject to two laws of thermodynamics

Thermodynamics = Study of energy transformations.

First Law of Thermodynamics = Energy can be transferred and transformed, but it cannot be created or destroyed (energy of the universe is constant).

Second Law of Thermodynamics = Every energy transfer or transformation makes the universe more disordered (every process increases the *entropy* of the universe).

Entropy = Quantitative measure of disorder that is proportional to randomness (designated by the letter S).

Closed system = Collection of matter under study which is isolated from its surroundings.

Open system = System in which energy can be transferred between the system and its surroundings.

NOTE: It is important to distinguish between open and closed systems, and to spend lecture time on the second law of thermodynamics. Be advised that the question often comes up: "How is the evolution of complex life forms possible since it violates the second law of thermodynamics?" (See question 1, at the end of the chapter under Science, Technology and Society.) Thoughtful preparation of an answer beforehand will be well worth the effort.

The entropy of a system may decrease, but the entropy of the system *plus its surroundings* must always increase. Highly ordered living organisms do not violate the second law because they are open systems. For example, animals:

- Maintain highly ordered structure at the expense of increased entropy of their surroundings.
- Take in complex high energy molecules as food and extract chemical energy to create and maintain order.
- Return to the surroundings simpler low energy molecules (CO_2 and water) and heat.

Energy can be transformed, but part of it is dissipated as heat which is largely unavailable to do work. Heat energy *can* perform work if there is a heat gradient resulting in heat flow from warmer to cooler.

Combining the first and second laws; the *quantity* of energy in the universe is constant, but its *quality* is not.

IV. Organisms live at the expense of free energy

A. Free Energy: A Criterion For Spontaneous Change

Not all of a system's energy is available to do work. The amount of energy that is available to do work is described by the concept of *free energy*. Free energy (G) is related to the system's total energy (H) and its entropy (S) in the following way:

$$G = H - TS$$

where:

G = Gibbs free energy (energy available to do work)
H = enthalpy or total energy
T = temperature in °K
S = entropy

<u>Free energy (G)</u> = Portion of a system's energy available to do work; is the difference between the total energy (*enthalpy*) and the energy *not* available for doing work (TS).

The maximum amount of useable energy that can be harvested from a particular reaction is the system's free energy change from the initial to the final state. This change in free energy (ΔG) is given by the Gibbs-Helmholtz equation at constant temperature and pressure:

$$\Delta G = \Delta H - T\Delta S$$

where:

ΔG = change in free energy
ΔH = change in total energy (enthalpy)
ΔS = change in entropy
T = absolute temperature in °K (which is °C + 273)

NOTE: To put these thermodynamic concepts in the context of chemical reactions, we also briefly discuss the other component of the Gibbs-Helmholtz equation — ΔH or change in enthalpy measured as the *heat of reaction*. Students should understand that during a chemical reaction, reactant molecules must absorb energy for their bonds to break, and that energy is released when bonds form between the rearranged atoms of the products. Consequently, the net energy consumed or released when reactants are converted to products is the *net difference* between the energy consumed to break chemical bonds of reactants and the energy released from the formation of the products.

Significance of Free Energy:

1. Indicates the maximum amount of a system's energy which is available to do work.
2. Indicates whether a reaction will occur spontaneously or not.

- A *spontaneous reaction* is one that will occur without additional energy.
- In a spontaneous process, ΔG or free energy of a system *decreases* (ΔG<0).
- A decrease in enthalpy ($-\Delta$H) and an increase in entropy ($+\Delta$S) reduce the free energy of a system and contribute to the spontaneity of a process.
- A higher temperature enhances the effect of an entropy change. Greater kinetic energy of molecules tends to disrupt order as the chances for random collisions increase.

- When enthalpy and entropy changes in a system have an opposite effect on free energy, temperature may determine whether the reaction will be spontaneous or not (e.g. protein denaturation by increased temperature).
- High energy systems, including high energy chemical systems, are unstable and tend to change to a more stable state with a lower free energy.

B.   Free Energy and Equilibrium

There is a relationship between chemical equilibrium and the free energy change (ΔG) of a reaction:

- As a reaction approaches equilibrium, the free energy of the system decreases (spontaneous and exergonic reaction).
- When a reaction is pushed away from equilibrium, the free energy of system increases (non-spontaneous and endergonic reaction).
- When a reaction reaches equilibrium, $\Delta G = 0$, because there is no net change in the system.

Metabolic Disequilibrium   Since many metabolic reactions are reversible, they have the potential to reach equilibrium.

- At equilibrium, $\Delta G = 0$, so the system can do no work.
- Metabolic disequilibrium is a necessity of life; a cell at equilibrium is dead.
- In the cell, these potentially reversible reactions are pulled forward away from equilibrium, because the products of some reactions become reactants for the next reaction in the metabolic pathway.
- For example, during cellular respiration a steady supply of high energy reactants such as glucose and removal of low energy products such as CO_2 and H_2O, maintain the disequilibrium necessary for respiration to proceed.

C.   Free Energy and Metabolism

Reactions can be classified based upon their free energy changes:

<u>Exergonic reaction</u> = A reaction that proceeds with a net loss of free energy.

<u>Endergonic reaction</u> = An energy-requiring reaction that proceeds with a net gain of free energy; a reaction that absorbs free energy from its surroundings.

Exergonic Reaction	Endergonic Reaction
Chemical products have <u>less</u> free energy than the reactant molecules.	Products store <u>more</u> free energy than reactants.
Reaction is energetically downhill.	Reaction is energetically uphill.
Spontaneous reaction.	Non-spontaneous reaction (requires energy input).
ΔG is negative.	ΔG is positive.
$-\Delta G$ is the maximum amount of work the reaction can perform.	$+\Delta G$ is the minimum amount of work required to drive the reaction.

If a chemical process is exergonic, the reverse process must be endergonic. For example:

- For each mole of glucose oxidized in the exergonic process of cellular respiration, 2870 kJ are released ($\Delta G = -2870$ kJ/mol or -686 kcal/mol).
- To produce a mole of glucose, the endergonic process of photosynthesis requires an energy input of 2870 kJ ($\Delta G = +2870$ kJ/mol or $+686$ kcal/mol).

NOTE: From this point on, the Campbell text uses joules and kilojoules as energy units and puts the caloric equivalent in parentheses. The *joule* (J) is the metric unit of energy; some handy conversions follow:

$$
\begin{aligned}
\text{joule } (J) &= 0.239 \text{ cal} \\
\text{Kilojoule } (kJ) &= 1000 \text{ J or } 0.239 \text{ kcal} \\
\text{calorie (cal)} &= 4.184 \text{ J}
\end{aligned}
$$

In cellular metabolism, endergonic reactions are driven by coupling them to reactions with a greater negative free energy (exergonic). ATP plays a critical role in this energy coupling.

V. ATP powers cellular work by coupling exergonic to endergonic reactions

ATP is the immediate source of energy that drives most cellular work, which includes:

1. *Mechanical work* such as beating of cilia, muscle contraction, cytoplasmic flow, and chromosome movement during mitosis and meiosis.
2. *Transport work* such as pumping substances across membranes.
3. *Chemical work* such as the endergonic process of polymerization.

A. The Structure and Hydrolysis of ATP

ATP (adenosine triphosphate) = Nucleotide with unstable phosphate bonds that the cell hydrolyzes for energy to drive endergonic reactions. ATP consists of:

- Adenine, a nitrogenous base.
- Ribose, a five-carbon sugar.
- Chain of three phosphate groups.

Unstable bonds between the phosphate groups can be hydrolyzed in an exergonic reaction that releases energy.

- When the terminal phosphate bond is hydrolyzed, a phosphate group is removed producing ADP (adenosine diphosphate).

$$\text{ATP} + \text{H}_2\text{O} \longrightarrow \text{ADP} + \textcircled{P}$$

- Under standard conditions in the laboratory, this reaction releases –31 kJ/mol (–7.3 kcal/mol).
- In a living cell, this reaction releases –55 kJ/mol (–13 kcal/mol) — about 77% more than under standard conditions.

The terminal phosphate bonds of ATP are unstable, so:
- The products of the hydrolysis reaction are more stable than the reactant.
- Hydrolysis of the phosphate bonds is thus exergonic as the system shifts to a more stable state.

B.   How ATP Performs Work

Exergonic hydrolysis of ATP is coupled with endergonic processes by transferring a phosphate group to another molecule.

- Phosphate transfer is enzymatically controlled.
- The molecule acquiring the phosphate (*phosphorylated* or *activated intermediate*) becomes more reactive.

For example, conversion of glutamic acid to glutamine: (See Campbell, Figure 6.7)

$$Glu \quad + \quad NH_3 \quad \longrightarrow \quad Gln \qquad \Delta G = +14.2 \text{ kJ/mol} (+3.4 \text{ kcal/mol})$$

glutamic acid   ammonia   glutamine   (endergonic)

Two step process of energy coupling with ATP hydrolysis:

(1) Hydrolysis of ATP and phosphorylation of glutamic acid.

$$Glu \quad + \quad ATP \quad \longrightarrow \quad Glu–\text{\textcircled{P}} \quad + \quad ADP$$

unstable
phosphorylated
intermediate

(2) Replacement of the phosphate with the reactant ammonia.

$$Glu–\text{\textcircled{P}} \quad + \quad NH_3 \quad \longrightarrow \quad Gln \quad + \quad \text{\textcircled{P}}$$

Overall ΔG:

$$Glu \quad + \quad NH_3 \quad \longrightarrow \quad Gln \qquad\qquad \Delta G = +14.2 \text{ kJ/mol}$$
$$ATP \quad \longrightarrow \quad ADP \quad + \quad \text{\textcircled{P}} \qquad \Delta G = \underline{-31.0 \text{ kJ/mol}}$$

Net ΔG = – 16.8 kJ/mol
(Overall process is exergonic.)

C.   The Regeneration of ATP

ATP is continually regenerated by the cell.

- Process is rapid (10^7 molecules used and regenerated/sec/cell).
- Reaction is endergonic.

$$ADP \quad + \quad \text{\textcircled{P}} \quad \longrightarrow \quad ATP \qquad \Delta G = +31 \text{ kJ/mol} (+7.3 \text{ kcal/mol})$$

- Energy to drive the endergonic regeneration of ATP comes from the exergonic process of cellular respiration.

VI. Enzymes speed up metabolic reactions by lowering energy barriers: *an overview*

Free energy change indicates whether a reaction will occur spontaneously, but does not give information about the speed of reaction.

- A chemical reaction will occur spontaneously if it releases free energy ($-\Delta G$), but it may occur too slowly to be effective in living cells.
- Biochemical reactions require *enzymes* to speed up and control reaction rates.

<u>Catalyst</u> = Chemical agent that accelerates a reaction without being permanently changed in the process, so it can be used over and over.

<u>Enzymes</u> = Biological catalysts, which are usually proteins.

Before a reaction can occur, the reactants must absorb energy to break chemical bonds. This initial energy investment is the *activation energy*.

<u>Free energy of activation</u> (*Activation energy*) = Amount of energy that reactant molecules must absorb to start a reaction (E_A).

<u>Transition state</u> = Unstable condition of reactant molecules that have absorbed sufficient free energy to react.

<u>Energy profile of an exergonic reaction:</u>

1. Reactants must absorb enough energy (E_A) to reach the transition state (uphill portion of the curve). Usually the absorption of thermal energy from the surroundings is enough to break chemical bonds.

2. Reaction occurs and energy is released as new bonds form (downhill portion of the curve).

3. ΔG for the overall reaction is the difference in free energy between products and reactants. In an exergonic reaction the free energy of the products is less than reactants.

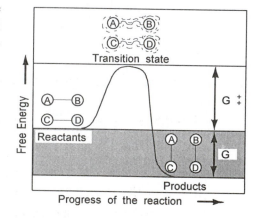

Even though a reaction is energetically favorable, there must be an initial investment of activation energy (E_A).

The breakdown of biological macromolecules is exergonic. However, these molecules react *very slowly* at cellular temperatures because they cannot absorb enough thermal energy to reach transition state.

In order to make these molecules reactive when necessary, cells use biological catalysts called *enzymes*, which:

- Are usually proteins.
- Lower E_A, so the transition state can be reached at cellular temperatures.
- Do *not change* the nature of a reaction (ΔG), but only speed up a reaction that would have occurred anyway.
- Are very selective for which reaction they will catalyze.

VII. **Enzymes are substrate-specific:** *a closer look*

Enzymes are specific for a particular *substrate*, and that specificity depends upon the enzyme's three-dimensional shape.

Substrate = The substance an enzyme acts on and makes more reactive.
- An enzyme binds to its substrate and catalyzes its conversion to product. The enzyme is released in original form.

    Substrate + enzyme ⟶ enzyme-substrate complex ⟶ product + enzyme
- The substrate binds to the enzyme's *active site*.

Active site = Restricted region of an enzyme molecule which binds to the substrate.
- Is usually a pocket or groove on the protein's surface.
- Formed with only a few of the enzyme's amino acids.
- Determines enzyme specificity which is based upon a compatible fit between the shape of an enzyme's active site and the shape of the substrate.
- Changes its shape in response to the substrate.
    ⇒ As substrate binds to the active site, it *induces* the enzyme to change its shape.
    ⇒ This brings its chemical groups into positions that enhance their ability to interact with the substrate and catalyze the reaction.

Induced fit = Change in the shape of an enzyme's active site, which is induced by the substrate.

VIII. **The active site is an enzyme's catalytic center:** *a closer look*

The entire enzymatic cycle is quite rapid. (See Campbell, Figure 6.12)

Steps in the Catalytic Cycle of Enzymes:
1. Substrate binds to the active site forming an *enzyme-substrate complex*. Substrate is held in the active site by weak interactions (e.g. hydrogen bonds and ionic bonds).
2. *Induced fit* of the active site around the substrate. Side chains of a few amino acids in the active site catalyze the conversion of substrate to product.
3. Product departs active site and the enzyme emerges in its original form. Since enzymes are used over and over, they can be effective in very small amounts.

Enzymes lower activation energy and speed up reactions by several mechanisms:
- Active site can hold two or more reactants in the proper position so they may react.
- Induced fit of the enzyme's active site may distort the substrate's chemical bonds, so less thermal energy (lower ΔG) is needed to break them during the reaction.
- Active site might provide a micro-environment conducive to a particular type of reaction (e.g. localized regions of low pH caused by acidic side chains on amino acids at the active site).
- Side chains of amino acids in the active site may participate directly in the reaction.

The initial substrate concentration partly determines the rate of an enzyme controlled reaction.
- The higher the substrate concentration, the faster the reaction - up to a limit.
- If substrate concentration is high enough, the enzyme becomes *saturated* with substrate. (The active sites of all enzymes molecules are engaged.)
- When an enzyme is saturated, the reaction rate depends upon how fast the active sites can convert substrate to product.
- When enzyme is saturated, reaction rate may be increased by adding more enzyme.

IX. A cell's chemical and physical environment affects enzyme activity: *a closer look*

Each enzyme has optimal environmental conditions that favor the most active enzyme conformation.

A. Effects of Temperature and pH

Optimal temperature allows the greatest number of molecular collisions without denaturing the enzyme.
- Enzyme reaction rate increases with increasing temperature. Kinetic energy of reactant molecules increases with rising temperature, which increases substrate collisions with active sites.
- Beyond the optimal temperature, reaction rate slows. The enzyme denatures when increased thermal agitation of molecules disrupts weak bonds that stabilize the active conformation.
- Optimal temperature range of most human enzymes is 35°– 40°C.

Optimal pH range for most enzymes is pH 6 – 8.
- Some enzymes operate best at more extremes of pH.
- For example, the digestive enzyme, pepsin, found in the acid environment of the stomach has an optimal pH of 2.

B. Cofactors

Cofactors = Small nonprotein molecules that are required for proper enzyme catalysis.
- May bind tightly to active site.
- May bind loosely to both active site and substrate.
- Some are inorganic (e.g. metal atoms of zinc, iron or copper).
- Some are organic and are called *coenzymes* (e.g. most vitamins).

C. **Enzyme Inhibitors**

Certain chemicals can selectively inhibit enzyme activity. (See Campbell, Figure 6.14)
- Inhibition may be *irreversible* if the inhibitor attaches by covalent bonds.
- Inhibition may be *reversible* if the inhibitor attaches by weak bonds.

Competitive inhibitors = Chemicals that resemble an enzyme's normal substrate and compete with it for the active site.
- Block active site from the substrate.
- If reversible, the effect of these inhibitors can be overcome by increased substrate concentration.

Noncompetitive inhibitors = Enzyme inhibitors that do not enter the enzyme's active site, but bind to another part of the enzyme molecule.
- Causes enzyme to change its shape so the active site cannot bind substrate.
- May act as metabolic poisons (e.g. DDT, many antibiotics).
- Selective enzyme inhibition is an essential mechanism in the cell for regulating metabolic reactions.

D. **Allosteric Regulation**

Allosteric site = Specific receptor site on some part of the enzyme molecule other than the active site.
- Most enzymes with allosteric sites have two or more polypeptide chains, each with its own active site. Allosteric sites are often located where the subunits join.
- Allosteric enzymes have two conformations, one catalytically active and the other inactive. (See Campbell, Figure 6.15)
- Binding of an *activator* to an allosteric site stabilizes the active conformation.
- Binding of an *inhibitor* (noncompetitive inhibitor) to an allosteric site stabilizes the inactive conformation.
- Enzyme activity changes continually in response to changes in the relative proportions of activators and inhibitors (e.g. ATP/ADP).
- Subunits may interact so that a single activator or inhibitor at one allosteric site will affect the active sites of the other subunits.

E. **Cooperativity**

Substrate molecules themselves may enhance enzyme activity.

Cooperativity = The phenomenon where substrate binding to the active site of one subunit induces a conformational change that enhances substrate binding at the active sites of the other subunits.

X. **Metabolic order emerges from the cell's regulatory systems and structural organization**

Metabolic pathways are regulated by controlling enzyme activity.

A. **Feedback Inhibition**

<u>Feedback inhibition</u> = Regulation of a metabolic pathway by its end product, which inhibits an enzyme within the pathway. For example:

Prevents the cell from wasting chemical resources by synthesizing more product than is necessary.

B. **Structural Order and Metabolism**

Cellular structure orders and compartmentalizes metabolic pathways.

- Some enzymes have fixed locations in the cell because they are incorporated into a membrane.
- Dissolved enzymes and their substrates may be localized within organelles such as chloroplasts and mitochondria .

REFERENCES

Atkins, P.W. *The Second Law*. New York, Oxford: W.H. Freeman and Company, 1984. A beautifully written, understandable description of the Second Law of Thermodynamics; addresses the role of the Second Law in life processes.

Campbell, N. *Biology*. 4th ed. Menlo Park, California: Benjamin/Cummings, 1996.

Lehninger, A.L., D.L. Nelson and M.M. Cox. *Principles of Biochemistry*. 2nd ed. New York: Worth, 1993.

CHAPTER 7

A TOUR OF THE CELL

OUTLINE

I. Microscopes provide windows to the world of the cell

II. Cell biologists can isolate organelles to study their function

III. A panoramic view of the cell
   A. Prokaryotic and Eukaryotic Cells
   B. Cell Size
   C. The Importance of Compartmental Organization

IV. The nucleus contains a cell's genetic library

V. Ribosomes build a cell's proteins

VI. Many organelles are related through the endomembrane system

VII. The endoplasmic reticulum manufactures membranes and performs many other biosynthetic functions
   A. Functions of Smooth ER
   B. Rough ER and Protein Synthesis
   C. Rough ER and Membrane Production

VIII. The Golgi apparatus finishes, sorts, and ships many products of the cell

IX. Lysomes are digestive compartments

X. Vacuoles have diverse functions in cell maintenance

XI. Peroxisomes consume oxygen in various metabolic functions

XII. Mitochondria and chloroplasts are the main energy transformers of cells

XIII. The cytoskeleton provides structural support and functions in cell motility
    A.   Microtubules
    B.   Microfilaments (Actin Filaments)
    C.   Intermediate Filaments

XIV. Plant cells are encased by cell walls

XV. The extracellular matrix (ECM) of animal cells functions in support, adhesion, movement and development

XVI. Intercellular junctions integrate cells into higher levels of structure and function

XVII. The cell is a living unit greater than the sum of its parts

OBJECTIVES

After reading this chapter and attending lecture, the student should be able to:

1. Describe techniques used to study cell structure and function.
2. Distinguish between magnification and resolving power.
3. Describe the principles, advantages and limitations of the light microscope, transmission electron microscope and the scanning electron microscope.
4. Describe the major steps of cell fractionation and explain why it is a useful technique.
5. Distinguish between prokaryotic and eukaryotic cells.
6. Explain why there are both upper and lower limits to cell size.
7. Explain why compartmentalization is important in eukaryotic cells.
8. Describe the structure and function of the nucleus, and briefly explain how the nucleus controls protein synthesis in the cytoplasm.
9. Describe the structure and function of a eukaryotic ribosome.
10. List the components of the *endomembrane system*, describe their structures and functions and summarize the relationships among them.
11. Explain how impaired lysosomal function causes the symptoms of storage diseases.
12. Describe the types of vacuoles and explain how their functions differ.
13. Explain the role of *peroxisomes* in eukaryotic cells.
14. Describe the structure of a *mitochondrion* and explain the importance of compartmentalization in mitochondrial function.
15. Distinguish among *amyloplast*, *chromoplast* and *chloroplast*.
16. Identify the three functional compartments of a chloroplast, and explain the importance of compartmentalization in chloroplast function.
17. Describe probable functions of the cytoskeleton.
18. Describe the structure, monomers and functions of microtubules, microfilaments and intermediate filaments.
19. Explain how the ultrastructure of cilia and flagella relates to their function.
20. Describe the development of plant cell walls.
21. Describe the structure and list some functions of the extracellular matrix in animal cells.
22. Describe the structure of intercellular junctions found in plant and animal cells, and relate their structure to function.

KEY TERMS

magnification
resolving power
light microscope
scanning electron
   microscope
transmission electron
   microscope
cell fractionation
cytoplasm
cytosol
nucleus
nuclear envelope
chromatin
chromosome
nucleolus
nucleolar organizer
ribosome
endomembrane system
vesicles

endoplasmic
   reticulum (ER)
smooth ER
rough ER
Golgi apparatus
cis face
trans face
lysosomes
phagocytosis
macrophages
food vacuole
contractile vacuole
central vacuole
tonoplast
peroxisome
mitochondria
intermembrane space
mitochondrial matrix

plastid
amyloplast
chromoplast
chloroplast
thylakoid
thylakoid space
grana
stroma
cytoskeleton
microtubules
microfilaments
intermediate filaments
motor molecule
centriole
cilia
flagella
tubulin
centrosome

dynein
actin
myosin
basal body
cytoplasmic
   streaming
cell wall
primary cell wall
secondary cell wall
extracellular matrix
collagen
proteoglycan
fibronectin
integrin
plasmodesmata
desmosomes
tight junctions
gap junctions

LECTURE NOTES

All organisms are made of cells, the organism's basic unit of structure and function.

The cell as a microcosm can be used to illustrate four themes integral to the text and course:

1. Theme of emergent properties. Life at the cellular level arises from interactions among cellular components.

2. Correlation of structure and function. Ordered cellular processes (e.g. protein synthesis, respiration, photosynthesis, cell-cell recognition, cellular movement, membrane production and secretion) are based upon ordered structures.

3. Interaction of organisms within their environment. Cells are excitable responding to environmental stimuli. In addition, cells are open systems that exchange materials and energy with their environment.

4. Unifying theme of evolution. Evolutionary adaptations are the basis for the correlation between structure and function.

NOTE: Students often find this material boring. A good set of micrographs and line drawings in the form of slides or transparencies will help. If the class size is small enough, a tour of an electron microscopy facility will help stimulate interest.

I. **Microscopes provide windows to the world of the cell**

The microscope's invention and improvement in the seventeenth century led to the discovery and study of cells.

In 1665, Robert Hooke described cells using a *light microscope*. Modern light microscopy is based upon the same principles as microscopy first used by Renaissance scientists.

- Visible light is focused on a specimen with a *condenser lens*.
- Light passing through the specimen is refracted with an *objective lens* and an *ocular lens*. The specimen's image is thus magnified and inverted for the observer.

Two important concepts in microscopy are *magnification* and *resolving power*.

<u>Magnification</u> = How much larger an object is made to appear compared to its real size.

<u>Resolving power</u> = Minimum distance between two points that can still be distinguished as two separate points.

- Resolution of a light microscope is limited by the wavelength of visible light. Maximum possible resolution of a light microscope is 0.2 µm.
- Highest magnification in a light microscope with maximum resolution is about 1000 times.
- By the early 1900's, optics in light microscopes were good enough to achieve the best resolution, so improvements since then have focused on improving contrast.

In the 1950's, researchers began to use the *electron microscope* which far surpassed the resolving power of the light microscope.

- Resolving power is inversely related to wavelength. Instead of light, electron microscopes use electron beams which have much shorter wavelengths than visible light.
- Modern electron microscopes have a resolving power of about 0.2nm.
- Enhanced resolution and magnification allowed researchers to clearly identify subcellular *organelles* and to study cell *ultrastructure*.
- Two types of electron microscopes are the *transmission electron microscope* (TEM) and the *scanning electron microscope*.

The *transmission electron microscope* (TEM) aims an electron beam at a thin section of specimen which may be stained with metals to absorb electrons and enhance contrast.

- Electrons *transmitted* through the specimen are focused and the image magnified by using electromagnetic lenses (rather than glass lenses) to bend the trajectories of the charged electrons.
- Image is focused onto a viewing screen or film.
- Used to study internal cellular ultrastructure.

The *scanning electron microscope* (SEM) is useful for studying the surface of a specimen.
- Electron beam *scans* the surface of the specimen usually coated with a thin film of gold.
- Scanning beam excites secondary electrons on the sample's surface.
- Secondary electrons are collected and focused onto a viewing screen.
- SEM has a great depth of field and produces a three-dimensional image.

Disadvantages of an Electron Microscope:
- Can usually only view dead cells because of the elaborate preparation required.
- May introduce structural artifacts.

NOTE: In our laboratory course students are given electron micrographs of organelles to identify and label. Many are disappointed when they view wet mounts of cells or prepared slides with their light microscopes and cannot find the detail seen in the micrographs. Clearly, some students have no conception of the resolution and magnifying power of an electron microscope. It would be helpful to indicate a size scale on micrographs you might use in lecture.

II. Cell biologists can isolate organelles to study their function

Modern cell biology integrates the study of cell structure (*cytology*) with the study of cell function. *Cell fractionation* is a technique that enables researchers to isolate organelles without destroying their function.

Cell fractionation = Technique which involves centrifuging disrupted cells at various speeds and durations to isolate components of different sizes, densities and shapes.
- Development of the *ultracentrifuge* made this technique possible.
- Ultracentrifuges can spin as fast as 80,000 rpm, applying a force of 500,000 g.

The process of cell fractionation involves the following:
- Homogenization of tissue and its cells using pistons, blenders or ultrasound devices.
- Centrifugation of the resulting homogenate at a slow speed. Nuclei and other larger particles settle at the bottom of the tube, forming a *pellet*.
- The unpelleted fluid or *supernatant* is decanted into another tube and centrifuged at a faster speed, separating out smaller organelles.
- The previous step is repeated, increasing the centrifugation speed each time to collect smaller and smaller cellular components from the pellet.
- Once the cellular components are separated and identified, their particular metabolic functions can be determined.

III.   A panoramic view of the cell

> **A.   Prokaryotic and Eukaryotic Cells**
>
> Living organisms are made of either prokaryotic or eukaryotic cells — two major kinds of cells, which can be distinguished by structural organization.

Prokaryotic (pro=before; karyon=kernel)	Eukaryotic (Eu=true; karyon=kernel)
Found only in the Kingdom Monera (bacteria and cyanobacteria).	Found in the Kingdoms Protista, Fungi, Plantae, and Animalia.
No true nucleus; lacks nuclear envelope.	True nucleus; bounded by nuclear envelope.
Genetic material in *nucleoid* region.	Genetic material within nucleus.
No membrane-bound organelles.	Contains cytoplasm with *cytosol* and membrane-bound *organelles*.

> Cytoplasm = Entire region between the nucleus and cell membrane.
>
> Cytosol = Semi-fluid medium found in the cytoplasm.

> **B.   Cell Size**
>
> Size ranges of cells:

Cell Type	Diameter
Mycoplasmas	0.1 – 1.0μm
Most bacteria	1.0 – 10.0μm
Most eukaryotic cells	10.0 – 100.0μm

> Range of cell size is limited by metabolic requirements. The lower limits are probably determined by the smallest size with:
> - Enough DNA to program metabolism.
> - Enough ribosomes, enzymes and cellular components to sustain life and reproduce.
>
> The upper limits of size are imposed by the surface area to volume ratio. As a cell increases in size, its volume grows proportionately more than its surface area. (See Campbell, Figure 7.5)
> - The surface area of the plasma membrane must be large enough for the cell volume, in order to provide an adequate exchange surface for oxygen, nutrients and wastes.

C. The Importance of Compartmental Organization

The average eukaryotic cell has a thousand times the volume of the average prokaryotic cell, but only a hundred times the surface area. Eukaryotic cells compensate for the small surface area to volume ratio by having internal membranes which:

- Partition the cell into compartments.
- Have unique lipid and protein compositions depending upon their specific functions.
- May participate in metabolic reactions since many enzymes are incorporated directly into the membrane.
- Provide localized environmental conditions necessary for specific metabolic processes.
- Sequester reactions, so they may occur without interference from incompatible metabolic processes elsewhere in the cell.

IV. The nucleus contains a cell's genetic library

Nucleus = A generally conspicuous membrane-bound cellular organelle in a eukaryote; contains most of the genes that control the entire cell.

- Averages about 5µm diameter.
- Enclosed by a *nuclear envelope*.

Nuclear envelope = A double membrane which encloses the nucleus in a eukaryotic cell.

- Is two lipid bilayer membranes separated by a space of about 20 to 40 nm. Each lipid bilayer has its own specific proteins.
- Attached to proteins on the envelope's nuclear side is a network of protein filaments, the *nuclear lamina*, which stabilizes nuclear shape.
- Is perforated by pores (100 nm diameter), which are ordered by an octagonal array of protein granules.
    ⇒ The envelope's inner and outer membranes are fused at the lip of each pore.
    ⇒ Pore complex regulates molecular traffic into and out of the nucleus.
- There is new evidence of an intranuclear framework of fibers, the *nuclear matrix*.

The nucleus contains most of the cell's DNA which is organized with proteins into a complex called *chromatin*.

Chromatin = Complex of DNA and histone proteins, which makes up *chromosomes* in eukaryotic cells; appears as a mass of stained material in nondividing cells.

Chromosomes = Long threadlike association of genes, composed of *chromatin* and found in the nucleus of eukaryotic cells.

- Each species has a characteristic chromosome number.
- Human cells have 46 chromosomes, except egg and sperm cells, which have half or 23.

The most visible structure within the nondividing nucleus is the *nucleolus*.

Nucleolus = Roughly spherical region in the nucleus of nondividing cells, which consists of *nucleolar organizers* and ribosomes in various stages of production.
- May be two or more per cell.
- Packages ribosomal subunits from:
  ⇒ rRNA transcribed in the nucleolus.
  ⇒ RNA produced elsewhere in the nucleus.
  ⇒ Ribosomal proteins produced and imported from the cytoplasm.
- Ribosomal subunits pass through nuclear pores to the cytoplasm, where their assembly is completed.

Nucleolar organizers = Specialized regions of some chromosomes, with multiple copies of genes for rRNA (ribosomal RNA) synthesis.

The nucleus controls protein synthesis in the cytoplasm:

Messenger RNA (mRNA) *transcribed* in the nucleus
from DNA instructions.

↓

Passes through nuclear pores into cytoplasm.

↓

Attaches to ribosomes where the genetic message
is *translated* into primary protein structure.

V.   Ribosomes build a cell's proteins

Ribosome = A cytoplasmic organelle which is the site for protein synthesis.
- Are complexes of RNA and protein.
- Constructed in the nucleolus in eukaryotic cells.
- Cells with high rates of protein synthesis have prominent nucleoli and many ribosomes (e.g. human liver cell has a few million).

NOTE: Since most organelles are membrane-bound, students frequently ask if the ribosome has a membrane. They can deductively answer the question themselves if they are reminded that prokaryotes have ribosomes as well.

Ribosomes function either free in the cytosol or bound to endoplasmic reticulum. Bound and free ribosomes are structurally identical and interchangeable.

<u>Free ribosomes</u> = Ribosomes suspended in the cytosol.
- Most proteins made by free ribosomes will function in the cytosol.

<u>Bound ribosomes</u> = Ribosomes attached to the outside of the endoplasmic reticulum.
- Generally make proteins that are destined for membrane inclusion or export.
- Cells specializing in protein secretion often have many bound ribosomes (e.g. pancreatic cells).

VI. Many organelles are related through the endomembrane system

Biologists now consider many membranes of the eukaryotic cell to be part of an *endomembrane system*.
- Membranes may be interrelated <u>directly</u> through physical contact.
- Membranes may be related <u>indirectly</u> through *vesicles*.

<u>Vesicles</u> = Membrane-enclosed sacs that are pinched off portions of membranes moving from the site of one membrane to another.

Membranes of the endomembrane system vary in structure and function, and the membranes themselves are dynamic structures changing in composition, thickness and behavior.

The endomembrane system includes:
- Nuclear envelope.
- Endoplasmic reticulum.
- Golgi apparatus.
- Lysosomes.
- Vacuoles.
- Plasma membrane (not actually an *endomembrane*, but related to endomembrane system).

VII. The endoplasmic reticulum manufactures membranes and performs many other biosynthetic functions

<u>Endoplasmic reticulum (ER)</u> = (Endoplasmic=within the cytoplasm; reticulum= network) Extensive membranous network of tubules and sacs (*cisternae*) which sequesters its internal lumen (*cisternal space*) from the cytosol.
- Most extensive portion of endomembrane system.
- Continuous with the outer membrane of the nuclear envelope; therefore, the space between the membranes of the nuclear envelope is continuous with cisternal space.

There are two distinct regions of ER that differ in structure and function: smooth ER and rough ER.

A. Functions of Smooth ER

Appears smooth in the electron microscope because its cytoplasmic surface lacks ribosomes. Smooth ER functions in diverse metabolic processes:

1. Synthesizes lipids, phospholipids and steroids.
   - For example, mammalian sex hormones and steroids secreted by the adrenal gland.
   - Cells that produce and secrete these products are rich in smooth ER (e.g. testes, ovaries, skin oil glands).

2. Participates in carbohydrate metabolism.
   - Smooth ER in liver contains an embedded enzyme that catalyzes the final step in the conversion of glycogen to glucose (removes the phosphate from glucose-phosphate).

3. Detoxifies drugs and poisons.
   - Smooth ER, especially in the liver, contains enzymes which detoxify drugs and poisons.
   - Enzymes catalyze the addition of hydroxyl groups to drugs and poisons. This makes them soluble in the cytosol, so they may be excreted from the body.
   - Smooth ER in liver cells proliferates in response to barbiturates, alcohol and other drugs. This, in turn, may increase drug tolerance.

4. Stores calcium ions necessary for muscle contraction.
   - In a muscle cell, the ER membrane pumps Ca^{++} from the cytosol into the cisternal space.
   - In response to a nerve impulse, Ca^{++} leaks from the ER back into the cytosol, which triggers muscle cell contraction.

B. Rough ER and Protein Synthesis

Rough ER:
- Appears rough under an electron microscope because the cytoplasmic side is studded with ribosomes.
- Is continuous with outer membrane of the nuclear envelope (which may also be studded with ribosomes on the cytoplasmic side).
- Manufactures secretory proteins and membrane.

Proteins destined for secretion are synthesized by ribosomes attached to rough ER:

Ribosomes attached to rough ER synthesize
secretory proteins.

↓

Growing polypeptide is threaded through ER
membrane into the lumen or *cisternal space*.

↓

Protein folds into its native conformation.

↓

If destined to be a *glycoprotein*, enzymes
localized in the ER membrane catalyze the
covalent bonding of an *oligosaccharide* to the
secretory protein.

↓

Protein departs in a *transport vesicle* pinched off
from *transitional ER* adjacent to the rough ER site of
production.

Glycoprotein = Protein covalently bonded to carbohydrate.

Oligosaccharide = Small polymer of sugar units.

Transport vesicle = Membrane vesicle in transit from one part of the cell to another.

C.   **Rough ER and Membrane Production**

Membranes of rough ER grow *in place* as newly formed proteins and phospholipids are assembled:

- Membrane proteins are produced by ribosomes. As a polypeptide grows, it is inserted directly into the rough ER membrane where it is anchored by hydrophobic regions of the proteins.
- Enzymes within the ER membrane synthesize phospholipids from raw materials in the cytosol.
- Newly expanded ER membrane can be transported as a vesicle to other parts of the cell.

VIII. The Golgi apparatus finishes, sorts, and ships many products of the cell

Many transport vesicles leave the ER and travel to the *Golgi apparatus.*

<u>Golgi apparatus</u> = Organelle made of stacked, flattened membranous sacs (*cisternae*), that modifies, stores and routes products of the endoplasmic reticulum. (See Campbell, Figure 7.12)

- Membranes of the cisternae sequester cisternal space from the cytosol.
- Vesicles may transport macromolecules between the Golgi and other cellular structures.
- Has a distinct polarity. Membranes of cisternae at opposite ends differ in thickness and composition.
- Two poles are called the *cis face* (forming face) and the *trans face* (maturing face).
- *Cis face*, which is closely associated with transitional ER, receives products by accepting transport vesicles from the ER. A vesicle fuses its membrane to the cis face of the Golgi and empties its soluble contents into the Golgi's cisternal space.
- *Trans face* pinches off vesicles from the Golgi and transports molecules to other sites.

Enzymes in the Golgi modify products of the ER in stages as they move through the Golgi stack from the *cis* to the *trans* face:

- Each cisternae between the *cis* and *trans* face contains unique combinations of enzymes.
- Golgi products in transit from one cisternae to the next, are carried in transport vesicles.

During this process, the Golgi:

- Alters some membrane phospholipids.
- Modifies the oligosaccharide portion of glycoproteins.
- Manufactures certain macromolecules itself (e.g. hyaluronic acid).
- Targets products for various parts of the cell.
  - ⇒ Phosphate groups or oligosaccharides may be added to Golgi products as molecular identification tags.
  - ⇒ Membranous vesicles budded from the Golgi may have external molecules that recognize docking sites on the surface of certain organelles.
- Sorts products for secretion. Products destined for secretion leave the *trans* face in vesicles which eventually fuse with the plasma membrane.

IX.   Lysosomes are digestive compartments

<u>Lysosome</u> = An organelle which is a membrane-enclosed bag of hydrolytic enzymes that digest all major classes of macromolecules.
- Enzymes include lipases, carbohydrases, proteases and nucleases.
- Optimal pH for lysosomal enzymes is about pH 5.
- Lysosomal membrane performs two important functions:
    1. Sequesters potentially destructive hydrolytic enzymes from the cytosol.
    2. Maintains the optimal acidic environment for enzyme activity by pumping H^+s inward from the cytosol to the lumen.
- Hydrolytic enzymes and lysosomal membrane are synthesized in the rough ER and processed further in the Golgi apparatus.
- Lysosomes probably pinch off from the *trans* face of the Golgi apparatus.

Functions of Lysosomes:
1. Intracellular digestion.

   <u>Phagocytosis</u> = (Phago=to eat; cyte=cell) Cellular process of ingestion, where the plasma membrane engulfs particulate substances and pinches off to form a particle-containing *vacuole*.
   - Lysosomes may fuse with food-filled vacuoles, and their hydrolytic enzymes digest the food.
   - Examples are *Amoeba* and other protists which eat smaller organisms or food particles.
   - Human cells called *macrophages* phagocytize bacteria and other invaders.
2. Recycle cell's own organic material.
   - Lysosomes may engulf other cellular organelles or part of the cytosol and digest them with hydrolytic enzymes (autophagy).
   - Resulting monomers are released into the cytosol where they can be recycled into new macromolecules.
3. Programmed cell destruction.

   Destruction of cells by their own lysosomes is important during metamorphosis and development.

Lysosomes and Human Disease:

Symptoms of inherited *storage diseases* result from impaired lysosomal function. Lack of a specific lysosomal enzyme causes substrate accumulation which interferes with lysosomal metabolism and other cellular functions.
- In Pompe's disease, the missing enzyme is a carbohydrase that breaks down glycogen. The resulting glycogen accumulation damages the liver.
- Lysosomal lipase is missing in Tay-Sachs disease, which causes lipid accumulation in the brain.

X.   Vacuoles have diverse functions in cell maintenance

Vacuole = Organelle which is a membrane-enclosed sac that is larger than a vesicle (transport vesicle, lysosome or microbody).

Vacuole Types and Functions:

Food vacuole = Vacuole formed by phagocytosis which is the site of intracellular digestion in some protists and macrophages.

Contractile vacuole = Vacuole, found in some fresh-water protozoa, that pumps excess water from the cell.

Central vacuole = Large vacuole found in most mature plant cells.

- Is enclosed by a membrane called the *tonoplast* which is part of the endomembrane system.
- Develops by the coalescence of smaller vacuoles derived from the ER and Golgi apparatus.
- Is a versatile compartment with many functions:
   ⇒ Stores organic compounds (e.g. protein storage in seeds).
   ⇒ Stores inorganic ions (e.g. K^+ and Cl^-).
   ⇒ Sequesters dangerous metabolic by-products from the cytoplasm.
   ⇒ Contains soluble pigments in some cells (e.g. red and blue pigments in flowers).
   ⇒ May protect the plant from predators by containing poisonous or unpalatable compounds.
   ⇒ Plays a role in plant growth by absorbing water and elongating the cell.
   ⇒ Contributes to the large ratio of membrane surface area to cytoplasmic volume. (There is only a thin layer of cytoplasm between the tonoplast and plasma membrane.)

A Summary of Relationships Among Endomembranes

Components of the endomembrane system are related through direct contact or through vesicles. (See Campbell, Figure 7.16)

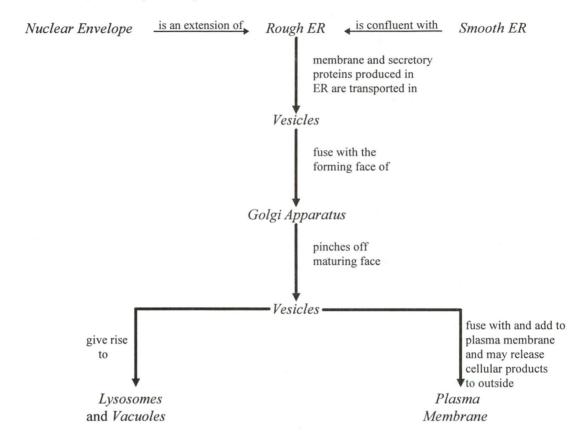

XI. Peroxisomes consume oxygen in various metabolic functions

<u>Peroxisomes</u> = Membrane-bound organelles that contain specialized teams of enzymes for specific metabolic pathways; all contain peroxide-producing oxidases.

- Bound by a single membrane.
- Found in nearly all eukaryotic cells.
- Often have a granular or crystalline core which is a dense collection of enzymes. (See Campbell, Figure 7.20)
- Contain peroxide-producing oxidases that transfer hydrogen from various substrates to oxygen, producing hydrogen peroxide.

$$RH_2 \ + \ O_2 \ \xrightarrow{\text{oxidase}} \ R \ + \ H_2O_2$$

- Contain catalase, an enzyme that converts toxic hydrogen peroxide to water.

$$2H_2O_2 \ \xrightarrow{\text{catalase}} \ 2H_2O \ + \ O_2$$

- Peroxisomal reactions have many functions, some of which are:
  - ⇒ Breakdown of fatty acids into smaller molecules (acetyl CoA). The products are carried to the mitochondria as fuel for cellular respiration.

⇒ Detoxification of alcohol and other harmful compounds. In the liver, peroxisomes enzymatically transfer H from poisons to O_2.

- Specialized peroxisomes (*glyoxysomes*) are found in heterotrophic fat-storing tissue of germinating seeds.

  ⇒ Contain enzymes that convert lipid to carbohydrate.

  ⇒ These biochemical pathways make energy stored in seed oils available for the germinating seedling.

- Current thought is that peroxisome biogenesis occurs by pinching off from preexisting peroxisomes. Necessary lipids and enzymes are imported from the cytosol.

XII. Mitochondria and chloroplasts are the main energy transformers of cells

Mitochondria and chloroplasts are organelles that transduce energy acquired from the surroundings into forms useable for cellular work.

- Enclosed by *double* membranes.
- Membranes are not part of endomembrane system. Rather than being made in the ER, their membrane proteins are synthesized by free ribosomes in the cytosol and by ribosomes located within these organelles themselves.
- Contain ribosomes and some DNA that programs a small portion of their own protein synthesis, though most of their proteins are synthesized in the cytosol programmed by nuclear DNA.
- Are semiautonomous organelles that grow and reproduce within the cell.

NOTE: We just *briefly* mention mitochondria and chloroplasts at this point in the course. Because structure is so closely tied to function, the organelle structure must be covered again in detail with cellular respiration and photosynthesis. In deference to time, we find it more practical to discuss it just once with the metabolism lectures.

A. Mitochondria

Mitochondria = Organelles which are the sites of cellular respiration, a catabolic oxygen-requiring process that uses energy extracted from organic macromolecules to produce ATP.

- Found in nearly all eukaryotic cells.
- Number of mitochondria per cell varies and directly correlates with the cell's metabolic activity.
- Are about 1 mm in diameter and 1-10 mm in length.
- Are dynamic structures that move, change their shape and divide.

Structure of the Mitochondrion:
- Enclosed by two membranes that have their own unique combination of proteins embedded in phospholipid bilayers. (See Campbell, Figure 7.18)

- Smooth *outer membrane* is highly permeable to small solutes, but it blocks passage of proteins and other macromolecules.
- Convoluted *inner membrane* contains embedded enzymes that are involved in cellular respiration. The membrane's many infoldings or *cristae* increase the surface area available for these reactions to occur.
- The inner and outer membranes divide the mitochondrion into two internal compartments:

  1. **Intermembrane Space**

     - Narrow region between the inner and outer mitochondrial membranes.
     - Reflects the solute composition of the cytosol, because the outer membrane is permeable to small solute molecules.

  2. **Mitochondrial Matrix**

     - Compartment enclosed by the inner mitochondrial membrane.
     - Contains enzymes that catalyze many metabolic steps of cellular respiration.
     - Some enzymes of respiration and ATP production are actually embedded in the inner membrane.

B.   Chloroplasts

Plastids = A group of plant and algal membrane-bound organelles that include *amyloplasts*, *chromoplasts* and *chloroplasts*.

Amyloplasts = (Amylo=starch) Colorless plastids that store starch; found in roots and tubers.

Chromoplasts = (Chromo=color) Plastids containing pigments other than chlorophyll; responsible for the color of fruits, flowers and autumn leaves.

Chloroplasts = (Chloro=green) Chlorophyll-containing plastids which are the sites of photosynthesis.

- Found in eukaryotic algae, leaves and other green plant organs.
- Are lens-shaped and measure about 2 mm by 5 mm.
- Are dynamic structures that change shape, move and divide.

Structure of the Chloroplast:

Chloroplasts are divided into three functional compartments by a system of membranes:

1. **Intermembrane Space.** The chloroplast is bound by a double membrane which partitions its contents from the cytosol. A narrow *intermembrane space* separates the two membranes.

2. **Thylakoid Space.** *Thylakoids* form another membranous system within the chloroplast. The thylakoid membrane segregates the interior of the chloroplast into two compartments: *thylakoid space* and *stroma*.

   <u>Thylakoid space</u> = Space inside the thylakoid.

   <u>Thylakoids</u> = Flattened membranous sacs inside the chloroplast.

   - Chlorophyll is found in the thylakoid membranes.
   - Thylakoids function in the steps of photosynthesis that initially convert light energy to chemical energy.
   - Some thylakoids are stacked into *grana*.

   <u>Grana</u> = (Singular, granum) Stacks of thylakoids in a chloroplast.

3. **Stroma.** Photosynthetic reactions that use chemical energy to convert carbon dioxide to sugar occur in the *stroma*.

   <u>Stroma</u> = Viscous fluid outside the thylakoids.

XIII. The cytoskeleton provides structural support and functions in cell motility

It was originally thought that organelles were suspended in a formless cytosol. Technological advances in both light and electron microscopy (e.g. high voltage E.M.) revealed a three-dimensional view of the cell, which showed a network of fibers throughout the cytoplasm — the *cytoskeleton*.

<u>Cytoskeleton</u> = A network of fibers throughout the cytoplasm that forms a dynamic framework for support and movement.

- Gives mechanical support to the cell and helps maintain its shape.
- Enables a cell to change shape.
- Associated with motility by interacting with specialized proteins called *motor molecules* (e.g. organelle movement, muscle contraction, and locomotor organelles).
- Constructed from at least three types of fibers: *microtubules* (thickest), *microfilaments* (thinnest) and *intermediate filaments* (intermediate in diameter). (See Campbell, Table 7.2)

A. **Microtubules**

Found in cytoplasm of all eukaryotic cells, *microtubules*:

- Are straight *hollow* fibers about 25 nm in diameter and 200 nm – 25 μm in length.
- Are constructed from globular proteins called *tubulin* that consists of one α-tubulin and one β-tubulin molecule.
- Begin as two-dimensional sheets of tubulin units, which roll into tubes.
- Elongate by adding tubulin units to its ends.
- May be disassembled and the tubulin units recycled to build microtubules elsewhere in the cell.

Functions of microtubules include:

1. Cellular support.
   - May radiate from the *centrosome*, a *microtubule-organizing center* near the nucleus of most animal cells, and form a framework for cellular support.
   - Microtubular bundles near plasma membrane reinforce cell shape.

2. Tracks for organelle movement. (See Campbell, Figure 7.21)
   - Protein *motor molecules* (e.g. kinesin) interact with microtubules to translocate organelles (e.g. vesicles from the Golgi to the plasma membrane).

3. Separation of chromosomes during cell division.

4. Make up *centrioles* in animal cells. (See Campbell, Figure 7.22)

   Centriole = Pair of cylindrical structures in animal cells, composed of nine sets of triplet microtubules arranged in a ring.
   - Are about 150 nm in diameter and are arranged at right angles to each other.
   - Pair of centrioles located within the centrosome, replicate during cell division.
   - May organize microtubule assembly during cell division, but must not be mandatory for this function since plants lack centrioles.

Cilia and Flagella.

Cilia and flagella = Locomotor organelles found in eukaryotes, which are formed from a specialized arrangement of microtubules.
   - May propel single-celled organisms (Protista) and motile sperm cells through an aquatic medium.
   - May function to draw fluid across the surface of stationary cells (e.g. ciliated cells lining trachea).

Cilia (singular, cilium)	Flagella (singular, flagellum)
Occur in large numbers on cell surface.	One or a few per cell.
Shorter; 2-20 mm in length.	Longer; 10-200 mm in length.
Work like oars, alternating power with recovery strokes. Creates force in a direction perpendicular to the axis of the cilium.	Undulating motion that creates force in the same direction as the axis of the flagellum.

Ultrastructure of Cilia and Flagella:

- Are extensions of plasma membrane with a core of microtubules. (See Campbell, Figure 7.24)
- Microtubular core is made of nine doublets of microtubules arranged in a ring with two single microtubules in the center (*9 + 2 pattern*).
- Each doublet is a pair of attached microtubules. One of the pair shares a portion of the other's wall.
- Each doublet is connected to the center of the ring by *radial spokes* that end near the central microtubules.
- Each doublet is attached to the neighboring doublet by a pair of *side arms*. Many pairs of sidearms are evenly spaced along the doublet's length.
- Structurally identical to centrioles, *basal bodies* anchor the microtubular assemblies.

9+2 Pattern in Cross Section

Basal Body = A cellular structure, identical to a centriole, that anchors the microtubular assembly of cilia and flagella.

- Can convert into a centriole and vice versa.
- May be a template for ordering tubulin into the microtubules of *newly* forming cilia or flagella. As cilia and flagella continue to grow, new tubulin subunits are added to the tips, rather than to the bases.

The unique ultrastructure of cilia and flagella is necessary for them to function:

- Sidearms are made of *dynein*, a large protein motor molecule that changes its conformation in the presence of ATP as an energy source.
- A complex cycle of movements caused by dynein's conformational changes, makes the cilium or flagellum bend: (See Campbell, Figure 7.25)

Sidearms of one doublet attach to the
adjacent doublet.

↓

Sidearms swing and the two doublets slide
past one another.

↓

Sidearms release.

↓

Sidearms reattach to the adjacent doublet
farther along its length.

↓

Cycle is repeated.

- In cilia and flagella, linear displacement of dynein sidearms is translated into a *bending* by the resistance of the *radial spokes*. Working against this resistance, the "dynein-walking" distorts the microtubules, causing them to bend.

B. Microfilaments (Actin Filaments)

Structure of Microfilaments:

- Solid rods about 7 nm in diameter.
- Built from globular protein monomers, *G-actin*, which are linked into long chains.
- Two actin chains are wound into a helix.

Actin subunit

7 nm

Function of Microfilaments:

1. Participate in muscle contraction.

   - Along the length of a muscle cell, parallel actin microfilaments are interdigitated with thicker filaments made of the protein *myosin*, a motor molecule.

   - With ATP as the energy source, a muscle cell shortens as the thin actin filaments slide across the myosin filaments. Sliding results from the swinging of myosin cross-bridges intermittently attached to actin.

2. Provide cellular support (e.g. bundles of microfilaments in the core of intestinal microvilli).

3. Responsible for localized contraction of cells. Small actin-myosin aggregates exist in some parts of the cell and cause localized contractions. Examples include:

   - Contracting ring of microfilaments pinches an animal cell in two during cell division. (See Campbell, Figure 11.9)

   - Elongation and contraction of *pseudopodia* during *amoeboid movement*.

   - Involved in *cytoplasmic streaming* (*cyclosis*) found in plant cells.

     Cytoplasmic streaming (cyclosis) = Flowing of the entire cytoplasm around the space between the vacuole and plasma membrane in a plant cell.

C. Intermediate Filaments

Structure of Intermediate Filaments:

- Filaments that are intermediate in diameter (8-12 nm) between microtubules and microfilaments.

- Diverse class of cytoskeletal elements that differ in diameter and composition depending upon cell type.

- Constructed from *keratin* subunits.

- More permanent than microfilaments and microtubules.

Fibrous subunits

10 nm

Function of Intermediate Filaments:

1. Specialized for bearing tension; may function as the framework for the cytoskeleton.
2. Reinforce cell shape (e.g. nerve axons).
3. Probably fix organelle position (e.g. nucleus).
4. Compose the nuclear lamina, lining the nuclear envelope's interior.

XIV. Plant cells are encased by cell walls

Most cells produce coats that are external to the plasma membrane.

A. Cell Walls

Plant cells can be distinguished from animal cells by the presence of a *cell wall*:

- Thicker than the plasma membrane (0.1 – 2 µm)..
- Chemical composition varies from cell to cell and species to species.
- Basic design includes strong *cellulose* fibers embedded in a matrix of other polysaccharides and proteins.
- Functions to protect plant cells, maintain their shape, and prevent excess water uptake.
- Has membrane-lined channels, *plasmodesmata*, that connect the cytoplasm of neighboring cells.

Plant cells develop as follows:

- Young plant cell secretes a thin, flexible *primary cell wall*. Between primary cell walls of adjacent cells is a *middle lamella* made of *pectins*, a sticky polysaccharide that cements cells together.
- Cell stops growing and strengthens its wall. Some cells:
    1. Secrete hardening substances into primary wall.
    2. Add a *secondary cell wall* between plasma membrane and primary wall.
- *Secondary cell wall* is often deposited in layers with a durable matrix that supports and protects the cell. (See Campbell, Figure 7.30)

XV. The extracellular matrix (ECM) of animal cells functions in support, adhesion, movement and development

Animal cells lack walls, but they do have an elaborate *extracellular matrix (ECM)*.

Extracellular matrix (ECM) = Meshwork of macromolecules outside the plasma membrane of animal cells. This ECM is:

- locally secreted by cells.
- composed mostly of glycoproteins, the most abundant of which is *collagen* that:
    ⇒ accounts for about half of the total protein in the vertebrate body.
    ⇒ forms strong extracellular fibers embedded in a meshwork of carbohydrate-rich glycoproteins called *proteoglycans*.

Some cells are attached:

- directly to the collagen and proteoglycan of their extracellular matrix.
- or to the ECM by another class of glycoproteins – *fibronectins*.

Fibronectins bind to transmembrane receptor proteins called *integrins* that:
- bond on their cytoplasmic side to microfilaments of the cytoskeleton.
- integrate cytoskeletal responses to ECM changes and vice versa.

The extracellular matrix:
- provides support and anchorage for cells.
- functions in a cell's dynamic behavior. For example, some embryonic cells migrate along specific pathways by orienting their intracellular microfilaments to the pattern of extracellular fibers in the ECM. (See Campbell, Figure 7.29)
- helps control gene activity in the cell's nucleus. Perhaps the transcription of specific genes is a response to chemical signals triggered by communication of mechanical stimuli across the plasma membrane from the ECM through integrins to the cytoskeleton.

XVI. Intercellular junctions integrate cells into higher levels of structure and function

Neighboring cells often adhere and interact through special patches of direct physical contact.

Intercellular Junctions in Plants:

Plasmodesmata (singular, plasmodesma) = Channels that perforate plant cell walls, through which cytoplasmic strands communicate between adjacent cells.
- Lined by plasma membrane. Plasma membranes of adjacent cells are continuous through a plasmodesma.
- Allows free passage of water and small solutes. This transport is enhanced by *cytoplasmic streaming*.

Intercellular Junctions in Animals: (See Campbell, Figure 7.30)

Tight junctions = Intercellular junctions that hold cells together tightly enough to block transport of substances through the intercellular space.
- Specialized membrane proteins in adjacent cells bond directly to each other allowing no space between membranes.
- Usually occur as belts all the way around each cell, that block intercellular transport.
- Frequently found in epithelial layers that separate two kinds of solutions.

Desmosomes = Intercellular junctions that rivet cells together into strong sheets, but still permit substances to pass freely through intracellular spaces. The desmosome is made of:
- Intercellular glycoprotein filaments that penetrate and attach the plasma membrane of both cells.
- A dense disk inside the plasma membrane that is reinforced by *intermediate filaments* made of *keratin* (a strong structural protein).

Gap junctions = Intercellular junctions specialized for material transport between the cytoplasm of adjacent cells.

- Formed by two connecting protein rings (*connexon*), each embedded in the plasma membrane of adjacent cells. The proteins protrude from the membranes enough to leave an intercellular gap of 2–4 nm.
- Have pores with diameters (1.5 nm) large enough to allow cells to share smaller molecules (e.g. inorganic ions, sugars, amino acids, vitamins), but not macromolecules such as proteins.
- Common in animal embryos and cardiac muscle where chemical communication between cells is essential.

REFERENCES

Alberts, B., D. Bray, J. Lewis, M. Raff, K. Roberts and J.D. Watson. *Molecular Biology of the Cell*. 2nd ed. New York: Garland, 1989.

Becker, W.M. and D.W. Deamer. *The World of the Cell*. 2nd ed. Redwood City, California: Benjamin/Cummings, 1991.

Campbell, N. *Biology*. 4th ed. Menlo Park, California: Benjamin/Cummings, 1996.

deDuve, C. *A Guided Tour of the Living Cell*. Volumes I and II. New York: Scientific American Books, 1984. Literally a guided tour of the cell with the reader as "cytonaut." This is an excellent resource for lecture material and enjoyable reading.

Kleinsmith, L.J. and V.M. Kish. *Principles of Cell Biology*. New York: Harper and Row, Publ., 1988.

CHAPTER **8**

MEMBRANE STRUCTURE AND FUNCTION

OUTLINE

I.   Membrane models have evolved to fit new data: *science as a process*

II. A membrane is a fluid mosaic of lipids, proteins and carbohydrates
   A.   The Fluid Quality of Membranes
   B.   Membranes as Mosaics of Structure and Function
   C.   Membrane Carbohydrates and Cell-Cell Recognition

III. A membrane's molecular organization results in selective permeability
   A.   Permeability of the Lipid Bilayer
   B.   Transport Proteins

IV. Passive transport is diffusion across a membrane

V.   Osmosis is the passive transport of water

VI. Cell survival depends on balancing water uptake and loss
   A.   Water Balance of Cells Without Walls
   B.   Water Balance of Cells With Walls

VII. Specific proteins facilitate the passive transport of selected solutes

VIII. Active transport is the pumping of solutes against their gradients

IX. Some ion pumps generate voltage across membranes

X.   In cotransport, a membrane protein couples the transport of one solute to another

XI. Exocytosis and endocytosis transport large molecules

XII. Specialized membrane proteins transmit extracellular signals to the inside of the cell

OBJECTIVES

After reading this chapter and attending lecture, the student should be able to:

1. Describe the function of the plasma membrane.
2. Explain how scientists used early experimental evidence to make deductions about membrane structure and function.
3. Describe the Davson-Danielli membrane model and explain how it contributed to our current understanding of membrane structure.
4. Describe the contribution J.D. Robertson, S.J. Singer and G.L. Nicolson made to clarify membrane structure.
5. Describe the fluid properties of the cell membrane and explain how membrane fluidity is influenced by membrane composition.
6. Explain how hydrophobic interactions determine membrane structure and function.
7. Describe how proteins are spatially arranged in the cell membrane and how they contribute to membrane function.
8. Describe factors that affect selective permeability of membranes.
9. Define diffusion; explain what causes it and why it is a spontaneous process.
10. Explain what regulates the rate of passive transport.
11. Explain why a concentration gradient across a membrane represents potential energy.
12. Define osmosis and predict the direction of water movement based upon differences in solute concentration.
13. Explain how bound water affects the osmotic behavior of dilute biological fluids.
14. Describe how living cells with and without walls regulate water balance.
15. Explain how transport proteins are similar to enzymes.
16. Describe one model for facilitated diffusion.
17. Explain how active transport differs from diffusion.
18. Explain what mechanisms can generate a membrane potential or electrochemical gradient.
19. Explain how potential energy generated by transmembrane solute gradients can be harvested by the cell and used to transport substances across the membrane.
20. Explain how large molecules are transported across the cell membrane.
21. Give an example of receptor-mediated endocytosis.
22. Explain how membrane proteins interface with and respond to changes in the extracellular environment.
23. Describe a simple signal-transduction pathway across the membrane including the roles of first and second messengers.

KEY TERMS

phospholipid bilayer	carrier-mediated	water potential	proton pump
amphipathic	transport	osmotic potential	cotransport
fluid mosaic model	permease	solution	exocytosis
J.F. Danielli	diffusion	solvent	endocytosis
H. Davson	osmosis	solute	phagocytosis
J.D. Robertson	dialysis	hypertonic	pinocytosis
S.J. Singer	facilitated diffusion	hypotonic	receptor-mediated
G.L. Nicolson	active transport	isotonic	endocytosis
integral proteins	concentration	membrane potential	signal-transduction
peripheral proteins	gradient	electrogenic pump	pathway
selective permeability	bulk flow	sodium-potassium pump	second messenger

LECTURE NOTES

The *plasma membrane* is the boundary that separates the living cell from its nonliving surroundings. It makes life possible by its ability to discriminate in its chemical exchanges with the environment. This membrane:

- Is about 8 nm thick.
- Surrounds the cell and controls chemical traffic into and out of the cell.
- Is *selectively permeable*; it allows some substances to cross more easily than others.
- Has a unique structure which determines its function and solubility characteristics.

NOTE: This is an opportune place to illustrate how form fits function. It is remarkable how much early models contributed to the understanding of membrane structure, since biologists proposed these models without the benefit of "seeing" a membrane with an electron microscope.

I.   **Membrane models have evolved to fit new data:** *science as a process*

Membrane function is determined by its structure. Early models of the plasma membrane were deduced from <u>indirect</u> evidence:

1. Evidence: Lipid and lipid soluble materials enter cells more rapidly than substances that are insoluble in lipids (C. Overton, 1895).

    Deduction: Membranes are made of lipids.

    Deduction: Fat-soluble substance move through the membrane by dissolving in it ("like dissolves like").

2. Evidence: *Amphipathic* phospholipids will form an artificial membrane on the surface of water with only the hydrophilic heads immersed in water (Langmuir, 1917).

    <u>Amphipathic</u> = Condition where a molecule has both a hydrophilic region and a hydrophobic region.

    Deduction: Because of their molecular structure, phospholipids can form membranes.

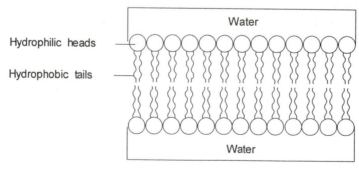

3. Evidence: Phospholipid content of membranes isolated from red blood cells is just enough to cover the cells with two layers (Gorter and Grendel, 1925).

   Deduction: Cell membranes are actually phospholipid bilayers, two molecules thick.

4. Evidence: Membranes isolated from red blood cells contain proteins as well as lipids.

   Deduction: There is protein in biological membranes.

5. Evidence: Wettability of the surface of an actual biological membrane is greater than the surface of an artificial membrane consisting only of a phospholipid bilayer.

   Deduction: Membranes are coated on both sides with proteins, which generally absorb water.

Incorporating results from these and other solubility studies, J.F. Danielli and H. Davson (1935) proposed a model of cell membrane structure:

- Cell membrane is made of a phospholipid bilayer sandwiched between two layers of globular protein.
- The polar (hydrophilic) heads of phospholipids are oriented towards the protein layers forming a hydrophilic zone.
- The nonpolar (hydrophobic) tails of phospholipids are oriented in between polar heads forming a hydrophobic zone.
- The membrane is approximately 8 nm thick.

In the 1950's, electron microscopy allowed biologists to visualize the plasma membrane for the first time and provided support for the Davson-Danielli model. Evidence from electron micrographs:

1. Confirmed the plasma membrane was 7 to 8 nm thick (close to the predicted size if the Davson-Danielli model was modified by replacing globular proteins with protein layers in pleated-sheets).
2. Showed the plasma membrane was trilaminar, made of two electron-dense bands separated by an unstained layer. It was assumed that the heavy metal atoms of the stain adhered to the hydrophilic proteins and heads of phospholipids and not to the hydrophobic core.
3. Showed internal cellular membranes that looked similar to the plasma membrane. This led biologists (J.D. Robertson) to propose that all cellular membranes were symmetrical and virtually identical.

Though the phospholipid bilayer is probably accurate, there are problems with the Davson-Danielli model:

1. Not all membranes are identical or symmetrical.
   - Membranes with different functions also differ in chemical composition and structure.
   - Membranes are bifacial with distinct inside and outside faces.

2. A membrane with an outside layer of proteins would be an unstable structure.

    - Membrane proteins are not soluble in water, and, like phospholipid, they are *amphipathic*.

    - Protein layer not likely because its hydrophobic regions would be in an aqueous environment, and it would also separate the hydrophilic phospholipid heads from water.

In 1972, S.J. Singer and G.L. Nicolson proposed the *fluid mosaic model* which accounted for the amphipathic character of proteins. They proposed:

- Proteins are individually embedded in the phospholipid bilayer, rather than forming a solid coat spread upon the surface.

Hydrophilic region of protein

Phospholipid bilayer

Hydrophobic region of protein

- Hydrophilic portions of both proteins and phospholipids are maximally exposed to water resulting in a stable membrane structure.

- Hydrophobic portions of proteins and phospholipids are in the nonaqueous environment inside the bilayer.

- Membrane is a mosaic of proteins bobbing in a fluid bilayer of phospholipids.

- Evidence from freeze fracture techniques have confirmed that proteins are embedded in the membrane. Using these techniques, biologists can delaminate membranes along the middle of the bilayer. When viewed with an electron microscope, proteins appear to penetrate into the hydrophobic interior of the membrane. (See Campbell, Methods Box)

II. A membrane is a fluid mosaic of lipids, proteins and carbohydrates

A. The Fluid Quality of Membranes

Membranes are held together by hydrophobic interactions, which are weak attractions. (See Campbell, Figure 8.3)

- Most membrane lipids and some proteins can drift laterally within the membrane.

- Molecules rarely flip transversely across the membrane, because hydrophilic parts would have to cross the membrane's hydrophobic core.

- Phospholipids move quickly along the membrane's plane, averaging 2 μm per second.

- Membrane proteins drift more slowly than lipids. The fact that proteins drift laterally was established experimentally by fusing a human and mouse cell (Frye and Edidin, 1970):

Membrane proteins of a human and mouse
cell were labeled with different green and
red fluorescent dyes.

↓

Cells were fused to form a hybrid cell with
a continuous membrane.

↓

Hybrid cell membrane had initially distinct
regions of green and red dye.

↓

In less than an hour, the two colors were
intermixed.

- Some membrane proteins are tethered to the cytoskeleton and cannot move far.

Membranes must be fluid to work properly. Solidification may result in permeability changes and enzyme deactivation.
- Unsaturated hydrocarbon tails enhance membrane fluidity, because kinks at the carbon-to-carbon double bonds hinder close packing of phospholipids.
- Membranes solidify if the temperature decreases to a critical point. Critical temperature is lower in membranes with a greater concentration of unsaturated phospholipids.
- Cholesterol, found in plasma membranes of eukaryotes, modulates membrane fluidity by making the membrane:
  ⇒ Less fluid at warmer temperatures (e.g. 37°C body temperature) by restraining phospholipid movement.
  ⇒ More fluid at lower temperatures by preventing close packing of phospholipids.
- Cells may alter membrane lipid concentration in response to changes in temperature. Many cold tolerant plants (e.g. winter wheat) increase the unsaturated phospholipid concentration in autumn, which prevents the plasma membranes from solidifying in winter.

B.   **Membranes as Mosaics of Structure and Function**

A membrane is a *mosaic* of different proteins embedded and dispersed in the phospholipid bilayer. These proteins vary in both structure and function, and they occur in two spatial arrangements:
  1. *Integral proteins*, which are inserted into the membrane so their hydrophobic regions are surrounded by hydrocarbon portions of phospholipids. They may be:
     - unilateral, reaching only partway across the membrane.
     - transmembrane, with hydrophobic midsections between hydrophilic ends exposed on both sides of the membrane.

2. *Peripheral proteins*, which are not embedded but attached to the membrane's surface.
- May be attached to integral proteins or held by fibers of the ECM.
- On cytoplasmic side, may be held by filaments of cytoskeleton.

Membranes are bifacial. The membrane's synthesis and modification by the ER and Golgi determines this asymmetric distribution of lipids, proteins and carbohydrates:
- Two lipid layers may differ in lipid composition.
- Membrane proteins have distinct directional orientation.
- When present, carbohydrates are restricted to the membrane's exterior.
- Side of the membrane facing the lumen of the ER, Golgi and vesicles is topologically the same as the plasma membrane's outside face. (See Campbell, Figure 8.7)
- Side of the membrane facing the cytoplasm has always faced the cytoplasm, from the time of its formation by the endomembrane system to its addition to the plasma membrane by the fusion of a vesicle.

C.   Membrane Carbohydrates and Cell-Cell Recognition

Cell-cell recognition = The ability of a cell to determine if other cells it encounters are alike or different from itself.

Cell-cell recognition is crucial in the functioning of an organism. It is the basis for:
- Sorting of an animal embryo's cells into tissues and organs.
- Rejection of foreign cells by the immune system.

The way cells recognize other cells is probably by keying on cell markers found on the external surface of the plasma membrane. Because of their diversity and location, likely candidates for such cell markers are membrane carbohydrates:
- Usually branched *oligosaccharides* (<15 monomers).
- Some covalently bonded to lipids (*glycolipids*).
- Most covalently bonded to proteins (*glycoproteins*).
- Vary from species to species, between individuals of the same species and among cells in the same individual.

III.   A membrane's molecular organization results in selective permeability

The *selectively permeable* plasma membrane regulates the type and rate of molecular traffic into and out of the cell.

Selective permeability = Property of biological membranes which allows some substances to cross more easily than others. The selective permeability of a membrane depends upon:
- membrane solubility characteristics of the phospholipid bilayer
- presence of specific integral transport proteins.

A.   **Permeability of the Lipid Bilayer**

The ability of substances to cross the hydrophobic core of the plasma membrane can be measured as the rate of transport through an artificial phospholipid bilayer:

1.   **Nonpolar (Hydrophobic) Molecules**

- Dissolve in the membrane and cross it with ease (e.g. hydrocarbons and O_2).
- If two molecules are equally lipid soluble, the smaller of the two will cross the membrane faster.

2.   **Polar (Hydrophilic) Molecules**

- Small, polar uncharged molecules (e.g. H_2O, CO_2) that are small enough to pass between membrane lipids, will easily pass through synthetic membranes.
- Larger, polar uncharged molecules (e.g. glucose) will *not* easily pass through synthetic membranes.
- All ions, even small ones (e.g. Na^+, H^+) have difficulty penetrating the hydrophobic layer.

B.   **Transport Proteins**

Water, CO_2 and nonpolar molecules rapidly pass through the plasma membrane as they do an artificial membrane.

Unlike artificial membranes, however, biological membranes *are* permeable to specific ions and certain polar molecules of moderate size. These hydrophilic substances avoid the hydrophobic core of the bilayer by passing through *transport proteins*.

Transport proteins = Integral membrane proteins that transport specific molecules or ions across biological membranes. (See Campbell, Figure 8.6)

- May provide a hydrophilic tunnel through the membrane.
- May bind to a substance and physically move it across the membrane.
- Are specific for the substance they translocate.

IV.   **Passive transport is diffusion across a membrane**

NOTE: Students have particular trouble with the concepts of *gradient* and *net* movement, yet their understanding of diffusion depends upon having a working knowledge of these terms.

Concentration gradient = Regular, graded concentration change over a distance in a particular direction.

Net directional movement = Overall movement away from the center of concentration, which results from random molecular movement in all directions.

Diffusion = The *net* movement of a substance down a *concentration gradient*.

- Results from the intrinsic kinetic energy of molecules (also called thermal motion, or heat).
- Results from random molecular motion, even though the *net* movement may be directional.

- Diffusion continues until a dynamic equilibrium is reached — the molecules continue to move, but there is no net directional movement.

In the absence of other forces, a substance will diffuse from where it is more concentrated to where it is less concentrated.

- A substance diffuses down its *concentration gradient*.
- Because it decreases free energy, diffusion is a spontaneous process ($-\Delta G$). It increases entropy of a system by producing a more random mixture of molecules.
- A substance diffuses down its *own* concentration gradient and is not affected by the gradients of other substances.

Much of the traffic across cell membranes occurs by diffusion and is thus a form of *passive transport*.

Passive transport = Diffusion of a substance across a biological membrane.

- Spontaneous process which is a function of a concentration gradient when a substance is more concentrated on one side of the membrane.
- Passive process which does not require the cell to expend energy. It is the potential energy stored in a concentration gradient that drives diffusion.
- Rate of diffusion is regulated by the permeability of the membrane, so some molecules diffuse more freely than others.
- Water diffuses freely across most cell membranes.

V.   Osmosis is the passive transport of water

Hypertonic solution = A solution with a greater solute concentration than that inside a cell.

Hypotonic solution = A solution with a lower solute concentration compared to that inside a cell.

Isotonic solution = A solution with an equal solute concentration compared to that inside a cell.

NOTE: These terms are clearly a source of confusion for students. It helps to point out that these are only relative terms used to compare the osmotic concentration of a solution to the osmotic concentration of a cell.

Osmosis = Diffusion of water across a selectively permeable membrane.

- Water diffuses down its concentration gradient.
- For example, if two solutions of different concentrations are separated by a selectively permeable membrane that is permeable to water but not the solute, water will diffuse from the hypoosmotic solution (solution with the lower osmotic concentration) to the hyperosmotic solution (solution with the higher osmotic concentration).

Hypoosmotic
solution

H_2O

Hyperosmotic
solution

Selectively
permeable membrane

- Some solute molecules can reduce the proportion of water molecules that can freely diffuse. Water molecules form a *hydration shell* around hydrophilic solute molecules, and this bound water cannot freely diffuse across a membrane.
- In dilute solutions including most biological fluids, it is the difference in the proportion of unbound water that causes osmosis, rather than the actual difference in water concentration.
- Direction of osmosis is determined by the difference in *total* solute concentration, regardless of the type or diversity of solutes in the solutions.
- If two *isosmotic* solutions are separated by a selectively permeable membrane, water molecules diffuse across the membrane in both directions at an equal rate. There is no *net* movement of water.

NOTE: Clarification of this point is often necessary. Students may need to be reminded that even though there is no *net* movement of water across the membrane (or osmosis), the water molecules do not stop moving. At equilibrium, the water molecules move in both directions at the same rate.

<u>Osmotic concentration</u> = Total solute concentration of a solution.

<u>Osmotic pressure</u> = Measure of the tendency for a solution to take up water when separated from pure water by a selectively permeable membrane.

- Osmotic pressure of pure water is zero.
- Osmotic pressure of a solution is proportional to its osmotic concentration. (The greater the solute concentration, the greater the osmotic pressure.)

Osmotic pressure can be measured by an *osmometer*:

- In one type of osmometer, pure water is separated from a solution by a selectively permeable membrane that is permeable to water but not solute.
- The tendency for water to move into the solution by osmosis is counteracted by applying enough pressure with a piston so the solution's volume will stay the same.
- The amount of pressure required to prevent net movement of water into the solution is the *osmotic pressure*.

Piston

Solute

VI.   Cell survival depends on balancing water uptake and loss

A.   Water Balance of Cells Without Walls

Since animal cells lack cell walls, they are not tolerant of excessive osmotic uptake or loss of water.

- In an isotonic environment, the volume of an animal cell will remain stable with no net movement of water across the plasma membrane.

- In a hypertonic environment, an animal cell will lose water by osmosis and *crenate* (shrivel).
- In a hypotonic environment, an animal cell will gain water by osmosis, swell and perhaps *lyse* (cell destruction).

Organisms without cell walls prevent excessive loss or uptake of water by:
- Living in an isotonic environment (e.g. many marine invertebrates are isosmotic with sea water).
- Osmoregulating in a hypo- or hypertonic environment. Organisms can regulate water balance (osmoregulation) by removing water in a hypotonic environment (e.g. *Paramecium* with contractile vacuoles in fresh water) or conserving water and pumping out salts in a hypertonic environment (e.g. bony fish in seawater).

B.   Water Balance of Cells With Walls

Cells of prokaryotes, some protists, fungi and plants have cell walls outside the plasma membrane.

- In a hypertonic environment, walled cells will lose water by osmosis and will *plasmolyze*, which is usually lethal.

    Plasmolysis = Phenomenon where a walled cell shrivels and the plasma membrane pulls away from the cell wall as the cell loses water to a hypertonic environment.

- In a hypotonic environment, water moves by osmosis into the plant cell, causing it to swell until internal pressure against the cell wall equals the osmotic pressure of the cytoplasm. A dynamic equilibrium is established (water enters and leaves the cell at the same rate and the cell becomes *turgid*).

    Turgid = Firmness or tension such as found in walled cells that are in a hypoosmotic environment where water enters the cell by osmosis.

    ⇒ Ideal state for most plant cells.

    ⇒ Turgid cells provide mechanical support for plants.

    ⇒ Requires cells to be hyperosmotic to their environment.

- In an isotonic environment, there is no net movement of water into or out of the cell.

    ⇒ Plant cells become *flaccid* or limp.

    ⇒ Loss of structural support from turgor pressure causes plants to wilt.

VII.   Specific proteins facilitate the passive transport of selected solutes

Facilitated diffusion = Diffusion of solutes across a membrane, with the help of transport proteins.

- Is passive transport because solute is transported down its concentration gradient.
- Helps the diffusion of many polar molecules and ions that are impeded by the membrane's phospholipid bilayer.

Transport proteins share some properties of enzymes:

- Transport proteins are *specific* for the solutes they transport. There is probably a specific binding site analogous to an enzyme's active site.
- Transport proteins can be *saturated* with solute, so the maximum transport rate occurs when all binding sites are occupied with solute.
- Transport proteins can be inhibited by molecules that resemble the solute normally carried by the protein (similar to competitive inhibition in enzymes).

Transport proteins differ from enzymes in they do not usually catalyze chemical reactions.

One Model for Facilitated Diffusion:

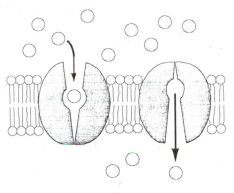

- Transport protein most likely remains in place in the membrane and translocates solute by alternating between two conformations.
- In one conformation, transport protein binds solute; as it changes to another conformation, transport protein deposits solute on the other side of the membrane.
- The solute's binding and release may trigger the transport protein's conformational change.

Other transport proteins are selective channels across the membrane.

- The membrane is thus permeable to specific solutes that can pass through these channels.
- Some selective channels (gated channels) only open in response to electrical or chemical stimuli. For example, binding of neurotransmitter to nerve cells opens gated channels, so that sodium ions can diffuse into the cell.

In some inherited disorders, transport proteins are missing or are defective (e.g. cystinuria, a kidney disease caused by missing carriers for cystine and other amino acids which are normally reabsorbed from the urine).

VIII. Active transport is the pumping of solutes against their gradients

Active transport = Energy-requiring process during which a transport protein pumps a molecule across a membrane, *against* its concentration gradient.

- Is energetically uphill ($+\Delta G$) and requires the cell to expend energy.
- Helps cells maintain steep ionic gradients across the cell membrane (e.g. Na^+, K^+, Mg^{++}, Ca^{++} and Cl^-).
- Transport proteins involved in active transport harness energy from ATP to pump molecules against their concentration gradients.

An example of an active transport system that translocates ions against steep concentration gradients is the *sodium-potassium pump*. Major features of the pump are:

1. The transport protein oscillates between two conformations:
   a. High affinity for Na^+ with binding sites oriented towards the cytoplasm.
   b. High affinity for K^+ with binding sites oriented towards the cell's exterior.
2. ATP phosphorylates the transport protein and powers the conformational change from Na^+ receptive to K^+ receptive.
3. As the transport protein changes conformation, it translocates bound solutes across the membrane.
4. Na^+K^+-pump translocates three Na^+ ions out of the cell for every two K^+ ions pumped into the cell. (Refer to Campbell, Figure 8.13 for the specific sequence of events.)

IX. Some ion pumps generate voltage across membranes

Because anions and cations are unequally distributed across the plasma membrane, all cells have voltages across their plasma membranes.

Membrane potential = Voltage across membranes.
- Ranges from -50 to -200 mv. As indicated by the negative sign, the cell's inside is negatively charged with respect to the outside.
- Affects traffic of charged substances across the membrane.
- Favors diffusion of cations into cell and anions out of the cell (because of electrostatic attractions).

Two forces drive passive transport of ions across membranes:
1. Concentration gradient of the ion.
2. Effect of membrane potential on the ion.

Electrochemical gradient = Diffusion gradient resulting from the combined effects of membrane potential and concentration gradient.
- Ions may not always diffuse down their concentration gradients, but they *always* diffuse down their electrochemical gradients.
- At equilibrium, the distribution of ions on either side of the membrane may be different from the expected distribution when charge is not a factor.
- Uncharged solutes diffuse down concentration gradients because they are unaffected by membrane potential.

Factors which contribute to a cell's membrane potential (net negative charge on the inside):
1. Negatively charged proteins in the cell's interior.
2. Plasma membrane's selective permeability to various ions. For example, there is a net loss of positive charges as K^+ leaks out of the cell faster than Na^+ diffuses in.
3. The sodium-potassium pump. This *electrogenic pump* translocates 3 Na^+ out for every 2 K^+ in - a net loss of one positive charge per cycle.

Electrogenic pump = A transport protein that generates voltage across a membrane.
- Na^+/K^+ ATPase is the major electrogenic pump in animal cells.
- A *proton pump* is the major electrogenic pump in plants, bacteria and fungi. Also, mitochondria and chloroplasts use a proton pump to drive ATP synthesis.
- Voltages created by electrogenic pumps are sources of potential energy available to do cellular work.

NOTE: This is a good place to emphasize that electrochemical gradients represent potential energy. Spending lecture time on cotransport and the proton pump will help prepare your students for the upcoming topic of chemiosmosis.

X.   In cotransport, a membrane protein couples the transport of one solute to another

Cotransport = Process where a single ATP-powered pump actively transports one solute and indirectly drives the transport of other solutes against their concentration gradients.

One mechanism of cotransport involves two transport proteins:

1. ATP-powered pump actively transports one solute and creates potential energy in the gradient it creates.
2. Another transport protein couples the solute's downhill diffusion as it leaks back across the membrane with a second solute's uphill transport against its concentration gradient.

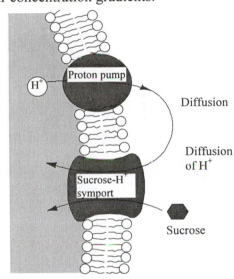

For example, plants use a proton pump coupled with sucrose-H^+ symport to load sucrose into specialized cells of vascular tissue. Both solutes, H^+ and sucrose, must bind to the transport protein for cotransport to take place.

XI.   Exocytosis and endocytosis transport large molecules

Water and small molecules cross membranes by:
1. Passing through the phospholipid bilayer.
2. Being translocated by a transport protein.

Large molecules (e.g. proteins and polysaccharides) cross membranes by the processes of *exocytosis* and *endocytosis*.

Exocytosis	Endocytosis
Process of exporting macromolecules from a cell by fusion of vesicles with the plasma membrane.	Process of importing macromolecules into a cell by forming vesicles derived from the plasma membrane.
Vesicle usually budded from the ER or Golgi and migrates to plasma membrane.	Vesicle forms from a localized region of plasma membrane that sinks inward; pinches off into the cytoplasm.
Used by secretory cells to export products (e.g. insulin in pancreas, or neuro-transmitter from neuron).	Used by cells to incorporate extracellular substances.

There are three types of endocytosis: (1) *phagocytosis*, (2) *pinocytosis* and (3) *receptor-mediated endocytosis*.

Phagocytosis = (cell eating) Endocytosis of solid particles.
- Cell engulfs particle with *pseudopodia* and pinches off a food vacuole.
- Vacuole fuses with a lysosome containing hydrolytic enzymes that will digest the particle.

Pinocytosis = (cell drinking) Endocytosis of fluid droplets.
- Droplets of extracellular fluid are taken into small vesicles.
- The process is not discriminating. The cell takes in all solutes dissolved in the droplet.

Receptor-mediated endocytosis = The process of importing specific macromolecules into the cell by the inward budding of vesicles formed from *coated pits*; occurs in response to the binding of specific *ligands* to receptors on the cell's surface.
- More discriminating process than pinocytosis.
- Membrane-embedded proteins with specific receptor sites exposed to the cell's exterior, cluster in regions called *coated pits*.
- A layer of *clathrin*, a fibrous protein, lines and reinforces the *coated pit* on the cytoplasmic side and probably helps deepen the pit to form a vesicle.
- A molecule that binds to a specific receptor site of another molecule is called a *ligand*.

Progressive Stages of Receptor-Mediated Endocytosis:

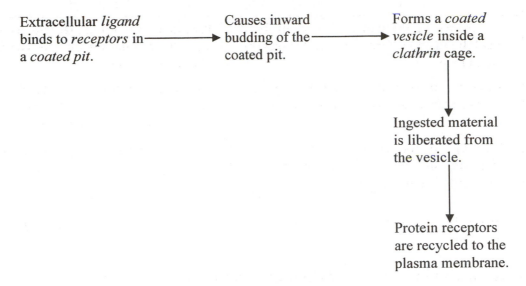

Extracellular *ligand* binds to *receptors* in a *coated pit.* → Causes inward budding of the coated pit. → Forms a *coated vesicle* inside a *clathrin* cage.

↓

Ingested material is liberated from the vesicle.

↓

Protein receptors are recycled to the plasma membrane.

Receptor-mediated endocytosis enables cells to acquire bulk quantities of specific substances, even if they are in low concentration in extracellular fluid. For example, cholesterol enters cells by receptor-mediated endocytosis.

- In the blood, cholesterol is bound to lipid and protein complexes called *low-density lipoproteins* (LDLs).
- These LDLs bind to LDL receptors on cell membranes, initiating endocytosis.
- An inherited disease call familial hypercholesterolemia is characterized by high cholesterol levels in the blood. The LDL receptors are defective, so cholesterol cannot enter the cells by endocytosis and thus accumulates in the blood, contributing to the development of atherosclerosis.

In a nongrowing cell, the amount of plasma membrane remains relatively constant.

- Vesicle fusion with the plasma membrane offsets membrane loss through endocytosis.
- Vesicles provide a mechanism to rejuvenate or remodel the plasma membrane.

XII. Specialized membrane proteins transmit extracellular signals to the inside of the cell.

In addition to their role in transport, membrane proteins interface with and respond to changes in the intracellular environment. For example,

- Specific integral proteins (*integrins*) transmit physical stimuli from the extracellular matrix to the cytoskeleton inside, influencing cell shape and movement.
- Some membrane proteins transduce chemical signals from the outside, beginning a cascade of responses inside the cell. Such *signal-transduction pathways* are often complex and begin with the binding of an extracellular molecule, such as a hormone, to a receptor. A simplified pathway might be:

Extracellular molecule (*first messenger*)
binds to membrane's receptor protein.

↓

Receptor protein activates
a relay protein in the membrane.

↓

Relay protein stimulates another membrane protein – the *effector*,
an enzyme which causes changes within the cell.

↓

The effector catalyzes the production of a cytoplasmic molecule
— the *second messenger*.

↓

Second messenger triggers metabolic and
structural responses within the cell.

REFERENCES

Alberts, B., D. Bray, J. Lewis, M. Raff, K. Roberts and J.D. Watson. *Molecular Biology of the Cell*. 2nd ed. New York: Garland, 1989.

Becker, W.M. and D.W. Deamer. *The World of the Cell*. 3rd ed. Redwood City, California: Benjamin/Cummings, 1996.

Campbell, N. *Biology*. 4th ed. Menlo Park, California: Benjamin/Cummings, 1996.

deDuve, C. *A Guided Tour of the Living Cell*. Volumes I and II. New York: Scientific American Books, 1984. Literally a guided tour of the cell with the reader as "cytonaut." This is an excellent resource for lecture material and enjoyable reading.

Kleinsmith, L.J. and V.M. Kish. *Principles of Cell Biology*. New York: Harper and Row, Publ., 1988.

CHAPTER 9

CELLULAR RESPIRATION:
HARVESTING CHEMICAL ENERGY

OUTLINE

XI.   Fermentation enables some cells to produce ATP without the help of oxygen
    A.   Comparison of Fermentation and Respiration
    B.   The Evolutionary Significance of Glycolysis

XII.   Glycolysis and the Krebs cycle connect to many other metabolic pathways
    A.   The Versatility of Catabolism
    B.   Biosynthesis (Anabolic Pathways)

XIII.   Feedback mechanisms control cellular respiration

OBJECTIVES

After reading this chapter and attending lecture, the student should be able to:

1. Diagram energy flow through the biosphere.
2. Describe the overall summary equation for cellular respiration.
3. Distinguish between substrate-level phosphorylation and oxidative phosphorylation.
4. Explain how exergonic oxidation of glucose is coupled to endergonic synthesis of ATP.
5. Define oxidation and reduction.
6. Explain how redox reactions are involved in energy exchanges.
7. Define coenzyme and list those involved in respiration.
8. Describe the structure of coenzymes and explain how they function in redox reactions.
9. Describe the role of ATP in coupled reactions.
10. Explain why ATP is required for the preparatory steps of glycolysis.
11. Describe how the carbon skeleton of glucose changes as it proceeds through glycolysis.
12. Identify where in glycolysis that sugar oxidation, substrate-level phosphorylation and reduction of coenzymes occur.
13. Write a summary equation for glycolysis and describe where it occurs in the cell.
14. Describe where pyruvate is oxidized to acetyl CoA, what molecules are produced and how it links glycolysis to the Krebs cycle.
15. Describe the location, molecules in and molecules out for the Krebs Cycle.
16. Explain at what point during cellular respiration glucose is completely oxidized.
17. Explain how the exergonic slide of electrons down the electron transport chain is coupled in the endergonic production of ATP by chemiosmosis.
18. Describe the process of chemiosmosis.
19. Explain how membrane structure is related to membrane function in chemiosmosis.
20. Summarize the net ATP yield from the oxidation of a glucose molecule by constructing an ATP ledger which includes coenzyme production during the different stages of glycolysis and cellular respiration.
21. Describe the fate of pyruvate in the absence of oxygen.
22. Explain why fermentation is necessary.
23. Distinguish between aerobic and anaerobic metabolism.
24. Describe how food molecules other than glucose can be oxidized to make ATP.
25. Describe evidence that the first prokaryotes produced ATP by glycolysis.
26. Explain how ATP production is controlled by the cell and what role the allosteric enzyme, phosphofructokinase, plays in this process.

KEY TERMS

fermentation
cellular respiration
phosphorylation
oxidative
    phosphorylation
substrate-level
    phosphorylation
oxidation
reduction
coenzyme
dinucleotide

nicotinamide adenine
    dinucleotide
flavin adenine
    dinucleotide
dehydrogenase
mitochondrion
matrix
cristae
intermembrane space
Sir Hans Krebs
electron carrier

cytochrome
heme group
electron transport
    chain
ubiquinone
proton pump
proton gradient
Peter Mitchell
chemiosmosis
electrochemical
    gradient

proton-motive force
ATP synthase
aerobic
anaerobic
fermentation
obligate aerobe
obligate anaerobe
facultative anaerobe
phosphofructokinase

LECTURE NOTES

As open systems, cells require outside energy sources to perform cellular work (e.g. chemical, transport and mechanical).

Energy flows into most ecosystems as sunlight.

Photosynthetic organisms trap a portion of the light energy and transform it into chemical bond energy of organic molecules. O_2 is released as a byproduct.

Cells use some of the chemical bond energy in organic molecules to make ATP–the energy source for cellular work.

Energy leaves living organisms as it dissipates as heat.

Light Energy

Metabolic reactions involve energy exchanges

Photosynthesis (anabolic pathways)

Chloroplast

CO_2 + H_2O

Organic molecules + O_2

Mitochondrion

Respiration (catabolic pathways)

ATP

Energy used in enzyme mediated reactions of metabolism

(Powers most cellular work)

Energy dissipated to the environment

Heat energy

The products of respiration (CO_2 and H_2O) are the raw materials for photosynthesis. Photosynthesis produces glucose and oxygen, the raw materials for respiration.

Chemical elements essential for life are recycled, but energy is not.

How do cells harvest chemical energy?

Complex organic molecules

Catabolic pathways

Simpler waste products with *less energy*

energy

Some energy used to do work & some energy dissipated as heat

I. **Cellular respiration and fermentation are catabolic (energy-yielding) pathways**

Fermentation = An ATP-producing catabolic pathway in which both electron donors and acceptors are organic compounds.

- Can be an anaerobic process.
- Results in a partial degradation of sugars.

Cellular respiration = An ATP-producing catabolic process in which the ultimate electron acceptor is an inorganic molecule, such as oxygen.

- Most prevalent and efficient catabolic pathway.
- Is an exergonic process ($\Delta G = -2870$ kJ/mol or -686 kcal/mol).
- Can be summarized as:

Organic compounds + Oxygen $\longrightarrow$ Carbon dioxide + Water + Energy
(food)

- Carbohydrates, proteins and fats can all be metabolized as fuel, but cellular respiration is most often described as the oxidation of glucose:

$$C_6H_{12}O_6 + 6 O_2 \longrightarrow 6 CO_2 + 6 H_2O + \text{Energy (ATP + Heat)}$$

II. **Cells must recycle the ATP they spend for work**

The catabolic process of respiration transfers the energy stored in food molecules to *ATP*.

ATP (adenosine triphosphate) = Nucleotide with unstable phosphate bonds that the cell hydrolyzes for energy to drive endergonic reactions.

- The cell taps energy stored in ATP by enzymatically transferring terminal phosphate groups from ATP to other compounds. (Recall that direct hydrolysis of ATP would release energy as heat, a form unavailable for cellular work. See Chapter 6.)
- The compound receiving the phosphate group from ATP is said to be *phosphorylated* and becomes more reactive in the process.
- The phosphorylated compound loses its phosphate group as cellular work is performed; inorganic phosphate and ADP are formed in the process. (See Campbell, Figure 9.2)
- Cells must replenish the ATP supply to continue cellular work. Respiration provides the energy to regenerate ATP from ADP and inorganic phosphate.

III. **Redox reactions release energy when electrons move closer to electronegative atoms**

A. **An Introduction to Redox Reactions**

Oxidation-reduction reactions = Chemical reactions which involve a partial or complete transfer of electrons from one reactant to another; called *redox reactions* for short.

Oxidation = Partial or complete loss of electrons.

Reduction = Partial or complete gain of electrons.

Generalized Redox Reaction:

Electron transfer requires both a donor and acceptor, so when one reactant is oxidized the other is reduced.

$$\text{Xe}^- + \text{Y} \longrightarrow \text{X} + \text{Ye}^-$$

(oxidation from Xe⁻ to X; reduction from Y to Ye⁻)

Where:

X = Substance being oxidized; acts as a *reducing agent* because it reduces Y.

Y = Substance being reduced; as as an *oxidizing agent* because it oxidizes X.

Not all redox reactions involve a complete transfer of electrons, but, instead, may just change the degree of sharing in covalent bonds. For example:

- Covalent electrons of methane are equally shared, because carbon and hydrogen have similar electronegativities.
- As methane reacts with oxygen to form carbon dioxide, electrons shift away from carbon and hydrogen to the more electronegative oxygen.

$$\underset{\text{methane}}{\text{CH}_4} + \underset{\text{oxygen}}{2\,\text{O}_2} \longrightarrow \underset{\substack{\text{carbon} \\ \text{dioxide}}}{\text{CO}_2} + \underset{\text{water}}{2\,\text{H}_2\text{O}} + \text{energy}$$

(oxidation; reduction)

- Since electrons *lose* potential energy when they shift *toward* more electronegative atoms, redox reactions that move electrons closer to oxygen release energy.
- Oxygen is a powerful oxidizing agent because it is so electronegative.

IV. Electrons "fall" from organic molecules to oxgyen during cellular respiration

Cellular respiration is a redox process that transfers hydrogen from sugar to oxygen.

$$\text{C}_6\text{H}_{12}\text{O}_6 + 6\,\text{O}_2 \longrightarrow 6\,\text{CO}_2 + 6\,\text{H}_2\text{O} + \text{energy (used to make ATP)}$$

(oxidation; reduction)

- Valence electrons of carbon and hydrogen lose potential energy as they shift toward electronegative oxygen.
- Released energy is used by cells to produce ATP.
- Carbohydrates and fats are excellent energy stores, because they are rich in C to H bonds.

Without the activation barrier, glucose would combine spontaneously with oxygen.
- Igniting glucose provides the activation energy for the reaction to proceed; a mole of glucose yields 686 kcal (2870 kJ) of heat when burned in air.
- Cellular respiration does not oxidize glucose in one explosive step, as the energy could not be efficiently harnessed in a form available to perform cellular work.
- Enzymes lower the activation energy in cells, so glucose can be slowly oxidized in a stepwise fashion during glycolysis and Krebs cycle.

V. The "fall" of electrons during respiration is stepwise, via NAD^+ and an electron transport chain

Hydrogens stripped from glucose are not transferred directly to oxygen, but are first passed to a special electron acceptor — NAD^+.

Nicotinamide adenine dinucleotide (NAD^+) = A *dinucleotide* that functions as a *coenzyme* in the redox reactions of metabolism. (See Campbell, Figure 9.4)
- Found in all cells.
- Assists enzymes in electron transfer during redox reactions of metabolism.

Coenzyme = Small nonprotein organic molecule that is required for certain enzymes to function.

Dinucleotide = A molecule consisting of two nucleotides.

During the oxidation of glucose, NAD^+ functions as an oxidizing agent by trapping energy-rich electrons from glucose or food. These reactions are catalyzed by enzymes called *dehydrogenases*, which:
- Remove a pair of hydrogen atoms (2 electrons and 2 protons) from substrate.
- Deliver the two electrons and *one* proton to NAD^+.
- Release the remaining proton into the surrounding solution.

Where:

$$R - \overset{\overset{\displaystyle H}{|}}{\underset{}{C}} - R' \quad = \quad \text{Substrate that is oxidized by enzymatic transfer of electrons to } NAD^+$$

NAD^+ = Oxidized coenzyme (net positive charge).

NADH = Reduced coenzyme (electrically neutral).

These high energy electrons transferred from substrate to NAD^+ are then passed down the *electron transport chain* to oxygen, powering ATP synthesis (*oxidative phosphorylation*).

NOTE: Our lecturers have found it difficult to drive this point home. Surprisingly, some students can recall the intermediate steps of glycolysis or the Krebs Cycle (which we do *not* require them to do), but cannot explain in general terms how energy from food is transferred to ATP. Figure 9.15 in the Campbell text can be used to give students an overview when respiration is introduced; it is useful to refer to here, so students can place the process you are describing in context; and it can be used later as a summary to bring closure to the topic.

Electron transport chains convert some of the chemical energy extracted from food to a form that can be used to make ATP. These transport chains:

- Are composed of electron-carrier molecules built into the inner mitochondrial membrane. Structure of this membrane correlates with its functional role (form fits function).

- Accept energy-rich electrons from reduced coenzymes (NADH and $FADH_2$); and during a series of redox reactions, pass these electrons down the chain to oxygen, the final electron acceptor. The electronegative oxygen accepts these electrons, along with hydrogen nuclei, to form water.

- Release energy from energy-rich electrons in a controlled stepwise fashion; a form that can be harnessed by the cell to power ATP production. If the reaction between hydrogen and oxygen during respiration occurred in a single explosive step, much of the energy released would be lost as heat, a form unavailable to do cellular work.

Electron transfer from NADH to oxygen is exergonic, having a free energy change of -222 kJ/mole (-53 kcal/mol).

- Since electrons lose potential energy when they shift toward a more electronegative atom, this series of redox reactions releases energy.

- Each successive carrier in the chain has a higher electronegativity than the carrier before it, so the electrons are pulled downhill towards oxygen, the final electron acceptor and the molecule with the highest electronegativity.

VI. Respiration is a cumulative function of glycolysis, the Krebs cycle and electron transport: *an overview*

There are three metabolic stages of cellular respiration:

1. Glycolysis
2. Krebs Cycle
3. Electron transport chain (ETC) and oxidative phosphorylation

Glycolysis is a catabolic pathway that:

- Occurs in the cytosol.
- Partially oxidizes glucose (6C) into two *pyruvate* (3C) molecules.

The *Krebs Cycle* is a catabolic pathway that:
- Occurs in the mitochondrial matrix.
- Completes glucose oxidation by breaking down a *pyruvate* derivative (acetyl CoA) into carbon dioxide.

Glycolysis and the *Krebs Cycle* produce:
- A *small* amount of ATP by *substrate-level phosphorylation*.
- NADH by transferring electrons from substrate to NAD^+. (Krebs Cycle also produces $FADH_2$ by transferring electrons to FAD.)

The *electron transport chain*:
- Is located at the inner membrane of the mitochondrion.
- Accepts energized electrons from reduced coenzymes (NADH and $FADH_2$) that are harvested during glycolysis and Krebs cycle. Oxygen pulls these electrons down the electron transport chain to a lower energy state.
- Couples this exergonic slide of electrons to ATP synthesis or *oxidative phosphorylation*. This process produces *most* (90%) of the ATP.

Oxidative phosphorylation = ATP production that is coupled to the exergonic transfer of electrons from food to oxygen.

A small amount of ATP is produced directly by the reactions of glycolysis and Krebs Cycle. This mechanism of producing ATP is called *substrate-level phosphorylation*.

Substrate-level phosphorylation = ATP production by direct enzymatic transfer of phosphate from an intermediate substrate in catabolism to ADP.

VII. **Glycolysis harvests chemical energy by oxidizing glucose to pyruvate:** *a closer look*

Glycolysis = (Glyco = sweet, sugar; lysis = to split) Catabolic pathway during which six-carbon glucose is split into two three-carbon sugars, which are then oxidized and rearranged by a step-wise process that produces two pyruvate molecules.
- Each reaction is catalyzed by specific enzymes dissolved in the cytosol.
- No CO_2 is released as glucose is oxidized to pyruvate; all carbon in glucose can be accounted for in the two molecules of pyruvate.
- Occurs whether or not oxygen is present.

The reactions of glycolysis occur in two phases:

Energy-investment phase. The cell uses ATP to phosphorylate the intermediates of glycolysis.

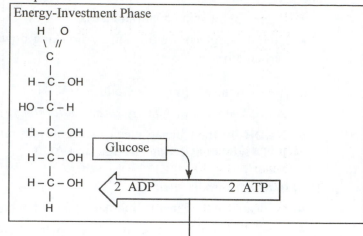

Energy-yielding phase. Two three-carbon intermediates are oxidized. For each glucose molecule entering glycolysis:

1. A net gain of two ATPs is produced by substrate-level phosphorylation.

2. Two molecules of NAD^+ are reduced to NADH. Energy conserved in the high-energy electrons of NADH can be used to make ATP by oxidative phosphorylation.

NOTE: We do not require students to memorize the structures or steps of glycolysis, but we do expect them to understand the process, where it occurs and the major molecules required and produced.   A table such as the one in the *Student Study Guide for Campbell's Biology* will help students organize the information. We have even found it helpful to summarize a lecture by filling in the table on an overhead transparency.

Energy-Investment Phase:

The *energy investment phase* includes five preparatory steps that split glucose in two. This process actually *consumes* ATP.

Step 1: Glucose enters the cell, and carbon six is phosphorylated. This ATP-coupled reaction:

- Is catalyzed by *hexokinase*. (*Kinase* is an enzyme involved in phosphate transfer.)

- Requires an initial <u>investment</u> of ATP.

- Makes glucose more chemically reactive.

- Produces glucose-6-phosphate. Since the plasma membrane is relatively impermeable to ions, addition of an electrically charged phosphate group traps the sugar in the cell.

Step 2: An *isomerase* catalyzes the rearrangement of glucose-6-phosphate to its isomer, fructose-6-phosphate.

Glucose 6–Phosphate   Fructose 6–Phosphate

Step 3: Carbon one of fructose-6-phosphate is phosphorylated. This reaction:

- Requires an investment of another ATP.

- Is catalyzed by *phosphofructokinase*, an allosteric enzyme that controls the rate of glycolysis.

Fructose 6–phosphate   Fructose 1, 6–diphosphate

Step 4: *Aldolase* cleaves the six-carbon sugar into two isomeric three-carbon sugars.

- This is the reaction for which glycolysis is named.

- For each glucose molecule that begins glycolysis, there are <u>two</u> product molecules for this and each succeeding step.

Fructose 1, 6–diphosphate   Dihydroxyacetone phosphate   Glyceraldehyde phosphate

Step 5: An isomerase catalyzes the reversible conversion between the two three-carbon sugars. This reaction:

- Never reaches equilibrium because only one isomer, *glyceraldehyde phosphate*, is used in the next step of glycolysis.

Dihydroxyyacetone phosphate   Glyceraldehyde phosphate

- Is thus pulled towards the direction of glyceraldehyde phosphate, which is removed as fast as it forms.

- Results in the net effect that, for each glucose molecule, <u>two</u> molecules of glyceraldehyde phosphate progress through glycolysis.

Energy-Yielding Phase:

The *energy-yielding phase* occurs after glucose is split into two three-carbon sugars. During these reactions, sugar is oxidized, and ATP and NADH are produced.

Step 6: An enzyme catalyzes two sequential reactions:

1. Glyceraldehyde phosphate is oxidized and NAD^+ is reduced to $NADH + H^+$.

   - This reaction is very exergonic and is coupled to the endergonic phosphorylation phase ($\Delta G = -10.3$ kcal/ mol).

   - For every glucose molecule, 2 NADH are produced.

2. Glyceraldehyde phosphate is phosphorylated on carbon one.

   - The phosphate source is inorganic phosphate, which is always present in the cytosol.

   - The new phosphate bond is a high energy bond with even more potential to transfer a phosphate group than ATP.

Step 7: ATP is produced by *substrate-level phosphorylation*.

   - In a very exergonic reaction, the phosphate group with the high energy bond is transferred from 1,3-diphosphoglycerate to ADP.

   - For each glucose molecule, two ATP molecules are produced. The ATP ledger now stands at zero as the initial debt of two ATP from steps one and three is repaid.

Step 8: In preparation for the next reaction, a phosphate group on carbon three is enzymatically transferred to carbon two.

Step 9: Enzymatic removal of a water molecule:

- Creates a double bond between carbons one and two of the substrate.
- Rearranges the substrate's electrons, which transforms the remaining phosphate bond into an unstable bond.

Step 10: ATP is produced by *substrate-level phosphorylation*.

- In a highly exergonic reaction, a phosphate group is transferred from PEP to ADP.
- For each glucose molecule, this step produces two ATP.

Summary Equation For Glycolysis:

$$\begin{aligned} &C_6H_{12}O_6 && 2\ C_3H_4O_3\ (\text{Pyruvate})\\ +\ &2\ NAD^+ && +\ 2\ NADH + 2\ H^+\\ +\ &2\ ADP\ +\ 2\ \text{(P)} \longrightarrow && +\ 2\ ATP\\ & && +\ 2\ H_2O \end{aligned}$$

- Glucose has been oxidized into two pyruvate molecules.
- The process is exergonic ($\Delta G = -140$ kcal/mol or -586 kJ/mol); most of the energy harnessed is conserved in the high-energy electrons of NADH and in the phosphate bonds of ATP.

VIII. The Krebs cycle completes the energy-yielding oxidation of organic molecules: *a closer look*

Most of the chemical energy originally stored in glucose still resides in the two pyruvate molecules produced by glycolysis. The fate of pyruvate depends upon the presence or absence of oxygen. If oxygen is present, pyruvate *enters the mitochondrion* where it is *completely oxidized* by a series of enzyme-controlled reactions.

Formation of Acetyl CoA. The junction between glycolysis and the Krebs Cycle is the oxidation of pyruvate to acetyl CoA:

- Pyruvate molecules are translocated from the cytosol into the mitochondrion by a carrier protein in the mitochondrial membrane.

- This step is catalyzed by a *multienzyme complex* which:

  1. Removes CO_2 from the carboxyl group of pyruvate, changing it from a three-carbon to a two-carbon compound. This is the first step where CO_2 is released.

  2. Oxidizes the two-carbon fragment to acetate, while reducing NAD^+ to NADH. Since glycolysis produces two pyruvate molecules per glucose, there are <u>two</u> NADH molecules produced.

  3. Attaches coenzyme A to the acetyl group, forming acetyl CoA. This bond is unstable, making the acetyl group very reactive.

Krebs Cycle: The Krebs Cycle reactions oxidize the remaining acetyl fragments of acetyl-CoA to CO_2. Energy released from this exergonic process is used to reduce coenzyme (NAD^+ and FAD) and to phosphorylate ATP (substrate-level phosphorylation).

- A German-British scientist, Hans Krebs, elucidated this catabolic pathway in the 1930's.

- The Krebs cycle, which is also known as the *citric acid cycle* or *TCA cycle*, has eight enzyme-controlled steps that occur in the *mitochondrial matrix*.

For every turn of Krebs Cycle:

- Two carbons enter in the acetyl fragment of acetyl CoA.

- Two different carbons are oxidized and leave as CO_2.

- Coenzymes are reduced; three NADH and one $FADH_2$ are produced.

- One ATP molecule is produced by substrate-level phosphorylation.

- Oxaloacetate is regenerated.

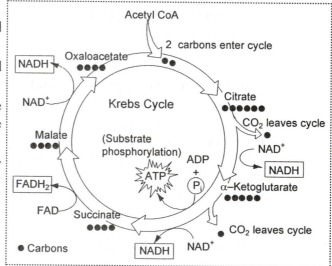

For every glucose molecule split during glycolysis:

- Two acetyl fragments are produced.
- It takes two turns of Krebs Cycle to complete the oxidation of glucose.

Steps of the Krebs Cycle: (This detailed description is now in the appendix of the Campbell text.)

Step 1: The unstable bond of acetyl CoA breaks, and the two-carbon acetyl group bonds to the four-carbon oxaloacetate to form six-carbon citrate.

Step 2: Citrate is isomerized to isocitrate.

Step 3: Two major events occur during this step:

- Isocitrate loses CO_2 leaving a *five-carbon* molecule.
- The five-carbon compound is oxidized and NAD^+ is reduced.

Step 4: A multienzyme complex catalyzes:

- Removal of CO_2.
- Oxidation of the remaining *four-carbon* compound and reduction of NAD^+.
- Attachment of CoA with a high energy bond to form succinyl CoA.

Step 5: Substrate-level phosphorylation occurs in a series of enzyme catalyzed reactions:

- The high energy bond of succinyl-CoA breaks, and some energy is conserved as CoA is displaced by a phosphate group.
- The phosphate group is transferred to GDP to form GTP and succinate.
- GTP donates a phosphate group to ADP to form ATP.

Step 6: Succinate is oxidized to fumarate and FAD is reduced.

- Two hydrogens are transferred to FAD to form FADH₂.
- The dehydrogenase that catalyzes this reaction is bound to the inner mitochondrial membrane.

Step 7: Water is added to fumarate which rearranges its chemical bonds to form malate.

Step 8: Malate is oxidized and NAD^+ is reduced.

- A molecule of NADH is produced.
- Oxaloacetate is regenerated to begin the cycle again.

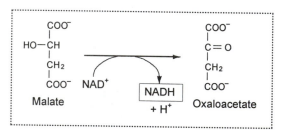

Two turns of the Krebs Cycle produces two ATPs by substrate-level phosphorylation. However, *most* ATP output of respiration results from *oxidative phosphorylation*.

- Reduced coenzymes produced by the Krebs Cycle (6 NADH and 2 FADH₂ per glucose) carry high energy electrons to the electron transport chain.
- The ETC couples electron flow down the chain to ATP synthesis.

IX. **The inner mitochondrial membrane couples electron transport to ATP synthesis:** *a closer look*

Only a few molecules of ATP are produced by substrate-level phosphorylation:

- 2 net ATPs per glucose from glycolysis.
- 2 ATPs per glucose from the Krebs Cycle.

Most molecules of ATP are produced by oxidative phosphorylation.

- At the end of the Krebs Cycle, most of the energy extracted from glucose is in molecules of NADH and $FADH_2$.
- These reduced coenzymes link glycolysis and the Krebs Cycle to oxidative phosphorylation by passing their electrons down the electron transport chain to oxygen. (Though the Krebs Cycle occurs only under aerobic conditions, it does not use oxygen directly. The ETC and oxidative phosphorylation require oxygen as the final electron acceptor.)
- This exergonic transfer of electrons down the ETC to oxygen is coupled to ATP synthesis.

A. **The Pathway of Electron Transport**

The *electron transport chain* is made of electron carrier molecules embedded in the inner mitochondrial membrane.

- Each successive carrier in the chain has a higher electronegativity than the carrier before it, so the electrons are pulled downhill towards oxygen, the final electron acceptor and the molecule with the highest electronegativity.
- Except for ubiquinone (Q), most of the carrier molecules are proteins and are tightly bound to *prosthetic groups* (nonprotein cofactors).
- Prosthetic groups alternate between reduced and oxidized states as they accept and donate electrons.

Protein Electron Carriers	Prosthetic Group
flavoproteins	flavin mononucleotide (FMN)
iron-sulfur proteins	iron and sulfur
cytochromes	heme group

Heme group = Prosthetic group composed of four organic rings surrounding a single iron atom.

Cytochrome = Type of protein molecule that contains a heme prosthetic group and that functions as an electron carrier in the electron transport chains of mitochondria and chloroplasts.

- There are several cytochromes, each a slightly different protein with a heme group.
- It is the iron of cytochromes that transfers electrons.

Sequence of Electron Transfers Along the Electron Transport Chain:

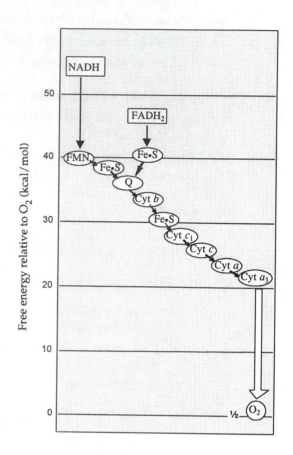

NADH is oxidized and *flavoprotein* is reduced as high energy electrons from NADH are transferred to FMN.

↓

Flavoprotein is oxidized as it passes electrons to an *iron-sulfur protein*, Fe•S.

↓

Iron-sulfur protein is oxidized as it passes electrons to *ubiquinone* (Q).

↓

Ubiquinone passes electrons on to a succession of electron carriers, most of which are cytochromes.

↓

Cyt a_3, the last cytochrome passes electrons to molecular oxygen, O_2.

↓

As molecular oxygen is reduced it also picks up two protons from the medium to form water. For every two NADHs, one O_2 is reduced to two H_2O molecules.

- $FADH_2$ also donates electrons to the electron transport chain, but those electrons are added at a lower energy level than NADH.
- The electron transport chain does not make ATP directly. It *generates a proton gradient across the inner mitochondrial membrane*, which stores potential energy that can be used to phosphorylate ADP.

B. Chemiosmosis: The Energy-Coupling Mechanism

The mechanism for coupling exergonic electron flow from the oxidation of food to the endergonic process of oxidative phosphorylation is *chemiosmosis*.

Chemiosmosis = The coupling of exergonic electron flow down an electron transport chain to endergonic ATP production by the creation of a proton gradient across a membrane. The proton gradient drives ATP synthesis as protons diffuse back across the membrane.

- Proposed by British biochemist, Peter Mitchell (1961).
- The term *chemiosmosis* emphasizes a coupling between (1) chemical reactions (phosphorylation) and (2) transport processes (proton transport).
- Process involved in oxidative phosphorylation and photophosphorylation.

The site of oxidative phosphorylation is the inner mitochondrial membrane, which has many copies of a protein complex, *ATP synthase*. This complex:

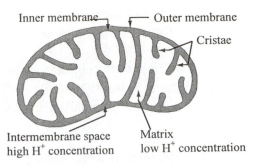

- Is an enzyme that makes ATP.
- Uses an existing *proton gradient* across the inner mitochondrial membrane to power ATP synthesis.

Cristae or infoldings of the inner mitochondrial membrane, increase the surface area available for chemiosmosis to occur.

Membrane structure correlates with the prominent functional role membranes play in chemiosmosis:

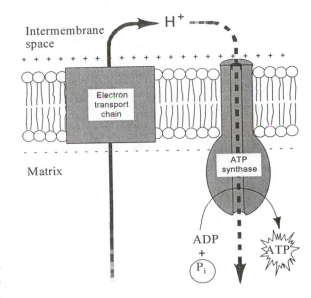

- Using energy from exergonic electron flow, the *electron transport chain* creates the proton gradient by pumping H$^+$s from the *mitochondrial matrix*, across the *inner membrane* to the *intermembrane space*.

- This proton gradient is maintained, because the membrane's phospholipid bilayer is impermeable to H$^+$s and prevents them from leaking back across the membrane by diffusion.

- *ATP synthases* use the potential energy stored in a proton gradient to make ATP by allowing H$^+$ to diffuse down the gradient, back across the membrane. Protons diffuse through the ATP synthase complex, which causes the phosphorylation of ADP.

How does the electron transport chain pump hydrogen ions from the matrix to the intermembrane space? The process is based on spatial organization of the electron transport chain in the membrane. Note that:

- Some electron carriers of the transport chain transport only electrons.
- Some electron carriers accept and release protons along with electrons. These carriers are spatially arranged so that protons are picked up from the matrix and are released into the intermembrane space.

Most of the electron carriers are organized into three complexes: 1) NADH dehydrogenase complex; 2) cytochrome b-c_1 complex; and 3) cytochrome oxidase complex. (See Campbell, Figure 9.14)

- Each complex is an asymmetric particle that has a specific orientation in the membrane.
- As complexes transport electrons, they also harness energy from this exergonic process to pump protons across the inner mitochondrial membrane.

Mobile carriers transfer electrons between complexes. These mobile carriers are:

1. Ubiquinone (Q). Near the matrix, Q accepts electrons from the NADH dehydrogenase complex, diffuses across the lipid bilayer, and passes electrons to the cytochrome b-c_1 complex.
2. Cytochrome c (Cyt c). Cyt c accepts electrons from the cytochrome b-c_1 complex and conveys them to the cytochrome oxidase complex.

When the transport chain is operating:

- The pH in the intermembrane space is one or two pH units lower than in the matrix.
- The pH in the intermembrane space is the same as the pH of the cytosol because the outer mitochondrial membrane is permeable to protons.

The H^+ gradient that results is called a *proton-motive force* to emphasize that the gradient represents potential energy.

Proton motive force = Potential energy stored in the proton gradient created across biological membranes that are involved in chemiosmosis.

- This force is an *electrochemical gradient* with two components:
    1. Concentration gradient of protons (chemical gradient).
    2. Voltage across the membrane because of a higher concentration of positively charged protons on one side (electrical gradient).
- It tends to drive protons across the membrane back into the matrix.

Chemiosmosis couples exergonic chemical reactions to endergonic H^+ transport, which creates the proton-motive force used to drive cellular work, such as:

- ATP synthesis in mitochondria (*oxidative phosphorylation*). The energy to create the proton gradient comes from the oxidation of glucose and the ETC.
- ATP synthesis in chloroplasts (*photophosphorylation*). The energy to create the proton gradient comes from light trapped during the energy-capturing reactions of photosynthesis.
- ATP synthesis, transport processes and rotation of flagella in bacteria. The proton gradient is created across the plasma membrane. Peter Mitchell first postulated chemiosmosis as an energy-coupling mechanism based on experiments with bacteria.

C. Biological Themes and Oxidative Phosphorylation

The working model of how mitochondria harvest the energy of food, illustrates many of the text's integrative themes in the study of life:

- Energy conversion and utilization.
- Emergent properties. Oxidative phosphorylation is an emergent property of the intact mitochondrion that uses a precise interaction of molecules.
- Correlation of structure and function. The chemiosmotic model is based upon the spatial arrangement of membrane proteins.
- Evolution. In an effort to reconstruct the origin of oxidative phosphorylation and the evolution of cells, biologists compare similarities in the chemiosmotic machinery of mitochondria to that of chloroplasts and bacteria.

X. Cellular respiration generates many ATP molecules for each sugar molecule it oxidizes: *a review*

During cellular respiration, *most* energy flows in this sequence:

Glucose $\Rightarrow$ NADH $\Rightarrow$ electron transport chain $\Rightarrow$ proton motive force $\Rightarrow$ ATP

The *net* ATP yield from the oxidation of one glucose molecule to six carbon dioxide molecules can be estimated by adding:

1. ATP produced directly by substrate-level phosphorylation during glycoloysis and the Krebs cycle.
    - A net of two ATPs is produced during glycolysis. The debit of two ATPs used during the investment phase is subtracted from the four ATPs produced during the energy-yielding phase.
    - Two ATPs are produced during the Krebs cycle.

2. ATP produced when chemiosmosis couples electron transport to oxidative phosphorylation.
    - The electron transport chain creates enough proton-motive force to produce a maximum of *three ATPs* for each electron pair that travels from NADH to oxygen. The average yield is actually between two and three ATPs per NADH (2.7).
    - $FADH_2$ produced during the Krebs Cycle is worth a maximum of only *two ATPs*, since it donates electrons at a lower energy level to the electron transport chain.
    - In most eukaryotic cells, the ATP yield is lower from an NADH produced during glycolysis. The mitochondrial membrane is impermeable to NADH, so its electrons must be carried across the membrane in another molecule. These electrons are received inside the mitochondrion by FAD, a process which downgrades the energy level of those electrons.

- Maximum ATP yield for each glucose oxidized during cellular respiration:

Process	ATP Produced Directly by Substrate-Level Phosphorylation	Reduced Coenzyme	ATP Produced by Oxidative Phosphorylation	Total
Glycolysis	*Net* 2 ATP	2 NADH	4 to 6 ATP	6-8
Oxidation of Pyruvate	———	2 NADH	6 ATP	6
Krebs Cycle	2 ATP	6 NADH 2 FADH$_2$	18 ATP 4 ATP	24
			Total	36-38

- This tally only <u>estimates</u> the ATP yield from respiration. Some variables that affect ATP yield include:

  ⇒ The proton-motive force may be used to drive other kinds of cellular work such as active transport.

  ⇒ The total ATP yield is inflated (~10%) by rounding off the number of ATPs produced per NADH to three.

Cellular respiration is remarkably efficient in the transfer of chemical energy from glucose to ATP.

- Estimated efficiency in eukaryotic cells is about 38%.
- Energy lost in the process is released as heat.

XI.   Fermentation enables some cells to produce ATP without the help of oxygen

Food can be oxidized under *anaerobic* conditions.

<u>Aerobic</u> = (Aer = air; bios = life) Existing in the presence of oxygen.

<u>Anaerobic</u> = (An = without; aer = air) Existing in the absence of free oxygen.

<u>Fermentation</u> = The anaerobic catabolism of organic nutrients.

Glycolysis oxidizes glucose to two pyruvate molecules, and the oxidizing agent for this process is NAD$^+$, *not* oxygen.

- Some energy released from the exergonic process of glycolysis drives the production of 2 net ATPs by substrate-level phosphorylation.
- Glycolysis produces a net of 2 ATPs whether conditions are aerobic or anaerobic.

1. *Aerobic conditions*: Pyruvate is *oxidized* further, and more ATP is made as NADH passes electrons removed from glucose to the electron transport chain. NAD^+ is regenerated in the process.

2. *Anaerobic conditions*: Pyruvate is *reduced*, and NAD^+ is regenerated. This prevents the cell from depleting the pool of NAD^+, which is the oxidizing agent necessary for glycolysis to continue. No additional ATP is produced.

Fermentation recycles NAD^+ from NADH. This process consists of anaerobic glycolysis plus subsequent reactions that regenerate NAD^+ by reducing pyruvate. Two of the most common types of fermentation are (1) alcohol fermentation and (2) lactic acid fermentation. (See Campbell, Figure 9.16)

1. Alcohol Fermentation

Pyruvate is converted to ethanol in two steps:

a. Pyruvate loses carbon dioxide and is converted to the two-carbon compound acetaldehyde.

b. NADH is oxidized to NAD^+ and acetaldehyde is reduced to ethanol.

Many bacteria and yeast carry out alcohol fermentation under anaerobic conditions.

2. Lactic Acid Fermentation

NADH is oxidized to NAD^+ and pyruvate is reduced to lactate.

- Commercially important products of lactic acid fermentation include cheese and yogurt.

- When oxygen is scarce, human muscle cells switch from aerobic respiration to lactic acid fermentation. Lactate accumulates, but it is gradually carried to the liver where it is converted back to pyruvate when oxygen becomes available.

A. Comparison of Fermentation and Respiration

The anaerobic process of fermentation and aerobic process of cellular respiration are similar in that both metabolic pathways:

- use glycolysis to oxidize glucose and other substrates to pyruvate, producing a net of 2 ATPs by substrate phosphorylation.
- use NAD^+ as the oxidizing agent that accepts electrons from food during glycolysis.

Fermentation and cellular respiration differ in:

- how NADH is oxidized back to NAD^+. Recall that the oxidized form, NAD^+, is necessary for glycolysis to continue.
  - ⇒ During fermentation, NADH passes electrons to pyruvate or some derivative. As pyruvate is reduced, NADH is oxidized to NAD^+. Electrons transferred from NADH to pyruvate or other substrates are not used to power ATP production.
  - ⇒ During cellular respiration, the stepwise electron transport from NADH to oxygen not only drives oxidative phosphorylation, but regenerates NAD^+ in the process.
- the final electron acceptor.
  - ⇒ In fermentation, the final electron acceptor is pyruvate (lactic acid fermentation), acetaldehyde (alcohol fermentation), or some other organic molecule.
  - ⇒ In cellular respiration, the final electron acceptor is oxygen.
- the amount of energy harvested.
  - ⇒ During fermentation, energy stored in pyruvate is unavailable to the cell.
  - ⇒ Cellular respiration yields 18 times more ATP per glucose molecule than does fermentation. The higher energy yield is a consequence of the Krebs Cycle which completes the oxidation of glucose and thus taps the chemical bond energy still stored in pyruvate at the end of glycolysis.
- their requirement for oxygen.
  - ⇒ Fermentation does not require oxygen.
  - ⇒ Cellular respiration occurs only in the presence of oxygen.

Organisms can be classified based upon the effect oxygen has on growth and metabolism.

<u>Strict (obligate) aerobes</u> = Organisms that require oxygen for growth and as the final electron acceptor for aerobic respiration.

<u>Strict (obligate) anaerobes</u> = Microorganisms that only grow in the absence of oxygen and are, in fact, poisoned by it.

Facultative anaerobes = Organisms capable of growth in either aerobic or anaerobic environments.

- Yeasts, many bacteria and mammalian muscle cells are facultative anaerobes.
- Can make ATP by fermentation in the absence of oxygen or by respiration in the presence of oxygen.
- Glycolysis is common to both fermentation and respiration, so pyruvate is a key juncture in catabolism.

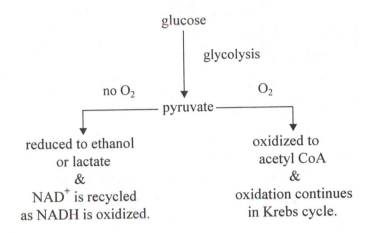

B. The Evolutionary Significance of Glycolsysis

The first prokaryotes probably produced ATP by glycolysis. Evidence includes the following:

- Glycolysis does not require oxygen, and the oldest known bacterial fossils date back to three-and-a-half billion years ago when oxygen was not present in the atmosphere.
- Glycolysis is the most widespread metabolic pathway, so it probably evolved early.
- Glycolysis occurs in the cytosol and does not require membrane-bound organelles. Eukaryotic cells with organelles probably evolved about two billion years after prokaryotic cells.

XII. Glycolysis and the Krebs cycle connect to many other metabolic pathways

A. The Versatility of Catabolism

Respiration can oxidize organic molecules other than glucose to make ATP. Organisms obtain most calories from fats, proteins, disaccharides and polysaccharides. These complex molecules must be enzymatically hydrolyzed into simpler molecules or monomers that can enter an intermediate reaction of glycolysis or the Krebs cycle. (See Campbell, Figure 9.19)

Glycolysis can accept a wide range of carbohydrates for catabolism.

- Starch is hydrolyzed to glucose in the digestive tract of animals.
- In between meals, the liver hydrolyzes glycogen to glucose.
- Enzymes in the small intestine break down disaccharides to glucose or other monosaccharides.

Proteins are hydrolyzed to amino acids.

- Organisms synthesize new proteins from some of these amino acids.
- Excess amino acids are enzymatically converted to intermediates of glycolysis and the Krebs cycle. Common intermediates are pyruvate, acetyl CoA and α-ketoglutarate.
- This conversion process deaminates amino acids, and the resulting nitrogenous wastes are excreted.

Fats are excellent fuels because they are rich in hydrogens with high energy electrons. Oxidation of one gram of fat produces twice as much ATP as a gram of carbohydrate.

- Fat sources may be from the diet or from storage cells in the body.
- Fats are digested into glycerol and fatty acids.
- Glycerol can be converted to glyceraldehyde phosphate, an intermediate of glycolysis.
- Most energy in fats is in fatty acids, which are converted into acetyl CoA by *beta oxidation*. The resulting two-carbon fragments can enter the Krebs cycle.

B. Biosynthesis (Anabolic Pathways)

Some organic molecules of food provide the carbon skeletons or raw materials for the synthesis of new macromolecules.

- Some organic monomers from digestion can be used *directly* in anabolic pathways.
- Some precursors for biosynthesis do not come directly from digested food, but instead come from glycolysis or Krebs cycle intermediates which are diverted into anabolic pathways.
- These anabolic pathways *consume* ATP produced by catabolic pathways of glycolysis and respiration.
- Glycolysis and the Krebs cycle are metabolic interchanges that can convert one type of macromolecule to another in response to the cell's metabolic demands.

XIII. Feedback mechanisms control cellular respiration

Cells respond to changing metabolic needs by controlling reaction rates.

- Anabolic pathways are switched off when their products are in ample supply. The most common mechanism of control is *feedback inhibition*. (See Campbell, Chapter 6.)
- Catabolic pathways, such as glycolysis and Krebs cycle, are controlled by regulating enzyme activity at strategic points.

A key control point of catabolism is the third step of glycolysis, which is catalyzed by an allosteric enzyme, *phosphofructokinase*.

- The ratio of ATP to ADP and AMP reflects the energy status of the cell, and phosphofructokinase is sensitive to changes in this ratio.
- Citrate (produced in Krebs cycle) and ATP are *allosteric inhibitors* of phosphofructokinase, so when their concentrations rise, the enzyme slows glycolysis. As the rate of glycolysis slows, Krebs cycle also slows since the supply of acetyl CoA is reduced. This synchronizes the rates of glycolysis and Krebs cycle.
- ADP and AMP are *allosteric activators* for phosphofructokinase, so when their concentrations relative to ATP rise, the enzyme speeds up glycolysis which speeds up the Krebs cycle.
- There are other allosteric enzymes that also control the rates of glycolysis and the Krebs cycle.

REFERENCES

Campbell, N. *Biology*. 4th ed. Menlo Park, California: Benjamin/Cummings, 1996.

Lehninger, A.L., D.L. Nelson and M.M. Cox. *Principles of Biochemistry*. 2nd ed. New York: Worth, 1993.

Matthews, C.K. and K.E. van Holde. *Biochemistry*. 2nd ed. Redwood City, California: Benjamin/Cummings, 1996.

CHAPTER 10

PHOTOSYNTHESIS

OUTLINE

OBJECTIVES

After reading this chapter and attending lecture, the student should be able to:
1. Distinguish between autotrophic and heterotrophic nutrition.
2. Distinguish between photosynthetic autotrophs and chemosynthetic autotrophs.
3. Describe the location and structure of the chloroplast.
4. Explain how chloroplast structure relates to its function.
5. Write a summary equation for photosynthesis.
6. Explain van Niel's hypothesis and describe how it contributed to our current understanding of photosynthesis.
7. Explain the role of REDOX reactions in photosynthesis.
8. Describe the wavelike and particlelike behaviors of light.
9. Describe the relationship between an action spectrum and an absorption spectrum.
10. Explain why the absorption spectrum for chlorophyll differs from the action spectrum for photosynthesis.
11. List the wavelengths of light that are most effective for photosynthesis.
12. Explain what happens when chlorophyll or accessory pigments absorb photons.
13. List the components of a photosystem and explain their function.
14. Trace electron flow through photosystems II and I.
15. Compare cyclic and noncyclic electron flow and explain the relationship between these components of the light reactions.
16. Summarize the light reactions with an equation and describe where they occur.
17. Describe important differences in chemiosmosis between oxidative phosphorylation in mitochondria and photophosphorylation in chloroplasts.
18. Summarize the carbon-fixing reactions of the Calvin cycle and describe changes that occur in the carbon skeleton of the intermediates.
19. Describe the role of ATP and NADPH in the Calvin cycle.
20. Describe what happens to rubisco when the O_2 concentration is much higher than CO_2.
21. Describe the major consequences of photorespiration.
22. Describe two important photosynthetic adaptations that minimize photorespiration.
23. Describe the fate of photosynthetic products.

KEY TERMS

autotrophic	C.B. van Niel	accessory pigment	cyclic electron flow
photoautotroph	electromagnetic	chlorophyll *b*	noncyclic electron
chemoautotroph	radiation	carotenoids	flow
heterotrophic	wavelength	ground state	photophosphorylation
chloroplast	electromagnetic	excited state	Calvin cycle
chlorophyll	spectrum	fluorescence	glyceraldehyde
mesophyll	photon	photosystem	phosphate
stomata	pigments	antenna assembly	rubisco
vascular bundles	spectrophotometer	reaction center	RuBP
stroma	photooxidation	primary electron	Melvin Calvin
thylakoid	absorption spectrum	acceptor	photorespiration
thylakoid membrane	action spectrum	P700	C_4 pathway
thylakoid space	chlorophyll *a*	P680	CAM pathway

LECTURE NOTES

Photosynthesis transforms solar light energy trapped by chloroplasts into chemical bond energy stored in sugar and other organic molecules. This process:

- Synthesizes energy-rich organic molecules from the energy-poor molecules, CO_2 and H_2O.
- Uses CO_2 as a carbon source and light energy as the energy source.
- Directly or indirectly supplies energy to most living organisms.

I. Plants and other autotrophs are the producers of the biosphere

Organisms acquire organic molecules used for energy and carbon skeletons by one of two nutritional modes: 1) *Autotrophic Nutrition* or 2) *Heterotrophic Nutrition*.

Autotrophic nutrition = (Auto = self; trophos = feed) Nutritional mode of synthesizing organic molecules from inorganic raw materials.

- Examples of autotrophic organisms are plants, which require only CO_2, H_2O and minerals as nutrients.
- Because autotrophic organisms produce organic molecules that enter an ecosystem's food store, autotrophs are also known as *producers*.
- Autotrophic organisms require an energy source to synthesize organic molecules. That energy source may be from light (*photoautotrophic*) or from the oxidation of inorganic substances (*chemoautotrophic*).

    Photoautotrophs = Autotrophic organisms that use light as an energy source to synthesize organic molecules. Examples are photosynthetic organisms such as plants, algae and some prokaryotes.

    Chemoautotrophs = Autotrophic organisms that use the oxidation of inorganic substances, such as sulfur or ammonia, as an energy source to synthesize organic molecules. Unique to some bacteria, this is a rarer form of autotrophic nutrition.

Heterotrophic nutrition = (Heteros = other; trophos = feed) Nutritional mode of acquiring organic molecules from compounds produced by other organisms; heterotrophs are unable to synthesize organic molecules from inorganic raw materials.

- Heterotrophs are also known as *consumers*.
- Examples are animals that eat plants or other animals.
- Examples also include *decomposers*, heterotrophs that decompose and feed on organic litter. Most fungi and many bacteria are decomposers.
- Most heterotrophs depend on photoautotrophs for food and oxygen (a by-product of photosynthesis).

II.   Chloroplasts are the sites of photosynthesis in plants

Although all green plant parts have chloroplasts, leaves are the major organs of photosynthesis in most plants.

- *Chlorophyll* is the green pigment in chloroplasts that gives a leaf its color and that absorbs the light energy used to drive photosynthesis.

LEAF CROSS SECTION

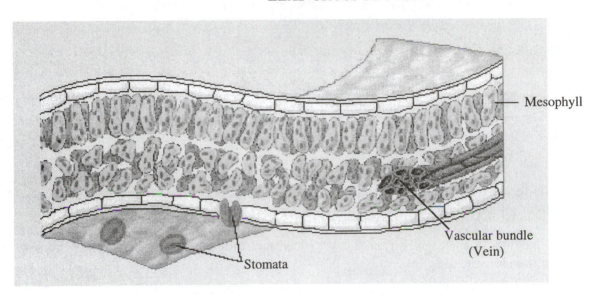

- Chloroplasts are primarily in cells of *mesophyll*, green tissue in the leaf's interior.
- CO_2 enters and O_2 exits the leaf through microscopic pores called *stomata*.
- Water absorbed by the roots is transported to leaves through veins or *vascular bundles* which also export sugar from leaves to nonphotosynthetic parts of the plant.

Chloroplasts are lens-shaped organelles measuring about 2 – 4 µm by 4 – 7 µm. These organelles are divided into three functional compartments by a system of membranes:

1. **Intermembrane Space** . The chloroplast is bound by a double membrane which partitions its contents from the cytosol. A narrow *intermembrane space* separates the two membranes.

2. **Thylakoid Space**. *Thylakoids* form another membranous system within the chloroplast. The thylakoid membrane segregates the interior of the chloroplast into two compartments: *thylakoid space* and *stroma*.

Thylakoids = Flattened membranous sacs inside the chloroplast.

- Chlorophyll is found in the thylakoid membranes.
- Thylakoids function in the steps of photosynthesis that initially convert light energy to chemical energy.

Thylakoid space = Space inside the thylakoid.

Grana = (Singular, granum) Stacks of thylakoids in a chloroplast.

3. **Stroma**. Reactions that use chemical energy to convert carbon dioxide to sugar occur in the *stroma*, viscous fluid outside the thylakoids.

Photosynthetic prokaryotes lack chloroplasts, but have chlorophyll built into the plasma membrane or into membranes of numerous vesicles within the cell.

- These membranes function in a manner similar to the thylakoid membranes of chloroplasts.
- Photosynthetic membranes of cyanobacteria are usually arranged in parallel stacks of flattened sacs similar to the thylakoids of chloroplasts.

III. **Evidence that chloroplasts split water molecules enabled researchers to track atoms through photosynthesis:** *science as a process*

Some steps in photosynthesis are not yet understood, but the following summary equation has been known since the early 1800's:

$$6\,CO_2 \; + \; 12\,H_2O \; + \; \text{light energy} \longrightarrow C_6H_{12}O_6 \; + \; 6\,O_2 \; + \; 6\,H_2O$$

- Glucose ($C_6H_{12}O_6$) is shown in the summary equation, though the main products of photosynthesis are other carbohydrates.
- Water is on both sides of the equation, because photosynthesis consumes 12 molecules and forms 6.

Indicating the net consumption of water simplifies the equation:

$$6\,CO_2 \; + \; 6\,H_2O \; + \; \text{light energy} \longrightarrow C_6H_{12}O_6 \; + \; 6\,O_2$$

- In this form, the summary equation for photosynthesis is the reverse of that for cellular respiration.
- Photosynthesis and cellular respiration both occur in plant cells, but plants do not simply reverse the steps of respiration to make food.

The simplest form of the equation is: $CO_2 \; + \; H_2O \longrightarrow CH_2O \; + \; O_2$

- CH_2O symbolizes the general formula for a carbohydrate.
- In this form, the summary equation emphasizes the production of a sugar molecule, one carbon at a time. Six repetitions produces a glucose molecule.

A.   The Splitting of Water

The discovery that O_2 released by plants is derived from H_2O and not from CO_2, was one of the earliest clues to the mechanism of photosynthesis.

- In the 1930's, C.B. van Niel from Stanford University challenged an early model, which predicted that:

    a. O_2 released during photosynthesis comes from CO_2.
    $$CO_2 \longrightarrow C + O_2$$

    b. CO_2 is split and water is added to the carbon.
    $$C + H_2O \longrightarrow CH_2O$$

- Van Niel studied bacteria that use hydrogen sulfide (H_2S) rather than H_2O for photosynthesis and that produce yellow sulfur globules as a by-product.
$$CO_2 + 2\,H_2S \longrightarrow CH_2O + H_2O + 2\,S$$

- Van Niel deduced that these bacteria split H_2S and use H to make sugar. He generalized that all photosynthetic organisms require hydrogen, but that the source varies:

general:	$CO_2 + 2\,H_2X \longrightarrow CH_2O + H_2O + 2\,X$
sulfur bacteria:	$CO_2 + 2\,H_2S \longrightarrow CH_2O + H_2O + 2\,S$
plants:	$CO_2 + 2\,H_2O \longrightarrow CH_2O + H_2O + O_2$

- Van Niel thus hypothesized that plants split water as a source of hydrogen and release oxygen as a by-product.

Scientists later confirmed van Niel's hypothesis by using a heavy isotope of oxygen (^{18}O) as a tracer to follow oxygen's fate during photosynthesis.

- If water was labeled with tracer, released oxygen was ^{18}O:

    Experiment 1: $CO_2 + H_2O^* \longrightarrow CH_2O + H_2O + O_2^*$

- If the ^{18}O was introduced to the plant as CO_2, the tracer did not appear in the released oxygen:

    Experiment 2: $CO_2^* + H_2O \longrightarrow CH_2O^* + H_2O^* + O_2$

An important result of photosynthesis is the extraction of hydrogen from water and its incorporation into sugar.

- Electrons associated with hydrogen have more potential energy in organic molecules than they do in water, where the electrons are closer to electronegative oxygen.
- Energy is stored in sugar and other food molecules in the form of these high-energy electrons.

B.   Photosynthesis as a Redox Process

Respiration is an exergonic redox process; energy is *released* from the oxidation of sugar.

- Electrons associated with sugar's hydrogens lose potential energy as carriers transport them to oxygen, forming water.
- Electronegative oxygen pulls electrons down the electron transport chain, and the potential energy released is used by the mitochondrion to produce ATP.

Photosynthesis is an endergonic redox process; energy is *required* to reduce carbon dioxide.

- Light is the energy source that boosts potential energy of electrons as they are moved from water to sugar.
- When water is split, electrons are transferred from the water to carbon dioxide, reducing it to sugar.

IV. The light reactions and the Calvin cycle cooperate in transforming light to the chemical energy of food: *an overview*

Photosynthesis occurs in two stages: the *light reactions* and the *Calvin cycle*.

Light reactions = In photosynthesis, the reactions that convert light energy to chemical bond energy in ATP and NADPH. These reactions:

- Occur in the thylakoid membranes of chloroplasts.
- Reduce $NADP^+$ to NADPH.
    - ⇒ Light absorbed by chlorophyll provides the energy to reduce $NADP^+$ to NADPH, which temporarily stores the energized electrons transferred from water.
    - ⇒ $NADP^+$ (nicotinamide adenine dinucleotide phosphate), a coenzyme similar to NAD^+ in respiration, is reduced by adding a pair of electrons along with a hydrogen nucleus, or H^+.
- Give off O_2 as a by-product from the splitting of water.
- Generate ATP. The light reactions power the addition of a phosphate group to ADP in a process called *photophosphorylation*.

Calvin cycle = In photosynthesis, the carbon-fixation reactions that assimilate atmospheric CO_2 and then reduce it to a carbohydrate; named for Melvin Calvin. These reactions:

- Occur in the stroma of the chloroplast.
- First incorporate atmospheric CO_2 into existing organic molecules by a process called *carbon fixation*, and then reduce fixed carbon to carbohydrate.

    Carbon fixation = The process of incorporating CO_2 into organic molecules.

The Calvin cycle reactions do not require light directly, but reduction of CO_2 to sugar requires the *products* of the light reactions:

- NADPH provides the reducing power.
- ATP provides the chemical energy.

Chloroplasts thus use light energy to make sugar by coordinating the two stages of photosynthesis. (See Campbell, Figure 10.4)

- Light reactions occur in the thylakoids of chloroplasts.
- Calvin cycle reactions occur in the stroma.
- As $NADP^+$ and ADP contact thylakoid membranes, they pick up electrons and phosphate respectively, and then transfer their high-energy cargo to the Calvin cycle.

V. **The light reactions transform solar energy to the chemical energy of ATP and NADPH:** *a closer look*

To understand how the thylakoids of chloroplasts transform light energy into the chemical energy of ATP and NADPH, it is necessary to know some important properties of light.

A. **The Nature of Sunlight**

Sunlight is *electromagnetic energy*. The quantum mechanical model of electromagnetic radiation describes light as having a behavior that is both wavelike and particlelike.

1. Wavelike properties of light.

- *Electromagnetic energy* is a form of energy that travels in rhythmic waves which are disturbances of electric and magnetic fields.

- A *wavelength* is the distance between the crests of electromagnetic waves.

- The electromagnetic spectrum ranges from wavelengths that are less than a nanometer (gamma rays) to those that are more than a kilometer (radio waves). (See Campbell, Figure 10.5)

- *Visible light*, which is detectable by the human eye, is only a small portion of the electromagnetic spectrum and ranges from about 380 to 750 nm. The wavelengths most important for photosynthesis are within this range of visible light.

2. Particlelike properties of light.

- Light also behaves as if it consists of discrete particles or quanta called *photons*.

- Each photon has a fixed quantity of energy which is *inversely* proportional to the wavelength of light. For example, a photon of violet light has nearly twice as much energy as a photon of red light.

The sun radiates the full spectrum of electromagnetic energy.

- The atmosphere acts as a selective window that allows visible light to pass through while screening out a substantial fraction of other radiation.

- The visible range of light is the radiation that drives photosynthesis.

- Blue and red, the two wavelengths most effectively absorbed by chlorophyll, are the colors most useful as energy for the light reactions.

B. **Photosynthetic Pigments: The Light Receptors**

Light may be reflected, transmitted or absorbed when it contacts matter. (See Campbell, Figure 10.6)

Pigments = Substances that absorb visible light.

- Different pigments absorb different wavelengths of light.

- Wavelengths that are absorbed disappear, so a pigment that absorbs all wavelengths appears black.

- When white light, which contains all the wavelengths of visible light, illuminates a pigment, the color you see is the color most reflected or transmitted by the pigment. For example, a leaf appears green because chlorophyll absorbs red and blue light but transmits and reflects green light.

Each pigment has a characteristic *absorption spectrum* or pattern of wavelengths that it absorbs. It is expressed as a graph of absorption versus wavelength.

- The absorption spectrum for a pigment in solution can be determined by using a *spectrophotometer*, an instrument used to measure what proportion of a specific wavelength of light is absorbed or transmitted by the pigment. (See Campbell Methods Box)

- Since chlorophyll *a* is the light-absorbing pigment that participates directly in the light reactions, the absorption spectrum of chlorophyll *a* provides clues as to which wavelengths of visible light are most effective for photosynthesis.

A graph of wavelength versus rate of photosynthesis is called an *action spectrum* and profiles the relative effectiveness of different wavelengths of visible light for driving photosynthesis.

- The action spectrum of photosynthesis can be determined by illuminating chloroplasts with different wavelengths of light and measuring some indicator of photosynthetic rate, such as oxygen release or carbon dioxide consumption.

- It is apparent from the action spectrum of photosynthesis that blue and red light are the most effective wavelengths for photosynthesis and green light is the least effective.

The *action spectrum* for photosynthesis does not exactly match the *absorption spectrum* for chlorophyll *a*.

- Since chlorophyll *a* is not the only pigment in chloroplasts that absorb light, the absorption spectrum for chlorophyll *a* underestimates the effectiveness of some wavelengths.

- Even though only special chlorophyll *a* molecules can participate directly in the light reactions, other pigments, called *accessory pigments*, can absorb light and transfer the energy to chlorophyll *a*.

The *accessory pigments* expand the range of wavelengths available for photosynthesis. These pigments include:

- *Chlorophyll b*, a yellow-green pigment with a structure similar to chlorophyll *a*. This minor structural difference gives the pigments slightly different absorption spectra.

- *Carotenoids*, yellow and orange hydrocarbons that are built into the thylakoid membrane with the two types of chlorophyll.

C. The Photoexcitation of Chlorophyll

What happens when chlorophyll or accessory pigments absorb photons?

- Colors of absorbed wavelengths disappear from the spectrum of transmitted and reflected light.
- The absorbed photon boosts one of the pigment molecule's electrons in its lowest-energy state (*ground state*) to an orbital of higher potential energy (*excited state*).

The only photons absorbed by a molecule are those with an energy state equal to the difference in energy between the ground state and excited state.

- This energy difference varies from one molecule to another. Pigments have unique absorption spectra because pigments only absorb photons corresponding to specific wavelengths.
- The photon energy absorbed is converted to potential energy of an electron elevated to the excited state.

The excited state is unstable, so excited electrons quickly fall back to the ground state orbital, releasing excess energy in the process. This released energy may be:

- Dissipated as heat.
- Reradiated as a photon of lower energy and longer wavelength than the original light that excited the pigment. This afterglow is called *fluorescence*.

Pigment molecules do not fluoresce when in the thylakoid membranes, because nearby *primary electron acceptor* molecules trap excited state electrons that have absorbed photons.

- In this redox reaction, chlorophyll is photooxidized by the absorption of light energy and the electron acceptor is reduced.
- Because no primary electron acceptor is present, *isolated* chlorophyll fluoresces in the red part of the spectrum and dissipates heat.

D.   Photosystems: Light-Harvesting Complexes of the Thylakoid Membrane

Chlorophyll *a*, chlorophyll *b* and the carotenoids are assembled into *photosystems* located within the thylakoid membrane. Each photosystem is composed of:

1. *antenna complex.*

   - Several hundred chlorophyll *a*, chlorophyll *b* and carotenoid molecules are light-gathering antennae that absorb photons and pass the energy from molecule to molecule. This process of resonance energy transfer is called *inductive resonance*.
   - Different pigments within the antennal complex have slightly different absorption spectra, so collectively they can absorb photons from a wider range of the light spectrum than would be possible with only one type of pigment molecule.

2. *reaction-center chlorophyll.*

- Only one of the many chlorophyll *a* molecules in each complex can actually *transfer* an excited electron to initiate the light reactions. This specialized chlorophyll *a* is located in the *reaction center*.

3. *primary electron acceptor.*

- Located near the reaction center, a primary electron acceptor molecule traps excited state electrons released from the reaction center chlorophyll.

- The transfer of excited state electrons from chlorophyll to primary electron acceptor molecules is the first step of the light reactions. The energy stored in the trapped electrons powers the synthesis of ATP and NADPH in subsequent steps.

Two types of photosystems are located in the thylakoid membranes, photosystem I and photosystem II.

- The reaction center of photosystem I has a specialized chlorophyll *a* molecule known as *P700*, which absorbs best at 700 nm (the far red portion of the spectrum).

- The reaction center of photosystem II has a specialized chlorophyll *a* molecule known as *P680*, which absorbs best at a wavelength of 680 nm.

- P700 and P680 are identical chlorophyll *a* molecules, but each is associated with a different protein. This affects their electron distribution and results in slightly different absorption spectra.

E. **Noncyclic Electron Flow**

There are two possible routes for electron flow during the light reactions: *noncyclic flow* and *cyclic flow*.

Both photosystem I and photosystem II function and cooperate in noncyclic electron flow which transforms light energy to chemical energy stored in the bonds of NADPH and ATP. This process:

- Occurs in the thylakoid membrane.
- Passes electrons continuously from water to $NADP^+$.
- Produces ATP by *noncyclic photophosphorylation*.
- Produces NADPH.
- Produces O_2.

Light excites electrons from P700, the reaction center chlorophyll in photosystem I. These excited state electrons do not return to the reaction center chlorophyll, but are ultimately stored in NADPH, which will later be the electron donor in the Calvin Cycle.

- Initially, the excited state electrons are transferred from P700 to the primary electron acceptor for photosystem I.

- The primary electron acceptor passes these excited state electrons to *ferredoxin* (Fd), an iron-containing protein.

- *NADP$^+$ reductase* catalyzes the redox reaction that transfers these electrons from ferredoxin to $NADP^+$, producing reduced coenzyme – NADPH.

- The oxidized P700 chlorophyll becomes an oxidizing agent as its electron "holes" must be filled; photosystem II supplies the electrons to fill these holes.

When the antenna assembly of photosystem II absorbs light, the energy is transferred to the P680 reaction center .

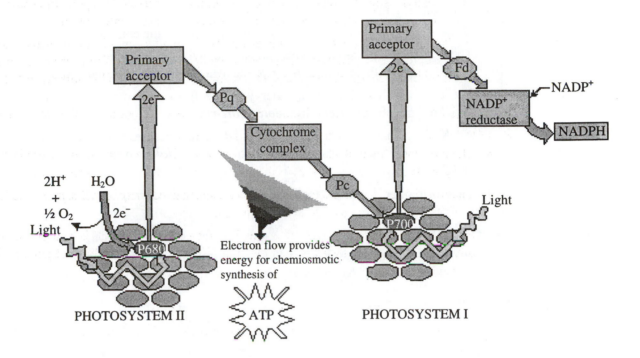

- Electrons ejected from P680 are trapped by the photosystem II primary electron acceptor.

- The electrons are then transferred from this primary electron acceptor to an electron transport chain embedded in the thylakoid membrane. The first carrier in the chain, *plastoquinone* (Pq) receives the electrons from the primary electron acceptor. In a cascade of redox reactions, the electrons travel from Pq to a complex of two cytochromes to plastocyanin (Pc) to P700 of photosystem I.

- As these electrons pass down the electron transport chain, they lose potential energy until they reach the ground state of P700.

- These electrons then fill the electron vacancies left in photosystem I when NADP$^+$ was reduced.

Electrons from P680 flow to P700 during noncyclic electron flow, restoring the missing electrons in P700. This, however, leaves the P680 reaction center of photosystem II with missing electrons; the oxidized P680 chlorophyll thus becomes a strong oxidizing agent.

- A water-splitting enzyme extracts electrons from water and passes them to oxidized P680, which has a high affinity for electrons.

- As water is oxidized, the removal of electrons splits water into two hydrogen ions and an oxygen atom.

- The oxygen atom immediately combines with a second oxygen atom to form O_2. It is this water-splitting step of photosynthesis that releases O_2.

As excited electrons give up energy along the transport chain to P700, the thylakoid membrane couples the exergonic flow of electrons to the endergonic reactions that phosphorylate ADP to ATP.

- This coupling mechanism is *chemiosmosis*.
- Some electron carriers can only transport electrons in the company of protons.
- The protons are picked up on one side of the thylakoid membrane and deposited on the opposite side as the electrons move to the next member of the transport chain.
- The electron flow thus stores energy in the form of a proton gradient across the thylakoid membrane – a *proton-motive force*.
- An ATP synthase enzyme in the thylakoid membrane uses the proton-motive force to make ATP. This process is called *photophosphorylation* because the energy required is light.
- This form of ATP production is called *noncyclic photophosphorylation*.

F.   **Cyclic Electron Flow**

Cyclic electron flow is the simplest pathway, but involves only photosystem I and generates ATP without producing NADPH or evolving oxygen.

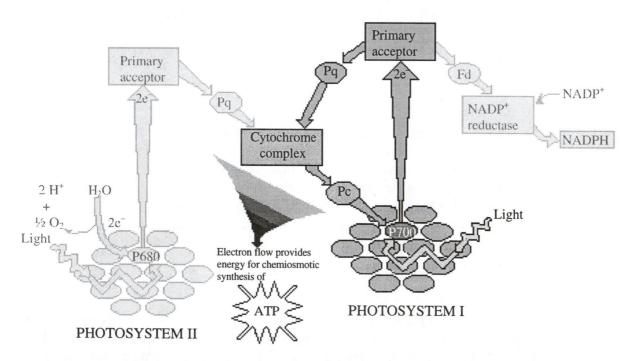

- It is cyclic because excited electrons that leave from chlorophyll *a* at the reaction center return to the reaction center.
- As photons are absorbed by Photosystem I, the P700 reaction center chlorophyll releases excited-state electrons to the primary electron acceptor; which, in turn, passes

them to ferredoxin. From there the electrons take an alternate path that sends them tumbling down the electron transport chain to P700. This is the same electron transport chain used in noncyclic electron flow.

- With each redox reaction along the electron transport chain, electrons lose potential energy until they return to their ground-state orbital in the P700 reaction center.

- The exergonic flow of electrons is coupled to ATP production by the process of chemiosmosis. This process of ATP production is called *cyclic photo-phosphorylation.*

- Absorption of another two photons of light by the pigments send a second pair of electrons through the cyclic pathway.

The function of the cyclic pathway is to produce additional ATP.

- It does so *without* the production of NADPH or O_2.

- Cyclic photophosphorylation supplements the ATP supply required for the Calvin cycle and other metabolic pathways. The noncyclic pathway produces approximately equal amounts of ATP and NADPH, which is not enought ATP to meet demand.

- NADPH concentration might influence whether electrons flow through cyclic or noncyclic pathways.

G.   **A Comparison of Chemiosmosis in Chloroplasts and Mitochondria**

Chemiosmosis = The coupling of exergonic electron flow down an electron transport chain to endergonic ATP production by the creation of an electrochemical proton gradient across a membrane. The proton gradient drives ATP synthesis as protons diffuse back across the membrane.

Chemiosmosis in chloroplasts and chemiosmosis in mitochondria are similar in several ways:

- An electron transport chain assembled in a membrane translocates protons across the membrane as electrons pass through a series of carriers that are progressively more electronegative.

- An ATP synthase complex built into the same membrane, couples the diffusion of hydrogen ions down their gradient to the phosphorylation of ADP.

- The ATP synthase complexes and some electron carriers (including quinones and cytochromes) are very similar in both chloroplasts and mitochondria.

Oxidative phosphorylation in mitochondria and photophosphorylation in chloroplasts differ in the following ways:

1.   Electron Transport Chain

- Mitochondria transfer chemical energy from food molecules to ATP. The high-energy electrons that pass down the transport chain are extracted by the oxidation of food molecules.

- Chloroplasts transform light energy into chemical energy. Photosystems capture light energy and use it to drive electrons to the top of the transport chain.

2. Spatial Organization

- The inner mitochondrial membrane pumps protons from the matrix out to the intermembrane space, which is a reservoir of protons that power ATP synthase.

- The chloroplast's thylakoid membrane pumps protons from the stroma into the thylakoid compartment, which functions as a proton reservoir. ATP is produced as protons diffuse from the thylakoid compartment back to the stroma through ATP synthase complexes that have catalytic heads on the membrane's stroma side. Thus, ATP forms in the stroma where it drives sugar synthesis during the Calvin cycle.

There is a large proton or pH gradient across the thylakoid membrane.

- When chloroplasts are illuminated, there is a thousand-fold difference in H^+ concentration. The pH in the thylakoid compartment is reduced to about 5 while the pH in the stroma increases to about 8.

- When chloroplasts are in the dark, the pH gradient disappears, but can be reestablished if chloroplasts are illuminated.

- Andre Jagendorf (1960's) produced compelling evidence for chemiosmosis when he induced chloroplasts to produce ATP in the dark by using artificial means to create a pH gradient. His experiments demonstrated that during photophosphorylation, the function of the photosystems and the electron transport chain is to create a proton-motive force that drives ATP synthesis.

A tentative model for the organization of the thylakoid membrane includes the following: (See Campbell, Figure 10.15)

- Proton pumping by the thylakoid membrane depends on an asymmetric placement of electron carriers that accept and release protons (H^+).

- There are three steps in the light reactions that contribute to the proton gradient across the thylakoid membrane:

    1. Water is split by Photosystem II on the thylakoid side, releasing protons in the process.

    2. As plastoquinone (Pq), a mobile carrier, transfers electrons to the cytochrome complex, it translocates protons from the stroma to the thylakoid space.

    3. Protons in the stroma are removed from solution as $NADP^+$ is reduced to NADPH.

- NADPH and ATP are produced on the side of the membrane facing the stroma where sugar is synthesized by the Calvin cycle.

NOTE: Students must be able to visualize the spatial arrangement of electron carriers in the membrane, since this arrangment is a crucial component of the chemiosmosis model. Figure 10.15 in the Campbell transparency set nicely illustrates this spatial arrangement.

Summary of the Light Reactions:

During *noncyclic electron flow*, the photosystems of the thylakoid membrane transform light energy to the chemical energy stored in NADPH and ATP. This process:

- Pushes low energy-state electrons from water to NADPH, where they are stored at a higher state of potential energy. NADPH, in turn, is the electron donor used to reduce carbon dioxide to sugar (Calvin Cycle).
- Produces ATP from this light driven electron current.
- Produces oxygen as a by-product.

During *cyclic electron flow*, electrons ejected from P700 reach ferredoxin and flow *back* to P700. This process:

- Produces ATP.
- Unlike noncyclic electron flow, does *not* produce NADPH or O_2.

VI. The Calvin cycle uses ATP and NADPH to convert CO_2 to sugar: *a closer look*

ATP and NADPH produced by the light reactions are used in the Calvin cycle to reduce carbon dioxide to sugar.

- The Calvin cycle is similar to the Krebs cycle in that the starting material is regenerated by the end of the cycle.
- Carbon enters the Calvin cycle as CO_2 and leaves as sugar.
- ATP is the energy source, while NADPH is the reducing agent that adds high-energy electrons to form sugar.
- The Calvin cycle actually produces a three-carbon sugar *glyceraldehyde 3-phosphate* (G3P).

NOTE: Students can easily follow the Calvin cycle if you use a diagram for reference, such as Figure 10.16 from the Campbell transparency set. This figure is especially helpful because you can go through the cycle twice; once to count carbons and once to follow the reactions pointing out where ATP and NADPH are used, where glyceraldehyde phosphate is produced and how RuBP is regenerated.

PHASE I: CARBON FIXATION

INPUT

3 ● CO_2

3 (P) ● ● ● ● — (P) Rubisco   6 ● ● ● — (P) 3-Phosphoglycerate

RuBP

6 ATP

6 ADP

3 ADP

CALVIN
CYCLE

6 (P) ● ● ● — (P)
1,3–Biphosphoglycerate

3 ATP

6 NADPH

6 $NADP^+$

5 ● ● ● — (P)

G3P

6 (P)

6 ● ● ● — (P)
Glyceraldehyde 3–phosphate
(G3P)

**PHASE 3: REGENERATION
OF CO_2 ACCEPTOR
(RuBP)**

PHASE 2: REDUCTION

1 ● ● ● — (P)

G3P
(a sugar)

Glucose and
other organic
compounds

OUTPUT

For the Calvin cycle to synthesize one molecule of sugar (G3P), three molecules of CO_2 must enter the cycle. The cycle may be divided into three phases:

Phase 1: Carbon Fixation. The Calvin cycle begins when each molecule of CO_2 is attached to a five-carbon sugar, *ribulose biphosphate* (*RuBP*).

- This reaction is catalyzed by the enzyme *RuBP carboxylase* (*rubisco*) – one of the most abundant proteins on Earth..

- The product of this reaction is an unstable six-carbon intermediate that immediately splits into two molecules of 3-phosphoglycerate.

- For every three CO_2 molecules that enter the Calvin cycle via rubisco, three RuBP molecules are carboxylated forming six molecules of 3-phosphoglycerate.

Phase 2: Reduction. This endergonic reduction phase is a two-step process that couples ATP hydrolysis with the reduction of 3-phosphoglycerate to glyceraldehyde phosphate.

| 3-phosphoglycerate | 1,3-bisphosphoglycerate | glyceraldehyde 3-phosphate (G3P) |

- An enzyme phosphorylates 3-phosphoglycerate by transferring a phosphate group from ATP. This reaction:
  - ⟹ produces 1, 3-bisphosphoglycerate.
  - ⟹ uses six ATP molecules to produce six molecules of 1,3-bisphosphoglycerate.
  - ⟹ primes 1,3-bisphosphoglycerate for the addition of high-energy electrons from NADPH.

- Electrons from NADPH reduce the carboxyl group of 1,3-bisphosphoglycerate to the aldehyde group of glyceraldehyde 3-phosphate (G3P).
  - ⟹ The product, G3P, stores more potential energy than the initial reactant, 3-phosphoglycerate.
  - ⟹ G3P is the same three-carbon sugar produced when glycolysis splits glucose.

- For every three CO_2 molecules that enter the Calvin cycle, six G3P molecules are produced, only one of which can be counted as net gain.
  - ⟹ The cycle begins with three five-carbon RuBP molecules – a total of 15 carbons.
  - ⟹ The six G3P molecules produced contain 18 carbons, a net gain of three carbons from CO_2.
  - ⟹ One G3P molecule exits the cycle; the other five are recycled to regenerate three molecules of RuBP.

Phase 3: Regeneration of Starting Material (RuBP). A complex series of reactions rearranges the carbon skeletons of five G3P molecules into three RuBP molecules.
- These reactions require three ATP molecules.
- RuBP is thus regenerated to begin the cycle again.

For the net synthesis of one G3P molecule, the Calvin cycle uses the products of the light reactions:
- 9 ATP molecules
- 6 NADPH molecules

G3P produced by the Calvin cycle is the raw material used to synthesize glucose and other carbohydrates.
- The Calvin cycle uses 18 ATP and 12 NADPH molecules to produce one glucose molecule.

VII.   Alternative mechanisms of carbon fixation have evolved in hot, arid climates

A.   Photorespiration: An evolutionary Relic?

A metabolic pathway called *photorespiration* reduces the yield of photosynthesis.

Photorespiration = In plants, a metabolic pathway that consumes oxygen, evolves carbon
dioxide, produces no ATP and decreases photosynthetic output.

- Occurs because the active site of rubisco can accept O_2 as well as CO_2.
- Produces no ATP molecules.
- Decreases photosynthetic output by reducing organic molecules used in the Calvin cycle.

When the O_2 concentration in the leaf's air spaces is higher than CO_2 concentration, rubisco
accepts O_2 and transfers it to RuBP. (The "photo" in photorespiration refers to the fact that
this pathway usually occurs in light when photosynthesis reduces CO_2 and raises O_2 in the
leaf spaces.)

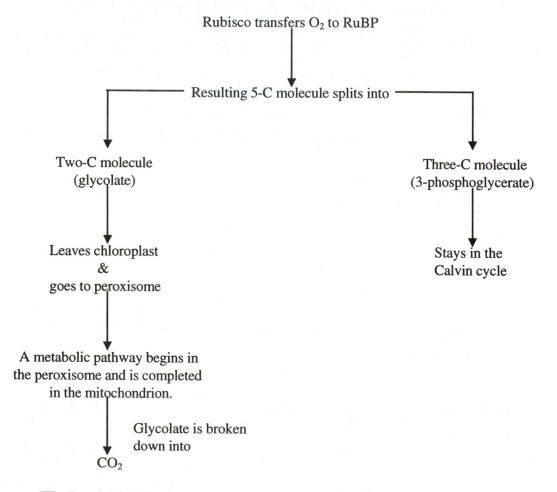

(The "respiration" in photorespiration refers to the fact that this process uses O_2 and releases
CO_2.)

Some scientists believe that photorespiration is a metabolic relic from earlier times when the atmosphere contained less oxygen and more carbon dioxide than is present today.

- Under these conditions, when rubisco evolved, the inability of the enzyme's active site to distinguish carbon dioxide from oxygen would have made little difference.
- This affinity for oxygen has been retained by rubisco and some photorespiration is bound to occur.

Whether photorespiration is beneficial to plants is not known.

- It is known that some crop plants (e.g. soybeans) lose as much as 50% of the carbon fixed by the Calvin cycle to photorespiration.
- If photorespiration could be reduced in some agricultural plants, crop yields and food supplies would increase.

Photorespiration is fostered by hot, dry, bright days.

- Under these conditions, plants close their stomata to prevent dehydration by reducing water loss from the leaf.
- Photosynthesis then depletes available carbon dioxide and increases oxygen within the leaf air spaces. This condition favors photorespiration.

Certain species of plants, which live in hot arid climates, have evolved alternate modes of carbon fixation that minimize photorespiration. C_4 and CAM are the two most important of these photosynthetic adaptations.

B. C_4 Plants

The Calvin cycle occurs in most plants and produces 3-phosphoglycerate, a three-carbon compound, as the first stable intermediate.

- These plants are called C_3 *plants*, because the first stable intermediate has three carbons.
- Agriculturally important C_3 plants include rice, wheat and soybeans.

Many plant species preface the Calvin cycle with reactions that incorporate carbon dioxide into four-carbon compounds.

- These plants are called C_4 *plants*.
- The C_4 pathway is used by several thousand species in at least 19 families including corn and sugarcane, important agricultural grasses.
- This pathway is adaptive, because it enhances carbon fixation under conditions that favor photorespiration, such as hot, arid environments.

Leaf anatomy of C_4 plants spatially segregates the Calvin cycle from the initial incorporation of CO_2 into organic compounds. There are two distinct types of photosynthetic cells:

1.   Bundle-sheath cells.

- Are arranged into tightly packed sheaths around the veins of the leaf.
- Thylakoids in the chloroplasts of bundle-sheath cells are not stacked into grana.
- The Calvin cycle is confined to the chloroplasts of the bundle sheath.

2.   Mesophyll cells.

- Are more loosely arranged in the area between the bundle sheath and the leaf surface.

The Calvin cycle of C4 plants is preceded by incorporation of CO_2 into organic compounds in the mesophyll. (See Campbell, Figure 10.18)

Step 1: CO_2 is added to phosphoenolpyruvate (PEP) to form oxaloacetate, a four-carbon product.

- *PEP carboxylase* is the enzyme that adds CO_2 to PEP. Compared to rubisco, it has a much greater affinity for CO_2 and has *no* affinity for O_2.
- Thus, PEP carboxylase can fix CO_2 efficiently when rubisco cannot — under hot, dry conditions that cause stomata to close, CO_2 concentrations to drop and O_2 concentrations to rise.

Step 2: After CO_2 has been fixed by mesophyll cells, they convert oxaloacetate to another four-carbon compound (usually malate).

Step 3: Mesophyll cells then export the four-carbon products (i.e. malate) through plasmodesmata to bundle-sheath cells.

- In the bundle-sheath cells, the four carbon compounds release CO_2, which is then fixed by rubisco in the Calvin cycle.
- Mesophyll cells thus pump CO_2 into bundle-sheath cells, minimizing photorespiration and enhancing sugar production by maintaining a CO_2 concentration sufficient for rubisco to accept CO_2 rather than oxygen.

C. CAM Plants

A second photosynthetic adaptation exists in succulent plants adapted to very arid conditions. These plants open their stomata primarily at night and close them during the day (opposite of most plants).

- This conserves water during the day, but prevents CO_2 from entering the leaves.
- When stomata are open at night, CO_2 is taken up and incorporated into a variety of organic acids. This mode of carbon fixation is called *crassulacean acid metabolism (CAM)*.
- The organic acids made at night are stored in vacuoles of mesophyll cells until morning, when the stomata close.
- During daytime, light reactions supply ATP and NADPH for the Calvin cycle. At this time, CO_2 is released from the organic acids made the previous night and is incorporated into sugar in the chloroplasts.

The CAM and C_4 pathways:

- are similar in that CO_2 is first incorporated into organic intermediates before it enters the Calvin cycle.
- differ in that the initial steps of carbon fixation in C_4 plants are structurally separate from the Calvin cycle; in CAM plants, the two steps occur at separate times.

Regardless of whether the plant uses a C_3, C_4 or CAM pathway, all plants use the Calvin cycle to produce sugar from CO_2.

VIII.   Photosynthesis is the biosphere's metabolic foundation: *a review*

On a global scale, photosynthesis makes about 160 billion metric tons of carbohydrate per year. No other chemical process on Earth is more productive or is as important to life.

- Light reactions capture solar energy and use it to:
  - ⇒ produce ATP
  - ⇒ transfer electrons from water to $NADP^+$ to form NADPH
- The Calvin cycle uses ATP and NADPH to fix CO_2 and produce sugar.

Photosynthesis transforms light energy to chemical bond energy in sugar molecules.

- Sugars made in chloroplasts supply the entire plant with chemical energy and carbon skeletons to synthesize organic molecules.
- Nonphotosynthetic parts of a plant depend on organic molecules exported from leaves in veins.
  - ⇒ The disaccharide *sucrose* is the transport form of carbohydrate in most plants.
  - ⇒ Sucrose is the raw material for cellular respiration and many anabolic pathways in nonphotosynthetic cells.
- Much of the sugar is *glucose* – the monomer linked to form *cellulose*, the main constituent of plant cell walls.

Most plants make more organic material than needed for respiratory fuel and for precursors of biosynthesis.

- Plants consume about 50% of the photosynthate as fuel for cellular respiration.
- Extra sugars are synthesized into starch and stored in storage cells of roots, tubers, seeds, and fruits.
- Heterotrophs also consume parts of plants as food.

Photorespiration can reduce photosynthetic yield in hot dry climates. Alternate methods of carbon fixation minimize photorespiration.

- C4 plants spatially separate carbon fixation from the Calvin cycle.
- CAM plants temporally separate carbon fixation from the Calvin cycle.

REFERENCES

Atkins, P.W. *Atoms, Electrons, and Change*. New York, Oxford: W.H. Freeman and Company, 1991. Chapter 9, "Light and Life" is a witty, imaginative description of photosynthesis. Though written for a lay audience, it is probably best appreciated by someone already familiar with photosynthesis.

Campbell, N. *Biology*. 4th ed. Menlo Park, California: Benjamin/Cummings, 1996.

Lehninger, A.L., D.L. Nelson and M.M. Cox. *Principles of Biochemistry*. 2nd ed. New York: Worth, 1993.

Matthews, C.K. and K.E. van Holde. *Biochemistry*. 2nd ed. Redwood City, California: Benjamin/Cummings, 1996.

CHAPTER 11
THE REPRODUCTION OF CELLS

OUTLINE

OBJECTIVES

After reading this chapter and attending lecture, the student should be able to:
1. Define genome and state what major events must occur during cell division for the entire genome to be passed on to daughter cells.
2. Describe the process of binary fission in prokaryotes.
3. Describe the composition of chromosomes and explain how chromosomal structure changes in preparation for cell division.
4. Describe how chromosome number changes throughout the human life cycle.
5. List the phases of the cell cycle and describe the sequence of events that occurs during each phase.

6. Describe what major events occur during the G_1, S and G_2 periods of interphase, and describe what characterizes a G_2 interphase cell.
7. Distinguish between interphase and mitosis proper.
8. List the phases of mitosis and describe the events characteristic of each phase.
9. Recognize the phases of mitosis from diagrams or micrographs.
10. Draw or describe the spindle apparatus including centrosomes, nonkinetochore microtubules, kinetochore microtubules, asters and centrioles (in animal cells).
11. Describe what characteristic changes occur in the spindle apparatus during each phase of mitosis.
12. Explain the current models for poleward chromosomal movement and elongation of the cell's polar axis.
13. Compare cytokinesis in animals and plants.
14. List several factors, identified from cell tissue-culture studies, which stimulate or inhibit cell growth.
15. Describe what point in the cell cycle determines whether a cell will divide and explain the role of volume-to-genome ratio.
16. Explain how MPF induces the changes that occur in mitosis and describe what causes the cyclical change in MPF concentration.
17. Explain how abnormal cell division of cancerous cells differs from normal cell division.

KEY TERMS

cell division	mitosis	mitotic spindle	cell plate
genome	cytokinesis	centrosome	density-dependent
binary fission	interphase	kinetochore	inhibition
somatic cell	S phase	microtubules	growth factors
chromosomes	G_1 phase	nonkinetochore	restriction point
chromatin	G_2 phase	microtubules	kinase
sister chromatids	prophase	dynein	cyclin
centromere	prometaphase	asters	Cdk
centriole	metaphase	metaphase plate	MPF
cell cycle	anaphase	cleavage	cancer
M phase	telophase	cleavage furrow	transformation

LECTURE NOTES

The ability to reproduce distinguishes living organisms from nonliving objects; this ability has a cellular basis.

"Where a cell exists, there must have been a preexisting cell, just as the animal arises only from an animal and the plant only from a plant." "*Omnis cellula e cellula*" or "All cells from cells." — Rudolf Virchow (1855).

I.   Cell division functions in reproduction, growth, and repair

The perpetuation of life is based on the reproduction of cells or *cell division*.

- In unicellular organisms, the division of one cell to form two reproduces an entire organism (e.g. *Amoeba*).
- In multicellular organisms, cell division allows:
  - ⇒ Growth and development from the fertilized egg.
  - ⇒ Replacement of damaged or dead cells.

Cell division is a complex process that faithfully passes along the *genome* from one generation of cells to the next. A dividing cell:

- Precisely replicates its DNA.
- Equally distributes the DNA to opposite ends of the cell.
- Separates into two identical daughter cells

Genome = Total endowment of DNA unique to each species.

II.   Bacteria reproduce by binary fission

Prokaryotes are smaller and simpler than eukaryotes. Prokaryotes:

- Contain most genes in a single circular chromosome composed of a double-stranded DNA molecule and associated proteins.
- Contain only about 1/1000 the DNA of eukaryotes, but prokaryotic chromosomes still contain a large amount of DNA relative to the small prokaryotic cell. Consequently, bacterial chromosomes are highly folded and packed within the cell.

Prokaryotes reproduce by *binary fission*, a process during which bacteria replicate their chromosomes and equally distribute copies between the two daughter cells.

- The chromosome is replicated; each copy remains attached to the plasma membrane at adjacent sites.
- Between the attachment sites the membrane grows and separates the two copies of the chromosome.
- The bacterium grows to about twice its initial size, and the plasma membrane pinches inward.
- A cell wall forms across the bacterium between the two chromosomes, dividing the original cell into two daughter cells.

III.   The genome of a eukaryotic cell is organized into multiple chromosomes

Dividing eukaryotic cells replicate and distribute their tens of thousands of genes – a process that is manageable because the genes are organized into multiple *chromosomes*.

<u>Chromosomes</u> = Threadlike structures in eukaryotic nuclei that are composed of DNA and protein.
- Each species has a characteristic chromosome number. (Human somatic cells have 46.)
- Gametes (sperm and ova) contain half the number of chromosomes of somatic cells. (Human gametes have 23.)

Chromatin, a DNA-protein complex, is organized into a long, thin fiber that is folded and coiled to form the chromosome. Each chromosome thus contains:
1. A long DNA molecule, with thousands of genes.
2. Various proteins that maintain chromosomal structure and help control gene activity.

<u>Mitosis</u> = (Mitos = thread) Nuclear division during which duplicated chromosomes are evenly distributed into two daughter nuclei; results in two daughter cells that are the genetic equivalent of the parent cell.
- Before mitosis, a cell copies its genome by duplicating every chromosome, each of which forms two identical *sister chromatids* joined at a specialized region called the *centromere*.
- During mitosis, sister chromatids are pulled apart forming two complete chromosome sets, one at each end of the cell.
- Mitosis (division of the nucleus) may be followed by *cytokinesis* (division of the cytoplasm).

<u>Cytokinesis</u> = Cytoplasmic division that forms two separate daughter cells, each containing a single nucleus.

HUMAN LIFE CYCLE

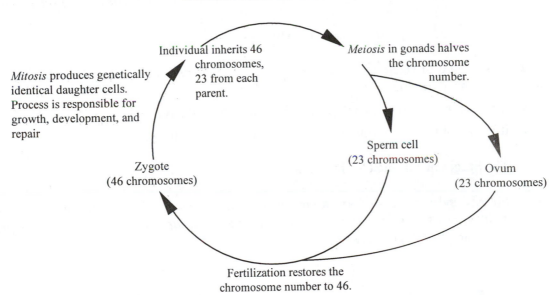

Mitosis produces genetically identical daughter cells. Process is responsible for growth, development, and repair

Individual inherits 46 chromosomes, 23 from each parent.

Meiosis in gonads halves the chromosome number.

Sperm cell (23 chromosomes)

Ovum (23 chromosomes)

Zygote (46 chromosomes)

Fertilization restores the chromosome number to 46.

IV.   Mitosis alternates with interphase in the cell cycle: *an overview*

<u>Cell cycle</u> = Well-ordered sequence of events between the time a cell divides to form two daughter cells and the time those daughter cells divide.

- Includes a doubling of a cell's cytoplasm, reproduction of cellular organelles, precise duplication of DNA, mitosis and cytokinesis.
- Duration varies with cell type. Some cells divide each hour, others take more than 24 hours.
- Nerve and muscle cells never or rarely divide once they are formed.

The *cell cycle* alternates between *M phase*, or dividing phase, and *Interphase*, the nondividing phase:

- *M phase*, the shortest part of the cell cycle and the phase during which the cell divides, includes:

     1. *Mitosis* - division of the nucleus.
     2. *Cytokinesis* - division of the cytoplasm.

- *Interphase*, the nondividing phase, includes most of a cell's growth and metabolic activities.
     ⇒ Is about 90% of the cell cycle.
     ⇒ Is a period of intense biochemical activity during which the cell grows and copies its chromosomes in preparation for cell division.
     ⇒ Consists of three periods:

     1. *G_1 phase* – first growth phase. (G stands for "gap".)
     2. *S phase* - synthesis phase when DNA is synthesized as chromosomes are duplicated. (S stands for "synthesis".)
     3. *G_2 phase* - second growth phase.

Mitosis is unique to eukaryotes and may be an evolutionary adaptation for distributing a large amount of genetic material.

- Details may vary, but overall process is similar in most eukaryotes.
- It is a reliable process with only one error per 100,000 cell divisions.

A.   The Stages of Mitotic Cell Division

Mitosis and cytokinesis form a continuum, but for ease of description, mitosis is usually divided into five stages: *prophase, prometaphase, metaphase, anaphase* and *telophase*. (See Campbell, Figure 11.6)

G₂ of Interphase

A G_2 cell is characterized by:

- A well-defined nucleus bounded by a nuclear envelope.
- One or more nucleoli.
- Two centrosomes adjacent to the nucleus (formed earlier by replication of a single centrosome).
- In animals, a pair of centrioles in each centrosome.
- In animals, a radial microtubular array (*aster*) around each pair of centrioles.
- Duplicated chromosomes that cannot be distinguished individually due to loosely packed chromatin fibers. (Chromosomes were duplicated earlier in S phase.)

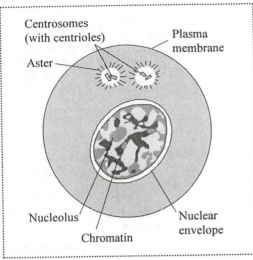

Prophase

In the nucleus:

- Nucleoli disappear.
- Chromatin fibers condense into discrete, observable chromosomes, composed of two identical sister chromatids joined at the centromere.

In the cytoplasm:

- Mitotic spindle forms. It is composed of microtubules between the two *centrosomes* or microtubule-organizing centers.
- Centrosomes move apart, apparently propelled along the nuclear surface by lengthening of the microtubule bundles between them.

Prometaphase

During prometaphase:

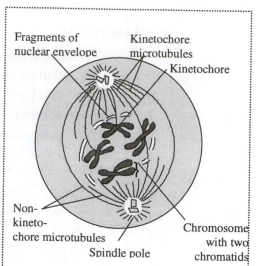

- Nuclear envelope fragments, which allows microtubules to interact with the highly condensed chromosomes.

- *Spindle fibers* (bundles of microtubules) extend from each pole toward the cell's equator.

- Each chromatid now has a specialized structure, the *kinetochore*, located at the centromere region.

- *Kinetochore microtubules* become attached to the kinetochores and put the chromosomes into agitated motion.

- *Nonkinetochore microtubules* radiate from each centrosome toward the metaphase plate without attaching to chromosomes. Nonkinetochore microtubules radiating from one pole overlap with those from the opposite pole.

Metaphase

During metaphase:

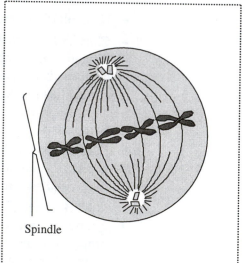

- Centrosomes are positioned at opposite poles of the cell.

- Chromosomes move to the *metaphase plate*, the plane equidistant between the spindle poles.

- Centromeres of all chromosomes are aligned on the metaphase plate.

- The long axis of each chromosome is roughly at a right angle to the spindle axis.

- Kinetochores of sister chromatids face opposite poles, so identical chromatids are attached to kinetochore fibers radiating from opposite ends of the parent cell.

- Entire structure formed by nonkinetochore microtubules plus kinetochore microtubules is called the *spindle*.

Anaphase

Anaphase is characterized by movement. It begins when paired centromeres of each chromosome move apart.

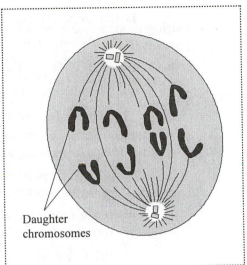

- Sister chromatids split apart into separate chromosomes and move towards opposite poles of the cell.
- Because kinetochore fibers are attached to the centromeres, the chromosomes move centromere first in a "V" shape.
- Kinetochore microtubules shorten at the kinetochore end as chromosomes approach the poles.
- Simultaneously, the poles of the cell move farther apart, elongating the cell.

Daughter chromosomes

At the end of anaphase, the two poles have identical collections of chromosomes.

Telophase and Cytokinesis

During telophase:

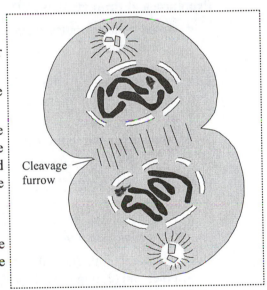

- Nonkinetochore microtubules further elongate the cell.
- Daughter nuclei begin to form at the two poles.
- Nuclear envelopes form around the chromosomes from fragments of the parent cell's nuclear envelope and portions of the endomembrane system.
- Nucleoli reappear.
- Chromatin fiber of each chromosome uncoils and the chromosomes become less distinct.

Cleavage furrow

By the end of telophase:

- Mitosis, the equal division of one nucleus into two genetically identical nuclei, is complete.
- Cytokinesis has begun and the appearance of two separate daughter cells occurs shortly after mitosis is completed.

NOTE: A lecture on mitosis may not last the entire period if it is limited to just a description of mitotic stages. Though it may be tempting to continue with meiosis during the same class period, it is not recommended. Students easily confuse the two processes because they are somewhat similar, so it helps to allow some time for students to assimilate the mitosis material, before discussing meiosis. It is effective to summarize with a comparison of the two processes after the topic of meiosis has been discussed.

V. **The mitotic spindle distributes chromosomes to daughter cells:** *a closer look*

During prophase, the mitotic spindle forms in the cytoplasm from *microtubules* and associated proteins.

- Microtubules of the cytoskeleton are partially disassembled during spindle formation.
    - ⇒ Spindle microtubules are aggregates of two proteins, α- and β-tubulin.
    - ⇒ Spindle microtubules elongate by the adding tubulin subunits at one end.
- Parallel microtubules form bundles called *spindle fibers* that are visible under a light microscope; their assembly begins in the *centrosome* or microtubule organizing center.
- In animal cells, a pair of centrioles is in the center of the centrosome, but there is evidence that they are not essential for cell division:
    - ⇒ If the centrioles of animal cells are destroyed with a laser microbeam, spindles still form and function during mitosis.
    - ⇒ Plant centrosomes lack centrioles.

Interphase: The centrosome replicates to form two centrosomes located just outside the nucleus.

Prophase and Prometaphase: The two centrosomes move farther apart.

- Spindle microtubules radiate from the centrosomes, elongating at the end away from their centrosome.

Late Prometaphase: By the end of prometaphase, the two centrosomes are at opposite poles and the chromosomes have moved to the cell's midline.

- Each chromatid of a replicated chromosome develops its own *kinetochore*, a protein structure on the centromere.
    - ⇒ Some spindle microtubules attach to the kinetochores and are called *kinetochore microtubules*.
    - ⇒ Some spindle microtubules extend from the centrosomes and overlap with those radiating from the cell's opposite pole. These are called *polar* or *nonkinetochore microtubules*.
- Kinetochore microtubules interact to: (1) arrange the chromosomes so kinetochores face the poles and (2) align the chromosomes at the cell's midline.
- The most stable arrangement occurs when sister kinetochores are attached by microtubules to opposite spindle poles.
    - ⇒ Initially, kinetochore microtubules from one pole may attach to a kinetochore, moving the chromosome toward that pole. This movement is checked when microtubules from the opposite pole attach to the chromosome's other kinetochore.
    - ⇒ The chromosome oscillates back and forth until it stabilizes and aligns at the cell's midline.
    - ⇒ Microtubules can remain attached to a kinetochore only if there is opposing tension from the other side. It is this opposing tension that stabilizes the microtubule-kinetochore connection and allows the proper alignment and movement of chromosomes at the cell's midline.

Metaphase: All the duplicated chromosomes align on the cell's midline, or *metaphase plate*.

Anaphase: The chromosome's centromeres split and the sister chromatids move as separate chromosomes toward opposite ends of the cell.

How do the kinetochore microtubules function in this poleward movement of chromosomes? Based on experimental evidence, the current model is that:

- Kinetochore microtubules shorten during anaphase by depolymerizing at their kinetochore ends; pulling the chromosomes poleward.

- The mechanism of this interaction between kinetochores and microtubules may involve microtubule-walking proteins similar to dynein that "walk" a chromosome along the shortening microtubules.

What is the function of the nonkinetochore microtubules?

- Nonkinetochore tubules elongate the whole cell along the polar axis during anaphase.

- These tubules overlap at the middle of the cell and slide past each other away from the cell's equator, reducing the degree of overlap.

- It is hypothesized that dynein cross-bridges may form between overlapping tubules to slide them past one another. Alternatively, motor molecules may link the microtubules to other cytoskeletal elements to drive the sliding.

- ATP provides the energy for this endergonic process.

Telophase: At the end of anaphase, the duplicate sets of chromosomes are clustered at opposite ends of the elongated parent cell.

- Nuclei reform during telophase.

- Cytokinesis usually divides the cell's cytoplasm during late telophase. In some exceptional cases, mitosis is not followed by cytokinesis (e.g. certain slime molds form multinucleated masses called *plasmodia*).

VI.   Cytokinesis divides the cytoplasm: *a closer look*

Cytokinesis, the process of cytoplasmic division, begins in telophase and is different in animal and plant cells. In animal cells, cytokinesis occurs as *cleavage*:

- First, a *cleavage furrow* forms as a shallow groove in the cell surface near the old metaphase plate. (See Campbell, Figure 11.9)

- A *contractile ring* of actin microfilaments forms on the cytoplasmic side of the furrow; this ring contracts until it pinches the parent cell in two.

- Finally, the remaining mitotic spindle breaks, and the two cells become completely separate.

In plant cells, cytokinesis occurs by *cell plate* formation across the parent cell's midline (old metaphase plate).

- Golgi-derived vesicles move along microtubules to the cell's center, where they fuse into a disc-like cell plate.

- Additional vesicles fuse around the edge of the plate, expanding it laterally until its membranes touch and fuse with the existing parent cell's plasma membrane.

- A new cell wall forms as cellulose is deposited between the two membranes of the cell plate.

VII.   External and internal cues control cell division

Normal growth, development and maintenance depend on the timing and rate of mitosis. Various cell types differ in their pattern of cell division; for example:

- Human skin cells divide frequently.

- Liver cells only divide in appropriate situations, such as wound repair.

- Nerve, muscle and other specialized cells do not divide in mature humans.

Using tissue culture, researchers have identified some factors that influence cell division:

1. *Contents of the culture medium.*

- If essential nutrients are left out of the culture medium, cells will not divide.

- Specific regulatory substances called *growth factors* are necessary for some cultured mammalian cells to divide. For example:

  ⇒ Binding of platelet-derived growth factor (PDGF) to cell membrane receptors, stimulates cell division in fibroblasts. This regulation probably occurs not only in cell culture, but in the animal's body as well – a response that helps heal wounds.

  ⇒ Other cell types may have membrane receptors for different growth factors or for different combinations of several growth factors.

2. *Cell density.*

- Crowding inhibits cell division in a phenomenon called *density-dependent inhibition*. Cultured cells stop dividing when they form a single layer on the container's inner surface. If some cells are removed, those bordering the open space divide again until the vacancy is filled.

- Density-dependent inhibition is apparently a consequence of the fact that:

  ⇒ Quantities of nutrients and growth regulators may be insufficient to support cell division, if cell density is too high.

  ⇒ For normal cell division, cells must adhere to a substratum – such as the surface of a culture dish. Cells will stop dividing if they become detached.

- Density-dependent inhibition probably occurs in the body's tissues as well as in cell culture. Cancer cells are abnormal and do not exhibit density-dependent inhibition.

3. *G_1 phase of the cell cycle.*
- Whether a cell is destined to divide is determined at the *restriction point* which occurs late in the G_1 phase of the cell cycle.
    - $\Rightarrow$ If a cell is destined to divide, it progresses beyond the restriction point into the S phase when DNA synthesis begins, and then proceeds through cell division.
    - $\Rightarrow$ If the cell is not destined to divide, it may exit from the cell cycle at the restriction point and switch to a nondividing state called the *G_0 phase*.
- Most human body cells are in the G_0 phase. Some most specialized cells, such as nerve and muscle, stay in G_0 and never divide. Other cells, such as liver cells, can be induced by environmental cues, such as injury, to continue through the cell cycle and divide.

4. *Cell size.*
- For actively dividing cells, the ratio of cytoplasmic volume to genome size is the most important indicator of whether a cell will pass the restriction point.
- Once a cell's volume-to-genome ratio reaches a critical threshold value, it may pass the restriction point and copy its DNA. Thus, a cell must add enough cytoplasm to attain a certain size, before DNA synthesis can begin. This prevents daughter cells from becoming progressively smaller with each cell cycle.

VIII. Cyclical changes in regulatory proteins function as a mitotic clock

Once the cell passes the restriction point, it is destined to divide.
- The onset of S phase commits the cell to continue through G_2 and M phases and divide.
- Biologists are only beginning to work out the switches that control the precise sequence of events in cell division. We do know that this sequence probably depends on the completion of each task before the cell can progress to the next stage.

The ordered sequence of cell cycle events is synchronized by rhythmic changes in the activity of regulatory proteins, some of which are *protein kinases*.
- Protein kinases are enzymes that catalyze the transfer of a phosphate group from ATP to a target protein.
- Phosphorylation, in turn, induces a conformational change that either activates or inactivates the target protein.

Cyclical changes in kinase activity are controlled by another class of regulatory proteins called *cyclins*.
- These regulatory proteins are named cyclins, because their concentrations change cyclically during the cell cycle.
- Protein kinases that regulate cell cycles are *cyclin-dependent kinases* (or *Cdks*); they are active only when attached to a particular cyclin.
- Even though Cdk concentration stays the same throughout the cell cycle, its *activity* changes in response to the changes in cyclin concentration.

An example of a cyclin-dependent kinase is *MPF (maturation promoting factor)*, which controls the cell's progress from late interphase (G_2) to mitosis.

- Active MPF phosphorylates chromatin proteins, causing chromosomes to condense during prophase.

- During prometaphase, the nuclear envelope disperses when some of its membrane proteins are phosphorylated.

Cyclin's rhythmic changes in concentration regulate MPF activity, and thus act as a mitotic clock that regulates the sequential changes in a dividing cell.

- Cyclin is produced at a uniform rate throughout the cell cycle, and it accumulates during interphase.

- Cyclin combines with Cdk to form active MPF, so as cyclin concentration rises and falls, the amount of active MPF changes in a similar way.

- Active MPF activates proteins that participate in mitosis and initiates prophase.

- Near the end of mitosis, cyclin is destroyed by an enzyme that is activated by MPF.

- The destruction of cyclin causes the decline in active MPF at the end of mitosis. Since MPF is activated by cyclin, it is indirectly responsible for its own decline.

- Continuing cyclin synthesis raises the concentration again during interphase. This newly synthesized cyclin binds to Cdk to form active MPF, and mitosis begins again.

Rhythmic changes in different cyclin-Cdk complexes regulate other cell cycle stages.

To summarize the control of the cell cycle:

- The G_1 phase of the cycle is the most variable phase in duration and in the variety of external and internal controls over cell division.

- Nutritional status, growth factors, cell density, and developmental state of the cell affect the length of G_1 and whether the cell will pass the restriction point and divide.

- A cell that does not pass the restriction point will enter the G_0 phase as a nondividing cell.

- If conditions favor cell division, the cell will proceed through the restriction point. When the cell acquires enough cytoplasm to reach the threshold volume-to-genome ratio, the cell will be irreversibly committed to divide.

- Rhythmic changes in cyclin concentration activate protein kinases at key regulatory points in the cell cycle, controlling the sequence of events during cell division.

IX. Cancer cells escape from the controls on cell division

Cancer cells do not respond normally to controls on cell division. They divide excessively, invade other tissues and, if unchecked, can kill the whole organism.

- Cancer cells in culture do not stop growing in response to cell density (density-dependent inhibition); they continue to grow until nutrients in the medium are exhausted.

- Cancer cells that stop dividing do so at random points in the cycle instead of at the restriction point in G_1.

- Cancer cells in culture are immortal in that they continue to divide indefinitely, as long as nutrients are available. Normal mammalian cells in culture divide only about 20 to 50 times before they stop.

Cells in tissue culture that have lost the normal controls on growth are said to be *transformed*.

<u>Transformation</u> = The conversion of eukaryotic cells in tissue culture to a condition of unregulated growth.

The immune system normally destroys abnormal cells that have converted from normal to cancer cells.

- If abnormal cells evade destruction, they may proliferate to form a *tumor*, an unregulated growing mass of cells within otherwise normal tissue.

- If the cells remain at this original site, the mass is *benign* and can be completely removed by surgery.

- A tumor is *malignant* if the cells have the ability to spread to other parts of the body. Only a malignant tumor is said to be cancer.

Besides having anomalous cell cycles, malignant cells are abnormal in other ways. For example, they may have:

- Unusual numbers of chromosomes.

- Aberrant metabolism.

- Lost attachments to neighboring cells and substratum – usually a consequence of abnormal cell surface changes.

Detached cancer cells may spread into other tissues surrounding the original tumor and may even enter the blood and lymph vessels of the circulatory system.

- These circulating cells can invade other parts of the body and proliferate to form more tumors.

- This spread of cancer cells beyond their original sites is called *metastasis*.

- If a tumor metastasizes, it is usually treated with radiation and chemotherapy, which is especially harmful to actively dividing cells.

Researchers are beginning to understand how a normal cell is transformed into a cancerous one. In most cases, it is probably caused by an alteration of genes that control cell division.

- Michael Bishop and Harold Varmus won the Nobel Prize for demonstrating that genetic alterations can cause cancer.

- Knowledge of how changes in the genome lead to the heritable abnormalities of cancer cells is still rudimentary and depends upon our increased understanding of how normal cells work.

REFERENCES

Alberts, B., D. Bray, J. Lewis, M. Raff, K. Roberts and J.D. Watson. *Molecular Biology of the Cell*. 2nd ed. New York: Garland, 1989.

Becker, W.M., J.B. Reece and M.F. Puente. *The World of the Cell*. 3rd ed. Redwood City, California: Benjamin/Cummings, 1996.

Campbell, N. *Biology*. 4th ed. Menlo Park, California: Benjamin/Cummings, 1996.

Kleinsmith, L.J. and V.M. Kish. *Principles of Cell Biology*. New York: Harper and Row, Publ., 1988.

Varmus, H. and R.A. Weinberg. *Genes and the Biology of Cancer*. New York: Scientific American Books, 1993.

CHAPTER 12

MEIOSIS AND
SEXUAL LIFE CYCLES

OUTLINE

   I.   Offspring acquire genes from parents by inheriting chromosomes

  II.   Like begets like, more or less: a comparison of asexual versus sexual reproduction

 III.   Fertilization and meiosis alternate in sexual life cycles: *an overview*

 IV.   Meiosis reduces chromosome number from diploid to haploid: *a closer look*
       A.   The Stages of Meiotic Cell Division
       B.   A Comparison of Mitosis and Meiosis

  V.   Sexual life cycles produce genetic variation among offspring
       A.   Independent Assortment of Chromosomes
       B.   Crossing Over
       C.   Random Fertilization

 VI.   Evolutionary adaptation depends on a population's genetic variation

OBJECTIVES

After reading this chapter and attending lecture, the student should be able to:

1. Explain why organisms only reproduce their own kind, and why offspring more closely resemble their parents than unrelated individuals of the same species.
2. Explain what makes heredity possible.
3. Distinguish between asexual and sexual reproduction.
4. Diagram the human life cycle and indicate where in the human body that mitosis and meiosis occur; which cells are the result of meiosis and mitosis; and which cells are haploid.
5. Distinguish among the life cycle patterns of animals, fungi and plants.
6. Distinguish sporophyte from gametophyte, and explain how plant life cycles differ from those of animals and fungi.

7. List the phases of meiosis I and meiosis II and describe the events characteristic of each phase.
8. Recognize the phases of meiosis from diagrams or micrographs.
9. Describe the process of synapsis during prophase I and explain how genetic recombination occurs.
10. Describe key differences between mitosis and meiosis; explain how the end result of meiosis differs from that of mitosis.
11. Compare the phases of meiosis I with mitosis and distinguish between the chromosomal arrangement during metaphase of both processes.
12. Distinguish between mitotic interphase and meiotic interkinesis.
13. Explain how independent assortment, crossing over and random fertilization contribute to genetic variation in sexually reproducing organisms.
14. Explain why inheritable variation was crucial to Darwin's theory of evolution.
15. List the sources of genetic variation.

KEY TERMS

heredity
genetics
variation
gene
chromosome
life cycle
somatic cell
asexual reproduction
sexual reproduction
karyotype

homologous
   chromosomes
locus
sex chromosome
autosome
diploid
haploid
gamete
fertilization

zygote
meiosis
meiosis I
interkinesis
meiosis II
synapsis
sister chromatid
tetrad
crossing over

chiasmata
synaptonemal
   complex
independent
   assortment
alternation of
   generations
sporophyte
gametophyte

LECTURE NOTES

Reproduction is an emergent property associated with life. The fact that organisms reproduce their own kind is a consequence of *heredity*.

Heredity = Continuity of biological traits from one generation to the next.

- Results from the transmission of hereditary units, or *genes*, from parents to offspring.
- Because they share similar genes, offspring more closely resemble their parents or close relatives than unrelated individuals of the same species.

Variation = Inherited differences among individuals of the same species.

- Though offspring resemble their parents and siblings, they also diverge somewhat as a consequence of inherited differences among them.
- The development of *genetics* in this century has increased our understanding about the mechanisms of variation and heredity.

Genetics = The scientific study of heredity and variation.

NOTE: Beginning students often compartmentalize their knowledge, which makes it difficult to transfer and apply information learned in one context to a new situation. Be forewarned that unless the instructor points it out, some students will never make the connection that meiosis, sexual reproduction and heredity are all aspects of the same process.

I.   Offspring acquire genes from parents by inheriting chromosomes

<u>DNA</u> = Type of nucleic acid that is a polymer of four different kinds of nucleotides.

<u>Genes</u> = Units of hereditary information that are made of *DNA* and are located on *chromosomes*.
- Have specific sequences of nucleotides, the monomers of DNA.
- Most genes program cells to synthesize specific proteins; the action of these proteins produce an organism's inherited traits.

Inheritance is possible because:
- DNA is precisely replicated producing copies of genes that can be passed along from parents to offspring.
- Sperm and ova carrying each parent's genes are combined in the nucleus of the fertilized egg.

The actual transmission of genes from parents to offspring depends on the behavior of *chromosomes*.

<u>Chromosomes</u> = Threadlike structures in eukaryotic nuclei that are made of DNA and protein.
- Consist of a single long DNA molecule that is highly folded and coiled along with proteins.
- Contain genetic information arranged in a linear sequence.
- Contain hundreds or thousands of genes, each of which is a specific region of the DNA molecule, or *locus*.

    <u>Locus</u> = Specific location on a chromosome that contains a gene.

- Each species has a characteristic chromosome number; humans have 46.

II.   Like begets like, more or less: a comparison of asexual versus sexual reproduction

Asexual Reproduction	Sexual Reproduction
Single individual is the sole parent.	Two parents give rise to offspring.
Single parent passes on *all* its genes to its offspring.	Each parent passes on *half* its genes, to its offspring.
Offspring are genetically identical to the parent.	Offspring have a unique combination of genes inherited from both parents.
Results in a *clone*, or genetically identical individual. Rarely, genetic differences occur as a result of *mutation*, a change in DNA.	Results in greater genetic variation; offspring vary genetically from their siblings and parents.

What generates this genetic variation during sexual reproduction? The answer lies in the process of *meiosis*.

III. **Fertilization and meiosis alternate in sexual life cycles:** *an overview*

The human *life cycle* follows the same basic pattern found in all sexually reproducing organisms; *meiosis* and *fertilization* result in alternation between the haploid and diploid condition.

Life cycle = Sequence of stages in an organism's reproductive history, from conception to production of its own offspring.

Somatic cell = Any cell other than a sperm or egg cell.
- Human somatic cells contain 46 chromosomes distinguishable by differences in size, position of the centromere, and staining or banding pattern.
- Using these criteria, chromosomes from a photomicrograph can be matched into *homologous* pairs and arranged in a standard sequence to produce a *karyotype*.

Karyotype = A display or photomicrograph of an individual's somatic-cell metaphase chromosomes that are arranged in a standard sequence. (See Campbell, Methods Box: *Preparation of a Karyotype*)
- Human karyotypes are often made with lymphocytes.
- Can be used to screen for chromosomal abnormalities.

Homologous chromosomes (homologues) = A pair of chromosomes that have the same size, centromere position and staining pattern.
- With one exception, homologues carry the same genetic loci.
- Homologous *autosomes* carry the same genetic loci; however, human *sex chromosomes* carry different loci even though they pair during prophase of Meiosis I.

Autosome = A chromosome that is not a sex chromosome.

Sex chromosome = Dissimilar chromosomes that determine an individual's sex.
- Females have a homologous pair of X chromosomes.
- Males have one X and one Y chromosome.
- Thus, humans have 22 pairs of autosomes and 1 pair of sex chromosomes.

Chromosomal pairs in the human karyotype are a result of our sexual origins.
- One homologue is inherited from each parent.
- Thus, the 46 somatic-cell chromosomes are actually two sets of 23 chromosomes; one a maternal set and the other a paternal set.
- Somatic cells in humans and most other animals are *diploid*.

Diploid = Condition in which cells contain two sets of chromosomes; abbreviated as $2n$.

Haploid = Condition in which cells contain one set of chromosomes; it is the chromosome number of *gametes* and is abbreviated as n.

<u>Gamete</u> = A haploid reproductive cell.

- *Sperm cells* and *ova* are gametes, and they differ from somatic cells in their chromosome number. Gametes only have one set of chromosomes.
- Human gametes contain a single set of 22 autosomes and one sex chromosome (either an X or a Y).
- Thus, the haploid number of humans is 23.

The diploid number is restored when two haploid gametes unite in the process of *fertilization*. Sexual intercourse allows a haploid sperm cell from the father to reach and fuse with an ovum from the mother.

<u>Fertilization</u> = The union of two gametes to form a *zygote*.

<u>Zygote</u> = A diploid cell that results from the union of two haploid gametes.

- Contains the maternal and parental haploid sets of chromosomes from the gametes and is diploid ($2n$).
- As humans develop from a zygote to a sexually mature adult, the zygote's genetic information is passed with precision to all somatic cells by mitosis.

Gametes are the only cells in the body that are <u>not</u> produced by mitosis.

- Gametes are produced in the ovaries or testes by the process of *meiosis*.
- *Meiosis* is a special type of cell division that produces haploid cells and compensates for the doubling of chromosome number that occurs at fertilization.
- Meiosis in humans produces sperm cells and ova which contain 23 chromosomes.
- When fertilization occurs, the diploid condition ($2n=46$) is restored in the zygote.

A. The Variety of Sexual Life Cycles

Alternation of meiosis and fertilization is common to all sexually reproducing organisms; however, the timing of these two events in the life cycle varies among species. There are three basic patterns of sexual life cycles:

Animal: In animals, including humans, gametes are the only haploid cells.

- Meiosis occurs during gamete production. The resulting gametes undergo no further cell division before fertilization.
- Fertilization produces a diploid zygote that divides by mitosis to produce a diploid multicellular animal.

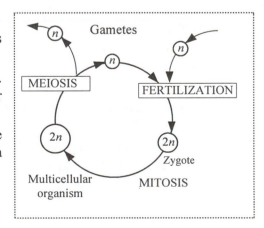

Fungi and Some Protists: In many fungi and some protists, the only diploid stage is the zygote.

- Meiosis occurs immediately after the zygote forms.
- Resulting haploid cells divide by *mitosis* to produce a haploid multicellular organism.
- Gametes are produced by *mitosis* from the already haploid organism.

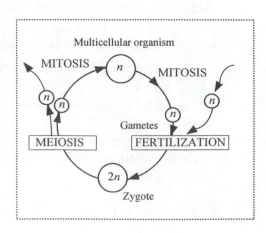

Plants and Some Algae: Plants and some species of algae alternate between multicellular haploid and diploid generations.

- This type of life cycle is called an *alternation of generations*.
- The multicellular diploid stage is called a *sporophyte*, or spore-producing plant. Meiosis in this stage produces haploid cells called *spores*.
- Haploid spores divide mitotically to generate a multicellular haploid stage called a *gametophyte*, or gamete-producing plant.
- Haploid gametophytes produce gametes by mitosis.
- Fertilization produces a diploid zygote which develops into the next sporophyte generation.

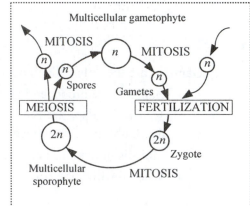

V. Meiosis reduces chromosome number from diploid to haploid: *a closer look*

Meiosis and sexual reproduction significantly contribute to genetic variation among offspring.

Meiosis includes steps that closely resemble corresponding steps in mitosis.

- Like mitosis, meiosis is preceded by replication of the chromosomes.
- Meiosis differs from mitosis in that this single replication is followed by two consecutive cell divisions: *Meiosis I* and *Meiosis II*.
- These cell divisions produce <u>four</u> daughter cells instead of two as in mitosis.
- The resulting daughter cells have *half* the number of chromosomes as the original cell; whereas, daughter cells of mitosis have the same number of chromosomes as the parent cell.

A. The Stages of Meiotic Cell Division

Interphase I: Interphase I precedes meiosis.

- Chromosomes replicate as in mitosis.
- Each duplicated chromosome consists of two identical sister chromatids attached at their centromeres.
- Centriole pairs in animal cells also replicate into two pairs.

Meiosis I: This cell division segregates the two chromosomes of each homologous pair and reduces the chromosome number by one-half. It includes the following four phases:

Prophase I. This is a longer and more complex process than prophase of mitosis.

- Chromosomes condense.
- *Synapsis* occurs. During this process, homologous chromosomes come together as pairs.
- Chromosomes condense further until they are distinct structures that can be seen with a microscope. Since each chromosome has two chromatids, each homologous pair in synapsis appears as a complex of four chromatids or a *tetrad*.
- In each tetrad, sister chromatids of the same chromosome are attached at their centromeres. Nonsister chromatids are linked by X-shaped *chiasmata*, sites where homologous strand exchange or *crossing-over* occurs.
- Chromosomes thicken further and detach from the nuclear envelope.

As prophase I continues, the cell prepares for nuclear division.

- Centriole pairs move apart and spindle microtubules form between them.
- Nuclear envelope and nucleoli disperse.
- Chromosomes begin moving to the metaphase plate, midway between the two poles of the spindle apparatus.
- Prophase I typically occupies more than 90% of the time required for meiosis.

Metaphase I. Tetrads are aligned on the metaphase plate.

- Each synaptic pair is aligned so that centromeres of homologues point towards opposite poles.

- Each homologue is thus attached to kinetochore microtubules emerging from the pole it faces, so that the two homologues are destined to separate in anaphase and move towards opposite poles.

Anaphase I. Homologues separate and are moved towards the poles by the spindle apparatus.

- Sister chromatids remain attached at their centromeres and move as a unit towards the same pole, while the homologue moves towards the opposite pole.

- This differs from mitosis during which chromosomes line up individually on the metaphase plate (rather than in pairs) and sister chromatids are moved apart towards opposite poles of the cell.

Telophase I and Cytokinesis. The spindle apparatus continues to separate homologous chromosome pairs until the chromosomes reach the poles.

- Each pole now has a haploid set of chromosomes that are each still composed of two sister chromatids attached at the centromere.

- Usually, cytokinesis occurs simultaneously with Telophase I, forming two *haploid* daughter cells. *Cleavage furrows* form in animal cells, and cell plates form in plant cells.

- In some species, nuclear membranes and nucleoli reappear, and the cell enters a period of *interkinesis* before meiosis II. In other species, the daughter cells immediately prepare for meiosis II.

- Regardless of whether a cell enters interkinesis, *no DNA replication occurs before meiosis I.*

Meiosis II: This second meiotic division separates sister chromatids of each chromosome.

Prophase II.
- If the cell entered interkinesis, the nuclear envelope and nucleoli disperse.
- Spindle apparatus forms and chromosomes move towards the metaphase II plate.

Metaphase II. Chromosomes align singly on the metaphase plate.
- Kinetochores of sister chromatids point towards opposite poles.

Anaphase II. Sister chromatids separate.
- Centromeres of sister chromatids separate.
- Sister chromatids of each pair (now individual chromosomes) move toward opposite poles of the cell.

Telophase II and Cytokinesis.

- Nuclei form at opposite poles of the cell.
- Cytokinesis occurs producing four haploid daughter cells.

Haploid daughter cells

B.   **A Comparison of Mitosis and Meiosis**

NOTE: Spending class time on a comparison of mitosis and meiosis is really worth the effort. It not only brings closure to the topic, but also provides an opportunity to check for understanding. One check is to ask students to identify unlabeled diagrams of various stages in mitosis and meiosis. The ability to distinguish metaphase of mitosis from metaphase of meiosis I, is particularly diagnostic of student understanding.

If you are fortunate enough to have video capability in your classroom, you can show moving sequences of mitosis and meiosis. The fact that these are dynamic processes involving chromosomal movement is not a trivial point, but is often lost in the course of a lecture where the only visuals are drawings or micrographs.

Though the processes of mitosis and meiosis are similar in some ways, there are some key differences:

- *Meiosis is a reduction division*. Cells produced by mitosis have the same number of chromosomes as the original cell, whereas cells produced by meiosis have <u>half</u> the number of chromosomes as the parent cell.
- *Meiosis creates genetic variation*. Mitosis produces *two* daughter cells genetically identical to the parent cell and to each other. Meiosis produces *four* daughter cells genetically different from the parent cell and from each other.
- *Meiosis is two successive nuclear divisions*. Mitosis, on the other hand, is characterized by just one nuclear division.

COMPARISON OF MEIOSIS I AND MITOSIS		
	<u>MEIOSIS I</u>	<u>MITOSIS</u>
Prophase	*Synapsis* occurs to form tetrads. *Chiasmata* appear as evidence that crossing over has occurred.	Neither synapsis nor crossing over occurs.
Metaphase	Homologous pairs (tetrads) align on the metaphase plate.	Individual chromosomes align on the metaphase plate.
Anaphase	Meiosis I separates pairs of chromosomes. Centromeres do *not* divide and sister chromatids stay together. Sister chromatids of each chromosome move to the *same* pole of the cell; only the homologues separate.	Mitosis separates sister chromatids of individual chromosomes. Centromeres divide and sister chromatids move to *opposite* poles of the cell.

Meiosis II is virtually identical in mechanism to mitosis, separating sister chromatids.

V. Sexual life cycles produce genetic variation among offspring

Meiosis and fertilization are the primary sources of genetic variation in sexually reproducing organisms. Sexual reproduction provides genetic variation by:

- independent assortment;
- crossing over during prophase I of meiosis;
- random fusion of gametes during fertilization.

A. Independent Assortment of Chromosomes

At Metaphase I, each homologous pair of chromosomes aligns on the metaphase plate. Each pair consists of one maternal and one paternal chromosome.

- The orientation of the homologous pair to the poles is random, so there is a fifty-fifty chance that a particular daughter cell produced by meiosis I will receive the maternal chromosome of a homologous pair, and a fifty-fifty chance that it will receive the paternal chromosome.
- Each homologous pair of chromosomes orients independently of the other pairs at metaphase I; thus, the first meiotic division results in *independent assortment* of maternal and paternal chromosomes. (See Campbell, Figure 12.8)
- A gamete produced by meiosis contains just one of all the possible combinations of maternal and paternal chromosomes.

<u>Independent assortment</u> = The random distribution of maternal and paternal homologues to the gametes. (In a more specific sense, assortment refers to the random distribution of genes located on different chromosomes.)

- Since each homologous pair assorts independently from all the others, the process produces 2^n possible combinations of maternal and paternal chromosomes in gametes, where n is the haploid number.
- In humans, the possible combinations would be 2^{23}, or about eight million.
- Thus, each human gamete contains one of eight million possible assortments of chromosomes inherited from that person's mother and father.
- Genetic variation results from this reshuffling of chromosomes, because the maternal and paternal homologues will carry different genetic information at many of their corresponding loci.

B.   Crossing Over

Another mechanism that increases genetic variation is the process of *crossing over*, during which homologous chromosomes exchange genes.

<u>Crossing over</u> = The exchange of genetic material between homologues; occurs during prophase of meiosis I. This process:

- Occurs when homologous portions of two nonsister chromatids trade places. During prophase I, X-shaped *chiasmata* become visible at places where this homologous strand exchange occurs.
- Produces chromosomes that contain genes from *both* parents.
- In humans, there is an average of two or three crossovers per chromosome pair.
- Synapsis during prophase I is precise, so that homologues align gene by gene. The exact mechanism of synapsis is still unknown, but involves the formation of the *synaptonemal complex*, a protein structure that brings the chromosomes into close association.

C.   Random Fertilization

Random fertilization is another source of genetic variation in offspring.

- In humans, an egg cell that is one of eight million different possibilities will be fertilized by a sperm cell that is also one of eight million possibilities. Thus, the resulting zygote can have one of 64 trillion possible diploid combinations.

To summarize, there are three important sources of genetic variability in sexually reproducing organisms:

1. Independent assortment of homologous chromosome pairs during meiosis I.
2. Crossing over between homologous chromosomes during prophase of meiosis I.
3. Random fusion of gametes.

VI.   Evolutionary adaptation depends on a population's genetic variation

Inheritable variation is the basis for Charles Darwin's theory that natural selection is the mechanism for evolutionary change. Natural selection:

- Increases the frequency of inheritable variations that favor the reproductive success of some individuals over others.
- Results in *adaptation*, the accumulation of inheritable variations that are favored by the environment.
- In the face of environmental change, genetic variation increases the likelihood that some individuals in a population will have inheritable variations that help them cope with the new conditions.

There are two sources of genetic variation:

1. Sexual reproduction: independent assortment in meiosis I, crossing over in prophase of meiosis I, and random fusion of gametes during fertilization.
2. Mutation, which is rare structural change in a gene.

REFERENCES

Alberts, B., D. Bray, J. Lewis, M. Raff, K. Roberts and J.D. Watson. *Molecular Biology of the Cell*. 2nd ed. New York: Garland, 1989.

Becker, W.M., J.B. Reece, and M.F. Puente. The World of the Cell. 3rd. ed. Redwood City, California: Benjamin/Cummings, 1996.

Campbell, N. *Biology*. 4th ed. Menlo Park, California: Benjamin/Cummings, 1996.

Kleinsmith, L.J. and V.M. Kish. *Principles of Cell Biology*. New York: Harper and Row, Publ., 1988.

CHAPTER **13**

MENDEL AND THE GENE IDEA

OUTLINE

I. Mendel brought an experimental and quantitative approach to genetics: *science as a process*

II. According to the law of segregation, the two alleles for a character are packaged into separate gametes
   A. Some Useful Genetic Vocabulary
   B. The Testcross

III. According to the law of independent assortment, each pair of alleles segregates into gametes independently

IV. Mendelian inheritance reflects rules of probability
   A. Rule of Multiplication
   B. Rule of Addition
   C. Using Rules of Probability to Solve Genetics Problems

V. Mendel discovered the particulate behavior of genes: *a review*

VI. The relationship between genotype and phenotype is rarely simple
   A. Incomplete Dominance
   B. What Is a Dominant Allele?
   C. Multiple Alleles
   D. Pleiotropy
   E. Epistasis
   F. Polygenic Inheritance
   G. Nature Versus Nurture: The Environmental Impact on Phenotype
   H. Integrating a Mendelian View of Heredity and Variation

VII. Pedigree analysis reveals mendelian patterns in human inheritance

VIII.   Many human disorders follow mendelian patterns of inheritance
   A.   Recessively Inherited Disorders
   B.   Dominantly Inherited Disorders
   C.   Multifactorial Disorders

IX.   Technology is providing new tools for genetic testing and counseling
   A.   Carrier Recognition
   B.   Fetal Testing
   C.   Newborn Screening

OBJECTIVES

After reading this chapter and attending lecture, the student should be able to:

1. Describe the favored model of heredity in the 19th century prior to Mendel, and explain how this model was inconsistent with observations.
2. Explain how Mendel's hypothesis of inheritance differed from the blending theory of inheritance.
3. List several features of Mendel's methods that contributed to his success.
4. List four components of Mendel's hypothesis that led him to deduce the law of segregation.
5. State, in their own words, Mendel's law of segregation.
6. Use a Punnett square to predict the results of a monohybrid cross and state the phenotypic and genotypic ratios of the F_2 generation.
7. Distinguish between genotype and phenotype; heterozygous and homozygous; dominant and recessive.
8. Explain how a testcross can be used to determine if a dominant phenotype is homozygous or heterozygous.
9. Define random event, and explain why it is significant that allele segregation during meiosis and fusion of gametes at fertilization are random events.
10. Use the rule of multiplication to calculate the probability that a particular F_2 individual will be homozygous recessive or dominant.
11. Given a Mendelian cross, use the rule of addition to calculate the probability that a particular F_2 individual will be heterozygous.
12. Describe two alternate hypotheses that Mendel considered for how two characters might segregate during gamete formation, and explain how he tested those hypothesis.
13. State, in their own words, Mendel's law of independent assortment.
14. Use a Punnett square to predict the results of a dihybrid cross and state the phenotypic and genotypic ratios of the F_2 generation.
15. Using the laws of probability, predict from a trihybrid cross between two individuals that are heterozygous for all three traits, what expected proportion of the offspring would be:
   a. Homozygous for the three dominant traits.
   b. Heterozygous for all three traits.
   c. Homozygous recessive for two specific traits and heterozygous for the third.
16. Give an example of incomplete dominance and explain why it is not evidence for the blending theory of inheritance.
17. Explain how the phenotypic expression of the heterozygote is affected by complete dominance, incomplete dominance and codominance.

18. Describe the inheritance of the ABO blood system and explain why the I^A and I^B alleles are said to be *codominant*.
19. Define and give examples of pleiotropy.
20. Explain, in their own words, what is meant by "one gene is epistatic to another."
21. Explain how epistasis affects the phenotypic ratio for a dihybrid cross.
22. Describe a simple model for polygenic inheritance, and explain why most polygenic characters are described in quantitative terms.
23. Describe how environmental conditions can influence the phenotypic expression of a character.
24. Given a simple family pedigree, deduce the genotypes for some of the family members.
25. Describe the inheritance and expression of cystic fibrosis, Tay-Sachs disease and sickle-cell disease.
26. Explain how a lethal recessive gene can be maintained in a population.
27. Explain why consanguinity increases the probability of homozygosity in offspring.
28. Explain why lethal dominant genes are much more rare than lethal recessive genes.
29. Give an example of a late-acting lethal dominant in humans and explain how it can escape elimination.
30. Explain how carrier recognition, fetal testing and newborn screening can be used in genetic screening and counseling.

KEY TERMS

blending theory	heterozygous	ABO blood system	consanguinity
Gregor Mendel	phenotype	pleiotropy	lethal dominant
true-breeding	genotype	epistasis	Huntington's disease
P generation	monohybrid	quantitative character	amniocentesis
F_1 generation	testcross	polygenic inheritance	chorionic villi
F_2 generation	dihybrid	norm of reaction	sampling
Mendel's law of	law of independent	pedigree	ultrasound
segregation	assortment	lethal recessive	fetoscopy
alleles	incomplete	carriers	PKU
dominant allele	dominance	cystic fibrosis	multifactorial
recessive allele	codominance	Tay-Sachs disease	disorder
homozygous	multiple alleles	sickle-cell disease	

LECTURE NOTES

NOTE: Listed below are a few suggestions for teaching Mendelian genetics:

1. There is a certain baseline working vocabulary that students need in order to follow your lecture, understand the text and solve problems. It is more economical to recognize this fact and just begin with some "definitions you should know". Once that is done, you can use the terms in context during the lecture and focus attention on the major points rather than on defining terms.

2. Demonstrating how to work a Punnett square and how to solve genetics problems is obviously necessary. But your students will learn best if they actively participate in the process. You can structure opportunities for students to solve problems during lecture and then let them participate in the explanation. If time does not allow this, it is highly recommended that there be an additional recitation or problem-solving session outside of class.

3. After Mendel's laws of segregation and independent assortment have been introduced, it is extremely useful to put up a transparency of meiosis and ask students to identify where in meiosis that segregation and assortment occur. Many students will not make this connection on their own.

Based upon their observations from ornamental plant breeding, biologists in the 19th century realized that both parents contribute to the characteristics of offspring. Before Mendel, the favored explanation of heredity was the *blending theory*.

Blending theory of heredity = Pre-Mendelian theory of heredity proposing that hereditary material from each parent mixes in the offspring; once blended like two liquids in solution, the hereditary material is inseparable and the offspring's traits are some intermediate between the parental types. According to this theory:
- Individuals of a population should reach a uniform appearance after many generations.
- Once hereditary traits are blended, they can no longer be separated out to appear again in later generations.

This blending theory of heredity was inconsistent with the observations that:
- Individuals in a population do not reach a uniform appearance; inheritable variation among individuals is generally preserved.
- Some inheritable traits skip one generation only to reappear in the next.

Modern genetics began in the 1860's when Gregor Mendel, an Augustinian monk, discovered the fundamental principles of heredity. Mendel's great contribution to modern genetics was to replace the blending theory of heredity with the *particulate theory of heredity*.

Particulate theory of heredity = Gregor Mendel's theory that parents transmit to their offspring discrete inheritable factors (now called genes) that remain as separate factors from one generation to the next.

I. **Mendel brought an experimental and quantitative approach to genetics:** *science as a process*

While attending the University of Vienna from 1851-1853, Mendel was influenced by two professors:

- Doppler, a physicist, trained Mendel to apply a *quantitative experimental* approach to the study of natural phenomena.
- Unger, a botanist, interested Mendel in the causes of inheritable variation in plants.

These experiences inspired Mendel to use key elements of the scientific process in the study of heredity. Unlike most nineteenth century biologists, he used a quantitative approach to his experimentation.

In 1857, Mendel was living in an Augustinian monastery, where he bred garden peas in the abbey garden. He probably chose garden peas as his experimental organisms because:

- They were available in many easily distinguishable varieties.
- Strict control over mating was possible to ensure the parentage of new seeds. Petals of the pea flower enclose the pistil and stamens, which prevents cross-pollination. Immature stamens can be removed to prevent self-pollination. Mendel hybridized pea plants by transferring pollen from one flower to another with an artist's brush.

Character = Detectable inheritable feature of an organism.

Trait = Variant of an inheritable character.

Mendel chose *characters* in pea plants that differed in a relatively clear-cut manner. He chose seven characters, each of which occurred in two alternative forms:

1. Flower color (purple or white)
2. Flower position (axial or terminal)
3. Seed color (yellow or green)
4. Seed shape (round or wrinkled)
5. Pod shape (inflated or constricted)
6. Pod color (green or yellow)
7. Stem length (tall or dwarf)

True breeding = Always producing offspring with the same *traits* as the parents when the parents are self-fertilized.

Mendel started his experiments with *true-breeding* plant varieties, which he hybridized (cross-pollinated) in experimental crosses.

- The true-breeding parental plants of such a *cross* are called the *P generation* (parental).
- The hybrid offspring of the P generation are the F_1 *generation* (first filial).
- Allowing F_1 generation plants to self-pollinate, produces the next generation, the F_2 *generation* (second filial).

Mendel observed the transmission of selected traits for at least three generations and arrived at two principles of heredity that are now known as the *law of segregation* and the *law of independent assortment*.

II. **According to the law of segregation, the two alleles for a character are packaged into separate gametes**

When Mendel crossed true-breeding plants with different character traits, he found that the traits did not blend.

- Using the scientific process, Mendel designed experiments in which he used large sample sizes and kept accurate quantitative records of the results.
- For example, a cross between true-breeding varieties, one with purple flowers and one with white flowers, produced F_1 *progeny* (offspring) with only purple flowers.

   Hypothesis: Mendel hypothesized that if the inheritable factor for white flowers had been lost, then a cross between F_1 plants should produce only purple-flowered plants.

   Experiment: Mendel allowed the F_1 plants to self-pollinate.

   Results: There were 705 purple-flowered and 224 white-flowered plants in the F_2 generation – a ratio of 3:1. The inheritable factor for white flowers was not lost, so the hypothesis was rejected.

   Conclusions: From these types of experiments and observations, Mendel concluded that since the inheritable factor for white flowers was not lost in the F_1 generation, it must have been masked by the presence of the purple-flower factor. Mendel's factors are now called *genes*; and in Mendel's terms, purple flower is the *dominant trait* and white flower is the *recessive trait*.

Mendel repeated these experiments with the other six characters and found similar 3:1 ratios in the F_2 generations. From these observations he developed a hypothesis that can be divided into four parts:

1. *Alternative forms of genes are responsible for variations in inherited characters.*
    - For example, the gene for flower color in pea plants exists in two alternative forms; one for purple color and one for white color.
    - Alternative forms for a gene are now called *alleles*.

2. *For each character, an organism inherits two alleles, one from each parent.*
    - Mendel deduced that each parent contributes one "factor," even though he did not know about chromosomes or meiosis.
    - We now know that Mendel's factors are genes. Each genetic locus is represented twice in diploid organisms, which have homologous pairs of chromosomes, one set for each parent. Homologous loci may have the same allele as in Mendel's true-breeding organisms or they may differ as in the F_1 hybrids.

3. *If the two alleles differ, one is fully expressed (dominant allele); the other is completely masked (recessive allele).*
    - Dominant alleles are designated by a capital letter: P = purple flower color.
    - Recessive alleles are designated by a lowercase letter: p = white flower color.

4. *The two alleles for each character segregate during gamete production.*
   - Without any knowledge of meiosis, Mendel deduced that a sperm cell or ovum carries only one allele for each inherited characteristic, because allele pairs separate (segregate) from each other during gamete production.
   - Gametes of true-breeding plants will all carry the same allele. If different alleles are present in the parent, there is a 50% chance that a gamete will receive the dominant allele, and a 50% chance that it will receive the recessive allele.
   - This sorting of alleles into separate gametes is known as Mendel's *law of segregation.*

<u>Mendel's law of segregation</u> = Allele pairs segregate during gamete formation (meiosis), and the paired condition is restored by the random fusion of gametes at fertilization.
   - This law predicts the 3:1 ratio observed in the F_2 generation of a monohybrid cross.
   - F_1 hybrids (Pp) produce two classes of gametes when allele pairs segregate during gamete formation. Half receive a purple-flower allele (P) and the other half the white-flower allele (p).
   - During self-pollination, these two classes of gametes unite randomly. Ova containing purple-flower alleles have equal chances of being fertilized by sperm carrying purple-flower alleles or sperm carrying white-flower alleles.
   - Since the same is true for ova containing white-flower alleles, there are four equally likely combinations of sperm and ova.

The combinations resulting from a genetic cross may be predicted by using a *Punnett Square*.

The F_2 progeny would include:

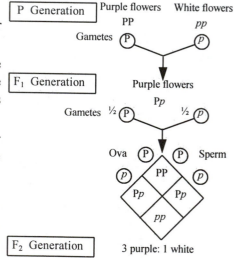

   - One-fourth of the plants with two alleles for purple flowers.
   - One-half of the plants with one allele for purple flowers and one allele for white flowers. Since the purple-flower allele is dominant, these plants have purple flowers.
   - One-fourth of the plants with two alleles for white flower color, which will have white flowers since no dominant allele is present.

The pattern of inheritance for all seven of the characteristics studied by Mendel was the same: one parental trait disappeared in the F_1 generation and reappeared in one-fourth of the F_2 generation.

A. **Some Useful Genetic Vocabulary**

   <u>Homozygous</u> = Having two identical alleles for a given trait (e.g. PP or pp).
   - All gametes carry that allele.
   - Homozygotes are *true-breeding*.

<u>Heterozygous</u> = Having two different alleles for a trait (e.g. Pp).

- Half of the gametes carries one allele (P) and the remaining half carries the other (p).
- Heterozygotes are not true-breeding.

<u>Phenotype</u> = An organism's expressed traits (e.g. purple or white flowers).

- In Mendel's experiment above, the F_2 generation had a 3:1 *phenotypic ratio* of plants with purple flowers to plants with white flowers.

<u>Genotype</u> = An organism's genetic makeup (e.g. PP, Pp or pp).

- The *genotypic ratio* of the F_2 generation was 1:2:1 (1 PP:2 Pp:1 pp).

B.   The Testcross

Because some alleles are dominant over others, the genotype of an organism may not be apparent. For example:

- A pea plant with purple flowers may be either homozygous dominant (PP) or heterozygous (Pp).

To determine whether an organism with a dominant phenotype (e.g. purple flower color) is homozygous dominant or heterozygous, you use a *testcross*.

<u>Testcross</u> = The breeding of an organism of unknown genotype with a homozygous recessive.

- For example, if a cross between a purple-flowered plant of unknown genotype (P___) produced only purple-flowered plants, the parent was probably homozygous dominant since a PP × pp cross produces all purple-flowered progeny that are heterozygous (Pp).

- If the progeny of the testcross contains both purple and white phenotypes, then the purple-flowered parent was heterozygous since a Pp × pp cross produces Pp and pp progeny in a 1:1 ratio.

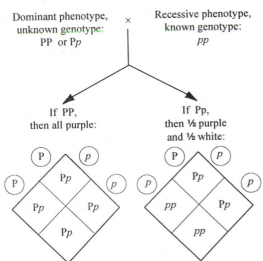

III. **According to the law of independent assortment, each pair of alleles segregates into gametes independently**

Mendel deduced the law of segregation from experiments with *monohybrid crosses*, breeding experiments that used parental varieties differing in a single trait. He then performed crosses between parental varieties that differed in two characters or *dihybrid crosses*.

<u>Dihybrid cross</u> = A mating between parents that are heterozygous for two characters (dihybrids).

- Mendel began his experiments by crossing true-breeding parent plants that differed in two characters such as seed color (yellow or green) and seed shape (round or wrinkled). From previous monohybrid crosses, Mendel knew that yellow seed (Y) was dominant to green (y), and that round (R) was dominant to wrinkled (r).
- Plants homozygous for round yellow seeds (RRYY) were crossed with plants homozygous for wrinkled green seeds (rryy).
- The resulting F_1 dihybrid progeny were heterozygous for both traits (RrYy) and had round yellow seeds, the dominant phenotypes.
- From the F_1 generation, Mendel could not tell if the two characters were inherited independently or not, so he allowed the F_1 progeny to self-pollinate. In the following experiment, Mendel considered two alternate hypotheses:

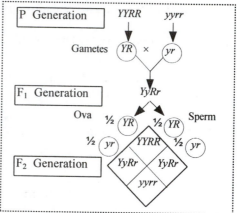

Hypothesis 1: If the two characters segregate *together*, the F_1 hybrids can only produce the same two classes of gametes (RY and ry) that they received from the parents, and the F_2 progeny will show a 3:1 phenotypic ratio.

Hypothesis 2: If the two characters segregate *independently*, the F_1 hybrids will produce four classes of gametes (RY, Ry, rY, ry), and the F_2 progeny will show a 9:3:3:1 ratio.

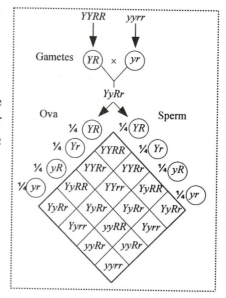

Experiment: Mendel performed a dihybrid cross by allowing self-pollination of the F_1 plants (RrYy × RrYy).

Results: Mendel categorized the F_2 progeny and determined a ratio of 315:108:101:32, which approximates 9:3:3:1.

These results were repeatable. Mendel performed similar dihybrid crosses with all seven characters in various combinations and found the same 9:3:3:1 ratio in each case.

He also noted that the ratio for each individual gene pair was 3:1, the same as that for a monohybrid cross.

Conclusions: The experimental results supported the hypothesis that *each allele pair segregates independently during gamete formation.*

This behavior of genes during gamete formation is referred to as *Mendel's law of independent assortment.*

<u>Mendel's law of independent assortment</u> = Each allele pair segregates independently of other gene pairs during gamete formation.

IV. Mendelian inheritance reflects rules of probability

Segregation and independent assortment of alleles during gamete formation and fusion of gametes at fertilization are random events. Thus, if we know the genotypes of the parents, we can predict the most likely genotypes of their offspring by using the simple *laws of probability*:

- The probability scale ranges from 0 to 1; an event that is certain to occur has a probability of 1, and an event that is certain *not* to occur has a probability of 0.
- The probabilities of all possible outcomes for an event must add up to 1.
- For example, when tossing a coin or rolling a six-sided die:

Event	Probability	
Tossing heads with a two-headed coin	1	
Tossing tails with a two-headed coin	0	$1 + 0 = 1$
Tossing heads with a normal coin	1/2	
Tossing tails with a normal coin	1/2	$1/2 + 1/2 = 1$
Rolling 3 on a six-sided die	1/6	
Rolling a number other than 3	5/6	$1/6 + 5/6 = 1$

Random events are *independent* of one another.

- The outcome of a random event is unaffected by the outcome of previous such events.
- For example, it is possible that five successive tosses of a normal coin will produce five heads; however, the probability of heads on the sixth toss is still 1/2.

Two basic rules of probability are helpful in solving genetics problems: the *rule of multiplication* and the *rule of addition*.

A. Rule of Multiplication

Rule of multiplication = The probability that independent events will occur simultaneously is the product of their individual probabilities. For example:

Question: In a Mendelian cross between pea plants that are heterozygous for flower color (Pp), what is the probability that the offspring will be homozygous recessive?

Answer:

Probability that an egg from the F_1 (Pp) will receive a p allele = 1/2.

Probability that a sperm from the F_1 will receive a p allele = 1/2.

The overall probability that two recessive alleles will unite at fertilization: $1/2 \times 1/2 = \underline{1/4}$.

This rule also applies to dihybrid crosses. For example:

Question: For a dihybrid cross, YyRr × YyRr, what is the probability of an F_2 plant having the genotype YYRR?

Answer:

Probability that an egg from a YyRr parent will receive the Y and R alleles = $1/2 \times 1/2 = 1/4$.

Probability that a sperm from a YyRr parent will receive a the Y and R alleles = $1/2 \times 1/2 = 1/4$.

The overall probability of an F_2 plant having the genotype YYRR: $1/4 \times 1/4 = \underline{1/16}$.

B. Rule of Addition

Rule of addition = The probability of an event that can occur in two or more independent ways is the sum of the separate probabilities of the different ways. For example:

Question: In a Mendelian cross between pea plants that are heterozygous for flower color (Pp), what is the probability of the offspring being a heterozygote?

Answer: There are two ways in which a heterozygote may be produced: the dominant allele (P) may be in the egg and the recessive allele (p) in the sperm, or the dominant allele may be in the sperm and the recessive in the egg. Consequently, the probability that the offspring will be heterozygous is the sum of the probabilities of those two possible ways:

Probability that the dominant allele will be in the egg with the recessive in the sperm is $1/2 \times 1/2 = 1/4$.

Probability that the dominant allele will be in the sperm and the recessive in the egg is $1/2 \times 1/2 = 1/4$.

Therefore, the probability that a heterozygous offspring will be produced is $1/4 + 1/4 = \underline{1/2}$.

C.   **Using Rules of Probability to Solve Genetics Problems**

The rules of probability can be used to solve complex genetics problems. For example, Mendel crossed pea varieties that differed in three characters (*trihybrid crosses*).

Question: What is the probability that a trihybrid cross between two organisms with the genotypes AaBbCc and AaBbCc will produce an offspring with the genotype aabbcc?

Answer: Because segregation of each allele pair is an independent event, we can treat this as three separate monohybrid crosses:

$Aa \times Aa$: probability for aa offspring $=$ $1/4$
$Bb \times Bb$: probability for bb offspring $=$ $1/4$
$Cc \times Cc$: probability for cc offspring $=$ $1/4$

The probability that these independent events will occur simultaneously is the product of their independent probabilities (rule of multiplication). So the probability that the offspring will be aabbcc is:

$1/4$ aa $\times$ $1/4$ bb $\times$ $1/4$ cc $=$ $\underline{1/64}$

For another example, consider a trihybrid cross of garden peas, where:

Character	Trait & Genotype	
Flower Color	Purple:	PP, Pp
	White:	pp
Seed Color	Yellow:	YY, Yy
	Green:	yy
Seed Shape	Round:	RR, Rr
	Wrinkled:	rr

Question: What fraction of offspring from the following cross of garden peas, would show recessive phenotypes for *at least* two of the three traits?

$$PpYyRr \times Ppyyrr$$

Answer: First list those genotypes that are homozygous recessive for at least two traits, (note that this includes the homozygous recessive for all three traits). Use the *rule of multiplication* to calculate the probability that offspring would be one of these genotypes. Then use the *rule of addition* to calculate the probability that two of the three traits would be homozygous recessive.

Genotypes with at least two homosygous recessives			Probability of genotype
ppyyRr	$1/4 \times 1/2 \times 1/2$	=	1/16
ppYyrr	$1/4 \times 1/2 \times 1/2$	=	1/16
Ppyyrr	$1/2 \times 1/2 \times 1/2$	=	2/16
PPyyrr	$1/4 \times 1/2 \times 1/2$	=	1/16
ppyyrr	$1/4 \times 1/2 \times 1/2$	=	<u>1/16</u>
		=	6/16 or <u>3/8</u> chance of two recessive traits

V.   Mendel discovered the particulate behavior of genes: *a review*

If a seed is planted from the F_2 generation of a monohybrid cross, we cannot predict with absolute certainty that the plant will grow to produce white flowers (pp). We *can* say that there is a 1/4 chance that the plant will have white flowers.

- Stated in statistical terms: among a large sample of F_2 plants, 25% will have white flowers.
- The larger the sample size, the closer the results will conform to predictions.

Mendel's quantitative methods reflect his understanding of this statistical feature of inheritance. Mendel's laws of segregation and independent assortment are based on the premise that:

- Inheritance is a consequence of discrete factors (genes) that are passed on from generation to generation.
- Segregation and assortment are random events and thus obey the simple laws of probability.

VI.   The relationship between genotype and phenotype is rarely simple

As Mendel described it, characters are determined by one gene with two alleles; one allele completely dominant over the other. There are other patterns of inheritance not described by Mendel, but his laws of segregation and independent assortment can be extended to these more complex cases.

A. Incomplete Dominance

In cases of *incomplete dominance*, one allele is not completely dominant over the other, so the heterozygote has a phenotype that is intermediate between the phenotypes of the two homozygotes.

Incomplete dominance = Pattern of inheritance in which the dominant phenotype is not fully expressed in the heterozygote, resulting in a phenotype intermediate between the homozygous dominant and homozygous recessive.

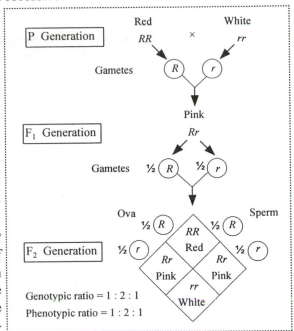

- For example, when red snapdragons (RR) are crossed with white snapdragons (rr), all F_1 hybrids (Rr) have pink flowers. (The heterozygote produces half as much red pigment as the homozygous red-flowered plant.)

- Since the heterozygotes can be distinguished from homozygotes by their phenotypes, the phenotypic and genotypic ratios from a monohybrid cross are the same – 1:2:1.

- Incomplete dominance is *not* support for the blending theory of inheritance, because alleles maintain their integrity in the heterozygote and segregate during gamete formation. Red and white phenotypes reappear in the F_2 generation.

B. What Is a Dominant Allele?

Dominance/recessiveness relationships among alleles vary in a continuum from *complete dominance* on one end of the spectrum to *codominance* on the other, with various degrees of incomplete dominance in between these extremes.

Complete dominance = Inheritance characterized by an allele that is fully expressed in the phenotype of a heterozygote and that masks the phenotypic expression of the recessive allele; state in which the phenotypes of the heterozygote and dominant homozygote are indistinguishable.

<u>Codominance</u> = Inheritance characterized by full expression of both alleles in the heterozygote.

- For example, the MN blood-group locus codes for the production of surface glycoproteins on the red blood cell. In this system, there are three blood types: M, N and MN.

Blood Type	Genotype
M	MM
N	NN
MN	MN

- The MN blood type is the result of full phenotypic expression of *both* alleles in the heterozygote; both molecules, M and N, are produced on the red blood cell.

Apparent dominance/recessiveness relationships among alleles reflect the level at which the phenotype is studied. For example:

- *Tay-Sachs disease* is a recessively inherited disease in humans; only children who are homozygous recessive for the Tay-Sachs allele have the disease.
- Brain cells of Tay-Sachs babies lack a crucial lipid-metabolizing enzyme. Thus, lipids accumulate in the brain, causing the disease symptoms and ultimately leading to death.
- At the *organismal level*, since heterozygotes are symptom free, it appears that the normal allele is completely dominant and the Tay-Sachs allele is recessive.
- At the *biochemical level*, inheritance of Tay-Sachs seems to be incomplete dominance of the normal allele, since there is an intermediate phenotype. Heterozygotes have an enzyme activity level that is intermediate between individuals homozygous for the normal allele and individuals with Tay-Sachs Disease.
- At the *molecular level*, the normal allele and the Tay-Sachs allele are actually codominant. Heterozygotes produce equal numbers of normal and dysfunctional enzymes. They lack disease symptoms, because half the normal amount of functional enzyme is sufficient to prevent lipid accumulation in the brain.

Dominance/recessiveness relationships among alleles:

- are a consequence of the mechanism that determines phenotypic expression, not the ability of one allele to subdue another at the level of the DNA.
- do not determine the relative abundance of alleles in a population.
  - ⇒ In other words, dominant alleles are not necessarily more common and recessive alleles more rare.
  - ⇒ For example, the allele for polydactyly is quite rare in the U.S. (1 in 400 births), yet it is caused by a dominant allele. (Polydactyly is the condition of having extra fingers or toes.)

C. Multiple Alleles

Some genes may have *multiple alleles*; that is, more than just two alternative forms of a gene. The inheritance of the ABO blood group is an example of a locus with three alleles.

Paired combinations of three alleles produce four possible phenotypes:

- Blood type A, B, AB, or O.
- A and B refer to two genetically determined polysaccharides (A and B antigens) which are found on the surface of red blood cells.

There are three alleles for this gene: I^A, I^B, and i.

- The I^A allele codes for the production of A antigen, the I^B allele codes for the production of B antigen, and the i allele codes for *no* antigen production on the red blood cell (neither A or B).
- Alleles I^A and I^B are *codominant* since both are expressed in heterozygotes.
- Alleles I^A and I^B are dominant to allele i, which is recessive.
- Even though there are three possible alleles, every person carries only two alleles which specify their ABO blood type; one allele is inherited from each parent.

Since there are three alleles, there are six possible genotypes:

Blood Type	Possible Genotypes	Antigens on the red blood cell	Antibodies in the serum
A	$I^A I^A$ I^Ai	A	anti-B
B	$I^B I^B$ I^Bi	B	anti-A
AB	$I^A I^B$	A, B	----
O	ii	----	anti-A anti-B

Foreign antigens usually cause the immune system to respond by producing *antibodies*, globular proteins that bind to the foreign molecules causing a reaction that destroys or inactivates it. In the ABO blood system:

- The antigens are located on the red blood cell and the antibodies are in the serum.
- A person produces antibodies against foreign blood antigens (those not possessed by the individual). These antibodies react with the foreign antigens causing the blood cells to clump or *agglutinate*, which may be lethal.
- For a blood transfusion to be successful, the red blood cell *antigens of the donor* must be compatible with the *antibodies of the recipient*.

D. Pleiotropy

Pleiotropy = The ability of a single gene to have multiple phenotypic effects.

- There are many hereditary diseases in which a single defective gene causes complex sets of symptoms (e.g. sickle-cell anemia).
- One gene can also influence a combination of seemingly unrelated characteristics. For example, in tigers and Siamese cats, the gene that controls fur pigmentation also influences the connections between a cat's eyes and the brain. A defective gene causes both abnormal pigmentation and cross-eye condition.

E. Epistasis

Different genes can interact to control the phenotypic expression of a single trait. In some cases, a gene at one locus alters the phenotypic expression of a second gene, a condition known as *epistasis*.

Epistasis = (Epi=upon; stasis=standing) Interaction between two nonallelic genes in which one modifies the phenotypic expression of the other.

- If one gene suppresses the phenotypic expression of another, the first gene is said to be *epistatic* to the second.
- If epistasis occurs between two nonallelic genes, the phenotypic ratio resulting from a dihybrid cross will deviate from the 9:3:3:1 Mendelian ratio.
- For example, in mice and other rodents, the gene for pigment deposition (C) is epistatic to the gene for pigment (melanin) production. In other words, whether the pigment can be deposited in the fur determines whether the coat color can be expressed. Homozygous recessive for pigment deposition (cc) will result in an albino mouse regardless of the genotype at the black/brown locus (BB, Bb or bb):

  CC, Cc = Melanin deposition

  cc = Albino

  BB, Bb = Black coat color

  bb = Brown coat color

- Even though both genes affect the same character (coat color), they are inherited separately and will assort independently during gamete formation. A cross between black mice that are heterozygous for the two genes results in a 9:3:4 phenotypic ratio:

  9 Black (B_C_)

  3 Brown (bbC_)

  4 Albino (__cc)

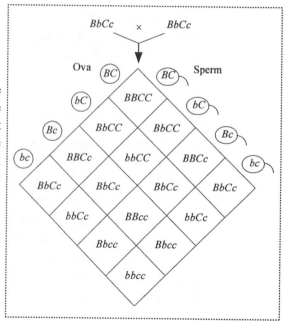

F. Polygenic Inheritance

Mendel's characters could be classified on an either-or basis, such as purple versus white flower. Many characters, however, are *quantitative characters* that vary in a continuum within a population.

Quantitative characters = Characters that vary by degree in a continuous distribution rather than by discrete (either-or) qualitative differences.

- Usually, continuous variation is determined not by one, but by many segregating loci or *polygenic inheritance*.

<u>Polygenic inheritance</u> = Mode of inheritance in which the additive effect of two or more genes determines a single phenotypic character.

For example, skin pigmentation in humans appears to be controlled by at least three separately inherited genes. The following is a simplified model for the polygenic inheritance of skin color:

- Three genes with the dark-skin allele (*A, B, C*) contribute one "unit" of darkness to the phenotype. These alleles are incompletely dominant over the other alleles (*a, b, c*).
- An *AABBCC* person would be very dark and an aabbcc person would be very light.
- An *AaBbCc* person would have skin of an intermediate shade.
- Because the alleles have a cumulative effect, genotypes *AaBbCc* and *AABbcc* make the same genetic contribution (three "units") to skin darkness. (See Campbell, Figure 13.13)
- Environmental factors, such as sun exposure, could also affect the phenotype.

G. Nature Versus Nurture: The Environmental Impact on Phenotype

Environmental conditions can influence the phenotypic expression of a gene, so that a single genotype may produce a range of phenotypes. This environmentally-induced phenotypic range is the *norm of reaction* for the genotype.

<u>Norm of reaction</u> = Range of phenotypic variability produced by a single genotype under various environmental conditions. Norms of reaction for a genotype:

- May be quite limited, so that a genotype only produces a specific phenotype, such as the blood group locus that determines ABO blood type.
- May also include a wide range of possibilities. For example, an individual's blood cell count varies with environmental factors such as altitude, activity level or infection.
- Are generally broadest for polygenic characters, including behavioral traits.

The expression of most polygenic traits, such as skin color, is *multifactorial*; that is, it depends upon many factors - a variety of possible genotypes, as well as a variety of environmental influences.

H. Integrating a Mendelian View of Heredity and Variation

These patterns of inheritance that are departures from Mendel's original description, can be integrated into a comprehensive theory of Mendelian genetics.

- Taking a holistic view, an organism's entire phenotype reflects its overall genotype and unique environmental history.
- Mendelism has broad applications beyond its original scope; extending the principles of segregation and independent assortment helps explain more complex hereditary patterns such as epistasis and quantitative characters.

VII. **Pedigree analysis reveals mendelian patterns in human inheritance**

Mendelian inheritance in humans is difficult to study because:

- The human generation time is about 20 years.
- Humans produce relatively few offspring compared to most other species.
- Well-planned breeding experiments are impossible.

Our understanding of Mendelian inheritance in humans is based on the analysis of family *pedigrees* or the results of matings that have already occurred.

<u>Pedigree</u> = A family tree that diagrams the relationships among parents and children across generations and that shows the inheritance pattern of a particular phenotypic character. By convention:

- Squares symbolize males and circles represent females.
- A horizontal line connecting a male and female indicates a mating; offspring are listed below in birth order, from left to right.
- Shaded symbols indicate individuals showing the trait being traced.

Following a dominant trait. For example, family members' genotypes can be deduced from a pedigree that traces the occurrence of widow's peak, the expression of a dominant allele.

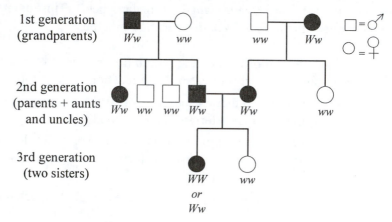

- If a widow's peak results from a dominant allele, *W*, then all individuals that do not have a widow's peak hairline must be homozygous recessive (*ww*). The genotypes of all recessives can be written on the pedigree.
- If widow's peak results from a dominant allele, W, then individuals that have a widow's peak hairline must be either homozygous dominant (*WW*) or heterozygous (*Ww*).
- If only some of the second generation offspring have a widow's peak, then the grandparents that show the trait must be heterozygous (*Ww*). (Note: if the grandparents with widow's peak were homozygous dominant, then all their respective offspring would show the trait.)
- Second generation offspring with widow's peaks must be heterozygous, because they are the result of *Ww* × *ww* matings.
- The third generation sister with widow's peak may be either homozygous dominant (*WW*) or heterozygous (*Ww*), because her parents are both heterozygous.

Following a recessive trait. For example, the same family can be used to trace a recessive trait such as attached ear lobes.

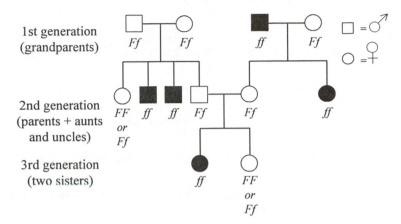

- If attached earlobes is due to a recessive allele (*f*), then all individuals with attached earlobes must be homozygous recessive (*ff*).
- Since attached earlobes appears in second generation offspring, the grandparents with free earlobes are heterozygous (*Ff*) since they must be capable of passing on a recessive allele (*f*).
- Since one of the third generation sisters has attached earlobes (ff), her parents are heterozygous; they have free earlobes (dominant trait) and yet must be able to contribute a recessive allele to their daughter. The other sister shows the dominant trait, so her genotype is unknown; it is possible that she may be either homozygous dominant or heterozygous.

Pedigree analysis can also be used to:

- Deduce whether a trait is determined by a recessive or dominant allele. Using the example above:

  ⇒ The first-born third generation daughter has attached earlobes. Since both parents *lack* the trait, it must not be determined by a dominant allele.

- Predict the occurrence of a trait in future generations. For example, if the second generation couple decide to have another child,

  ⇒ What is the probability the child will have a widow's peak? From a mating of *Ww* × *Ww*:

  probability of a child being *WW* = 1/4
  probability of a child being *Ww* = 2/4

  probability of widow's peak = 3/4

  ⇒ What is the probability the child will have attached earlobes? From a mating of *Ff* × *Ff*:

  probability of a child being *ff* = 1/4

  ⇒ What is the probability the child will have a widow's peak and attached earlobes? From a cross of *WwFf* × *WwFf*, use the rule of multiplication:

  3/4 (probability of widow's peak) × 1/4 (probability of attached earlobes) = 3/16

This type of analysis is important to geneticists and physicians, especially when the trait being analyzed can lead to a disabling or lethal disorder.

VIII. Many human disorders follow mendelian patterns of inheritance

A. Recessively Inherited Disorders

Recessive alleles that cause human disorders are usually defective versions of normal alleles.

- Defective alleles code for either a malfunctional protein or no protein at all.
- Heterozygotes can be phenotypically normal, if one copy of the normal allele is all that is needed to produce sufficient quantities of the specific protein.

Recessively inherited disorders range in severity from nonlethal traits (e.g. albinism) to lethal diseases (e.g. cystic fibrosis). Since these disorders are caused by recessive alleles:

- The phenotypes are expressed only in homozygotes (aa) who inherit one recessive allele from each parent.
- Heterozygotes (Aa) can be phenotypically normal and act as *carriers*, possibly transmitting the recessive allele to their offspring.

Most people with recessive disorders are born to normal parents, both of whom are carriers.

- The probability is 1/4 that a mating of two carriers (Aa × Aa) will produce a homozygous recessive zygote.
- The probability is 2/3 that a normal child from such a mating will be a heterozygote, or a carrier.

Human genetic disorders are not usually evenly distributed among all racial and cultural groups due to the different genetic histories of the world's people. Three examples of such recessively inherited disorders are *cystic fibrosis*, *Tay-Sachs disease* and *sickle-cell disease*.

Cystic fibrosis, the most common lethal genetic disease in the United States, strikes 1 in every 2,500 Caucasians (it is much rarer in other races).

- Four percent of the Caucasian population are carriers.
- The dominant allele codes for a membrane protein that controls chloride traffic across the cell membrane. Chloride channels are defective or absent in individuals that are homozygous recessive for the cystic fibrosis allele.
- Disease symptoms result from the accumulation of thickened mucus in the pancreas and lungs.

Tay-Sachs disease occurs in 1 out of 3,600 births. The incidence is about 100 times higher among Ashkenazic (central European) Jews than among Sephardic (Mediterranean) Jews and non-Jews.

- Brain cells of babies with this disease are unable to metabolize gangliosides (a type of lipid), because a crucial enzyme does not function properly.
- As lipids accumulate in the brain, the infant begins to suffer seizures, blindness and degeneration of motor and mental performance. The child usually dies after a few years.

Sickle-cell disease is the most common inherited disease among African-Americans. It affects 1 in 400 African-Americans born in the United States.

- The disease is caused by a single amino acid substitution in hemoglobin.
- The abnormal hemoglobin molecules tend to link together and crystallize, especially when blood oxygen content is lower than normal. This causes red blood cells to deform from the normal disk-shape to a sickle-shape.
- The sickled cells clog tiny blood vessels, causing the pain and fever characteristic of a sickle-cell crisis.

About 1 in 10 African-Americans are heterozygous for the sickle-cell allele and are said to have the *sickle-cell trait*.

- These carriers are usually healthy, although some suffer symptoms after an extended period of low blood oxygen levels.
- Carriers can function normally because the two alleles are codominant (heterozygotes produce not only the abnormal hemoglobin but also normal hemoglobin).
- The high incidence of heterozygotes is related to the fact that in tropical Africa where malaria is endemic, heterozygotes have enhanced resistance to malaria compared to normal homozygotes. Thus, heterozygotes have an advantage over both homozygotes – those who have sickle cell disease and those who have normal hemoglobin.

The probability of inheriting the same rare harmful allele from both parents, is greater if the parents are closely related.

Consanguinity = A genetic relationship that results from shared ancestry.

- The probability is higher that consanguinous matings will result in homozygotes for harmful recessives, since parents with recently shared ancestry are more likely to inherit the same recessive alleles than unrelated persons.
- It is difficult to accurately assess the extent to which human consanguinity increases the incidence of inherited diseases, because embryos homozygous for deleterious mutations are affected so severely that most are spontaneously aborted before birth.
- Most cultures forbid marriage between closely related adults. This may be the result of observations that stillbirths and birth defects are more common when parents are closely related.

B. Dominantly Inherited Disorders

Some human disorders are dominantly inherited.

- For example, *achondroplasia* (a type of dwarfism) affects 1 in 10,000 people who are heterozygous for this gene.
- Homozygous dominant condition results in spontaneous abortion of the fetus, and homozygous recessives are of normal phenotype (99.9% of the population).

Lethal dominant alleles are much rarer than lethal recessives, because they:

- Are always expressed, so their effects are not masked in heterozygotes.
- Usually result from new genetic mutations that occur in gametes and later kill the developing embryo.

Late-acting lethal dominants can escape elimination if the disorder does not appear until an advanced age after afflicted individuals may have transmitted the lethal gene to their children. For example,

- *Huntington's disease*, a degenerative disease of the nervous system, is caused by a late-acting lethal dominant allele. The phenotypic effects do not appear until 35 to 40 years of age. It is irreversible and lethal once the deterioration of the nervous system begins.
- Molecular geneticists have recently located the gene for Huntington's near the tip of chromosome #4.
- Children of an afflicted parent have a 50% chance of inheriting the lethal dominant allele. A newly developed test can detect the Huntington's allele before disease symptoms appear.

C. Multifactorial Disorders

Not all hereditary diseases are simple Mendelian disorders; that is, diseases caused by the inheritance of certain alleles at a single locus. More commonly, people are afflicted by *multifactorial* disorders, diseases that have both genetic and environmental influences.

- Examples include heart disease, diabetes, cancer, alcoholism and some forms of mental illness.
- The hereditary component is often polygenic and poorly understood.
- The best public-health strategy is to educate people about the role of environmental and behavioral factors that influence the development of these diseases.

IX. Technology is providing new tools for genetic testing and counseling

Genetic counselors in many hospitals can provide information to prospective parents concerned about a family history for a genetic disorder.

- This preventative approach involves assessing the risk that a particular genetic disorder will occur.
- Risk assessment includes studying the family history for the disease using Mendel's law of segregation to deduce the risk.

For example, a couple is planning to have a child, and both the man and woman had siblings who died from the same recessively inherited disorder. A genetic counselor could deduce the risk of their first child inheriting the disease by using the laws of probability:

Question: What is the probability that the husband and wife are each carriers?

Answer: The genotypic ratio from an Aa × Aa cross is 1 AA:2 Aa:1 aa. Since the parents are normal, they have a <u>2/3</u> of being carriers.

Question: What is the chance of two carriers having a child with the disease?

Answer: 1/2 (mother's chance of passing on the gene) × 1/2 (father's chance of passing on the gene) = <u>1/4</u>

Question: What is the probability that their firstborn will have the disorder?

Answer: (Chance that the father is a carrier) × (chance that mother is a carrier) × (chance of two carriers having a child with the disease).

$$2/3 \times 2/3 \times 1/4 = \underline{1/9}$$

If the first child is born with the disease, what is the probability that the second child will inherit the disease?

- If the first child is born with the disease, then it is certain that both the man and the woman are carriers. Thus, the probability that other children produced by this couple will have the disease is <u>1/4</u>.
- The conception of each child is an independent event, because the genotype of one child does not influence the genotype of the other children. So there is a 1/4 chance that any additional child will inherit the disease.

A. Carrier Recognition

Several tests are available to determine if prospective parents are carriers of genetic disorders.

- Tests are currently available that can determine heterozygous carriers for the Tay-Sachs allele, cystic fibrosis, and sickle-cell disease.
- Tests such as these enable people to make informed decisions about having children, but they could also be abused. Ethical dilemmas about how this information should be used points to the immense social implications of such technological advances.

B. Fetal Testing

A couple that learns they are both carriers for a genetic disease and decide to have a child can determine if the fetus has the disease. Between the fourteenth and sixteenth weeks of pregnancy, *amniocentesis*. can be done to remove amniotic fluid for testing.

- During amniocentesis, a physician inserts a needle into the uterus and extracts about 10 milliliters of amniotic fluid.
- The presence of certain chemicals in amniotic fluid indicate some genetic disorders.
- Some tests (including one for Tay-Sachs) are performed on cells grown in culture from fetal cells sloughed off in the amniotic fluid. These cells can also be karyotyped to identify chromosomal defects.

Chorionic villus sampling (CVS) is a newer technique during which a physician suctions off a small amount of fetal tissue from the chorionic villi of the placenta.

- These rapidly dividing embryonic cells can be karyotyped immediately, usually providing results in 24 hours – a major advantage over amniocentesis which may take several weeks. (Amniocentesis requires that the cells must first be cultured before karyotyping can be done.)
- Another advantage of CVS is that it can be performed at only 8 to 10 weeks of pregnancy.

Other techniques such as *ultrasound* and *fetoscopy* allow physicians to examine a fetus for major abnormalities.

- Ultrasound is a non-invasive procedure which uses sound waves to create an image of the fetus.
- Fetoscopy involves inserting a thin fiber-optic scope into the uterus.

Amniocentesis and fetoscopy have a 1% risk of complication such as maternal bleeding or fetal death. Thus, they are used only when risk of genetic disorder or birth defect is relatively high.

C.   Newborn Screening

In most U.S. hospitals, simple tests are routinely performed at birth, to detect genetic disorders such as *phenylketonuria* (PKU).

- PKU is recessively inherited and occurs in about 1 in 15,000 births in the United States.
- Children with this disease cannot properly break down the amino acid phenylalanine.
- Phenylalanine and its by-product (phenylpyruvic acid) can accumulate in the blood to toxic levels, causing mental retardation.
- Fetal screening for PKU can detect the deficiency in a newborn and retardation can be prevented with a special diet (low in phenylalanine) that allows normal development.

REFERENCES

Campbell, N. *Biology*. 4th ed. Menlo Park, California: Benjamin/Cummings, 1996.

Griffith, A.J.E., J.H. Miller, D.T. Suzuki, R.C. Lewontin, and W.M. Gelbart. *An Introduction to Genetic Analysis*. 5th ed. New York: W.H. Freeman, 1993.

Kowles, R.V. *Genetics, Society and Decisions*. 1st ed. Columbus, Ohio: Charles E. Merrill, 1985. Though the target audience for this book is non-majors, it can be a useful lecture supplement for contrasting controversial social issues.

Russell, P.J. *Genetics*. 2nd ed. Glenview, Illinois: Scott, Foresman, 1990.

CHAPTER **14**

THE CHROMOSOMAL
BASIS OF INHERITANCE

OUTLINE

OBJECTIVES

After reading this chapter and attending lecture, the student should be able to:

1. Explain how the observations of cytologists and geneticists provided the basis for the chromosome theory of inheritance.
2. Describe the contributions that Thomas Hunt Morgan, Walter Sutton and A.H. Sturtevant made to current understanding of chromosomal inheritance.
3. Explain why *Drosophila melanogaster* is a good experimental organism.
4. Define linkage and explain why linkage interferes with independent assortment.
5. Distinguish between parental and recombinant phenotypes.
6. Explain how crossing over can unlink genes.
7. Map a linear sequence of genes on a chromosome using given recombination frequencies from experimental crosses.
8. Explain what additional information cytological maps provide over crossover maps.
9. Distinguish between a heterogametic sex and a homogametic sex.
10. Describe sex determination in humans.
11. Describe the inheritance of a sex-linked gene such as color-blindness.
12. Explain why a recessive sex-linked gene is always expressed in human males.
13. Explain how an organism compensates for the fact that some individuals have a double dosage of sex-linked genes while others have only one.
14. Distinguish among nondisjunction, aneuploidy and polyploidy; explain how these major chromosomal changes occur and describe the consequences.
15. Distinguish between trisomy and triploidy.
16. Distinguish among deletions, duplications, translocations and inversions.
17. Describe the effects of alterations in chromosome structure, and explain the role of position effects in altering the phenotype.
18. Describe the type of chromosomal alterations implicated in the following human disorders: Down syndrome, Klinefelter syndrome, extra Y, triple-X syndrome, Turner syndrome, *cri du chat* syndrome and chronic myelogenous leukemia.
19. Define genomic imprinting and provide evidence to support this model.
20. Explain how the complex expression of a human genetic disorder, such as fragile-X syndrome, can be influenced by triplet repeats and genomic imprinting.
21. Give some exceptions to the chromosome theory of inheritance, and explain why cytoplasmic genes are not inherited in a Mendelian fashion.

KEY TERMS

Thomas Hunt Morgan
Drosophila
    melanogaster
wild type
mutant phenotype
linkage
sex linkage
Walter Sutton
genetic
    recombination
recombinant
parental type

locus (pl. loci)
chromosome map
map unit
cytological map
point mutation
chromosomal
    mutation
chromosomal
heterogametic sex
homogametic sex
autosome
sex chromosome

hemizygous
gene dosage
    compensation
Barr body
Mary F. Lyon
nondisjunction
aneuploidy
polyploidy
trisomy
triploidy
tetraploidy

deletion
duplication
translocation
inversion
position effect
amniocentesis
Down Syndrome
genomic imprinting
triplet repeats
Fragile-X syndrome
Huntington's disease

LECTURE NOTES

I. **Mendelian inheritance has its physical basis in the behavior of chromosomes during sexual life cycles**

Genetics	Cytology
1860's: Mendel proposed that discrete inherited factors segregate and assort independently during gamete formation	
	1875: Cytologists worked out process of mitosis
	1890: Cytologists worked out process of meiosis
1900: Three botanists (Correns, de Vries and von Seysenegg) independently rediscovered Mendel's principles of segregation and independent assortment	
1902: Cytology and genetics converged as Walter Sutton, Theodor Boveri and others noticed parallels between the behavior of Mendel's factors and the behavior of chromosomes. For example: • Chromosomes and genes are both paired in diploid cells. • Homologous chromosomes separate and allele pairs segregate during meiosis. • Fertilization restores the paired condition for both chromosomes and genes.	

Based upon these observations, biologists developed the *chromosome theory of inheritance*. According to this theory:

- Mendelian factors or genes are located on chromosomes.
- It is the chromosomes that segregate and independently assort.

II. Morgan traced a gene to a specific chromosome: *science as a process*

Thomas Hunt Morgan from Columbia University performed experiments in the early 1900's which provided convincing evidence that Mendel's inheritable factors are located on chromosomes.

Morgan selected the fruit fly, *Drosophila melanogaster*, as the experimental organism because these flies:

- Are easily cultured in the laboratory.
- Are prolific breeders.
- Have a short generation time.
- Have only four pairs of chromosomes which are easily seen with a microscope.

There are three pairs of autosomes (II, III and IV) and one pair of sex chromosomes. Females have two *X* chromosomes, and males have one *X* and one *Y* chromosome.

A. A Note on Genetic Symbols

Morgan and his colleagues used genetic symbols that are now convention. For a particular character:

- A gene's symbol is based on the first *mutant*, non-wild type discovered.
  - ⇒ If the mutant is recessive, the first letter is lowercase. (e.g. *w* = white eye allele in *Drosophila*.)
  - ⇒ If the mutant is dominant, the first letter is capitalized. (e.g. *Cy* = "curly" allele in *Drosophila* that causes abnormal, curled wings.)
- Wild-type trait is designated by a superscript +. (e.g. Cy^+ = allele for normal, straight wings.)

<u>Wild type</u> = Normal or most frequently observed phenotype.

<u>Mutant phenotypes</u> = Phenotypes that are alternatives to the wild type and which are due to mutations in the wild-type gene.

B.   Discovery of a Sex-Linked Gene

After a year of breeding *Drosophila* to find variant phenotypes, Morgan discovered a single male fly with white eyes instead of the wild-type red. Morgan mated this mutant white-eyed male with a red-eyed female. The cross is outlined below.

w = white-eye allele

w^+ = red-eye or wild-type allele

Drosophila geneticists symbolize a recessive mutant allele with one or more lower case letters. The corresponding wild-type allele has a superscript plus sign.

P generation:

w^+w^+
red-eyed ♀ × w white-eyed ♂

F$_1$ generation:

w^+w
red-eyed ♀ × w^+ red-eyed ♂

The fact that all the F$_1$ progeny had red eyes, suggested that the wild-type allele was dominant over the mutant allele.

F$_2$ generation:

w^+w^+ red-eyed ♀ ww^+ red-eyed ♀

w^+ red-eyed ♂ w white-eyed ♂

White-eyed trait was expressed only in the male, and all the F$_2$ females had red eyes.

Morgan deduced that eye color is linked to sex and that the gene for eye color is located only on the X chromosome. Premises for his conclusions were:

- If eye color is located only on the X chromosome, then females (XX) carry two copies of the gene, while males (XY) have only one.
- Since the mutant allele is recessive, a white-eyed female must have that allele on both X chromosomes which was impossible for F$_2$ females in Morgan's experiment.
- A white-eyed male has no wild-type allele to mask the recessive mutant allele, so a single copy of the mutant allele confers white eyes.

Sex-linked genes = Genes located on sex chromosomes. The term is commonly applied only to genes on the X chromosome.

III. Linked genes tend to be inherited together because they are located on the same chromosome

Genes located on the same chromosome tend to be linked in inheritance and do not assort independently.

<u>Linked genes</u> = Genes that are located on the same chromosome and that tend to be inherited together.

- Linked genes do not assort independently, because they are on the same chromosome and move together through meiosis and fertilization.
- Since independent assortment does not occur, a dihybrid cross following two linked genes will not produce an F_2 phenotypic ratio of 9:3:3:1.

T.H. Morgan and his students performed a dihybrid testcross between flies with autosomal recessive mutant alleles for black bodies and vestigial wings and wild-type flies heterozygous for both traits. (A more detailed description follows in a later section.)

$$b = \text{black body} \qquad vg = \text{vestigial wings}$$
$$b^+ = \text{gray body} \qquad vg^+ = \text{wild-type wings}$$

$$\begin{array}{ccc} b^+bvg^+vg & \times & bbvgvg \\ \text{gray, normal wings} & & \text{black, vestigial wings} \end{array}$$

- Resulting phenotypes of the progeny did not occur in the expected 1:1:1:1 ratio for a dihybrid testcross.
- A disproportionately large number of flies had the phenotypes of the parents: gray with normal wings and black with vestigial wings.
- Morgan proposed that these unusual ratios were due to linkage. The genes for body color and wing size are on the same chromosome and are usually thus inherited together.

IV. Independent assortment of chromosomes and crossing over cause genetic recombination

<u>Genetic recombination</u> = The production of offspring with new combinations of traits different from those combinations found in the parents; results from the events of meiosis and random fertilization.

A. The Recombination of Unlinked Genes: Independent Assortment of Chromosomes

Mendel discovered that some offspring from dihybrid crosses have phenotypes unlike either parent. An example is the following test cross between pea plants:

YY, Yy	=	yellow seeds	RR, Rr	=	round seeds
yy	=	green seeds	rr	=	wrinkled seeds

P generation:

YyRr	×	yyrr
yellow round		green wrinkled

Testcross progeny:

¼ YyRr	¼ yyrr	Parental types
yellow, round	green, wrinkled	(50%)
¼ yyRr	¼ Yyrr	Recombinant types
green, round	yellow, wrinkled	(50%)

<u>Parental types</u> = Progeny that have the same phenotype as one or the other of the parents.

<u>Recombinants</u> = Progeny whose phenotypes differ from either parent.

In this cross, seed shape and seed color are unlinked.

- One-fourth of the progeny have round yellow seeds, and one fourth have wrinkled green seeds. Therefore, one-half of the progeny are *parental types*.
- The remaining half of the progeny are *recombinants*. One fourth are round green and one fourth are wrinkled yellow – phenotypes not found in either parent.
- When half the progeny are recombinants, there is a 50% frequency of recombination.
- A 50% frequency of recombination usually indicates that the two genes are on different chromosomes, because it is the expected result if the two genes assort randomly.
- The genes for seed shape and seed color assort independently of one another because they are located on different chromosomes which randomly align during metaphase of meiosis I.

B. The Recombination of Linked Genes: Crossing Over

If genes are totally linked, some possible phenotypic combinations should not appear. Sometimes, however, the unexpected recombinant phenotypes do appear.

As described earlier, T.H. Morgan and his students performed the following dihybrid testcross between flies with autosomal recessive mutant alleles for black bodies and vestigial wings and wild-type flies heterozygous for both traits.

$$b = \text{black body} \qquad vg = \text{vestigial wings}$$
$$b^+ = \text{gray body} \qquad vg^+ = \text{wild-type wings}$$

$$b^+bvg^+vg \qquad \times \qquad bbvgvg$$
$$\text{gray, normal wings} \qquad \text{black, vestigial wings}$$

Phenotypes	Genotypes	Expected Results If Genes Are Unlinked	Expected Results If Genes Are Totally Linked	Actual Results
Black body, normal wings	$\dfrac{b\ vg^+}{b\ vg}$	575	—	206
Gray body, normal wings	$\dfrac{b^+vg^+}{b\ vg}$	575	1150	965
Black body, vestigial wings	$\dfrac{b\ vg}{b\ vg}$	575	1150	944
Gray body, vestigial wings	$\dfrac{b^+vg}{b\ vg}$	575	—	185

$$\text{Recombination Frequency} = \frac{391 \text{ recombinants}}{2300 \text{ total offspring}} \times 100 = 17\%$$

Morgan's results from this dihybrid testcross showed that the two genes were neither unlinked nor totally linked.

- If wing type and body color genes were unlinked, they would assort independently, and the progeny would show a 1:1:1:1 ratio of all possible phenotypic combinations.

- If the genes were completely linked, expected results from the testcross would be a 1:1 phenotypic ratio of *parental types only.*

- Morgan's testcross did not produce results consistent with unlinkage or total linkage. The high proportion of parental phenotypes suggested linkage between the two genes.

- Since 17% of the progeny were recombinants, the linkage must be incomplete. Morgan proposed that there must be some mechanism that occasionally breaks the linkage between the two genes.

- It is now known that *crossing over* during meiosis accounts for the recombination of linked genes. The exchange of parts between homologous chromosomes breaks linkages in parental chromosomes and forms recombinants with new allelic combinations.

V. Geneticists can use recombination data to map a chromosome's genetic loci

Scientists used recombination frequencies between genes to *map* the sequence of linked genes on particular chromosomes.

Morgan's *Drosophila* studies showed that some genes are linked more tightly than others.
- For example, the recombination frequency between the *b* and *vg* loci is about 17%.
- The recombination frequency is only 9% between *b* and *cn*, a third locus on the same chromosome. (The cinnabar gene, *cn*, for eye color has a recessive allele causing "cinnabar eyes.")

A.H. Sturtevant, one of Morgan's students, assumed that if crossing over occurs randomly, the probability of crossing over between two genes is directly proportional to the distance between them.
- Sturtevant used recombination frequencies between genes to assign them a linear position on a chromosome *map*.
- He defined one *map unit* as 1% recombination frequency. (Map units are now called *centimorgans*, in honor of Morgan.)

Using crossover data, a map may be constructed as follows:

Loci	Recombination Frequency	Approximate Map Units
b vg	17.0%	18.5*
cn b	9.0%	9.0
cn vg	9.5%	9.5

1. Establish the relative distance between those genes farthest apart or with the highest recombination frequency.

    b ——————————— vg
    ←——————— 17 ———————→

2. Determine the recombination frequency between the third gene (*cn*) and the first (*b*).

    ←—— 9 ——→
    cn ——————— b

3. Consider the two possible placements of the third gene:

    ←—— 9 ——→
    cn ——————— b ——————————— vg
    ←——————— 17 ———————→

    ←—— 9 ——→
    b ——————— cn ——————— vg
    ←——————— 17 ———————→

4. Determine the recombination frequency between the third gene (*cn*) and the second (*vg*) to eliminate the incorrect sequence.

    ←—— 9 ——→←—— 9.5 ——→
    b ——————— cn ——————— vg
    ←——————— 17 ———————→

So, the correct sequence is *b–cn–vg*.

* Note that there are actually 18.5 map units between *b* and *vg*. This is higher than that predicted from the recombination frequency of 17.0%. Because *b* and *vg* are relatively far apart, double crossovers occur between these loci and cancel each other out, leading us to underestimate the actual map distance.

If linked genes are so far apart on a chromosome that the recombination frequency is 50%, they are indistinguishable from unlinked genes that assort independently.

- Linked genes that are far apart can be mapped, if additional recombination frequencies can be determined between intermediate genes and each of the distant genes.

Sturtevant and his coworkers extended this method to map other *Drosophila* genes in linear arrays. (See Campbell, Figure 14.7)

- The crossover data allowed them to cluster the known mutations into four major linkage groups.
- Since *Drosophila* has four sets of chromosomes, this clustering of genes into four linkage groups was further evidence that genes are on chromosomes.

Maps based on crossover data only give information about the relative position of linked genes on a chromosome. Another technique, *cytological mapping* pinpoints the actual location of genes and the real distance between them.

- *Cytological mapping* involves screening offspring for mutant phenotypes and associating mutants with chromosomal defects visible by direct microscopic examination.
- The location of loci derived from maps based on crossover data differs from the spacing derived from cytological mapping, because the frequency of crossing over is not the same for all chromosomal regions.

VI. The chromosomal basis of sex produces unique patterns of inheritance

In most species, sex is determined by the presence or absence of special chromosomes. As a result of meiotic segregation, each gamete has one sex chromosome to contribute at fertilization.

<u>Heterogametic sex</u> = The sex that produces two kinds of gametes and determines the sex of the offspring.

<u>Homogametic sex</u> = The sex that produces one kind of gamete.

A.   The Chromosomal Basis of Sex in Humans

Mammals, including humans, have an *X-Y* mechanism that determines sex at fertilization.

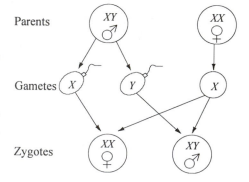

- There are two chromosomes, *X* and *Y*. Each gamete has one sex chromosome, so when sperm cell and ovum unite at fertilization, the zygote receives one of two possible combinations: *XX* or *XY*.
- Males are the heterogametic sex (*XY*). Half the sperm cells contain an *X* chromosome, while the other half contain a *Y* chromosome.
- Females are the homogametic sex (*XX*); all ova carry an *X* chromosome.

Whether an embryo develops into a male or female depends upon the presence of a *Y* chromosome.

- A British research team has identified a single gene, *Sry* (sex-determining region), on the *Y* chromosome that is responsible for triggering the complex series of events that lead to normal testicular development.
- *Sry* probably codes for a protein that regulates other genes.

B. Sex-Linked Disorders in Humans

Some genes on sex chromosomes play a role in sex determination, but these chromosomes also contain genes for other traits.

In humans, the term *sex-linked traits* usually refers to *X*-linked traits.

- The human *X*-chromosome is much larger than the *Y*. Thus, there are more *X*-linked than *Y*-linked traits.
- Most *X*-linked genes have no homologous loci on the *Y* chromosome.
- Most genes on the *Y* chromosome not only have no *X* counterparts, but they encode traits found only in males (e.g. testis-determining factor).
- Examples of sex-linked traits in humans are color blindness, *Duchenne muscular dystrophy* and hemophilia.

Fathers pass *X*-linked alleles to *only* and *all* of their daughters.

- Males receive their *X* chromosome only from their mothers.
- Fathers cannot, therefore, pass sex-linked traits to their sons.

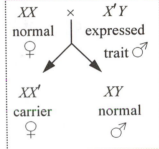

Mothers can pass sex-linked alleles to *both* sons and daughters.

- Females receive two *X* chromosomes, one from each parent.
- Mothers pass on one *X* chromosome (either maternal or paternal homologue) to every daughter and son.

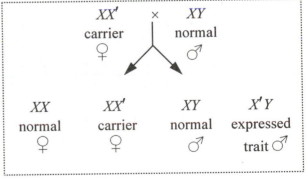

If a sex-linked trait is due to a recessive allele, a female will express the trait only if she is homozygous.

- Females have two *X* chromosomes, therefore they can be either homozygous or heterozygous for sex-linked alleles.
- There are fewer females with sex-linked disorders than males, because even if they have one recessive allele, the other dominant allele is the one that is expressed. A female that is heterozygous for the trait can be a *carrier*, but not show the recessive trait herself.

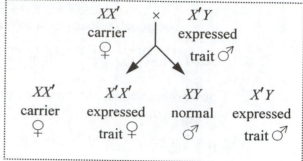

- A carrier that mates with a normal male will pass the mutation to half her sons and half her daughters.
- If a carrier mates with a male who has the trait, there is a 50% chance that each child born to them will have the trait, regardless of sex.

Because males have only one *X*-linked locus, any male receiving a mutant allele from his mother will express the trait.

- Far more males than females have sex-linked disorders.
- Males are said to be *hemizygous*.

Hemizygous = A condition where only one copy of a gene is present in a diploid organism.

C. *X*-Inactivation in Females

How does an organism compensate for the fact that some individuals have a double dosage of sex-linked genes while others have only one?

In female mammals, most diploid cells have only one fully functional *X* chromosome.

- The explanation for this process is known as the *Lyon hypothesis*, proposed by the British geneticist Mary F. Lyon.
- In females, each of the embryonic cells inactivates one of the two *X* chromosomes.
- The inactive *X* chromosome contracts into a dense object called a *Barr body*.

Barr body = Inside the nuclear envelope, a densely staining object that is an inactivated *X* chromosome in female mammalian cells.

- Most Barr body genes are not expressed.
- Are reactivated in gonadal cells that undergo meiosis to form gametes.

Female mammals are a *mosaic* of two types of cells – those with an active maternal *X* and those with an active paternal *X*.

- Which of the two *X*s will be inactivated is determined randomly in embryonic cells.
- After an *X* is inactivated, all mitotic descendants will have the same inactive *X*.
- As a consequence, if a female is heterozygous for a sex-linked trait, about half of her cells will express one allele and the other cells well express the alternate allele.
- Examples of this type of mosaicism are coloration in calico cats and normal sweat gland development in humans.

X chromosome inactivation is associated with DNA methylation.

- Methyl groups ($-CH_3$) attach to cytosine, one of DNA's nitrogenous bases.
- Barr bodies are highly methylated compared to actively transcribed DNA.

What determines which of the two *X* chromosomes will be methylated?

- A recently discovered gene, *XIST* is active *only* on the Barr body.
- The product of the *XIST* gene, *X-inactive specific transcript*, is an RNA that interacts with the *X* chromosome and maintains its inactivation.

Many questions are yet to be answered.

- How does *XIST* initiate *X*-inactivation?
- What determines which *X* chromosome in each of a female's cells will have an active *XIST* gene and become a Barr body?

VII. Alterations of chromosome number or structure cause some genetic disorders

Meiotic errors and mutagens can cause major chromosomal changes such as altered chromosome numbers or altered chromosomal structure.

A. Alterations of Chromosome Number: Aneuploidy and Polyploidy

Nondisjunction = Meiotic or mitotic error during which certain homologous chromosomes or sister chromatids fail to separate.

- Meiotic nondisjunction:
  ⇒ May occur during Meiosis I so that a homologous pair does not separate. (See Campbell, Figure 14.11a)
  ⇒ May occur during Meiosis II when sister chromatids do not separate. (See Campbell, Figure 14.11b)
  ⇒ Results in one gamete receiving two of the same type of chromosome and another gamete receiving no copy. The remaining chromosomes may be distributed normally.
- Mitotic nondisjunction:
  ⇒ Also results in abnormal number of certain chromosomes.
  ⇒ If it occurs in embryonic cells, mitotic division passes this abnormal chromosome number to a large number of cells, and can thus have a large effect.

Aneuploidy = Condition of having an abnormal number of certain chromosomes.

- Aneuploid offspring may result if a normal gamete unites with an aberrant one produced as a result of *nondisjunction*.
- An aneuploid cell that has a chromosome in triplicate is said to be *trisomic* for that chromosome.
- An aneuploid with a missing chromosome is said to be *monosomic* for that chromosome.
- When an aneuploid zygote divides by mitosis, it transmits the chromosomal anomaly to all subsequent embryonic cells.
- Abnormal gene dosage in aneuploids causes characteristic symptoms in survivors. An example is Down's syndrome which results from trisomy of chromosome 21.

Polyploidy = A chromosome number that is more than two complete chromosome sets.

- *Triploidy* is a polyploid chromosome number with three haploid chromosome sets (3N).
- *Tetraploidy* is polyploidy with four haploid chromosome sets (4N).
- Triploids may be produced by fertilization of an abnormal diploid egg produced by nondisjunction of all chromosomes.
- Tetraploidy may result if a diploid zygote undergoes mitosis without cytokinesis. Subsequent normal mitosis would produce a 4N embryo.
- Polyploidy is common in plants and important in plant evolution.
- Polyploids occur rarely among animals, and they are more normal in appearance than aneuploids. Mosaic polyploids, with only patches of polyploid cells, are more common than complete polyploid animals.

B.   Alterations of Chromosome Structure

Chromosome breakage can alter chromosome structure in four ways:

- Chromosomes which lose a fragment lacking a centromere will have a deficiency or *deletion*.
- Fragments without centromeres are usually lost when the cell divides, or they may:
  ⇒ Join to a homologous chromosome producing a *duplication*.
  ⇒ Join to a nonhomologous chromosome (*translocation*).
  ⇒ Reattach to the original chromosome in reverse order (*inversion*).

Crossing-over error is another source of deletions and duplications.

- Crossovers are normally reciprocal, but sometimes one sister chromatid gives up more genes than it receives in an unequal crossover.
- A nonreciprocal crossover results in one chromosome with a deletion and one chromosome with a duplication.

Alterations of chromosome structure, can have various effects:

- Homozygous deletions, including a single X in a male, are usually lethal.
- Duplications and translocations tend to have deleterious effects.
- Even if all genes are present in normal dosages, reciprocal translocations between nonhomologous chromosomes and inversions can alter the phenotype because of subtle *position effects*.

Position effect = Influence on a gene's expression because of its location among neighboring genes.

C. Human Disorders Due to Chromosomal Alterations

Chromosomal alterations are associated with some serious human disorders.

Aneuploidy, resulting from meiotic nondisjunction during gamete formation, usually prevents normal embryonic development and often results in spontaneous abortion.

- Some types of aneuploidy cause less severe problems, and aneuploid individuals may survive to birth and beyond with a set of characteristic symptoms or *syndrome*.
- Aneuploid conditions can be diagnosed before birth by *fetal testing*.

Down syndrome, an aneuploid condition, affects 1 out of 700 U.S. children.

- Is usually the result of trisomy 21.
- Includes characteristic facial features, short stature, heart defects, mental retardation, susceptibility to respiratory infections, and a proneness to developing leukemia and Alzheimer's disease.
- Though most are sexually underdeveloped and sterile, a few women with Down syndrome have had children.
- The incidence of Down syndrome offspring correlates with maternal age.
  ⇒ May be related to the long time lag between the first meiotic division during the mother's fetal life and the completion of meiosis at ovulation.
  ⇒ May be that older women have less chance of miscarrying a trisomic embryo.

Other rarer disorders caused by autosomal aneuploidy are:

- *Patau syndrome* (trisomy 13)
- *Edwards syndrome* (trisomy 18).

Sex chromosome aneuploidies result in less severe conditions than those from autosomal aneuploidies. This may be because:

- The *Y* chromosome carries few genes.
- Copies of the *X* chromosome become inactivated as Barr bodies.

The basis of sex determination in humans is illustrated by sex chromosome aneuploidies.

- A single *Y* chromosome is sufficient to produce maleness.
- The absence of *Y* is required for femaleness.

Examples of sex chromosome aneuploidy in males are:

Klinefelter Syndrome

Genotype: Usually *XXY*, but may be associated with *XXYY*, *XXXY*, *XXXXY*, *XXXXXY*.

Phenotype: Male sex organs with abnormally small testes; sterile; feminine body contours and perhaps breast enlargement; usually of normal intelligence.

Extra Y

Genotype: *XYY*.

Phenotype: Normal male; usually taller than average; normal intelligence and fertility.

Abnormalities of sex chromosome number in females include:

Triple-X Syndrome

Genotype: *XXX*.

Phenotype: Usually fertile; can show a normal phenotype.

Turner Syndrome

Genotype: *XO* (only known viable human monosomy).

Phenotype: Short stature; at puberty, secondary sexual characteristics fail to develop; internal sex organs do not mature; sterile.

Structural chromosomal alterations such as deletions and translocations can also cause human disorders.

- Deletions in human chromosomes cause severe defects even in the heterozygous state. For example,
  ⇒ *Cri du chat* syndrome is caused by a deletion on chromosome 5. Symptoms are mental retardation, a small head with unusual facial features and a cry that sounds like a mewing cat.
- Translocations associated with human disorders include:
  ⇒ Certain cancers such as *chronic myelogenous leukemia* (CML). A portion of chromosome 22 switches places with a small fragment from chromosome 9.
- Some cases of Down syndrome. A third chromosome 21 translocates to chromosome 15, resulting in two normal chromosomes 21 plus the translocation.

VIII. The phenotypic effects of some genes depend on whether they were inherited from the mother or father

A. **Genomic Imprinting**

The expression of some traits may depend upon which parent contributes the alleles for those traits.

- For example, two genetic disorders, *Prader-Willi syndrome* and *Angelman syndrome*, are caused by the same deletion on chromosome 15. The symptoms differ depending upon whether the gene was inherited from the mother or from the father.
- Prader-Willi syndrome is caused by a deletion from the *paternal* version of chromosome 15. The syndrome is characterized by mental retardation, obesity, short stature, and unusually small hands and feet.
- Angelman syndrome is caused by a deletion from the *maternal* version of chromosome 15. This syndrome is characterized by uncontrollable spontaneous laughter, jerky movements, and other motor and mental symptoms.
- The Prader-Willi/Angelman syndromes imply that the deleted genes normally behave differently in offspring, depending on whether they belong to the maternal or the paternal homologue.
- In other words, homologous chromosomes inherited from males and females are somehow differently *imprinted*, which causes them to be functionally different in the offspring.

Genomic imprinting = Process that induces intrinsic changes in chromosomes inherited from males and females; causes certain genes to be differently expressed in the offspring depending upon whether the alleles were inherited from the ovum or from the sperm cell.

- According to this hypothesis, certain genes are imprinted in some way each generation, and the imprint is different depending on whether the genes reside in females or in males.
- The same alleles may have different effects on offspring depending on whether they are inherited from the mother or the father.
- In the new generation, both maternal and paternal imprints can be reversed in gamete-producing cells, and all the chromosomes are re-coded according to the sex of the individual in which they now reside.
- *DNA methylation* may be one mechanism for genomic imprinting. (See chapter 18 for further discussion.)

B. **Fragile-*X* and Triplet Repeats**

Triplet repeats = Sections of DNA where a specific triplet of nucleotides is repeated many times.

- Occur normally in many places within the human genome.
- Progressive addition of triplet repeats can lead to genetic disorders such as *Fragile-X syndrome* and *Huntington's disease*.

Affecting about one in every 1500 males and one in every 2500 females, *Fragile-X syndrome* is the most common genetic cause of mental retardation.

- The "fragile *X*" is an abnormal *X* chromosome, the tip of which hangs on the rest of the chromosome by a thin DNA thread. This altered region in fragile-*X*, as well as the comparable region in normal *X* chromosomes, contains triplet repeats.
- The triplet repeat, CGG (cytosine, guanine, guanine), is repeated up to 50 times on one tip of a normal *X* chromosome, but is repeated more than 200 times in a fragile *X* chromosome.
- Abnormal addition of triplet repeats occurs incrementally over generations, so there is a "prefragile-*X*" condition between 50 to 200 CGG repeats; these individuals are phenotypically normal. Eventually, as enough repeats accrue from one generation to the next, *fragile-X syndrome* appears.

Fragile-X syndrome's complex expression may be a consequence of maternal genomic imprinting.

- The syndrome is more likely to appear if the abnormal *X* chromosome is inherited from the mother rather than the father, and pre-fragile *X* chromosomes in ova producing cells are more likely to acquire new CGG triplets than chromosomes in sperm producing cels.
- In the female parent, the site of triplet repeats on the *X* chromosome is imprinted by DNA methylation. Excessive methylation may inactivate one or more genes and prevent their normal expression offspring.
- Maternal imprinting explains why fragile-*X* disorder is more common in males. Males (XY) inherit the fragile *X* chromosome only from their mothers – maternally imprinted versions of the abnormal chromosome.
- Females can inherit the fragile *X* chromosome from either parent, but only the maternal version is imprinted and causes expression of the syndrome. Heterozygous carriers for this recessively inherited trait, have partial protection from the normal *X* chromosome and are usually only mildly retarded.

Huntington's disease is another example of how extended triplet repeats and genomic imprinting can influence the expression of a human genetic disorder.

- The Huntington's locus, near the tip of chromosome #4, has a CAG extended triplet repeat.
- Genomic imprinting influences the expression of the gene. The triplet repeat at the Huntington's locus is more likely to extend if the allele is inherited from the father, rather than the mother.

IX. Extranuclear genes exhibit a non-mendelian pattern of inheritance

There are some exceptions to the chromosome theory of inheritance.

- Extranuclear genes are found in cytoplasmic organelles such as plastids and mitochondria.
- These cytoplasmic genes are not inherited in Mendelian fashion, because they are not distributed by segregating chromosomes during meiosis.

In plants, a zygote receives its plastids from the ovum, not from pollen. Consequently, offspring receive only maternal cytoplasmic genes.

- Cytoplasmic genes in plants were first described by Karl Corens (1909) when he noticed that plant coloration of an ornamental species was determined by the seed bearing plants and not by the pollen producing plants.
- It is now known that maternal plastid genes control variegation of leaves.

In mammals, inheritance of mitochondrial DNA is also exclusively maternal.

- Since the ovum contributes most of the cytoplasm to the zygote, the mitochondria are all maternal in origin.

REFERENCES

Campbell, N. *Biology*. 4th ed. Menlo Park, California: Benjamin/Cummings, 1996.

Griffith, A.J.E., J.H. Miller, D.T. Suzuki, R.C. Lewontin, and W.M. Gelbart. *An Introduction to Genetic Analysis*. 5th ed. New York: W.H. Freeman, 1993.

Kowles, R.V. *Genetics, Society and Decisions*. 1st ed. Columbus, Ohio: Charles E. Merrill, 1985.

CHAPTER 15

THE MOLECULAR BASIS
OF INHERITANCE

OUTLINE

I.   The search for the genetic material led to DNA: *science as a process*
   A.   Evidence That DNA Can Transform Bacteria
   B.   Evidence That Viral DNA Can Program Cells
   C.   Additional Evidence That DNA Is the Genetic Material of Cells

II. Watson and Crick discovered the double helix by building models to conform to X-ray data: *science as a process*

III. During DNA replication, base-pairing enables existing DNA strands to serve as templates for new complementary strands

IV. A team of enzymes and other proteins functions in DNA replication
   A.   Getting Started: Origins of Replication
   B.   Elongating a New DNA Strand

V.   Enzymes proofread DNA during its replication and repair damage to existing DNA

OBJECTIVES

After reading this chapter and attending lecture, the student should be able to:

1.   Explain why researchers originally thought protein was the genetic material.
2.   Summarize experiments performed by the following scientists, which provided evidence that DNA is the genetic material:
   a.   Frederick Griffith.
   b.   Alfred Hershey and Martha Chase.
   c.   Erwin Chargaff.
3.   List the three components of a nucleotide.
4.   Distinguish between deoxyribose and ribose.

5. List the nitrogen bases found in DNA and distinguish between pyrimidine and purine.
6. Explain how Watson and Crick deduced the structure of DNA and describe what evidence they used.
7. Explain the "base-pairing rule" and describe its significance.
8. Describe the structure of DNA and explain what kind of chemical bond connects the nucleotides of each strand and what type of bond holds the two strands together.
9. Explain, in their own words, semiconservative replication and describe the Meselson-Stahl experiment.
10. Describe the process of DNA replication and explain the role of helicase, single strand binding protein, DNA polymerase, ligase and primase.
11. Explain what energy source drives endergonic synthesis of DNA.
12. Define antiparallel and explain why continuous synthesis of both DNA strands is not possible.
13. Distinguish between the leading strand and the lagging strand.
14. Explain how the lagging strand is synthesized when DNA polymerase can add nucleotides only to the 3′ end.
15. Explain the role of DNA polymerase, ligase and repair enzymes in DNA proofreading and repair.

KEY TERMS

Frederick Griffith	deoxyribose	semiconservative	single-strand binding
transformation	ribose	replication	protein
bacteriophage	nitrogen base	conservative	primer
Alfred Hershey	purine	replication	primase
Martha Chase	pyrimidine	dispersive replication	DNA polymerase
Erwin Chargaff	adenine	ultracentrifugation	replication complex
Chargaff's rules	thymine	Matthew Meselson	Okazaki fragment
James Watson	quanine	Franklin Stahl	leading strand
Francis Crick	cytosine	origin of replication	lagging strand
Rosalind Franklin	nucleoside	replication fork	ligase
nucleotide	triphosphate	replication bubble	mismatch repair
nucleic acid	antiparallel	helicase	excision repair

LECTURE NOTES

Deoxyribonucleic acid or DNA is the genetic material – Mendel's heritable factors and Morgan's genes on chromosomes. Inheritance has its molecular basis in the precise replication and transmission of DNA from parent to offspring.

I. **The search for the genetic material led to DNA:** *science as a process*

By the 1940's, scientists knew that chromosomes carry hereditary material and consist of DNA and protein. Most researchers thought protein was the genetic material because:

- Proteins are macromolecules with great heterogeneity and functional specificity.
- Little was known about nucleic acids.
- The physical and chemical properties of DNA seemed too uniform to account for the multitude of inherited traits.

A. **Evidence That DNA Can Transform Bacteria**

In 1928, Frederick Griffith performed experiments which provided evidence that genetic material is a specific molecule.

Griffith was trying to find a vaccine against *Streptococcus pneumoniae*, a bacterium that causes pneumonia in mammals. He knew that:

- There are two distinguishable strains of the pneumococcus: one produces smooth colonies (S) and the other rough colonies (R).
- Cells of the smooth strain are encapsulated with a polysaccharide coat and cells of the rough strain are not.
- These alternative phenotypes (S and R) are inherited.

Griffith performed four sets of experiments: (See Campbell, Figure 15.1)

Experiment: Griffith injected live S strain of *Streptococcus pneumoniae* into mice.
    Results: Mice died of pneumonia.
    Conclusions: Encapsulated strain is pathogenic.

Experiment: Mice were injected with live R strain.
    Results: Mice survived and were healthy.
    Conclusions: The bacterial strain lacking the polysaccharide coat was non-pathogenic.

Experiment: Mice were injected with heat-killed S strain of pneumococcus.
    Results: Mice survived and were healthy.
    Conclusions: Polysaccharide coat does not cause pneumonia because it is still present in heat-killed bacteria which proved to be non-pathogenic.

Experiment: Heat-killed S cells mixed with live R cells were injected into mice.
    Results: Mice developed pneumonia and died. Blood samples from dead mice contained live S cells.
    Conclusions: R cells had acquired from the dead S cells the ability to make polysaccharide coats. Griffith cultured S cells from the dead mice. Since the dividing bacteria produced encapsulated daughter cells, he concluded that this newly acquired trait was inheritable. This phenomenon is now called *transformation*.

<u>Transformation</u> = The assimilation of external genetic material by a cell.

What was the chemical nature of the transforming agent?

- Griffith was unable to answer this question, but other scientists continued the search.
- Griffith's experiments hinted that protein is not the genetic material. Heat denatures protein, yet it did not destroy the transforming ability of the genetic material in the heat-killed S cells.
- In 1944, after a decade of research, Oswald Avery, Maclyn McCarty and Colin MacLeod discovered that the transforming agent had to be DNA.
- The discovery by Avery and his coworkers was met with skepticism by other scientists, because they still believed protein was a better candidate for the genetic material and so little was known about DNA.

B.   Evidence That Viral DNA Can Program Cells

More evidence that DNA is the genetic material came from studies of *bacteriophages*.

<u>Bacteriophage (phage)</u> = Virus that infects bacteria.

In 1952, Alfred Hershey and Martha Chase discovered that DNA is the genetic material of a phage known as T2. They knew that T2:

- Is one of many phages to infect the enteric bacterium *Escherichia coli* (*E. coli*).
- Like other viruses, is little more than DNA enclosed by a protein coat.
- Can quickly reprogram an *E. coli* cell to produce T2 phages and release the viruses when the cell lyses.

What Hershey and Chase did not know is which viral component – DNA or protein – was responsible for reprogramming the host bacterial cell. They answered this question by performing the following experiment: (See Campbell, Figure 15.2)

Experiment:

Step 1: Viral protein and DNA were tagged with different radioactive isotopes.

⇒ *Protein Tagging*: T2 and *E. coli* were grown in media with radioactive sulfur (^{35}S) which incorporated only into the phage protein.

⇒ *DNA Tagging*: T2 and *E. coli* were grown in media containing radioactive phosphorus (^{32}P) which was incorporated only into the phage DNA.

Step 2: Protein-labeled and DNA-labeled T2 phages were allowed to infect separate samples of nonradioactive *E. coli* cells.

Step 3: Cultures were agitated to shake loose phages that remained outside the bacterial cells.

Step 4: Mixtures were centrifuged forcing the heavier bacterial cells into a pellet on the bottom of the tubes. The lighter viruses remained in the supernatant.

Step 5: Radioactivity in the pellet and supernatant was measured and compared.

Results:

1. In tubes with *E. coli* infected with protein-labeled T2, most of the radioactivity was in the supernatant with viruses.

2. In tubes with *E. coli* infected with DNA-labeled T2, most of the radioactivity was in the pellet with the bacterial cells.

3. When the bacteria containing DNA-labeled phages were returned to culture medium, the bacteria released phage progeny which contained ^{32}P in their DNA.

Conclusions:

1. Viral proteins remain outside the host cell.

2. Viral DNA is injected into the host cell.

3. Injected DNA molecules cause cells to produce additional viruses with more viral DNA and proteins.

4. These data provided evidence that nucleic acids rather than proteins are the hereditary material.

C. Additional Evidence That DNA Is the Genetic Material of Cells

Hershey and Chase's experiments provided evidence that DNA is the hereditary material in viruses. Additional evidence pointed to DNA as the genetic material in eukaryotes as well.

Some circumstantial evidence was:

- A eukaryotic cell doubles its DNA content prior to mitosis.
- During mitosis, the doubled DNA is equally divided between two daughter cells.
- An organism's diploid cells have twice the DNA as its haploid gametes.

Experimental evidence for DNA as the hereditary material in eukaryotes came from the laboratory of Erwin Chargaff. In 1947, he analyzed the DNA composition of different organisms.

Using paper chromatography to separate nitrogenous bases, Chargaff reported the following:

- DNA composition is species-specific.
  - ⇒ The amount and ratios of nitrogenous bases vary from one species to another.
  - ⇒ This source of molecular diversity made it more credible that DNA is the genetic material.

- In every species he studied, there was a regularity in base ratios.
  - ⇒ The number of adenine (A) residues approximately equaled the number of thymines (T), and the number of guanines (G) equaled the number of cytosines (C).
  - ⇒ The A=T and G=C equalities became known later as *Chargaff's rules*. The explanation for these rules came with Watson and Crick's structural model for DNA.

II. Watson and Crick discovered the double helix by building models to conform to X-ray data: *science as a process*

NOTE: Sometime before this discussion, it is helpful to review material from chapter 7, such as: components of nucleotides; nitrogenous bases in DNA; difference between purine and pyrimidine; what kind of chemical bond connects nucleotides of each DNA strand; and what type of bond holds the two strands together. Because so much time elapses between the lectures on organic molecules and molecular genetics, students may forget crucial information necessary to understand this material. For this reason, some of our lecturers find it best to defer discussion of nucleic acids (from chapter 7) until this point in the course.

By the 1950's, DNA was accepted as the genetic material, and the covalent arrangement in a nucleic acid polymer was well established. The three dimensional structure of DNA, however, was yet to be discovered. (See Campbell, Figure 15.3)

Among scientists working on the problem were:

Linus Pauling – California Institute of Technology

Maurice Wilkins and Rosalind Franklin – King's College in London

James D. Watson (American) and Francis Crick – Cambridge University

James Watson went to Cambridge to work with Francis Crick who was studying protein structure with *X-ray crystallography*.

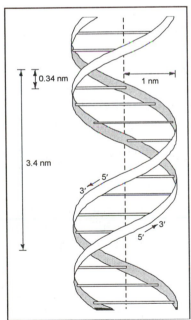

Watson saw an X-ray photo of DNA produced by Rosalind Franklin at King's College, London. Watson and Crick deduced from Franklin's X-ray data that:

a. DNA is a helix with a uniform width of 2 nm. This width suggested that it had two strands.

b. Purine and pyrimidine bases are stacked .34 nm apart.

c. The helix makes one full turn every 3.4 nm along its length.

d. There are ten layers of nitrogenous base pairs in each turn of the helix.

Watson and Crick built scale models of a double helix that would conform to the X-ray data and the known chemistry of DNA.

- One of their unsuccessful attempts placed the sugar-phosphate chains inside the molecule.

- Watson next put the sugar-phosphate chains on the outside which allowed the more hydrophobic nitrogenous bases to swivel to the interior away from the aqueous medium.

- Their proposed structure is a ladder-like molecule twisted into a spiral, with sugar-phosphate backbones as uprights and pairs of nitrogenous bases as rungs.
- The two sugar-phosphate backbones of the helix are *antiparallel*; that is, they run in opposite directions.

Watson and Crick finally solved the problem of DNA structure by proposing that there is a specific pairing between nitrogenous bases. After considering several arrangements, they concluded:

- To be consistent with a 2 nm width, a purine on one strand must pair (by hydrogen bonding) with a pyrimidine on the other.
- Base structure dictates which pairs of bases can hydrogen bond. The base pairing rule is that adenine can only pair with thymine, and guanine with cytosine.

Purines	Pyrimidines	Possible Base Pairs	Number of Hydrogen Bonds
Adenine (A)	Thymine (T)	A – T	2
Guanine (G)	Cytosine (C)	G – C	3

ADENINE (A)　　THYMINE (T)　　　　GUANINE (G)　　CYTOSINE (C)

The base-pairing rule is significant because:

- It explains Chargaff's rules. Since A must pair with T, their amounts in a given DNA molecule will be about the same. Similarly, the amount of G equals the amount of C.
- It suggests the general mechanisms for DNA replication. If bases form specific pairs, the information on one strand complements that along the other.
- It dictates the combination of complementary base pairs, but places no restriction on the linear sequence of nucleotides along the length of a DNA strand. The sequence of bases can be highly variable which makes it suitable for coding genetic information.
- Though hydrogen bonds between paired bases are weak bonds, collectively they stabilize the DNA molecule. Van der Waals forces between stacked bases also help stabilize DNA.

III.   **During DNA replication, base-pairing enables existing DNA strands to serve as templates for new complementary strands**

In April 1953, Watson and Crick's new model for DNA structure, the double helix, was published in the British journal *Nature*.   This model of DNA structure suggested a *template* mechanism for DNA replication.

- Watson and Crick proposed that genes on the original DNA strand are copied by a specific pairing of complementary bases, which creates a complementary DNA strand.
- The complementary strand can then function as a template to produce a copy of the original strand.

In a second paper, Watson and Crick proposed that during DNA replication:

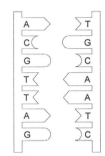

1.   The two DNA strands separate.

2.   Each strand is a template for assembling a complementary strand.

3.   Nucleotides line up singly along the template strand in accordance with the base-pairing rules (A-T and G-C).

4.   Enzymes link the nucleotides together at their sugar-phosphate groups.

Watson and Crick's model is a *semiconservative model* for DNA replication.

- They predicted that when a double helix replicates, each of the two daughter molecules will have one old or *conserved* strand from the parent molecule and one newly created strand.
- In the late 1950's, Matthew Meselson and Franklin Stahl provided the experimental evidence to support the semiconservative model of DNA replication.   A brief description of the experimental steps follows. (See Campbell, Figure 15.8.)

   Hypotheses:

   There were three alternate hypotheses for the pattern of DNA replication:

   a. If DNA replication is *conservative*, then the parental double helix should remain intact and the second DNA molecule should be constructed as entirely new DNA.

   b. If DNA replication is *semiconservative*, then each of the two resulting DNA molecules should be composed of one original or conserved strand (template) and one newly created strand.

    c. If DNA replication is *dispersive*, then both strands of the two newly produced DNA molecules should contain a mixture of old and new DNA.

***Experiment*:**

    Step 1: Labeling DNA strands with ^{15}N.

    E. coli was grown for several generations in a medium with heavy nitrogen (^{15}N). As *E. coli* cells reproduced, they incorporated the ^{15}N into their nitrogenous bases.

    Step 2: Transfer of *E. coli* to a medium with ^{14}N.

    E. coli cells grown in heavy medium were transferred to a medium with light nitrogen, ^{14}N. There were three experimental classes of DNA based upon times after the shift from heavy to light medium:

    a. Parental DNA from *E. coli* cells grown in heavy medium.

    b. First-generation DNA extracted from *E. coli* after one generation of growth in the light medium.

    c. Second-generation DNA extracted from *E. coli* after two replications in the light medium.

    Step 3: Separation of DNA classes based upon density differences.

    Meselson and Stahl used a new technique to separate DNA based on density differences between ^{14}N and ^{15}N.

Isolated DNA was mixed with a CsCl
solution and placed in an ultracentrifuge.

↓

DNA-CsCl solution was centrifuged at high
speed for several days.

↓

Centrifugal force created a CsCl gradient
with increased concentration toward the
bottom of the centrifuge tube.

↓

DNA molecules moved to a position in the
tube where their density equaled that of the
CsCl solution. Heavier DNA molecules were
closer to the bottom and lighter DNA
molecules were closer to the top where the
CsCl solution was less dense.

Results:

DNA from *E. coli* grown with ¹⁵N was heavier than DNA containing the more common, lighter isotope, ¹⁴N.	First-generation DNA after one generation of bacterial growth, was all of intermediate density.	Second generation DNA after two generations of bacterial growth in light medium was of intermediate and light density.

Conclusions:

Their results supported the hypothesis of semiconservative replication for DNA.

- The first generation DNA was all hybrid DNA containing one heavy parental strand and one newly synthesized light strand.

- These results were predicted by the semiconservative model. If DNA replication were conservative, two classes of DNA would be produced. The heavy parental DNA would be conserved and newly synthesized DNA would be light. There would be no intermediate density hybrid DNA.

- The results for first-generation DNA eliminated the possibility of conservative replication, but did not eliminate the possibility of dispersive replication.

- Consequently, Meselson and Stahl examined second-generation DNA. The fact that there were two bands, one light and one hybrid band supported the semiconservative model and allowed the investigators to rule out dispersive replication. If DNA replication were dispersive, there would have been only one band intermediate in density between light and hybrid DNA.

IV. A team of enzymes and other proteins functions in DNA replication

The general mechanism of DNA replication is conceptually simple, but the actual process:

- Is complex. The helical molecule must untwist while it copies its two antiparallel strands simultaneously. This requires the cooperation of over a dozen enzymes and other proteins.

- Is extremely rapid. In prokaryotes, up to 500 nucleotides are added per second. It takes only a few hours to copy the 6 billion bases of a single human cell.

- Is accurate. Only about one in a billion nucleotides is incorrectly paired.

A. Getting Started: Origins of Replication

DNA replication begins at special sites called *origins of replication* that have a specific sequence of nucleotides.

- Specific proteins required to initiate replication bind to each origin.
- The DNA double helix opens at the origin and *replication forks* spread in both directions away from the central initiation point creating a *replication bubble*.
- Bacterial or viral DNA molecules have only one replication origin.
- Eukaryotic chromosomes have hundreds or thousands of replication origins. The many replication bubbles formed by this process, eventually merge forming two continuous DNA molecules.

<u>Replication forks</u> = The Y-shaped regions of replicating DNA molecules where new strands are growing.

B. Elongating a New DNA Strand

1. Strand Separation

There are two types of proteins involved in the separation of parental DNA strands:

- a. *Helicases* are enzymes which catalyze unwinding of the parental double helix to expose the template.
- b. *Single-strand binding proteins* are proteins which keep the separated strands apart and stabilize the unwound DNA until new complementary strands can be synthesized.

2. Synthesis of the New DNA Strands

Enzymes called *DNA polymerases* catalyze synthesis of a new DNA strand.

- According to base-pairing rules, new nucleotides align themselves along the templates of the old DNA strands.
- *DNA polymerase* links the nucleotides to the growing strand. These strands grow in the 5′ → 3′ direction since new nucleotides are added only to the 3′ end of the growing strand.

Hydrolysis of *nucleoside triphosphates* provides the energy necessary to synthesize the new DNA strands.

<u>Nucleoside triphosphate</u> = Nucleotides with a triphosphate (three phosphates) covalently linked to the 5′ carbon of the pentose.

- Recall that the pentose in DNA is deoxyribose, and the pentose in RNA is ribose.
- Nucleoside triphosphates that are the building blocks for DNA, lose two phosphates (pyrophosphate group) when they form covalent linkages to the growing chain.
- Exergonic hydrolysis of this phosphate bond drives the endergonic synthesis of DNA; it provides the required energy to form the new covalent linkages between nucleotides.

Continuous synthesis of both DNA strands at a replication fork is not possible, because:

- The sugar phosphate backbones of the two complementary DNA strands run in opposite directions; that is, they are *antiparallel*.
- Recall that each DNA strand has a distinct polarity. At one end (3′ end), a hydroxyl group is attached to the 3′ carbon of the terminal deoxyribose; at the other end (5′ end), a phosphate group is attached to the 5′ carbon of the terminal deoxyribose.
- DNA polymerase can only elongate strands in the 5′ to 3′ direction.

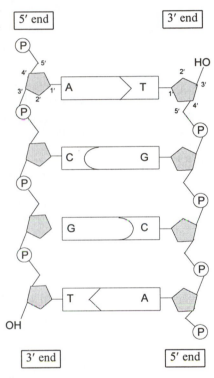

The problem of antiparallel DNA strands is solved by the continuous synthesis of one strand (*leading strand*) and discontinuous synthesis of the complementary strand (*lagging strand*).

Leading strand = The DNA strand which is synthesized as a single polymer in the 5′ → 3′ direction towards the replication fork.

Lagging strand = The DNA strand that is discontinuously synthesized against the overall direction of replication.

- Lagging strand is produced as a series of short segments called *Okazaki fragments* which are each synthesized in the 5′ → 3′ direction.

- Okazaki fragments are 1000 – 2000 nucleotides long in bacteria and 100 to 200 nucleotides long in eukaryotes.

- The many fragments are ligated by *DNA ligase*, a linking enzyme that catalyzes the formation of a covalent bond between the 3′ end of each new Okazaki fragment to the 5′ end of the growing chain.

3. Priming DNA Synthesis

Before new DNA strands can form, there must be small pre-existing *primers* to start the addition of new nucleotides.

Primer = Short RNA segment that is complementary to a DNA segment and that is necessary to begin DNA replication.

- Primers are short segments of RNA polymerized by an enzyme called *primase*.

- A portion of the parental DNA serves as a template for making the primer with a complementary base sequence that is about 10 nucleotides long in eukaryotes.

- Primer formation must precede DNA replication, because DNA polymerase can only add nucleotides to a polynucleotide that is already correctly base-paired with a complementary strand.

Only one primer is necessary for replication of the leading strand, but many primers are required to replicate the lagging strand.

- An RNA primer must initiate the synthesis of each Okazaki fragment.

- The many Okazaki fragments are ligated in two steps to produce a continuous DNA strand:

    1. DNA polymerase removes the RNA primer and replaces it with DNA.

    2. DNA ligase catalyzes the linkage between the 3′ end of each new Okazaki fragment to the 5′ end of the growing chain.

V.   Enzymes proofread DNA during its replication and repair damage to existing DNA

DNA replication is highly accurate, but this accuracy is not solely the result of base-pairing specificity.

- Initial pairing errors occur at a frequency of about one in ten thousand, while errors in a complete DNA molecule are only about one in one billion.
- DNA can be repaired as it is being synthesized (e.g. *mismatch repair*) or after accidental changes in existing DNA (e.g. *excision repair*).

Mismatch repair, corrects mistakes when DNA is synthesized.

- In bacteria, DNA polymerase proofreads each newly added nucleotide against its template. If polymerase detects an incorrectly paired nucleotide, the enzyme removes and replaces it before continuing with synthesis.
- In eukaryotes, additional proteins as well as polymerase participate in mismatch repair. A hereditary defect in one of these proteins has been associated with a form of colon cancer. Apparently, DNA errors accumulate in the absence of adequate proofreading.

Excision repair, corrects accidental changes that occur in existing DNA.

- Accidental changes in DNA can result from exposure to reactive chemicals, radioactivity, X-rays and ultraviolet light.
- There are more than fifty different types of DNA repair enzymes that repair damage. For example, in *excision repair*:
  ⇒ The damaged segment is excised by one *repair enzyme* and the remaining gap is filled in by base-pairing nucleotides with the undamaged strand.
  ⇒ *DNA polymerase* and *DNA ligase* are enzymes that catalyze the filling-in process.

REFERENCES

Alberts, B., D. Bray, J. Lewis, M. Raff, K. Roberts and J.D. Watson. *Molecular Biology of the Cell*. 2nd ed. New York: Garland, 1989.

Becker, W.M. and D.W. Deamer. *The World of the Cell*. 3rd ed. Redwood City, California: Benjamin/Cummings, 1996.

Campbell, N. *Biology*. 4th ed. Menlo Park, California: Benjamin/Cummings, 1996.

Culotta, E. and D.E. Koshland, Jr. "DNA Repair Works its Way to the Top." *Science*. December 23, 1994.

Watson, J.D., N.H. Hopkins, J.W. Roberts, J.A. Steitz and A.M. Weiner. *Molecular Biology of the Gene*. 4th ed. Menlo Park, California: Benjamin/Cummings, 1987.

CHAPTER 16

FROM GENE TO PROTEIN

OUTLINE

VIII. Eukaryotic cells modify RNA after transcription
    A. Alteration of mRNA Ends
    B. Split Genes and RNA Splicing
    C. Ribozymes
    D. The Functional and Evolutionary Importance of Introns

IX. A point mutation can affect the function of a protein
    A. Types of Point Mutations
    B. Mutagens

OBJECTIVES

After reading this chapter and attending lecture, the student should be able to:

1. Give early experimental evidence that implicated proteins as the links between genotype and phenotype.
2. Describe Beadle and Tatum's experiments with *Neurospora*, and explain the contribution they made to our understanding of how genes control metabolism.
3. Distinguish between "one gene-one enzyme" hypothesis and "one gene-one polypeptide," and explain why the original hypothesis was changed.
4. Explain how RNA differs from DNA.
5. In their own words, briefly explain how information flows from gene to protein.
6. Distinguish between transcription and translation.
7. Describe where transcription and translation occur in prokaryotes and in eukaryotes; explain why it is significant that, in eukaryotes, transcription and translation are separated in space and time.
8. Define codon, and explain what relationship exists between the linear sequence of codons on mRNA and the linear sequence of amino acids in a polypeptide.
9. List the three stop codons and the one start codon.
10. Explain in what way the genetic code is redundant and unambiguous.
11. Explain the evolutionary significance of a nearly universal genetic code.
12. Explain the process of transcription including the three major steps of initiation, elongation and termination.
13. Describe the general role of RNA polymerase in transcription.
14. Explain how RNA polymerase recognizes where transcription should begin.
15. Specifically, describe the primary functions of RNA polymerase II.
16. Distinguish among mRNA, tRNA and rRNA.
17. Describe the structure of tRNA and explain how the structure is related to function.
18. Given a sequence of bases in DNA, predict the corresponding codons transcribed on mRNA and the corresponding anticodons of tRNA.
19. Describe the wobble effect.
20. Explain how an aminoacyl-tRNA synthetase matches a specific amino acid to its appropriate tRNA; describe the energy source that drives this endergonic process.
21. Describe the structure of a ribosome, and explain how this structure relates to function.
22. Describe the process of translation including initiation, elongation and termination and explain what enzymes, protein factors and energy sources are needed for each stage.

23. Explain what determines the primary structure of a protein and describe how a polypeptide must be modified before it becomes fully functional.
24. Describe what determines whether a ribosome will be free in the cytosol or attached to rough ER.
25. Explain how proteins can be targeted for specific sites within the cell.
26. Describe the difference between prokaryotic and eukaryotic mRNA.
27. Explain how eukaryotic mRNA is processed before it leaves the nucleus.
28. Describe some biological functions of introns and gene splicing.
29. Explain why base-pair insertions or deletions usually have a greater effect than base-pair substitutions.
30. Describe how mutagenesis can occur.

KEY TERMS

Archibald Garrod
George Beadle
Edward Tatum
Neurospora
auxotroph
"one gene–one enzyme"
"one gene–one polypeptide"
transcription
codon
messenger RNA (mRNA)
RNA polymerase
promoter

terminator
transfer RNA (tRNA)
anticodon
triphosphate nucleotide
aminoacyl-tRNA synthetases
ribosome
ribosomal RNA (rRNA)
P site
A site
initiation factors
start codon
reading frame

elongation factor
peptidyl transferase
translocation
termination codon
release factor
wobble effect
redundance
heterogeneous nuclear RNA (hnRNA)
intron
exon
RNA splicing
snRNPS
small nuclear RNA (sn RNA)

spliceosome
ribozyme
CAP (mRNA)
Poly-A tail
leader sequence
trailer sequence
point mutation
base-pair substitution
missense mutation
nonsense mutation
insertion
deletion
frameshift mutation
mutagenesis
mutagens

LECTURE NOTES

Inherited instructions in DNA direct protein synthesis. Thus, proteins are the links between genotype and phenotype, since proteins are directly involved in the expression of specific phenotypic traits.

I. **The study of metabolic defects provided evidence that genes specify proteins:** *science as a process*

Archibald Garrod was the first to propose the relationship between genes and proteins (1909).
- He suggested that genes dictate phenotypes through enzymes that catalyze reactions.
- As a physician, Garrod was familiar with inherited diseases which he called "inborn errors in metabolism." He hypothesized that such diseases reflect the patient's inability to make particular enzymes.
- One example he studied was *alkaptonuria*, which causes the afflicted person's urine to turn black.
  ⇒ People with alkaptonuria accumulate alkapton in their urine, causing it to darken on contact with air.
  ⇒ Garrod reasoned that alkaptonurics, unlike normal individuals, lack the enzyme that breaks down alkapton.

A. **How Genes Control Metabolism**

Garrod's hypothesis was confirmed several decades later by research which determined that specific genes direct production of specific enzymes.
- Biochemists found that cells synthesize and degrade organic compounds via metabolic pathways, with each sequential step catalyzed by a specific enzyme.
- Geneticists George Beadle and Boris Ephrussi (1930's) studied eye color in *Drosophila*. They speculated that mutations affecting eye color block pigment synthesis by preventing enzyme production at certain steps in the pigment synthesis pathway.

George Beadle and Edward Tatum were later able to demonstrate the relationship between genes and enzymes by studying mutants of a bread mold, *Neurospora crassa*.
- Wild-type *Neurospora* in laboratory colonies can survive on *minimal medium*. All other molecules needed by the mold are produced by its own metabolic pathways from this minimal nutrient source.
- Beadle and Tatum searched for mutants or *auxotrophs* that could not survive on minimal medium because they lacked the ability to synthesize essential molecules.
- Mutants were identified by transferring fragments of growing fungi (in complete medium) to vials containing minimal medium. Fragments that didn't grow were identified as auxotrophic mutants.

Auxotroph = (Auxo = to augment; troph = nourishment) Nutritional mutants that can only be grown on *minimal medium* augmented with nutrients not required by the wild type.

<u>Minimal medium</u> = Support medium that is mixed only with molecules required for the growth of wild-type organisms.

- Minimal medium for *Neurospora* contains inorganic salts, sucrose and the vitamin, biotin.

- Nutritional mutants cannot survive only on minimal medium.

<u>Complete growth medium</u> = Minimal medium supplemented with all 20 amino acids and some other nutrients.

- Nutritional mutants can grow on complete growth medium, since all essential nutrients are provided.

Beadle and Tatum then identified specific metabolic defects (from mutations) by transferring fragments of auxotrophic mutants growing on complete growth medium to vials containing minimal medium each supplemented with only *one* additional nutrient.

- Vials where growth occurred indicated the metabolic defect, since the single supplement provided the necessary component.

- For example, if a mutant grew on minimal medium supplemented with only arginine, it could be concluded that the mutant was defective in the arginine synthesis pathway.

Experiment:

Beadle and Tatum experimented further to more specifically describe the defect in the multistep pathway that synthesizes the amino acid arginine.

- Arginine synthesis requires three steps each catalyzed by a specific enzyme:

- They distinguished between three classes of arginine auxotrophs by adding either arginine, citrulline, or ornithine to the medium and seeing if growth occurred.

Results:

Some mutants required arginine, some either arginine or citrulline, and others could grow when any of the three were added.

	Minimal Medium (MM)	MM plus Ornithine	MM plus Citrulline	MM plus Arginine
Wild Type	+	+	+	+
Class I Mutants	−	+	+	+
Class II Mutants	−	−	+	+
Class III Mutants	−	−	−	+

+ = growth, − = no growth

Conclusions:

Beadle and Tatum deduced from their data that the three classes of mutants each lacked a different enzyme and were thus blocked at different steps in the arginine synthesis pathway.

- Class I mutants lacked enzyme A; Class II mutants lacked enzyme B; and Class III mutants lacked enzyme C.
- Assuming that each mutant was defective in a single gene, they formulated the *one gene-one enzyme* hypothesis, which states that the function of a gene is to dictate the production of a specific enzyme.

B.   One Gene-One Polypeptide

Beadle and Tatum's one gene-one enzyme hypothesis has been slightly modified:

- While most enzymes are proteins, many proteins are not enzymes. Proteins that are not enzymes are still, nevertheless, gene products.
- Also, many proteins are comprised of two or more polypeptide chains, each chain specified by a different gene (e.g. globulin chains of hemoglobin).

As a result of this new information, Beadle and Tatum's hypothesis has been restated as *one gene-one polypeptide*.

II.   Transcription and translation are the two main steps from gene to protein: *an overview*

Ribonucleic acid (RNA) links DNA's genetic instructions for making proteins to the process of protein synthesis. It copies or transcribes the message from DNA and then translates that message into a protein.

- RNA, like DNA, is a nucleic acid or polymer of nucleotides.
- RNA structure differs from DNA in the following ways:
  - ⇒ The five-carbon sugar in RNA nucleotides is *ribose* rather than deoxyribose.
  - ⇒ The nitrogenous base *uracil* is found in place of thymine.

The linear sequence of nucleotides in DNA ultimately determines the linear sequence of amino acids in a protein.

- Nucleic acids are made of four types of nucleotides which differ in their nitrogenous bases. Hundreds or thousands of nucleotides long, each gene has a specific linear sequence of the four possible bases.
- Proteins are made of twenty types of amino acids linked in a particular linear sequence (the protein's primary structure).
- Information flows from gene to protein through two major processes, *transcription* and *translation*.

Transcription = The synthesis of RNA using DNA as a template.

- A gene's unique nucleotide sequence is transcribed from DNA to a complementary nucleotide sequence in messenger RNA (mRNA).
- The resulting mRNA carries this transcript of protein-building instructions to the cell's protein-synthesizing machinery.

Translation = Synthesis of a polypeptide, which occurs under the direction of messenger RNA (mRNA).

- During this process, the linear sequence of bases in mRNA is translated into the linear sequence of amino acids in a polypeptide.
- Translation occurs on *ribosomes*, complex particles composed of ribosomal RNA (rRNA) and protein that facilitate the orderly linking of amino acids into polypeptide chains.

Prokaryotes and eukaryotes differ in how protein synthesis is organized within their cells.

- Prokaryotes lack nuclei, so DNA is not segregated from ribosomes or the protein-synthesizing machinery. Thus, transcription and translation occur in rapid succession.
- Eukaryotes have nuclear envelopes that segregate transcription in the nucleus from translation in the cytoplasm; mRNA, the intermediary, is modified before it moves from the nucleus to the cytoplasm where translation occurs. This *RNA processing* occurs only in eukaryotes.

III.   **In the genetic code, a particular triplet of nucleotides specifies a certain amino acid:** *a closer look*

There is not a one-to-one correspondence between the nitrogenous bases and the amino acids they specify, since there are only 4 nucleotides and 20 amino acids.

- A two-to-one correspondence of bases to amino acids would only specify 16 (4^2) of the 20 amino acids.
- A three-to-one correspondence of bases to amino acids would specify 64 (4^3) amino acids.

Researchers have verified that the flow of information from a gene to a protein is based on a triplet code.

- Triplets of nucleotides are the smallest units of uniform length to allow translation into all 20 amino acids with plenty to spare.
- These three-nucleotide "words" are called *codons*.

Codon = A three-nucleotide sequence in mRNA that specifies which amino acid will be added to a growing polypeptide or that signals termination; the basic unit of the genetic code.

Genes are not directly translated into amino acids, but are first transcribed as codons into mRNA.

- For each gene, only one of the two DNA strands (the *template strand*) is transcribed.
- The complementary nontemplate strand is the parental strand for makine a new template when DNA replicates.
- The same DNA strand can be the template strand for some genes and the nontemplate strand for others.

An mRNA is complementary to the DNA template from which it is transcribed.

- For example, if the triplet nucleotide sequence on the template DNA strand is CCG; GGC, the codon for glycine, will be the complementary mRNA transcript.
- Recall that according to the base-pairing rules, uracil (U) in RNA is used in place of thymine (T); uracil thus base pairs with adenine (A).

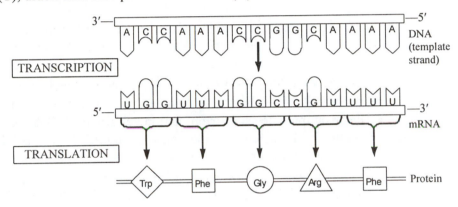

During translation, the linear sequence of codons along mRNA is translated into the linear sequence of amino acids in a polypeptide.

- Each mRNA codon specifies which one of 20 amino acids will be incorporated into the corresponding position in a polypeptide.
- Because codons are base triplets, the number of nucleotides making up a genetic message is three times the number of amino acids making up the polypeptide product.

A.   Cracking the Genetic Code

The first codon was deciphered in 1961 by Marshall Nirenberg of the National Institutes of Health.

- He synthesized an mRNA by linking only uracil-bearing RNA nucleotides, resulting in UUU codons.
- Nirenberg added this "poly U" to a test-tube mixture containing the components necessary for protein synthesis. The artificial mRNA (poly U) was translated into a polypeptide containing a string of only one amino acid, phenylalanine.
- Nirenberg concluded that the mRNA codon UUU specifies the amino acid phenylalanine.
- These same techniques were used to determine amino acids specified by the codons AAA, GGG, and CCC.

More elaborate techniques allowed investigators to determine all 64 codons by the mid-1960's. (See Campbell, Figure 16.5)

- 61 of the 64 triplets code for amino acids.
- The triplet AUG has a *dual* function — it is the start signal for translation and codes for methionine.
- Three codons do not code for amino acids, but signal termination (UAA, UAG and UGA).

There is redundancy in the genetic code, but no ambiguity.

- *Redundancy* exists since two or more codons differing only in their third base can code for the same amino acid (UUU and UUC both code for phenylalanine).
- *Ambiguity* is absent, since codons code for only one amino acid.

The correct ordering and grouping of nucleotides is important in the molecular language of cells. This ordering is called the *reading frame*.

<u>Reading frame</u> = The correct grouping of adjacent nucleotide triplets into codons that are in the correct sequence on mRNA.

- For example, the sequence of amino acids – Trp – Phe – Gly – Arg – Phe – can be assembled in the correct order only if the mRNA codons UGGUUUGGCCGUUUU are read in the correct sequence and groups.
- The cell reads the message in the correct frame as a series of <u>nonoverlapping</u> three-letter words: UGG – UUU – GGC – CGU – UUU.

B.   The Evolutionary Significance of a Common Genetic Language

The genetic code is shared nearly universally among living organisms.

- For example, the RNA codon CCG is translated into proline in all organisms whose genetic codes have been examined.
- The technology exists to transfer genes from one species to another. For example, the human gene for insulin can be inserted into bacteria where it is successfully expressed.

There are some exceptions to this universality:

- Several ciliates (e.g. *Paramecium* and *Tetrahymena*) depart from standard code; codons UAA and UAG are not stop signals, but code for glutamine.
- Mitochondria and chloroplasts have their own DNA that codes for some proteins.
- Mitochondrial genetic codes vary even among organisms; for example, CUA codes for threonine in yeast mitochondria and leucine in mammalian mitochondria.

The fact that the genetic code is shared nearly universally by all organisms indicates that this code was established very early in life's history.

IV. **Transcription is the DNA-directed synthesis of RNA:** *a closer look*

Transcription of messenger RNA (mRNA) from template DNA is catalyzed by *RNA polymerases*, which:

- Separate the two DNA strands and link RNA nucleotides as they base-pair along the DNA template.
- Add nucleotides only to the 3′ end; thus, mRNA molecules grow in the 5′ to 3′ direction.

There are several types of RNA polymerase.

- Prokaryotes have only one type of RNA polymerase that synthesizes all types of RNA – mRNA, rRNA and tRNA.
- Eukaryotes have three RNA polymerases that transcribe genes. *RNA polymerase II* is the polymerase that catalyzes mRNA synthesis; it transcribes genes that will be translated into proteins.

Specific DNA nucleotide sequences mark where transcription of a gene begins (*initiation*) and ends (*termination*). Initiation and termination sequences plus the nucleotides in between are called a *transcription unit*.

Transcription unit = Nucleotide sequence on the template strand of DNA that is transcribed into a single RNA molecule by RNA polymerase; it includes the initiation and termination sequences, as well as the nucleotides in between.

- In eukaryotes, a transcription unit contains a single gene, so the resulting mRNA codes for synthesis of only one polypeptide.
- In prokaryotes, a transcription unit can contain several genes, so the resulting mRNA may code for different, but functionally related, proteins.

Transcription occurs in three key steps: a) polymerase binding and initiation; b) elongation; and c) termination.

A. **RNA Polymerase Binding and Initiation of Transcription**

RNA polymerases bind to DNA at regions called *promoters*.

Promoter = Region of DNA that includes the site where RNA polymerase binds and where transcription begins (*initiation site*). In eukaryotes, the promoter is about 100 nucleotides long and consists of:

   a. The initiation site, where transcription begins.
   b. A few nucleotide sequences recognized by specific DNA-binding proteins (transcription factors) that help initiate transcription.

In eukaryotes, RNA polymerases cannot recognize the promoter without the help of *transcription factors*.

Transcription factors = DNA-binding proteins that bind to specific DNA nucleotide sequences at the promoter and that help RNA polymerase recognize and bind to the promoter region, so that transcription can begin.

- RNA polymerase II, the enzyme that synthesizes mRNA in eukaryotes, usually cannot recognize a promoter unless a specific transcription factor binds to a region on the promoter called a *TATA box*.

  TATA box = A short nucleotide sequence at the promoter that is rich in thymine (T) and adenine (A) and that is located about 25 nucleotides upstream from the initiation site.

- RNA polymerase II recognizes the complex between the bound TATA transcription factor and the DNA binding site.

- Once RNA polymerase recognizes and attaches to the promoter region, it probably associates with other transcription factors before RNA synthesis begins.

When active RNA polymerase binds to a promoter, the enzyme separates the two DNA strands at the initiation site, and transcription begins.

B. Elongation of the RNA Strand

Once transcription begins, RNA polymerase II moves along DNA and performs two primary functions:

1. It untwists and opens a short segment of DNA exposing about ten nucleotide bases; one of the exposed DNA strands is the template for base-pairing with RNA nucleotides.

2. It links incoming RNA nucleotides to the 3′ end of the elongating strand; thus, RNA grows one nucleotide at a time in the 5′ to 3′ direction.

During transcription, mRNA grows about 30 – 60 nucleotides per second. As the mRNA strand elongates:

- It peels away from its DNA template.

- The nontemplate strand of DNA re-forms a DNA – DNA double helix by pairing with the template strand.

Following in series, several molecules of RNA polymerase II can simultaneously transcribe the same gene.

- Cells can thus produce particular proteins in large amounts.

- The growing RNA strands hang free from each polymerase. The length of each strand varies and reflects how far the enzyme has traveled from the initiation site on template DNA.

C.   **Termination of Transcription**

Transcription proceeds until RNA polymerase reaches a *termination site* on the DNA.

Terminator sequence = DNA sequence that signals RNA polymerase to stop transcription and to release the RNA molecule and DNA template.
- Additional proteins may cooperate with RNA polymerase in termination.
- In eukaryotes, the most common terminator sequence is AATAAA.

Prokaryotic mRNA is ready for translation as soon as it leaves the DNA template. Eukaryotic mRNA, however, must be processed before it leaves the nucleus and becomes functional.

V.   **Translation is the RNA-directed synthesis of a polypeptide:** *a closer look*

During translation, proteins are synthesized according to a genetic message of sequential codons along mRNA.
- *Transfer RNA* (*tRNA*) is the interpreter between the two forms of information – base sequence in mRNA and amino acid sequence in polypeptides.
- tRNA aligns the appropriate amino acids to form a new polypeptide. To perform this function, tRNA must:
  ⇒ Transfer amino acids from the cytoplasm's amino acid pool to a ribosome.
  ⇒ Recognize the correct codons in mRNA.

Molecules of tRNA are specific for only one particular amino acid. Each type of tRNA associates a distinct mRNA codon with one of the 20 amino acids used to make proteins.
- One end of a tRNA molecule attaches to a specific amino acid.
- The other end attaches to an mRNA codon by base pairing with its *anticodon*.

Anticodon = A nucleotide triplet in tRNA that base pairs with a complementary nucleotide triplet (codon) in mRNA.

tRNAs decode the genetic message, codon by codon. For example,
- The mRNA codon UUU is translated as the amino acid phenylalanine. (See Campbell, Figure 16.5)
- The tRNA that transfers phenylalanine to the ribosome has an anticodon of AAA.
- When the codon UUU is presented for translation, phenylalanine will be added to the growing polypeptide.
- As tRNAs deposit amino acids in the correct order, ribosomal enzymes link them into a chain.

A.   **The Structure and Function of Transfer RNA**

All types of eukaryotic RNA, including tRNA, are transcribed from template DNA located within the nucleus.

- tRNA must travel from the nucleus to the cytoplasm, where translation occurs.
- Once in the cytoplasm, each tRNA molecule can be used repeatedly.

The ability of tRNA to carry specific amino acids and to recognize the correct codons depends upon its structure; its form fits function.

- tRNA is a single-stranded RNA only about 80 nucleotides long.
- The strand is folded, forming several double-stranded regions where short base sequences hydrogen bond with other complementary base sequences.
- A single-plane view reveals a clover leaf shape.

The three-dimensional structure is roughly L-shaped.

- A loop protrudes at one end of the L and has a specialized sequence of three bases called the *anticodon*.
- At the other end of the L protrudes the 3′ end of the tRNA molecule - the attachment site for an amino acid.

There are only about 45 distinct types of tRNA. However, this is enough to translate the 64 codons, since some tRNAs recognize two or three mRNA codons specifying the same amino acid.

- This is possible because the base-pairing rules are relaxed between the third base of an mRNA codon and the corresponding base of a tRNA anticodon.
- This exception to the base-pairing rule is called *wobble*.

<u>Wobble</u> = The ability of one tRNA to recognize two or three different mRNA codons; occurs when the third base (5′ end) of the tRNA anticodon has some play or wobble, so that it can hydrogen bond with more than one kind of base in the third position (3′ end) of the codon.

- For example, the base U in the wobble position of a tRNA anticodon can pair with either A or G in the third position of an mRNA codon.
- Some tRNAs contain a modified base called inosine (I), which is in the anticodon's wobble position and can base pair with U, C or A in the third position of an mRNA codon.
- Thus, a single tRNA with the anticodon CCI will recognize three mRNA codons: GGU, GGC or GGA - all of which code for glycine.

B. Aminoacyl-tRNA Synthetases

The correct linkage between tRNA and its designated amino acid must occur before the anticodon pairs with its complementary mRNA codon. This process of correctly pairing a tRNA with its appropriate amino acid is catalyzed by an *aminoacyl-tRNA synthetase*.

<u>Aminoacyl-tRNA synthetase</u> = A type of enzyme that catalyzes the attachment of an amino acid to its tRNA.

- Each of the 20 amino acids has a specific aminoacyl-tRNA synthetase.
- In an endergonic reaction driven by the hydrolysis of ATP, the proper synthetase attaches an amino acid to its tRNA in two steps:
  1. *Activation of the amino acid with AMP*. The synthetase's active site binds the amino acid and ATP; the ATP loses two phosphate groups and attaches to the amino acid as AMP (adenosine monophosphate).
  2. *Attachment of the amino acid to tRNA*. The appropriate tRNA covalently bonds to the amino acid, displacing AMP from the enzyme's active site.
- The aminoacyl-tRNA complex releases from the enzyme and transfers its amino acid to a growing polypeptide on the ribosome.

C. Ribosomes

Ribosomes coordinate the pairing of tRNA anticodons to mRNA codons.

- Ribosomes have two subunits (small and large) which are separated when not involved in protein synthesis. (See Campbell, Figure 16.12)
- Ribosomes are composed of about 60% *ribosomal RNA (rRNA)* and 40% protein.

The large and small subunits of eukaryotic ribosomes are:

- Constructed in the nucleolus.
- Dispatched through nuclear pores to the cytoplasm.
- Once in the cytoplasm, are assembled into functional ribosomes only when attached to an mRNA.

Compared to eukaryotic ribosomes, prokaryotic ribosomes are smaller and have a different molecular composition.

- Selection of effective drug therapies against bacterial pathogens capitalizes on this difference.
- For example, the antibiotics tetracycline and streptomycin can be used to combat bacterial infections, because they inhibit bacterial protein synthesis without afffecting the ribosomes of the eukaryotic host.

In addition to an mRNA binding site, each ribosome has two tRNA binding sites (P and A).

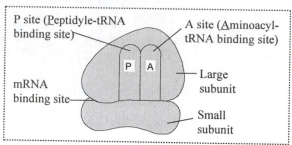

- The *P site* holds the tRNA carrying the growing polypeptide chain.
- The *A site* holds the tRNA carrying the next amino acid to be added.

As the ribosome holds the tRNA and mRNA molecules together, enzymes transfer the new amino acid from its tRNA to the carboxyl end of the growing polypeptide.

D. Building a Polypeptide

The building of a polypeptide, or translation, occurs in three stages: 1) initiation, 2) elongation, and 3) termination.

- All three stages require enzymes and other protein factors.
- Initiation and elongation also require energy provided by GTP (a molecule closely related to ATP).

1. Initiation

Initiation must bring together the mRNA, the first amino acid attached to its tRNA, and the two ribosomal subunits.

An initiation complex is assembled as mRNA and a special initiator tRNA bind to a small ribosomal subunit .

- In eukaryotes, the small ribosomal subunit binds first to an initiator tRNA with the anticodon UAC. This tRNA carries methionine, the first amino acid to be added.
- The small ribosomal subunit next binds to the 5′ end of mRNA. mRNA has a specific ribosome-

recognition sequence at the 5′ upstream end that base pairs with a complementary sequence on rRNA.

- With help from the small ribosomal subunit, bound initiator tRNA finds and base pairs with the initiation or start codon on mRNA. This start codon, AUG, marks the place where translation will begin and is located just downstream from the ribosome-recognition sequence.

- Assembly of the initiation complex – small ribosomal subunit, initiator tRNA and mRNA – requires:

    ⇒ Protein *initiation factors* that are bound to the small ribosomal subunit.

    ⇒ One GTP molecule that probably stabilizes the binding of initiation factors and upon hydrolysis drives the attachment of the large ribosomal subunit.

In the second step, a large ribosomal subunit binds to the small one to form a functional ribosome.

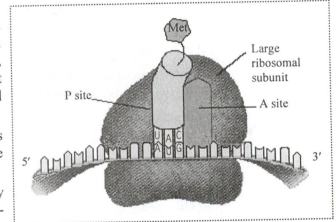

- Initiation factors attached to the small ribosomal subunit are released, allowing the large subunit to bind with the small subunit.

- The initiator tRNA fits into the P site on the ribosome.

- The vacant A site is ready for the next aminoacyl-tRNA.

2. Elongation

Several proteins called *elongation factors* take part in this three-step cycle which adds amino acids one by one to the initial amino acid.

a. *Codon recognition.* The mRNA codon in the A site of the ribosome forms hydrogen bonds with the anticodon of an entering tRNA carrying the next amino acid in the chain.

- An elongation factor directs tRNA into the A site.

- Hydrolysis of GTP provides energy for this step.

b. *Peptide bond formation.* An enzyme, *peptidyl transferase*, catalyzes the formation of a peptide bond between the polypeptide in the P site and the new amino acid in the A site.

- Peptidyl transferase is part of the large ribosomal subunit and consists of ribosomal proteins and rRNA.

- The polypeptide separates from its tRNA and is transferred to the new amino acid carried by the tRNA in the A site.

c. *Translocation.* The tRNA in the P site releases from the ribosome, and the tRNA in the A site is translocated to the P site.

- During this process, the codon and anticodon remain bonded, so the mRNA and the tRNA move as a unit, bringing the next codon to be translated into the A site.
- The mRNA is moved through the ribosome only in the 5′ to 3′ direction.
- GTP hydrolysis provides energy for each translocation step.

NOTE: Some students have trouble visualizing translocation, especially how the tRNA and mRNA move as a unit, exposing a new codon in the A site. Showing the class an animated sequence would no doubt solve the problem. However, if you do not have a monitor or video-projection capability, a paper simulation is just as effective. Paper models can be tacked to a large bulletin board or small cutouts can be moved on an overhead projector. Students may also actively participate by practicing their own simulations.

3. Termination

Each iteration of the elongation cycle takes about 60 milliseconds and is repeated until synthesis is complete and a *termination codon* reaches the ribosome's A site.

Termination codon (stop codon) = Base triplet (codon) on mRNA that signals the end of translation.

- Stop codons are UAA, UAG and UGA.
- Stop codons do not code for amino acids.

NOTE: Students often confuse terminator sequence (on DNA) that signals the end of transcription with termination or stop codons (on mRNA) that signal the end of translation.

When a stop codon reaches the ribosome's A site, a protein *release factor* binds to the codon and initiates the following sequence of events:

- Peptidyl transferase hydrolyzes the bond between the completed polypeptide and the tRNA in the P site.
- This frees the polypeptide and tRNA, so they can both release from the ribosome.
- The two ribosomal subunits dissociate from mRNA and separate back into a small and a large subunit.

E. Polyribosomes

Single ribosomes can make average-sized polypeptides in less than a minute; usually, however, clusters of ribosomes simultaneously translate an mRNA.

Polyribosome = A cluster of ribosomes simultaneously translating an mRNA molecule.

- Once a ribosome passes the initiation codon, a second ribosome can attach to the mRNA.
- Several ribosomes may translate an mRNA at once, making many copies of a polypeptide.

F. **From Polypeptide to Functional Protein**

The biological activity of proteins depends upon a precise folding of the polypeptide chain into a native three-dimensional conformation.

- Genes determine *primary structure*, the linear sequence of amino acids.
- Primary structure determines how a polypeptide chain will spontaneously coil and fold to form a three-dimensional molecule with *secondary* and *tertiary structure*.

Some proteins must undergo *post-translational modification* before they become fully functional in the cell.

- Sugars, lipids, phosphate groups, or other additives may be attached to some amino acids.
- One or more amino acids may be enzymatically cleaved from the leading (amino) end of the polypeptide chain.
- Single polypeptide chains may be divided into two or more pieces (e.g. insulin).
- Two or more polypeptides may join as subunits of a protein that has quaternary structure.

VI. **Some polypeptides have signal sequences that target them to specific destinations in the cell**

Eukaryotic ribosomes function either free in the cytosol or bound to endomembranes.

- Bound and free ribosomes are structurally identical and interchangeable.
- Most proteins made by free ribosomes will function in the cytosol.
- Attached to the outside of the endoplasmic reticulum, bound ribosomes generally make proteins that are:
  ⇒ destined for membrane inclusion in endomembrane system (e.g. nuclear envelope, ER, Golgi, lysosomes, vacuoles and plasma membrane).
  ⇒ secretory proteins destined for export.

There is only one type of ribosome, and synthesis of all proteins *begins* in the cytosol. *What determines whether a ribosome will be free in the cytosol or attached to rough ER?*

- Messenger RNA for secretory proteins codes for an initial *signal sequence* of 16 – 20 hydrophobic amino acids at the amino end of the forming polypeptide.
- When a ribosome begins to synthesize a protein with a signal sequence, it moves to the ER membrane by a mechanism that involves two other components.
  ⇒ *Signal recognition particle (SRP)*. This protein complex moves between the ER membrane and the cytosol. It is an adaptor that attaches both to the signal sequence of a growing polypeptide and to a receptor protein in the ER membrane (SRP receptor).
  ⇒ *SRP receptor*. This receptor protein is built into the ER membrane. The signal recognition particle docks with the receptor, and the ribosome thus becomes bound to the ER membrane.

- The ribosome continues protein synthesis and the leading end of the new polypeptide (N-terminus) threads into the cisternal space.
- The signal sequence is removed by an enzyme.
- Newly formed polypeptide is released from the ribosome and folds into its native conformation.
- If an mRNA does not code for a signal sequence, the ribosome remains free and synthesizes its protein in the cytosol.

Different signal sequences may also dispatch proteins to specific sites other than the ER. For example, newly formed proteins may be targeted for mitochondria or chloroplasts.

VII. Comparing protein synthesis in prokaryotes and eukaryotes: *a review*

Transcription and translation are similar in prokaryotes and eukaryotes, but protein synthesis is organized differently within their cells.

- Prokaryotes lack nuclei, so transcription is not segregated from translation; consequently, translation may begin as soon as the 5′ end of mRNA peels away from template DNA, even before transcription is complete.

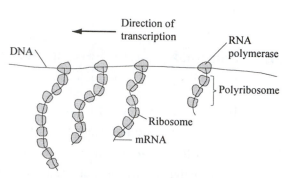

- The significance of a eukaryotic cell's compartmental organization is that transcription and translation are segregated by the nuclear envelope. This allows mRNA to be modified before it moves from the nucleus to the cytoplasm. Such *RNA processing* occurs only in eukaryotes.

VIII. Eukaryotic cells modify RNA after transcription

Before eukaryotic mRNA is exported from the nucleus, it is processed in two ways: a) both ends are covalently altered and b) intervening sequences are removed and the remainder spliced together.

A. Alteration of mRNA Ends

During mRNA processing, both the 5′ and 3′ ends are covalently modified.

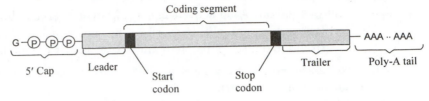

5′ cap = Modified guanine nucleotide (guanosine triphosphate) that is added to the 5′ end of mRNA shortly after transcription begins; has two important functions:

- Protects the growing mRNA from degradation by hydrolytic enzymes.
- Helps small ribosomal subunits recognize the attachment site on mRNA's 5′ end. A *leader* segment of mRNA may also be part of the ribosome recognition signal.

    Leader sequence = Noncoding (untranslated) sequence of mRNA from the 5′ end to the start codon.

The 3′ end, which is transcribed last, is modified by enzymatic addition of a *poly-A tail*, before the mRNA exits the nucleus.

Poly-A tail = Sequence of about 200 adenine nucleotides added to the 3′ end of mRNA before it exits the nucleus.

- May inhibit degradation of mRNA in the cytoplasm.
- May regulate protein synthesis by facilitating mRNA's export from the nucleus to the cytoplasm.
- Is not attached directly to the stop codon, but to an untranslated *trailer* segment of mRNA.

    Trailer sequence = Noncoding (untranslated) sequence of mRNA from the stop codon to the poly-A tail.

B. Split Genes and RNA Splicing

The original RNA transcript accurately reflects the complementary base sequence of the gene in template DNA; however, it is much longer than the mRNA that functions in the cytoplasm.

- The original transcript, or precursor mRNA, is *heterogeneous nuclear RNA*.

    Heterogeneous nuclear RNA (hnRNA) = Pool of RNA in the nucleus that contains molecules of widely varied sizes; includes primary mRNA transcripts or precursor mRNA (pre-mRNA).

- After RNA processing, only a small portion of the original transcript (pre-mRNA) leaves the nucleus as mRNA.

Genes that code for proteins in eukaryotes are not continuous sequences.

- Coding sequences of a gene are interrupted by noncoding segments of DNA called intervening sequences, or *introns*.

  Introns = Noncoding sequences in DNA that intervene between coding sequences (exons); are initially transcribed, but not translated, because they are excised from the transcript before mature RNA leaves the nucleus.

- Coding sequences of a gene are called *exons*, because they are eventually expressed (translated into protein).

  Exons = Coding sequences of a gene that are transcribed and expressed.

- In 1977, Richard Roberts and Philip Sharp independently found evidence for "split genes"; they won the 1993 Nobel Prize for their discovery.

Introns and exons are both transcribed to form hnRNA, but the introns are subsequently removed and the remaining exons linked together during the process of *RNA splicing*.

RNA splicing = RNA processing that removes introns and joins exons from eukaryotic hnRNA; produces mature mRNA that will move into the cytoplasm from the nucleus.

- Enzymes excise introns and splice together exons to form an mRNA with a continuous coding sequence.
- RNA splicing also occurs during post-transcriptional processing of tRNA and rRNA.

Though there is much left to be discovered, some details of RNA splicing are now known.

- Each end of an intron has short boundary sequences that accurately signal the RNA splicing sites.
- *Small nuclear ribonucleoproteins (snRNPs)*, play a key role in RNA splicing.

Small nuclear ribonucleoproteins (snRNPs) = Complexes of proteins and small nuclear RNAs that are found only in the nucleus; some participate in RNA splicing. (SnRNPs is pronounced "snurps".)

- These small nuclear particles are composed of:
  1. *Small nuclear RNA (snRNA)*. This small RNA molecule has less than 300 nucleotides – much shorter than mRNA.
  2. *Protein*. Each snRNP has seven or more proteins.
- There are various types of snRNPs with different functions; those involved in RNA splicing are part of a larger, more complex assembly called a *spliceosome*.

Spliceosome = A large molecular complex that catalyzes RNA splicing reactions; composed of small nuclear ribonucleoproteins (snRNPs) and other proteins. (See Campbell, Figure 16.21)

- As the spliceosome is assembled, one type of snRNP base pairs with a complementary sequence at the 5' end of the intron.
- The spliceosome precisely cuts the RNA transcript at specific splice sites at either end of the intron, which is excised as a lariat-shaped loop.
- The intron is released and the adjacent exons are immediately spliced together by the spliceosome.

C.   Ribozymes

Other kinds of RNA transcripts, such as tRNA and rRNA, are spliced differently; however, as with mRNA splicing, RNA is often involved in catalyzing the reactions.

Ribozymes = RNA molecules that can catalyze reactions by breaking and forming covalent bonds; are called ribozymes to emphasize their enzymelike catalytic activity.
- Ribozymes were first discovered in *Tetrahymena*, a ciliated protozoan that has self-splicing rRNA. That is, intron rRNA catalyzes its own splicing, completely without proteins or extra RNA molecules.
- Since RNA is acting as an enzyme, it can no longer be said that "All enzymes are proteins".
- It has since been discovered that rRNA also functions as an enzyme during translation.

Three-dimensional conformations vary among the types of RNA. These differences in shape give RNA its ability to perform a variety of functions, such as:
1. *Information carrier.* Messenger RNA (mRNA) carries genetic information from DNA to ribosomes; this genetic message specifies a protein's primary structure.
2. *Adaptor molecule.* Transfer RNA (tRNA) acts as an adaptor in protein synthesis by translating information from one form (mRNA nucleotide sequence) into another (protein amino acid sequence).
3. *Catalyst and structural molecule.* During translation, ribosomal RNA (rRNA) plays structural and probably enzymatic roles in ribosomes. Small nuclear RNA (snRNA) in snRNP particles also plays structural and enzymatic roles within spliceosomes that catalyze RNA splicing reactions.
4. *Viral genomes.* Some viruses use RNA as their genetic material.

D.   The Functional and Evolutionary Importance of Introns

What are the biological functions of introns and gene splicing?

Introns may play a regulatory role in the cell.
- Intron DNA sequences may control gene activity.
- The splicing process itself may regulate the export of mRNA to the cytoplasm.

Introns may allow a single gene to direct the synthesis of different proteins.
- This can occur if the same RNA transcript is processed differently among various cell types in the same organism.
- For example, all introns may be removed from a particular transcript in one case; but in another, one or more of the introns may be left in place to be translated. Thus, the resulting proteins in each case would be different.

Introns play an important role in the evolution of protein diversity; they increase the probability that recombination of exons will occur between alleles.

- In split genes, coding sequences can be separated by long distances, so they have higher recombination frequencies than continuously coded genes without introns.
- Exons of a "split gene" may code for different *domains* of a protein that have specific functions, such as, an enzyme's active site or a protein's binding site.

  Protein domains = Continuous polypeptide sequences that are structural and functional units in proteins with a modular architecture.

- Genetic recombination can occur in just one exon resulting in the synthesis of a novel protein with only one altered domain.

IX. A point mutation can affect the function of a protein

Knowing how genes are translated into proteins, scientists can give a molecular description of heritable changes that occur in organisms.

Mutation = A permanent change in DNA that can involve large chromosomal regions or a single nucleotide pair.

Point mutation = A mutation limited to about one or two nucleotides in a single gene.

A. Types of Point Mutations

There are two categories of point mutations: 1) base-pair substitutions and 2) base-pair insertions or deletions.

1. Substitutions

Base-pair substitution = The replacement of one base pair with another; occurs when a nucleotide and its partner from the complementary DNA strand are replaced with another pair of nucleotides according to base-pairing rules.

Depending on how base-pair substitutions are translated, they can result in little or no change in the protein encoded by the mutated gene.

- Redundancy in the genetic code is why some substitution mutations have no effect. A base pair change may simply transform one codon into another that codes for the same amino acid.
- Even if the substitution alters an amino acid, the new amino acid may have similar properties to the one it replaces, or it may be in a part of the protein where the exact amino acid sequence is not essential to its activity.

Some base-pair substitutions result in readily detectable changes in proteins.

- Alteration of a single amino acid in a crucial area of a protein will significantly alter protein activity.
- On rare occasions, such a mutation will produce a protein that is improved or has capabilities that enhance success of the mutant organism and its descendants.
- More often, such mutations produce a less active or inactive protein that impairs cell function.

Base-pair substitutions are usually *missense mutations* or *nonsense mutations*.

<u>Missense mutation</u> = Base-pair substitution that alters an amino acid codon (sense codon) to a new codon that codes for a different amino acid.

- Altered codons make sense (are translated), but not necessarily that originally intended.
- Base-pair substitutions are usually missense mutations.

<u>Nonsense mutation</u> = Base-pair substitution that changes an amino acid codon (sense codon) to a chain termination codon, or *vice versa*.

- Nonsense mutations can result in premature termination of translation and the production of a shorter than normal polypeptide.
- Nearly all nonsense mutations lead to nonfunctional proteins.

2. Insertions or Deletions

Base-pair insertions or *deletions* usually have a greater negative effect on proteins than substitutions.

<u>Base-pair insertion</u> = The insertion of one or more nucleotide pairs into a gene.

<u>Base-pair deletion</u> = The deletion of one or more nucleotide pairs from a gene.

Because mRNA is read as a series of triplets during translation, insertion or deletion of nucleotides may alter the reading frame (triplet grouping) of the genetic message. This type of *frameshift mutation* will occur whenever the number of nucleotides inserted or deleted is not 3 or a multiple of 3.

<u>Frameshift mutation</u> = A base-pair insertion or deletion that causes a shift in the reading frame, so that codons beyond the mutation will be the wrong grouping of triplets and will specify the wrong amino acids.

- A frameshift mutation causes the nucleotides following the insertion or deletion to be improperly grouped into codons.
- This results in extensive missense, which will sooner or later end in nonsense (premature termination).
- Frameshift will produce a nonfunctional protein unless the insertion or deletion is very near the end of the gene.

B. Mutagens

<u>Mutagenesis</u> = The creation of mutations.

- Mutations can occur as errors in DNA replication, repair or recombination that result in base-pair substitutions, insertions or deletions.
- Mutagenesis may be a naturally occurring event causing *spontaneous mutations*, or mutations may be caused by exposure to *mutagens*.

Mutagen = Physical or chemical agents that interact with DNA to cause mutations.

- Radiation is the most common physical mutagen in nature and has been used in the laboratory to induce mutations.

- Several categories of chemical mutagens are known including *base analogues*, which are chemicals that mimic normal DNA bases, but base pair incorrectly.

- The Ames test, developed by Bruce Ames, is one of the most widely used tests for measuring the mutagenic strength of various chemicals. Since most mutagens are carcinogenic, this test is also used to screen for chemical carcinogens. (See Campbell, Chapter 16 Methods Box)

REFERENCES

Alberts, B., D. Bray, J. Lewis, M. Raff, K. Roberts and J.D. Watson. *Molecular Biology of the Cell*. 2nd ed. New York: Garland, 1989.

Campbell, N. *Biology*. 4th ed. Menlo Park, California: Benjamin/Cummings, 1996.

Tijan, R. "Molecular Machines that Control Genes." *Scientific American*, February, 1995.

Watson, J.D., N.H. Hopkins, J.W. Roberts, J.A. Steitz and A.M. Weiner. *Molecular Biology of the Gene*. 4th ed. Menlo Park, California: Benjamin/Cummings, 1987.

CHAPTER 17

MICROBIAL MODELS: THE GENETICS OF VIRUSES AND BACTERIA

OUTLINE

X. Genetic recombination and transposition produce new bacterial strains
   A. Transformation
   B. Transduction
   C. Conjugation and Plasmids
   D. Transposons

XI. The control of gene expression enables individual bacteria to adjust their metabolism to environmental change
   A. Operons: The Basic Concept
   B. Repressible Versus Inducible Enzymes: Two Types of Negative Gene Regulation
   C. An Example of Positive Gene Regulation

OBJECTIVES

After reading this chapter and attending lecture, the student should be able to:

1. Recount the history leading up to the discovery of viruses and include the contributions of A. Mayer, D. Ivanowsky, Martinus Beijerinck and Wendell Stanley.
2. List and describe structural components of viruses.
3. Explain why viruses are obligate parasites.
4. Describe three patterns of viral genome replication.
5. Explain the role of reverse transcriptase in retroviruses.
6. Describe how viruses recognize host cells.
7. Distinguish between lytic and lysogenic reproductive cycles using phage T_4 and phage λ as examples.
8. Outline the procedure for measuring phage concentration in a liquid medium.
9. Describe several defenses bacteria have against phage infection.
10. Using viruses with envelopes and RNA viruses as examples, describe variations in replication cycles of animal viruses.
11. Explain how viruses may cause disease symptoms and describe some medical weapons used to fight viral infections.
12. List some viruses that have been implicated in human cancers and explain how tumor viruses transform cells.
13. Distinguish between horizontal and vertical routes of viral transmission in plants.
14. List some characteristics that viruses share with living organisms and explain why viruses do not fit our usual definition of life.
15. Provide evidence that viruses probably evolved from fragments of cellular nucleic acid.
16. Describe the structure of a bacterial chromosome.
17. Describe the process of binary fission in bacteria and explain why replication of the bacterial chromosome is considered to be semiconservative.
18. List and describe the three natural processes of genetic recombination in bacteria.
19. Distinguish between general transduction and specialized transduction.
20. Explain how the F plasmid controls conjugation in bacteria.
21. Explain how bacterial conjugation differs from sexual reproduction in eukaryotic organisms.
22. For donor and recipient bacterial cells, predict the consequences of conjugation between an F^+ and F^- cell; Hfr and F^- cell.

23. Explain how a geneticist can use several features of conjugation to roughly map a bacterial chromosome.
24. Define transposon and describe two essential types of nucleotide sequences found in transposon DNA.
25. Distinguish between an insertion sequence and a complex transposon.
26. Describe the role of transposase and DNA polymerase in the process of transposition.
27. Explain how transposons can generate genetic diversity.
28. Briefly describe two main strategies cells use to control metabolism.
29. Explain why grouping genes into an operon can be advantageous.
30. Using the *trp* operon as an example, explain the concept of an operon and the function of the operator, repressor and corepressor.
31. Distinguish between structural and regulatory genes.
32. Describe how the *lac* operon functions and explain the role of the inducer, allolactose.
33. Explain how repressible and inducible enzymes differ and how these differences reflect differences in the pathways they control.
34. Distinguish between positive and negative control, and give examples of each from the *lac* operon.
35. Explain how CAP is affected by glucose concentration.
36. Describe how *E. coli* uses the negative and positive controls of the *lac* operon to economize on RNA and protein synthesis.

KEY TERMS

A. Mayer
D. Ivanowsky
Martinus Beijerinck
Wendell Stanley
virus
tobacco mosaic virus
virion
capsid
envelope
bacteriophage
T-even phage
obligate parasite
reverse transcriptase
reverse transcription
host range
lytic cycle
lysogenic cycle
virulent
restriction enzymes
temperate virus

phage λ
prophage
lysogenic conversion
horizontal
   transmission
vertical
   transmission
viroids
Herpes virus
retrovirus
HIV virus
provirus
vaccine
oncogene
nucleoid region
plasmid
binary fission
transformation
   (in bacteria)

transduction
conjugation
F factor
F^+ cell
F^- cell
Hfr cell
pili
conjugation tube
episome
nonepisomal plasmid
R plasmid
transposon
Barbara McClintock
transposase
inverted repeat (IR)
insertion sequence (IS)
complex transposon
direct repeat
feedback inhibition

gene repression
structural gene
regulatory gene
operon
polycistronic mRNA
operator
repressor
corepressor
inducer
Francois Jacob
Jacques Monod
lac operon
trp operon
negative control
positive control
CAP (catabolite
   activator protein)
cAMP

LECTURE NOTES

Scientists discovered the role of DNA in heredity by studying the simplest of biological systems — viruses and bacteria. Most of the molecular principles discovered through microbe research applies to higher organisms, but viruses and bacteria also have unique genetic features.

- Knowledge of these unique genetic features has helped scientists understand how viruses and bacteria cause disease.
- Techniques for gene manipulation emerged from studying genetic peculiarities of microorganisms.

I. Researchers discovered viruses by studying a plant disease: *science as a process*

The discovery of viruses resulted from the search for the infectious agent causing tobacco mosaic disease. This disease stunts the growth of tobacco plants and gives their leaves a mosaic coloration.

1883: A. Mayer, a German scientist demonstrated that the disease was contagious and proposed that the infectious agent was an unusually small bacterium that could not be seen with a microscope.

- He successfully transmitted the disease by spraying sap from infected plants onto the healthy ones.
- Using a microscope, he examined the sap and was unable to identify a microbe.

1890's: D. Ivanowsky, a Russian scientist proposed that tobacco mosaic disease was caused by a bacterium that was either too small to be trapped by a filter or that produced a filterable toxin.

- To remove bacteria, he filtered sap from infected leaves.
- Filtered sap still transmitted disease to healthy plants.

1897: Martinus Beijerinck, a Dutch microbiologist proposed that the disease was caused by a reproducing particle much smaller and simpler than a bacterium.

- He ruled out the theory that a filterable toxin caused the disease by demonstrating that the infectious agent in filtered sap could reproduce.

Plants were sprayed with filtered sap from
diseased plant.

↓

Sprayed plants developed tobacco mosaic
disease.

↓

Sap from newly infected plants was used to
infect others.

- This experiment was repeated for several generations. He concluded that the pathogen must be reproducing because its ability to infect was undiluted by transfers from plant to plant.
- He also noted that unlike bacteria, the pathogen:
  ⇒ Reproduced only within the host it infected.
  ⇒ Could not be cultured on media.
  ⇒ Could not be killed by alcohol.

1935: Wendell M. Stanley, an American biologist, crystallized the infectious particle now known as *tobacco mosaic virus* (*TMV*).

II. Most viruses consist of a genome enclosed in a protein shell

In the 1950's, TMV and other viruses were finally observed with electron microscopes. Viral structure appeared to be unique from the simplest of cells.
- The smallest viruses are only 20 nm in diameter.
- The virus particle, or *virion*, is just nucleic acid enclosed by a protein coat.

A. Viral Genomes

Depending upon the virus, viral genomes:
- May be double-stranded DNA, single-stranded DNA, double-stranded RNA or single-stranded RNA.
- Are organized as single nucleic acid molecules that are either linear or circular.
- May have as few as four genes or as many as several hundred.

B. Capsids and Envelopes

Capsid = Protein coat that encloses the viral genome.
- Its structure may be rod-shaped, polyhedral or complex.
- Composed of many *capsomeres* – protein subunits made from only one or a few types of protein.

Envelope = Membrane that cloaks some viral capsids.
- Helps viruses infect their host.
- Derived from host cell membrane which is usually virus-modified and contains proteins and glycoproteins of viral origin.

The most complex capsids are found among *bacteriophages* or bacterial viruses.

- Of the first phages studied, seven infected *E. coli*. These were named types 1 – 7 (T1, T2, T3 ... T7).
- The T-even phages – T2, T4 and T6 – are structurally very similar.
  ⇒ The icosohedral head encloses the genetic material.
  ⇒ The protein tailpiece with tail fibers attaches the phage to its bacterial host and injects its DNA into the bacterium.

Structure of the
T-Even Phage

III. Viruses can only reproduce within a host cell

Viral reproduction differs markedly from cellular reproduction, because viruses are *obligate intracellular parasites* which can express their genes and reproduce only within a living cell. Each virus has a specific *host range*.

Host range = Limited number or range of host cells that a parasite can infect.
- Viruses recognize host cells by a complementary fit between external viral proteins and specific cell surface *receptor sites*.
- Some viruses have broad host ranges which may include several species (e.g. swine flu and rabies).
- Some viruses have host ranges so narrow that they can:
  ⇒ infect only one species (e.g. phages of *E. coli*)
  ⇒ infect only a single tissue type of one species (e.g. human cold virus infects only cells of the upper respiratory tract; AIDS virus binds only to specific receptors on certain white blood cells).

There are many patterns of viral life cycles, but they all generally include:
- Infecting the host cell with viral genome.
- Coopting host cell's resources to:
  ⇒ replicate the viral genome.
  ⇒ manufacture capsid protein.
- Assembling newly produced viral nucleic acid and capsomeres into the next generation of viruses.

There are several mechanisms used to infect host cells with viral DNA.

- For example, T-even phages use an elaborate tailpiece to inject DNA into the host cell.
- Once the viral genome is inside its host cell, it commandeers the host's resources and reprograms the cell to copy the viral genes and manufacture capsid protein.

There are three possible patterns of viral genome replication:

1. DNA → DNA. If viral DNA is double-stranded, DNA replication resembles that of cellular DNA, and the virus uses DNA polymerase produced by the host.

2. RNA → RNA. Since host cells lack the enzyme to copy RNA, most RNA viruses contain a gene that codes for *RNA replicase*, an enzyme that uses viral RNA as a template to produce complementary RNA.

3. RNA → DNA → RNA. Some RNA viruses encode *reverse transcriptase*, an enzyme that transcribes DNA from an RNA template.

Regardless of how viral genomes replicate, all viruses divert host cell resources for viral production.

- Viral genes use the host cell's enzymes, ribosomes, tRNAs, amino acids, ATP and other resources to make copies of the viral genome and produce viral capsid proteins.
- These viral components – nucleic acid and capsids – are assembled into hundreds or thousands of virions, which leave to parasitize new hosts.

Viral nucleic acid and capsid proteins assemble spontaneously into new virus particles, a process called *self-assembly*.

- Since most viral components are held together by weak bonds (e.g. hydrogen bonds and Van der Waals forces), enzymes are not usually necessary for assembly.
- For example, TMV can be disassembled in the laboratory. When mixed together, the RNA and capsids spontaneously reassemble to form complete TMV virions.

IV. Phages exhibit two reproductive cycles, the lytic and lysogenic cycles

Bacteriophages are the best understood of all viruses, and many of the important discoveries in molecular biology have come from bacteriophage studies.

- In the 1940's, scientists determined how the T phages reproduce within a bacterium; this research:
  ⇒ Demonstrated that DNA is the genetic material.
  ⇒ Established the phage-bacterium system as an important experimental tool.
- Studies on lambda (λ) phage of *E. coli* showed that double-stranded DNA viruses reproduce by two alternative mechanisms: the *lytic cycle* and the *lysogenic cycle*.

A. The Lytic Cycle

Virulent bacteriophages reproduce by a *lytic* replication cycle.

<u>Virulent phages</u> = Phages that lyse their host cells.

<u>Lytic cycle</u> = A viral replication cycle that results in the death or lysis of the host cell.

The lytic cycle of phage T4 illustrates this type of replication cycle:

1. Phage attaches to cell surface.
   - T4 recognizes a host cell by a complementary fit between proteins on the virion's tail fibers and specific receptor sites on the outer surface of an *E. coli* cell.

2. Phage contracts sheath and injects DNA.
   - ATP stored in the phage tailpiece is the energy source for the phage to: a) pierce the *E. coli* wall and membrane, b) contract its tail sheath, and c) inject its DNA.
   - The genome separates from the capsid leaving a capsid "ghost" outside the cell.

3. Hydrolytic enzymes destroy host cell's DNA.
   - The *E. coli* host cell begins to transcribe and translate the viral genome.
   - One of the first viral proteins produced is an enzyme that degrades host DNA. The phage's own DNA is protected, because it contains modified cytosine not recognized by the enzyme.

4. Phage genome directs the host cell to produce phage components: DNA and capsid proteins.
   - Using nucleotides from its own degraded DNA, the host cell makes many copies of the phage genome.
   - The host cell also produces three sets of capsid proteins and assembles them into phage tails, tail fibers and polyhedral heads.
   - Phage components spontaneously assemble into virions.

5. Cell lyses and releases phage particles.
   - Lysozymes specified by the viral genome digest the bacterial cell wall.
   - Osmotic swelling lyses the cell which releases hundreds of phages from their host cell.

- Released virions can infect nearby cells.
- Lytic cycle takes only 20 – 30 minutes at 37°C. In that period, a T4 population can increase a hundredfold.

Bacteria have several defenses against destruction by phage infection.

- Bacterial mutations can change receptor sites used by phages for recognition, and thus avoid infection.
- Bacterial *restriction enzymes* recognize and cut up foreign DNA, including certain phage DNA. Bacterial DNA is chemically altered, so it is not destroyed by the cell's own restriction enzymes.

Restriction Enzymes = Naturally occurring bacterial enzymes that protect bacteria against intruding DNA from other organisms; catalyze restriction – the process of cutting foreign DNA into small segments.

Bacterial hosts and their viral parasites are continually *coevolving*.

- Most successful bacteria have effective mechanisms for preventing phage entry or reproduction.
- Most successful phages have evolved ways around bacterial defenses.
- Many phages check their own destructive tendencies and may coexist with their hosts.

B. The Lysogenic Cycle

Some viruses can coexist with their hosts by incorporating their genome into the host's genome.

Temperate viruses = Viruses that can integrate their genome into a host chromosome and remain latent until they initiate a lytic cycle.

- They have two possible modes of reproduction, the lytic cycle and the *lysogenic cycle*.
- An example is phage λ, discovered by E. Lederberg in 1951. (See Campbell, Figure 17.5)

Lysogenic cycle = A viral replication cycle that involves the incorporation of the viral genome into the host cell genome.

Details of the lysogenic cycle were discovered through studies of phage λ life cycle:

1. Phage λ binds to the surface of an *E. coli* cell.
2. Phage λ injects its DNA into the bacterial host cell.
3. λ DNA forms a circle and either begins a lytic cycle or a lysogenic cycle.
4. During a lysogenic cycle, λ DNA inserts by genetic recombination into a specific site on the bacterial chromosome and becomes a *prophage*.

<u>Prophage</u> = A phage genome that is incorporated into a specific site on the bacterial chromosome.

- Most prophage genes are inactive.
- One active prophage gene codes for the production of *repressor protein* which switches off most other prophage genes.
- Prophage genes are copied along with cellular DNA when the host cell reproduces. As the cell divides, both prophage and cellular DNA are passed on to daughter cells.
- A prophage may be carried in the host cell's chromosomes for many generations.

Occasionally, a prophage may leave the bacterial chromosome.

- This may be spontaneous or caused by environmental factors (e.g. radiation).
- The excision process may begin the phage's lytic reproductive cycle.
- Virions produced during the lytic cycle may begin either a lytic or lysogenic cycle in their new host cells.

<u>Lysogenic cell</u> = Host cell carrying a prophage in its chromosome.

- It is called lysogenic because it has the potential to lyse.
- Some prophage genes in a lysogenic cell may be expressed and change the cell's phenotype in a process called *lysogenic conversion.*
- Lysogenic conversion occurs in bacteria that cause diphtheria, botulism and scarlet fever. Pathogenicity results from toxins coded for by prophage genes.

V. Animal viruses are diverse in their modes of infection and mechanisms of replication

A. Reproductive Cycles of Animal Viruses

Replication cycles of animal viruses may show some interesting variations from those of other viruses. Two examples are the replication cycles of: 1) viruses with envelopes and 2) viruses with RNA genomes. (See Campbell, Table 17.1 for families of animal viruses grouped by type of nucleic acid.)

1. Viruses with Envelopes

Some animal viruses are surrounded by a membranous envelope, which is unique to several groups of animal viruses. This envelope is:

- Outside the capsid and helps the virus enter host cells.
- A lipid bilayer with glycoprotein spikes protruding from the outer surface.

Enveloped viruses have replication cycles characterized by: (See Campbell, Figure 17.6)

    a. *Attachment*. Glycoprotein spikes protruding from the viral envelope attach to receptor sites on the host's plasma membrane.

    b. *Entry*. As the envelope fuses with the plasma membrane, the entire virus (capsid and genome) is transported into the cytoplasm by receptor-mediated endocytosis.

    c. *Uncoating*. Cellular enzymes uncoat the genome by removing the protein capsid from viral RNA.

    d. *Viral RNA and Protein Synthesis*. Viral enzymes are required to replicate the RNA genome and to transcribe mRNA.

- Some viral RNA polymerase is packaged in the virion.
- Viral RNA polymerase (transcriptase) replicates the viral genome and transcribes viral mRNA. Note that the viral genome is a strand complementary to mRNA.
- Viral mRNA is translated into viral proteins including:
  ⇒ Capsid proteins synthesized in the cytoplasm by free ribosomes.
  ⇒ Viral-envelope glycoproteins synthesized by ribosomes bound to rough ER. Glycoproteins produced in the host's ER are sent to the Golgi apparatus for further processing. Golgi vesicles transport the glycoproteins to the plasma membrane, where they cluster at exit sites for the virus.

    e. *Assembly and Release*. New capsids surround viral genomes. Once assembled, the virions envelop with host plasma membrane as they bud off from the cell's surface. The viral envelope is derived from:

- Host cell's plasma membrane lipid.
- Virus-specific glycoprotein.

Some viral envelopes are not derived from host plasma membrane. For example herpesviruses are double-stranded DNA viruses which:

- Contain envelopes derived from the host cell's nuclear envelope rather than from the plasma membrane.
- Reproduce within the host cell's nucleus.
- Use both viral and cellular enzymes to replicate and transcribe their genomic DNA.
- May integrate their DNA into the cell's genome as a *provirus*. Evidence comes from the nature of herpes infections, which tend to recur. After a period of latency, physical or emotional stress may cause the proviruses to begin a productive cycle again.

Provirus = Viral DNA that inserts into a host cell chromosome.

2. RNA Viruses

All possible types of viral genomes are represented among animal viruses. Since mRNA is common to all types, DNA and RNA viruses are classified according to the relationship of their mRNA to the genome. In this classification:

- mRNA or the strand that corresponds to mRNA is the *plus (+) strand*; it has the nucleotide sequence that codes for proteins.
- The *minus (–) strand* is a template for synthesis of a plus strand; it is complementary to the sense strand or mRNA.

Animal RNA viruses are classified as following:

- *Class III RNA viruses*. Double-stranded RNA genome; the minus strand is the template for mRNA. (Reoviruses)
- *Class IV RNA viruses*. Single plus strand genome; the plus strand can function directly as mRNA, but also is a template for synthesis of minus RNA. (Minus RNA is a template for synthesis of additional plus strands.) Viral enzymes are required for RNA synthesis from RNA templates. (Picornavirus, Togavirus)
- *Class V RNA viruses*. Single minus strand genome; mRNA is transcribed directly from this genomic RNA. (Rhabdovirus, Paramyxovirus, Orthomyxovirus)
- *Class VI RNA viruses*. Single plus strand genome; the plus strand is a template for complementary DNA synthesis. *Reverse transcriptase* catalyzes this reverse transcription from RNA to DNA. mRNA is then transcribed from a DNA template. (*Retroviruses*)

Retrovirus = (Retro = backward) RNA virus that uses *reverse transcriptase* to transcribe DNA from the viral RNA genome.

- *Reverse transcriptase* is a type of DNA polymerase that transcribes DNA from an RNA template.
- *HIV* (*human immunodeficiency virus*), the virus that causes *AIDS* (*acquired immunodeficiency syndrome*) is a retrovirus.

RNA viruses with the most complicated reproductive cycles are the retroviruses, because retroviruses must first carry out *reverse transcription*: (See Campbell, Figure 17.7)

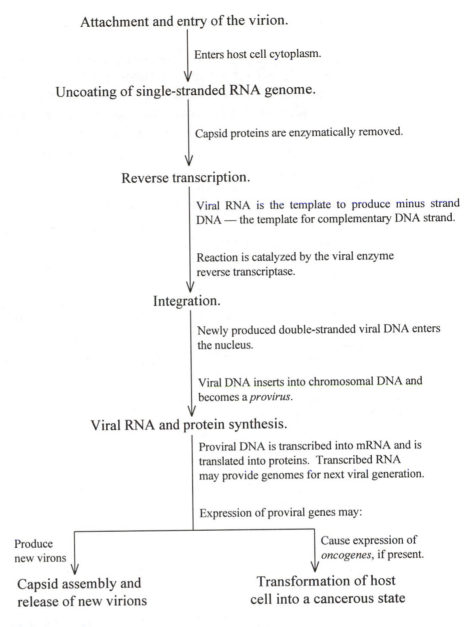

Attachment and entry of the virion.

Enters host cell cytoplasm.

Uncoating of single-stranded RNA genome.

Capsid proteins are enzymatically removed.

Reverse transcription.

Viral RNA is the template to produce minus strand DNA — the template for complementary DNA strand.

Reaction is catalyzed by the viral enzyme reverse transcriptase.

Integration.

Newly produced double-stranded viral DNA enters the nucleus.

Viral DNA inserts into chromosomal DNA and becomes a *provirus*.

Viral RNA and protein synthesis.

Proviral DNA is transcribed into mRNA and is translated into proteins. Transcribed RNA may provide genomes for next viral generation.

Expression of proviral genes may:

Produce new virons

Capsid assembly and release of new virions

Cause expression of *oncogenes*, if present.

Transformation of host cell into a cancerous state

B. Important Viral Diseases in Animals

It is often unclear how certain viruses cause disease symptoms. Viruses may:

- Damage or kill cells. In response to a viral infection, lysosomes may release hydrolytic enzymes.
- Be toxic themselves or cause infected cells to produce toxins.
- Cause varying degrees of cell damage depending upon regenerative ability of the infected cell. We recover from colds because infected cells of the upper respiratory tract can regenerate by cell division. Poliovirus, however, causes permanent cell damage because the virus attacks nerve cells which cannot divide.
- Be indirectly responsible for disease symptoms. Fever, aches and inflammation may result from activities of the immune system.

Medical weapons used to fight viral infections include *vaccines* and *antiviral drugs*.

<u>Vaccines</u> = Harmless variants or derivatives of pathogenic microbes that mobilize a host's immune mechanism against the pathogen.

- Edward Jenner developed the first vaccine (against smallpox) in 1796. According to the WHO, a vaccine has almost completely eradicated smallpox.

- Effective vaccines now exist for polio, rubella, measles, mumps and many other viral diseases.

While vaccines can prevent some viral illnesses, little can be done to cure a viral disease once it occurs. Some *antiviral drugs* have recently been developed.

- Several are analogs of purine nucleosides that interfere with viral nucleic acid synthesis (e.g. adenine arabinoside and acyclovir).

C. Emerging Viruses

Emerging viruses are viruses that make an apparent sudden appearance. In reality, they are not likely to be new viruses, but rather existing ones that have expanded their host territory. Emerging viral diseases can arise if an existing virus:

1. *Evolves and thus causes disease in individuals who have immunity only to the ancestral virus.* (e.g. influenza virus)

2. *Spreads from one host species to another.*
   - For example, monkeypox virus spread from African to Asian monkeys in the laboratory (1950's); and, in Zaire (1970's), spread to humans from other mammals that harbored the virus.

3. *Disseminates from a small population to become more widespread.*
   - For example, the 1993 hantavirus outbreak in New Mexico was the result of a population explosion in deer mice that are the viral reservoirs. Humans became infected by inhaling airborne hantavirus that came from the excreta of deer mice.
   - AIDS, once a rare disease, has become a global epidemic. Technological and social factors influenced the spread of AIDS virus.

Environmental disturbances can increase the viral traffic responsible for emerging diseases. For example:

- Traffic on newly cut roads through remote areas can spread viruses among previously isolated human populations.

- Deforestation activities brings humans into contact with animals that may host viruses capable of infecting humans.

D. **Viruses and Cancer**

Some tumor viruses cause cancer in animals.
- When animal cells grown in tissue culture are infected with tumor viruses, they *transform* to a cancerous state.
- Examples are members of the retrovirus, papovavirus, adenovirus and herpesvirus groups.
- Certain viruses are implicated in human cancers:

Viral Group	Examples/Diseases	Cancer Type
Retrovirus	HTLV-1/adult leukemia	Leukemia
Herpesvirus	Epstein-Barr/infectious mononucleosis	Burkitt's lymphoma
Papovavirus	Papilloma/human warts	Cervical cancer
Hepatitis B virus	Chronic hepatitis	Liver cancer

Tumor viruses transform cells by inserting viral nucleic aids into host cell DNA.
- This insertion is permanent as the provirus never excises.
- Insertion for DNA tumor viruses is straightforward.

Several viral genes have been identified as *oncogenes*.

Oncogenes = Genes found in viruses or as part of the normal eukaryotic genome, that trigger transformation of a cell to a cancerous state.
- Code for cellular growth factors or for proteins involved in the function of growth factors.
- Are not unique to tumor viruses, but are found in the normal cells of many species. In fact, some tumor viruses transform cells by activating cellular oncogenes.

More than one oncogene must usually be activated to completely transform a cell.
- Indications are that tumor viruses are effective only in combination with other events such as exposure to carcinogens.
- Carcinogens probably also act by turning on cellular oncogenes.

VI. **Plant viruses are serious agricultural pests**

As serious agricultural pests, many of the plant viruses:
- Stunt plant growth and diminish crop yields.
- Are RNA viruses.
- Have rod-shaped capsids with *capsomeres* arranged in a spiral.

Capsomere = Complex capsid subunit consisting of several identical or different protein molecules.

Plant viruses spread from plant to plant by two major routes: *horizontal transmission* and *vertical transmission*.

Horizontal transmission = Route of viral transmission in which an organism receives the virus from an external source.

- Plants are more susceptible to viral infection if their protective epidermal layer is damaged.
- Insects may be *vectors* that transmit viruses from plant to plant and can inject the virus directly into the cytoplasm.
- By using contaminated tools, gardeners and farmers may transmit plant viruses.

Vertical transmission = Route of viral transmission in which an organism inherits a viral infection from its parent.

- Can occur in asexual propagation of infected plants (e.g. by taking cuttings).
- Can occur in sexual reproduction via infected seeds.

Once a plant is infected, viruses reproduce and spread from cell to cell by passing through plasmodesmata.

Most plant viral diseases have no cure, so current efforts focus on reducing viral propagation and breeding resistant plant varieties.

VII. Viroids and prions are infectious agents even simpler than viruses

Another class of plant pathogens called *viroids* are smaller and simpler than viruses.
- They are small naked RNA molecules with only several hundred nucleotides.
- It is likely that viroids disrupt normal plant metabolism, development and growth by causing errors in regulatory systems that control gene expression.
- Viroid diseases affect many commercially important plants such as coconut palms, chrysanthemums, potatoes and tomatoes.

Some scientists believe that viroids originated as escaped introns.
- Nucleotide sequences of viroid RNA are similar to self-splicing introns found within some normal eukaryotic genes, including rRNA genes.
- An alternative hypothesis is that viroids and self-splicing introns share a common ancestral molecule.

As nucleic acids, viroids self-direct their replication and thus are not diluted during transmission from host to host. Molecules other than nucleic acids can be infectious agents even though they cannot self-replicate.

- For example, *prions* are pathogens that are proteins.
    - ⇒ Cause scrapie in sheep.
    - ⇒ May cause degenerative diseases of the nervous system in humans.
- *How can a protein which cannot replicate itself, be an infectious pathogen?* According to one hypothesis:
    - ⇒ Prions are defective versions of normally occurring cellular proteins.
    - ⇒ When prions infect normal cells, they catalyze conversion of normal protein to the prion version.
    - ⇒ Prions could thus trigger chain reactions that increase their numbers and allow them to spread through a host population without dilution.

VIII. Viruses may have evolved from other mobile genetic elements

Viruses do not fit our usual definitions of living organisms. They cannot reproduce independently, yet they:

- Have a genome with the same genetic code as living organisms.
- Can mutate and evolve.

Viruses probably evolved after the first cells, from fragments of cellular nucleic acid that were mobile genetic elements. Evidence to support this includes:

- Genetic material of different viral families is more similar to host genomes than to that of other viral families.
- Some viral genes are identical to cellular genes (e.g. oncogenes in retroviruses).
- Viruses of eukaryotes are more similar in genomic structure to their cellular hosts than to bacterial viruses.
- Viral genomes are similar to certain cellular genetic elements such as plasmids and transposons; they are all *mobile* genetic elements.

IX. The short generation span of bacteria facilitates their evolutionary adaptation to changing environments

The average bacterial genome is larger than a viral genome, but much smaller than a typical eukaryotic genome.

- Though prokaryotes contain only about 1/1000 the DNA of eukaryotes, prokaryotic chromosomes still contain a large amount of DNA relative to the small prokaryotic cell.
- Bacterial chromosomes, consequently, are highly folded and packed within the cell.

Most DNA in a bacterium is found in a single circular *bacterial chromosome* (*genophore*) that is:

- Composed of double-stranded DNA.
- Structurally simpler and has fewer associated proteins than a eukaryotic chromosome.
- Found in the *nucleoid* region. Since this region is not separated from the rest of the cell (by a membrane), transcription and translation can occur simultaneously.

Many bacteria also contain extrachromosomal DNA in *plasmids*.

Plasmid = A small double-stranded ring of DNA that carries extrachromosomal genes in some bacteria.

Most bacteria can rapidly reproduce by *binary fission*, which is preceded by DNA replication.

- Semi-conservative replication of the bacterial chromosome begins at a single origin of replication.
- The two replication forks move bidirectionally until they meet and replication is complete. (See Campbell, Figure 17.9)
- Under optimal conditions, some bacteria can divide in twenty minutes. Because of this rapid reproductive rate, bacteria are useful for genetic studies.

Binary fission is asexual reproduction that produces clones - daughter cells that are genetically identical to the parent.

- Though mutations are rare events, they can impact genetic diversity in bacteria because of their rapid reproductive rate.
- Though mutation can be a major source of genetic variation in bacteria, it is *not* a major source in more slowly reproducing organisms (e.g. humans). In most higher organisms, genetic recombination from sexual reproduction is responsible for most of the genetic diversity within populations.

X. Genetic recombination and transposition produce new bacterial strains

There are three natural processes of genetic recombination in bacteria: *transformation*, *transduction* and *conjugation*. These mechanisms of gene transfer occur separately from bacterial reproduction; and, in addition to mutation, are another major source of genetic variation in bacterial populations.

A. Transformation

Transformation = Process of gene transfer during which a bacterial cell assimilates foreign DNA from the surroundings.

- Some bacteria can take up naked DNA from the surroundings. (Refer to Avery's experiments with *Streptococcus pneumoniae* in Chapter 15.)
- Assimilated foreign DNA may be integrated into the bacterial chromosome by recombination (crossing over).
- Progeny of the recipient bacterium will carry a new combination of genes.

Many bacteria have surface proteins that recognize and import naked DNA from closely related bacterial species.

- Though lacking such proteins, *E. coli* can be artificially induced to take up foreign DNA by incubating the bacteria in a culture medium that has a high concentration of calcium ions.
- This technique of artificially inducing transformation is used by the biotechnology industry to introduce foreign genes into bacterial genomes, so that bacterial cells can produce proteins characteristic of other species (e.g. insulin and human growth hormone).

B.   Transduction

Transduction = Gene transfer from one bacterium to another by a bacteriophage.

Generalized transduction = Transduction that occurs when random pieces of host cell DNA are packaged within a phage capsid during the lytic cycle of a phage.

- This process can transfer almost any host gene and little or no phage genes.
- When the phage particle infects a new host cell, the donor cell DNA can recombine with the recipient cell DNA.

Specialized transduction = Transduction that occurs when a prophage excises from the bacterial chromosome and carries with it some host genes adjacent to the excision site. Also known as *restricted transduction*.

- Carried out only by temperate phages.
- Differs from general transduction in that:
  - ⇒ Specific host genes and most phage genes are packed into the same virion.
  - ⇒ Transduced bacterial genes are restricted to specific genes adjacent to the prophage insertion site. In general transduction, host genes are randomly selected and almost any host gene can be transferred.

C.   Conjugation and Plasmids

Conjugation = The direct transfer of genes between two cells that are temporarily joined.

- Discovered by Joshua Lederberg and Edward Tatum.
- Conjugation in *E. coli* is one of the best-studied examples:

A DNA-donating *E. coli* cell extends
external appendages called *sex pili*.

↓

Sex pili attach to a DNA-receiving cell.

↓

A cytoplasmic bridge forms through which
DNA transfer occurs.

The ability to form sex pili and to transfer DNA is conferred by genes in a plasmid called the *F plasmid*.

1. **General Characteristics of Plasmids**

   Plasmid = A small double-stranded ring of DNA that carries extrachromosomal genes in some bacteria.
   - Plasmids have only a few genes, and they are not required for survival and reproduction.
   - Plasmid genes can be beneficial in stressful environments. Examples include the F plasmid, which confers ability to conjugate; and the R plasmid, which confers antibiotic resistance.

   These small circular DNA molecules replicate independently:
   - Some plasmids replicate in synchrony with the bacterial chromosome, so only a few are present in the cell.
   - Some plasmids under more relaxed control can replicate on their own schedule, so the number of plasmids in the cell at any one time can vary from only a few to as many as 100.

   Some plasmids are *episomes* that can reversibly incorporate in the cell's chromosome.

   Episomes = Genetic elements that can replicate either independently as free molecules in the cytoplasm or as integrated parts of the main bacterial chromosome.
   - Examples include some plasmids, as well as, temperate viruses such as lambda phage.
   - Temperate phage genomes replicate separately in the cytoplasm during a lytic cycle, and as an integral part of the host's chromosome during a lysogenic cycle.

   While plasmids and viruses can both be episomes, they differ in that:
   - Plasmids, unlike viruses, lack an extracellular stage.
   - Plasmids are generally beneficial to the cell, while viruses are parasites that usually harm their hosts.

2. **The F Plasmid and Conjugation**

   The F plasmid (F for fertility) has about 25 genes, most of which are involved in the production of sex pili.
   - Bacterial cells that contain the F factor and can donate DNA ("male") are called F^+ cells.
   - The F factor replicates in synchrony with chromosomal DNA, so the F^+ factor is inheritable; that is, division of an F^+ cell results in two F^+ daughter cells.
   - Cells without the F factor are designated F^- ("female").

During conjugation between an F^+ and an F^- bacterium:

- The F factor replicates by *rolling circle replication*. The 5' end of the copy peels off the circular plasmid and is transferred in linear form.
- The F^+ cell transfers a copy of its F factor to the F^- partner, and the F^- cell becomes F^+.
- The donor cell remains F^+, with its original DNA intact.

The F factor is an episome and occasionally inserts into the bacterial chromosome.

- Integrated F factor genes are still expressed.
- Cells with integrated F factors are called *Hfr cells* (high frequency of recombination).

Conjugation can occur between an Hfr and an F^- bacterium.

- As the integrated F factor of the Hfr cell transfers to the F^- cell, it pulls the bacterial chromosome behind its leading end.
- The F factor always opens up at the same point for a particular Hfr strain. As rolling circle replication proceeds, the sequence of chromosomal genes behind the leading 5' end is always the same.
- The conjugation bridge usually breaks before the entire chromosome and tail end of the F factor can be transferred. As a result:
  ⇒ Only some bacterial genes are donated.
  ⇒ The recipient F^- cell does not become an F^+ cell, because only part of the F factor is transferred.
  ⇒ The recipient cell becomes a partial diploid.
  ⇒ Recombination occurs between the Hfr chromosomal fragment and the F^- cell. Homologous strand exchange results in a *recombinant F^- cell*.
  ⇒ Asexual reproduction of the recombinant F^- cell produces a bacterial colony that is genetically different from both original parental cells.

3. **Interrupting Conjugation to Map Bacterial Chromosomes**

Scientists have mapped gene sequence around the *E. coli* chromosome by using several features of conjugation.

- A specific strain of Hfr bacteria always transfers genes in the same sequence.
- The sequence of gene transfer results from where the F factor is inserted and how it is oriented in the bacterial chromosome. (See Campbell, Figure 17.12c.)
- The duration of conjugation determines how many chromosomal genes will be transferred.

By artificially interrupting conjugation at different time intervals, geneticists can roughly determine gene location on the bacterial chromosome. The experimental steps are outlined below.

1. Liquid cultures of an Hfr strain and an F⁻ strain with different alleles are mixed together.
2. At successive time intervals, a sample is taken and agitated in a blender to disrupt conjugating pairs.
3. Bacteria are then cultured and genetic analysis of the recombinants indicates which genes were transferred during the time period allowed for conjugation.
4. Gene sequence and relative distances between genes are deduced from the above information.

4. R Plasmids and Antibiotic Resistance

One class of nonepisomal plasmids, the *R plasmids* (for resistance), carry genes that confer resistance to certain antibiotics.

- Some carry up to 10 genes for resistance to antibiotics.
- During conjugation, some mobilize their own transfer to nonresistant cells.
- Increased antibiotic use has selected for antibiotic resistant bacterial strains carrying the R plasmid.
- Additionally, R plasmids can transfer resistance genes to bacteria of different species including pathogenic strains. As a consequence, resistant strains of pathogens are becoming more common.

D. Transposons

Pieces of DNA called *transposons*, or *transposable genetic elements*, can actually move from one location to another in a cell's genome.

Transposons = DNA sequences that can move from one chromosomal site to another.
- Occur as natural agents of genetic change in both prokaryotic and eukaryotic organisms.
- Were first proposed in the 1940's by Barbara McClintock, who deduced their existence in maize. Decades later, the importance of her discovery was recognized; in 1983, at the age of 81, she received the Nobel Prize for her work.

There are two patterns of transposition: a) *conservative transposition* and b) *replicative transposition*.

Conservative transposition = Movement of preexisting genes from one genomic location to another; the transposon's genes are not replicated before the move, so the number of gene copies is conserved.

Replicative transposition = Movement of gene copies from their original site of replication to another location in the genome, so the transposon's genes are inserted at some new site without being lost from the original site.

Transposition is fundamentally different from all other mechanisms of genetic recombination, because transposons may scatter certain genes throughout the genome with no apparent single, specific target.

- All other mechanisms of genetic recombination depend upon homologous strand exchange: meiotic crossing over in eukaryotes; and transformation, transduction, and conjugation in prokaryotes.
- Insertion of episomic plasmids into chromosomes is also site specific, even though it does not require an extensive stretch of DNA homologous to the plasmid.

1. Insertion Sequences

The simplest transposons are *insertion sequences*.

Insertion sequences (IS) = The simplest transposons, which contain only the genes necessary for the process of transposition. Insertion sequence DNA includes two essential types of nucleotide sequences:

    a. Nucleotide sequence coding for *transposase*.

    b. *Inverted repeats*.

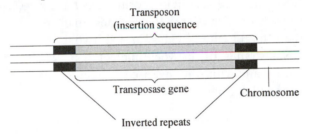

Transposase = Enzyme that catalyzes insertion of transposons into new chromosomal sites.

- The transposase gene in an insertion sequence is flanked by *inverted repeats*.

Inverted repeats (IR) = Short noncoding nucleotide sequences of DNA that are repeated in reverse order on opposite ends of a transposon. For example:

DNA strand #1	...A T C C G G T...	...A C C G G A T...
DNA strand #2	...A T C C G G T...	...A C C G G A T...

Note that each base sequence (IR) is repeated in reverse, on the DNA strand *opposite* the inverted repeat at the other end. Inverted repeats:

- Contain only 20 to 40 nucleotide pairs.
- Are recognition sites for transposase.

Transposase catalyzes the recombination by:

- Binding to the inverted repeats and holding them close together.
- Cutting and resealing DNA required for insertion of the transposon at a new site.

Insertion of transposons also requires other enzymes, such as DNA polymerase. For example,

- At the target site, transposase makes staggered cuts in the two DNA strands, leaving short segments of unpaired DNA at each end of the cut.
- Transposase inserts the transposon into the open target site.

- DNA polymerase helps form *direct repeats*, which flank transposons in their target site. Gaps in the two DNA strands fill in when nucleotides base pair with the exposed single-stranded regions.

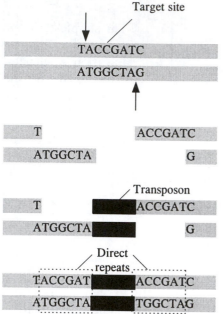

<u>Direct repeats</u> = Two or more identical DNA sequences in the same molecule.

- The transposition process creates direct repeats that flank transposons in their target site.

Transposed insertion sequences are likely to somehow alter the cell's phenotype; they may:

- Cause mutations by interrupting coding sequences for proteins.
- Increase or decrease a protein's production by inserting within regulatory regions that control transcription rates.

Transposition of insertion sequences probably plays a significant role in bacterial evolution as a source of genetic variation.

- Though insertion sequences only rarely cause mutations (about one in every 10^6 generations), the mutation rate from transpositions is about the same as the mutation rate from extrinsic causes, such as radiation and chemical mutagens.

2. Complex Transposons and R Plasmids

<u>Complex transposons</u> = Transposons which include additional genetic material besides that required for transposition; consist of one or more genes flanked by insertion sequences.

- The additional DNA may have any nucleotide sequence.
- Can insert into almost any stretch of DNA since their insertion is not dependent upon DNA sequence homology.
- Generate genetic diversity in bacteria by moving genes from one chromosome, or even one species, to another. This diversity may help bacteria adapt to new environmental conditions.
- An example is a transposon that carries a bacterial gene for antibiotic resistance.

Examples of genetic elements that contain one or more complex transposons include:

- F factor.
- R plasmids. (See Campbell, Figure 17.18c)
- DNA version of the retrovirus genome.

XI. The control of gene expression enables individual bacteria to adjust their metabolism to environmental change

NOTE: This material is difficult to teach. The problem is that there are so many components to track, and the students are not familiar enough with the vocabulary. Figures 17.17 and 17.18, which are available as transparencies in the Campbell set, are good teaching aids for the *trp* and *lac* operons. It is also helpful to construct flow charts where there are decision points (e.g. glucose present – glucose absent), so that students can visually follow alternate paths and their respective consequences.

Genes switch on and off as conditions in the intracellular environment change. Bacterial cells have two main ways of controlling metabolism:

1. *Regulation of enzyme activity.* The catalytic activity of many enzymes increases or decreases in response to chemical cues.

- For example, the end product of an anabolic pathway may turn off its own production by inhibiting activity of an enzyme at the beginning of the pathway (*feedback inhibition*).
- Useful for immediate short-term response.

2. *Regulation of gene expression.* Enzyme concentrations may rise and fall in response to cellular metabolic changes that switch genes on or off.

- For example, accumulation of product may trigger a mechanism that inhibits transcription of mRNA production by genes that code for an enzyme at the beginning of the pathway (*gene repression*).
- Slower to take effect than feedback inhibition, but is more economical for the cell. It prevents unneeded protein synthesis for enzymes, as well as, unneeded pathway product.

An example illustrating regulation of a metabolic pathway is the tryptophan pathway in *E. coli.* (See Campbell, Figure 17.16) Mechanisms for gene regulation were first discovered for *E. coli*, and current understanding of such regulatory mechanisms at the molecular level is still limited to bacterial systems.

A.   Operons: The Basic Concept

Regulated genes can be switched on or off depending on the cell's metabolic needs. From their research on the control of lactose metabolism in *E. coli*, Francois Jacob and Jacques Monod proposed a mechanism for the control of gene expression, the *operon* concept.

Structural gene = Gene that codes for a polypeptide.

Operon = A regulated cluster of adjacent *structural genes* with related functions.
- Common in bacteria and phages.
- Has a single promoter region, so an RNA polymerase will transcribe all structural genes on an all-or-none basis.
- Transcription produces a single *polycistronic* mRNA with coding sequences for all enzymes in a metabolic pathway (e.g. tryptophan pathway in *E. coli*).

Polycistronic mRNA = A large mRNA molecule that is a transcript of several genes.
- Is translated into separate polypeptides.
- Contains stop and start codons for the translation of each polypeptide.

Grouping structural genes into operons is advantageous because:
- Expression of these genes can be coordinated. When a cell needs the product of a metabolic pathway, all the necessary enzymes are synthesized at one time.
- The entire operon can be controlled by a single *operator*.

Operator = A DNA segment between an operon's promoter and structural genes, which controls access of RNA polymerase to structural genes.
- Sometimes overlaps the transcription starting point for the operon's first structural gene.
- Acts as an on/off switch for movement of RNA polymerase and transcription of the operon's structural genes.

What determines whether an operator is in the "on" or "off" mode? By itself, the operator is on; it is switched off by a protein *repressor*.

<u>Repressor</u> = Specific protein that binds to an operator and blocks transcription of the operon.
- Blocks attachment of RNA polymerase to the promoter.
- Is similar to an enzyme, in that it:
  - ⇒ Has an active site with a specific conformation, which discriminates among operators. Repressor proteins are specific only for operators of certain operons.
  - ⇒ Binds *reversibly* to DNA.
  - ⇒ May have an allosteric site in addition to its DNA-binding site.
- Repressors are encoded by *regulatory genes*.

<u>Regulatory genes</u> = Genes that code for repressor or regulators of other genes.
- Are often located some distance away from the operons they control.
- Are involved in switching on or off the transcription of structural genes by the following process:

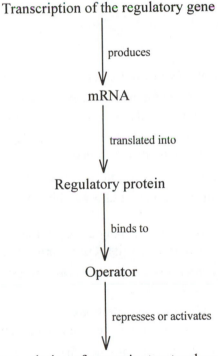

Transcription of the regulatory gene

↓ produces

mRNA

↓ translated into

Regulatory protein

↓ binds to

Operator

↓ represses or activates

Transcription of operon's structural genes

Regulatory genes are continually transcribed, so their activity depends upon how efficient their promoters are in binding RNA polymerase.
- They produce repressor molecules continuously, but slowly.
- Operons are still expressed even though repressor molecules are always present, because repressors are not always capable of blocking transcription; they alternate between inactive and active conformations.

A repressor's activity depends upon the presence of key metabolites in the cell.

- Regulation of the *trp* operon in *E. coli* is an example of how a metabolite cues a repressor: (See Campbell, Figure 17.17)
- Repressible enzymes catalyze the anabolic pathway that produces tryptophan, an amino acid.
- Tryptophan accumulation represses synthesis of the enzymes that catalyze its production.

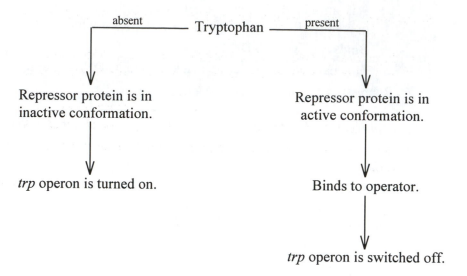

How does tryptophan activate the repressor protein?

- The repressor protein, which normally has a low affinity for the operator, has a DNA binding site plus an allosteric site specific for tryptophan.
- When tryptophan binds to the repressor's allosteric site, it activates the repressor causing it to change its conformation.
- The activated repressor binds to the operator, which switches the *trp* operon off.
- Tryptophan functions in this regulatory system as a *corepressor*.

Corepressor = A molecule, usually a metabolite, that binds to a repressor protein, causing the repressor to change into its active conformation.

- Only the *repressor-corepressor complex* can attach to the operator and turn off the operon.
- When tryptophan concentrations drop, it is less likely to be bound to repressor protein. The *trp* operon, once free from repression, begins transcription.
- As concentrations of tryptophan rise, it turns off its own production by activating the repressor.
- Enzymes of the tryptophan pathway are said to be *repressible*.

B.   Repressible Versus Inducible Enzymes: Two Types of Negative Gene Regulation

Repressible enzymes = Enzymes which have their synthesis inhibited by a metabolite (e.g. tryptophan).

Inducible enzymes = Enzymes which have their synthesis stimulated or induced by specific metabolites.

Some operons can be switched on or *induced* by specific metabolites (e.g. *lac* operon in *E. coli*).

- *E. coli* can metabolize the disaccharide lactose. Once lactose is transported into the cell, β-galactosidase cleaves lactose into glucose and galactose:

$$\text{lactose} \xrightarrow{\text{β-galactosidase}} \text{glucose + galactose}$$
$$\text{(disaccharide)} \qquad\qquad\qquad \text{(monosaccharides)}$$

- When *E. coli* is in a lactose-free medium, it only contains a few β-galactosidase molecules.
- When lactose is added to the medium, *E. coli* increases the number of mRNA molecules coding for β-galactosidase. These mRNA molecules are quickly translated into thousands of β-galactosidase molecules.
- Lactose metabolism in *E. coli* is programmed by the *lac* operon which has three structural genes:
    1. *lac Z* – Codes for β-galactosidase which hydrolyzes lactose.
    2. *lac Y* – Codes for a permease, a membrane protein that transports lactose into the cell.
    3. *lac A* – Codes for transacetylase, an enzyme that has no known role in lactose metabolism.
- The *lac* operon has a single promoter and operator. The *lac* repressor is innately active, so it attaches to the operon without a corepressor.
- Allolactose, an isomer of lactose, acts as an *inducer* to turn on the *lac* operon: (See Campbell, Figure 17.18)

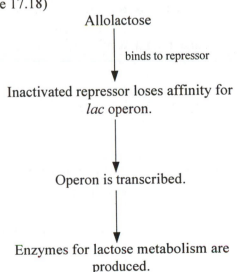

Allolactose

↓ binds to repressor

Inactivated repressor loses affinity for *lac* operon.

↓

Operon is transcribed.

↓

Enzymes for lactose metabolism are produced.

Differences between repressible and inducible operons reflect differences in the pathways they control.

Repressible Enzymes	Inducible Enzymes
Their genes are switched on until a specific metabolite activates the repressor.	Their genes are switched off until a specific metabolite inactivates the repressor.
Generally function in anabolic pathways.	Function in catabolic pathways.
Pathway end product switches off its own production by repressing enzyme synthesis.	Enzyme synthesis is switched on by the nutrient the pathway uses.

Repressible and inducible operons share similar features of gene regulation. In both cases:
- Specific repressor proteins control gene expression.
- Repressors can assume an active conformation that blocks transcription and an inactive conformation that allows transcription.
- Which form the repressor assumes depends upon cues from a metabolite.

Both systems are thus examples of *negative control*.
- Binding of active repressor to an operator always turns off structural gene expression.
- The *lac* operon is a system with negative control, because allolactose does not interact directly with the genome. The derepression allolactose causes is indirect, by freeing the *lac* operon from the repressor's negative effect.

Positive control of a regulatory system occurs only if an activator molecule interacts directly with the genome to turn on transcription.

C.   An Example of Positive Gene Regulation

The *lac* operon is under dual regulation which includes negative control by repressor protein and positive control by *catabolite activator protein*.

CAP (catabolite activator protein) = A protein that binds within an operon's promoter region and enhances the promoter's affinity for RNA polymerase.
- It is necessary for the normal expression of the *lac* operon. (Even if allolactose is present to inactivate the repressor, transcription proceeds slowly because the promoter has such a low affinity for RNA polymerase.)
- It is a *positive regulator* because it *directly* interacts with the genome to stimulate gene expression.
- It can bind to the promoter *only if glucose is absent* from the cell.

E. coli preferentially uses glucose over lactose as a substrate for glycolysis. So, normal expression of the lac operon requires:

- Presence of lactose.
- *Absence* of glucose.

How is CAP affected by the absence or presence of glucose?

- When glucose is missing, the cell accumulates *cyclic AMP* (*cAMP*), a nucleotide derived from ATP. cAMP activates CAP so that it can bind to the *lac* promoter.
- When glucose concentration rises, glucose catabolism decreases the intracellular concentration of cAMP. Thus, cAMP releases CAP.

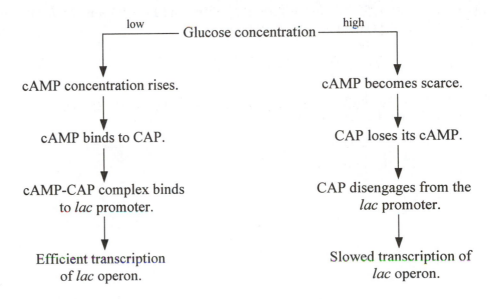

In this dual regulation of the *lac* operon:

- Negative control by the repressor determines whether or not the operon will transcribe the structural genes.
- Positive control by CAP determines the rate of transcription.

E. coli economizes on RNA and protein synthesis with the help of these negative and positive controls.

- CAP is an activator of several different operons that program catabolic pathways.
- Glucose's presence deactivates CAP. This, in turn, slows synthesis of those enzymes a cell needs to use catabolites other than glucose.
- *E. coli* preferentially uses glucose as its primary carbon and energy source, and the enzymes for glucose catabolism are coded for by unregulated genes that are continuously transcribed (constitutive).
- Consequently, when glucose is present, CAP does not work and the cell's systems for using secondary energy sources are inactive.

When glucose is absent, the cell metabolizes alternate energy sources.

- The cAMP level rises, CAP is activated and transcription begins of operons that program the use of alternate energy sources (e.g. lactose).
- Which operon is actually transcribed depends upon which nutrients are available to the cell. For example, if lactose is present, the *lac* operon will be switched on as allolactose inactivates the repressor.

REFERENCES

Alberts, B., D. Bray, J. Lewis, M. Raff, K. Roberts and J.D. Watson. *Molecular Biology of the Cell*. 3rd ed. New York: Garland, 1994.

Campbell, N. *Biology*. 4th ed. Menlo Park, California: Benjamin/Cummings, 1996.

Watson, J.D., N.H. Hopkins, J.W. Roberts, J.A. Steitz and A.M. Weiner. *Molecular Biology of the Gene*. 4th ed. Menlo Park, California: Benjamin/Cummings, 1987.

CHAPTER **18**

GENOME ORGANIZATION AND EXPRESSION IN EUKARYOTES

OUTLINE

I. Each cell of a multicellular eukaryote expresses only a small fraction of its genome

II. The structural organization of chromatin sets coarse controls on gene expression
   A. Nucleosomes, or "Beads on a String"
   B. Higher Levels of DNA Packing

III. Noncoding sequences and gene duplications account for much of a eukaryotic genome
   A. Repetitive Sequences
   B. Multigene Families

IV. The control of gene expression can occur at any step in the pathway from gene to functional protein
   A. Organization of a Typical Eukaryotic Gene: A Review
   B. Transcriptional Control
   C. Posttranscriptional Control
   D. Translational and Posttranslational Control
   E. The Arrangement of Coordinately Controlled Genes
   F. Gene Expression and Differentiation

V. Chemical signals that help control gene expression include hormones
   A. Chromosome Puffs: Evidence for the Regulatory Role of Steroid Hormones in Insects
   B. The Action of Steroid Hormones in Vertebrates

VI. Chemical modification or relocation of DNA within a genome can alter gene expression
   A. Gene Amplification and Selective Gene Loss
   B. Rearrangements in the Genome
   C. DNA Methylation

VII. Cancer can result from the abnormal expression of genes that regulate cell growth and division

OBJECTIVES

After reading this chapter and attending lecture, the student should be able to:

1. Compare the organization of prokaryotic and eukaryotic genomes.
2. Describe the current model for progressive levels of DNA packing.
3. Explain how histones influence folding in eukaryotic DNA.
4. Distinguish between heterochromatin and euchromatin.
5. Using the Barr body as an example, describe the function of heterochromatin in interphase cells.
6. Describe where satellite DNA is found and what role it may play in the cell.
7. Describe the role of telomeres in solving the end-replication problem with the lagging DNA strand.
8. Using the genes for rRNA as an example, explain how multigene families of identical genes can be advantageous for a cell.
9. Using α-globin and β-globin genes as examples, describe how multigene families of nonidentical genes probably evolve, including the role of transposition.
10. Explain the potential role that promoters and enhancers play in transcriptional control.
11. Explain why the nuclear envelope in eukaryotes offers a level of post-transcriptional control beyond that found in prokaryotes.
12. Explain why the ability to rapidly degrade mRNA can be an adaptive advantage for prokaryotes.
13. Describe the importance of mRNA degradation in eukaryotes, describe how it can be prevented.
14. Explain how gene expression may be controlled at the translational and post-translational level.
15. Compare the arrangement of coordinately controlled genes in prokaryotes and eukaryotes.
16. Explain how eukaryotic genes can be coordinately expressed and give some examples of coordinate gene expression in eukaryotes.
17. Provide evidence from studies of polytene chromosomes, that eukaryotic gene expression is controlled at transcription and that gene regulation responds to chemical signals such as steroid hormones.
18. Describe the key steps of steroid hormone action on gene expression in vertebrates.
19. In general terms, explain how genome plasticity can influence gene expression.
20. Describe the effects of gene amplification, selective gene loss and DNA methylation.
21. Explain how rearrangements in the genome can activate or inactivate genes.
22. Explain the genetic basis for antibody diversity.
23. Explain how DNA methylation may be a cellular mechanism for long-term control of gene expression and how it can influence early development.
24. Describe the normal control mechanisms that limit cell growth and division.
25. Briefly describe the four mechanisms that can convert proto-oncogenes to oncogenes.
26. Explain how changes in tumor-suppressor genes can be involved in transforming normal cells into cancerous cells.
27. Explain how oncogenes are involved in virus-induced cancers.

KEY TERMS

cellular differentiation

chromatin

histones

nucleosome

30-nm chromatin fiber

looped domains

heterochromatin

euchromatin

satellite DNA

telomere

multigene family

pseudogene

enhancer

gene amplification

chromosome diminution

DNA methylation

immunoglobulins

transcription factor

steroid-receptor protein

inhibitory protein

carcinogens

oncogene

proto-oncogenes

tumor-suppressor genes

LECTURE NOTES

Eukaryotic gene regulation is more complex than in prokaryotes, because eukaryotes:
- Have larger, more complex genomes. This requires that eukaryotic DNA be more complexly organized than prokaryotic DNA.
- Require cell specialization or *differentiation*.

I.   Each cell of a multicellular eukaryote expresses only a small fraction of its genome

Cellular differentiation = Divergence in structure and function of different cell types, as they become specialized during an organism's development.
- Cell differentiation requires that gene expression must be regulated on a long-term basis.
- Highly specialized cells, such as muscle or nerve, express only a small percentage of their genes, so transcription enzymes must locate the right genes at the right time.
- Uncontrolled or incorrect gene action can cause serious imbalances and disease, including cancer. Thus, eukaryotic gene regulation is of interest in medical as well as basic research.

DNA-binding proteins regulate gene activity in all organisms – prokaryotes as well as eukaryotes.
- Usually, it is DNA transcription that is controlled.
- Eukaryotes have more complex chromosomal structure, gene organization and cell structure than prokaryotes, which offer added opportunities for controlling gene expression.

II.   The structural organization of chromatin sets coarse controls on gene expression

Prokaryotic and eukaryotic cells both contain double-stranded DNA, but their genomes are organized differently.

Prokaryotic DNA is:

- Usually circular.

- Much smaller than eukaryotic DNA; it makes up a small nucleoid region only visible with an electron microscope.

- Associated with only a few protein molecules.

- Less elaborately structured and folded than eukaryotic DNA; bacterial chromosomes have some additional structure as the DNA-protein fiber forms loops that are anchored to the plasma membrane.

Eukaryotic DNA is:

- Complexed with a large amount of protein to form *chromatin*.

- Highly extended and tangled during interphase.

- Condensed into short, thick, discrete *chromosomes* during mitosis; when stained, chromosomes are clearly visible with a light microscope.

Eukaryotic chromosomes contain an enormous amount of DNA, which requires an elaborate system of DNA packing to fit all of the cell's DNA into the nucleus.

A.   Nucleosomes, or "Beads on a String"

Histone proteins associated with DNA are responsible for the first level of DNA packing in eukaryotes.

Histones = Small proteins that are rich in basic amino acids and that bind to DNA, forming chromatin.

- Contain a high proportion of positively charged amino acids (arginine and lysine), which bind tightly to the negatively charged DNA.

- Are present in approximately equal amounts to DNA in eukaryotic cells.

- Are similar from one eukaryote to another, suggesting that histone genes have been highly conserved during evolution. There are five types of histones in eukaryotes.

Partially unfolded *chromatin* (DNA and its associated proteins) resembles beads spaced along the DNA string. Each beadlike structure is a *nucleosome*. (See Campbell, Figure 18.1a).

Nucleosome = The basic unit of DNA packing; it is formed from DNA wound around a protein core that consists of two copies each of four types of histone (H2A, H2B, H3, H4). A fifth histone (H1) may be present on DNA next to the nucleosome.

- Nucleosomes may control gene expression by controlling access of transcription proteins to DNA.

- Nucleosome heterogeneity may also help control gene expression; nucleosomes may differ in the extent of amino acid modification and in the type of nonhistone proteins present.

B.   **Higher Levels of DNA Packing**

The *30-nm chromatin fiber* is the next level of DNA packing. (See Campbell, Figure 18.1b)

- This structure consists of a tightly wound coil with six nucleosomes per turn.
- Molecules of histone H1 pull the nucleosomes into a cylinder 30nm in diameter.

In the next level of higher-order packing, the 30-nm chromatin fiber forms *looped domains*, which:

- Are attached to a nonhistone protein scaffold.
- Contain 20,000 to 100,000 base pairs.
- Coil and fold, further compacting the chromatin into a mitotic chromosome characteristic of metaphase.

Interphase chromatin is much less condensed than mitotic chromatin, but it still exhibits higher-order packing.

- Its nucleosome string is usually coiled into a 30-nm fiber, which is folded into looped domains.
- Interphase looped domains attach to a scaffolding inside the nuclear envelope (nuclear lamina); this helps organize areas of active transcription.
- Chromatin fibers of different chromosomes do not become entangled as they occupy restricted areas within the nucleus.

Portions of some chromosomes remain highly condensed throughout the cell cycle, even during interphase. Such *heterochromatin* is not transcribed.

Heterochromatin = Chromatin that remains highly condensed during interphase and that is not actively transcribed.

Euchromatin = Chromatin that is less condensed during interphase and is actively transcribed; euchromatin becomes highly condensed during mitosis.

What is the function of heterochromatin in interphase cells?

- Since most heterochromatin is not transcribed, it may be a coarse control of gene expression.
- For example, Barr bodies in mammalian cells are *X* chromosomes that are mostly condensed into heterochromatin. In female somatic cells, one *X* chromosome is a Barr body, so the other *X* chromosome is the only one transcribed.

III. Noncoding sequences and gene duplications account for much of a eukaryotic genome

DNA in eukaryotic genomes is organized differently from that in prokaryotes.

- In prokaryotes, most DNA codes for protein (mRNA), tRNA or rRNA, and coding sequences are uninterrupted. Small amounts of noncoding DNA consist mainly of control sequences, such as promoters.

- In eukaryotes, most DNA does *not* encode protein or RNA, and coding sequences may be interrupted by long stretches of noncoding DNA (introns). Certain DNA sequences may be present in multiple copies.

A. Repetitive Sequences

About 10–25% of total DNA in higher eukaryotes is *satellite DNA* that consists of short (5 to 10 nucleotides) sequences that are tandemly repeated thousands of times.

Satellite DNA = In eukaryotic chromosomes, highly repetitive DNA consisting of short unusual nucleotide sequences that are tandemly repeated thousands of times.

- Called satellite DNA because its unusual nucleotide ratio gives it a density different from the rest of the cell's DNA. Thus, during ultracentrifugation, satellite DNA separates out in a cesium chloride gradient as a "satellite" band separate from the rest of the DNA.

- Is not transcribed and its function is not known. Since most satellite DNA in chromosomes is located at the tips and the centromere, scientists speculate that it plays a structural role during chromosome replication and chromatid separation in mitosis and meiosis.

It is known that short tandem repeats called *telomeres* – at the ends of eukaryotic chromosomes – are important in maintaining the integrity of the lagging DNA strand during replication.

Telomere = Series of short tandem repeats at the ends of eukaryotic chromosomes; prevents chromosomes from shortening with each replication cycle.

- Before an Okazaki fragment of the lagging DNA strand can be synthesized, RNA primers must be produced on a DNA template ahead of the sequence to be replicated.

- Since such a template is not possible for the end of a linear DNA molecule, there must be a mechanism to prevent DNA strands from becoming shorter with each replication cycle.

- This end-replication problem is solved by the presence of special repeating telomeric sequences on the ends of linear chromosomes.

- To compensate for the loss of telomeric nucleotides that occurs each replication cycle, the enzyme *telomerase* periodically restores this repetitive sequence to the ends of DNA molecules.

- Telomeric sequences are similar among many organisms and contain a block of G nucleotides. For example, human chromosomes have 250–1500 repetitions of the base sequence TTAGGG (AATCCC on the complementary strand).

There are other highly repetitive sequences in eukaryotic genomes. For example,

- Some are transposons; generally regarded as nonfunctional, they are associated with some diseases (e.g. neurofibromatosis-1 or elephant man's disease and some cancers).
- Mutations can extend the repetitive sequences normally found within the boundary of genes and cause them to malfunction. (e.g. fragile X syndrome and Huntington's disease.)

B. Multigene Families

Most eukaryotic genes are *unique sequences* present as single copies in the genome, but some genes are part of a *multigene family*.

Multigene family = A collection of genes that are similar or identical in sequence and presumably of common ancestral origin; such genes may be clustered or dispersed in the genome.

Families of *identical* genes:

- Probably arise from a single ancestral gene that has undergone repeated duplication. Such *tandem gene duplication* results from mistakes made during DNA replication and recombination.
- Are usually clustered and almost exclusively genes for RNA products. (One exception is the gene family coding for histone proteins.)
- Include genes for the major rRNA molecules; huge tandem repeats of these genes enable cells to make millions of ribosomes during active protein synthesis. (See Campbell, Figure 18.2)

Families of *nonidentical* genes:

- Arise over time from mutations that accumulate in duplicated genes.
- Can be clustered on the same chromosome or scattered throughout the genome. (Note that, for various reasons, gene sequences in tandem arrays on the same chromosome tend to stay very similar to one another. Transposition events that translocate variants of duplicated genes to different chromosomes, help stabilize their differences and thus promote diversity.)
- May include pseudogenes or nonfunctional versions of the duplicated gene.

    Pseudogene = Nonfunctional gene that has a DNA sequence similar to a functional gene; but, as a consequence of mutation, lacks sites necessary for expression.

A good example of how multigene families can evolve from a single ancestral gene is the globin gene family – actually two related families of genes that encode globins, the α and β polypeptide subunits of hemoglobin. Based on amino acid homologies, the evolutionary history has been reconstructed as follows:

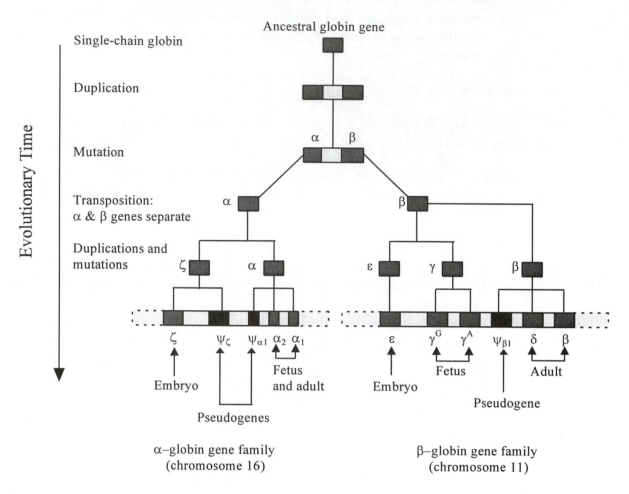

- The original α and β genes evolved from duplication of a common ancestral globin gene. Gene duplication was followed by mutation.
- Transposition separated the α globin and β globin families, so they exist on different chromosomes.
- Subsequent episodes of gene duplication and mutation resulted in new genes and pseudogenes in each family.

The consequence is that each globin gene family consists of a group of similar, but not identical genes clustered on a chromosome.

- In each gene family, the genes are arranged in order of their expression. During development, genes are turned on or off in response to the organism's changing environment as it develops from an embryo into a fetus, and then into an adult.
- At all times during development, functional hemoglobin consists of two α-like and two β-like polypeptides.
  - ⇒ During embryogenesis, the ζ and ε forms predominate.
  - ⇒ About 10 weeks after fertilization, the products of the α-globin genes replace that of the ζ gene, and the γ-chain gene products – γ^G and the γ^A globins – become more prevalent.
  - ⇒ Prior to birth, the β-gene product begins to replace the γ^G and γ^A globins, so by six months of age, the adult β-like globins, δ and β, are present.
- In humans, embryonic and fetal hemoglobins have a higher affinity for oxygen than the adult forms, allowing efficient oxygen exchange between mother and developing fetus.

IV.   The control of gene expression can occur at any step in the pathway from gene to functional protein

Chromatin organization:
- Packages DNA into a compact form that can be contained by the cell's nucleus.
- Controls which DNA regions are available for transcription.
  - ⇒ Condensed heterochromatin is not expressed.
  - ⇒ A gene's location relative to nucleosomes and to scaffold attachment sites influences its expression.

Availability for transcription is only the first level of control of gene expression in eukaryotic cells. Additional levels of control allow cells to alter gene expression in response to changing conditions and needs.

Complexities in chromosome structure, gene organization and cell structure provide opportunities for the control of gene expression in eukaryotic cells. The steps of gene expression where gene regulation can occur are outlined below.

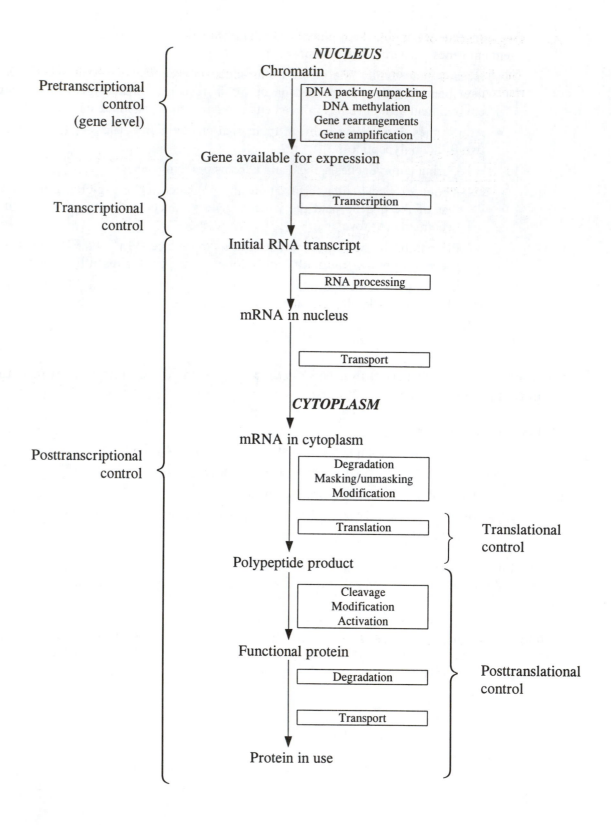

A. Organization of a Typical Eukaryotic Gene: A Review

The following is a brief review of the molecular anatomy of a eukaryotic gene and its transcript.

Eukaryotic genes:

- Contain introns, noncoding sequences that intervene within the coding sequence.
- Contain a promoter sequence at the 5′ upstream end; RNA polymerase attaches to a promoter sequence and transcribes introns along with the coding sequences, or exons.
- May be regulated by *enhancers*, other noncoding control sequences that can be located thousands of base pairs away from the promoter.

  Enhancer = Noncoding DNA control sequence that enhances a gene's transcription and that is located thousands of bases away from the gene's promoter.

The primary RNA transcript (hnRNA) is processed into mature mRNA by:

- Removal of introns.
- Addition of a modified guanosine triphosphate cap at the 5′ end.
- Addition of a poly-A tail at the 3′ end.

B.   Transcriptional Control

In both prokaryotes and eukaryotes, transcription requires that RNA polymerase recognize and bind to DNA at the promoter. However, transcription in eukaryotes differs in two important ways, offering additional opportunities for transcriptional control:

- Eukaryotic RNA polymerase cannot recognize the promoter without the help of *general transcription factors* – proteins that must assemble on DNA at the promoter before transcription can begin.

    ⇒ For example, recall from Chapter 16 that most eukaryotic genes are transcribed by RNA polymerase II, which cannot recognize a promoter unless a specific general transcription factor is bound to the TATA box.

    ⇒ Gene regulatory proteins can influence the rate of transcription by speeding up or slowing down this assembly process at the promoter. (See discussion below.)

- *Enhancer sequences*, noncoding sequences thousands of nucleotide pairs away from the promoter, influence gene expression in eukaryotes.

    Enhancer sequence = Noncoding DNA control sequence to which transcription factors (gene regulatory proteins) bind, controlling transcription of a structural gene; may be thousands of nucleotide pairs away from the promoter of the gene it influences.

NOTE: Transcription factors and the TATA box were discussed with the material in Chapter 16. This is another point in the course where students, if they compartmentalize their learning, may not relate your discussion of transcriptional control to what they have already learned about transcription. We encourage you to experiment in organizing the presentation of this material. It may be more prudent to cover transcriptional control of gene expression (both prokaryotic and eukaryotic) with the lectures on protein synthesis. If not, you should probably begin your lecture with a review of transcription, especially RNA polymerase binding and initiation of transcription.

Associations between transcription factors and enhancer sites are important transcriptional controls in eukaryotes. *How do enhancers stimulate transcription of specific genes?*

- One hypothesis is that a hairpin loop forms in DNA, bringing the transcription factor bound to enhancer into contact with transcription factors and polymerase at the promoter. (See Campbell, Figure 18.6)

- Diverse enhancer sequences and their transcription factors may selectively activate gene expression at appropriate stages in cell development.

Control of transcription thus depends on selective binding of gene regulatory proteins to specific DNA sequences and proteins. This highly selective binding depends on molecular shape.

- There must be a complementary fit between the surfaces of a transcription factor and its specific DNA-binding site.

- Hundreds of transcription factors have been discovered; and though each of these proteins is unique, many recognize their DNA-binding sites with only one of a few possible structural motifs or *domains* containing α helices or β sheets.

C. Posttranscriptional Control

Transcription produces a primary transcript, but gene expression – the production of protein, tRNA, or rRNA – may be stopped or enhanced at any posttranscriptional step.

1. RNA Processing and Export

Because eukaryotic cells have a nuclear envelope, translation is segregated from transcription. This offers additional opportunities for controlling gene expression.

Before leaving the nucleus, the initial mRNA transcript (pre-mRNA) is processed in two ways:
- Both ends are covalently altered; a 5′ cap and poly-A tail are added.
- Introns are removed and the remaining exons are spliced together.

The processed mRNA transcript moves to the cytoplasm through pores in the nuclear envelope.
- Once in the cytoplasm, the mRNA interacts with specific proteins and may associate with ribosomes to begin translation.
- Little is known about how these posttranscriptional steps are regulated.

2. Regulation of mRNA Degradation

Protein synthesis is also controlled by mRNA's lifespan in the cytoplasm.
- Prokaryotic mRNA molecules are degraded by enzymes after only a few minutes. Thus, bacteria can quickly alter patterns of protein synthesis in response to environmental change.
- Eukaryotic mRNA molecules can exist for several hours or even weeks.
- The longevity of a mRNA affects how much protein synthesis it directs. Those that are viable longer can produce more of their protein.
- For example, long-lived mRNAs for hemoglobin are repeatedly translated in developing vertebrate red blood cells.

In some cases, mRNA molecules accumulate and a mass translation occurs after some control signal is received.
- For example, prior to fertilization, the ovum produces and stores inactive mRNA that will be used later in development.
- In cases such as these, the stored mRNA must not only be blocked from translation, but must be protected from degradation as well. mRNA degradation can be prevented by:
  - ⇒ *RNA-binding proteins.* Some mRNA-binding proteins that repress translation, also prevent mRNA degradation.
  - ⇒ *Selective poly-A addition or removal.* There is a minimum poly-A tail length required for mRNA stability and a longer length required for translation.

D.   Translational and Posttranslational Control

Gene expression can also be regulated by mechanisms that control translation of mRNA into protein. Most of these translational controls repress initiation; for example.

- Binding of translation repressor protein to the 5′-end of a particular mRNA can prevent ribosome attachment.
- Translation of all mRNAs can be blocked by the inactivation of certain initiation factors. Such global translational control occurs during early embryonic development of many animals.
  - ⇒ Prior to fertilization, the ovum produces and stores inactive mRNA to be used later during the first embryonic cleavage.
  - ⇒ The inactive mRNA is stored in the ovum's cytosol until fertilization, when the sudden activation of an initation factor triggers translation.
  - ⇒ Delayed translation of stockpiled mRNA allows cells to respond quickly with a burst of protein synthesis when it is needed.

Posttranslational control is the last level of control for regulating gene expression.

- Many eukaryotic polypeptides must be modified or transported before becoming biologically active. Such modifications include:
  - ⇒ Cleavage of the polypeptide into smaller molecules (e.g. insulin).
  - ⇒ Adding chemical groups, such as sugars.
  - ⇒ Dispatching proteins targeted by signal sequences for specific sites.
- Selective degradation of particular proteins and regulation of enzyme activity are also control mechanisms of gene expression.

E.   The Arrangement of Coordinately Controlled Genes

Coordinately controlled genes are arranged differently in a eukaryotic chromosome than in prokaryotic genomes.

- Prokaryotic genes that are turned on and off together are often clustered into operons; these adjacent genes share regulatory sites located at one end of the cluster. All genes of the operon are transcribed into one mRNA molecule and are translated together.
- Eukaryotic genes coding for enzymes of a metabolic pathway are often scattered over different chromosomes. Even functionally related genes on the same chromosome have their own promoters and are individually transcribed.

Eukaryotic genes can be coordinately expressed, even though they may be scattered throughout the genome.

- Coordinately controlled genes are each associated with specific regulatory DNA sequences or enhancers. These sequences are recognized by a single type of transcription factor that activates or represses a group of genes in synchrony.

- Examples of coordinate gene expression in eukaryotes includes:
  ⇒ *Heat shock response.* Exposure to high temperature activates genes coding for heat shock proteins, which help stabilize and repair heat-denatured proteins in the cell.
  ⇒ *Steroid hormone action.* Steroids activate protein receptors, and the protein-receptor complex, in turn, activates genes. In a secondary response, proteins produced this way can activate another group of genes.
  ⇒ *Cellular differentiation.* During cellular differentiation, coordinately controlled genes producing particular sets of proteins are switched on and off.

F. Gene Expression and Differentiation

In multicellular organisms, cells differentiate into specialized cell types which become organized into tissues and are characterized by tissue-specific gene expression.

- Developmental fate of embryonic cells is determined by cytoplasmic content and cell position within the embryo; these factors influence what developmental chemical signals will be present.
- Affecting only certain cells, these chemical signals activate transcription factors, which result in gene expression for other regulatory proteins.
- Cascades of regulatory proteins, in turn, cause batteries of tissue-specific genes to be expressed.

V. Chemical signals that help control gene expression include hormones

Small molecules in the cell's external environment can influence gene expression. Animal steroid hormones, for example, help control transcription by binding to regulatory proteins, altering their conformation.

- As chemical messengers, hormones such as steroids are produced in one part of the body and then travel in the bloodstream to affect target organs elsewhere, far removed from the site of production.
- The effect of steroids on gene expression in target cells was recognized first in insects.

A. Chromosome Puffs: Evidence for the Regulatory Role of Steroid Hormones in Insects

Evidence for transcriptional control and the influence of steroids on gene expression have come from studies of polytene chromosomes from the salivary gland and other tissues of insects.

- Polytene chromosomes are single giant chromosomes produced by repeated DNA replication without cell division; rather than separating out as individual chromosomes, the hundreds of homologous chromosome copies remain next to each other.
- At particular points of larval development, *chromosome puffs* – loops of decondensed DNA – appear at specific sites on the polytene chromosomes.
- Growing RNA transcripts detected by autoradiography, indicate that intense transcription occurs at puffs.

As a larva develops, puff location along the chromosome changes.

- At molting, some puffs disappear and new puffs form at other sites.
- Shifting puffs indicate that selective switching on and off of specific genes occurs during development.
- *Ecdysone* (molting hormone) can induce changes in puff patterns, demonstrating that gene regulation responds to specific chemical signals.

B. The Action of Steroid Hormones in Vertebrates

Steroid hormones are chemical signals that can activate gene expression in target cells of vertebrates. The key steps of steroid hormone action are outlined below.

- Since steroids are lipid soluble, they readily diffuse across the plasma membrane into the cytoplasm.
- From the cytoplasm, the steroid enters the nucleus, where it binds to a soluble, *steroid-receptor* protein – a DNA-binding protein that can activate transcription of a particular gene. (Some steroids bind to receptors in the cytoplasm, and the hormone-receptor complex moves as a unit into the nucleus.)

- In absence of steroid, an *inhibitory protein* binds to the steroid receptor and blocks its DNA-binding domain, preventing the receptor from binding to DNA.

Hormone-receptor site

Inhibitor protein

- When steroid is present, its binding to the receptor causes release of inhibitory protein and activates the steroid receptor, so it can attach to DNA at *enhancer* sequences that control steroid-responsive genes.

Hormone

DNA binding survace

Summary: The Control of Gene Expression in Eukaryotes

- Various cell types of multicellular organisms express different genes.
- Chromatin organization makes certain genes available for expression and other genes unavailable.
- Opportunities for the control of gene expression exist at each step in the pathway from gene to functional protein.
- Transcriptional control is important in determining which genes are expressed; selective binding of regulatory proteins to enhancer sequences in DNA, stimulates transcription of specific genes.
- Regulatory activity of some DNA-binding proteins is affected by certain hormones and other chemical cues.

VI. **Chemical modification or relocation of DNA within a genome can alter gene expression**

An organism's genome is plastic, or changeable, in ways that affect the availability of specific genes for expression.

- Genes may be available for expression in some cells and not others, or at some time in the organism's development and not others.
- Genes may, under some conditions, be amplified or made more available than usual.
- Changes in the physical arrangement of DNA, such as levels of DNA packing, affect gene expression. For example, genes in heterochromatin and mitotic chromosomes are not expressed.

The structural organization of an organism's genome is also somewhat plastic; movement of DNA within the genome and chemical modification of DNA influence gene expression.

A. **Gene Amplification and Selective Gene Loss**

Gene amplification may temporarily increase the number of gene copies at certain times in development.

Gene amplification = Selective synthesis of DNA, which results in multiple copies of a single gene. For example,

- Amphibian rRNA genes are selectively amplified in the oocyte, which:
  ⇒ Results in a million or more additional copies of the rRNA genes that exist as extrachromosomal circles of DNA in the nucleoli.
  ⇒ Permits the oocyte to make huge numbers of ribosomes that will produce the vast amounts of proteins needed when the egg is fertilized.
- Gene amplification occurs in cancer cells exposed to high concentrations of chemotherapeutic drugs.
  ⇒ Some cancer cells survive chemotherapy, because they contain amplified genes conferring drug resistance.
  ⇒ Increased drug resistance can be created experimentally by exposing a cell population to increasing drug doses and artificially selecting for surviving cells that have amplified drug-resistance genes.

Genes may also be selectively lost in certain tissues by elimination of chromosomes.

Chromosome diminution = Elimination of whole chromosomes or parts of chromosomes from certain cells early in embryonic development.

- For example, chromosome diminution occurs in gall midges during early development; all but two cells lose 32 of their 40 chromosomes during the first mitotic division after the 16-cell stage.
- The two cells that retain the complete genome are germ cells that will produce gametes in the adult. The other 14 cells become somatic cells with only 8 chromosomes.

B. Rearrangements in the Genome

Rearrangements of DNA in somatic cells can activate or inactivate specific genes.

1. Transposons

All organisms probably have transposons that move DNA from one location to another within the genome. (See Campbell, Chapter 17) Transposons can rearrange the genome by:

- Inserting into the middle of a coding sequence of another gene; it can prevent the interrupted gene from functioning normally.

- Inserting within a sequence that regulates transcription; the transposition may increase or decrease a protein's production.

- Inserting its own gene just downstream from an active promoter that activates its transcription.

2. Immunoglobulin Genes

During cellular differentiation in mammals, permanent rearrangements of DNA segments occur in those genes that encode antibodies, or *immunoglobulins*.

Immunoglobulins = A class of proteins (antibodies) produced by B lymphocytes that specifically recognize and help combat viruses, bacteria, and other invaders of the body. Immunoglobulin molecules consist of: (See Campbell, Figure 18.10)

- Four polypeptide chains held together by disulfide bridges.

- Each chain has two major parts:

  ⇒ A *constant region*, which is the same for all antibodies of a particular class.

  ⇒ A *variable region*, which gives an antibody the ability to recognize and bind to a specific foreign molecule.

B lymphocytes, which produce immunoglobulins, are a type of white blood cell found in the mammalian immune system.

- The human immune system contains millions of subpopulations of B lymphocytes that produce different antibodies.

- B lymphocytes are very specialized; each differentiated cell and its descendants produce only one specific antibody.

Antibody specificity and diversity are properties that emerge from the unique organization of the antibody gene, which is formed by a rearrangement of the genome during B cell development.

- As an unspecialized cell differentiates into a B lymphocyte, its antibody gene is pieced together randomly from several DNA segments that are physically separated in the genome.

  ⇒ In the genome of an embryonic cell, there is an intervening DNA sequence between the sequence coding for an antibody's constant region and the site containing hundreds of coding sequences for the variable regions.

  ⇒ As a B cell differentiates, the intervening DNA is deleted, and the DNA sequence for a variable region connects with the DNA sequence for a constant region, forming a continuous nucleotide sequence that will be transcribed.

⇒ The primary RNA transcript is processed to form mRNA that is translated into one of the polypeptides of an antibody molecule.

- Antibody variation results from:

⇒ Different combinations of variable and constant regions in the polypeptides.

⇒ Different combinations of polypeptides.

C. DNA Methylation

DNA methylation = The addition of methyl groups ($-CH_3$) to bases of DNA, after DNA synthesis.

- Most plant and animal DNA contains methylated bases (usually cytosine); about 5% of the cytosine residues are methylated.
- May be a cellular mechanism for long-term control of gene expression. When researchers examine the same genes from different types of cells, they find:

⇒ Genes that are not expressed (e.g. Barr bodies) are more heavily methylated than those that are expressed.

⇒ Drugs that inhibit methylation can induce gene reactivation, even in Barr bodies.

- In vertebrates, DNA methylation reinforces developmental decisions made earlier by other mechanisms.

⇒ For example, genes must be selectively turned on or off for normal cell differentiation to occur. DNA methylation insures that once a gene is turned off, it stays off.

⇒ DNA methylation patterns are inherited and thus perpetuated as cells divide; clones of a cell lineage forming specialized tissues have a chemical record of regulatory events that occurred during early development.

VII. Cancer can result from the abnormal expression of genes that regulate cell growth and division

Cancer is a variety of diseases in which cells escape from the normal controls on growth and division, and it can result from mutations that alter normal gene expression in somatic cells. These mutations:

- Can be random and spontaneous.
- Most likely occur as a result of environmental influences, such as:

⇒ Infection by certain viruses.

⇒ Exposure to *carcinogens*.

Carcinogens = Physical agents such as X-rays and chemical agents that cause cancer by mutating DNA.

Whether cancer is caused by physical agents, chemicals or viruses, the mechanism is the same – the activation of *oncogenes* that are either native to the cell or introduced in viral genomes.

<u>Oncogene</u> = Gene responsible for a cell becoming cancerous.
- Discovered during the study of tumors induced by specific viruses.
- Harold Varmus and Michael Bishop won a Nobel Prize for their discovery of oncogenes in RNA viruses (retroviruses) that cause uncontrolled growth of infected cells in culture.

Researchers later discovered that some animal genomes, including human, contain genes that closely resemble viral oncogenes. These *proto-oncogenes* normally regulate growth, division and adhesion in cells.

<u>Proto-oncogenes</u> = Gene that normally codes for regulatory proteins controlling cell growth, division and adhesion, and that can be transformed by mutation into an oncogene. Four types of mutations can convert proto-oncogenes to oncogenes:

1. *Gene amplification.* Sometimes more copies of oncogenes are present in a cell than is normal.

2. *Chromosome translocation.* Malignant cells frequently contain chromosomes that have broken and rejoined, placing pieces of different chromosomes side-by-side and possibly separating the oncogene from its normal control regions. In its new position, an oncogene may be next to active promoters or other control sequences that enhance transcription.

3. *Gene transposition.* Abnormal expression of an oncogene may occur if the oncogene or its regulatory gene are transposed to another locus. For example, a transposon may move:
    - An oncogene to a new locus that has a highly active promoter.
    - A promoter or other regulatory sequence to the upstream end of the oncogene.

4. *Point mutation.* A slight change in the nucleotide sequence might produce a growth-stimulating protein that is more active or more resistant to degradation than the normal protein.

In addition to mutations affecting growth-stimulating proteins, changes in *tumor-suppressor genes* coding for proteins that normally *inhibit* growth can also promote cancer. For example,
- Children who are homozygous for the retinoblastoma allele develop malignant eye tumors.
    - ⇒ The mutated genes do not produce normal tumor-suppressor protein.
    - ⇒ When a normal allele coding for tumor suppressor is inserted into cells with mutated genes, abnormal growth of cancerous cells in culture is stopped.
- Mutant tumor-suppressor genes are associated with inherited forms of colorectal cancer, Wilm's tumor and breast cancer.

More than one somatic mutation is probably needed to transform normal cells into cancerous cells. One of the best understood examples is colorectal cancer.

- Development of metastasizing colorectal cancer is gradual, and the first sign is unusually rapid cell division of apparently normal cells in the colon lining; a benign tumor (polyp) appears, and eventually a malignant tumor may develop.
- During this process, mutations in oncogenes and tumor-suppressor genes gradually accumulate. After a number of genes have changed, the tumor becomes malignant.
- In most cases studied, four changes occur at the DNA level: the activation of an oncogene (*ras*) and the inactivation of three tumor-suppressor genes on chromosomes 5, 18, and 17. (See Campbell, Figure 18.12)

Breast cancer, the second most common type of cancer in women, is associated with somatic mutations of tumor-suppressor genes.

- Inherited breast cancer accounts for 5–10% of all breast cancer cases.
  - ⇒ More than half of these are associated with mutations of a tumor-suppressor gene called BRCA1 (for BReast CAncer) on chromosome #17.
  - ⇒ Another locus, BRCA2, on chromosome #13, accounts for most of the remaining breast cancer cases linked to family history.
- In breast cancer patients with no family history for the disease, new somatic mutations BRCA1 and BRCA2 are usually present.
- The study of such genes associated with breast cancer may lead to early diagnosis and treatment.

How are oncogenes involved in virus-induced cancers, which probably account for 15% of human cancer worldwide?

- Viruses might add oncogenes to cells or disrupt DNA, affecting proto-oncogenes or a tumor-suppressor gene.
- However, viral-induced DNA changes are probably not all that is required to cause cancer; additional DNA changes needed may be triggered by environmental agents.

REFERENCES

Alberts, B., D. Bray, J. Lewis, M. Raff, K. Roberts and J.D. Watson. *Molecular Biology of the Cell.* 3rd ed. New York: Garland, 1994.

Campbell, N. *Biology.* 4th ed. Menlo Park, California: Benjamin/Cummings, 1996.

Cavenee, W.K., and R.L. White. "The Genetic Basis of Cancer." *Scientific American,* March, 1995.

Watson, J.D., N.H. Hopkins, J.W. Roberts, J.A. Steitz and A.M. Weiner. *Molecular Biology of the Gene.* 4th ed. Menlo Park, California: Benjamin/Cummings, 1987.

CHAPTER **19**

DNA TECHNOLOGY

OUTLINE

I.   DNA technology makes it possible to clone genes for basic research and commercial applications: *an overview*

II. The toolkit for DNA technology includes restriction enzymes, DNA vectors, and host organisms
    A.   Restriction Enzymes
    B.   Vectors
    C.   Host Organisms

III. Recombinant DNA technology provides a means to transplant genes from one species into the genome of another
    A.   Steps for Using Bacteria and Plasmids to Clone Genes
    B.   Sources for Cloning
    C.   Inserting DNA Into Cells
    D.   Selection: Finding a Gene of Interest
    E.   Achieving Expression of Cloned Genes

IV. Additional methods for analyzing and cloning nucleotide sequences increase the power of DNA technology
    A.   Gel Electrophoresis
    B.   DNA Synthesis and Sequencing
    C.   The Polymerase Chain Reaction (PCR)
    D.   Hybridization
    E.   RFLP Analysis

V.   DNA technology is catalyzing progress in many fields of biology

VI. The Human Genome Project is an enormous collaborative effort to map and sequence DNA

VII.   DNA technology is reshaping the medical and pharmaceutical industries
    A.   Diagnosis of Diseases
    B.   Human Gene Therapy
    C.   Vaccines and Other Pharmaceutical Products

VIII.   DNA technology offers forensic, environmental, and agricultural applications
    A.   Forensic Use of DNA Technology
    B.   Environmental Uses of DNA Technology
    C.   Agricultural Uses of DNA Technology

IX.   DNA technology raises important safety and ethical questions

OBJECTIVES

After reading this chapter and attending lecture, the student should be able to:

1.   Explain how advances in recombinant DNA technology have helped scientists study the eukaryotic genome.
2.   Describe the natural function of restriction enzymes.
3.   Describe how restriction enzymes and gel electrophoresis are used to isolate DNA fragments.
4.   Explain how the creation of sticky ends by restriction enzymes is useful in producing a recombinant DNA molecule.
5.   Outline the procedures for producing plasmid and phage vectors.
6.   Explain how vectors are used in recombinant DNA technology.
7.   List and describe the two major sources of genes for cloning.
8.   Describe the function of reverse transcriptase in retroviruses and explain how they are useful in recombinant DNA technology.
9.   Describe how "genes of interest" can be identified with the use of a probe.
10.   Explain the importance of DNA synthesis and sequencing to modern studies of eukaryotic genomes.
11.   Outline the Sanger method for sequencing DNA.
12.   Describe how bacteria can be induced to produce eukaryotic gene products.
13.   List some advantages for using yeast in the production of gene products.
14.   Describe some practical applications of recombinant DNA technology in biological research.
15.   List and describe four complementary approaches used to map the human genome.
16.   Explain how RFLP analysis and PCR can be applied to the Human Genome Project.
17.   Describe how recombinant DNA technology can have medical applications such as diagnosis of genetic disease, development of gene therapy, vaccine production, and development of pharmaceutical products.
18.   Describe how gene manipulation has practical applications for agriculture.
19.   Describe how plant genes can be manipulated using the Ti plasmid carried by *Agrobacterium* as a vector.
20.   Explain how foreign DNA may be transferred into monocotyledonous plants.
21.   Describe how recombinant DNA studies and the biotechnology industry are regulated with regards to safety and policy matters.

KEY TERMS

recombinant DNA technology	DNA ligase	restriction fragment length polymor-	polymerase chain reaction (PCR)
biotechnology	gel electrophoresis	phisms (RFLPs)	subunits
human genome project	plasmid	reverse transcriptase	gene-splicing
phage	plasmid DNA	retroviruses	vaccinia
restriction enzymes	transformation	probe	erythropoietin (EPO)
restriction enzyme cleavage site	vector	Sanger Method	transgenic animals
restriction	recombinant plasmid	dideoxyribo-	Ti plasmid
modification	gene cloning	nucleotide	T DNA
sticky ends	bacteriophage lambda	in *situ* hybridization	*Agrobacterium tumefaciens*
restriction fragments	genomic library	genetic (linkage) mapping	electroporation
	complementary DNA (cDNA)	physical mapping	microinjection

LECTURE NOTES

Recombinant DNA technology refers to the set of techniques for recombining genes from different sources *in vitro* and transferring this recombinant DNA into a cell where it may be expressed.

- These techniques were first developed around 1975 for basic research in bacterial molecular biology, but this technology has also led to many important discoveries in basic eukaryotic molecular biology.

- Such discoveries resulted in the appearance of the *biotechnology* industry. Biotechnology refers to the use of living organisms or their components to do practical tasks such as:
  ⇒ The use of microorganisms to make wine and cheese.
  ⇒ Selective breeding of livestock and crops.
  ⇒ Production of antibiotics from microorganisms.
  ⇒ Production of monoclonal antibodies.

Using recombinant DNA techniques, modern biotechnology is a more precise and systematic process than earlier research methods.

- It is also a powerful tool since it allows genes to be moved across the species barrier.

- Using these techniques, scientists have advanced our understanding of eukaryotic molecular biology.

- The *human genome project* is an important application of this technology. This project's goal is to transcribe and translate the entire human genome in order to better understand the human organism.

- A variety of applications are possible for this technology, and the practical goal is the improvement of human health and food production.

I. DNA technology makes it possible to clone genes for basic research and commercial applications: *an overview*

Prior to the discovery of recombinant DNA techniques, procedures for altering the genes of organisms were constrained by the need to find and propagate desirable mutants.

- Geneticists relied on either natural processes, mutagenic radiation, or chemicals to induce mutations.
- In a laborious process, each organism's phenotype was checked to determine the presence of the desired mutation.
- Microbial geneticists developed techniques for screening mutants. For example, bacteria was cultured on media containing an antibiotic to isolate mutants which were antibiotic resistant.

Before 1975, transferring genes between organisms was accomplished by cumbersome and nonspecific breeding procedures. The only exception to this was the use of bacteria and their phages.

- Genes can be transferred from one bacterial strain to another by the natural processes of transformation, conjugation or transduction.
- Geneticists used these processes to carry out detailed molecular studies on the structure and functioning of prokaryotic and phage genes.
- Ideal for laboratory experiments, bacteria and phages are relatively small, have simple genomes, and are easily propagated.
- Although the technique was available to grow plant and animal cells in culture, the workings of their genomes could not be examined using existing methods.

Recombinant DNA technology now makes it possible for scientists to examine the structure and function of the eukaryotic genome, because it contains several key components:

- Biochemical tools that allow construction of recombinant DNA.
- Methods for purifying DNA molecules and proteins of interest.
- Vectors for carrying recombinant DNA into cells and replicating it.
- Techniques for determining nucleotide sequences of DNA molecules.

II. The toolkit for DNA technology includes restriction enzymes, DNA vectors, and host organisms

A. Restriction Enzymes

Restriction enzymes are major tools in recombinant DNA technology.

- First discovered in the late 1960's, these enzymes occur naturally in bacteria where they protect the bacterium against intruding DNA from other organisms.
- This protection involves *restriction*, a process in which the foreign DNA is cut into small segments.
- Most restriction enzymes only recognize short, specific nucleotide sequences called *recognition sequences*. They only cut at specific points within those sequences.

Bacterial cells protect their own DNA from restriction through *modification* or methylation of DNA.

- Methyl groups are added to nucleotides within the recognition sequences.
- Modification is catalyzed by separate enzymes that recognize these same DNA sequences.

There are several hundred restriction enzymes and about a hundred different specific recognition sequences.

- Recognition sequences are symmetric in that the same sequence of four to eight nucleotides is found on both strands, but run in opposite directions.
- Restriction enzymes usually cut phosphodiester bonds of both strands in a staggered manner, so that the resulting double-stranded DNA fragments have single-stranded ends, called *sticky ends*.
- The single-stranded short extensions form hydrogen-bonded base pairs with complementary single-stranded stretches on other DNA molecules.

Sticky ends of *restriction fragments* are used in the laboratory to join DNA pieces from different sources (cells or even different organisms).

- These unions are temporary since they are only held by a few hydrogen bonds.
- These unions can be made permanent by adding the enzyme *DNA ligase*, which catalyzes formation of covalent phosphodiester bonds.

The outcome of this process is the same as natural genetic recombination, the production of recombinant DNA – a DNA molecule carrying a new combination of genes.

B.   Vectors

Most DNA technology procedures use carriers or vectors for moving DNA from test tubes back into cells.

Cloning vector = Bacterial plasmid or virus used to move DNA into cells.

- Two most often used types of vectors are bacterial plasmids and viruses.
- Restriction fragments of foreign DNA can be spliced into a bacterial plasmid without interfering with its ability to replicate within the bacterial cell. Isolated recombinant plasmids can be introduced into bacterial cells by transformation.

Bacteriophages, such as lambda phage, can also be used as vectors.

- The middle of the linear genome, which contains nonessential genes, is deleted by using restriction enzymes.
- Restriction fragments of foreign DNA are then inserted to replace the deleted area.
- The recombinant phage DNA is introduced into an *E. coli* cell.
- The phage replicates itself inside the bacterial cell.
- Each new phage particle carries the foreign DNA "passenger."

Sometimes it is necessary to clone DNA in eukaryotic cells rather than in bacteria. Under the right conditions, yeast and animal cells growing in culture can also take up foreign DNA from the medium.

- If the new DNA becomes incorporated into chromosomal DNA or *can* replicate itself, it can be cloned with the cell.
- Since yeast cells have plasmids, scientists can construct recombinant plasmids that combine yeast and bacterial DNA and that can replicate in either cell type.
- Viruses can also be used as vectors with eukaryotic cells. For example, retroviruses used as vectors in animal cells can integrate DNA directly into the chromosome.

C. Host Organisms

Bacteria are commonly used hosts in genetic engineering because:

- DNA can be easily isolated from and reintroduced into bacterial cells
- Bacterial cultures grow quickly, rapidly cloning the inserted foreign genes.

Some disadvantages to using bacterial host cells are that:

- Bacterial cells may not be able to use the information in a eukaryotic gene, since eukaryotes and prokaryotes use different enzymes and regulatory mechanisms during transcription and translation.
- Bacterial cells cannot make the posttranslational modifications required to produce some eukaryotic proteins. (e.g. addition of lipid or carbohydrate groups)

Eukaryotic cells may be used as hosts.

- Yeast cells can be used as hosts to overcome some of the limitations of using bacterial cells.
- Cultured plant and animal cells can be hosts for foreign DNA, but it is often difficult to get such cells to take up engineered DNA.

III. **Recombinant DNA technology provides a means to transplant genes from one species into the genome of another**

A. **Steps for Using Bacteria and Plasmids to Clone Genes**

Recombinant DNA molecules are only useful if they can be made to replicate and produce a large number of copies. A typical gene-cloning procedure includes the following steps: (See Campbell, Figure 19.3)

Step 1: *Isolation of two kinds of DNA.*
- Bacterial plasmids and foreign DNA containing the gene of interest are isolated.
- In this example, the foreign DNA is human, and the plasmid is from *E. coli* and has two genes:
  ⇒ *amp*^{R} that confers antibiotic resistance to ampicillin.
  ⇒ *lacZ* that codes for β-galactosidase, the enzyme that catalyzes the hydrolysis of lactose
- Note that the recognition sequence for the restriction enzyme used in this example is *within* the *lacZ* gene.

Step 2: *Treatment of plasmid and foreign DNA with the same restriction enzyme.*
- The restriction enzyme cuts plasmid DNA at the *restriction site*, disrupting the *lacZ* gene.
- The foreign DNA is cut into thousands of fragments by the same restriction enzyme; one of the fragments contains the gene of interest.
- When the restriction enzyme cuts, it produces *sticky ends* on both the foreign DNA fragments and the plasmid.

Step 3: *Mixture of foreign DNA with clipped plasmids.*
- Sticky ends of the plasmid base pair with complementary sticky ends of foreign DNA fragments.

Step 4: *Addition of DNA ligase.*
- DNA ligase catalyzes the formation of covalent bonds, joining the two DNA molecules and forming a new plasmid with recombinant DNA.

Step 5: *Introduction of recombinant plasmid into bacterial cells.*
- The naked DNA is added to a bacterial culture.
- Some bacteria will take up the plasmid DNA by transformation.

Step 6: *Production of multiple gene copies by gene cloning and selection process for transformed cells.*
- Bacteria with the recombinant plasmid are allowed to reproduce, cloning the inserted gene in the process.
- Recombinant plasmids can be identified by the fact that they are ampicillin resistant and will grow in the presence of ampicillin.

Step 7: *Final screening for transformed cells.*

- X-gal, a modified sugar added to the culture medium, turns blue when hydrolyzed by β-galactosidase. It is used as an indicator that cells have been transformed by plasmids containing the foreign insert.
- Since the foreign DNA insert disrupts the *lacZ* gene, bacterial colonies that have successfully acquired the foreign DNA fragment will be white. Those bacterial colonies lacking the DNA insert will have a complete *lacZ* gene that produces β-galactosidase and will turn blue in the presence of X-gal.

B. Sources of Genes for Cloning

There are two major sources of DNA which can be inserted into vectors and cloned:
1. DNA isolated directly from an organism.
2. Complementary DNA made in the laboratory from mRNA templates.

DNA isolated directly from an organism contains all genes including the gene of interest.
- Restriction enzymes are used to cut this DNA into thousands of pieces which are slightly larger than a gene.
- All of these pieces are then inserted into plasmids or viral DNA.
- These vectors containing the foreign DNA are introduced into bacteria.
- This produces the *genomic library*: thousands of DNA segments from a genome, each of which is carried by a plasmid or phage.

One problem with this method is that eukaryotic genes of interest may be too large to clone easily because they contain introns (noncoding regions). A solution is to make *complementary DNA (cDNA)* in the laboratory, by using mRNA molecules as templates. (See Campbell, Figure 19.5)
- The enzyme *reverse transcriptase* (from retroviruses) is used to catalyze DNA synthesis from an RNA template. A second DNA strand is synthesized using DNA polymerase and the first DNA strand as a template.
- Even if the gene of interest contains introns, the cDNA does not, because they have been removed from the template during RNA processing.
- Without introns, the cDNA gene is more manageable in size than the original gene and also has the potential of being translated into protein by bacterial cells which lack RNA-processing capabilities.
- To be transcribed in bacteria, the cDNA is attached to other DNA containing a promoter and other essential transcription signals.

The cDNA method produces partial genomic libraries because mRNA molecules for a particular gene usually cannot be distinguished from other RNA in a cell.
- By using cells from specialized tissues or a cell culture used exclusively for making one gene product, the majority of mRNA produced is for the gene of interest.
- For example, most of the mRNA in precursors of mammalian erythrocytes is for the protein hemoglobin.

C.   **Inserting DNA Into Cells.**

Which method is used to insert DNA into cells depends upon the vector and type of host cell.

- Most commonly, DNA is transferred into bacterial cells by transformation, the absorption of DNA from the surrounding solution
- Recombinant phage vectors package foreign DNA inside the protein coat and then infect host cells.
- Yeast cells can take up recombinant plasmids by transformation. They can also take up linear DNA, which becomes incorporated into a chromosome by recombination.

There are more aggressive techniques for inserting foreign DNA into eukaryotic cells:

- In *electroporation*, a brief electric pulse applied to a cell solution causes temporary holes in the plasma membrane, through which DNA can enter.
- With thin needles, DNA can be injected directly into a eukaryotic cell.
- DNA attached to microscopic metal particles can be fired into plant cells with a gun.

D.   **Selection: Finding a Gene of Interest**

Identifying the bacterial clone containing the gene of interest is difficult regardless of which approach (genomic library or cDNA) is used.

- Screening for a protein can be used if the clones translate the gene into a product. Detection may be based on activity (for enzymes) or structure (using antibodies that combine with the protein).
- Screening for the gene itself is most often used. These methods use base pairing between the gene and a complementary sequence on another nucleic acid (RNA or DNA) — a process called *hybridization*.
  - ⇒ Nucleic acid segments can be synthesized and used to identify a gene if a part of the gene's nucleotide sequence is known or inferred by its protein product.
  - ⇒ The synthesized nucleic acid used to find the gene of interest is called a *probe*.
  - ⇒ The location of a probe, which will hydrogen bond to a specific gene, is traced by labeling it with a radioactive isotope or fluorescent tag. (See Campbell, Figure 19.6)
- When a bacterial clone carrying the gene of interest is identified, the gene can be easily isolated in large amounts and used for further studies or be used as probes themselves to identify similar or identical genes.

E.   **Achieving Expression of Cloned Genes**

Yeast genes, as well as those of more complex eukaryotes, can be expressed in bacteria. This possibility surprised scientists because:

- Prokaryotes have different signals that control gene transcription and translation.
- The enzymes that recognize these signals are also different.

Bacteria are the organisms of choice for large scale, commercial production of gene products, because:

- They can be grown rapidly and cheaply in large fermenters.
- Their genomes are relatively simple and easy to manipulate.
- Bacteria can be manipulated to secrete a protein as it is made, simplifying purification of the protein.

Expression of eukaryotic genes in bacterial cells can be maximized by several genetic and biochemical methods. They include:

- Using plasmid vectors that produce many copies per cell.
- Changing the promoter controlling the gene's expression to a highly active one.
- Attaching a eukaryotic gene to the initial portion of a bacterial gene for a protein that is usually produced in large quantities.

Eukaryotic cells must be used to study eukaryotic mechanisms of gene function and regulation.

- While eukaryotic genes are initially cloned in bacteria, they are later returned to eukaryotic cells such as yeast for expression and study.
- Yeast have advantages for use in the production of gene products (commercially or for research).
  1. They are easy to grow.
  2. They are very simple eukaryotes.
  3. It is simple to reinsert genes into yeast since they can be made to take up DNA by transformation and have plasmids for the process.

Bacteria and yeast are not suitable for every purpose. For certain applications, plant or animal cell cultures must be used.

- Cells of more complex eukaryotes carry out certain biochemical processes not found in yeast (e.g. only animal cells produce antibodies).

IV. Additional methods for analyzing and cloning nucleotide sequences increase the power of DNA technology

A. Gel Electrophoresis

Gel electrophoresis is used to separate either nucleic acids or proteins based upon molecular size, charge and other physical properties. Using this technique:

- Viral DNA, plasmid DNA, and segments of chromosomal DNA can be identified by their characteristic banding patterns after being cut with various restriction enzymes.
- DNA fragments containing genes of interest can be isolated, purified and then recovered from the gel with full biological activity.

B.   DNA Synthesis and Sequencing

It is possible to synthesize artificial genes in the laboratory (e.g. genes for insulin).

- This once tedious task can now be done with machines that can rapidly produce genes several hundred nucleotides long.
- This approach can only be used for short genes with known nucleotide sequences.

The most common method for determining the sequence of a DNA molecule is the Sanger Method. (See Campbell, Methods Box, for a description of DNA sequencing by the Sanger Method.) The method is based on:

- Use of restriction enzymes to cut DNA into discrete, reproducible fragments with unique sequences.
- Synthesizing *in vitro* DNA strands complementary to a strand from the restriction fragment being sequenced.
- Incorporation of a modified nucleotide that blocks further DNA synthesis (*dideo*xyribonucleotide, which lacks 2 OH groups).

Restriction enzymes are used to cut the DNA into restriction fragments. Each fragment is processed in the following manner:

Step 1: A preparation containing one of the strands of the DNA fragment is divided into four portions.

- Each portion is incubated with a radioactively labeled primer, DNA polymerase, and the four deoxyribonucleotide triphosphates (ingredients necessary for complementary strand synthesis).
- Each mixture also contains <u>one</u> of the four nucleotides in the modified dideoxy (dd) form.

Step 2: New strand synthesis begins with the radioactive primer and continues until a dideoxyribonucleotide is incorporated, where synthesis stops.

- Since both deoxy and dideoxy forms of one nucleotide are present, they compete for positions.
- A set of radioactive strands of varying lengths are eventually produced.

Step 3: The new DNA strands are separated by polyacrylamide gel electrophoresis, which can separate strands differing by one nucleotide in length.

- The sequence in the new strands is then read from the radioactive bands, and the sequence of the original template is deduced.

DNA sequencing techniques have enabled scientists to collect thousands of DNA sequences in computer data banks.

- This knowledge greatly improves the understanding of genes and genetic control elements.
- Using a computer, scientists can scan long sequences for shorter sequences known to be protein recognition sites and control sequences. They can also scan for similarities to known sequences of other genes or other organisms.
- Nucleotide sequences can also be automatically translated into amino acid sequences.

C. The Polymerase Chain Reaction (PCR)

PCR is another promising new technique which allows any piece of DNA to be quickly amplified (copied many times) in vitro. (See Campbell, Methods Box, page 380)
- DNA is incubated under appropriate conditions with special primers and DNA polymerase molecules.
- Billions of copies of the DNA are produced in just a few hours.
- PCR is highly specific; primers determine the sequence to be amplified.
- Only minute amounts of DNA are needed.

PCR is presently being applied in many ways for analysis of DNA from a wide variety of sources:
- Ancient DNA fragments from a woolly mammoth; DNA is a stable molecule and can be amplified by PCR from sources thousands, even millions, of years old.
- DNA from tiny amounts of tissue or semen found at crime scenes.
- DNA from single embryonic cells for prenatal diagnosis.
- DNA of viral genes from cells infected with difficult to detect viruses such as HIV.

Amplification of DNA by PCR is being used in the human genome project to produce linkage maps without the need for large family pedigree analysis.
- DNA from sperm of a single donor can be amplified to analyze the immediate products of meiotic recombination.
- This process eliminates the need to rely on chance that offspring will be produced with a particular type of recombinant chromosome.
- It makes it possible to study genetic markers that are extremely close together.

D. Hybridization

The technique of hybridization is used to determine the presence of a specific nucleotide sequence.
- Labeled probes complementary to the gene of interest are allowed to bind to DNA from cells being tested.
- Variations of this technique allows researchers to:
  ⇒ determine if a gene is present in various organisms.
  ⇒ determine not only whether a sequence is present, but how many sequences there are and the size of the restriction fragments containing these sequences (*Southern hybridization*). (See Campbell's Methods Box.)
  ⇒ determine whether a gene is made into mRNA, how much of that mRNA is present, and whether the amount of that mRNA changes at different stages of development or in response to certain regulatory signals. (*Northern hybridization*)

E.   RFLP Analysis

Gel electrophoresis of restriction fragments results in a characteristic banding pattern.

- Each band corresponds to a DNA restriction fragment of a certain length.
- DNA segments carrying different alleles of a gene can result in dissimilar banding patterns, since numbers and locations of restriction sites may not be the same in the different nucleotide sequences. (See Campbell, Figure 19.7)
- Similar differences in banding patterns result when *noncoding* segments of DNA are used as starting material.

Differences in restriction fragment length that reflect variations in homologous DNA sequences are called *restriction fragment length polymorphisms* (*RFLPs*).

- DNA sequence differences on homologous chromosomes that result in RFLPs are scattered abundantly throughout genomes, including the human genome.
- RFLPs are not only abundant, but can easily be detected whether or not they affect the organism's phenotype; they can be located in an exon, intron, or any noncoding part of the genome.
- RFLP analysis involves techniques often used in recombinant DNA technology:
  ⇒ Restriction enzyme digestion of DNA.
  ⇒ Gel electrophoresis.
  ⇒ Southern blotting.
  ⇒ Use of radioactive DNA probes.
  ⇒ Autoradiography.
- Because RFLPs can be readily detected with recombinant DNA methods, they are extremely useful as *genetic markers* for making linkage maps.
- A RFLP marker is often found in numerous variants in a population. RFLPs have provided many markers for mapping the human genome since geneticists are no longer limited to genetic variations that lead to phenotypic differences or protein products.

RFLPs are proving useful in several areas.

- Disease genes are being located by examining known RFLPs for linkage to them. RFLP markers inherited at a high frequency with a disease are probably located close to the defective gene on the chromosome.
- An individual's RFLP markers provide a "genetic fingerprint" which can be used in forensics, since there is a very low probability that two people would have the same set of RFLP markers.

V. DNA technology is catalyzing progress in many fields of biology

Recombinant DNA technology is a refined research tool which allows scientists to study more specific questions in almost all fields of biology.

- Study of the molecular details of eukaryotic gene structure and function is possible with the use of identified clones as probes.
  - ⇒ A radioactively labeled probe can be used to detect similar DNA segments in the same or different genomes by hydrogen bonding to complementary nucleotide sequences.
  - ⇒ These methods do not depend on gene expression, so geneticists do not have to infer genotype from phenotype.
- *In vitro mutagenesis* is a technique that can be used to determine the function of a protein product from cDNA cloning of an mRNA.
  - ⇒ Mutations are induced into the sequence of a cloned gene.
  - ⇒ The mutated gene is returned to the host cell.
  - ⇒ If the mutation alters the function of the protein product, it may be possible to determine the function of the protein by examining what changes occur in cell physiology or developmental pattern.
- Finding DNA segments with a probe is helpful, because it can:
  - ⇒ Provide information about evolutionary relationships between the gene of interest within or among species.
  - ⇒ Be used to examine the natural form of a gene, allowing the study of regulatory and other noncoding sequences adjacent to the gene.
  - ⇒ Increase the understanding of gene organization and control.

DNA probes can be used to map genes on eukaryotic chromosomes.

- *In situ hybridization* uses a radioactive DNA probe that base pairs with complementary sequences on intact chromosomes.
- Autoradiography and chromosome staining reveal to which band of which chromosome the probe has attached.

Recombinant DNA technology allows for the production of large quantities of proteins which are usually present in small amounts.

- Traditional techniques cannot purify and characterize many scarce molecules that control cell metabolism and development.
- The production of large amounts of proteins will allow scientists to catalog all proteins produced by particular organs such as the brain.

VI. **The Human Genome Project is an enormous collaborative effort to map and sequence DNA**

Four complementary approaches are being used in an attempt to map the entire human genome.

1. *Genetic mapping (linkage mapping) of the human genome.*
   - At least 3000 genes or other identifiable loci on the DNA (genetic markers) are to be identified initially. Many of the markers will be RFLPs.
   - This initial map will allow researchers to more easily map other genes by testing for genetic linkage to known markers.

2. *Physical mapping of the human genome.*
   - Each chromosome will be broken into a number of identifiable fragments.
   - The actual order of fragments in the chromosome will then be determined.

3. *Sequencing the human genome.*
   - The exact order of nucleotide pairs of each chromosome will be determined.
   - This will be the most time consuming part of the project as each haploid set of human chromosomes contains about 3 *billion* nucleotide pairs.

4. *Analyzing the genomes of other species.*
   - Scientists will also map the genomes of species that are particularly useful for genetic research.
   - Such species will include *E. coli*, yeast, the plant *Arabadopsis* and the mouse.

These four approaches will be used to completely map the human genome and provide an understanding of how the human genome compares to those of other organisms. Potential benefits include:

- Identification and mapping of genes responsible for genetic diseases will aid diagnosis, treatment and prevention.
- Detailed knowledge of the genomes of humans and other species will give insight into genome organization, control of gene expression, cellular growth and differentiation, and evolutionary biology.

Methods used to construct the human genetic map combine:

- Classical pedigree analysis of large families.
- Many molecular and recombinant DNA techniques.

Two valuable techniques used in the human genome project include: 1) *chromosome walking* (overlap hybridization) and 2) *PCR amplification*.

1. *Chromosome walking* (*overlap hybridization*). This process combines several techniques to order numerous restriction fragments that result from treating long eukaryotic DNA molecules with restriction enzymes. (See Campbell, Figure 19.8)
   - Overlapping DNA segments are created by cutting the original DNA with two different restriction enzymes; two libraries are produced with the resulting fragments.

- Beginning with a probe for an already known DNA sequence, the researcher "walks" along the chromosome from this point. From overlaps between DNA segments, one can deduce the order in which the segments occur in the original DNA.

2. *PCR amplification.*
   - PCR can amplify specific portions of DNA from individual sperm cells – the immediate products of meiotic recombination.
   - Researchers can amplify DNA in amounts sufficient for study, and analyze samples as large as thousands of sperm.
   - Based on the crossover frequencies between genes, researchers can deduce human linkage maps without having to find large families for pedigree analysis.

Achieving the goals of the Human Genome Project in a timely way will come from advances in automation and utilization of the latest electronic technology.

VII. DNA technology is reshaping the medical and pharmaceutical industries

Modern biotechnology has resulted in significant advances in many areas of medicine.

A. Diagnosis of Diseases

Medical scientists currently use DNA technology, in particular PCR and labeled DNA probes, to detect genetic diseases and identify heterozygote carriers.

- This allows early disease detection and identification of carriers for potentially harmful recessive mutations – even before the onset of symptoms.
- Genes have been cloned for many genetic disorders including hemophilia, phenylketonuria, cystic fibrosis, and Duchenne's muscular dystrophy.
- Gene cloning permits direct detection of gene mutations. A cloned normal gene can be used as a probe to find the corresponding gene in cells being tested; the alleles are compared with normal and mutant standards usually by RFLP analysis.

When the normal gene has not been cloned, a closely linked RFLP marker may indicate the presence of an abnormal allele if the RFLP marker is frequently co-inherited with the disease.

- Blood samples from relatives can be used to determine which RFLP marker is linked to the abnormal allele and which is linked to the normal allele.
- The RFLP markers must be different for the normal and abnormal alleles.
- Under these conditions, the RFLP marker variant found in the person being tested can reveal whether the normal or abnormal allele is likely to be present.
- Alleles for cystic fibrosis and Huntington's disease can be detected in this manner.

B.   Human Gene Therapy

Traceable genetic disorders in individuals may eventually be correctable.

- Theoretically, it should be possible to replace or supplement defective genes with functional normal genes using recombinant DNA techniques.
- Correcting somatic cells of individuals with well-defined, life-threatening genetic defects will be the starting place.

The principle behind human gene therapy is that normal genes are introduced into a patient's own somatic cells.   For this therapy to be effective, there are several requirements:

- The few cells receiving the normal gene must actively reproduce, so the normal gene will be replicated.
- The cells' normal protein products can correct the disorder.   Gene therapy is especially suited for single-enzyme deficiency diseases.

For example, several children with an immunodeficiency disease are currently being treated successfully by gene therapy; their disorder results from a lack of the enzyme adenosine deaminase (ADA).

- Some of the patient's T lymphocytes (a type of white blood cell) are removed and cultured in vitro with harmless retroviral vectors carrying the normal ADA allele.
- The recombinant retroviruses infect the cells, inserting the normal ADA allele (along with a DNA version of the viral genome).
- The patient's own engineered cells carrying a normal ADA allele are then returned to the patient's body by intravenous injection.

A newer approach using undifferentiated bone marrow cells could offer a broader, more long-lasting treatment. (See Campbell, Figure 19.10)

- The normal ADA allele is inserted into a patient's undifferentiated bone marrow cells rather than into lymphocytes.
- The bone marrow cells multiply and give rise to all the cells of the immune system. Also, the bone marrow cells continue to reproduce throughout life.

Many technical questions have been raised by this possibility:

- Can the proper genetic control mechanisms be made to operate on the transferred gene?
- At what stage of biological development is intervention most effective?
- When is intervention too late?
- Would undifferentiated or differentiated cells be best to use?
- How can the correct gene or gene product be placed in the tissues where it is required?
- How can developmental diseases be treated when a gene's product is only needed during a brief period in development?

Gene therapy raises difficult social and ethical questions:
- Who will have access to gene therapy? The procedures are very expensive and only available in major medical centers.
- Should we try to treat germ cells with the hope of correcting a defect in future generations?

Treating germ cells is possible and has been used in mice for some time.
- Mice defective in a gene for a crucial blood polypeptide have been treated with a corresponding human gene.
- Recipient mice and their descendants contain the active human gene not only in the proper location (erythrocytes) but also during the proper stage of development.
- Expression of foreign genes in other animal experiments has not been so successful.

Some critics feel that tampering with human genes is wrong regardless of the situation, because it will lead to eugenics. Others say that genetic engineering of somatic cells is no different than other conventional medical interventions used to save lives.

C. Vaccines and Other Pharmaceutical Products

Developing vaccines for prevention of infectious diseases is also possible using recombinant DNA techniques. Prevention by vaccine is the only way to fight diseases for which no treatment exists.

Traditional vaccines for viral diseases are of two types:
- Particles of virulent virus that have been inactivated by chemical or physical means.
- Active virus particles of an attenuated viral strain.

Since the particles in both cases are similar to active virus, both types of vaccine will trigger an animal's immune system to produce antibodies, which react very specifically against invading pathogens.

Biotechnology is being used in several ways to modify current vaccines and to produce new ones. Recombinant DNA techniques can be used to:
- Produce large amounts of specific protein molecules (*subunits*) from the protein coat of disease causing viruses, bacteria, or other microbes. If these protein subunits cause immune responses to the pathogens, they can be used as vaccines.
- Modify genomes of pathogens to directly attenuate them. Vaccination with live, attenuated organisms is more effective than a subunit vaccine. Small amounts of material trigger greater immune response, and pathogens attenuated by gene-splicing may also be safer than using natural mutants.
- Alter a virus to produce a vaccine against several diseases. For example, vaccinia, the virus used to produce smallpox vaccine, may have its genes that induce immunity to smallpox replaced with genes that induce immunity to other diseases. The vaccinia virus can even carry genes required to vaccinate for several diseases at once.

Recombinant DNA techniques use bacteria to produce hormones and other mammalian proteins for use in treating human patients. Insulin and growth hormone have already been approved.

Treatment of diabetics in the United States has been improved by genetically engineering bacteria to produce insulin.

- Insulin produced this way is chemically identical to that made by the human pancreas, so it causes fewer adverse reactions than previous insulin extracted from pig and cattle pancreas.

The growth hormone molecule is much larger than insulin (almost 200 amino acids long) and more species specific.

- Thus, growth hormone from other animals is not an effective growth stimulator in humans.
- Hormone produced by genetically engineered bacteria was approved in 1985 for the clinical treatment of children with hypopituitarism (pituitary dwarfism).
- Previously, these individuals were treated with growth hormone obtained from human cadavers.
- Genetically engineered growth hormone will eventually be used in other treatments (to promote healing in burn victims and individuals with bone fractures).

Erythropoietin (EPO) was produced in 1989 using recombinant DNA methods.

- This protein is normally produced by the kidney and stimulates erythrocyte production in the bone marrow.
- EPO produced by biotechnology will provide a source of the protein to treat anemia from various causes including kidney damage.

Disease-causing viruses are less susceptible to drug treatment than bacterial or fungal pathogens, so recent pharmaceutical research has focused on novel ways to fight viral infection. For example, two new approaches using biotechnology include the production of: a) *antisense nucleic acid* and b) *surface-receptor drugs*.

a. *Antisense nucleic acid*. Drugs may soon be produced that prevent viral mRNAs from being translated in an infected cell.

- For example, single-stranded antisense nucleic acid (DNA or RNA) could be used to base pair with viral mRNA and block its translation.

b. *Surface-receptor drugs*. Recombinant DNA techniques can be used to make drugs that either block or mimic surface receptors on cell membranes.

- For example, a new experimental drug substitutes for a receptor protein that HIV binds to when it attacks white blood cells. The premise is that the drug will decoy the AIDS virus and prevent it from infecting white blood cells.

VIII.   DNA technology offers forensic, environmental, and agricultural applications

A.   Forensic Uses of DNA Technology

Forensic labs can determine blood or tissue type from blood, small fragments of other tissue or semen left at the scene of violent crime. These tests, however, have limitations:

- They require fresh tissue in sufficient amounts for testing.
- This approach can exclude a suspect but is not evidence of guilt, because many people have the same blood type or tissue type.

Theoretically, DNA testing can identify an individual with certainty, since everyone's DNA base sequence is unique (except for identical twins). The most commonly used DNA technology for this forensic application is the RFLP autoradiograph. (See Campbell, Figure 19.11)

- This method is used to compare DNA samples from the suspect, the victim, and a small amount of semen, blood or other tissue found at the scene of the crime.
- Restriction fragments from the DNA samples are separated by electrophoresis; radioactive probes mark the bands containing RFLP markers.
- Even a small set of RFLP markers from an individual can provide a *DNA fingerprint* that is of forensic use; the probability that two individuals would have the same RFLP markers is quite low.

How reliable is DNA fingerprinting?

- Though each individual's DNA fingerprint is unique, most forensic tests do not analyze the entire genome but focus on tiny regions known to be highly variable from one person to another.
- Even though the probability is minute (one in a million) that two people will have matching DNA fingerprints for the few regions tested, some courts have thrown out DNA evidence because of uncertainty about its probability of being correct.

As with most new technology, forensic applications of DNA fingerprinting raises important ethical questions such as:

- Once collected, what happens to the DNA data?
- Should DNA fingerprints be filed or destroyed? Some states now save DNA data from convicted criminals.

B.   Environmental Uses of DNA Technology

Scientists are engineering metabolic capabilities of organisms so they can transform chemicals and thus help solve environmental problems. For example:

- Some microorganisms can extract heavy metals (i.e. copper, lead, and nickel) from their environments and incorporate them into recoverable compounds such as copper sulfate or lead sulfate.
- As metal reserves are depleted, genetically engineered microbes may perhaps be used in mining and in cleaning up mining waste.

Metabolic diversity of microbes is used in the recycling of wastes and detoxification of toxic chemicals

- Sewage treatment plants use microorganisms to degrade organic compounds into non-toxic forms.
- Biologically and chemically active compounds that cannot be easily degraded are often released into the environment.
    ⇒ The intent is to engineer microorganisms that can degrade these compounds and can be used in waste water treatment plants.
    ⇒ Such microbes might be incorporated directly into the manufacturing process, preventing toxic chemicals from ever being released as waste in the first place.

C. Agricultural Uses of DNA Technology

Recombinant DNA techniques are being used to study plants and animals of agricultural importance to improve their productivity.

1. Animal Husbandry

Products produced by recombinant DNA methods, such as vaccines, antibodies and growth hormone, are already used in animal husbandry. For example,

- *Bovine growth hormone* (*BGH*). Made by *E. coli*, BGH is injected into milk cows to enhance milk production and into beef cattle to increase weight gain.
- *Cellulase*. Also produced by *E. coli*, this enzyme hydrolyzes cellulose making all plant parts useable for animal feed.

Transgenic animals, those that contain DNA from other species, have been commercially produced by injecting foreign DNA into egg nuclei or early embryos.

- Transgenic beef and dairy cattle, hogs, sheep and several species of commercially raised fish have been produced for potential agricultural use.

2. Manipulating Plant Genes

Plant cells are easier to engineer than animal cells, because an adult plant can be produced from a single cell growing in tissue culture.

- This is important since many types of genetic manipulation are easier to perform and assess on single cells than on whole organisms.
- Asparagus, cabbage, citrus fruits, sunflowers, carrots, alfalfa, millet, tomatoes, potatoes, and tobacco are all commercial plants that can be grown from single somatic cells.

The best developed DNA vector for plant cells is the *Ti plasmid* (tumor inducing), carried by the normally pathogenic bacterium *Agrobacterium tumefaciens*.

- Ti plasmid usually induces tumor formation in infected plants by integrating a segment of its DNA (called T DNA) into the chromosomes of the host plant cell.
- Researchers have turned this plasmid into a useful vector by eliminating its disease-causing ability without interfering with its potential to move genetic material into infected plants.

With recombinant DNA methods, foreign genes can be inserted into Ti plasmid.

- The recombinant plasmid can either be put back into *Agrobacterium tumefaciens*, which is used to infect plant cells in culture, or it can be introduced directly into plant cells.
- Individual modified plant cells are then used to regenerate whole plants that contain, express and pass on to their progeny the foreign gene. (See Campbell, Figure 19.12)

Using Ti plasmid as a vector has one major drawback, only dicotyledons are susceptible to infection by *Agrobacterium*; important commercial plants such as corn and wheat are monocotyledons and cannot be infected.

- Newer methods that allow researchers to overcome this limitation are *electroporation* and use of the *DNA particle gun*.
- Electroporation uses high-voltage jolts of electricity to open temporary pores in the cell membrane; foreign DNA can enter the cells through these pores.
- The DNA gun shoots tiny DNA-coated metal pellets through the cell walls into the cytoplasm, where the foreign DNA becomes integrated into the host cell DNA.

Though cloning plant DNA is straightforward, plant molecular geneticists still face several technical problems:

- Identifying genes of interest may be difficult.
- Many important plant traits (such as crop yield) are polygenic.

3. **Some Early Successes of Genetic Engineering in Plants**

Genetic engineering of plants has yielded positive results in cases where useful traits are determined by single genes. For example,

- A bacterial gene that makes plants resistant to glyphosate (a powerful herbicide) has been successfully introduced into several crop plants; glyphosate-resistant plants makes it easier to grow crops and destroy weeds simultaneously.
- The first gene-spliced fruits approved by the FDA for human consumption were tomatoes engineered with antisense RNA to suppress ripening and retard spoilage. The general process is outlined below:
   a. Researchers clone the tomato gene coding for enzymes responsible for ripening.
   b. The complementary (antisense) gene is cloned.
   c. The antisense gene is spliced into the tomato plant's DNA, where it transcribes mRNA complementary to the ripening genes' mRNA.
   d. When the ripening gene produces a normal mRNA transcript, the antisense mRNA binds to it, blocking synthesis of the ripening enzyme.

Crop plants are being engineered to resist pathogens and pest insects. For example,

- Tomato and tobacco plants can be engineered to carry certain genes of viruses that normally infect and damage plants; expression of these versions of viral genes confer resistance to viral attack.

- Some plants have been engineered to resist insect attack, reducing the need to apply chemical insecticides to crops. Genes for an insecticidal protein have been successfully transferred from the bacterium, *Bacillus thuringiensis*, to corn, cotton and potatoes; field tests show that these engineered plants are insect resistant.

In the near future, crop plants developed with recombinant DNA techniques are likely to:

- Be made more productive by enlarging agriculturally valuable parts - whether they be roots, leaves, flowers, or stems.

- Have an enhanced food value. For example, corn and wheat might be engineered to produce mixes of amino acids optimal for the human diet.

4. The Nitrogen Fixation Challenge

Nitrogen fixation is the conversion of atmospheric, gaseous nitrogen (N_2) into nitrogen-containing compounds.

- Gaseous nitrogen is useless to plants.

- Nitrogen-fixing bacteria live in the soil or are symbiotic within plant roots.

- The nitrogen-containing compounds produced by nitrogen fixation are taken up from the soil by plants and used to make organic molecules such as amino acids and nucleotides.

- A major limiting factor in plant growth and crop yield is availability of useable nitrogen compounds. This is why nitrogenous fertilizers are used in agriculture.

Recombinant DNA technology can possibly be used to increase biological nitrogen fixation of bacterial species living in the soil or in association with plants.

IX. DNA technology raises important safety and ethical questions

When scientists realized the potential power of DNA technology, they also became concerned that recombinant microorganisms could create hazardous new pathogens, which might escape from the laboratory.

- In response to these concerns, scientists developed and agreed upon a self-monitoring approach, which was soon formalized into federal regulatory programs.

- Today, governments and regulatory agencies worldwide are monitoring the biotechnology industry - promoting potential industrial, medical, and agricultural benefits, while ensuring that new products are safe.

- In the U.S., the FDA, National Institutes of Health (NIH) Recombinant DNA Advisory Committee, Department of Agriculture (USDA), and Environmental Protection Agency (EPA) set policies and regulate new developments in genetic engineering.

While genetic engineering holds enormous potential for improving human health and increasing agricultural productivity, new developments in DNA technology raise ethical concerns. For example, mapping the human genome will contribute to significant advances in gene therapy, but:

- Who should have the right to examine someone else's genes?
- Should a person's genome be a factor in their suitability for a job?
- Should insurance companies have the right to examine an applicant's genes?
- How do we weigh the benefits of gene therapy against assurances that the gene vectors are safe?

For environmental problems such as oil spills, genetically engineered organisms may be part of the solution, but what is their potential impact on native species?

For new medical products, what is the potential for harmful side effects, both short-term and long-term?

- New medical products must pass exhaustive tests before the FDA approves it for general marketing.
- Currently awaiting federal approval are hundreds of new genetically engineered diagnostic products, vaccines, and drugs - including some to treat AIDS and cancer.

For agricultural products, what are the potential dangers of introducing new genetically engineered organisms into the environment?

- Some argue that producing transgenic organisms is only an extension of traditional hybridization and should not be treated differently from the production of other hybrid crops or animals. The FDA holds that if products of genetic engineering are not significantly different from products already on the market, testing is not required.
- Others argue that creating transgenic organisms by splicing genes from one species to another is radically different from hybridizing closely related species of plants or animals.
- Some concerns are that genetically altered food products may contain new proteins that are toxic or cause severe allergies; genetically engineered crop plants could become superweeds resistant to herbicides, disease and insect pests; and engineered crop plants may hybridize with native plants and pass their new genes to closely related plants in the wild.

REFERENCES

Alberts, B., D. Bray, J. Lewis, M. Raff, K. Roberts and J.D. Watson. *Molecular Biology of the Cell*. 3rd ed. New York: Garland, 1994.

Campbell, N. *Biology*. 4th ed. Menlo Park, California: Benjamin/Cummings, 1996.

Watson, J.D., N.H. Hopkins, J.W. Roberts, J.A. Steitz and A.M. Weiner. *Molecular Biology of the Gene*. 4th ed. Menlo Park, California: Benjamin/Cummings, 1987.

CHAPTER 20

DESCENT WITH MODIFICATION:
A DARWINIAN VIEW OF LIFE

OUTLINE

I. Western culture resisted evolutionary views of life
- A. The Scale of Life and Natural Theology
- B. Cuvier, Fossils, and Catastrophism

II. Theories of geological gradualism helped clear the path for evolutionary biologists

III. Lamarck placed fossils in an evolutionary context

IV. Field research helped Darwin frame his view of life: *science as a process*
- A. The Voyage of the Beagle
- B. Darwin Focuses on Adaptation

V. *The Origin of Species* developed two main points: the occurrence of evolution and natural selection as its mechanism
- A. Descent with Modification
- B. Natural Selection and Adaptation

VI. Evidence from many fields of biology validates the evolutionary view of life
- A. Biogeography
- B. The Fossil Record
- C. Comparative Anatomy
- D. Comparative Embryology
- E. Molecular Biology

VII. What is theoretical about the Darwinian view of life?

OBJECTIVES

After reading this chapter and attending lecture, the student should be able to:
1. State the two major points Darwin made in *The Origin of Species* concerning the Earth's biota.
2. Compare and contrast Plato's philosophy of idealism and Aristotle's *scala naturae*.
3. Describe Carolus Linnaeus' contribution to Darwin's theory of evolution.
4. Describe Georges Cuvier's contribution to paleontology.
5. Explain how Cuvier and his followers used the concept of catastrophism to oppose evolution.
6. Explain how the principle of gradualism and Charles Lyell's theory of uniformitarianism influenced Darwin's ideas about evolution.
7. Describe Jean Baptiste Lamarck's model for how adaptations evolve.
8. Describe how Charles Darwin used his observations from the voyage of the HMS *Beagle* to formulate and support his theory of evolution.
9. Describe how Alfred Russel Wallace influenced Charles Darwin.
10. Explain what Darwin meant by the principle of common descent and "descent with modification".
11. Explain what evidence convinced Darwin that species change over time.
12. State, in their own words, three inferences Darwin made from his observations, which led him to propose natural selection as mechanism for evolutionary change.
13. Explain why variation was so important to Darwin's theory.
14. Explain how Reverend Thomas Malthus' essay influenced Charles Darwin.
15. Distinguish between artificial selection and natural selection.
16. Explain why the population is the smallest unit that can evolve.
17. Using some contemporary examples, explain how natural selection results in evolutionary change.
18. Explain why the emergence of population genetics was an important turning point for evolutionary theory.
19. Describe the lines of evidence Charles Darwin used to support the principle of common descent.
20. Describe how molecular biology can be used to study the evolutionary relationships among organisms.
21. Explain the problem with the statement that Darwinism is "just a theory".
22. Distinguish between the scientific and colloquial use of the word "theory".

KEY TERMS

On the Origin of Species by Means of Natural Selection
Charles Darwin
evolution
Plato
Aristotle
idealism (essentialism)
scala naturae
natural theology
Carolus Linnaeus
taxonomy
fossils

catastrophism
paleontology
Georges Cuvier
gradualism
James Hutton
Charles Lyell
uniformitarianism
Jean Baptiste Lamarck
inheritance of acquired characteristics
Reverend John Henslow
Captain Robert FitzRoy

HMS *Beagle*
Principles of Geology
Galapagos islands
natural selection
Alfred Wallace
principle of common descent
descent with modification
Ernst Mayr
Reverend Thomas Malthus
artificial selection

population
modern synthesis
Gregor Mendel
mutations
orthogenesis
population genetics
biogeography
homologous structures
homology
vestigial organs
ontogeny
phylogeny
cytochrome *c*

LECTURE NOTES

Evolution, the unifying theme woven throughout the text and course, refers to the processes that have transformed life on earth from its earliest forms to the enormous diversity that characterizes it today.

The first convincing case for evolution was published in a book by Charles Darwin on November 24, 1859. In this book, *On the Origin of Species by Means of Natural Selection*, Darwin:

- Synthesized seemingly unrelated facts into a conceptual framework that accounts for both the unity and diversity of life.

- Discussed important biological issues about organisms, such as why there are so many kinds of organisms, their origins and relationships, similarities and differences, geographic distribution, and adaptations to their environment.

- Made two major points:

  1. Species evolved from ancestral species and were not specially created.

  2. *Natural selection* is a mechanism that could result in this evolutionary change.

I. Western culture resisted evolutionary views of life

The impact of Darwin's ideas partially depended upon historical and social context. (See Campbell, Figure 20.1)

- Darwin's view of life contrasted sharply with the accepted viewpoint: the Earth was only a few thousand years old and was populated by unchanging life forms made by the Creator during a single week.

- Thus, *On the Origin of Species by Means of Natural Selection* not only challenged prevailing scientific views, but also challenged the roots of Western culture.

A. The Scale of Life and Natural Theology

Many Greek philosophers believed in the gradual evolution of life. However, the two that influenced Western culture most, Plato (427 – 347 B.C.) and his student Aristotle (384 – 322 B.C.), held opinions which were inconsistent with a concept of evolution.

- Plato, whose philosophy is known as *idealism* (*essentialism*), believed that there were two coexisting worlds: an ideal eternal real world and an illusionary imperfect world that humans perceive with their senses. To Plato,

  ⇒ Variations in plant and animal populations were merely imperfect representatives of ideal forms; only these perfect ideal forms are real.

  ⇒ Evolution would be counterproductive in a world where ideal organisms are already perfectly adapted to their environments.

- Aristotle questioned the Platonic philosophy of dual worlds, but his beliefs also excluded evolution.

  ⇒ Recognizing that organisms range from simple to complex, he believed that they could be placed on a scale of increasing complexity (*scala naturae*); on this ladder of life, each form had its allotted rung and each rung was occupied.

  ⇒ In this view of life, species are fixed and do not evolve.

  ⇒ The *scala naturae* view of life prevailed for over 2000 years.

The *creationist–essentialist* dogma that species were individually created and fixed became embedded in Western thought as the Old Testament account of creation from the Judeo–Christian culture fortified prejudice against evolution.

- *Natural Theology*, a philosophy that the Creator's plan could be revealed by studying nature, dominated European and American biology even as Darwinism emerged.
- For natural theologians, adaptations of organisms were evidence that the Creator had designed every species for a particular purpose.
- Natural theology's major objective was to classify species revealing God's created steps on the ladder of life.

Carolus Linnaeus (1707 – 1778), a Swedish physician and botanist, sought order in the diversity of life *ad majorem Dei gloriam* (for the greater glory of God).

- Known as the father of *taxonomy* – the naming and classifying of organisms – he developed the system of *binomial nomenclature* still used today.
- He adopted a system for grouping species into categories and ranking the categories into a hierarchy. For example, similar species are grouped into a genus; similar genera are grouped into the same order; etc.

Linnaeus found order in the diversity of life with his hierarchy of taxonomic categories.

- The clustering of species in taxonomic groups did not imply evolutionary relationships to Linnaeus, since he believed that species were permanent creations.
- Linnaeus, a natural theologian, developed his classification scheme only to reveal God's plan and even stated *Deus creavit, Linnaeus disposuit* ("God creates, Linnaeus arranges").

B.   Cuvier, Fossils, and Catastrophism

Fossils = Relics or impressions of organisms from the past, preserved in rock.

- Most fossils are found in *sedimentary rocks*, which:
    ⇒ Form when new layers of sand and mud settle to the bottom of seas, lakes, and marshes, covering and compressing older layers into rock (e.g. sandstone and shale).
    ⇒ May be deposited in many layers (*strata*) in places where shorelines repeatedly advance and retreat.
    ⇒ Later erosion can wear away the upper (younger) strata, revealing older strata which had been buried.
- The fossil record thus provides evidence that Earth has had a succession of flora and fauna.

The study of fossils, *paleontology*, was founded by the French anatomist Georges Cuvier (1769-1832) who:

- Realized life's history was recorded in fossil-containing strata and documented the succession of fossil species in the Paris Basin.

- Noted each stratum was characterized by a unique set of fossil species and that the older (deeper) the stratum, the more dissimilar the flora and fauna from modern life forms.
- Understood that extinction had been a common occurrence in the history of life since, from stratum to stratum, new species appeared and others disappeared.

Even with paleontological evidence, Cuvier was an effective opponent to the evolutionists of his day.

- He reconciled the fossil evidence with his belief in the fixity of species by speculating that boundaries between fossil strata corresponded in time to catastrophic events, such as floods or droughts.
- This view of Earth's history is known as *catastrophism*.

  Catastrophism = Theory that major changes in the Earth's crust are the result of catastrophic events rather than from gradual processes of change.

Cuvier explained the appearance of new species in younger rock that were absent from older rock by proposing that:

- Periodic localized catastrophes resulted in mass extinctions.
- After the local flora and fauna had become extinct, the region would be repopulated by foreign species immigrating from other areas.

II. Theories of geological gradualism helped clear the path for evolutionary biologists

In the late 18th century, a new theory of geological *gradualism* gained popularity among geologists that would greatly influence Darwin.

Gradualism = Principle that profound change is the cumulative product of slow, continuous processes.

- Competed with Cuvier's theory of catastrophism.
- Proposed by James Hutton (1975), a Scottish geologist. He proposed that it was possible to explain the various land forms by looking at mechanisms currently operating in the world.
- For example, canyons form by erosion from rivers, and fossil-bearing sedimentary rocks form from particles eroded from the land and carried by rivers to the sea.

Charles Lyell, a leading geologist of Darwin's time expanded Hutton's gradualism into the theory known as *uniformitarianism*.

Uniformitarianism = Theory that geological processes are uniform and have operated from the origin of the Earth to the present.

- Was Lyell's extreme idea that geological processes are so uniform that their rates and effects must balance out through time.
- For example, processes that build mountains are eventually balanced by the erosion of mountains.

Darwin rejected uniformitarianism, but was greatly influenced by conclusions that followed directly from the observations of Hutton and Lyell:

- The Earth must be ancient. If geological change results from slow, gradual processes rather than sudden events, then the Earth must be much older than the 6000 years indicated by many theologians on the basis of biblical inference.
- Very slow and subtle processes persisting over a great length of time can cause substantial change.

III. Lamarck placed fossils in an evolutionary context

Several 18th century naturalists suggested that life had evolved along with Earth's changes. Only Jean Baptiste Lamarck (1744-1829) developed and published (1809) a comprehensive model which attempted to explain how life evolved.

In charge of the invertebrate collection at the Natural History Museum in Paris, Lamarck:

- Compared modern species to fossil forms and, in the process, identified several lines of descent composed of a chronological series of older fossils to younger fossils to modern species.
- Envisioned many ladders of life which organisms could climb (as opposed to Aristotle's single ladder without movement).
  - ⇒ The bottom rungs were occupied by microscopic organisms which were continually generated spontaneously from inanimate material.
  - ⇒ At the top of the ladders were the most complex plants and animals.

Lamarck believed that evolution was driven by an innate tendency toward increasing complexity, which he equated with perfection.

- As organisms attained perfection, they became better and better adapted to their environments.
- Thus, Lamarck believed that evolution responded to organisms' *sentiments interieurs* ("felt needs").

Lamarck proposed a mechanism by which specific adaptations evolve, which included two related principles:

1. *Use and disuse.* Those body organs used extensively to cope with the environment become larger and stronger while those not used deteriorate.
2. *Inheritance of acquired characteristics.* The modifications an organism acquired during its lifetime could be passed along to its offspring.

Although his mechanism of evolution was in error, Lamarck deserves credit for proposing that:

- Evolution is the best explanation for both the fossil record and the extant diversity of life.
- The Earth is ancient.
- Adaptation to the environment is a primary product of evolution.

IV. Field research helped Darwin frame his view of life: *science as a process*

At the beginning of the 19th century, natural theology still dominated the European and American intellectual climate. In 1809, the same year Lamarck published his theory of evolution, Charles Darwin was born in Shrewsbury, England.

- Though interested in nature, Charles (at 16) was sent by his physician father to the University of Edinburgh to study medicine, which he found boring and distasteful.
- He left Edinburgh without a degree and enrolled at Christ College, Cambridge University to prepare for the clergy.
    - ⇒ Nearly all naturalists and other scientists were clergymen, and a majority held to the philosophy of natural theology.
    - ⇒ Charles studied under the Reverend John Henslow, a botany professor at Cambridge, and received his B.A. degree in 1831.
    - ⇒ Professor Henslow recommended him to Captain Robert FitzRoy who was preparing the survey ship HMS *Beagle* for an around the world voyage.

A. The Voyage of the Beagle

The HMS *Beagle*, with Darwin aboard, sailed from England in December 1831.

- The voyage's mission was to chart the poorly known South American coastline.
- While the ship's crew surveyed the coast, Darwin spent most of his time ashore collecting specimens of the exotic and diverse flora and fauna.

While the ship worked its way around the continent, Darwin observed the various adaptations of plants and animals that inhabited the diverse environments of South America: Brazilian jungles, grasslands of the Argentine pampas, desolate islands of Tierra del Fuego, and the Andes Mountains. Darwin noted the following:

- The South American flora and fauna from different regions were distinct from the flora and fauna of Europe.
- Temperate species were taxonomically closer to species living in tropical regions of South America than to temperate species of Europe.
- The South American fossils he found (while differing from modern species) were distinctly South American in their resemblance to the living plants and animals of that continent.

Geographical distribution was particularly confusing in the case of the fauna of the Galapagos, recently formed volcanic islands which lie on the equator about 900 km west of South America.

- Most animal species on the Galapagos are unique to those islands, but resemble species living on the South American mainland.
- Darwin collected 13 types of finches from the Galapagos, and although they were similar, they seemed to be different species.
    - ⇒ Some were unique to individual islands.
    - ⇒ Others were found on two or more islands that were close together.

By the time the *Beagle* left the Galapagos, Darwin had read Lyell's *Principles of Geology*, and was influenced by Lyell's ideas.

- Darwin had begun to doubt the church's position that the Earth was static and had been created only a few thousand years before.
- When Darwin acknowledged that the Earth was ancient and constantly changing, he had taken an important step toward recognizing that life on Earth had also evolved.

B.   Darwin Focuses on Adaptation

Darwin was not sure whether the 13 types of finches he collected on the Galapagos were different species or varieties of the same species.

- After he returned to England in 1836, an ornithologist indicated that they were actually different species.
- He reassessed observations made during the voyage and in 1837 began the first notebook on the origin of species.

Darwin perceived the origin of new species and adaptation as closely related processes; new species could arise from an ancestral population by gradually accumulating adaptations to a different environment. For example,

- Two populations of a species could be isolated in different environments and diverge as each adapted to local conditions.
- Over many generations, the two populations could become dissimilar enough to be designated separate species.
- This is apparently what happened to the Galapagos finches; their different beaks are adaptations to specific foods available on their home islands. (See Campbell, Figure 20.4)

By the early 1840's, Darwin had formed his theory of natural selection as the mechanism of adaptive evolution, but delayed publishing it.

- Reclusive and in poor health, Darwin was well known as a naturalist from the specimens and letters he had sent to Britain from the voyage on the *Beagle*.
- He frequently corresponded and met with Lyell, Henslow, and other scientists.

In 1844, Darwin wrote a long essay on the origin of species and natural selection.

- He realized the importance and subversive nature of his work, but did not publish the information because he wished to gather more evidence in support of his theory.
- Evolutionary thinking was emerging at this time, and Lyell admonished Darwin to publish on the subject before someone else published it first.

In June 1858, Darwin received a letter from Alfred Wallace who was working as a specimen collector in the East Indies.

- Accompanying the letter was a manuscript detailing Wallace's own theory of natural selection which was almost identical to Darwin's.
- The letter asked Darwin to evaluate the theory and forward the manuscript to Lyell if it was thought worthy of publication.
- Darwin did so, although he felt that his own originality would be "smashed."

- Lyell and a colleague presented Wallace's paper along with excerpts from Darwin's unpublished 1844 essay to the Linnaean Society of London on July 1, 1858.

Darwin finished *The Origin of Species* and published it the next year.

- Darwin is considered the main author of the idea since he developed and supported natural selection much more extensively than Wallace.
- Darwin's book and its proponents quickly convinced the majority of biologists that biodiversity is a product of evolution.
- Darwin succeeded where previous evolutionists had failed not only because science was moving away from natural theology, but because he convinced his readers with logic and evidence.

V. *The Origin of Species* developed two main points: the occurrence of evolution and natural selection as its mechanism

A. Descent with Modification

Darwin used the phrase "descent with modification," not evolution, in the first edition of *The Origin of Species*.

- He perceived a unity in life with all organisms related through descent from some unknown ancestral population that lived in the remote past.
- Diverse modifications (*adaptations*) accumulated over millions of years, as descendants from this common ancestor moved into various habitats.

Darwin's metaphor for the history of life was a branching tree with multiple branching from a common trunk to the tips of living twigs, symbolic of the diversity of contemporary organisms.

- At each fork or branch point is an ancestral population common to all evolutionary lines of descent branching from that fork.
- Species that are very similar share a common ancestor at a recent branch point on the phylogenetic tree.
- Less closely related organisms share a more ancient common ancestor at an earlier branch point.
- Most branches of evolution are dead ends since about 99% of all species that ever lived are extinct.

To Darwin, Linnaeus' taxonomic scheme reflected the branching genealogy of the tree of life.

- It recognized that the diversity of organisms could be ordered into "groups subordinate to groups" with organisms at the different taxonomic levels related through descent from common ancestors.
- Classification alone does not confirm the principle of common descent, but when combined with other lines of evidence, the relationships are clear.
- For example, genetic analysis of species that are thought to be closely related on the basis of anatomical features and other criteria reveals a common hereditary background.

B. Natural Selection and Adaptation

Darwin's book focused on the role of natural selection in adaptation. Ernst Mayr of Harvard University dissected the logic of Darwin's theory into three inferences based on five observations:

Observation 1: All species have such great fertility that their population size would increase exponentially if all individuals that are born would reproduce successfully.

Observation 2: Most populations are normally stable in size except for seasonal fluctuations.

Observation 3: Natural resources are limited.

> Inference 1: Production of more individuals than the environment can support leads to a struggle for existence among individuals of a population, with only a fraction of offspring surviving each generation.

Observation 4: Individuals of a population vary extensively in their characteristics; no two individuals are exactly alike.

Observation 5: Much of this variation is heritable.

> Inference 2: Survival in the struggle for existence is not random, but depends in part on the hereditary constitution of the surviving individuals. Those individuals whose inherited characteristics fit them best to their environment are likely to leave more offspring than less fit individuals.

> Inference 3: This unequal ability of individuals to survive and reproduce will lead to a gradual change in a population, with favorable characteristics accumulating over the generations.

Natural selection is this differential success in reproduction, and its product is adaptation of organisms to their environment.

- Natural selection occurs from the interaction between the environment and the inherent variability in a population.
- Variations in a population arise by chance, but natural selection is not a chance phenomenon, since environmental factors set definite criteria for reproductive success.

Darwin was already aware of the struggle for existence caused by overproduction, when he read an essay on human population written by the Reverend Thomas Malthus (1798).

- In this essay, Malthus held that much of human suffering was a consequence of human populations growing faster than the food supply.
- This capacity for overproduction is common to all species, and only a fraction of new individuals complete development and leave offspring of their own; the rest die or are unable to reproduce.

Variation and overproduction in populations make natural selection possible.

- On the average, the most fit individuals pass their genes on to more offspring than less fit individuals.
- This results from environmental editing, which favors some variations over others.

From his experiences with *artificial selection*, Darwin inferred that natural selection could cause substantial change in populations.

- Through the breeding of domesticated plants and animals, humans have modified species over many generations by selecting individuals with desired traits as breeding stock.

- The plants and animals we grow for food show little resemblance to their wild ancestors.

- Darwin reasoned that if such change could be achieved by artificial selection in a relatively short period of time, then natural selection should be capable of considerable modifications of species over hundreds of thousands of generations.

Gradualism is fundamental to the Darwinian view of evolution. Darwin reasoned that:

- Life did not evolve suddenly by quantum leaps, but instead by a gradual accumulation of small changes.

- Natural selection operating in differing contexts over vast spans of time could account for the diversity of life.

Summarizing Darwin's view of evolution:

- The diverse forms of life have arisen by descent with modification from ancestral species.

- The mechanism of modification has been natural selection working gradually over long periods of time.

1. Some Subtleties of Natural Selection

Populations are important in evolutionary theory, since a population is the smallest unit that can evolve.

Population = A group of interbreeding individuals belonging to a particular species and sharing a common geographic area.

Natural selection is a consequence of interactions between individual organisms and their environment, but individuals do not evolve.

- Evolution can only be measured as change in relative proportions of variations in a population over several generations.

- Natural selection can only amplify or diminish heritable variations.

- Organisms can adapt to changes in their immediate environment and can be otherwise modified by life experiences, but these acquired characteristics cannot be inherited.

- Evolutionists must distinguish between adaptations an organism acquires during its lifetime and those inherited adaptations that evolve in a population over many generations as a result of natural selection.

Specifics of natural selection are situational.

- Environmental factors vary from area to area and from time to time.

- An adaptation under one set of conditions may be useless or detrimental in different circumstances.

2. Natural Selection in Action: Two Examples

In an effort to test Darwin's hypothesis that the beaks of Galapagos finches are evolutionary adaptations to different food sources, Peter and Rosemary Grant of Princeton University have been conducting a long-term study on medium ground finches (*Geospiza fortis*) on Daphne Major, a tiny Galapagos island. They have discovered that:

- Average beak depth (an inherited trait) oscillates with rainfall.
  - ⇒ In wet years, birds preferentially feed on small seeds, and average beak depth decreases.
  - ⇒ In dry years, small seeds are less plentiful, so survival depends on the finches being able to crack the less preferred larger seeds. Average beak depth increases during dry years.
- It can be inferred that the change in beak depth is an adaptive response to the relative availability of small seeds from year to year.

This study illustrates some important points about adaptive change:

- *Natural selection is situational.* What works in one environmental context may not work in another.
- *Beak evolution on Daphne Major does not result from inheritance of acquired characteristics.* The environment did not *create* beaks specialized for large or small seeds, but only acted on inherited variations already present in the population. The proportion of thicker-beaked finches increased during dry periods because, on average, thicker-beaked birds transmitted their genes to more offspring than did thinner-beaked birds.

Michael Singer and Camille Parmesan of the University of Texas, have documented rapid evolutionary adaptation in a butterfly population (Edith's checkerspot) living in a meadow near Carson City, Nevada.

- In only a decade, this butterfly population apparently adapted to changing vegetation by inherited changes in reproductive behavior.
- Females lay eggs preferentially on certain plants which provide food for the larvae after they hatch. In 1983, checkerspots laid about 80% of their eggs on a native plant, *Collinsia parviflora*.
- By 1993, the butterflies were laying about 70% of their eggs on *Plantago lanceolata*, an invading weed from surrounding cattle ranches.
- The researchers demonstrated that the switch in plant preference is genetic; daughters of butterflies that deposited eggs on *Plantago* inherited the taste for that plant, choosing it over *Collinsia* when they laid their eggs.

There are hundreds of examples of natural selection in laboratory populations of *Drosophila*. A few other examples of natural selection in action include:

- Antibiotic resistance in bacteria.
- Body size of guppies exposed to different predators. (See Campbell, Chapter 1)

VI.   **Evidence from many fields of biology validates the evolutionary view of life**

Darwin used several lines of evidence to support his principle of common descent, an evolutionary change. Recent discoveries, including those from molecular biology, lend support to his evolutionary view of life.

A.   **Biogeography**

It was *biogeographical* evidence that first suggested common descent to Darwin, because the biogeographical patterns he observed only made sense in the light of evolution.

Biogeography = The geographical distribution of species.

Islands have many endemic species which are closely related to species on the nearest mainland or neighboring island. Some logical questions follow:

- Why are two islands with similar environments in different parts of the world not populated by closely related species, but rather by species more closely related to those from the nearest mainland even when that environment is quite different?

- Why are South American tropical animals more closely related to South American desert animals than to African tropical animals?

- Why does Australia have such a diversity of marsupial animals and very few placental animals even though the environment can easily support placentals?

B.   **The Fossil Record**

Darwin was troubled by the absence of transitional fossils linking modern life to ancestral forms.

- Even though the fossil record is still incomplete, paleontologists continue to find important new fossils, and many key links are no longer missing.

- For example, fossilized whales link these aquatic mammals to their terrestrial predecessors.

Although still incomplete, the fossil record provides information that supports other types of evidence about the major branches of the phylogenetic tree. For example:

- Prokaryotes are placed as the ancestors of all life by evidence from cell biology, biochemistry, and molecular biology.

- Fossil evidence shows the chronological appearance of the vertebrates as being sequential with fishes first, followed by amphibians, reptiles and then birds and mammals. This sequence is also supported by many other types of evidence.

C. Comparative Anatomy

Anatomical similarities among species grouped in the same taxonomic category are a reflection of their common descent.

- The skeletal components of mammalian forelimbs are a good example. (See Campbell, Figure 20.10)
    - ⇒ Although the limbs are used for different functions, it is obvious that the same skeletal elements are present.
    - ⇒ It is logical that whether the forelimb is a foreleg, wing, flipper, or arm, the basic similarity is the consequence of descent from a common ancestor and that the limbs have been modified for different functions. They are *homologous* structures.

Homologous structures = Structures that are similar because of common ancestry.

- Other evidence from comparative anatomy supports that evolution is a remodeling process in which ancestral structures that functioned in one capacity have become modified as they take on new functions.
- Some homologous structures are *vestigial organs*.

Vestigial organs = Rudimentary structures of marginal or no use to an organism.

- Vestigial organs are remnants of structures that had important functions in ancestral forms but are no longer essential.
- An example of vestigial organs are the remnants of pelvic and leg bones in snakes. They show descent from a walking ancestor, but have no function in the snake.

D. Comparative Embryology

Closely related organisms go through similar stages in their embryonic development.

Vertebrate embryos (fishes, amphibians, reptiles, birds, mammals) go through an embryonic stage in which they possess gill slits on the sides of their throats.

- As development progresses, the gill slits develop into divergent structures characteristic of each vertebrate class.?
- In fish, the gill slits form gills; in humans, they form the eustachian tubes that connect the middle ear with the throat.
- Development of diverse organs with different functions from the gill slits common to all vertebrate embryos supports the conclusion that all vertebrates descended from aquatic ancestors with gills.

In the late nineteenth century, embryologists developed the view that "ontogeny recapitulates phylogeny."

- This view held that the embryonic development of an individual organism (*ontogeny*) is a replay of the evolutionary history of the species (*phylogeny*).
- This is an extreme view; what does occur is a series of similar *embryonic* stages that exhibit the same characteristics, not a sequence of adult-like stages.
- These characteristics then become modified, possibly by natural selection, since even embryonic processes can ultimately affect the fitness of the adult organism.

Comparative embryology can often establish homology among structures that are so altered in later development that their common origin can not be determined by comparing their fully developed forms.

E.   Molecular Biology

An organism's hereditary background is reflected in its genes and their protein products.
- Siblings have greater similarity in their DNA and proteins than do two unrelated organisms of the same species.
- Likewise, two species considered to be closely related by other criteria should have a greater proportion of their DNA and proteins in common than more distantly related species.

Molecular taxonomists use a variety of modern techniques to measure the degree of similarity among DNA nucleotide sequences of different species.
- The closer two species are taxonomically, the higher the percentage of common DNA; this evidence supports common descent.
- Common descent is also supported by the fact that closely related species also have proteins of similar amino acid sequence (resulting from inherited genes).
- If two species have many genes and proteins with sequences of monomers that match closely, the sequences must have been copied from a common ancestor.

Molecular biology has also substantiated Darwin's idea that all forms of life are related to some extent through branching descent from the earliest organisms.
- Even taxonomically distant organisms (bacteria and mammals) have some proteins in common.
- For example, cytochrome c (the respiratory protein) is found in all aerobic species. Cytochrome c molecules of all species are very similar in structure and function even though mutations have substituted amino acids in some areas of the protein during the course of evolution.
- Additional evidence for the unity of life is the common genetic code. This mechanism has been passed through all branches of evolution since its beginning in an early form of life.

VII.   What is theoretical about the Darwinian view of life?

Dismissing Darwinism as "just a theory" is flawed because:
- Darwin made *two* claims:
    1. Modern species evolved from ancestral forms.
    2. The mechanism for evolution is natural selection.
- The conclusion that species change or evolve is based on historical fact.

What then is theoretical about evolution?

- Theories are conceptual frameworks with great explanatory power used to interpret facts.
- That species can evolve is fact, but the *mechanism* Darwin proposed for that change – natural selection – is a theory. Darwin used this theory of natural selection to explain facts of evolution documented by fossils, biogeography, and other historical evidence.

In science, "theory" is very different from the colloquial use of the word, which comes closer to what scientists mean by a hypothesis, or educated guess.

- Unifying concepts do not become scientific theories, unless their predictions stand up to thorough and continuous testing by experiment and observation.
- Good scientists, however, do not allow theories to become dogma; many evolutionary biologists now question whether natural selection alone can account for evolutionary history observed in the fossil record.

REFERENCES

Campbell, N. *Biology*. 4th ed. Menlo Park, California: Benjamin/Cummings, 1996.

Futuyma, D.J. *Evolutionary Biology*. 2nd ed. Sunderland, Massachusetts: Sinauer, 1986. Because it is broad in scope, this advanced textbook is a good reference for background information.

Mayr, E. *The Growth of Biological Thought: Diversity, Evolution and Inheritance*. Cambridge, Massachusetts: Harvard University Press, 1982. This is an excellent reference for historical perspective and could be a useful companion to the text for supplementing lecture material on many topics.

CHAPTER 21

THE EVOLUTION OF POPULATIONS

OUTLINE

X. Genetic variation is the substrate for natural selection
   A. How Extensive *Is* Genetic Variation Within and Between Populations?
   B. How *Is* Genetic Variation Generated?
   C. How *Is* Genetic Variation Preserved?
   D. Does All Genetic Variation Affect Survival and Reproductive Success?

XI. Natural selection is the mechanism of adaptive evolution
   A. Fitness
   B. What Does Selection Act On?
   C. Modes of Natural Selection
   D. Sexual Selection

XII. Does evolution fashion perfect organisms?

OBJECTIVES

After reading this chapter and attending lecture, the student should be able to:
1. Explain what is meant by the "modern synthesis".
2. Explain how microevolutionary change can affect a gene pool.
3. In their own words, state the Hardy-Weinberg theorem.
4. Write the general Hardy-Weinberg equation and use it to calculate allele and genotype frequencies.
5. Explain the consequences of Hardy-Weinberg equilibrium.
6. Demonstrate, with a simple example, that a disequilibrium population requires only one generation of random mating to establish Hardy-Weinberg equilibrium.
7. Describe the usefulness of the Hardy-Weinberg model to population geneticists.
8. List the conditions a population must meet in order to maintain Hardy-Weinberg equilibrium.
9. Explain how genetic drift, gene flow, mutation, nonrandom mating and natural selection can cause microevolution.
10. Explain the role of population size in genetic drift.
11. Distinguish between the bottleneck effect and the founder effect.
12. Explain why mutation has little quantitative effect on a large population.
13. Describe how inbreeding and assortative mating affect a population's allele frequencies and genotype frequencies.
14. Explain, in their own words, what is meant by the statement that natural selection is the only agent of microevolution which is adaptive.
15. Describe the technique of electrophoresis and explain how it has been used to measure genetic variation within and between populations.
16. List some factors that can produce geographical variation among closely related populations.
17. Explain why even though mutation can be a source of genetic variability, it contributes a negligible amount to genetic variation in a population.
18. Give the cause of nearly all genetic variation in a population.
19. Explain how genetic variation may be preserved in a natural population.
20. In their own words, briefly describe the neutral theory of molecular evolution and explain how changes in gene frequency may be nonadaptive.
21. Explain what is meant by "selfish" DNA.
22. Explain the concept of relative fitness and its role in adaptive evolution.

23. Explain why the rate of decline for a deleterious allele depends upon whether the allele is dominant or recessive to the more successful allele.
24. Describe what selection acts on and what factors contribute to the overall fitness of a genotype.
25. Give examples of how an organism's phenotype may be influenced by the environment.
26. Distinguish among stabilizing selection, directional selection and diversifying selection.
27. Define sexual dimorphism and explain how it can influence evolutionary change.
28. Give at least four reasons why natural selection cannot breed perfect organisms.

KEY TERMS

species	inbreeding	sexual recombination	relative fitness
gene flow	assortative mating	balanced	selection coefficient
gene pool	polygenic traits	polymorphism	pleiotropy
fixed allele	morphs	heterozygote	coadapted gene
microevolution	polymorphic	advantage	complex
Hardy-Weinberg	electrophoresis	sickle-cell anemia	norm of reaction
equilibrium	geographical	frequency-dependent	stabilizing selection
nonadaptive	variation	selection	directional selection
genetic drift	cline	Batesian mimicry	diversifying selection
sampling error	step cline	neutral variation	sexual dimorphism
bottleneck effect	point-mutation	selfish DNA	sexual selection
founder effect	supergenes	adaptive evolution	

LECTURE NOTES

Natural selection works on individuals, but it is the *population* that evolves. Darwin understood this, but was unable to determine its genetic basis.

I.   **The modern evolutionary synthesis integrated Darwinism and Mendelism:** *science as a process*

Shortly after the publication of *The Origin of Species*, most biologists were convinced that species evolved. Darwin was less successful in convincing them that natural selection was the mechanism for evolution, because little was known about inheritance.

- An understanding about inheritance was necessary to explain:
  ⇒ How chance variations arise in populations.
  ⇒ How these variations are precisely transmitted from parents to offspring.
- Though Gregor Mendel was a contemporary of Darwin's, Mendel's principles of inheritance went unnoticed until the early 1900's.

For Darwin, the raw material for natural selection was variation in quantitative characters that vary along a continuum in a population.

- We now know that continuous variation is usually determined by many segregating loci (polygenic inheritance).
- As did Mendel, geneticists in the early 1900's recognized only discrete characters inherited on an either-or basis. Thus, for them, there appeared to be no genetic basis for the subtle variations that were central to Darwin's theory.

During the 1920's, genetic research focused on mutations, and a widely accepted alternative to Darwin's theory was that evolution occurs in rapid leaps as a result of radical phenotypic changes caused by mutations.

- This idea contrasted with Darwin's view of gradual evolution due to environmental selection acting on continuous variation among individuals of a population.
- Another popular theory was *orthogenesis*, the idea that evolution has been a predictable progression to more and more elite forms of life.

In the 1930's, the science of *population genetics* emerged, which:

- Emphasized genetic variation within populations and recognized the importance of quantitative characters.
- Was an important turning point for evolutionary theory, because it reconciled Mendelian genetics with Darwinian evolution.

In the 1940's, the genetic basis of variation and natural selection was worked out, and *the modern synthesis* was formulated. This comprehensive theory:

- Integrated discoveries from different fields (i.e. paleontology, taxonomy, biogeography, and population genetics).
- Was collectively developed by many scientists including:
  - ⇒ Theodosius Dobzhansky – geneticist
  - ⇒ Ernst Mayr – biogeographer and systematist
  - ⇒ George Gaylord Simpson – paleontologist
  - ⇒ G. Ledyard Stebbins – botanist
- Emphasized the following:
  - ⇒ Importance of populations as units of evolution.
  - ⇒ The central role of natural selection as the primary mechanism of evolutionary change.
  - ⇒ Gradualism as the explanation of how large changes can result from an accumulation of small changes occurring over long periods of time.

Most of Darwin's ideas persist in the modern synthesis although many evolutionists are challenging some generalizations of the modern synthesis.

- This debate focuses on the rate of evolution and on the relative importance of evolutionary mechanisms other than natural selection.
- These debates do not question the fact of evolution, only what mechanisms are most important in the process.
- Such disagreements indicate that the study of evolution is very lively and that it continues to develop as a science.

II.   A population has a genetic structure defined by its gene pool's allele and genotype frequencies

Population = Localized group of organisms which belong to the same *species*.

Species = Groups of actually or potentially interbreeding natural populations, which are reproductively isolated from other such groups.

Most species are not evenly distributed over a geographical range, but are concentrated in several localized population centers.
- Each population center is isolated to some extent from other population centers with only occasional gene flow among these groups.
- Obvious examples are isolated populations found on widely separated islands or in unconnected lakes.
- Some populations are not separated by such sharp boundaries.
    ⇒ For example, a species with two population centers may be connected by an intermediate sparsely populated range.
    ⇒ Even though these two populations are not absolutely isolated, individuals are more likely to interbreed with others from their population center. Gene flow between the two population centers is thus reduced by the intermediate range.

Gene pool = The total aggregate of genes in a population at any one time.
- Consists of all the alleles at all gene loci in all individuals of a population. Alleles from this pool will be combined to produce the next generation.
- In a diploid species, an individual may be homozygous or heterozygous for a locus since each locus is represented twice.
- An allele is said to be *fixed* in the gene pool if all members of the population are homozygous for that allele.
- Normally there will be two or more alleles for a gene, each having a relative frequency in the gene pool.

III.   The Hardy-Weinberg Theorem describes a nonevolving population

NOTE: The Hardy-Weinberg model is so much easier to teach if the students calculate gene frequencies along with the instructor. This means that you must pause frequently to allow plenty of time for students to actively process the information and practice the calculations.

In the absence of other factors, the segregation and recombination of alleles during meiosis and fertilization will not alter the overall genetic makeup of a population.
- The frequencies of alleles in the gene pool will remain constant unless acted upon by other agents; this is known as the *Hardy-Weinberg Theorem*.
- The Hardy-Weinberg model describes the genetics of nonevolving populations. This theorem can be tested with theoretical population models.

To test the Hardy-Weinberg theorem, imagine an isolated population of wildflowers with the following characteristics: (See Campbell, Figure 21.3)

- It is a diploid species with both pink and white flowers.
- The population size is 500 plants: 480 plants have pink flowers, 20 plants have white flowers.
- Pink flower color is coded for by the dominant allele "A," white flower color is coded for by the recessive allele "a."
- Of the 480 pink-flowered plants, 320 are homozygous (AA) and 160 are heterozygous (Aa). Since white color is recessive, all white flowered plants are homozygous aa.
- There are 1000 genes for flower color in this population, since each of the 500 individuals has two genes (this is a diploid species).
- A total of 320 genes are present in the 160 heterozygotes (Aa): half are dominant (160 A) and half are recessive (160 a).
- 800 of the 1000 total genes are dominant.
- The frequency of the A allele is 80% or 0.8 (800/1000).

Genotypes	# of plants		# of A alleles per individual		Total # A alleles
AA plants	320	×	2	=	640
Aa plants	160	×	1	=	160
					800

- 200 of the 1000 total genes are recessive.
- The frequency of the a allele is 20% or 0.2 (200/1000).

Genotypes	# of plants		# of A alleles per individual		Total # A alleles
aa plants	20	×	2	=	40
Aa plants	160	×	1	=	160
					200

Assuming that mating in the population is completely random (all male-female mating combinations have equal chances), the frequencies of A and a will remain the same in the next generation.

- Each gamete will carry one gene for flower color, either A or a.
- Since mating is random, there is an 80% chance that any particular gamete will carry the A allele and a 20% chance that any particular gamete will carry the a allele.

The frequencies of the three possible genotypes of the next generation can be calculated using the rule of multiplication: (See Campbell, Chapter 13)

- The probability of two A alleles joining is $0.8 \times 0.8 = 0.64$; thus, 64% of the next generation will be AA.
- The probability of two a alleles joining is $0.2 \times 0.2 = 0.04$; thus, 4% of the next generation will be aa.
- Heterozygotes can be produced in two ways, depending upon whether the sperm or ovum contains the dominant allele (Aa or aA). The probability of a heterozygote being produced is thus $(0.8 \times 0.2) + (0.2 \times 0.8) = 0.16 + 0.16 = 0.32$.

The frequencies of possible genotypes in the next generation are 64% AA, 32% Aa and 4% aa.

- The frequency of the A allele in the new generation is $0.64 + (0.32/2) = 0.8$, and the frequency of the a allele is $0.04 + (0.32/2) = 0.2$. Note that the alleles are present in the gene pool of the new population at the *same* frequencies they were in the original gene pool.
- Continued sexual reproduction with segregation, recombination and random mating would *not alter* the frequencies of these two alleles: the gene pool of this population would be in a state of equilibrium referred to as *Hardy-Weinberg equilibrium*.
- If our original population had not been in equilibrium, only one generation would have been necessary for equilibrium to become established.

From this theoretical wildflower population, a general formula, called the *Hardy-Weinberg equation*, can be derived to calculate allele and genotype frequencies.

- The Hardy-Weinberg equation can be used to consider loci with three or more alleles.
- By way of example, consider the simplest case with only two alleles with one dominant to the other.
- In our wildflower population, let p represent allele A and q represent allele a, thus $p = 0.8$ and $q = 0.2$.
- The sum of frequencies from all alleles must equal 100% of the genes for that locus in the population: $p + q = 1$.
- Where only two alleles exist, only the frequency of one must be known since the other can be derived:

$$1 - p = q \quad \text{or} \quad 1 - q = p$$

When gametes fuse to form a zygote, the probability of producing the AA genotype is p^2; the probability of producing aa is q^2; and the probability of producing an Aa heterozygote is $2pq$ (remember heterozygotes may be formed in two ways).

- The sum of these frequencies must equal 100%, thus:

$$\underset{\substack{\text{Frequency} \\ \text{of AA}}}{p^2} \quad + \quad \underset{\substack{\text{Frequency} \\ \text{of Aa}}}{2pq} \quad + \quad \underset{\substack{\text{Frequency} \\ \text{of aa}}}{q^2} \quad = \quad 1$$

The Hardy-Weinberg equation permits the calculation of allelic frequencies in a gene pool, if the genotype frequencies are known. Conversely, the genotype can be calculated from known allelic frequencies.

For example, the Hardy-Weinberg equation can be used to calculate the frequency of inherited diseases in humans (e.g. phenylketonuria):

- 1 of every 10,000 babies in the United States is born with phenylketonuria (PKU) which can result in mental retardation if untreated.
- The allele for PKU is recessive, so babies with this disorder are homozygous recessive = q^2.
- Thus $q^2 = 0.0001$, with $q = 0.01$ (the square root of 0.0001).
- The frequency of p can be determined since $p = 1 - q$:

$$p = 1 - 0.01 = 0.99$$

- The frequency of carriers (heterozygotes) in the population is 2pq.

$$2pq = 2(0.99)(0.01) = 0.0198$$

- Thus, about 2% of the U.S. population are carriers for PKU.

IV. Microevolution is a generation–to–generation change in a population's allele or genotype frequencies: *an overview*

The Hardy-Weinberg equilibrium is important to the study of evolution since it tells us what will happen in a *nonevolving* population.

- This equilibrium model provides a base line from which evolutionary departures take place.
- It provides a reference point with which to compare the frequencies of alleles and genotypes of natural populations whose gene pools may be changing.

For Hardy-Weinberg equilibrium to be maintained, five conditions *must* be met:

1. *Very large population size*.
2. *Isolation from other populations*. There is no migration of individuals into or out of the population.
3. *No mutations*.
4. *Random mating*.
5. *No natural selection*. All genotypes are equal in survival and reproductive success. Differential reproductive success can alter gene frequencies.

In real populations, several factors can upset Hardy-Weinberg equilibrium and cause *microevolutionary* change.

Microevolution = Small scale evolutionary change represented by a generational shift in a population's relative allelic frequencies.

- Microevolution can be caused by *genetic drift, gene flow, mutation, nonrandom mating,* and *natural selection;* each of these conditions is a deviation from the criteria for Hardy-Weinberg equilibrium.
- Of these five possible agents for microevolution, only natural selection generally leads to an accumulation of favorable adaptations in a population.
- The other four are nonadaptive and are usually called non-Darwinian changes.

V. Genetic drift can cause evolution via chance fluctuation in a small population's gene pool: *a closer look*

Genetic drift = Changes in the gene pool of a small population due to chance.

- If a population is small, stochastic events have a greater impact on gene frequencies.
- Chance events may cause the frequencies of alleles to drift randomly from generation to generation, since the existing gene pool may not be accurately represented in the next generation.

For example, assume our theoretical wildflower population contains only 25 plants, and the genotypes for flower color occur in the following numbers: 16 AA, 8 Aa and 1 aa. In this case, a chance event could easily change the frequencies of the two alleles for flower color.

- A rock slide or passing herbivore which destroys three AA plants would immediately change the frequencies of the alleles from A = 0.8 and a = 0.2, to A = 0.77 and a = 0.23.
- Although this change does not seem very drastic, the frequencies of the two alleles were changed by a chance event.

The larger the population, the less important is the effect of genetic drift.

- Even though natural populations are not infinitely large (in which case genetic drift could be completely eliminated as a cause of microevolution), most are so large that the effect of genetic drift is negligible.
- However, some populations are small enough that genetic drift can play a major role in microevolution, especially when the population has less than 100 individuals.

Two situations which result in populations small enough for genetic drift to be important are the *bottleneck effect* and the *founder effect.*

A. Bottleneck Effect

The size of a population may be reduced drastically by such natural disasters as volcanic eruptions, earthquakes, fires, floods, etc. which kill organisms nonselectively.

- The small surviving population is unlikely to represent the genetic makeup of the original population.
- Genetic drift which results from drastic reduction in population size is referred to as the *bottleneck effect.*

- By chance some individuals survive. In the small remaining population, some alleles may be overrepresented, some under-represented, and some alleles may be totally absent.
- Genetic drift which has occurred may continue to affect the population for many generations, until it is large enough for random drift to be insignificant.

The bottleneck effect reduces overall genetic variability in a population since some alleles may be entirely absent.

- For example, a population of northern elephant seals was reduced to just 20 individuals by hunters in the 1890's.
  - ⇒ Since this time, these animals have been protected and the population has increased to about 30,000 animals.
  - ⇒ Researchers have found that <u>no</u> variation exists in the 24 loci examined from the present population. A single allele has been fixed at each of the 24 loci due to genetic drift by the bottleneck effect.
  - ⇒ This contrasts sharply with the large amount of genetic variation found in southern elephant seal populations which did not undergo the bottleneck effect.
- A lack of genetic variation in South African cheetahs may also have resulted from genetic drift, since the large population was severely reduced during the last ice age and again by hunting to near extinction.

B. Founder Effect

When a few individuals colonize a new habitat, genetic drift is also likely to occur. Genetic drift in a new colony is called the *founder effect*.

- The smaller the founding population, the less likely its gene pool will be representative of the original population's genetic makeup.
- The most extreme example would be when a single seed or pregnant female moves into a new habitat.
- If the new colony survives, random drift will continue to affect allele frequencies until the population reaches a large enough size for its influence to be negligible.
- No doubt, the founder effect was instrumental in the evolutionary divergence of the Galapagos finches.

The founder effect probably resulted in the high frequency of *retinitis pigmentosa* (a progressive form of blindness that affects humans homozygous for this recessive allele) in the human population of Tristan da Cunha, a group of small Atlantic islands.

- This area was colonized by 15 people in 1814, and one must have been a carrier.
- The frequency of this allele is much higher on this island than in the populations from which the colonists came.

Although inherited diseases are obvious examples of the founder effect, this form of genetic drift can alter the frequencies of any alleles in the gene pool.

VI. **Gene flow can cause evolution by transferring alleles between populations:** *a closer look*

Gene flow = The migration of fertile individuals, or the transfer of gametes, between populations.

- Natural populations may gain or lose alleles by gene flow, since they do not have gene pools which are closed systems required for Hardy-Weinberg equilibrium.
- Gene flow tends to reduce between-population differences which have accumulated by natural selection or genetic drift.
- An example of gene flow would be if our theoretical wildflower population was to begin receiving wind blown pollen from an all white-flower population in a neighboring field. This new pollen could greatly increase the frequency of the white flower allele, thus also altering the frequency of the red flower allele.
- Extensive gene flow can eventually group neighboring populations into a single population.

VII. **Mutations can cause evolution by substituting one allele for another in a gene pool:** *a closer look*

A new mutation which is transmitted in gametes immediately changes the gene pool of a population by substituting one allele for another.

In our theoretical wildflower population, if a mutation in a white flowered plant caused that plant to begin producing gametes which carried a red flower allele, the frequency of the white flower allele is reduced and the frequency of the red flower allele is increased.

Mutation itself has little quantitative effect on large populations in a single generation, since mutation at any given locus is very rare.

- Mutation rates of one mutation per 10^5 to 10^6 gametes are typical, but vary depending on the species and locus.
- An allele with a 0.50 frequency in the gene pool that mutates to another allele at a rate of 10^{-5} mutations per generation would take 2000 generations to reduce the frequency of the original allele from 0.50 to 0.49.
- The gene pool is effected even less, since most mutations are reversible.
- If a new mutation increases in frequency, it is because individuals carrying this allele are producing a larger percentage of offspring in the population due to genetic drift or natural selection, not because mutation is producing the allele in abundance.

Mutation is important to evolution since it is the original source of genetic variation, which is the raw material for natural selection.

VIII. Nonrandom mating can cause evolution by shifting the frequencies of genotypes in a gene pool: *a closer look*

Nonrandom mating increases the number of homozygous loci in a population, but does not in itself alter frequencies of alleles in a population's gene pool. There are two kinds of nonrandom mating: *inbreeding* and *assortative mating*.

A. Inbreeding

Individuals of a population usually mate with close neighbors rather than with more distant members of a population, especially if the members of the population do not disperse widely.

- This violates the Hardy-Weinberg criteria that an individual must choose its mate at random from the population.
- Since neighboring individuals of a large population tend to be closely related, inbreeding is promoted.
- Self-fertilization, which is common in plants, is the most extreme example of inbreeding.

Inbreeding results in relative genotypic frequencies (p^2, $2pq$, q^2) that deviate from the frequencies predicted for Hardy-Weinberg equilibrium, but does not in itself alter frequencies of alleles (p and q) in the gene pool.

Self-fertilization in our theoretical wildflower population would increase the frequencies of homozygous individuals and reduce the frequency of heterozygotes.

- Selfing of AA and aa individuals would produce homozygous plants.
- Selfing of Aa plants would produce half homozygotes and half heterozygotes.
- Each new generation would see the proportion of heterozygotes decrease, while the proportions of homozygous dominant and homozygous recessive plants would increase.
- Inbreeding without selfing would also result in a reduction of heterozygotes, although it would take much longer.

One effect of inbreeding is that the frequency of homozygous recessive phenotypes increases.

An interesting thing to note is that even if the phenotypic and genotypic ratios change, the values of p and q do not change in these situations, only the way they are combined. A smaller proportion of recessive alleles are masked by the heterozygous state.

B. Assortative Mating

Assortative mating is another type of nonrandom mating which results when individuals mate with partners that are like themselves in certain phenotypic characters. For example:

- Snow geese occur in a blue variety and a white variety, with the blue color allele being dominant. Birds prefer to mate with those of their own color, blue with blue and white with white; this results in a lower frequency of heterozygotes than predicted by Hardy-Weinberg.
- Blister beetles (*Lytla magister*) in the Sonoran Desert usually mate with a same-size individual.

IX.   Natural selection can cause evolution via differential reproductive success among varying members of a population: *a closer look*

The Hardy-Weinberg equilibrium condition that all individuals in a population have equal ability to produce viable, fertile offspring is probably never met.

- In any sexually reproducing population, variation among individuals exists and some variants leave more offspring than others.
- *Natural selection* is this differential success in reproduction.

Due to selection, alleles are passed on to the next generation in disproportionate numbers relative to their frequencies in the present generation.

- If in our theoretical wildflower population, white flowers are more visible to herbivores than pink flowers, plants with pink flowers (both AA and Aa) would leave more offspring on the average.
- Genetic equilibrium would be disturbed and the frequency of allele A would increase and the frequency of the a allele would decrease.

Natural selection is the only agent of microevolution which is adaptive, since it accumulates and maintains favorable genotypes.

- Environmental change would result in selection favoring genotypes present in the population which can survive the new conditions.
- Variability in the population makes it possible for natural selection to occur.

X.   Genetic variation is the substrate for natural selection

Members of a population may vary in subtle or obvious ways. It is the genetic basis of this variation that makes natural selection possible.

A.   How Extensive *Is* Genetic Variation Within and Between Populations?

Darwin considered the slight differences between individuals of a population as raw material for natural selection.

While we are more conscious of the variation among humans, an equal if not greater amount of variation exists among the many plant and animals species.

- Phenotypic variation is a product of inherited genotype and numerous environmental influences.
- Only the genetic or inheritable component of variation can have adaptive impact as a result of natural selection.

Polygenic characters which vary quantitatively within a population are responsible for much of the inheritable variation.

- For example, the height of the individuals in our theoretical wildflower population may vary from very short to very tall with all sorts of intermediate heights.

Discrete characters which are determined by only one locus vary categorically, such as flower color in our wildflowers, without intermediates.

- In our wildflower population, the red and white flowers would be referred to as different *morphs* (contrasting forms of a Mendelian character).
- A population is referred to as *polymorphic* for a character if two or more morphs are present in noticeable frequencies. (See Campbell, Figure 21.8)
- Polymorphism is found in human populations not only in physical characters (e.g. presence or absence of freckles) but also in biochemical characters (e.g. ABO blood group).

Darwin did not realize the extent of genetic variation in populations, since much of the genetic variation can only be determined with biochemical methods.

- Electrophoresis has been used to determine genetic variation among individuals of a population. This technique allows researchers to identify variations in protein products of specific gene loci.
- Electrophoretic studies show that in *Drosophila* populations the gene pool has two or more alleles for about 30% of the loci examined, and each fly is heterozygous at about 12% of its loci.
- Thus, there are about 700 – 1200 heterozygous loci in each fly. Any two flies in a *Drosophila* population will differ in genotype at about 25% of their loci.
- Electrophoretic studies also show comparable variation in the human population.

Note that electrophoresis underestimates genetic variation:

- Proteins produced by different alleles may vary in amino acid composition and still have the same overall charge, which makes them indistinguishable by electrophoresis.
- Also, DNA variation not expressed as protein is not detected by electrophoresis.

Geographical variation in allele frequencies exists among populations of most species.

- Natural selection can contribute to geographical variation, since at least some environmental factors are different between two locales. For example, one population of our wildflowers may have a higher frequency of white flowers because of the prevalence in that area of pollinators that prefer white flowers.
- Genetic drift may cause chance variations among different populations.
- Also, subpopulations may appear within a population due to localized inbreeding resulting from a "patchy" environment.

<u>Cline</u> = One type of geographical variation that is a graded change in some trait along a geographic transect.

- Clines may result from a gradation in some environmental variable.
- It may be a graded region of overlap where individuals of neighboring populations interbreed.
- For example, the average body size of many North American mammal species gradually increases with increasing latitude. It is presumed that the reduced surface area to volume ratio associated with larger size helps animals in cold environments conserve body heat.
- Studies of geographical variation confirm that genetic variation affects spatial differences of phenotypes in some clines. For example, yarrow plants are shorter at higher elevations, and some of this phenotypic variation has a genetic basis. (See Campbell, Figure 21.9)

B.   How *Is* Genetic Variation Generated?

Genetic variation results from mutation and sexual recombination.

1.   Mutation

Mutations produce new alleles. They are rare and random events which usually occur in somatic cells and are thus not inheritable.

- Only mutations that occur in cell lines which will produce gametes can be passed to the next generation.
- Geneticists estimate that only an average of one or two mutations occur in each human gamete-producing cell line.

<u>Point mutation</u> = Mutation affecting a single base in DNA.

- Much of the DNA in eukaryotes does not code for proteins, and it is uncertain how a point mutation in these regions affect an organism.
- Point mutations in structural genes may cause little effect, partly due to the redundancy of the genetic code.

Mutations that alter a protein enough to affect the function are more often harmful than beneficial, since organisms are evolved products shaped by selection and a chance change is unlikely to improve the genome.

- Occasionally, a mutant allele is beneficial, which is more probable when environmental conditions are changing.
- The mutation which allowed house flies to be resistant to DDT was present in the population and under normal conditions resulted in reduced growth rate. It became beneficial to the house fly population only after a new environmental factor (DDT) was introduced and tipped the balance in favor of the mutant alleles.

Chromosomal mutations usually affect many gene loci and tend to disrupt an organism's development.

- On rare occasions, chromosomal rearrangement may be beneficial. These instances (usually by translocation) may produce a cluster of genes with cooperative functions when inherited together.

Duplication of chromosome segments is nearly always deleterious.

- If the repeated segment does not severely disrupt genetic balance, it may persist for several generations and provide an expanded genome with extra loci.
- These extra loci may take on new functions by mutation while the original genes continue to function.
- Shuffling of exons within the genome (single locus or between loci) may also produce new genes.

Mutation can produce adequate genetic variation in bacteria and other microorganisms which have short generation times.

- Some bacteria reproduce asexually by dividing every 20 minutes, and a single cell can produce a billion descendants in only 10 hours.
- With this type of reproduction, a beneficial mutation can increase in frequency in a bacterial population very rapidly.
- A bacterial cell with a mutant allele which makes it antibiotic resistant could produce an extremely large population of clones in a short period, while other cells without that allele are eliminated.
- Although bacteria reproduce primarily by asexual means, most increase genetic variation by occasionally exchanging and recombining genes through processes such as conjugation, transduction and transformation.

2. Recombination

The contribution of mutations to genetic variation is negligible.

- Mutations are so infrequent at a single locus that they have little effect on genetic variation in a large gene pool.
- Although mutations produce new alleles, nearly all genetic variation in a population results from new combinations of alleles produced by sexual recombination.

Gametes from each individual vary extensively due to crossing over and random segregation during meiosis.

- Thus, each zygote produced by a mating pair possesses a unique genetic makeup.
- Sexual reproduction produces new combinations of old alleles each generation.

Plants and animals depend almost entirely on sexual recombination for genetic variation which makes adaptation possible.

C.   **How *Is* Genetic Variation Preserved?**

Natural selection tends to produce genetic uniformity in a population by eliminating unfavorable genotypes. This tendency is opposed by several mechanisms that preserve or restore variation.

1.   **Diploidy**

Diploidy hides much genetic variation from selection by the presence of recessive alleles in heterozygotes.

- Since recessive alleles are not expressed in heterozygotes, less favorable or harmful alleles may persist in a population.

- This variation is only exposed to selection when two heterozygotes mate and produce offspring homozygous for the recessive allele.

- If a recessive allele has a frequency of 0.01 and its dominant counterpart 0.99, then 99% of the recessive allele copies will be protected in heterozygotes. Only 1% of the recessive alleles will be present in homozygotes and exposed to selection.

- The more rare the recessive allele, the greater its protection by heterozygosity. That is, a greater proportion are hidden in heterozygotes by the dominant allele.

- This type of protection maintains a large pool of alleles which may be beneficial if conditions change.

2.   **Balanced Polymorphism**

Selection may also preserve variation at some gene loci.

Balanced polymorphism = The ability of natural selection to maintain diversity in a population.

One mechanism by which selection preserves variation is *heterozygote advantage.*

- Natural selection will maintain two or more alleles at a locus if heterozygous individuals have a greater reproductive success than any type of homozygote.

- An example is the recessive allele that causes sickle-cell anemia in homozygotes. The locus involved codes for one chain of hemoglobin.

- Homozygotes for this recessive allele develop sickle-cell anemia which is often fatal.

- Heterozygotes are resistant to malaria. Heterozygotes thus have an advantage in tropical areas where malaria is prevalent, since homozygotes for the dominant allele are susceptible to malaria and homozygous recessive individuals are incapacitated by the sickle-cell condition.

- In some African tribes from areas where malaria is common, 20% of the hemoglobin loci in the gene pool is occupied by the recessive allele.

Other examples of heterozygote advantage are found in crop plants (e.g. corn) where inbred lines become homozygous at more loci and show stunted growth and sensitivity to diseases.

- Crossbreeding different inbred varieties often produces hybrids which are more vigorous than the parent stocks.
- This *hybrid vigor* is probably due to:
  1. Segregation of harmful recessives that were homozygous in the inbred varieties.
  2. Heterozygote advantage at many loci in the hybrids.

Balanced polymorphism can also result from patchy environments where different phenotypes are favored in different subregions of a populations geographic range.

Frequency-dependent selection also causes balanced polymorphism.

- In this situation the reproductive success of any one morph declines if that phenotype becomes too common in the population.
- For example, in *Papilio dardanus*, an African swallowtail butterfly, males have similar coloration but females occur in several morphs.
- The female morphs resemble other butterfly species which are noxious to bird predators. *Papilio* females are not noxious, but birds avoid them because they look like distasteful species.
- This type of protective coloration (*Batesian mimicry*) would be less effective if all the females looked like the same noxious species, because birds would encounter good-tasting mimics as often as noxious butterflies and would not associate a particular color pattern with bad taste.

D. **Does All Genetic Variation Affect Survival and Reproductive Success?**

Some genetic variations found in populations confer no selective advantage or disadvantage. They have little or no impact on reproductive success. This type of variation is called *neutral variation*.

- Much of the protein variation found by electrophoresis is adaptively neutral.
- For example, 99 known mutations affect 71 of 146 amino acids in the beta hemoglobin chain in humans. Some, like the sickle-cell anemia allele, affect the reproductive potential of an individual, while others have no obvious effect.
- The *neutral theory* of molecular evolution states than many variant alleles at a locus may confer no selective advantage or disadvantage.
- Natural selection would not affect the relative frequencies of neutral variations. Frequency of some neutral alleles will increase in the gene pool and others will decrease *due the chance effects of genetic drift*.

Variation in DNA which does not code for proteins may also be nonadaptive.

- Most eukaryotes contain large amounts of DNA in their genomes which have no known function. Such noncoding DNA can be found in varying amounts in closely related species.

- Some scientists speculate that noncoding DNA has resulted from the inherent capacity of DNA to replicate itself and has expanded to the tolerance limits of the each species. The entire genome could exist as a consequence of being self-replicating rather than by providing an adaptive advantage to the organism.
- Transposons might fit this definition of "selfish DNA," although the degree of influence these sequences have on the evolution of genomes is not known.

Evolutionary biologists continue to debate how much variation, or even whether variation, is neutral.

- It is easy to show that an allele is deleterious to an organism.
- It is not easily shown that an allele provides no benefits, since such benefits may occur in immeasurable ways.
- Also, a variation may appear to be neutral under one set of environmental conditions and not neutral under other conditions.

We cannot know how much genetic variation is neutral, but if even a small portion of a population's genetic variation significantly affects the organisms, there is still a tremendous amount of raw material for natural selection and adaptive evolution.

XI. Natural selection is the mechanism of adaptive evolution

Adaptive evolution results from a combination of both:

- Chance events that produce new genetic variation (e.g. mutation and sexual recombination).
- Natural selection that favors propagation of some variations over others.

A. Fitness

Darwinian fitness is measured by the relative contribution an individual makes to the gene pool of the next generation.

- It is not a measure of physical and direct confrontation, but of the success of an organism in producing progeny.
- Organisms may produce more progeny because they are more efficient feeders, attract more pollinators (as in our wildflowers), avoid predators, etc.

Survival does not guarantee reproductive success, since a sterile organism may outlive fertile members of the population.

- A long life span may increase fitness if the organism reproduces over a longer period of time (thus leaving more offspring) than other members of the population.
- Even if all members of a population have the same life span, those that mature early and thus have a longer reproductive time span, have increased their fitness.
- Every aspect of survival and fecundity are components of fitness.

<u>Relative fitness</u> = The contribution of a genotype to the next generation compared to the contributions of alternative genotypes for the same locus.

- For example, if pink flower plants (AA and Aa) in our wildflower population produce more offspring than white flower plants (aa), then AA and Aa genotypes have a higher relative fitness.

Statistical estimates of fitness can be produced by the relative measure of selection *against* an inferior genotype. This measure is called the *selection coefficient.*

- For comparison, relative fitness of the most fecund variant (AA or Aa in our wildflower population) is set at 1.0.
- If white flower plants produce 80% as many progeny on average, then the white variant relative fitness is 0.8.
- The selection coefficient is the difference between these two values ($1.0 - 0.8 = 0.2$).
- The more disadvantageous the allele, the greater the selection coefficient.
- Selection coefficents can range up to 1.0 for a lethal allele.

The rate of decline in relative frequencies of deleterious alleles in a population depends on the magnitude of the selection coefficient working against it and whether the allele is dominant or recessive to the more successful allele.

- Deleterious recessives are normally protected from elimination by heterozygote protection.
- Selection against harmful dominant alleles is faster since they are expressed in heterozygotes.

The rate of increase in relative frequencies of beneficial alleles is also affected by whether it is a dominant or recessive.

- New recessive mutations spread slowly in a population (even if beneficial) because selection can not act in its favor until the mutation is common enough for homozygotes to be produced.
- New dominant mutations that are beneficial increase in frequency faster since even heterozygotes benefit from the allele's presence (for example, the mutant dark color producing allele in peppered moths).

Most new mutations, whether harmful or beneficial, probably disappear from the gene pool early due to genetic drift.

B.   What Does Selection Act On?

Selection acts on phenotypes, indirectly adapting a population to its environment by increasing or maintaining favorable genotypes in the gene pool.

- Since it is the phenotype (physical traits, metabolism, physiology, and behavior) which is exposed to the environment, selection can only act indirectly on genotypes.

The connection between genotype and phenotype may not be as simple and definite as with our wildflower population where pink was dominant to white.

- *Pleiotropy* (the ability of a gene to have multiple effects) often clouds this connection. The overall fitness of a genotype depends on whether its beneficial effects exceed any harmful effects on the organism's reproductive success.

- Polygenic traits also make it difficult to distinguish the phenotype-genotype connection. Whenever several loci influence the same characteristic, the members of the population will not fit into definite categories, but represent a continuum along a range.

An organism is an integrated composite of many phenotypic features, and the fitness of a genotype at any one locus depends upon the entire genetic context. A number of genes may work cooperatively to produce related phenotypic traits.

C. Modes of Natural Selection

The frequency of a heritable characteristic in a population may be affected in one of three different ways by natural selection, depending on which phenotypes are favored. (See Campbell, Figure 21.12)

1. Stabilizing Selection

Stabilizing selection favors intermediate variants by selecting against extreme phenotypes.

- The trend is toward reduced phenotypic variation and greater prevalence of phenotypes best suited to relatively stable environments.

- For example, human birth weights are in the 3 – 4 kg range. Much smaller and much higher birth weight babies have a greater infant mortality.

2. Directional Selection

Directional selection favors variants of one extreme. It shifts the frequency curve for phenotypic variations in one direction toward rare variants which deviate from the average of that trait.

- This is most common when members of a species migrate to a new habitat with different environmental conditions or during periods of environmental change.

- For example, fossils show the average size of European black bears increased after periods of glaciation, only to decrease during warmer interglacial periods.

3. Diversifying Selection

In diversifying selection, opposite phenotypic extremes are favored over intermediate phenotypes.

- This occurs when environmental conditions are variable in such a way that extreme phenotypes are favored.

- For example, balanced polymorphism of *Papilio* where butterflies with characteristics between two noxious model species (thus not favoring either) gain no advantage from their mimicry.

D. **Sexual Selection**

Sexual dimorphism = Distinction between the secondary sexual characteristics of males and females.

- Often seen as differences in size, plumage, lion's manes, deer antlers, or other adornments in males.
- In vertebrates it is usually the male that is the "showier" sex.
- In some species, males use their secondary sexual characteristics in direct competition with other males to obtain female mates (especially where harem building is common). These males may defeat other males in actual combat, but more often they use ritualized displays to discourage male competitors.

Darwin viewed sexual selection as a separate selection process leading to sexual dimorphism.

- These enhanced secondary sexual characteristics usually have no adaptive advantage other than attracting mates.
- However, if these adornments increase a males ability to attract more mates, his reproductive success is increased and he contributes more to the gene pool of the next generation.

The evolutionary outcome is usually a compromise between natural selection and sexual selection.

- In some cases the line between these two types of selection is not distinct, as in male deer.
- A stag may use his antlers to defend himself from predators and also to attract females.

XII. **Does Evolution Fashion Perfect Organisms?**

Natural selection <u>cannot</u> breed perfect organisms because:

1. *Organisms are locked into historical constraints.* Each species has a history of descent with modification from ancestral forms.
   - Natural selection modifies existing structures and adapts them to new situations, it does not rebuild organisms.
   - For example, back problems suffered by some humans are in part due to the modification of a skeleton and musculature from the anatomy of four-legged ancestors which are not fully compatible to upright posture.

2. *Adaptations are often compromises.*
   - Each organism must be versatile enough to do many different things.
   - For example, seals spend time in the water and on rocks; they could walk better with legs, but swim much better with flippers.
   - Prehensile hands and flexible limbs allow humans to be very versatile and athletic, but they also make us prone to sprains, torn ligaments, and dislocations. Structural reinforcement would prevent many of these disabling occurrences but would limit agility.

3. *Not all evolution is adaptive.*
   - Genetic drift probably affects the gene pool of populations to a large extent.
   - Alleles which become fixed in small populations formed by the founder effect may not be better suited for the environment than alleles that are eliminated.
   - Similarly, small surviving populations produced by bottleneck effect may be no better adapted to the environment or even less well adapted than the original population.

4. *Selection can only edit variations that exist.*
   - These variations may not represent ideal characteristics.
   - New genes are not formed by mutation on demand.

These limitations thus allow natural selection to operate on a "better than" basis and subtle imperfections are the best evidence for evolution.

REFERENCES

Campbell, N. *Biology.* 4th ed. Menlo Park, California: Benjamin/Cummings, 1996.

Doolittle, W.F. and C. Sapienza. "Selfish Genes, the Phenotype Paradigm and Genome Evolution." *Nature*, 284: 601-603, 1980.

Mettler, L.E., T.G. Gregg, and H.E. Schaffer. *Population Genetics and Evolution.* 2nd ed. Englewood Cliffs, New Jersey: Prentice Hall, 1988.

Milkman, R. *Perspectives on Evolution.* Sinauer, Massachusetts: Sunderland, 1982.

CHAPTER 22

THE ORIGIN OF SPECIES

OUTLINE

I. The biological species concept emphasizes reproductive isolation

II. Reproductive barriers separate species
   A. Prezygotic Barriers
   B. Postzygotic Barriers
   C. Introgression

III. Geographical isolation can lead to the origin of species: allopatric speciation
   A. Conditions Favoring Allopatric Speciation
   B. Adaptive Radiation on Island Chains

IV. A new species can originate in the geographical midst of the parent species: sympatric speciation

V. Population genetics can account for speciation
   A. Speciation by Adaptive Divergence
   B. Speciation by Shifts in Adaptive Peaks
   C. Hybrid Zones and the Cohesion Concept of Species
   D. How Much Genetic Change Is Required For Speciation?

VI. The Theory of Punctuated Equilibrium has stimulated research on the tempo of speciation

OBJECTIVES

After reading this chapter and attending lecture, the student should be able to:
1. Distinguish between anagenesis and cladogenesis.
2. Define morphospecies and explain how this concept can be useful to biologists.
3. Define biological species (E. Mayr).
4. Describe some limitations of the biological species concept.
5. Explain how gene flow between closely related species can be prevented.

6. Distinguish between prezygotic and postzygotic isolating mechanisms.
7. Describe five prezygotic isolating mechanisms and give an example of each.
8. Explain why many hybrids are sterile.
9. Explain, in their own words, how hybrid breakdown maintains separate species even if gene flow occurs.
10. Explain how introgression affects the gene pool of a species.
11. Distinguish between allopatric and sympatric speciation.
12. Explain, in their own words, the allopatric speciation model and describe the role of intraspecific variation and geographical isolation.
13. Explain why peripheral isolates are susceptible if geographic barriers arise.
14. Describe the adaptive radiation model and use it to describe how it might be possible to have many sympatric closely related species even if geographic isolation is necessary for them to evolve.
15. Define sympatric speciation and explain how polyploidy can cause reproductive isolation.
16. Distinguish between autopolyploidy and allopolyploidy.
17. List some points of agreement and disagreement between the two schools of thought about the tempo of speciation (gradualism vs punctuated equilibrium).

KEY TERMS

anagenesis	mechanical isolation	adaptive radiation	adaptive landscape
cladogenesis	gametic isolation	sympatric speciation	adaptive peak
Ernst Mayr	postzygotic	peripheral isolate	peak shift
morphospecies	reduced hybrid	polyploidy	hybrid zone
biological species	viability	autopolyploidy	cohesion concept of
reproductive barrier	reduced hybrid	tetraploid	species
prezygotic	fertility	Hugo DeVries	gradualism
habitat isolation	hybrid breakdown	*Oenothera*	punctuated equilibrium
temporal isolation	introgression	allopolyploidy	Niles Eldredge
behavioral isolation	allopatric speciation	Sewell Wright	Stephen J. Gould

LECTURE NOTES

Evolutionary theory must explain:

- How adaptations evolve in populations.
- The origin of new species which results in biological diversity.

The fossil record provides evidence for two patterns of speciation: *anagenesis* and *cladogenesis*.

Anagenesis (phyletic evolution) = The transformation of an unbranched lineage of organisms, sometimes to a state different enough from the ancestral population to justify renaming it as a new species.

Cladogenesis (branching evolution) = The budding of one or more new species from a parent species that continues to exist; is more important than anagenesis in life's history, because it is more common and can promote biological diversity.

I. **The biological species concept emphasizes reproductive isolation**

Species exist in nature as discrete units, usually distinguishable from other species.

- Trained taxonomists often give scientific names to flora and fauna which are already recognized by native populations. For example:
  - ⇒ In 1927, Ernst Mayr traveled to the remote Arafak mountains of New Guinea to study its diverse animal fauna.
  - ⇒ Using morphological differences, he identified 138 species of birds – 137 of which the Papuan natives had already named.
- Although species are discrete in nature, it is difficult to devise a formal definition of a species.

Linnaeus (founder of modern taxonomy) described species in terms of their physical form (morphology). This *morphospecies* concept is still the most common method used for describing species.

Morphospecies = Species defined by their anatomical features.

- It is difficult to apply to some situations.
  - ⇒ Sometimes, it is difficult to determine if a set of organisms represents multiple species or a single species with extensive phenotypic variation.
  - ⇒ Two populations morphologically almost indistinguishable may be different species based on other criteria.
- Can be useful in the field, since its basis is observable and measurable anatomy.
- Does not address the discontinuity that exists between species from the standpoint of evolutionary theory.

In 1942, Ernst Mayr proposed the *biological species concept* as an alternative to the morphospecies.

Biological species = Groups of interbreeding natural populations that are reproductively isolated from other such groups.

- Is circumscribed by reproductive barriers that preserve its integrity by preventing genetic mixing with other species.
- Is the largest unit of population in which gene flow is possible.
- Is defined by reproductive isolation from other species in *natural* environments (hybrids may be possible between two species in the laboratory).

The biological species concept cannot be applied to:

- Organisms that are completely asexual in their reproduction. Some protists and fungi, some commercial plants (bananas), and many bacteria are exclusively asexual.
  - ⇒ Asexual reproduction effectively produces a series of clones which, genetically speaking, represent a single organism.
  - ⇒ Asexual organisms can be assigned to species only by grouping clones with the same morphology and biochemical characteristics.
- Extinct organisms represented only by fossils. These must be classified by the morphospecies concept.

- Population which are geographically segregated. Can they potentially interbreed in nature even if they are similar enough to be placed into the same species based on morphological characteristics?

In some cases, unambiguous determination of species is not possible, even though the populations are sexual, contemporaneous and contiguous.

- Four phenotypically distinct populations of the deer mouse (*Peromyscus maniculatus*) found in the Rocky Mountains are geographically isolated and referred to as *subspecies*. (See Campbell, Figure 22.3)
- These populations overlap at certain locations and some interbreeding occurs in these areas of cohabitation, which indicates they are the same species by the biological species criteria.
- Two subspecies (*P. m. artemisiae* and *P. m. nebrascensis*) are an exception, since they do not interbreed in the area of cohabitation. However, their gene pools are not completely isolated since they freely interbreed with other neighboring populations.
- This circuitous route could only produce a very limited gene flow, but the route is open and possible between the populations of *P. m. artemisiae* and *P. m. nebrascensis* through the other populations.
- If this route was closed by extinction or geographic isolation of the neighboring populations, then *P. m. artemisiae* and *P. m. nebrascensis* could be named separate species without reservation.

More examples are being discovered where there is a blurry distinction between populations with limited gene flow and full biological species with segregated gene pools.

- If two populations cannot interbreed when in contact, they are clearly distinct species.
- When there is gene flow (even very limited) between two populations that are in contact, it is difficult to apply the biological species concept.
- This is equivalent to finding two populations at different stages in their evolutionary descent from a common ancestor, which is to be expected if new species arise by gradual divergence of populations.

Other species concepts have been developed in an effort to accommodate the dynamic, quantitative aspects of speciation; however, the species problem may never be completely resolved as it is unlikely that a single definition of species will apply to all cases.

II. Reproductive barriers separate species

<u>Reproductive barrier</u> = Any factor that impedes two species from producing fertile hybrids, thus contributing to reproductive isolation.

- Most species are genetically sequestered from other species by more than one type of reproductive barrier.
- Only intrinsic biological barriers to reproduction will be considered here. Geographic segregation (even though it prevents interbreeding) will not be considered.
- Reproductive barriers prevent interbreeding between closely related species.

The various reproductive barriers which isolate the gene pools of species are classified as either prezygotic or postzygotic depending on whether they function before or after the formation of zygotes.

- Prezygotic barriers impede mating between species or hinder fertilization of the ova should members of different species attempt to mate.
- In the event fertilization does occur, postzygotic barriers prevent the hybrid zygote from developing into a viable, fertile adult.

A. Prezygotic Barriers

1. Habitat Isolation

Two species living in different habitats within the same area may encounter each other rarely if at all, even though they are not technically geographically isolated.

- For example, two species of garter snakes (*Thamnophis*) occur in the same areas but for intrinsic reasons, one species lives mainly in water and the other is mainly terrestrial.
- Since these two species live primarily in separate habitats, they seldom come into contact as they are ecologically isolated.

2. Temporal Isolation

Two species that breed at different times of the day, seasons, or years cannot mix their gametes.

- For example, brown trout and rainbow trout cohabit the same streams, but brown trout breed in the fall and rainbow trout breed in the spring.
- Since they breed at different times of the year, their gametes have no opportunity to contact each other and reproductive isolation is maintained.

3. Behavioral Isolation

Species-specific signals and elaborate behavior to attract mates are important reproductive barriers among closely related species. For example,

- Male fireflies of different species signal to females of the same species by blinking their lights in a characteristic pattern; females discriminate among the different signals and respond only to flashes of their own species by flashing back and attracting the males.

Many animals recognize mates by sensing pheromones (distinctive chemical signals).

- Female Gypsy moths attract males by emitting a volatile compound to which the olfactory organs of male Gypsy moths are specifically tuned: when a male detects this pheromone, it follows the scent to the female.
- Males of other moth species do not recognize this chemical as a sexual attractant.

Other factors may also act as behavioral isolating mechanisms:

- Eastern and western meadowlarks are almost identical in morphology and habitat, and their ranges overlap in the central United States.
- They retain their species integrity partly because of the difference in their songs, which enables them to recognize potential mates as members of their own kind.

4. Mechanical Isolation

Anatomical incompatibility may prevent sperm transfer when closely related species attempt to mate.

- For example, male dragonflies use a pair of special appendages to clasp females during copulation; when a male tries to mount a female of a different species, he is unsuccessful because his clasping appendages do not fit the female's form well enough to grip securely.
- In plants that are pollinated by insects or other animals, the floral anatomy is often adapted to a specific pollinator that transfers pollen only among plants of the same species.

5. Gametic Isolation

Gametes of different species that meet rarely fuse to form a zygote.

- For animals that use internal fertilization, the sperm of one species may not be able to survive the internal environment of the female reproductive tract of a different species.
- Cross-specific fertilization is also uncommon for animals that utilize external fertilization due to a lack of gamete recognition.

Gamete recognition may be based on the presence of specific molecules on the coats around the egg which adhere only to complementary molecules on sperm cells of the same species.

- Similar mechanisms of molecular recognition enables a flower to discriminate between pollen of the same species and pollen of different species.

B.   Postzygotic Barriers

When prezygotic barriers are crossed and a hybrid zygote forms, one of several postzygotic barriers may prevent development of a viable, fertile hybrid.

1.   Reduced Hybrid Viability

Genetic incompatibility between the two species may abort development of the hybrid at some embryonic stage.

- For example, several species of frogs in the genus *Rana* live in the same regions and habitats.
- They occasionally hybridize but the hybrids generally do not complete development, and those that do are frail and soon die.

2.   Reduced Hybrid Fertility

If two species mate and produce hybrid offspring that are viable, reproductive isolation is intact if the hybrids are sterile because genes cannot flow from one species' gene pool to the other.

- One cause of this barrier is that if chromosomes of the two parent species differ in number or structure, meiosis cannot produce normal gametes in the hybrid.
- The most familiar case is the mule which is produced by crossing a donkey and a horse; very rarely are mules able to backbreed with either parent species.

3.   Hybrid Breakdown

When some species cross-mate, the first generation hybrids are viable and fertile, but when these hybrids mate with one another or with either parent species, offspring of the next generation are feeble or sterile.

- For example, different cotton species can produce fertile hybrids, breakdown occurs in the next generation when progeny of the hybrids die in their seeds or grow into weak defective plants.

C.   Introgression

Introgression = The transplantation of alleles between species.

- Introgression occurs when alleles occasionally seep through all reproductive barriers and pass between the gene pools of closely related species when fertile hybrids mate successfully with one of the parent species.

For example, corn (*Zea mays*) contains some alleles traceable to the closely related wild grass teosinte (*Zea mexicana*).

- Introgressions occur when the two species hybridize and a small number of the hybrids manage to cross with corn plants.
- The transplant of alleles increases the genetic variation that can be exploited by breeders attempting to produce new corn varieties by artificial selection.
- Occasional hybridization does not erase the boundary between corn and teosinte as only a very rare introgression occurs, the isolation of the two gene pools is not seriously breached and the two species remain distinct.

III.   Geographical isolation can lead to the origin of species: allopatric speciation

Reproductive barriers form boundaries around species, and the evolution of these barriers is the key biological event in the origin of new species.

- An essential episode in the origin of a species occurs when the gene pool of a population is separated from other populations of the parent species.
- This genetically isolated splinter group can then follow its own evolutionary course as changes in allele frequencies caused by selection, genetic drift, and mutations occur undiluted by gene flow from other populations.

Speciation episodes can be classified into two modes based on the geographical relationship of a new species to its ancestral species: *allopatric speciation* and *sympatric speciation*.

Allopatric speciation = Speciation that occurs when the initial block to gene flow is a geographical barrier that physically isolates the population.

- Geological processes can fragment a population into two or more allopatric populations (having separate ranges).
    - ⇒ Such occurrences include emergence of mountain ranges, movement of glaciers, formation of land bridges, subsidence of large lakes.
    - ⇒ Also small populations may become geographically isolated when individuals from the parent population travel to a new location.
- The extent of development of a geographical barrier necessary to isolate two populations depends on the ability of the organisms to disperse due to the mobility of animals or the dispersibility of spores, pollen and seeds of plants.
    - ⇒ For example, the Grand Canyon is an impassable barrier to small rodents but is easily crossed by birds.

An example of how geographic isolation can result in allopatric speciation is the pupfish.

- 50,000 years ago, during an ice age, the Death Valley region of California and Nevada had a rainy climate and a system of interconnecting lakes and rivers.
- 10,000 years ago a drying trend began and by 4,000 years ago, the region had become a desert.
- Presently, isolated springs in deep clefts between rocky walls are the only remnants of the lake and river networks. Living in many of these isolated springs are small pupfishes (*Cyprinodon* spp.).
- Each inhabited spring contains its own species of pupfish which is adapted to that pool and found nowhere else in the world.
- The endemic pupfish species probably descended from a single ancestral species whose range was fragmented when the region became arid, thus isolating several small populations that diverged in their evolution as they adapted to their spring's environment.

A.   **Conditions Favoring Allopatric Speciation:**

When populations become allopatric, speciation can potentially occur as the isolated gene pools accumulate differences by microevolution that may cause the populations to diverge in phenotype.

- A small isolated population is more likely to change substantially enough to become a new species than is a large isolated population.
- The geographic isolation (peripheral isolate) of a small population usually occurs at the fringe of the parent population's range.
- As long as the gene pools are isolated from the parental population, *peripheral isolates* are good candidates for speciation for three reasons:

    1. *The gene pool of the peripheral isolate probably differs from that of the parent population* initially, since fringe inhabiters usually represent the extremes of any genotypic and phenotypic clines in an original sympatric population. With a small peripheral isolate, there will be a founder effect with chance resulting in a gene pool that is not representative of the gene pool of the parental population.
    2. *Genetic drift* will continue to cause chance changes in the gene pool of the small peripheral isolate until a large population is formed. New mutations or combinations of alleles that are neutral in adaptive value may become fixed in the population by chance alone, causing phenotypic divergence from the parent population.
    3. *Evolution caused by selection* is likely to take a different direction in the peripheral isolate than in the parental population. Since the peripheral isolate inhabits a frontier with a somewhat different environment, it will probably be exposed to different selection pressures than those encountered by the parental population.

- Due to the severity of a fringe environment, most peripheral isolates do not survive long enough to speciate.

Although most peripheral isolates become extinct, evolutionary biologists agree that a small population can accumulate enough genetic change to become a new species in only hundreds to thousands of generations.

B.   **Adaptive Radiation on Island Chains**

Allopatric speciation occurs on island chains where new populations, which stray or are passively dispersed from their ancestral populations, evolve in isolation.

Adaptive radiation = The evolution of many diversely adapted species from a common ancestor.

Examples of adaptive radiation are the endemic species of the Galapagos Islands which descended from small populations which floated, flew, or were blown from South America to the islands. Darwin's finches can be used to illustrate a model for such adaptive radiation on island chains.

- A single dispersal event may have seeded one island with a peripheral isolate of the ancestral finch which diverged as it underwent allopatric speciation.

- A few individuals of this new species may have reached neighboring islands, forming new peripheral isolates which also speciated.

- After diverging on the island it invaded, a new species could re-colonize the island from which its founding population emigrated and coexist with the ancestral species or form still another species.

- Multiple invasions of islands could eventually lead to coexistence of several species on each island since the islands are distant enough from each other to permit geographic isolation, but near enough for occasional dispersal.

Similar evolutionary events have occurred on the Hawaiian Archipelago. These volcanic islands are 3500 km from the nearest continent.

- Hawaii is the youngest (<one million years old), largest island and has active volcanos.

- The islands grow progressively older in a northwesterly direction away from Hawaii.

- As each island was formed and cooled, flora and fauna carried by ocean and wind currents from other islands and continents became established.

- The physical diversity of each island provided many environmental opportunities for evolutionary divergence by natural selection.

- Multiple invasions and allopatric speciations have permitted such a degree of adaptive radiations that there are thousands of endemic species on the archipelago which are found no where else on Earth.

In contrast to the Hawaiian Archipelago, islands such as the Florida Keys are close enough to a mainland to allow free movement from the island to the mainland.

- Such islands are not characterized by endemic species since there is no long-term isolation of founding populations.
- Intrinsic reproductive barriers that block gene flow do not develop due to a steady influx of immigrants from the mainland parental populations.

IV. A new species can originate in the geographical midst of the parent species: sympatric speciation

Sympatric speciation = Formation of new species within the range of parent populations.

- Reproductive isolation evolves without geographical isolation.
- Can occur quickly (in one generation) if a genetic change results in a reproductive barrier between the mutants and the parent population.

Many plant species have originated from improper cell division that results in extra sets of chromosomes — a mutant condition called *polyploidy*. (See Campbell, Figure 22.10)

Depending on the origin of the extra set of chromosomes, polyploids are classified in two forms: *autopolyploids* and *allopolyploids*.

Autopolyploid = An organism that has more than two chromosome sets, all derived from a single species. For example,

- Nondisjunction in the germ cell line (in either mitosis or meiosis) results in diploid gametes.
- Self-fertilization would double the chromosome number to the tetraploid state.
- Tetraploids can self-pollinate or mate with other tetraploids.
- The mutants cannot interbreed with diploids of the parent population because hybrids would be triploid ($3n$) and sterile due to impaired meiosis from unpaired chromosomes.
- An instantaneous special genetic event would thus produce a postzygotic barrier which isolates the gene pool of the mutant in just one generation.
- Sympatric speciation by autopolyploidy was first discovered by Hugo De Vries in the early 20th century while working with *Oenothera*, the evening primrose.

Allopolyploid = A polyploid hybrid resulting from contributions by two different species.

- More common than autopolyploidy.
- Potential evolution of an allopolyploid begins when two different species interbreed and a hybrid is produced. (See Campbell, Figure 22.10b)
- Such interspecific hybrids are usually sterile, because the haploid set of chromosomes from one species cannot pair during meiosis with the haploid set of chromosomes from the second species.
- These sterile hybrids may actually be more vigorous than the parent species and propagate asexually.

At least two mechanisms can transform sterile alloployploid hybrids into fertile polyploids:

1. During the history of the hybrid clone, mitotic nondisjunction in the reproductive tissue may double the chromosome number. (See Campbell, Figure 22.10b)

    *   The hybrid clone will then be able to produce gametes since each chromosome will have a homologue to synapse with during meiosis.
    *   Gametes from this fertile tetraploid could unite and produce a new species of interbreeding individuals, reproductively isolated from both parent species.

2. Meiotic nondisjunction in one species produces an unreduced (diploid) gamete.

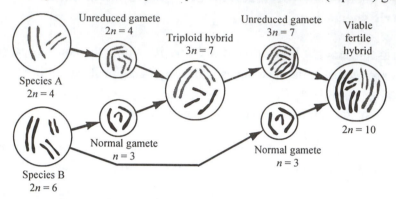

*   This abnormal gamete fuses with a normal haploid gamete of a second species and produces a triploid hybrid.
*   The triploid hybrid will be sterile, but may propagate asexually.
*   During the history of this sterile triploid clone, meiotic nondisjunction again produces an unreduced gamete (triploid).
*   Combination of this triploid gamete with a normal haploid gamete from the second parent species would result in a fertile hybrid with homologous pairs of chromosomes.
*   This allopolyploid would have a chromosome number equal to the sum of the chromosome numbers of the two ancestral species (as in 1 above).

Speciation of polyploids (especially allopolyploids) has been very important in plant evolution.

*   Some allopolyploids are very vigorous because they contain the best qualities of both parent species.
*   The accidents required to produce these new plant species (interspecific hybridization coupled with nondisjunction) have occurred often enough that between 25% and 50% of all plant species are polyploids.

Some of these species have originated and spread in relatively recent times and many others are of importance to humans.

*   *Spartina angelica* is a species of salt-marsh grass which evolved as an allopolyploid in the 1870s.

    ⇒ It is derived from a European species (*Spartina maritima*) and an American species (*Spartina alternaflora*).

$\Rightarrow$ In addition to being morphologically distinct and reproductively isolated from its parent species, *S. angelica* has a chromosome number ($2n = 122$) indicative of its mechanism of speciation (*S. maritima*, $2n =$, *S. alternaflora*, $2n = 62$).

- *Triticum aestivum*, bread wheat, is a 42 chromosome allopolyploid that is believed to have originated about 8000 years ago as a hybrid of a 28 chromosome cultivated wheat and a 14 chromosome wild grass.
- Other important polyploid species include oats, cotton, potatoes, and tobacco.
- Plant geneticists are presently inducing these genetic accidents to produce new polyploids which will combine high yield and disease resistance.

Sympatric speciation may also occur in animal evolution through different mechanisms.

- A group of animals may become isolated within the range of a parent population if genetic factors cause them to become fixed on resources not used by the parent population as a whole. For example,
  - $\Rightarrow$ A particular species of wasp pollinates each species of figs. The wasps mate and lay their eggs in the figs.
  - $\Rightarrow$ A genetic change causing wasps to select a different fig species would segregate mating individuals of the new phenotype from the parental population.
  - $\Rightarrow$ Divergence could then occur after such an isolation.
  - $\Rightarrow$ The great diversity of cichlid fishes in Lake Victoria may have arisen from isolation due to exploitation of differend food sources and other resources in the lake.
- Sympatric speciation could also result from a balanced polymorphism combined with assortative mating. For example,
  - $\Rightarrow$ If birds in a population that is dimorphic for beak size began to selectively mate with birds of the same morph, speciation could occur over time.

While both allopatric speciation and sympatric speciation have important roles in plant evolution, allopatric speciation is far more common in animals.

V.   Population genetics can account for speciation

Classifying modes of speciation as allopatric or sympatric emphasizes biogeographical factors but does not emphasize the actual genetic mechanisms. An alternative method which takes genetic mechanism into account, groups speciation into two categories: *speciation by adaptive divergence* and *speciation by shifts in adaptive peaks*.

A.   Speciation by Adaptive Divergence

Two populations which adapt to different environments accumulate differences in the frequencies of alleles and genotypes.

- During this gradual adaptive divergence of the two gene pools, reproductive barriers may evolve between the two populations.
- Evolution of reproductive barriers would differentiate these populations into two species.

A key point in evolution by divergence is that reproductive barriers can arise without being favored directly by natural selection.

- Divergence of two populations is due to their adaptation to separate environments, with reproductive isolation being a secondary development.
- Postzygotic barriers may be pleiotropic effects of interspecific differences in those genes that control development. For example,
  ⇒ Hybrids may be inviable if both sets of genes for rRNA synthesis are not active (e.g. hybrids between *D. melanogaster* and *D. simulans*).
- Gradual genetic divergence of two populations may also result in the evolution of prezygotic barriers. For instance,
  ⇒ An ecological barrier to inbreeding may secondarily result from the adaptation of an insect population to a new host plant different from the original population's host.

In some isolated populations, reproductive isolation has evolved more directly from sexual selection. For example,

- In *Drosophila heteroneura*, the male's wide head enhances reproductive success with females of the same species while reducing the probability that a male *D. heteroneura* will mate with females of other species.
- Sexual selection, in this case, probably evolved as an adaptation for enhanced reproductive success. A secondary consequence is that it prevents interbreeding with other *Drosophila* species.
- Since reproductive barriers usually evolve when populations are allopatric, they do not function directly to isolate the gene pools of populations.
  ⇒ For this reason, the emphasis on reproductive isolating mechanisms is one criticism of the "biological species concept".
- An alternative to the biological species concept is the *recognition concept of species*.

The recognition concept of species assumes that the reproductive adaptations of a species consists of a set of characteristics that maximize successful mating with members of the same population.

- Characteristics include the molecular, morphological, and behavioral characteristics that permit individuals to recognize a mate of the same species.
- Reproductive isolation from other populations would thus be a side effect.
- Focuses on characteristics which are actually subject to natural selection in isolated populations that are in the process of speciation.

B.   Speciation by Shifts in Adaptive Peaks

The concept of the "adaptive landscape" was proposed by Sewell Wright in the 1930's.

- The landscape includes many *adaptive peaks* separated by valleys.
- Each adaptive peak represents an equilibrium state in which allelic frequencies optimize the population's success in that environment.
- The landscape could include many adaptive peaks, even under stable environmental conditions, but natural selection would tend to maintain the population at a single peak.

For a population to reach an alternate peak by a change in its gene pool, it must pass through a valley where genetic combinations are of low average fitness.

- Thus, if a slight change in allele frequencies at one or more loci pushes a population from its adaptive peak, natural selection will tend to push that population back to its original adaptive peak.

Environmental changes redefine the landscape, making new adaptive peaks possible.

- For a population to survive the new conditions, it must reach a new adaptive peak through microevolution of its gene pool.

Population geneticists use the term *peak shift* for speciation caused by non-adaptive changes in the genetic system.

- Peak shifts may be caused by founder effect or by bottlenecks.
- Genetic drift can remove a small population from its original adaptive peak and, if the gene pool is sufficiently destabilized, new adaptive peaks may be reached.
- Such a population that survives will be pushed to a new adaptive peaks by natural selection in succeeding generations.
- Adaptive evolution thus plays a major role in peak shift, but genetic drift is the mechanism that makes the shift possible.
- Peak shifts can also occur in bottlenecked populations, even if there is no environmental change, due to the many adaptive peaks possible under the same environmental conditions.

A small, genetically destabilized population that is dislodged from its adaptive peak may undergo a generation-to-generation climb to a new adaptive peak due to natural selection.

- With founder effect, the splinter population is affected not only by genetic drift moving the gene pool randomly over the adaptive landscape, but also by new environmental conditions which form a new set of adaptive peaks.
- The combination of genetic drift followed by natural selection under new environmental conditions may be responsible for the rapid radiation of island species (e.g. Hawaiian *Drosophila*).

C. Hybrid Zones and the Cohesion Concept of Species

Three possible outcomes are possible when two closely related populations that have been allopatric for some time come back into contact:

- The two populations may interbreed freely.
    - ⇒ The gene pools would become incorporated into a single pool indicating that speciation had not occcured during their time of geographical isolation.
- The two populations may not interbreed due to reproductive barriers.
    - ⇒ The gene pools would remain separate due to the evolutionary divergence which occurred during the time of geographical isolation. Speciation has taken place.
- A *hybrid zone* may be established.

Hybrid zone = A region where two related populations that diverged after becoming geographically isolated make secondary contact and interbreed where their geographical ranges overlap. For example,

- The red-shafted flicker of western North America and the yellow-shafted flicker of central North America are two phenotypically distinct populations of woodpeckers that interbreed in a hybrid zone stretching from southern Alaska to the Texas panhandle.
- The two populations came into renewed contact a few centuries ago after being separated during the ice ages.
- The hybrid zone is relatively stable and not expanding.
    - ⇒ The introgression of alleles between the populations has not penetrated far beyond the hybrid zone, although the two populations have been interbreeding for at least two hundred years.
    - ⇒ The genotypic and pheontypic frequencies that distinguish the two populations form steep clines into the hybrid zone.
- Away from the hybrid zone, the two populations remain distinct.

Should two populations which form a hybrid zone be considered subspecies or separate species?

- Some researchers who support species status for such populations, recognize that the presence of stable hybrid zones creates a problem for the biological species concept.
    - ⇒ If the taxonomic identity of two species is maintained, even though they hybridize, there must be cohesive forces other than reproductive isolation that maintain the species and prevent their merging into a single species.
- These researchers favor an alternative known as the *cohesion concept of species*.

The cohesion concept of species holds that the cohesion may involve a distinctive, integrated set of adaptations that has been refined during the evolutionary history of a population.

- Phenotypic variation would be restricted by stabilizing selection to a range narrow enough to define the species as separate from other species.
- In the adaptive landscape view:
  ⇒ The red-shafted and yellow-shafted flickers are clustered around different adaptive peaks.
  ⇒ Specific combinations of alleles and specific linkages between gene loci on chromosomes may form a genetic basis for the cohesion of phenotypes.
  ⇒ The clinal change of genetic structure and phenotype noted in the hybrid zone may be correlated with transitions in environmental factors that help shape the two distinct populations.

D. How Much Genetic Change Is Required For Speciation?

No generalizations can be made about genetic distance between closely related species. Reproductive isolation may result from changes in many loci or only in a few.

- Two species of Hawaiian *Drosophila* (*D. silvestris* and *D. heteroneura*) differ at only one locus which determines head shape, an important factor in mate recognition.
  ⇒ The phenotypic effect of different alleles at this locus is multiplied by epistasis involving at least ten other loci.
  ⇒ Thus, no more than one mutation was necessary to differentiate the two species.
- Changes in one gene in a coadapted gene complex can substantially impact the development of an organism.

VI. The Theory of Punctuated Equilibrium has stimulated research on the tempo of speciation

Traditional evolutionary trees diagram the descent of species from ancestral forms as branches that gradually diverge with each new species evolving continuously over long spans of time. (See Campbell, Figure 22.14a)

- The theory behind such a tree is the extrapolation of microevolutionary processes (allele frequency changes in the gene pool) to the divergence of species.
- Big changes thus occur due to the accumulation of many small changes.

Paleontologists rarely find gradual transitions of fossil forms but often observe species appearing as new forms suddenly in the rock layers.

- These species persist virtually unchanged and then disappear as suddenly as they appeared.
- Even Darwin, who believed species from a common ancestral stock evolve differences gradually, was perplexed by the lack of transitional forms in the fossil record.

Advocates of *punctuated equilibrium* have redrawn the evolutionary tree to represent fossil evidence for evolution occurring in spurts of relatively rapid change instead of gradual divergence. (See Campbell, Figure 22.14b)

- This theory was proposed by Niles Eldredge and Stephen Jay Gould in 1972.
- It depicts species undergoing most of their morphological modification as they first separate from the parent species then showing little change as they produce additional species.
- In this theory gradual change is replaced with long periods of stasis punctuated with episodes of speciation.
- The origin of new polyploid plants through genome changes is one mechanism of sudden speciation.
- Allopatric speciation of a splinter population separated from its parent population by geographical barriers may also be rapid.
  ⇒ For a population facing new environmental conditions, genetic drift and natural selection can cause significant change in only a few hundred or thousand generations.

A few thousand generations is considered rapid in reference to the geologic time scale.

- The fossil record indicates that successful species survive for a few million years on average.
- If a species survives for five million years and most of its morphological changes occur in the first 50,000 years, then the speciation episode occurred in just 1% of the species' lifetime.
- With this time scale, a species will appear suddenly in rocks of a certain age, linger relatively unchanged for millions of years, then become extinct.
- While forming, the species may have gradually accumulated modifications, but with reference to its overall history, its formation was sudden.
- An evolutionary spurt preceding a longer period of morphological stasis would explain why paleontologists find so few transitions in the fossils record of a species.

Because "sudden" can refer to thousands of years on the geological time scale, differing opinions of punctuationalists and gradualists about the rate of speciation may be more a function of time perspective than conceptual difference. There is clear disagreement, however, over how much a species changes after its origin.

- In a species adapted to an environment that stays the same, natural selection would counter changes in the gene pool.
  ⇒ Once selection during speciation produces new complexes of coadapted genes, mutations are likely to impose disharmony on the genome and disrupt the development of the organism.
- Stabilizing selection would thus hold a population at one adaptive peak to produce long periods of stasis.

Some gradualists feel that stasis is an illusion since many species may continue to change, after they have diverged from the parent population, in ways undetectable in fossils.

- Changes in internal anatomy, physiology and behavior would go unnoticed by paleontologists as fossils only show external anatomy and skeletons.
- Population geneticists also point out that many microevolutionary changes occur at the molecular level without affecting morphology.

Some biologists even contest the long periods of morphological stasis in the history of species. Peter Sheldon has analyzed approximately 15,000 trilobite fossils which represent an unusually complete record.

- Paleontologists have arranged the fossils into several evolutionary lineages and classified the youngest and oldest as different species based on morphological differences.
- Based on his extensive study, Sheldon concludes that specific species boundaries cannot be drawn among these trilobite fossils.
    ⇒ He found a gradual change in the morphological characteristics used to name the new species as he examined successive rock layers.
- Sheldon suggests that giving fossil species names can lead to an artificial punctuated interpretation of a gradually changing fossil series.

Alan Cheetham's research on fossilized bryozoans supports punctuated equilibrium.

- His studies have found long periods of morphological stasis in the bryozoans which are punctuated by the relatively rapid appearance of new species.

It is obvious that additional extensive studies of fossil morphology where specific lineages are preserved should be carried out to assess the relative importance of gradual and punctuated tempos in the origin of new species.

REFERENCES

Campbell, N. *Biology*. 4th ed. Menlo Park, California: Benjamin/Cummings, 1996.

Futuyma, D.J. *Evolutionary Biology*. 2nd ed. Sunderland, Massachusetts: Sinauer, 1986.

Mayr, E. *Populations, Species and Evolution: An Abridgement of Animal Species and Evolution*. Cambridge, Massachusetts: Harvard University Press, 1970.

CHAPTER **23**

TRACING PHYLOGENY: MACROEVOLUTION, THE FOSSIL RECORD, AND SYSTEMATICS

OUTLINE

I. The fossil record documents macroevolution
    A.   How Fossils Form
    B.   Limitations of the Fossil Record

II. Paleontologists use a variety of methods to date fossils
    A.   Relative Dating
    B.   Absolute Dating

III. What are the major questions about macroevolution: *an overview*

IV. Some evolutionary novelties are modified versions of older structures

V. Genes that control development play a major role in evolutionary novelty

VI. Recognizing trends in the fossil record does not mean that macroevolution is goal-oriented

VII. Macroevolution has a biogeographical basis in continental drift

VIII. The history of life is punctuated by mass extinctions followed by adaptive radiations of the survivors
    A.   Examples of Major Adaptive Radiations
    B.   Examples of Mass Extinctions

IX. Systematics connects biological diversity to phylogeny
    A.   Taxonomy and Hierarchical Classification
    B.   The Relationship of Classification to Phylogeny
    C.   Sorting Homology from Analogy

X. Molecular biology provides powerful new tools for systematics
    A. Protein Comparison
    B. DNA Comparison
    C. Analysis of Fossilized DNA
    D. Molecular Clocks

XI. Cladistics highlights the phylogenetic significance of systematics
    A. Phenetics
    B. Cladistics
    C. Classical Evolutionary Systematics

XII. Is a new evolutionary synthesis necessary?

OBJECTIVES

After reading this chapter and attending lecture, the student should be able to:

1. Explain the importance of the fossil record to the study of evolution.
2. Describe how fossils form.
3. Distinguish between relative dating and absolute dating.
4. Explain how isotopes can be used in absolute dating.
5. Explain how preadaptation can result in macroevolutionary change.
6. Explain how modification of regulatory genes can result in macroevolutionary change.
7. Describe the importance of evolutionary trends to macroevolution.
8. Explain how continental drift may have played a role in macroevolutionary change.
9. Describe how radiation into new adaptive zones could result in macroevolutionary change.
10. Explain how mass extinctions could occur and affect evolution of surviving forms.
11. Distinguish between systematics and taxonomy.
12. Describe the contribution Carolus Linnaeus made to biology.
13. Distinguish between a taxon and a category.
14. List the major taxonomic categories from the most to least inclusive.
15. Explain why it is important when constructing a phylogeny to distinguish between homologous and analogous character traits.
16. Distinguish between homologous and analogous structures.
17. Describe three techniques used in molecular systematics and explain what information each provides.
18. Distinguish between a monophyletic and a polyphyletic group, and explain what is meant by a "natural taxon."
19. Distinguish among the three schools of taxonomy: phenetics, cladistics, and classical evolutionary taxonomy.
20. Distinguish between plesiomorphic and apomorphic characters.
21. Describe how a new synthesis incorporates both the evolutionists' and punctuationalists' views of evolution.

KEY TERMS

macroevolution
fossil
paleontologists
sedimentary rocks
sedimentation
trace fossils
petrification
geological time scale
relative dating
index fossils
absolute dating
radiometric dating
half-life
carbon-14
racemization
preadaptation

allometric growth
paedomorphosis
heterochrony
homeosis
Steven Stanley
species selection
biogeography
continental drift
Pangaea
adaptive zone
adaptive radiation
Walter and Luis
     Alvarez
phylogeny
phylogenetic tree

systematics
taxonomy
Carolus Linnaeus
binomial nomenclature
genus
specific epithet
taxon
monophyletic
polyphyletic
paraphyletic
homology
analogy
convergent evolution
DNA-DNA
     hybridization

restriction mapping
DNA sequencing
molecular clocks
divergence
convergence
phenetics
cladistics
cladogram
classical evolutionary
     taxonomy
shared primitive
     character
synapomorphies
modern synthesis
historical contingency

LECTURE NOTES

<u>Macroevolution</u> = The origin of taxonomic groups higher than the species level (genera, families, etc).

- Macroevolution is concerned with major events in the history of life as found in the fossil record.

- Encompasses the origin of new designs such as: feathers and wings of birds, upright posture of humans, increasing brain size in mammals, diversification of certain groups of organisms following evolutionary breakthroughs, and extinction.

- Macroevolution extends the study of evolution to include global environmental and biological change.

Is macroevolutionary change primarily a cumulative product of microevolution working gradually over time, or is it the product of mechanisms other than the gradual modification of populations by natural selection?

I.   The fossil record documents macroevolution

Biologists reconstruct evolutionary history by studying the succession of organisms in the fossil record. Fossils are collected and interpreted by *paleontologists*.

A.   How Fossils Form

Fossil = Any preserved remnant or impression left by an organism that lived in the past.

Sedimentary rocks are the richest sources of fossils.

- These rocks form from deposits of sand and silt which have weathered or eroded from the land and are carried by rivers to seas and swamps.
- Aquatic organisms and some terrestrial forms which were swept into seas and swamps became trapped in sediments when they died.
- New deposits pile on and compress older sediments below into rock. Sand is compressed into sandstone and mud into shale.
- A small proportion of the organims left a fossil record.

Fossils usually form from mineral rich hard parts of organisms (bones, teeth, shells of invertebrates) since most organic substances usually decay rapidly.

- Paleontologists usually find parts of skulls, bone fragments, or teeth; although nearly complete skeletons of dinosaurs and other forms have been found.
- Many of the parts found have been hardened by *petrification*, which occurs when minerals dissolved in groundwater seep into the tissues of dead organisms and replace organic matter.

Some fossils, found as thin films pressed between layers of sandstone or shale, retain organic material.

- Paleontologists have found leaves millions of years old that are still green with chlorophyll and preserved well enough that their organic composition and ultrastructure could be analyzed. (See Campbell, Figure 23.1d)
- One research team was even able to clone a very small sample of DNA from an ancient magnolia leaf.

Other fossils found by paleontologists are replicas formed in molds left when corpses were covered by mud or sand. (See Campbell, Figure 23.1e)

- Minerals from the water which filled the mold eventually crystallized in the shape of the organism.

Trace fossils form in footprints, animal burrows, and other impressions left in sediments by animal activity. These can provide a great deal of information.

- Dinosaur tracks can provide information about the animals gait, stride length, and speed.

Rarely, an entire organism has been fossilized, which could only have happened if the individual was buried in a medium that prevented bacteria and fungi from decomposing the body. (See Campbell, Figure 23.1g)

B.   **Limitations of the Fossil Record**

A fossil represents a sequence of improbable events:
- An organism had to die in the right place and at the proper time for burial conditions favoring fossilization.
- The rock layer containing the fossil had to escape geologic events (erosion, pressure, extreme heat) which would have distorted or destroyed the rock.
- The fossil had to be exposed and not destroyed.
- Someone who knew what they were doing had to find the fossil.

Paleontologists thus work with an incomplete record for many reasons.
- A large fraction of species that have lived probably left no fossils.
- Most fossils that were formed have probably been destroyed.
- Only a small number of existing fossils have been discovered.
- Consequently, the fossil record is comprised primarily of species that lived a long time, were abundant and widespread, and had shells or hard skeletons.

Even though it is incomplete, the fossil record provides the outline of macroevolution, but the evolutionary relationships between modern organisms must be studied to provide the details.

II.   **Paleontologists use a variety of methods to date fossils**

Several methods are used to determine the age of fossils, which makes them useful in studies of macroevolution.

A.   **Relative Dating**

Sedimentation may occur when the sea-level changes or lakes and swamps dry and refill.
- The rate of sedimentation and the types of particles that sediment vary with time when a region is submerged.
- The different periods of sedimentation resulted in formation of rock layers called *strata*.
- Younger strata are superimposed on top of older ones.
- The succession of fossil species chronicles macroevolution, since fossils in each layer represent organisms present at the time of sedimentation.

Strata from different locations can often be correlated by the presence of similar fossils, known as *index fossils*.
- The shells of widespread marine organisms are the best index fossils for correlating strata from different areas.
- Gaps in the sequence may appear in an area if it was above sea level (which prevents sedimentation) or if it was subjected to subsequent erosion.

Geologists have formulated a sequence of geological periods by comparing many different sites. This sequence is known as the *geological time scale*. (See Campbell, Table 23.1)

- These periods are grouped into four eras with boundaries between the eras marking major transitions in the life forms fossilized in the rocks.
- Periods within each era are subdivided into shorter intervals called *epochs*.
- The divisions are not arbitrary, but are associated with boundaries that correspond to times of change.

This record of the rocks presents a chronicle of the *relative* ages of fossils, showing the order in which species groups evolved.

B.   Absolute Dating

Absolute dating is not errorless, but it does give the age in years rather than in relative terms (e.g. before, after).

The most common method for determining the age of rocks and fossils on an absolute time scale is *radiometric dating*.

- Fossils contain isotopes of elements that accumulated in the living organisms.
- Since each radioactive isotope has a fixed *half-life*, it can be used to date fossils by comparing the ratio of certain isotopes (e.g. ^{14}C and ^{12}C) in a living organism to the ratio of the same isotopes in the fossil.

  half-life = the number of years it takes for 50% of the original sample to decay.
- The half-life of an isotope is not affected by temperature, pressure or other environmental variables.

Carbon-14 has a half-life of 5600 years, meaning that one-half of the carbon-14 in a specimen will be gone in 5600 years; half of the remaining carbon-14 would disappear from the specimen in the next 5600 years; this would continue until all of the carbon-14 had disappeared.

- Thus a sample beginning with 8 g of carbon-14 would have 4 g left after 5600 years and 2 g after 11,200 years.
- Carbon-14 is useful in dating fossils less than 50,000 years old due to its relatively short half-life.

Paleontologists use other radioactive isotopes with longer half-lives to date older fossils. For example,

- Uranium-238 has a half-life of 4.5 billion years and is reliable for dating rocks (and fossils within those rocks) hundreds of millions of years old.
- This isotope was used to place the oldest fossil-containing rocks in the Cambrian period.
- An error of < 10% is present with radioactive dating.

Other methods may also be used to date some fossils.
- Amino acids can have either left-handed (L-form) or right-handed (D-form) symmetry.
- Living organisms only synthesize L-form amino acids to incorporate into proteins.
- After an organism dies, L-form amino acids are slowly converted to D-form.
- The ratio of L-form to D-form amino acids can be measured in fossils.
- Knowing the rate of chemical conversion (*racemization*) allows this ratio to be used in determining how long the organism has been dead.
- This method is most reliable in environments where the climate has not changed significantly, since the conversion is temperature sensitive.

The dating of rocks and fossils they contain has enabled researchers to determine the geological periods. (See Campbell, Table 23.1)

III.   What are the major questions about macroevolution? *an overview*

Evolutionary biologists interested in macroevolution try to determine the processes involved in large-scale evolutionary changes that can be traced through paleontology and taxonomy.

Some of the questions to which they are looking for answers are:
- How do novel features which define taxonomic groups above the species level arise?
- What accounts for the appearantly progressive evolutionary trends found in the fossil record?
- How has macroevolution been affected by global geological changes?
- How can we explain the major fluctuations in biological diversity evident in the fossil record?

IV.   Some evolutionary novelties are modified versions of older structures

Higher taxa such as families and classes are defined by evolutionary novelties. For example:
- Birds evolved from dinosaurs, and their wings are homologous to the forelimbs of modern reptiles.
- Birds are adapted to flight, yet their ancestors were earthbound.

How could these new designs evolve?
- One mechanism is the gradual refinement of existing structures for new functions.
- Most biological structures have an evolutionary plasticity that makes alternative functions possible.

Preadaptation is a term applied to a structure that evolved in one context and became co-opted for another function.

- Natural selection cannot anticipate the future but can improve an existing structure in context of its current utility. For example,
  - ⇒ The honeycombed bones and feathers of birds did not evolve as adaptations for flight.
  - ⇒ They must have been beneficial (reduction of weight, gathering food, courtship) to the bipedal reptilian ancestors of birds and later, through modification, become functional for flying.
- Preadaptation cannot be proven, but provides an explanation for how novel designs can arise gradually through a series of intermediate stages, each having some function in the organism.
- The evolution of novelties by remodeling of old structures for new functions reflects the Darwinian tradition of large changes being crafted by natural selection through an accumulation of many small changes.

V. Genes that control development play a major role in evolutionary novelty

The evolution of complex structures (e.g. wings) requires such large modifications that changes at many gene loci are probably involved.

- In other cases, relatively few changes in the genome can cause major modifications in morphology (e.g. humans vs chimpanzees).
- Thus, slight genetic divergence can become magnified into major differences.

In animal development, a system of regulatory genes coordinates activities of structural genes to guide the rate and pattern of development from zygote to adult.

- A slight alteration of development becomes compounded in its effect on adult *allometric growth* (differences in relative rates of growth of various parts of the body) which helps to shape an organism.
- A slight change in these relative rates of growth will result in a substantial change in the adult. (See Campbell, Figure 23.2a)
- Thus, altering the parameters of allometric growth is one way relatively small genetic differences can have major morphological impact. (See Campbell, Figure 23.2b)

Changes in developmental dynamics, both *temporal* and *spatial*, have played a major role in macroevolution.

Temporal changes in development that create evolutionary novelties are called *heterochrony*.
Heterochrony = Evolutionary changes in the timing or rate of development.

- Genetic changes that alter the timing of development can also produce novel organisms.
  Paedomorphosis = Retention of ancestral juvenile structures in a sexually mature adult organism.
  - ⇒ In some species of salamanders, sexually mature adults retain certain larval features. (See Campbell, Figure 23.4)

⇒ A slight change in timing that retards the development of some organs in comparison to others produces a different kind of animal.

- Changes in developmental chronology may have contributed to human evolution.
  ⇒ Humans and chimpanzees are closely related through descent from a common ancestor.
  ⇒ They are much more similar as fetuses than as adults.
  ⇒ Different allometric properties and variations result in the human brain being proportionally larger than that in chimpanzees.
  ⇒ The human brain continues to grow several years longer than the chimpanzee brain.
  ⇒ Thus, the genetic changes responsible for humanness are not great, but have profound effects.

Equally important in evolution is the alteration of the spatial pattern of development or *homeosis*.

Homeosis = Alteration in the placement of different body parts.

- For example, to the arrangement of different kinds of appendages in animals or the placement of flower parts on a plant.

Since each regulatory gene may influence hundreds of structural genes, there is a potential for evolutionary novelties that define higher taxa to arise much faster than would occur by the accumulation of changes in only structural genes.

VI. Recognizing trends in the fossil record does not mean that macroevolution is goal-oriented

In most cases, a correct single evolutionary progression cannot be produced from the fossil record.

- This is a result of divergence which may produce organisms that seem to fit a single progression, but actually represent related but divergent forms.
- Modern horses (*Equus*) are believed to have evolved from *Hyracotherium*.
- A single evolutionary progression would have to include:
  ⇒ An increase in size.
  ⇒ Reduction of four toes to one.
  ⇒ Modification of browsing teeth to grazing teeth.

Selection of certain fossils can produce an apparent single evolutionary progression of intermediate organisms between *Hyracotherium* and modern horses (*Equus*). (See Campbell, Figure 23.5)

- This is a misrepresentation as a more complete phylogeny shows that *Equus* is the only survivor of a much more complicated evolutionary tree.
- Examination of the complete fossil sequence also shows that the transition from *Hyracotherium* to *Equus* was not a smooth gradation but involved a number of transitional steps which included a series of speciation episodes and several adaptive radiations.

Evolution has produced many genuine trends. For example,

- Punctuated equilibrium seems to have produced a trend toward larger size in the titanotheres. (See Campbell, Figure 23.6)
- The titanotheres were about as large as elephants and the fossil record shows that they evolved from a mouse-sized ancestral mammal.
- Although each species in the various lineages of titanotheres remain the same size, there was a succession of progressively larger species.

Branching evolution (cladogenesis) can produce a trend even if some new species counter the trend.

- There was an overall trend in reptilian evolution toward large size during the Mesozoic era which eventually produced the dinosaurs.
- This trend was sustained even though some new species were smaller than their parental species.

Even with an equal large-to-larger and large-to-smaller ratio, the trend would be maintained if the larger groups speciated at a greater rate than smaller groups or lasted longer before becoming extinct.

- Steven Stanley originated this view of macroevolution which holds that species are analogous to individuals with speciation being their birth and extinction their death.
- According to the Stanley model, an evolutionary trend is produced by *species selection* which is analogous to the production of a trend within a population by natural selection.
- The species that live longest and generate the greatest number of new species determine the direction of major evolutionary trends.
- Differential speciation thus may play a role in macroevolution similar to the role of differential reproduction in microevolution.

Qualities unrelated to the success of organisms in a specific environment may be equally important in species selection.

- The ability of a species to disperse to new habitats may result in development of new "daughter species" as organisms adapt to new conditions.
- A criticism of species selection is the argument that gradual modification of populations in response to environmental change is the most common stimulus to evolutionary trends.

No intrinsic drive toward a preordained state of being is indicated by the presence of an evolutionary trend.

- Evolution is a response to interactions between organisms and their current environments.
- An evolutionary trend may cease or reverse itself under changing conditions. For example, conditions of the Mesozoic era favored giant reptiles, but by the end of that era, smaller species prevailed.

VII.   Macroevolution has a biogeographical basis in continental drift

Macroevolution has dimension in space as well as time.

- Biogeography was a major influence on Darwin and Wallace in developing their views on evolution.

- Drifting of continents is the major geographical factor correlated with the spatial distribution of life. (See Campbell, Figure 23.8a)

Continental drift results from the movement of great plates of crust and upper mantle that float on the Earth's molten core.

- The relative positions of two land masses to each other changes unless they are embedded on the same plate.

- North America and Europe are drifting apart at a rate of 2 cm per year.

- Where two plates meet (boundaries), many important geological phenomena occur: mountain building, volcanism, and earthquakes. (See Campbell, Figure 23.8b)

  ⇒ Volcanism, in turn, forms volcanic islands (e.g. Galapagos), which opens new environments for founders and adaptive radiation.

Plate movements continually rearrange geography, however, two occurrences had important impacts on life: the formation of *Pangaea* and the subsequent breakup of Pangaea.

At the end of the Paleozoic era (250 million years ago), plate movements brought all land masses together into a super-continent called *Pangaea*.

- Species evolving in isolation where brought together and competition increased.

- Total shoreline was reduced and the ocean basins became deeper (draining much of the remaining shallow coastal seas).

- Marine species (which inhabit primarily the shallow coastal areas) were greatly affected by reduction of habitat.

- Terrestrial organisms were affected as continental interior habitats (and their harsher environments) increased in size.

- Changes in ocean currents would have affected both terrestrial and marine organisms.

- Overall diversity was thus impacted by extinctions and increased opportunities for surviving species.

During the early Mesozoic era (about 180 million years ago) Pangaea began to breakup due to continuing continental drift.

- This isolated the fauna and flora occupying different plates.

- The biogeographical realms were formed and divergence of organisms in the different realms continued.

Many biogeographical puzzles are explained by the pattern of continental separations. For example,

- Matching fossils recovered from widely separated areas.
  - ⇒ Although Ghana and Brazil are separated by 3000 km of ocean, matching fossils of Triassic reptiles have been recovered from both areas.
- Why Australia has such a unique fauna and flora.
  - ⇒ Australian marsupials are very diverse and occupy the same ecological roles as placental mammals on other continents.
  - ⇒ Marsupials probably evolved on the portion of Pangaea that is now North America and migrated into the area that would become Australia.
  - ⇒ The breakup of Pangea isolated Australia (and its marsupial populations) 50 million years ago, while placental mammals evolved and diversified on the other continents.

VIII. The history of life is punctuated by mass extinctions followed by adaptive radiations of survivors

The evolution of modern life has included long, relatively quiescent periods punctuated by briefer intervals of more extensive turnover in species composition.

- These intervals of extensive turnover included explosive adaptive radiations of major taxa as well as mass extinctions.

A. Examples of Major Adaptive Radiations

The evolution of some novel characteristics opened the way to new *adaptive zones* allowing many taxa to diversify greatly during their early history. For example,

- Evolution of wings allowed insects to enter an adaptive zone with abundant new food sources and adaptive radiation resulted in hundreds of thousands of variations on the basic insect body plan.
- A large increase in the diversity of sea animals occurred at the boundary between the Precambrian and Paleozoic eras. This was a result, in part, of the origin of shells and skeletons in a few key taxa.
  - ⇒ Precambrian rock contains the oldest animals (700 million years old) which were shell-less invertebrates that differed significantly from their successors found in Paleozoic rock.
  - ⇒ Nearly all the extant animal phyla and many extinct phyla evolved in less than 10 million years during the mid-Cambrian (early Paleozoic era).
  - ⇒ Shells and skeletons opened a new adaptive zone by making many new complex body designs possible and altering the basis of predator-prey relationships.
  - ⇒ It is possible that genes controlling development evolved during this time, resulting in a potential for increased morphological complexity and diversity.

An empty adaptive zone can be exploited only if the appropriate evolutionary novelties arise. For example,

- Flying insects existed for 100 million years before the appearance of flying reptiles and birds that fed on them.

Conversely, an evolutionary novelty cannot enable organisms to exploit adaptive zones that are occupied or that do not exist.

- Mass extinctions have often opened adaptive zones and allowed new adaptive radiations.
- For example, mammals existed 75 million years before their first large adaptive radiation in the early Cenozoic. This may have resulted from the ecological void created with the extinction of the dinosaurs.

B.   Examples of Mass Extinctions

Extinction is inevitable in a changing world. The average rate of extinction has been between 2.0 and 4.6 families (each family may include many species) per million years.

Extinctions may be caused by habitat destruction or by unfavorable environmental changes.

- Many very well adapted marine species would become extinct if the ocean's temperature fell only a few degrees.
- Changes in biological factors may cause extinctions even if physical factors remain stable.
- Since many species coexist in each community, an evolutionary change in one species will probably impact other species. For example,
  ⇒ The evolution of shells by some Cambrian animals may have contributed to the extinction of some shell-less forms.

There have been periods of global environmental change which greatly disrupted life and resulted in mass extinctions.

- During these periods, the rate of extinction escalated to as high 19.3 families per million years.
- These mass extinctions are recognized primarily from the decimation of hard-bodied animals in shallow seas which have the most complete fossil record.
- Two (of about a dozen) mass extinction episodes have been studied extensively by paleontologists.

The Permian extinctions (the boundary between the Paleozoic and Mesozoic eras) eliminated over 90% of the species of marine animals about 250 million years ago.

- Terrestrial life was probably also affected greatly. For example, 8 of the 27 orders of Permian insects did not survive into the Triassic.
- This mass extinction took place in less than 5 million years and probably resulted from several factors.
  ⇒ Occurred about the time Pangaea was formed by the merging of continents which disturbed many habitats and altered the climate.
  ⇒ A period of extreme vulcanism and resulting volcanic debris in the atmosphere may have altered the global temperature.

The Cretaceous extinction (the boundary between the Mesozoic and Cenozoic eras) occurred about 65 million years ago.

- More than 50% of the marine species and many terrestrial plants and animals (including dinosaurs) were eliminated.
- During this time the climate was cooling and many shallow seas receded from continental lowlands.
- Increased volcanic activity during this time may have contributed to the cooling by releasing materials into the atmosphere and blocking the sunlight.

Evidence also indicates that an asteroid or comet struck the Earth (*impact hypothesis*) while the Cretaceous extinctions were in progress.

- Iridium, an element rare on earth but common in meteorites, is found in large quantities in the clay layer separating Mesozoic and Cenozoic sediments.
- Walter and Luis Alvarez (and colleagues), after studying this clay layer, proposed that it is fallout from a huge cloud of dust ejected into the atmosphere when an asteroid collided with the Earth.
- This cloud would have both blocked the sunlight and severely disturbed the climate for several months.
- Although the asteroid hit the earth during this time, some researchers feel it did not cause the mass extinction of this period.

The impact hypothesis consisted of two parts: a large asteroid or comet collided with the Earth and the collision caused the Cretaceous extinctions.

- Many forms of evidence support the idea that a large comet or small asteroid collided with Earth 65 million years ago.
  ⇒ Many craters have been found and indicate that a large number of objects have fallen to the Earth's surface.
  ⇒ A large crater (≈ 300 km in diameter) located beneath sediments on the Yucatan coast of Mexico has been located.
- Questions about the impact hypothesis are now concentrated on the second part: the collision caused the Cretaceous extinctions.
- Advocates of the impact hypothesis point to several items in their support:
  ⇒ The large size of the impact would darken the Earth for several years and the reduction in photosynthesis output would be sufficient to cause food chains to collapse.
  ⇒ Severe acid precipitation would result from the increased mineral content of the atmosphere.
  ⇒ The content of sediments at the upper Cretaceous boundry indicated global fires were occuring and smoke from these fires would increase the atmospheric effects of the impact.
- Opponents of the impact hypothesis hold that the impact occurred within the period of mass extinction, but that the two occurrences are not a cause and effect event.
  ⇒ Many paleontologists and geologists believe that the climatic changes which occurred were due to continental drift, increased vulcanism, and other processes.
  ⇒ They also feel these events were sufficient to cause the mass extinction.

Many paleontologists are now trying to determine how sudden and uniform the Cretaceous extinctions were (on a geological time scale).

- Disappearance of diverse groups (from microscopic marine plankton to dinosaurs) during a short time span would support the impact hypothesis.
- A gradual decline with different groups disappearing at different rates, would support hypotheses emphasizing terrestrial causes.
- It is possible that the impact was a final, sudden event in environmental changes that were affecting the biota of the late Cretaceous.

Mass extinctions, whatever the cause, profoundly affect biological diversity.

- Not only are many species eliminated, but those that survive are able to undergo new adaptive radiations into the vacated adaptive zones and produce new diversity.

IX.   Systematics connects biological diversity to phylogeny

Phylogeny = The evolutionary history of a species or group of related species.

- Phylogeny is usually diagrammed as phylogenetic trees that trace inferred evolutionary relationships.

Systematics = The study of biological diversity in an evolutionary context.

- Biological diversity reflects past episodes of speciation and macroevolution.
- Encompasses *taxonomy* in its search for evolutionary relationships.
    Taxonomy = Identification and classification of species; is a component of systematics.

A.   Taxonomy and Hierarchical Classification

The taxonomic system used today was developed by Linnaeus in the eighteenth century. This system had two main features: the assignment of a bionomial to each species and a filing sytem for grouping species.

The *binomial* (two part Latin name) assigned to each species is unique to that species.

- The first word of the binomial is the *genus* (pl. genera); the second word is the *specific epithet* of the species.
- The scientific name of a species combines the genus and specific epithet.
- Each genus can include many species of related organisms. For example, *Felis silvestris* is the domestic cat; *Felis lynx* is the lynx.
- Use of the scientific name defines the organism referred to and removes ambiguity.

The filing system for grouping species into a hierarchy of increasingly general categories formalizes the grouping of organisms.

- Binomial nomenclature is the first step in grouping: similar species are grouped in the same genus.
- The system then progresses into broader categories:
  ⇒ Similar genera are grouped into the same family.
  ⇒ Families are grouped into orders.
  ⇒ Orders are grouped into classes.
  ⇒ Classes are grouped into phyla.
  ⇒ Phyla are grouped into kingdoms.
- Each taxonomic level is more inclusive than the one below. The more closely related two species are, the more levels they share: (See Campbell, Appendix Two)

Category	Domestic Cat	Bobcat	Lion	Dog
specific epithet	silvestris	rufus	leo	familiaris
genus	Felis	Felis	Panthera	Canis
family	Felidae	Felidae	Felidae	Canidae
order	Carnivora	Carnivora	Carnivora	Carnivora
class	Mammalia	Mammalia	Mammalia	Mammalia
phylum	Chordata	Chordata	Chordata	Chordata
kingdom	Animalia	Animalia	Animalia	Animalia

The two main objectives of taxonomy are to sort out and identify closely related species and to order species into the broader taxonomic categories.

- In sorting, closely related organisms are assigned to separate species (with the proper binomial) and described using the diagnostic characteristics which distinguish the species from one another.
- In categorizing, the species are grouped into broader categories from genera to kingdoms.
- In some cases, intermediate categories (i.e. subclasses; between orders and classes) are also used.
- The named taxonomic unit at any level is called a *taxon* (pl. taxa).
- Rules of nomenclature have been established: the genus name and specific epithet are italicized, all taxa from the genus level and higher are capitalized.

B. The Relationship of Classification to Phylogeny

The goal of systematics is to have classification reflect the evolutionary affinities of species.

- Groups subordinate to other groups in the taxonomic hierarchy should represent finer and finer branching of phylogenetic trees. (See Campbell, Figure 23.12)

- A *monophyletic taxon* is one where a single ancestor gave rise to all species in that taxon and to no species placed in any other taxon. For example, Family Ursidae evolved from a common ancestor. (Taxon 1)

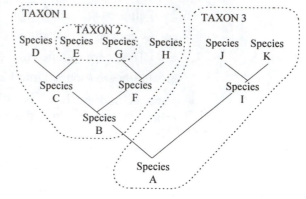

- A *polyphyletic taxon* is one whose members are derived from two or more ancestral forms not common to all members. For example, Kingdom Plantae includes both vascular plants and mosses which evolved from different algal ancestors. (Taxon 2)

- A *paraphyletic taxon* is one that excludes species that share a common ancestor that gave rise to the species included in the taxon. For example, lass Reptilia excludes the Class Aves although a reptilian ancestor common to all reptiles is shared. (Taxon 3)

The ideal in systematics is for each taxon to be monophyletic, creating a classification reflecting the evolutionary history of the organisms.

C. Sorting Homology from Analogy

Systematists classify species into higher taxa based on the extent of similarities in morphology and other characteristics.

<u>Homology</u> = Likeness attributed to shared ancestry.

- The forelimbs of mammals are homologous, they share a similarity in the skeletal support that has a genealogical basis.
- Homology must be distinguished from analogy in evolutionary trees.

<u>Analogy</u> = Similarities due to *convergent evolution*, not common ancestry.

<u>Convergent evolution</u> = Acquisition of similar characteristics in species from different evolutionary branches due to sharing similar ecological roles with natural selection shaping analogous adaptations.

The distinction between homology and analogy is some times relative. For example,

- The wings of birds and bats are modifications of the vertebrate forelimb.
  - ⇒ The appendages are thus homologous.
  - ⇒ As wings they are analogous since they evolved independently from the forelimbs of different flightless ancestors.
- Insect wings and bird wings are analogous.
  - ⇒ They evolved independently and are constructed from entirely different structures.
- Convergent evolution has produced analogous similarities between Australian marsupials and placental mammals on other continents.

Homology must be sorted from analogy to reconstruct phylogenetic trees on the basis of homologous similarities.

- Generally, the greater the amount of homology, the more closely related the species and this should be reflected in their classification.
- Adaptation and convergence often obscure homologies, although studies of embryonic development can expose homology that is not apparent in mature structures.
- Additionally, the more complex two similar structures are, the less likely it is that they have evolved independently. For example,
  - ⇒ The skulls of humans and chimpanzees are composed of many bones which are fused together and match almost perfectly. It is unlikely such a complex structure would have evolved indepently in separate groups.

X. Molecular biology provides powerful new tools for systematics

Molecular comparisons of proteins and DNA have added another useful method for studying the evolutionary relationships between species.

- Inherited nucleotide sequences in DNA program the corresponding sequences of amino acids in proteins.
- Examination of these macromolecules provide much information about evolutionary relationships.

A. Protein Comparison

The primary structure of proteins is genetically programmed and a similarity in the amino acid sequence of two proteins from different species indicates that the genes for those proteins evolved from a common gene present in a shared ancestor.

The advantages of molecular comparisons are:

- They are objective and quantitative.
- They can be used to assess relationships between species so distantly related that no morphological similarities exist.

Studies of cytochrome c (an ancient protein common to all aerobic organisms) have been used to compare many diverse species.

- The amino acid sequence has been determined for species ranging from bacteria to complex plants and animals.
- The sequence in cytochrome c is identical in chimpanzees and humans.
- The sequence in humans and chimpanzees differs at only one of the 104 amino acid positions in the rhesus monkeys.
- Chimpanzees, humans, and rhesus monkeys belong to the Order Primates.
    - ⇒ Comparing these sequences with nonprimate species shows greater differences (13 with the dog and 20 with the rattlesnake).
- Phylogenetic trees based on cytochrome c are consistent with evidence from comparative anatomy and the fossil record.

B. DNA Comparison

The most direct measure of common inheritance from shared ancestors is a comparison of the genes or genomes of two species.

Three methods can be used for DNA comparison: *DNA-DNA hybridization, restriction mapping*, and *DNA sequencing*.

DNA-DNA hybridization can compare whole genomes by measuring the degree of hydrogen bonding between single-stranded DNA obtained from two sources.

- DNA is extracted from different species and the complementary strands separated by heating.
- The single-stranded DNA from the two species is mixed and cooled to allow double-stranded DNA reformation which results from hydrogen bonding.
- The hybrid DNA is then reheated to separate the double strands.
- The temperature necessary to separate the hybrid DNA is indicative of the similarity in the DNA from the two species.
    - ⇒ The temperature correlation is based on the degree of bonding between the strands of the two species with more bonding occurring with greater similarity.
    - ⇒ The more extensive the pairing, the more heat is needed to separate the hybrid strand.
- Evolutionary trees constructed through DNA-DNA hybridization usually agree with those based on other methods, however, this technique is very beneficial in settling taxonomic debates that have not been finalized by other methods.

Restriction mapping provides precise information about the match-up of specific DNA nucleotide sequences.

- Restriction enzymes are used to cut DNA into fragments which can be separated by electrophoresis and compared to restriction fragments of other species.
- Two samples of DNA with similar maps for the locations of restriction sites will produce similar collections of fragments.
- The greater the divergence of two species from the common ancestor, the greater the differences in restriction sites and less similarity of the restriction fragments.
- This method works best when comparing small fragments of DNA.
- Mitochondrial DNA (mtDNA) is best suited for this type of comparison since it is smaller than nuclear DNA (produces smaller fragments) and mutates about ten times faster than nuclear DNA.
- The faster mutation rate of mtDNA allows it to be used to determine phylogenetic relationships between not only closely related species, but also populations of the same species.
    ⇒ mtDNA was used to establish the close relationship among the Pima, Mayan, and Yanomami groups of Native Americans.
    ⇒ The results supported linguistic evidence that these groups descended from the first wave of immigrants to cross the Bering land bridge from Asia during the late Pleistocene.

DNA sequencing is the most precise method of comparing DNA as it determines the actual nucleotide sequence of a DNA segment.

- Uses polymerase chain reaction (PCR) technology to clone traces of DNA.
- PCR is coupled with automated sequencing to provide a simpler and faster method of collecting sequence data.
- DNA sequencing and comparisons show exactly how much divergence there has been in the evolution of two genes derived from the same ancestral gene.
- Ribosomal RNA (rRNA) sequencing is a similar technique which can provide information about some of the earliest branching in phylogenetic relationships since DNA coding for rRNA changes very slowly.
- rRNA sequencing has been very useful in examining the relationships among bacteria.

C.   **Analysis of Fossilized DNA**

The nucleotide sequences in DNA traces recovered from fossils that retain organic material can be analyzed by using PCR.

- Fossilized DNA was first successfully analyzed in 1990. Using PCR, researchers amplified DNA extracted from 17 million-year-old magnolia leaves.
- A short piece of DNA from this sample was compared to homologous DNA from modern magnolias.
- Since 1990, DNA fragments have been sequenced from a frozen mammoth (40,000 years old), an insect fossilized in amber (40 million years old), and a frozen Stone Age man (5000 years old).

- Mary Schweitzer found DNA traces in the bone marrow of a 65-million-year-old *Tyrannosaurus rex* from Montana.
  - ⇒ Analysis is currently underway to eliminate the possibility that the trace is a result of contamination from a fungus or other organism.

It is obvious that the techniques from molecular biology will make even more important contributions to the fields of paleontology, systematics, and anthropology.

D.   Molecular Clocks

Different proteins evolve at different rates, although each type of protein evolves at a relatively constant rate over time.

When comparing homologous proteins from taxa that are known to have diverged from common ancestors, the number of amino acid substitutions is proportional to the elapsed time since divergence. For example,

- Homologous proteins of bats and dolphins are more similar than those of sharks and tuna.
- This is consistent with fossil evidence showing that tuna and sharks have been separated much longer than bats and dolphins.

DNA comparisons may be even more reliable than protein comparisons.

- Phylogenetic branching based on nucleotide substitutions in DNA generally approximates dates determined from the fossil record.
- The difference in DNA between two taxa is more closely correlated with the time since divergence than is morphological difference.

Molecular clocks (DNA and protein) are calibrated by graphing the number of nucleotide or amino acid differences against the times for a series of evolutionary branch points known from the fossil record.

- The graph can then be used to determine the time of divergence between taxa for which no substantial fossil record is available.

The assumption that mutation rates for genes (and their protein products) are relatively constant is the basis for using molecular clocks in evolutionary biology.

- This assumption is relatively solid when comparing groups of closely related species.
- Molecular clocks are less reliable when comparing more distantly related groups since differences in generation times and metabolice rates affect mutation rates.

Among closely related species, the constant mutation rate for specific genes implies that the accumulation of selectively neutral mutations, changes the genome more than adaptive mutations.

- Many evolutionary biologists doubt the prevalence of neutral variation, so they also question the use of molecular clocks to accurately date the time of divergence.
- There is less skepticism about the value of molecular clocks for determining the relative sequence of branch points in phylogeny.
- Modern systematists use available molecular data along with all other evidence to reconstruct phylogeny.

XI. Cladistics highlights the phylogenetic significance of systematics

The two significant features of a phylogenetic tree are the location of branch points along the tree and the degree of divergence between branches.

- The locations of branch points along the tree symbolize the relative times of origin for different taxa.
- The degree of divergence between branches represents how different two taxa have become since branching from a common ancestor.

The question of which property of phylogenetic trees should be most important in grouping species into taxa has divided taxonomy into three schools: *phenetics*, *cladistics*, and *classical evolutionary taxonomy*.

A. Phenetics

Phenetics makes no evolutionary assumptions and decides taxonomic affinities entirely on the basis of measurable similarities and differences.

- A comparison is made of as many characters (anatomical characteristics) as possible without attempting to sort homology from analogy.
- Pheneticists feel that the contribution of analogy to overall similarity will be overridden by the degree of homology if enough characters are compared.
- Critics of phenetics argue that overall phenotypic similarity is not a reliable index of phylogenetic proximity.
- While supported by few systematists, the emphasis of phenetics on multiple quantitative comparisons has made important contributions to systematics.
    ⇒ Especially useful for analyzing DNA sequence data and other molecular comparisons between species.

B. Cladistics

Cladistics classifies organisms according to the order in time that branches arise along a phylogenetic tree, without considering the degree of divergence.

- This produces a cladogram, a dichotomous tree that branches repeatedly.
    - ⇒ Each branch point is defined by novel homologies unique to the various species on that branch.

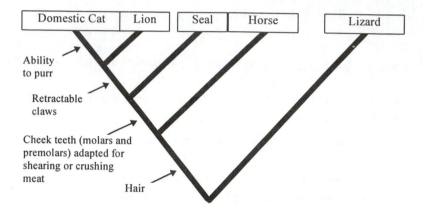

Each species in the cladogram has a mixture of primitive characters that existed in the common ancestor and characters that evolved more recently.

- The sharing of primitive characters indicates nothing about the pattern of evolutionary branching from a common ancestor.
    - ⇒ The presence of five separate toes can not be used to divide these vertebrates among evolutionary branches.
- Fossil evidence indicates the distant common ancestor to all of these species had five toes, thus this homology is a *shared primitive character*.
    - ⇒ Seals and horses lost this trait independently while the other species retained the trait.
- A major difficulty in cladistics is finding characters that are appropriate for each branch point.
- The cladogram is based on *synapomorphies* which are shared derived characters.
    - ⇒ These are homologies which evolved in an ancestor common to all species on one branch of the cladogram but not common to the other branch.
- Hair and mammary glands are synapomorphies for the branching point leading to the mammalian limb and the lizard on the other limb.
- Branching points along the mammalian limb can also be based on synapomorphies.
    - ⇒ Dental and skeletal modifications can be used to define the horse branch from the cat, lion, and seal.
    - ⇒ Retractable claws and other synapomorphic characters separate the seal from the cat and lion branch.
    - ⇒ The ability to purr is used to define the final branching between the cat and lion.

Some taxonomic surprises are produced by cladistic systematics:

- The branch point between crocodiles and birds is more recent than between crocodiles and other reptiles (a fact also supported by the fossil record).

    ⇒ Crocodiles and birds have synapomorphies not present in lizards and snakes.

- In a strict cladistic analysis, the Class Aves and the Class Reptilia, as we know them now, would be eliminated.

    ⇒ The birds would be included in a cladogram of the animals we know as reptiles.

- Birds are deemed superficially different because of the morphological changes associated with flight which have developed since their divergence from reptilian ancestors.

C. Classical Evolutionary Systematics

Classical evolutionary systematics attempts to balance the extent of divergence and the branching sequence.

- Predates both phenetics and cladistics, but now incorporates some ideas from both.
- In cases of systematic conflict, a subjective judgement is made about which type of information receives the highest priority.

In reference to crocodiles and birds, classical systematists recognize a closer genealogical relationship between crocodiles and birds than between crocodiles and lizards.

- However, they combine crocodiles and lizards in a taxon (Class Reptilia) that excludes birds because the ability to fly was an evolutionary advancement which allowed birds to enter a new adaptive zone.
- The resulting divergence of birds was so extensive that they are placed in a separate class (Aves).

XII. Is a new evolutionary synthesis necessary?

No biological theory has produced as much debate and controversy as the Theory of Evolution proposed by Charles Darwin.

- The debate began when Darwin published *The Origin of Species* in 1859 and continues today.
- The closest thing to a consensus has been the modern synthesis which has dominated evolutionary theory for the past 50 years.
- This view of evolution is called a synthesis since the ideas presented were drawn from several disciplines (paleontology, biogeography, systematics, population genetics) and continues to incorporate discoveries from new fields such as molecular biology.

Darwin's view of life was reaffirmed by the modern synthesis and updated by applying the principles of genetics.

- Its view is distinctly gradualistic in that it holds large-scale evolutionary changes are the accumulations of many minute changes occurring over vast spans of time.
- Microevolution (changes in allele frequencies in populations) is extrapolated to explain most macroevolution.

The classical version of the modern synthesis holds that natural selection is the major cause of evolution at all levels.

- Populations adapt by natural selection.
- New species arise when isolated populations diverge as different adaptations evolve.
- Continued divergence due to natural selection differentiates the higher taxa.
  - ⇒ That mechanisms such as genetic drift and chromosomal mutations can cause rapid, nonadaptive evolution is recognized by the modern synthesis; but major emphasis of the synthesis is on gradualism and natural selection.

Many evolutionary biologists disagree with the view that evolution recorded in the fossil record can be explained by extrapolating the processes of microevolution. The debate focuses in part on the pace of evolution.

- Many transitions in the fossil record are punctuational, not gradual.
- Gradualists hold that the apparent abruptness is in part derived from the imperfection of the fossil record and in part is a semantic issue confused by the vastness of geological time (e.g. is a change occurring over 10,000 years sudden or gradual?).
- Punctuationalists state that the imperfection of the fossil record is not enough to account for the rarity of transitional forms if speciation and the origin of higher taxa were primarily gradual extensions of microevolution.
- The debate also includes differing opinions about the degree to which microevolution compounded over time is sufficient to explain macroevolution.

The hierarchical theory favored by many evolutionary biologists gives mechanisms different levels of importance at different levels of evolution.

- In this theory, natural selection is the key to adaptive evolution of populations, plays a smaller role in speciation, and plays an even smaller role in macroevolution.
- Most new species begin as small populations isolated from their parent populations (geographical or genetic isolation).
- These isolates can evolve relatively rapidly, divergence from the parent population being due to genetic drift as much as to selection.
- Chance may cause speciation to begin before selection has produced new adaptations. Also, some major new adaptations may evolve with minimal genetic change if regulatory genes are involved.
- Continental drift and mass extinctions have probably had at least as much effect on the history of biological diversity as gradual adaptation caused by selection influencing gene pools at the population level. The historical contingency is gaining favor with researchers holding this view.

  Historical contingency = The occurrence of unforeseen events.

- In a hierarchical theory of evolution, most evolutionary trends progress by species selection not by phyletic transition due to an accumulation of microevolutionary changes.
- Species selection being the differential survival and branching of separate species that change little after they come into existence.

All groups in this debate recognize that natural selection is the mechanism of adaptation and should be the centerpiece of evolutionary biology.

- Selection adapts a population to its environment with generation-to-generation changes in the gene pool that are adaptive.

- It is also natural selection that refines unique adaptations when a new species or higher taxon comes into existence.

- Although events leading to speciation and episodes of macroevolution may have little influence on adaptation, new species only persist long enough to be entered into the fossil record only if they have adapted to their environment through natural selection.

The modern synthesis has never proposed that evolution is always smooth and gradual or that processes other than changes in gene pools due to selection are unimportant.

- The questions are not so much about the nature of evolutionary mechanisms as about their relative importance.

- Debate about how life evolved is indicative that evolutionary biology is an active science and the debate will continue.

REFERENCES

Campbell, N. *Biology*. 4th ed. Menlo Park, California: Benjamin/Cummings, 1996.

Gould, S.J. "The Wheel of Fortune and the Wedge of Progress." *Natural History*, March 1989.

Stanley, S. *Extinction*. New York: Scientific American Library, 1987.

CHAPTER 24

EARLY EARTH AND THE ORIGIN OF LIFE

OUTLINE

OBJECTIVES

After reading this chapter and attending lecture, the student should be able to:
1. Provide at least two lines of evidence for the antiquity of life.
2. Describe the contributions that A.I. Oparin, J.B.S. Haldane, Stanley Miller and Harold Urey made towards developing a model for abiotic synthesis of organic molecules.

3. Provide plausible evidence to support the hypothesis that chemical evolution resulting in life's origin occurred in four stages:
   a. Abiotic synthesis of organic monomers.
   b. Abiotic synthesis of polymers.
   c. Formation of protobionts.
   d. Origin of genetic information.
4. Describe the basis for Whittaker's five-kingdom system.
5. List and distinguish among the five kingdoms.
6. Give examples of representatives from each kingdom.
7. Describe three alternatives to the five-kingdom system and explain the rationale for each.

KEY TERMS

stromatolites	protobionts	photosynthetic	Protista
A.I. Oparin	liposomes	chemosynthetic	Plantae
J.B.S. Haldane	coacervates	heterotrophic	Fungi
Stanley Miller	ribozyme	autotrophic	Animalia
Harold Urey	prokaryotic	decomposer	Julius Rebek
polymers	eukaryotic	Robert H. Whittaker	panspermia
proteinoids	absorptive nutrition	Lynn Margulis	Thomas Cech
Sidney Fox	ingestive nutrition	Monera	domain

LECTURE NOTES

The history of living organisms and the history of Earth are inextricably linked. For example:
- Formation and subsequent breakup of Pangaea affected biotic diversity.
- The first photosynthetic organisms released oxygen into the air and altered Earth's atmosphere.
- *Homo sapiens* has changed the land, water and air on a scale and at a rate unprecedented for a single species.

In order to reconstruct life's history, scientists use evidence from:
- The fossil record, which is less complete the older the strata studied. In fact, there is no fossil record for the seminal episode of the origin of Earth's life.
- Contemporary organisms, which in their molecules and anatomy carry traces of their evolutionary histories.

I. Life on Earth originated between 3.5 and 4.0 billion years ago

Life appeared relatively early in the Earth's history. Precambrian fossils indicate the earliest organisms were bacteria that appeared at least 2 billion years before the oldest eukaryotic fossils.

- Fossils similar to spherical and filamentous prokaryotes have been recovered from *stromatolites* 3.5 billion years old in western Australia and southern Africa.

    Stromatolites = Banded domes of sediment similar to the layered mats constructed by colonies of bacteria and cyanobacteria currently living in salty marshes.

- The western Australian fossils appear to be of photosynthetic organisms, which indicates life evolved before these organisms lived — perhaps some 4 billion years ago.

- Other fossils similar to the prokaryotes have been recovered from the Fig Tree Chert rock formation in southern Africa which date to 3.4 billion years.

II. The first cells may have originated by chemical evolution on a young Earth: *an overview*

Life originated between 3.5 and 4.0 billion years ago. During this timespan the Earth's crust began to solidify (4.1 billion) and bacteria advanced enough to build stromatolites (3.5 billion).

The origin of life was possible in Earth's ancient environment, which was different from today:
- There was little atmospheric oxygen.
- Lightning, volcanic activity, meteorite bombardment, and ultraviolet radiation were more intense.

One hypothesis about the first living organisms is that they were the products of a chemical evolution that occurred in four stages:

1. Abiotic synthesis and accumulation of monomers, small organic molecules that are the building blocks for more complex molecules.
2. Joining of monomers into polymers (e.g. proteins and nucleic acids).
3. Formation of *protobionts* – droplets which formed from aggregates of abiotically produced molecules and which differed chemically from their surroundings.
4. Origin of heredity during or before protobiont appearance.

III. Abiotic synthesis of organic monomers is a testable hypothesis: *science as a process*

In the 1920's, A.I. Oparin and J.B.S. Haldane independently postulated that the reducing atmosphere and greater UV radiation on primitive Earth favored reactions that built complex organic molecules from simple monomers as building blocks. This is not possible today because:

- Oxygen in Earth's oxidizing environment attacks chemical bonds, removing electrons. An important characteristic of the early atmosphere must have been the rarity of oxygen.

- The modern atmosphere has a layer of ozone that screens UV radiation, so the energy required to abiotically synthesize organic molecules is not available. On primitive Earth, energy was available from frequent lightning and intense UV. radiation that penetrated the atmosphere.

Stanley Miller and Harold Urey tested the Oparin/Haldane hypothesis. They simulated conditions on early Earth by constructing an apparatus containing H_2O, H_2, CH_4 and NH_3.

- Their simulated environment produced some amino acids and other organic molecules.
- Now we know the atmosphere of early Earth probably included CO, CO_2, and N_2, and was less reducing than the Miller-Urey model and, thus, less favorable to formation of organic compounds.
- Additional experiments have produced all 20 amino acids, ATP, some sugars, lipids and purine and pyrimidine bases of RNA and DNA.

IV. Laboratory simulations of early Earth conditions have produced organic polymers

The forming of complex organic molecules, or *polymers*, from simpler building-block molecules may have been inevitable on the primitive Earth.

Polymers = Chains of similar building blocks or monomers; are synthesized by *dehydration* (condensation) reactions. For example:

$$\square—H + OH—\square \longrightarrow \square—\square + H_2O$$

- H and OH groups are removed from the monomers.
- H_2O is produced as a by-product.

Abiotic polymerization reactions in early-earth conditions must have occurred:

- Without the help of enzymes.
- With dilute solutions of monomers. (Spontaneous dehydration reactions that produce water would be unlikely in already dilute solutions.)

Abiotic polymerization does occur with dilute solutions of monomers under certain laboratory conditions:

- Dilute solutions of organic monomers are dripped onto hot sand, clay, or rock. Water vaporizes and concentrates the monomers on the substrate.
- Sidney Fox (University of Miami) used this method to abiotically produce polypeptides called *proteinoids*.

Clay may have been an important substrate for abiotic synthesis of polymers since:

- Monomers bind to charged sites in clay, concentrating amino acids and other monomers.
- Metal ions (e.g. iron and zinc) could catalyze dehydration reactions.
- The binding sites on clay could have brought many monomers close together and assisted in forming polymers.
- Pyrite (iron and sulfur) may also have been an important substrate. It has a charged surface and electrons freed during its formation could support bonding between molecules.

V. Protobionts can form by self-assembly

Living cells may have been preceded by *protobionts*.

> Protobionts = Aggregates of abiotically produced molecules able to maintain an internal environment different from their surroundings and exhibiting some life properties such as metabolism and excitability.

There is experimental evidence for the spontaneous formation of protobionts:

- When mixed with cool water, proteinoids self-assemble into *microspheres* surrounded by a selectively permeable protein membrane. These microspheres:
  ⇒ Undergo osmotic swelling and shrinking.
  ⇒ Have potential energy in the form of a membrane potential.
- *Liposomes* can form spontaneously when phospholipids form a bilayered membrane similar to those of living cells.
- *Coacervates* (colloidal drops of polypeptides, nucleic acids and polysaccharides) self-assemble.

VI. RNA was probably the first genetic material

Today's cells transcribe DNA into RNA, which is then translated into proteins. This chain of command must have evolved from a simpler mechanism of heritable control.

- One hypothesis is that before DNA, there existed a primitive mechanism for aligning amino acids along RNA molecules, which were the first genes. Evidence to support this hypothesis includes:
  ⇒ RNA molecules may have been able to self-replicate. Short polymers of ribonucleotides that can base pair (5 – 10 bases without enzyme, up to 40 bases with zinc added as catalyst) have been produced abiotically in test tubes.
  ⇒ RNA is autocatalytic, as indicated by *ribozymes* (RNA that acts as a catalyst to remove introns, or catalyze synthesis of mRNA, tRNA or rRNA).
- RNA folds uniquely depending on sequence (unlike DNA), thereby providing raw materials for natural selection — different molecular shapes (phenotypes) varying in stability and catalytic properties. Replication errors (mutations) probably created additional variation within families of closely related sequences.
- In addition to molecular competition, molecular cooperation probably evolved as RNA-directed protein synthesis produced short polypeptides that catalyzed RNA replication.
- Once this simple machinery for replication and translation of genetic information became sequestered into membrane-bound protobionts, molecular cooperation could be refined as natural selection acted on the level of the entire protobiont.

VII. The origin of hereditary information made Darwinian evolution possible

Perhaps this hypothetical membrane-bound protobiont:
- Incorporated genetic information.
- Selectively accumulated monomers from its surroundings.
- Used enzymes programmed by genes to make polymers and carry out other chemical reactions.
- Grew and split, distributing copies of its genes to offspring.

If these cell precursors could also grow, divide and distribute genes to offspring, the descendant protobionts would vary because of errors in the copying of RNA (mutations).
- The variation among related protobionts would be subject to natural selection.
- Evolution in the Darwinian sense - differential reproductive success - presumably accumulated refinements to primitive metabolism and inheritance, including the appearance of DNA as the hereditary material.
  ⇒ Initially, RNA could have provided the template to produce DNA.
  ⇒ Because it is more stable, DNA would have replaced RNA as the store of genetic information.
  ⇒ RNA's role would change as it became an intermediate in translation.

VIII. Debate about the origin of life abounds

No one knows how life actually began on Earth. The chemical evolution described and supporting lab simulations indicate key steps that could have occurred.

Several alternatives have been proposed.
- *Panspermia* - Some organic compounds may have reached Earth by way of meteorites and comets. Organic compounds (e.g. amino acids) have been recovered from modern meteorites. These extraterrestrial organic compounds may have contributed to the pool of molecules which formed early life.
- Most researchers believe life first appeared in shallow water or moist sediments. Some now feel the first organisms developed on the sea floor due to the harsh conditions on the surface during that time. This position was strengthened in the 1970's by discovery of the deep sea vents. Hot water and minerals emitted from such vents may have provided the energy and chemicals necessary for early protobionts.
- Simpler hereditary systems may have preceded nucleic acid genes. Julius Rebek synthesized a simple organic molecule in 1991. The importance of this molecule was that it served as a template for self-replication. This discovery supported the idea held by some biologists that RNA strands are too complicated to be the first self-replicating molecules.

IX. Arranging the diversity of life into kingdoms is a work in progress

The two kingdom system (animals and plants) long prevailed, but was not suitable as biologists learned more about the structures and life histories of different organisms.

The five kingdom system proposed by Robert H. Whittaker (1969) and modified by Lynn Margulis is widely accepted today.

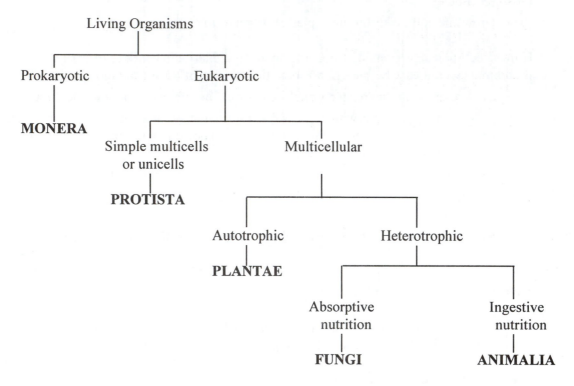

Plants, fungi, and animals are also distinguished by unique characteristics of structure and development.

Defining the kingdoms of life is a work in progress that reflects our increased understanding of the phylogeny of living organisms.

- Using the tools of molecular systematics, biologists have gathered new data that leads them to challenge the traditional five-kingdom system.

- This new information has reopened issues of biological diversity at the highest taxonomic levels. Three alternative classification systems are outlined below:

Six-kingdom system. The prokaryotes are split into two kingdoms based on molecular evidence for an early evolutionary divergence between eubacteria and archaebacteria.

Eu-bacteria	Archae-bacteria	Protista	Plantae	Fungi	Animalia

Three-domain system. This scheme assigns more significance to the ancient evolutionary split between eubacteria and archaebacteria by using a superkingdom taxon called the *domain*. The domain Eucarya includes four kingdoms of eukaryotic organisms.

Bacteria (Eubacteria)	Archaea (Archaebacteria)	Eucarya (Eukaryotes)

Eight-kingdom system. In addition to two separate prokaryotic kingdoms, this system also splits the protists into three kingdoms.

Eubacteria	Archaebacteria	Archaezoa	Chromista	Protista	Plantae	Fungi	Animalia

REFERENCES

Campbell, N. *Biology*. 4th ed. Menlo Park, California: Benjamin/Cummings, 1996.

Margulis, L., and K.V. Schwartz. *Five Kingdoms: An Illustrated Guide to the Phyla of Life on Earth.* 2nd ed. New York: Freeman, 1987.

Orgel, L.E. "*The Origin of Life on Earth.*" Scientific American, October 1994.

Whittaker, R.H. "New Concepts of Kingdoms of Organisms." *Science* 163: 150-160, 1969.

CHAPTER 25

PROKARYOTES AND THE ORIGINS OF METABOLIC DIVERSITY

OUTLINE

VII.   Prokaryotes continue to have an enormous ecological impact
    A.   Prokaryotes and Chemical Cycles
    B.   Symbiotic Bacteria
    C.   Bacteria and Disease
    D.   Putting Bacteria to Work

OBJECTIVES

After reading this chapter and attending lecture, the student should be able to:
1.   List unique characteristics that distinguish the archaebacteria from the eubacteria.
2.   Describe the three-domain system of classification and explain how it differs from previous systems.
3.   Using a diagram or micrograph, distinguish among the three most common shapes of prokaryotes.
4.   Describe the structure and functions of prokaryotic cell walls.
5.   Distinguish between the structure and staining properties of gram-positive and gram-negative bacteria.
6.   Explain why disease-causing gram-negative bacterial species are generally more pathogenic than disease-causing gram-positive bacteria.
7.   Describe three mechanisms motile bacteria use to move.
8.   Explain how prokaryotic flagella work and why they are not considered to be homologous to eukaryotic flagella.
9.   Indicate where photosynthesis and cellular respiration take place in prokaryotic cells.
10.   Explain how organization of the prokaryotic genome differs from that in eukaryotic cells.
11.   Explain, in their own words, what is meant by geometric growth.
12.   List the mechanisms that are sources of genetic variation in prokaryotes and indicate which one is the major source.
13.   Distinguish between autotrophs and heterotrophs.
14.   Describe four modes of bacterial nutrition and give examples of each.
15.   Distinguish among obligate aerobes, facultative anaerobes and obligate anaerobes.
16.   Describe, with supporting evidence, plausible scenarios for the evolution of metabolic diversity including:
    a.   Nutrition of early prokaryotes.   c.   Origin of photosynthesis.
    b.   Origin of electron transport chains.   d.   Origin of aerobic respiration.
17.   Explain how molecular systematics has been used in developing a moneran classification.
18.   List the three main groups of archaebacteria, describe distinguishing features among the groups and give examples of each.
19.   List the major groups of eubacteria, describe their mode of nutrition, some characteristic features and representative examples.
20.   Explain how endospores are formed and why endospore-forming bacteria are important to the food-canning industry.
21.   Explain how the presence of *E. coli* in public water supplies can be used as an indicator of water quality.
22.   State which organism is responsible for the most common sexually transmitted disease in the United States.
23.   Describe how mycoplasmas are unique from other prokaryotes.
24.   Explain why all life on earth depends upon the metabolic diversity of prokaryotes.
25.   Distinguish among mutualism, commensalism and parasitism.

26. List Koch's postulates that are used to substantiate a specific pathogen as the cause of a disease.
27. Distinguish between exotoxins and endotoxins.
28. Describe how humans exploit the metabolic diversity of prokaryotes for scientific and commercial purposes.
29. Describe how *Streptomyces* can be used commercially.

KEY TERMS

prokaryotes	magnetotactic	purple sulfur bacteria	*Chlamydia trachomatis*
bacteria	genophore	bacteriochlorophyll	nongonococcal
archaebacteria	nucleoid region	molecular systematics	urethritis (NGU)
eubacteria	plasmids	signature sequences	endospore
Carl Woese	endospore	methanogens	*Clostridium botulinum*
Domain Archea	transformation	extreme halophiles	botulism
Domain Bacteria	conjugation	extreme thermophiles	enteric
Domain Eukarya	transduction	bacteriorhodopsin	myxobacteria
cocci	nutrition	Proteobacteria	decomposers
bacilli	photoautotrophs	*Rhizobium*	symbiosis
spirilla	chemoautotrophs	*Escherichia coli*	symbiont host
spirochetes	photoheterotrophs	*Salmonella typhi*	mutualism
peptidoglycan	chemoheterotrophs	mycoplasmas	commensalism
gram stain	saprobes	*Mycoplasma*	parasitism
gram positive	parasites	*pneumoniae*	opportunistic
gram negative	non-biodegradable	actinomycetes	*Streptococcus pneumoniae*
capsule	obligate aerobes	*Streptomyces*	Louis Pasteur
pili	facultative anaerobes	spirochetes	Joseph Lister
flagella	obligate anaerobes	*Treponema pallidum*	Robert Koch
flagellin	cyanobacteria	*Borrelia burgdorferi*	Koch's postulates
taxis	heterocysts	Lyme disease	exotoxins
phototactic	nitrogen fixation	chlamydias	endotoxins
chemotactic	green sulfur bacteria		

LECTURE NOTES

Appearing about 3.5 billion years ago, prokaryotes (*bacteria*) were the earliest living organisms and the only forms of life for 2 billion years.

I. **They're (almost) everywhere!** *an overview of prokaryotic life*

Prokaryotes dominate the biosphere; they are the most numerous organisms and can be found in all habitats.

- Approximately 4,000 species are currently recognized, however, estimates of the actual diversity range from 400,000 – four million species
- Are structurally and metabolically diverse.

Prokaryotic cells differ from eukaryotic cells in several ways:

- Prokaryotes are smaller and lack membrane-bound organelles.
- Most have cell walls but the composition and structure differ from those found in plants, fungi and protists.
- Prokaryotes have simpler genomes. They also differ in genetic replication, protein synthesis and recombination.

Prokaryotes, while very small, have a tremendous impact on the Earth.

- A small percentage cause disease.
- Some are decomposers, key organisms in life-sustaining chemical cycles.
- Many form symbiotic relationships with other prokaryotes and eukaryotes. Mitochondria and chloroplasts may have evolved from such symbioses.

II.   Archaea and Bacteria are the two main branches of prokaryotic evolution

The traditional five-kingdom system recognizes one kingdom of prokaryotes (Monera) and four kingdoms of eukaryotes (Protista, Plantae, Fungi, and Animalia).

- This system emphasizes the structural differences between prokaryotic and eukaryotic cells.

Recent research in systematics has resulted in questions about the placement of a group as diverse as the prokaryotes in a single kingdom. Two major branches of prokaryotic evolution have been indicated by comparing ribosomal RNA and other genetic products:

- One branch is called the *archaebacteria*.
  ⇒ Believed to have evolved from the earliest cells.
  ⇒ Inhabit extreme environments which may resemble the Earth's early habitats (hot springs and salt ponds).
- The second branch is called the *eubacteria*.
  ⇒ Considered the more "modern" prokaryotes, having evolved later in Earth's history.
  ⇒ More numerous than archaebacteria.
  ⇒ Differ from archaebacteria in structural, biochemical, and physiological characters.

This recently acquired molecular data has led to new proposals for the systematic relationships of organisms.

- Carl Woese has proposed a six-kingdom system that includes:
  ⇒ Two prokaryotic kingdoms.
  ⇒ Four eukaryotic kingdoms.
- An eight-kingdom system has also been proposed. (Covered in more detail in Chapter 26.)
  ⇒ This system also contains two kingdoms of prokaryotic organisms.

In addition to the kingdoms systems proposed, many systematists now favor an organization of the diversity of life which includes three *domains*, with the domain being a taxonomic level higher than kingdom.

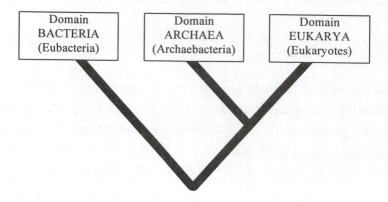

- Prokaryotes comprise two of the three domains, Archaea and Bacteria, while eukaryotes fill the third domain, Eukarya.
    - ⇒ The domain Archaea includes the archaebacteria, the Bacteria includes the eubacteria, and the Eukarya includes all eukaryotic organisms.
    - ⇒ The Eukarya and Archaea shared a common ancestor which lived more recently than the ancestor shared between the Archaea and Bacteria.
- Although systematic debates continue, the archaebacteria and eubacteria are structurally organized at the prokaryotic level while differing on structural, genetic, and metabolic levels.

III. The success of prokaryotic life is based on diverse adaptations of form and function

A. Morphological Diversity of Prokaryotes

A majority of prokaryotes are single-celled, although some aggregate into two-celled to several celled groups. Others form true, permanent aggregates and some bacterial species have a simple multicellular form with a division of labor between specialized cells.

- Cells have a diversity of shapes, the most common being spheres (*cocci*), rods (*bacilli*), and helices (*spirilla* and *spirochetes*).
- One rod-shaped species measures ≈0.5 mm in length, and is the largest prokaryotic cell known.
- Most have diameters of 1–5mm (compared to eukaryotic diameters of 10 – 100mm).

B. The Cell Surface of Prokaryotes

A majority of prokaryotes have external cell walls that:
- Maintain the cell shape.
- Protect the cell.
- Prevent the cell from bursting in a hypotonic environment.
- In eubacteria, contain *peptidoglycan*; archaebacteria lack peptidoglycan in their cell walls.

Peptidoglycan = Modified sugar polymers cross-linked by short polypeptides.

- Exact composition varies among species.
- Some antibiotics work by preventing formation of the cross-links in peptidoglycan, thus preventing the formation of a functional cell wall.

Gram stain = A stain used to distinguish two groups of eubacteria by virtue of a structural difference in their cell walls.

1. Gram-positive bacteria.
   - Have simple cell walls with large amounts of peptidoglycan.
   - Stain blue.

2. Gram-negative bacteria.
   - Have more complex cell walls with smaller amounts of peptidoglycan.
   - An outer lipopolysaccharide-containing membrane covers the cell wall.
   - Stain pink.
   - These cells are more often disease-causing (*pathogenic*) than gram-positive bacteria.
   - Lipopolysaccharides are often toxic and the outer membrane helps protect these bacteria from host defense systems.
   - Lipopolysaccharides also impede entry of drugs into the cells, making gram-negative bacteria more resistant to antibiotics.

Capsule = A gelatinous secretion of some prokaryotes which provides cells with additional protection, helps them adhere to hosts and helps form aggregates.

Pili = Surface appendages used for adherence to a host (in the case of a pathogen), or for transferring DNA when bacteria conjugate.

C.   The Motility of Prokaryotes

Motile bacteria ($\approx$ 50% of known species) use one of three mechanisms to move:

1. Flagella.
   - Prokaryotic flagella differ from eukaryotic flagella in that they are:
     ⇒ Unique in structure and function. Prokaryotic flagella lack the "9 + 2" microtubular structure and rotate rather than whip back and forth like eukaryotic flagella.
     ⇒ Not covered by an extension of the plasma membrane.
     ⇒ One-tenth the width of eukaryotic flagella.
   - Filaments, composed of chains of the protein *flagellin*, are attached to another protein hook which is inserted into the basal apparatus.
   - The basal apparatus consist of 35 different proteins arranged in a system of rings which sit in the various cell wall layers.
   - Their rotation is powered by the diffusion of protons into the cell. The proton gradient is maintained by an ATP-driven proton pump.

2. Filaments which are characteristic of spirochetes, helical-shaped bacteria.
   - Several filaments spiral around the cell inside the cell wall.
   - Similar to prokaryotic flagella in structure, axial filaments are attached to basal motors at either end of the cell. Filaments attached at opposite ends move relative to each other, rotating the cell like a corkscrew.

3. Gliding.
   - Some bacteria move by gliding through a layer of slimy chemicals secreted by the organism.
   - The movement may result from flagellar motors that lack flagellar filaments.

Prokaryotic movement is fairly random in homogenous environments but may become directional in a heterogenous environment.

Taxis = Movement to or away from a stimulus.
- The stimulus can be light (phototaxis), a chemical (chemotaxis), or a magnetic field (magnetotaxis).
- Movement toward a stimulus is a positive taxis (i.e. positive phototaxis = toward light) while movement away from a stimulus is a negative taxis (i.e. negative phototaxis = away from light).

During taxis (directed movement), bacteria move by *running* and *tumbling* movements:
- Enabled by rotation of flagella either counterclockwise or clockwise respectively.
- Caused by flagella moving coordinately about each other (for a run), or in separate and randomized movements (for a tumble).

D. Internal Membranous Organization

Prokaryotes lack the diverse internal membranes characteristic of eukaryotes. Some prokaryotes, however, do have specialized membranes, formed by invaginations of the plasma membranes.
- Infoldings of the plasma membrane function in cellular respiration of aerobic bacteria.
- Cyanobacteria have thylakoid membranes that contain chlorophyll and that function in photosynthesis.

E. Prokaryotic Genomes

The prokaryotic genome has only 1/1000 as much DNA as the eukaryotic genome.

Genophore = The bacterial chromosome, usually one double-stranded, circular DNA molecule.
- This DNA is concentrated in the nucleoid region, and is not surrounded by a membrane; therefore, there is no true nucleus.
- Has very little protein associated with the DNA.

Many bacteria also have *plasmids*.

Plasmid = Smaller rings of DNA having supplemental (usually not essential) genes for functions such as antibiotic resistance or metabolism of unusual nutrients.
- Replicate independently of the genophore.
- Can be transferred between partners during conjugation.

While prokaryotic and eukaryotic DNA replication and translation are similar, there are some differences. For example,
- Bacterial ribosomes are smaller and have different protein and RNA content than eukaryotic ribosomes.
  ⇒ This difference permits some antibiotics (i.e. tetracycline) to block bacterial protein synthesis while not inhibiting the process in eukaryotic cells.

F.   **Growth, Reproduction, and Gene Exchange**

Neither mitosis nor meiosis occur in the prokaryotes.
- Reproduction is asexual by *binary fission*.
- DNA synthesis is almost continuous.

Growth in the numbers of cells is geometric in an environment with unlimited resources.
- Generation time is usually 1 – 3 hours, although some can be 20 minutes in optimal environments.
- At high concentrations of cells, growth slows due to accumulation of toxic wastes, lack of nourishment, etc.
- Competition in natural environments is reduced by the release of antibiotic chemicals which inhibit the growth of other species.
- Optimal growth requirements vary depending upon the species.

Some bacteria survive adverse environmental conditions and toxins by producing *endospores*.

Endospore = Resistant cell formed by some bacteria; contains one chromosome copy surrounded by a thick wall.
- When endospores form, the original cell replicates its chromosome and surrounds one copy with a durable wall. The original surrounding cell disintegrates, releasing the resistant endospore.
- Since some endospores can survive boiling water for a short time, home canners and food canning industry must take special precautions to kill endospores of dangerous bacteria.
- May remain dormant for many years until proper environmental conditions return.

Although meiosis and syngamy do not occur in prokaryotes, genetic recombination can take place through three mechanisms that transfer variable amounts of DNA:

<u>Transformation</u> = The process by which external DNA is incorporated by bacterial cells.

<u>Conjugation</u> = The direct transfer of genes from one bacterium to another.

<u>Transduction</u> = The transfer of genes between bacteria by viruses.

Short generation times allow prokaryotic populations to adapt to rapidly changing environmental conditions.

- New mutations and genomes (from recombination) are screened by natural selection very quickly.

- Has resulted in the current diversity and success of prokaryotes as well as the variety of nutritional and metabolic mechanisms found in this group.

IV.   All major types of nutrition and metabolism evolved among prokaryotes

The prokaryotes exhibit some unique modes of nutrition as well as every type of nutrition found in eukaryotes.

A.   Major Modes of Nutrition

Prokaryotes exhibit a great diversity in how they obtain the necessary resources (energy and carbon) to synthesize organic compounds.

- Some obtain energy from light (*phototrophs*), while others use chemicals taken from the environment (*chemotrophs*).

- Many can utilized CO_2 as a carbon source (*autotrophs*) and others require at least one organic nutrient as a carbon source (*heterotrophs*).

Depending upon the energy source and the carbon source, prokaryotes have four possible nutritional modes:

1. *Photoautotrophs*: Use light energy to synthesize organic compounds from CO_2. Includes the cyanobacteria. (Actually all photosynthetic eukaryotes fit in this category.)

2. *Chemoautotrophs*: Require only CO_2 as a carbon source and obtain energy by oxidizing inorganic compounds such as H_2S, NH_3 and Fe^{2+}. This mode of nutrition is unique to certain prokaryotes (i.e. archaebacteria of the genus *Sulfobolus*).

3. *Photoheterotrophs*: Use light to generate ATP from an organic carbon source. This mode of nutrition is unique to certain prokaryotes.

4. *Chemoheterotrophs*: Must obtain organic molecules for energy and as a source of carbon. Found in many bacteria as well as most eukaryotes.

B.   Nutritional Diversity Among Chemoheterotrophs

Most bacteria are chemoheterotrophs and can be divided into two subgroups: *saprobes* and *parasites*.

- Saprobes are decomposers that absorb nutrients from dead organic matter.
- Parasites are bacteria that absorb nutrients from body fluids of living hosts.

The chemoheterotrophs are a very diverse group, some have very strict requirements while others are extremely versatile.

- *Lactobacillus* will grow well only when the medium contains all 20 amino acids, several vitamins, and other organic compounds.
- *E. coli* will grow on a medium which contains only a single organic ingredient (i.e. glucose or some other substitute).

Almost any organic molecule can serve as a carbon source for some species.

- Some bacteria are capable of degrading petroleum and are used to clean oil spills.
- Those compounds that cannot be used as a carbon source by bacteria are considered *non-biodegradable* (e.g. some plastics).

C.   Nitrogen Metabolism

While eukaryotes can only use some forms of nitrogen to produce proteins and nucleic acid, prokaryotes can metabolize most nitrogen compounds.

Prokaryotes are extremely important to the cycling of nitrogen through ecosystems.

- Some chemoautotrophic bacteria (*Nitrosomonas*) convert $NH_3 \rightarrow NO_2^-$.
- Other bacteria, such as *Pseudomonas*, denitrify NO_2^- or NO_3^- to atmospheric N_2.
- Nitrogen fixation ($N_2 \rightarrow NH_3$) is unique to certain prokaryotes (cyanobacteria) and is the only mechanism that makes atmospheric nitrogen available to organisms for incorporation into organic compounds.
- The nitrogen fixing cyanobacteria are very self-sufficient, they need only light energy, CO_2, N_2, water and a few minerals to grow.

D.   Metabolic Relationships to Oxygen

Prokaryotes differ in their growth response to the presence of oxygen.

Obligate aerobes = Prokaryotes needing O_2 for cellular respiration.

Facultative anaerobes = Prokaryotes that use O_2 when present, but in its absence can grow using fermentation.

Obligate anaerobes = Prokaryotes that are poisoned by oxygen.

- Some species live exclusively by fermentation.
- Other species use inorganic molecules (other than O_2) as electron acceptors during anaerobic respiration.

V.   The evolution of prokaryotic metabolism was both cause and effect of changing environments on Earth

Prokaryotes evolved all forms of nutrition and most metabolic pathways eons before eukaryotes arose.

- Evolution of these new metabolic capabilities were a response to the changing environment of the early atmosphere.
- As these new capabilities evolved, they changed the environment for subsequent prokaryotic communities.

Information from molecular systematics, comparisons of energy metabolism, and geological studies about Earth's early atmosphere have resulted in many hypotheses about the evolution of prokaryotes and their metabolic diversity.

A.   The Origin of Glycolysis

The first prokaryotes, which evolved 3.5 billion years ago, were probably chemoheterotrophs that absorbed free organic compounds (including ATP) generated by abiotic synthesis.

The universal role of ATP implies that prokaryotes used that molecule for energy very early in their evolution.

- As ATP supplies were depleted, natural selection favored those prokaryotes that could regenerate ATP from ADP, leading to step by step evolution of glycolysis and other catabolic pathways.

Glycolysis is the only metabolic pathway common to all modern organisms and does not require O_2 (which was not abundant on early Earth).

- Some extant archaebacteria and other obligate anaerobes that live by fermentation have forms of nutrition believed to be similar to those of the original prokaryotes.

B.   The Origin of Electron Transport Chains and Chemiosmosis

Chemiosmotic ATP synthesis probably evolved in early prokaryotes as it is a common mechanism in all three domains.

- Early prokaryotes may have used the transmembrane pumps to help regulate their internal pH by expelling hydrogen ions produced by fermentation. Energy (ATP) would have been necessary to drive these pumps.
- ATP may have been saved by the first electron transport chains by coupling oxidation of organic acids to the transport of H^+ out of the cell.
- Some bacteria may have evolved electron transport chains so efficient that more H^+ was extruded than was necessary for pH regulation. These cells could then utilize the influx of H^+ to reverse the proton pump and generate ATP.
  - ⇒ Some modern bacteria use this form of energy metabolism (= anaerobic respiration).
  - ⇒ For example, members of the genus *Pseudomonas* pass electrons down transport chains from organic substrates to NO_3^-.

C. **The Origin of Photosynthesis**

As the supply of free ATP and abiotically produced organic molecules was depleted, natural selection may have favored organisms that could make their own organic molecules from inorganic resources.

Light absorbing pigments in the earliest prokaryotes may have provided protection to the cells by absorbing excess light energy, especially ultraviolet, that could be harmful.

- These energized pigments may have then been coupled with electron transport systems to power ATP synthesis.
- Bacteriorhodopsin, the light-energy capturing pigment in the membrane of extreme halophiles (a group of archaebacteria), uses light energy to pump H^+ out of the cell to produce a gradient of hydrogen ions. This gradient provides the power for production of ATP.
- This mechanism is being studied as a model system of solar energy conversion.

Components of electron transport chains that functioned in anaerobic respiration in other prokaryotes may have been co-opted to also provide reducing power. For example, H_2S could be used as a source of electrons and hydrogen for fixing CO_2.

- The nutritional modes of modern purple and green sulfur bacteria are believed the most similar to early prokaryotes.
- The colors of these bacteria are due to bacteriochlorophyll, their main photosynthetic pigment.

D. **Cyanobacteria, the Oxygen Revolution, and the Origins of Cellular Respiration**

Eventually, some prokaryotes evolved that could use H_2O as the electron source. Thus evolved cyanobacteria, which released oxygen.

- Cyanobacteria evolved between 2.5 and 3.4 billion years ago.
- They lived with other bacteria in colonies that resulted in the formation of the stromatolites. (See Campbell, Figure 24.3.)

Oxygen released by photosynthesis may have first reacted with dissolved iron ions to precipitate as iron oxide (supported by geological evidence of deposits), preventing accumulation of free O_2.

- Precipitation of iron oxide would have eventually depleted the supply of dissolved iron and O_2 would have accumulated in the seas.
- As seas became saturated with O_2, the gas was released to the atmosphere.
- As O_2 accumulated, many species became extinct while others survived in anaerobic environments (including some archaebacteria) and others evolved with antioxidant mechanisms.
- Aerobic respiration may have originated as a modification of electron transport chains used in photosynthesis. The purple non-sulfur bacteria are photoheterotrophs which still use a hybrid electron transport system between a photosynthetic and respiratory system.
- Other bacterial lineages reverted to chemoheterotrophic nutrition with electron transport chains adapted only to aerobic respiration.

All major forms of nutrition evolved among prokaryotes before the first eukaryotes arose.

VI. Molecular systematics is leading to a phylogenetic classification of prokaryotes

The use of molecular systematics (especially ribosomal RNA comparisons) has shown that prokaryotes diverged into the archaebacteria and eubacteria lineages very early in prokaryotic evolution.

- Studies of ribosomal RNA indicate the presence of *signature sequences*.

    Signature sequences = Domain-specific base sequences at comparable locations in ribosomal RNA or other nucleic acids.

- Numerous other characteristics differentiate these two domains. (See Campbell, Table 25.2.)

- A somewhat surprising result of these types of studies has been the realization that the archaebacteria have at least as much in common with the eukaryotes as they do with the eubacteria.

A. Domain Archaea (Archaebacteria)

Some unique characteristics of archaebacteria include:

- Cell walls lack peptidoglycan.
- Plasma membranes have a unique lipid composition.
- RNA polymerase and ribosomal protein are more like those of eukaryotes than of eubacteria.

The archaebacteria inhabit the most extreme environments of the Earth. Studies of these organisms have identified three main groups:

1. *Methanogens* are named for their unique form of energy metabolism.
    - Use H_2 to reduce CO_2 to CH_4 and are strict anaerobes.
    - Some species are important decomposers in marshes and swamps (form marsh gas) and some are used in sewage treatment.
    - Other species are important digestive system symbionts in termites and herbivores that subsist on cellulose diets.

2. *Extreme halophiles* inhabit high salinity (15–20%) environments (e.g. Dead Sea).
    - Some species simply tolerate extreme salinities while others require such conditions.
    - They have the pigment bacteriorhodopsin in their plasma membrane which absorbs light to pump H^+ ions out of the cell.
    - This pigment is also responsible for the purple-red color of the colonies.

3. *Extreme thermophiles* inhabit hot environments.
    - Live in habitats of $60 - 80°C$.
    - One sulfur-metabolizing thermophile inhabits water of $105°C$ near deep sea hydrothermal vents.

B. **Domain Bacteria (Eubacteria)**

The major groups of eubacteria include a very diverse assemblage of organisms. Among the thousands of known species are forms which exhibit every known mode of nutrition and energy metabolism.

Molecular systematics has provided an increased understanding of the once hazy relationships among members of this taxon. While most prokaryotic systematists recognize a dozen groups of eubacteria, the following figure shows the relationship of five with the other domains.

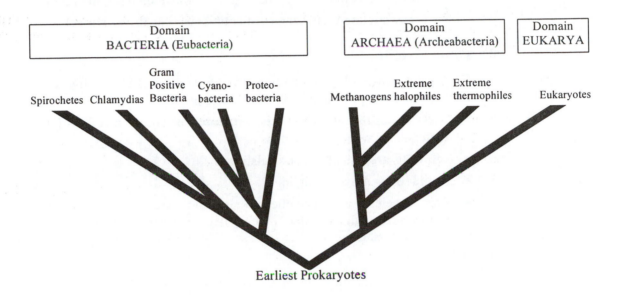

1. **Proteobacteria**

The most diverse group of bacteria, containing 3 main subgroups:

- Purple bacteria
    ⇒ Photoauto- and photoheterotrophic organisms with bacteriochlorophylls built into plasma membrane invaginations.
    ⇒ Extract electrons from molecules other than H_2O (i.e. H_2S), thus they release no oxygen.
    ⇒ Most are obligate anaerobes found in sediments of ponds, lakes, and mud flats.
    ⇒ Many species are flagellated.
- Chemoautotrophic proteobacteria
    ⇒ Includes both free-living and symbiotic species.
    ⇒ Many play key roles in the nitrogen cycle, including nitrogen fixation.
    ⇒ Example: *Rhizobium*
- Chemoheterotrophic proteobacteria
    ⇒ Enteric bacteria found in the intestinal tracts of animals.
    ⇒ Most are rod-shaped facultative anaerobes.
    ⇒ Examples: *E. coli* and *Salmonella*.

2. **Gram-positive eubacteria**

   Most are gram-positive while a few are gram-negative. Includes many photosynthetic members, but most are chemoheterotrophs. Many form endospores.
   - Mycoplasmas
     ⇒ Smallest of all known cells, 0.10 – 0.25 μm.
     ⇒ Only eubacteria that lack cell walls.
     ⇒ Common in soil and some are pathogenic (i.e. *Mycoplasma pneumoniae*)
   - Actinomycetes
     ⇒ Soil bacteria that form branching colonies which resemble fungi.
     ⇒ Many are important sources of antibiotics (i.e. *Streptomyces*).

3. **Cyanobacteria**

   Photoautotrophs with plantlike photosynthesis; possess chlorophyll *a* and use two photosystems to split water and yield O_2 as a product.
   - Most species inhabit fresh water, but some are marine and others form symbiotic relationships with fungi (lichens).
   - Cell walls are often thick and gelatinous.
   - Motile forms move by gliding.
   - Many species are single-celled forms, others are colonial, and some are truly multicellular with a division of labor between specialized cells.

4. **Spirochetes**

   Helical cells which are sometimes very long (up to 0.25 mm) but very thin.
   - Internal flagellar filaments function in corkscrew-like movements.
   - Chemoheterotrophs which include both free-living species and pathogens.
     ⇒ Includes *Treponema pallidum* (causes syphilis) and *Borrelia burgdorferi* (causes Lyme disease).

5. **Chlamydias**

   Obligate intracellular parasites of animals.
   - Obtain all of their ATP from host cells.
   - Have gram-negative walls but lack peptidogylcan found in other eubacteria.
   - *Chlamydia trachomatis* is the most common cause of blindness in the world and causes the most common sexually transmitted disease (non-gonococcal urethritis) in the U.S.

VII. Prokaryotes continue to have an enormous ecological impact

A. Prokaryotes and Chemical Cycles

Prokaryotes are critical links in the recycling of chemical elements between the biological and physical components of ecosystems — a critical element in the continuation of life.

Decomposers = Prokaryotes that decompose dead organisms and waste of live organisms.
- Return elements such as carbon and nitrogen to the environment in inorganic forms needed for reassimilation by other organisms, many of which are also prokaryotes.

Autotrophic bacteria = Bacteria that fix CO_2, thus supporting food chains through which organic nutrients pass.

Cyanobacteria supplement plants in restoring oxygen to the atmosphere as well as fixing nitrogen into nitrogenous compounds used by other organisms.

Other prokaryotes also support cycling of nitrogen, sulfur, iron and hydrogen.

B. Symbiotic Bacteria

Most prokaryotes form associations with other organisms; usually with other bacterial species possessing complementary metabolisms.

Symbiosis = Ecological relationships between organisms of different species that are in direct contact.
- Usually the smaller organism, the *symbiont*, lives within or on the larger *host*.

Three Categories of Symbiosis:

Mutualism = Symbiosis in which both symbionts benefit.
- For example, nitrogen-fixing bacteria in root nodules of certain plants fix nitrogen to be used by the plant, which in turn furnishes sugar and other nutrients to the bacteria.

Commensalism = Symbiosis in which one symbiont benefits while neither helping nor harming the other symbiont.

Parasitism = Symbiosis in which one symbiont (the *parasite*) benefits at the expense of the host.

Symbiosis is believed to have played a major role not only in the evolution of prokaryotes, but also in the origin of early eukaryotes.

C. Bacteria and Disease

About 1/2 of human disease is caused by bacteria. To cause a disease, the bacteria must invade the host, evade or resist the host's internal defenses long enough to grow, and harm the host.

Some pathogens are *opportunistic*.

<u>Opportunistic</u> = Normal inhabitants of the body that become pathogenic only when defenses are weakened by other factors such as poor nutrition or other infections.

- For example, *Streptococcus pneumoniae* lives in the throat of most healthy humans, but can cause pneumonia if the host's defenses are weakened.

While Louis Pasteur, Joseph Lister and others began linking disease to pathogenic microbes in the late 1800s, Robert Koch was the first to determine a direct connection between specific bacteria and certain diseases.

- Koch identified the bacteria responsible for anthrax and tuberculosis, and his methods established the four criteria (Koch's postulates) used as guidelines in medical microbiology.

<u>Koch's postulates</u> = Four criteria to substantiate a specific pathogen as the cause for a disease are:

1. Find the same pathogen in each diseased individual.
2. Isolate the pathogen from a diseased subject and grow it in a pure culture.
3. Use cultured pathogen to induce the disease in experimental animals.
4. Isolate the same pathogen in the diseased experimental animal.

Some pathogens cause disease by growth and invasion of tissues which disrupts the physiology of the host while others cause disease by production of a *toxin*. Two major types of toxins have been found:

<u>Exotoxins</u> = Proteins secreted by bacterial cells.

- Can cause disease without the organism itself being present; the toxin is enough.
- Among the most potent poisons known.
- Elicits specific symptoms.
- For example, botulism toxin from *Clostridium botulinum* and cholera toxin from *Vibrio cholerae*.

<u>Endotoxins</u> = Toxic component of outer membranes of some gram-negative bacteria.

- All induce the general symptoms of fever and aches.
- Examples are *Salmonella typhi* (typhoid fever) and other species of *Salmonella* which cause food poisoning.

Improved sanitation measures and development of antibiotics have greatly reduced mortality due to bacterial diseases over the last half of the century.

- Many of the antibiotics now in use are produced naturally by members of the genus *Streptomyces*. In its natural habitat (soil) such materials would reduce competition from other prokaryotes.
- Although beneficial, the excessive and improper use of antibiotics has resulted in the evolution of many antibiotic-resistant bacterial species which now pose a major health problem.

D.   Putting Bacteria to Work

Humans use the metabolic diversity of bacteria for a multitude of purposes. The range of these purposes has increased through the application of recombinant DNA technology.

- Pharmaceutical companies use cultured bacteria to make vitamins and antibiotics. More than half of the antibiotics used to treat bacterial diseases come from cultures of various species of *Streptomyces* maintained by pharmaceutical companies.
- As simple models of life to learn about metabolism and molecular biology. (*E. coli* is the best understood of all organisms.)
- Methanogens are used to digest organic wastes at sewage treatment plants.
- Some species of pseudomonads are used to decompose pesticides and other synthetic compounds.
- Industry uses bacterial cultures to produce products such acetone and butanol.
- The food industry uses bacteria to convert milk into yogurt and cheese.

REFERENCES

Campbell, N. *Biology*. 4th ed. Menlo Park, California: Benjamin/Cummings, 1996.

Tortora, G.J. et al. *Microbiology: An Introduction.*   5th. ed.   Redwood City, California: Benjamin/Cummings, 1995.

Woese, Carl R. "Archaebacteria." *Scientific American*, June 1981.

CHAPTER 26

THE ORIGINS OF EUKARYOTIC DIVERSITY

OUTLINE

I. Eukaryotes originated by symbiosis among prokaryotes

II. Archezoans provide clues to the early evolution of eukaryotes

III. The diversity of protists represents different "experiments" in the evolution of eukaryotic organization

IV. Protistan taxonomy is in a state of flux

V. Diverse modes of locomotion and feeding evolved among protozoa
   A.   Rhizopoda (Amoebas)
   B.   Actinopoda (Heliozoans and Radiozoans)
   C.   Foraminifera (Forams)
   D.   Apicomplexa (Sporozoans)
   E.   Zoomastigophora (Zooflagellates)
   F.   Ciliophora (Ciliates)

VI. Funguslike protists have morphological adaptations and life cycles that enhance their ecological role as decomposers
   A.   Myxomycota (Plasmodial Slime Molds)
   B.   Acrasiomycota (Cellular Slime Molds)
   C.   Oomycota (Water Molds)

VII. Eukaryotic algae are key producers in most aquatic ecosystems
   A.   Dinoflagellata (Dinoflagellates)
   B.   Bacillariophyta (Diatoms)
   C.   Chrysophyta (Golden Algae)
   D.   Phaeophyta (Brown Algae)
   E.   Rhodophyta (Red Algae)
   F.   Chlorophyta (Green Algae)

VIII. Systematists continue to refine their hypotheses of eukaryotic phylogeny

IX. Multicellularity originated independently many times

OBJECTIVES

After reading this chapter and attending lecture, the student should be able to:

1. List characteristics that distinguish protists from organisms in the other four kingdoms.
2. Explain why some biologists prefer to use the term *undulipodia* for eukaryotic flagella and cilia.
3. Briefly summarize and compare the two major models of eukaryotic origins, the autogenous hypothesis and the endosymbiotic hypothesis.
4. Provide three major lines of evidence for the endosymbiotic hypothesis.
5. Explain why some critics are skeptical about the eubacterial origins for chloroplasts and mitochondria.
6. Explain why modern biologists recommend expanding the original boundaries of the Kingdom Protista.
7. Explain what is meant by the statement that the Kingdom Protista is a polyphyletic group.
8. List the six major protozoan phyla and distinguish among them based upon locomotor structures, reproduction (asexual and sexual) and habitat.
9. Describe amoeboid movement.
10. Explain why the Phylum Apicomplexa is so named.
11. Outline the life cycle of *Plasmodium*.
12. Indicate the organism that causes African sleeping sickness and explain how it is spread and why it is difficult to control.
13. Describe the function of contractile vacuoles in freshwater ciliates.
14. Distinguish between macronuclei and micronuclei.
15. Using diagrams, describe conjugation in *Paramecium caudatum*.
16. Explain how accessory pigments can be used to classify algae and determine phylogenetic relationships among divisions.
17. Distinguish among the following algal divisions based upon pigments, cell wall components, storage products, reproduction, number and position of flagella and habitat:
    a. Dinoflagellata
    b. Bacillariophyta
    c. Chrysophyta
    d. Phaeophyta
    e. Rhodophyta
    f. Chlorophyta
18. Describe three possible evolutionary trends that led to multicellularity in the Chlorophyta.
19. Outline the life cycles of *Chlamydomonas*, *Ulva* and *Laminaria* and indicate whether the stages are haploid or diploid.
20. Distinguish between isogamy and oogamy; sporophyte and gametophyte; and isomorphic and heteromorphic generations.
21. Compare the life cycles of plasmodial and cellular slime molds and describe the major differences between them.
22. Provide evidence that the oomycetes are not closely related to true fungi.
23. Give examples of oomycetes and describe their economic importance.
24. Explain the most widely accepted hypothesis for the evolution of multicellularity.

KEY TERMS

Anton von Leeuwenhoek
endosymbiotic theory
endosymbiont
Lynn Margulis
archezoa
Giardia intestinalis
diplomonads
acritarchs
plankton
Robert Whittaker
protozoa
Rhizopoda
amoebas
pseudopodia
Actinopoda
axopodia
heliozoans
radiolarians
Foraminifera
Apicomplexa

sporozoite
Anopheles
Plasmodium
Zoomastigophora
Trypanosoma
Ciliophora
cilia
cirri
Stentor
macronucleus
micronucleus
conjugation
Myxomycota
plasmodial slime
    molds
plasmodium
sporangia
Acrasiomycota
cellular slime molds
Oomycota

white rusts
downy mildew
algae
chlorophyll *a*
carotenoids
xanthophylls
phycobilins
Dinoflagellata
red tides
peridinin
Bacillariophyta
diatoms
Chrysophyta
laminarin
Phaeophyta
fucoxanthin
seaweeds
thallus
holdfast
blades

algin
agar
carageenan
alternation of
    generations
Laminaria
sporophyte
gametophyte
Ulva
heteromorphic
isomorphic
Rhodophyta
Chlorophyta
lichen
Volvox
Chlamydomonas
isogamy
anisogamy
oogamy
Chromista

LECTURE NOTES

Using lenses he developed, Anton von Leeuwenhoek (17th century) was the first to describe the diversity of microscopic protists.

There are about 60,000 extant species of protists.
- Most are unicellular, but colonial forms and even some simple multicellular forms exist.
- Their eukaryotic structure makes even the simplest more complex than prokaryotes.
- Primal eukaryotes not only gave rise to current protists, but also to plants, fungi, and animals.

I. **Eukaryotes originated by symbiosis among prokaryotes**

There is a greater difference between prokaryotic and eukaryotic cells than between the cells of plants and animals.

The cellular structures and process unique arose during the genesis of the protists:
- A membrane-bound nucleus.
- Mitochondria, chloroplasts, and the endomembrane system.
- A cytoskeleton.
- 9 + 2 flagella.
- Multiple chromosomes consisting of linear DNA molecules compactly arranged with proteins.
- Diploid life cycle stages.
- Mitosis.
- Meiosis.
- Sex.

The small size and simpler construction of the prokaryotic cell has many advantages but also imposes a number of limitations. For example,
- The number of metabolic activities that can occur at one time is smaller.
- The smaller size of the prokaryotic genome limits the number of genes which code for enzymes controlling these activities.

While prokaryotes are extremely successful, natural selection resulted in increasing complexity in some groups. Three trends were:
- Toward multicellular forms such as the cyanobacteria which have different cells types with specialized functions.
- The evolution of complex bacterial communities where each species benefits from the metabolic activities of other species.
- The compartmentalization of different functions within single cells.
  - ⇒ The first eukaryotes resulted from this solution.

The evolution of the compartmentalized nature of eukaryotic cells may have resulted from two processes.
- Specialization of plasma membrane invaginations.
  - ⇒ Invaginations and subsequent specializations may have given rise to the nuclear envelope, endoplasmic reticulum, Golgi apparatus, and other components of the endomembrane system.
- Endosymbiotic associations of prokaryotes may have resulted in the appearance of some organelles.
  - ⇒ Mitochondria, chloroplasts, and some other organelles evolved from prokaryotes living within other prokaryotic cells.

The *endosymbiotic theory* proposes that certain prokaryotic species, called *endosymbionts* lived within larger prokaryotes. This theory was extensively developed by Lynn Margulis. (See Campbell, Figure 26.1b)

- Focuses mainly on mitochondria and chloroplasts.
- Chloroplasts are descended from endosymbiotic photosynthesizing prokaryotes, such as cyanobacteria, living in larger cells.
- Mitochondria are postulated to be descendants of prokaryotic aerobic heterotrophs.
  - ⇒ May have been parasites or undigested prey of larger prokaryotes.
  - ⇒ The association progressed from parasitism or predation to mutualism.
  - ⇒ As the host and endosymbiont became more interdependent, they integrated into a single organism.
- Many extant organisms are involved in endosymbiotic relationships.

Evidence for the endosymbiotic origin of mitochondria and chloroplasts includes the similarities between these organelles and eubacteria.

- Are of appropriate size to be descendants of eubacteria.
- Have inner membranes containing several enzymes and transport systems similar to those on prokaryotic plasma membranes.
- Replicate by splitting processes similar to binary fission present in prokaryotes.
- Have DNA which is circular and not associated with histones or other proteins, as in prokaryotes.
- Contain their own tRNA, ribosomes, and other components for DNA transcription and translation into proteins.
- Chloroplasts have ribosomes more similar to prokaryotic ribosomes (with regards to size, biochemical characteristics, and antibiotic sensitivity) than to eukaryotic ribosomes.
- Mitochondrial ribosomes vary, but are also more similar to prokaryotic ribosomes.

Molecular systematics lends even more evidence to support the endosymbiotic theory.

- The rRNA of chloroplasts is more similar in base sequence to RNA from certain photosynthetic eubacteria than to rRNA in eukaryotic cytoplasm.
  - ⇒ Chloroplast rRNA is transcribed from genes in the chloroplast while eukaryotic rRNA is transcribed from nuclear DNA.
- Mitochondrial rRNA also has a base sequence which supports a eubacterial origin.

A comprehensive theory for the origin of eukaryotic cells must also include the evolution of:

- 9 + 2 flagella and cilia which are analogous to prokaryotic flagella.
- The origins of mitosis and meiosis which also utilize microtubules.
  - ⇒ Mitosis made it possible for large eukaryotic genomes to be reproduced.
  - ⇒ Meiosis is essential to sexual reproduction.
  - ⇒ Protists have the most varied sexual life histories of the eukaryotes.

II.   **Archezoans provide clues to the early evolution of eukaryotes**

An ancient lineage of eukaryotes branched away from the eukaryotic tree very early, perhaps as long as two billion years ago. This group is referred to as the *archezoa* and contains only a few phyla.

- These organisms lack mitochondria and plastids, and have relatively simple cytoskeletons.
- Their ribosomes have some characteristics more closely aligned with prokaryotes than with eukaryotes.
  ⇒ rRNA sequencing indicates a closer relationship with prokaryotes than eukaryotes.

Giardia intestinalis is a modern representative of the archezoa.

- It is a flagellated unicellular eukaryotic that is parasitic in the human intestine.
- It is most commonly transmitted in the cyst form through water contaminated with human feces.

Giardia's importance to evolutionary biologists is related more to its (and other diplomonad species) characteristics than to its role as a human parasite.

- Diplomonads have two separate haploid nuclei which produce a "face-like" appearance.
  ⇒ Dual nuclei may be a vestige of early eukaryotic evolution.
- Prokaryotes have haploid genomes and some researchers postulate that early eukaryotes had a single haploid nucleus bounded by a nuclear envelope.
- In most modern eukaryotes, the diploid stages in the life cycle result from the fusion of haploid nuclei which form the diploid nucleus.
  ⇒ Diplomonads may represent an early mechanism in the evolution of diploidy in eukaryotes.

If the diplomonads diverged from the eukaryotic lineage before the process of nuclear fusion and meiosis evolved, their dual nuclei may be a clue to the past.

- This coupled with the absence of mitochondria in this group and other archezoans is consistent with an origin occurring before the endosymbiotic relationships that gave rise to mitochondria in aerobic species.

III.   **The diversity of protists represents different "experiments" in the evolution of eukaryotic organization**

Precambrian rock dated to about 2.1 billion years of age contain *acritarchs*, the oldest commonly accepted fossils of protists.

- Remnants of the proper size and structure to be ruptured coats of cysts similar to those of extant protists.
- Adaptive radiation produced a diversity of protists over the next billion years.
- The variations present in these organisms were representative of the structure and function possible in eukaryotic cells.

Protists are found in almost all moist environments, the seas, freshwater systems, and moist terrestrial habitats.

- They are important components of marine and freshwater *plankton*.

    Plankton = Communities of organisms, mostly microscopic, that drift passively or swim weakly near the surface of oceans, ponds, and lakes.

- Many are bottom dwellers in freshwater and marine habitats where they attach to rocks or live in the sand and silt.

- Photosynthetic species form mats at the still-water edges of lakes and ponds where they provide a food source for other protists.

- Damp soil, leaf litter, and other moist terrestrial habitats also contain large numbers of protists.

- Symbiotic species are found in the body fluids, tissues, and cells of the host.

    ⇒ These associations range from mutualism to parasitism and many are important pathogens.

Almost all protists are aerobic, using mitochondria for cellular respiration.

- Anaerobic forms lack mitochondria and live in anaerobic environments or have mutualistic respiring bacteria.

- May be photoautotrophic, heterotrophic, or mixotrophic (combining photosynthesis and heterotrophy).

    ⇒ The different modes of nutrition are used to separate protists into non-technical categories: photosynthetic forms are typically referred to as algae, ingestive forms as protozoa, and absorptive protists.

Most protists have flagella or cilia (not homologous to prokaryotic flagella) at some time in life cycle.

- Eukaryotic cilia and flagella are extensions of the cytoplasm.

- These cilia and flagella have the same basic 9 + 2 microtubular ultrastructure.

    ⇒ Differ in that cilia are shorter and more numerous.

Cell division and nuclear structure are very variable in the protists.

- Unique mitotic divisions occur in many groups.

- All can reproduce asexually; some can also reproduce sexually or at least use syngamy to trade genes between asexual reproductive episodes.

- Some form resistant cysts when stressed by harsh environments.

Protists are considered the simplest eukaryotic organisms because most are unicellular.

- At the cellular level, protists are extremely complex.

- The unicellular protist is not analogous to a single plant or animal cell, but is a complete organism.

    ⇒ The single cell of a protist must perform all the basic functions carried out by the specialized cells of plants and animals.

IV.   Protistan taxonomy is in a state of flux

Classification schemes and the phylogeny they reflect are based on available information.

- These presentations are tentative and often change as additional information becomes available.

In 1969, Robert H. Whittaker popularized the five-kingdom taxonomic system and placed only unicellular eukaryotes in the kingdom Protista.

- During the 1970s and 1980s, the Kingdom Protista was expanded to include some multicellular organisms earlier classified as either plants or fungi.
- Studies of cell ultrastructure and life cycle details formed the basis for such taxonomic transfers.
  ⇒ Seaweeds were found to exhibit characteristics which indicated a closer relationship with certain algae than to plants.
  ⇒ Slime molds and water molds were found to be more closely related to certain protozoans than to fungi.
- The tendency was to place all eukaryotes that could not comfortably be fitted into the plants, fungi, or animals into this kingdom. (Since many multicellular forms were included, it was suggested the kingdom be named Protoctista.)

Molecular systematics, especially rRNA comparisons, have stimulated three main trends in eukaryotic systematics and taxonomy over the last decade.

1. Reassessment of the number and membership of protistan phyla.
   - The phylum Sarcodina once housed all unicellular organisms that possessed pseudopodia, current classification splits these organisms into several phyla.
2. Arrangement of the phyla into a cladogram based largely on what molecular methods and cell structure comparisons reveal about evolutionary relationships of protists.
3. Reevaluation of the five-kingdom system and debate about the addition of new kingdoms.
   - The kingdom Archezoa has been proposed as well as other kingdoms for forms previously classified as protists.

The proposal of additional kingdoms challenges the current view of biological diversity at the highest taxonomic levels.

V.   Diverse modes of locomotion and feeding evolved among protozoa

The term *protozoa* is an informal reference to a very diverse group of heterotrophic protists.

- These organisms seek and consume bacteria, other protists, and *detritus* (dead organic mater).
- Some are symbiotic and many are pathogens.
- The subdivision of this group into different phyla is based in part on how they feed and move.
- As protistan systematics continues to be reevaluated, some of the following phyla may be split or combined.

A.   **Rhizopoda (Amoebas)**

The phylum Rhizopoda (= rootlike feet) includes the amoebas and their relatives.
  • Simplest of protists; all are unicellular.
  • No flagellated stages in life cycle.
  • *Pseudopodia* form as cellular extensions and function in feeding and movement. (See Campbell, Figure 26.5)
     ⇒ The cytoskeleton of microtubules and microfilaments functions in amoeboid movement.
  • All reproduction is by asexual mechanisms: no meiosis or sexual reproduction are known to occur.
  • During mitosis, spindle fibers form, but typical stages of mitosis are not apparent in most species.
     ⇒ The nuclear envelope persists during cell division in many genera.
  • Rhizopods inhabit freshwater, marine and soil habitats.
  • Most are free-living, although some are parasitic.

B.   **Actinopoda (Heliozoans and Radiozoans)**

Actinopoda (= ray feet) possess *axopodia*, a slender form of pseudopodia.

Axopodia = Projections reinforced by bundles of microtubules thinly covered by cytoplasm.

Axopodia increase the surface area that comes into contact with the surrounding water.
  • Helps the organisms float and function in feeding.
  • Small protist and other microorganisms which stick to the axopodia are phagocytized by the thin layer of cytoplasm and moved to the main portion of the cell by cytoplasmic streaming.

The two main groups of the Actinopoda are the *heliozoans* and the *radiozoans*.
  • Most are planktonic.
  • *Heliozoans* live primarily in fresh water.
  • *Radiozoans* are primarily marine and have delicate shells, usually made of silica.

C.   **Foraminifera (Forams)**

Forams have porous, multi-chambered shells of organic material hardened by calcium carbonate.
  • Forams are exclusively marine with most living in the sand or attached to algae and rocks; some are planktonic.
  • Cytoplasmic strands extend through the shell's pores and function in swimming, feeding and shell formation.
  • Many have symbiotic algae living beneath the shell which provide nourishment through photosynthesis.
  • 90% of the described species are fossils.
  • Foram shells are an important component of sediments and sedimentary rocks.

D. Apicomplexa (Sporozoans)

All species of the phylum Apicomplexa are parasites of animals.
- The infectious cells produced in the life cycle are called *sporozoites*.
- The apex of sporozoites has organelles for penetrating host cells and tissues; it is for these apical organelles that the phylum is named.
- Life cycles are intricate having both sexual and asexual reproduction; often requiring two or more different host species.

Several species of *Plasmodium* cause malaria. (See Campbell, Figure 26.8)
- *Anopheles* mosquitoes serve as the intermediate host and humans as the final host.
- The incidence of malaria was greatly reduced by the use of insecticides against mosquitoes in the 1960s.
- More recently, incidence of malaria has increased due to insecticide resistant strains of mosquitoes and drug resistant strains of *Plasmodium*.
  - ⇒ This is a relatively common ($\approx$ 300,000 new cases per year) and potentially fatal tropical disease resulting in about two million deaths each year.

There has been little success in developing a vaccine against *Plasmodium*.
- The human immune system has little effect on the parasite.
  - ⇒ *Plasmodium* spends most of its life cycle in blood cells or liver cells.
  - ⇒ *Plasmodium* also has the ability to alter its surface proteins.
- The most promising vaccine is a mixture of synthetic proteins which mimic several *Plasmodium* membrane vaccines.

E. Zoomastigophora (Zooflagellates)

All species within the phylum Zoomastigophora are heterotrophs that absorb organic molecules or phagocytize prey.
- Use whip-like flagella to move.
- Most are solitary cells although some are colonial.
- Many are free-living although a large number are symbiotic.
  - ⇒ Symbiotic forms which live in the gut of termites digest the cellulose in wood eaten by the host.

Species of *Trypanosoma* cause African sleeping sickness and are spread by the bite of the tsetse fly.
- There is a large natural reservoir for the disease since many tropical African mammals harbor the parasite.
- Trypanosomes can evade the host's immune system by rearranging genes coding for the molecular composition of coat proteins.

Molecular systematics has recently confirmed that the zooflagellates are closely related to the group of flagellated protists which includes photosynthetic forms such as *Euglena*.

F.   **Ciliophora (Ciliates)**

Species within the phylum Ciliophora use cilia to move and feed.
- Most ciliates exist as solitary cells in fresh water.
- Cilia are relatively short and beat in synchrony.
- The cilia are associated with a submembranous system that coordinates the movement of thousands of cilia.
- Cilia may be dispersed over surface, or clustered in fewer rows or tufts.
- Some species move on leg-like *cirri* (many cilia bonded together).
- Other species have rows of tightly packed cilia that function together as locomotor membranelles (i.e., *Stentor*).

Ciliates possess two types of nuclei: one large *macronucleus* and from one to several small *micronuclei*.
- Characteristics of the macronucleus:
  ⇒ It is large and has over 50 copies of the genome.
  ⇒ Genes are packaged in a large number of small units, each with hundreds of copies of just a few genes.
  ⇒ It controls everyday functions of the cell by synthesizing RNA.
  ⇒ It is also necessary for asexual reproduction during binary fission. The macronucleus elongates and splits instead of undergoing mitosis.
- Characteristics of the micronucleus:
  ⇒ It is small and may number from 1 to 80 micronuclei, depending on the species.
  ⇒ It does *not* function in growth, maintenance or asexual reproduction.
  ⇒ Functions in *conjugation*, a sexual process which produces genetic variation. (See Campbell, Figure 26.11)
- Note that meiosis and syngamy are separate from reproduction

VI.   **Funguslike protists have morphological adaptations and life cycles that enhance their ecological role as decomposers**

The resemblance of slime molds and water molds to true fungi is a result of convergent evolution of filamentous body structure.
- A filamentous body structure increases exposure to the environment and enhances their roles as decomposers.
- Slime molds differ from true fungi in their cellular organization, reproduction, and life cycles.
  ⇒ Slime molds are more closely related to amoeboid protists than to true fungi.
- Molecular comparisons indicate water molds are related to certain algae although they lack chloroplasts.

A. Myxomycota (Plasmodial Slime Molds)

The phylum Myxomycota consists of the plasmodial slime molds which are all heterotrophs and many are brightly pigmented.

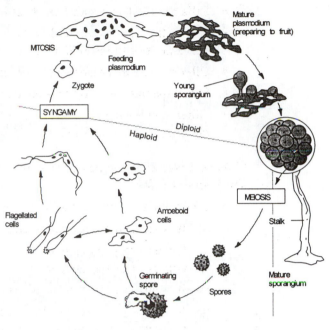

Plasmodium = Feeding stage of life cycle consisting of an amoeboid, coenocytic (multi-nucleated cytoplasm undivided by membranes) mass.

- In most species, the nuclei of plasmodia are diploid and exhibit synchronous mitotic divisions.

- Cytoplasmic streaming within the plasmodium helps distribute nutrients and oxygen.

- Engulfs food by phagocytosis as it grows by extending pseudopodia.

- Live in moist soil, leaf mulch and rotting logs.

- When stressed by drying or lack of food, the plasmodium ceases growth and forms sexually reproductive structures called fruiting bodies, or *sporangia*.

B. Acrasiomycota (Cellular Slime Molds)

The phylum Acrasiomycota includes the cellular slime molds.

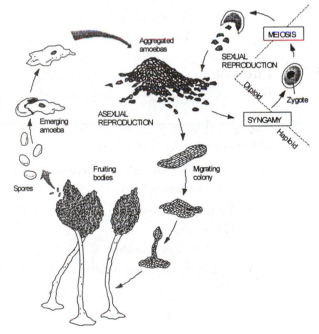

- Feeding stage of life cycle consists of individual, solitary haploid cells.

- When the food supply is depleted, cells aggregate to form a mass similar to those of myxomycota but cells remain separate (not coenocytic).

- Fruiting bodies function in *asexual* reproduction (unlike plasmodial slime molds).

- Only a few have flagellated stages.

C.   Oomycota

The phylum Oomycota includes water molds, white rusts, and downy mildews.
- Have coenocytic hyphae (fine, branching filaments) that are analogous to fungal hyphae.
- Cell walls are made of cellulose rather than the chitin found in true fungi.
- Diploid condition in the life cycle prevails in most species.
- Biflagellated cells are present in the life cycles, while fungi lack flagellated cells.

In water molds:
- A large egg is fertilized by a smaller sperm cell to form a resistant zygote. (See Campbell, Figure 26.14)
- These organisms are usually decomposers which grow on dead algae and animals in fresh water.
- Some are parasitic, growing on injured tissue but may also grow on the skin and gills of fish.

White rusts and downy mildews:
- Usually are parasitic on terrestrial plants.
- Disperse by windblown spores but also form flagellated zoospores at some point in their life cycle.
- Some of the most important plant pathogens are members of this phylum.

VII.   Eukaryotic algae are key producers in most aquatic ecosystems

A majority of the eukaryotic algae are photosynthetic with only a few of the phyla having heterotrophic or mixotrophic members.

Algae = relatively simple photoautotrophic aquatic organisms.
- These organisms were classified in the kingdom Protista in the five-kingdom system.
- Algae are of great ecological significance:
  ⇒ Account for about 50% of the global photosynthetic production of organic material.
  ⇒ Forms include fresh water plankton, marine plankton, and intertidal seaweeds that form the basis of aquatic food webs.

All algae have chlorophyll *a* (like plants). The phylogenetic relationships among algal phyla have been determined by differences in:
- Accessory pigments such as carotenoids, xanthophylls, phycobilins, and other forms of chlorophyll.
- Chloroplast structure.
- Cell wall chemistry.
- Number, type and position of flagella.
- Food storage product.

A.   Dinoflagellata (Dinoflagellates)

Dinoflagellates are components of phytoplankton which provides the foundation of most marine food chains.

- May cause *red tides* by explosive growth (bloom).
    - ⇒ These dinoflagellates produce a toxin that is concentrated by invertebrates, including shellfish.
    - ⇒ The toxin is dangerous to humans consuming shellfish and causes the condition known as paralytic shellfish poisoning.
- Most are unicellular, some are colonial.
- Cell surface is reinforced by cellulose plates with flagella in perpendicular grooves, creating its whirling movement and resulting in a characteristic shape.   (See Campbell, Figure 26.15)
- Some live as photosynthetic symbionts of the cnidarians that build coral reefs.
- Some lack chloroplasts and live as parasites; a few carnivorous species are known.
- Have brownish plastids containing chlorophyll *a*, chlorophyll *c* and a mix of carotenoids, including *peridinin* (found only in this phylum).
- Food is stored as starch.
- Chromosomes lack histones and are always condensed.
- Has no mitotic stages.
- Kinetochores are attached to the nuclear envelope and chromosomes distributed to daughter cells by the splitting of the nucleus.

Dinoflagellates are probably more closely related to zooflagellates than to any phylum of the algae.

B.   Bacillariophyta (Diatoms)

The phylum Bacillariophyta includes the diatoms which are yellow or brown in color due to the presence of brown plastids.

- Many have a gliding movement produced by chemical secretions.
- Usually reproduce asexually; sexual stages (egg and sperm production) are rare.
- Some produce resistant cysts.
- Mostly unicellular organisms with overlapping glasslike walls of hydrated silica in an organic matrix. (See Campbell, Figure 26.16)
- Have the same photosynthetic pigments as in Chrysophyta.
- Components of freshwater and marine plankton.
- Store food in a form of oil, which also makes cells buoyant.

C.   Chrysophyta (Golden Algae)

The phylum Chrysophyta includes the golden algae. (See Campbell, Figure 26.17)

- Plastids have chlorophyll *a*, chlorophyll *c*, yellow and brown carotenoids, and xanthophyll.
- Live among freshwater plankton; most are colonial.
- Have flagellated cells with both flagella attached near one end of the cell.

- Store carbohydrates in the form of *laminarin*, a polysaccharide.
- Survive environmental stress by forming resistant cysts.
  ⇒ Microfossils resembling these cysts have been found in Precambrian rocks.

D.   Phaeophyta (Brown Algae)

Phaeophyta (brown algae):
- Largest and most complex of the algae.
- All are multicellular and most are marine inhabitants.
- Have chlorophyll *a*, chlorophyll *c*, and the carotenoid pigment *fucoxanthin*.
- Store carbohydrate food reserves in the form of laminarin.
- Cell walls made of cellulose and *algin*.

1.   Evolutionary Adaptations of Seaweed

Seaweeds are large, multicellular marine algae which are found in the intertidal and subtidal zones of coastal waters.

- A diverse group of algae including members of the Phaeophyta (brown algae), Rhodophyta (red algae), and Chlorophyta (green algae).
- The following emphasizes adaptations found in the red algae, however, many of these adaptations also apply to the brown algae and green algae seaweeds.

The habitat of seaweeds, particularly the intertidal zone, poses several challenges to the survival of these organisms.

- Movement of the water due to wave action and winds produces a physically active habitat.
- Tidal rhythms result in the seaweeds being alternately covered by seawater and exposed to direct sunlight and the drying conditions of the air.

Seaweeds have evolved several unique structural and biochemical adaptations to survive the conditions of their habitats.

Structural adaptations found in seaweeds are a result of their complex multicellular anatomy. Some forms have differentiated tissues and organs analogous to those of plants.

- The body of a seaweed is called a *thallus*. It is plantlike in appearance but has no true roots, stems, or leaves.
- A thallus consists of a rootlike *holdfast* (maintains position), a stemlike *stipe* (supports the blades), and leaflike *blades* (large surfaces for photosynthesis).
- Floats which help suspend blades near the water surface are present in some brown algae.
- Brown algae known as giant kelp occur beyond the intertidal zone where less harsh conditions exist and may have stipes which may reach a length of 100 meters.

Biochemical adaptations in some seaweeds reinforce the anatomical adaptations and enhance survival.

- Cellulose cell walls also contain gel-forming polysaccharides (algin in brown algae; carageenan in red algae) which cushion the thalli against wave action and prevent desiccation during low tide.
- Some red algae retard grazing by marine invertebrates by incorporating large amounts of calcium carbonate into their cell walls.

Seaweeds are used by humans in a variety of ways:

- Brown and red alga are used as food in many parts of the Orient.
- Marine algae are used as nutrient supplements due their high content of iodine and other minerals.
- Algin, agar, and carageenan are extracted and used as thickeners for processed foods and lubricants in oil drilling.
- Agar is also used as a microbiological culture media.

2. **Alternation of Generations in the Life Cycles of Some Algae**

A variety of life cycles in the brown algae, the most complex involve *alternation of generations*.

> Alternation of generations = Alternation between multicellular haploid forms and multicellular diploid forms in a life history.

- Alteration of generations also evolved in certain groups of red algae and green algae.

The life cycle of *Lamanaria* is an example of a complex life cycle with an alteration of generations. (See Campbell, Figure 26.20)

- The diploid individual is called a *sporophyte* because it produces reproductive cells called spores.
- The haploid individual is called the *gametophyte* because it produces gametes.
- The sporophyte and gametophyte generations of the life cycle take turns producing one another.
  - ⇒ Spores released from the sporophyte develop into gametophytes.
  - ⇒ Gametophytes produce gametes which fuse (fertilization) to form a diploid zygote that develops into a sporophyte.
- In *Laminaria* the sporophyte and gametophyte generations are said to be *heteromorphic* because they are morphologically different.
- In *Ulva*, a green algae exhibiting alteration of generations, the generations are referred to as *isomorphic* because they look alike.

E.   **Rhodophyta (Red Algae)**

Rhodophyta (red algae) are primarily warm, tropical marine inhabitants, although some are found in fresh water and soil.

- Contain chlorophyll *a*, carotenoids, phycobilins, and chlorophyll *d* in some.
- Red color of plastids due to the accessory pigment, phycoerythrin.
    ⇒ Phycoerythrin is a phycobilin, a pigment found only in red algae and cyanobacteria.
- Color of the thallus may vary (even in a single species) with depth as pigmentation changes to optimize photosynthesis.
    ⇒ Deep water forms are almost black, moderate depth forms are red, and shallow water forms are green.
    ⇒ One species has been discovered near the Bahams at a depth of 260 meters.
    ⇒ Some tropical species lack pigmentation and survive as parasites on other red algae.
- Carbohydrate food reserves stored as floridean starch (similar to glycogen).
- Cell walls are cellulose with agar and carageenan.
- Most red algae are multicellular and are known as seaweeds.
- Most thalli are filamentous and are often branched forming an interwoven lacy network.

All red algae reproduce sexually:
- Have no flagellated stages, unlike other algal protists.
- Alternation of generations is common.

F.   **Chlorophyta (Green Algae)**

Members of the Chlorophyta (green algae) contain plant-like chloroplasts and are believed to be related to the ancestors of the plant kingdom.
- At least 7000 species are known with most being freshwater; fewer are marine.
- Many unicellular types live as plankton, inhabit damp soil, coat snow surfaces, or are symbionts with protozoa or invertebrates.
- When living mutualistically with fungi they form the association known as *lichens*.
- Colonial forms are often filamentous ("pond scum").
- Multicellular forms may have large, complex structures resembling true plants and comprise a group of seaweeds.

Evolutionary trends that probably produced colonial and multicellular forms from flagellated unicellular ancestors are:
- Formation of colonies of individual cells (as seen in *Volvox*).
- Repeated division of nuclei with no cytoplasmic division (as in *Bryopsis*).
- Formation of true multicellular forms as in *Ulva*.

Most chlorophytes have complex life histories involving sexual and asexual reproductive stages. (See Campbell, Figure 26.23)

- Some are conjugating algae (i.e. *Spirogyra*) which produce amoeboid gametes.
- A majority produce biflagellated gametes.

The life cycle of *Chlamydomonas* is a good example of the life history of a unicellular chlorophyte. Note, a mature *Chlamydomonas* is a single haploid cell. (See Campbell, Figure 26.24)

- During asexual reproduction, the flagella are resorbed and the cell divides twice by mitosis to form four cells (more in some species).
  - ⇒ The daughter cells develop and emerge as swimming zoospores. Zoospore development includes formation of flagella and cell walls.
  - ⇒ Zoospores grow into mature cells, thus completing asexual reproduction.
- Sexual reproduction is stimulated by environmental stress from such things as a shortage of nutrients, drying of the pond, or some other factor.
  - ⇒ During sexual reproduction, many gametes are produced by mitotic division within the wall of the parent cell. The gametes escape the parent cell wall.
  - ⇒ Gametes of opposite mating strains (+ and −) pair off and cling together by the tips of their flagella.
  - ⇒ The gametes are morphologically indistinguishable and their fusion is known as *isogamy*.
  - ⇒ The slow fusion of the gametes forms a diploid zygote which secretes a resistant coat that protects it from harsh environmental conditions.
  - ⇒ When dormancy of the zygote is broken, four haploid individuals (two of each mating type) are produced by meiosis.
  - ⇒ These haploid cells emerge from the coat and develop into mature cells, thus completing the sexual life cycle.

Many features of *Chlamydomonas* sex are believed to have evolved early in the chlorophyte lineage. Using this basic life cycle, many refinements that evolved among the chlorophytes have been identified.

- Some green algae produce gametes that differ from vegetative cells and, in some species, the male gamete differs in size or morphology from the female gamete (*anisogamy*).
- Many species exhibit *oogamy*, a type of anisogamy in which a flagellated sperm fertilizes a nonmotile egg.
- Some multicellular species also exhibit alternation of generations.
  - ⇒ *Ulva* produces isomorphic thalli for its diploid sphorophyte and haploid gametophyte.

VIII.   Systematists continue to refine their hypotheses of eukaryotic phylogeny

New evidence from fossils, cell ultrastructure, and molecular comparisons have stimulated much debate about eukaryotic relationships among protists.

- Many systematists feel the kingdom Protista of the five-kingdom system is obsolete since it includes all of the groups covered in this chapter.
- Current debate is in part concerned with subdivison of the eukaryotes into phyla:
   ⇒ How many phyla should there be, what are their boundaries, and what to name the new phyla.
- Another component of the debate concerns higher taxonomic levels and many researchers feel the five-kingdom system does not reflect the current understanding of phylogeny and should be abandoned.

A hypothetical phylogeny of eukaryotes based on molecular systematics and other evidence is shown in Figure 26.26.

- The tree relates some of the phyla discussed in this chapter to the origin of eukaryotes by serial endosymbiosis among prokaryotic ancestors and the later diversification of other prokaryotes.
- Note that chloroplasts probably evolved from cyanobacterial endosymbionts relatively early but that several lineages of heterotrophic eukaryotes lost their chloroplasts during their subsequent evolution.

The proposed eight-kingdom system differs from the five-kingdom system in several ways.

- Prokaryotes are split into separate kingdoms for archaebacteria and eubacteria.
- The early branching of archaezoans from the eukaryotic tree is also recognized by inclusion of the kingdom Archaezoa.
- The kingdom Protista is retained for most of the species covered in this chapter, however, the brown algae and some related phyla are separated into the kingdom Chromista.
- Organisms in the kingdom Chromista are distinguished from those in other kingdoms by:
   ⇒ The presence of unusual chloroplasts that have two additional membranes outside of the chloroplast envelope.
   ⇒ A small amount of cytoplasm.
   ⇒ The presence of a vestigial nucleus.
   ⇒ Recent evidence indicates these chloroplasts are descended from eukaryotic endosymbionts (probably red algae).
   ⇒ The Oomycota (water molds) are also placed on this branch even though they lost their chloroplasts.
- The green algae (which have an obviously close relationship with plants) and red algae are moved to the plant kingdom.

A consensus has not been reached among biologists about whether to replace the five-kingdom system with an alternative classification system.

IX. Multicellularity originated independently many times

Early eukaryotes were more complex than prokaryotes and this increase in complexity allowed for greater morphological variations to evolve.

- Extant protists are more complex in structure and show a greater diversity of morphology than the simpler prokaryotes.
- The ancestral stock which gave rise to new waves of adaptive radiations were the protists with multicellular bodies.

Multicellularity evolved several times among the early eukaryotes and gave rise to the multicellular algae, plants, fungi, and animals.

Most researchers believe that the earliest multicellular forms arose from unicellular ancestors as colonies or loose aggregates of interconnected cells.

- Multicellular algae, plants, fungi, and animals probably evolved from several lineages of protists that formed by amalgamations of individual cells.
- Evolution of multicellularity from colonial aggregates involved cellular specialization and division of labor.
    ⇒ The earliest specialization may have been locomotor capabilities provided by flagella.
    ⇒ As cells became more interdependent, some lost their flagella and performed other functions.
- Further division of labor may have separated sex cells from somatic cells.
    ⇒ This type of specialization and cooperation is seen today in colonial species such as *Volvox* (a green alga).
    ⇒ Gametes specialized for reproduction are dependent on somatic cells while developing.
- Many additional steps were involved in the evolution of specialized somatic cells capable of performing all the nonreproductive function in a multicellular organism.
    ⇒ Extensive division of labor exists among the different tissues that comprise the thalli of seaweeds.
- Multicellular forms more complex than filamentous algae appeared approximately 700 million years ago.
- A variety of animal fossils has been found in late Precambrian strata and many new forms evolved in the Cambrian period (about 570 million years ago).
- Seaweeds and other complex algae were also abundant during the Cambrian period.
- Primitive plants are believed to have evolved from certain green algae living in shallow waters about 400 million years ago.

REFERENCES

Bold, Harold C., and Michael J. Wynne. *Introduction to the Algae: Structure and Reproduction.* 2nd ed. Englewood Cliffs, New Jersey: Prentice-Hall, Inc., 1985.

Campbell, N. *Biology.* 4th ed. Menlo Park, California: Benjamin/Cummings, 1996.

Lee, John J., Seymour H. Hutner and Eugene C. Bovee. *An Illustrated Guide to the Protozoa.* Lawrence, Kansas: Allen Press, Inc., 1985.

CHAPTER 27

PLANTS AND THE COLONIZATION OF LAND

OUTLINE

I. Structural and reproductive adaptations made the colonization of land possible: *an overview of plant evolution*
    A. General Characteristics of Plants
    B. The Embryophyte Condition
    C. Alternation of Generations: *A Review*
    D. Some Highlights of Plant Phylogeny
    E. Classification of Plants

II. Plants probably evolved from green algae called charophytes
    A. The Origin of Alternation of Generations in Plants
    B. Adaptations to Shallow Water as Preadaptations for Living on Land

III. Bryophytes are embryophytes that generally lack vascular tissue and require environmental water to reproduce

IV. The origin of vascular tissue was an evolutionary breakthrough in the colonization of land
    A. The Earliest Vascular Plants

V. Ferns and other seedless vascular plants dominated the Carboniferous "coal forests"
    A. Division Psilophyta
    B. Division Lycophyta (Lycopods)
    C. Division Sphenophyta (Horsetails)
    D. Division Pterophyta (Ferns)
    E. The Coal Forests

VI. Reproductive adaptations catalyzed the success of the seed plants

VII. Gymnosperms began to dominate landscapes as climates became drier at the end of the Paleozoic era
    A.   Division Coniferophyta (Conifers)
    B.   The Life History of a Pine
    C.   The History of Gymnosperms

VIII. The evolution of flowers and fruits contributed to the radiation of angiosperms
    A.   The Flower
    B.   The Fruit
    C.   Life Cycle of an Angiosperm
    D.   The Rise of Angiosperms
    E.   Relationships Between Angiosperms and Animals
    F.   Angiosperms and Agriculture

IX. Plant diversity is a nonrenewable resource

OBJECTIVES

After reading this chapter and attending lecture, the student should be able to:
1. List characteristics that distinguish plants from organisms in the other four kingdoms.
2. Diagram a generalized plant life cycle indicating which generation is the sporophyte/ gametophyte, which individuals are haploid/diploid, where meiosis occurs and where mitosis occurs.
3. Describe four major periods of plant evolution that opened new adaptive zones on land.
4. Distinguish between the categories division and phylum.
5. Using the classification scheme presented in the text, list the ten plant divisions; give the common name for each; and categorize them into nonvascular, vascular seedless and vascular seed plants.
6. Provide evidence to defend the position that plants evolved from green algae.
7. Describe two adaptations that made the bryophytes' move onto land possible.
8. Explain how bryophytes are still tied to water.
9. List and distinguish among three division of Bryophyta.
10. Diagram the life cycle of a moss including gamete production, fertilization and spore production.
11. Compare environmental conditions faced by algae in an aquatic environment and plants in a terrestrial environment.
12. Provide evidence that suggests the division Bryophyta is a phylogenetic branch separate from vascular plants.
13. Describe adaptations of vascular plants, including modifications of the life cycle and modifications of the sporophyte, that have contributed to their success on land.
14. List and distinguish among the four extant divisions of seedless vascular plants.
15. Distinguish between homosporous and heterosporous.
16. Distinguish among spore, sporophyte, sporophyll and sporangium.
17. Diagram the life cycle of a fern including spore production, gamete production and fertilization.
18. Point out the major life cycle differences between mosses and ferns.
19. Describe how coal is formed and during which geological period the most extensive coal beds were produced.
20. List and distinguish among the four divisions of gymnosperms.
21. Describe and distinguish between the structures of ovulate and pollen cones of a pine.

22. Describe the life history of a pine and indicate which structures are part of the gametophyte generation and which are part of the sporophyte generation.
23. Point out the major life cycle differences between ferns and pines.
24. Distinguish between pollination and fertilization.
25. Explain how pollination and seed formation are adaptive for terrestrial environments.
26. Describe a pine seed and indicate which structures are old sporophyte, gametophyte and new sporophyte.
27. Describe how the needle-shaped leaves of pines and firs are adapted to dry conditions.
28. List and give examples of the two classes of Anthophyta.
29. Compare the life cycles of mosses, ferns, conifers and flowering plants in terms of:
    a. Dominant life cycle stage (gametophyte/sporophyte).
    b. Whether they are homosporous or heterosporous.
    c. Mechanism of gamete transfer.
30. Describe some refinements in vascular tissue that occurred during angiosperm evolution.
31. Explain how evolution of the flower enhanced the reproductive efficiency of angiosperms.
32. Identify the following floral structures and describe a function for each:
    a. Sepals.   d. Carpels.   g. Stigma.
    b. Petals.   e. Filament.   h. Style.
    c. Stamens.   f. Anther.   i. Ovary.
33. Describe four commonly recognized evolutionary trends in floral structure found in various angiosperm lineages.
34. Define fruit and explain how fruits are modified in ways that help disperse seeds.
35. Diagram the generalized life cycle of an angiosperm, identify which structures are haploid, and explain how it differs from the life cycle of a pine.
36. Explain the process of double fertilization and describe the fate of the polyploid nucleus.
37. Explain how an angiosperm seed differs from that of a pine.
38. Explain why paleobotanists have difficulty piecing together the origin of angiosperms and describe some current theories on how flowering plants may have evolved.
39. Explain how animals may have influenced the evolution of terrestrial plants, and vice versa.

KEY TERMS

cuticle
stomata
gametangia
embryophyte
charophyte
alternation of
    generation
gametophyte
sporophyte
heteromorphic
Silurian
Devonian
vascular tissue
seed
gymnosperm
Cretaceous
angiosperm
division
Bryophyta
antheridium
archegonium
mosses
liverworts

hornworts
rhizoids
elaters
gemmae
Psilophyta
Lycophyta
Sphenophyta
Pterophyta
Coniferophyta
Cycadophyta
Ginkgophyta
Gnetophyta
Anthophyta
lignin
xylem
phloem
Lycophyta
club mosses
sporophylls
strobili
Lycopodium
homosporous
Selaginella

heterosporous
megaspores
microspores
Equisetum
microphylls
megaphylls
fronds
sori
Ordovician
Silurian
Devonian
Carboniferous
seed
Cycadophyta
Ginkgophyta
Gnetophyta
Coniferophyta
wood
conifers
fibers
tracheids
cone
pollen cones

ovulate cones
Anthophyta
Monocotyledons
Dicotyledons
vessel elements
flower
sepal
petal
stamen
carpel
stigma
ovary
fruit
aggregate fruit
multiple fruit
pollen grains
ovules
embryo sac
double fertilization
cross pollination
cotyledons
endosperms
coevolution

LECTURE NOTES

Plants appeared on land about 425 million years ago, and the evolutionary history of the plant kingdom reflects increasing adaptation to the terrestrial environment.

I. **Structural and reproductive adaptations made the colonization of land possible:** *an overview of plant evolution*

    A. **General Characteristics of Plants**

Plants are multicellular eukaryotes that are photosynthetic autotrophs. They share the following characteristics with their green algal ancestors:

- Chloroplasts with the photosynthetic pigments: chlorophyll *a*, chlorophyll *b*, and carotenoids.
- Cell walls containing cellulose.
- Food reserve is starch that is stored in plastids.

As plants adapted to terrestrial life, they evolved complex bodies with cell specialization for different functions.

- Aerial plant parts are coated with a waxy *cuticle* that helps prevent desiccation.
- Though gas exchange cannot occur across the waxy cuticle, CO_2 and O_2 can diffuse between the leaf's interior and the surrounding air through *stomata*, microscopic pores on the leaf's surface.

B.   The Embryophyte Condition

With the move from an aquatic to terrestrial environment, a new mode of reproduction was necessary to solve two problems:

1. *Gametes must be dispersed in a nonaquatic environment.* Plants produce gametes within *gametangia*, organs with protective jackets of sterile (nonreproductive) cells that prevent gametes from drying out. The egg is fertilized within the female organ.

2. *Embryos must be protected against desiccation.* The zygote develops into an embryo that is retained for awhile within the female gametangia's jacket of protective cells. Emphasizing this terrestrial adaptation, plants are often referred to as *embryophytes*.

C.   Alternation of Generations: *A Review*

Most plants reproduce sexually, and most are also capable of asexual propagation. All plants have life cycles with an *alternation of generations*.

- A haploid *gametophyte* generation produces and alternates with a diploid *sporophyte* generation. The sporophyte, in turn, produces gametophytes.

- The life cycles are *heteromorphic*; that is, sporophytes and gametophytes differ in morphology.

- The sporophyte is larger and more noticeable than the gametophyte in all plants but mosses and their relatives.

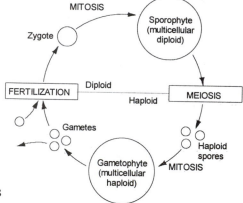

A comparison of life cycles among plant divisions is instructive because:

- It points to an important trend in plant evolution: reduction of the haploid gametophyte generation and dominance of the diploid sporophyte.

- Certain life cycle features are adaptations to a terrestrial environment; for example the replacement of flagellated sperm by pollen.

D.   Some Highlights of Plant Phylogeny

There are four major periods of plant evolution that opened new adaptive zones on land:

1. Origin of plants from aquatic ancestors (probably green algae) in the Silurian (425 million years ago).

    - Cuticle and jacketed gametangia evolved that protected gametes and embryos.
    - *Vascular tissue* evolved with conducting cells that transport water and nutrients throughout the plant.

2. Diversification of seedless vascular plants, such as ferns, during the early Devonian about 400 mya.

3. Origin of the seed near the end of the Devonian about 360 mya.

   <u>Seed</u> = Plant embryo packaged with a store of food within a resistant coat.

   - Early seed plants bore seeds as naked structures and evolved into gymnosperms including conifers.
   - Conifers and ferns coexisted in the landscape for more than 200 million years.

4. Emergence of flowering plants during the early Cretaceous, about 130 mya.

   - Unlike gymnosperms, flowering plants bear seeds within the flower's protective ovaries.
   - Most contemporary plants are flowering plants or *angiosperms*.

E. Classification of Plants

The major taxonomic category of plants is the *division*; it is comparable to phylum, the highest category in the animal kingdom.

A CLASSIFICATION OF PLANTS

	Common Name	Approximate Number of Extant Species
Nonvascular Plants		
Division Bryophyta	Mosses	10,000
Division Hepatophyta	Liverworts	6,500
Division Anthocerophyta	Hornworts	100
Vascular Seedless Plants		
Division Psilophyta	Whiskferns	10 to 13
Division Lycophyta	Club mosses	1,000
Division Sphenophyta	Horsetails	15
Division Pterophyta	Ferns	12,000
Vascular Seed Plants		
Division Coniferophyta	Conifers	550
Division Cycadophyta	Cycads	100
Division Ginkgophyta	Ginkgo	1
Division Gnetophyta	Gnetae	70
Division Anthophyta	Flowering plants	235,000

II.   Plants probably evolved from green algae called charophytes

The green algae are likely the photosynthetic protists most closely related to plants. This conclusion is based on homologies in:

- Cell wall composition.
- Structure and pigmentation of chloroplasts.

Available evidence supports the hypothesis that plants and green algae called *charophytes* both evolved from a common ancestor. Researchers have found the following homologies between charophytes and plants:

1. *Homologous chloroplasts.*

   - Green algae and plants both have the accessory pigments, chlorophyll *b* and beta-carotene.

   - Green algae and plants both have chloroplasts with thylakoid membranes stacked as grana.

   - Compared to chloroplast DNA of various green algae, plant chloroplast DNA most closely matches that of charophytes.

2. *Biochemical similarity.*

   - Most green algae and plants contain cellulose in their cell walls. Charophytes are the most plantlike in wall composition with cellulose making up 20% to 26% of the wall material.

   - Charophyte peroxisomes are the only algal peroxisomes with the same enzyme composition as plant peroxisomes.

3. *Similarity in mitosis and cytokinesis.* During cell division in charophytes and plants:
   - The nuclear envelope completely disperses during late prophase.
   - The mitotic spindle persists until cytokinesis begins.
   - Cell plate formation during cytokinesis involves cooperation of microtubules, actin microfilaments, and vesicles.

4. *Similarity in sperm ultrastructure.* Charophyte sperm ultrastructure is more similar to certain plants than to other green algae.

5. *Genetic relationship.* DNA and rRNA similarities in charophytes and plants provides additional evidence for the hypothesis that charophytes are the closest relatives of plants.

A.   The Origin of Alternation of Generations in Plants

The alternation of haploid and diploid generations apparently evolved independently among various groups of algae.

- Since alternation of generations does not occur among modern charophytes, it is presumed that alternation of generations in plants has had a separate origin from alternation of generations in other algal groups.

- Its appearance in plants is thus *analogous*, not homologous, to the alternation of generations observed in various groups of algae.

How did alternation of generations evolve in plant ancestors? *Coleochaete*, a modern charophyte, holds some clues:

- The *Coleochaete* thallus is haploid.
    - ⇒ In contrast to most algae, the parental thallus of *Coleochaete* retains the eggs, and after fertilization, the zygotes remain attached to the parent.
    - ⇒ Nonreproductive cells of the thallus grow around each zygote, which enlarges, undergoes meiosis, and releases haploid swimming spores.
    - ⇒ Haploid spores develop into new individuals.
    - ⇒ The only diploid stage is the zygote; there is no alternation of multicellular diploid and haploid generations.
- If an ancestral charophyte delayed meiosis until after the zygote divided mitotically, there would be a multicellular diploid generation (sporophyte) still attached to the haploid parent (gametophyte). Such a life cycle would be an alternation of generations.
- If specialized gametophyte cells formed protective layers around a tiny sporophyte, this hypothetical ancestor would also be a primitive embryophyte.

What would be the adaptive advantage of delaying meiosis and forming a mass of diploid cells? It may maximize the production of haploid spores.

- If the zygote undergoes meiosis directly, each fertilization event results in only a few haploid spores.
- Mitotic division of the zygote to form a multicellular sporophyte amplifies the sexual product. Many diploid cells can undergo meiosis producing a large number of haploid spores, enhancing the chances of survival in unfavorable environments.

B.   Adaptations to Shallow Water as Preadaptations for Living on Land

Some adaptations for life in shallow water could also have been adaptive for life on land.

- Many modern charophytes live in shallow water, and some ancient charophytes may have also lived in shallow-water habitats subject to occasional drying.
- About 440 million years ago, during the transition from Ordivician to Silurian, repeated glaciation and climatic changes caused fluctuations in the water levels of lakes and ponds.
- Natural selection may have favored shallow-water plants tolerant to periodic drying. Adaptations to shallow water may also have been preadaptive for terrestrial life. For example:
    - ⇒ Waxy cuticles.
    - ⇒ Protection of gametes.
    - ⇒ Protection of developing embryos.
- Eventually, accumulated adaptations made it possible for ancestral plants to live permanently above the water line, opening a new adaptive zone with:
    - ⇒ Sunlight unfiltered by water and algae.
    - ⇒ Soil rich in minerals.
    - ⇒ Absence of terrestrial herbivores.

III. Bryophytes are embryophytes that generally lack vascular tissue and require environmental water to reproduce

The *bryophytes* include plants found in three divisions:

- Bryophyta (mosses)
- Hepatophyta (liverworts)
- Anthocerophyta (hornworts)

Bryophytes have two adaptations that made the move onto land possible:

- A waxy cuticle that prevents desiccation.
- Gametangia that protect developing gametes.
    a. *Antheridium*, or male gametangium, produces flagellated sperm cells.
    b. *Archegonium*, or female gametangium, produces a single egg; fertilization occurs within the archegonium, and the zygote develops into an embryo within the protective jacket of the female organ (embryophyte condition).

Bryophytes are not totally free from their ancestral aquatic habitat.

- They need water to reproduce. Their flagellated sperm cells must swim from the antheridium to the archegonium to fertilize the egg.
- Most have no vascular tissue to carry water from the soil to aerial plant parts; they imbibe water and distribute it throughout the plant by the relatively slow processes of diffusion, capillary action, and cytoplasmic streaming.

Bryophytes lack woody tissue and cannot support tall plants on land; they may sprawl horizontally as mats, but always have a low profile.

1. Mosses (Division Bryophyta)

A tight pack of many moss plants forms a spongy mat that can absorb and retain water.

- Each plant grips the substratum with *rhizoids*, elongate cells or cellular filaments.
- Photosynthesis occurs mostly in the small stemlike and leaflike structures found in upper parts of the plant; these structures are not homologous with stems and leaves in vascular plants.

There is an alternation of haploid and diploid generations in the moss life cycle. (See Campbell, Figure 27.7)

- The sporophyte (2n) produces haploid spores by meiosis in a *sporangium*; the spores divide by mitosis to form new gametophytes.
- Contrary to the life cycles of vascular plants, the haploid gametophyte is the dominant generation in mosses and other bryophytes. Sporophytes are generally smaller and depend on the gametophyte for water and nutrients.

2. Liverworts (Division Hepatophyta)

Less conspicuous than mosses, liverworts:

- Sometimes have bodies divided into lobes.
- Have a life cycle similar to mosses. Their sporangia have *elaters*, coil-shaped cells that spring out of the capsule and disperse spores.
- Can also reproduce asexually from *gemmae* (small bundles of cells that can bounce out of cups on the surface of the gametophyte when hit by rainwater).

3. **Hornworts (Division Anthocerophyta)**

    Hornworts:
    - Resemble liverworts, but sporophytes are horn-shaped, elongated capsules that grow from the mat-like gametophyte.
    - Their photosynthetic cells have only one large chloroplast, unlike the many smaller ones of other plants.

IV.   The origin of vascular tissue was an evolutionary breakthrough in the colonization of land

In addition to cuticles and jacketed sex organs, other adaptations for terrestrial life evolved in vascular plants as they colonized land:
- *Regional specialization of the plant body.* Unlike aquatic environments, terrestrial environments spatially segregate the resources of water and light. This problem was solved as plants evolved subterranean roots that absorb water and minerals from the soil and an aerial shoot system of stems and leaves to make food.
- *Structural support.* In aquatic environments, the denser medium of water buoys plants up toward the light, but in terrestrial environments plants must have structural support to stand upright in air. Such support was provided as the hard material lignin was embedded into the cellulose matrix of cell walls.
- *Vascular system.* Regional specialization of the plant body presented the problem of transporting substances between the root and shoot systems. This problem was solved as a vascular system evolved with two types of conducting tissues:

    Xylem = Complex, plant vascular tissue that conducts water and minerals from the roots to the rest of the plant.
    - ⇒ Composed of dead, tube-shaped cells that form a microscopic water-pipe system.
    - ⇒ Cell walls are usually lignified, giving the plant structural support.

    Phloem = Plant vascular tissue that conducts food throughout the plant.
    - ⇒ Composed of living cells arranged into tubules.
    - ⇒ Distributes sugars, amino acids and other organic nutrients.
- *Pollen.* Pollination eliminated the need for water to transport gametes.
- *Seeds.*
- *Increased dominance of the diploid sporophyte.*

A.   The Earliest Vascular Plants

Oldest fossilized vascular plant is *Cooksonia* (late Silurian):
- Discovered in both European and North American Silurian rocks; North America and Europe were probably connected during the late Silurian, about 408 million years ago.
- Simple plant with dichotomous branching and bulbous terminal sporangia on some stems.
- True roots and leaves were absent; the largest species was about 50 cm tall.
- Grew in dense stands around marshes.
- As vascular plants became more widespread, new species appeared.

V. **Ferns and other seedless vascular plants dominated the Carboniferous "coal forests"**

The earliest vascular plants were seedless and they dominated the Carboniferous forests. Modern flora includes four divisions of seedless vascular plants.

A. **Division Psilophyta**

Psilophyta consists of only two genera: *Psilotum* (whiskferns) and *Tmesipteris*. Whiskferns are the most well known and share the following characteristics:
- True roots and leaves are absent, subterranean rhizomes are covered with hair-like rhizoids, and shoots have scales which lack vascular tissue.
- Dichotomous branching.
- The gametophytes are subterranean and lack chlorophyll, depending on symbiotic soil fungi for food.
- Flagellated sperm swim through the soil from antheridia to the archegonium of the gametophyte.
- The sporophyte emerges from the gametophyte, which then dies.

B. **Division Lycophyta (Lycopods)**

The Division Lycophyta include the club mosses and ground pines.
- Survived through the Devonian period and dominated land during the Carboniferous Period (340–280 million years ago).
- Some are temperate, low-growing plants with rhizomes and true leaves.
- Some species of *Lycopodium* are *epiphytes*, plants that use another organism as a substratum but are not parasites.
  ⇒ The sporangia of *Lycopodium* are borne on *sporophylls*, leaves specialized for reproduction. In some, sporophylls are clustered at branch tips into club-shaped *strobili* – hence the name club moss.
  ⇒ Spores develop into inconspicuous gametophytes. The non-photosynthetic gametophytes are nurtured by symbiotic fungi.
- Most are *homosporous* (making only one type of spore which develops into a bisexual gametophyte).
- Genus *Selaginella* is *heterosporous*, having *megaspores* which develop into gametophytes bearing archegonia, and *microspores* which develop into gametophytes with antheridia. The gametophytes are unisexual, either male or female.

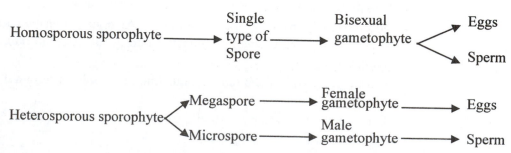

C. **Division Sphenophyta (Horsetails)**

The division Sphenophyta includes the horsetails; it survived through the Devonian and reached its zenith during the Carboniferous period.

The only existing genus is *Equisetum*, which:

- Lives in damp locations and has flagellated sperm.
- Is homosporous.
- Has a conspicuous sporophyte generation.
- Has photosynthetic, free-living gametophytes (not dependent on the sporophyte for food).

D. **Division Pterophyta (Ferns)**

Appearing in the Devonian, ferns radiated into diverse species that coexisted with tree lycopods and horsetails in the great Carboniferous forests.

- Most diverse in the tropics, ferns are the most well represented seedless plants in modern floras; there are more than 12,000 existing species of ferns.
- Fern leaves are generally much larger than those of lycopods and probably evolved in a different way.
  - ⇒ Lycopods have *microphylls*, small leaves that probably evolved as emergences from the stem that contained a single strand of vascular tissue.
  - ⇒ Ferns have *megaphylls*, leaves with a branched system of veins. Megaphylls probably evolved from webbing formed between separate branches growing close together.

Most ferns have *fronds*, compound leaves that are divided into several leaflets.

- The emerging frond is coiled into a *fiddlehead* that unfurls as it grows.
- Leaves may sprout directly from a prostrate stem (bracken and sword ferns) or from upright stems many meters tall (tropical tree ferns).

Ferns are *homosporous* and the conspicuous leafy fern plant is the sporophyte. (See Campbell, Figure 27.15)

- Specialized sporophylls bear sporangia on their undersides; many ferns have sporangia arranged in clusters called *sori* and are equipped with springlike devices that catapult spores into the air, where they can be blown by the wind far from their origin.
- The spore is the dispersal stage.
- The free-living gametophyte is small and fragile, requiring a moist habitat.
- Water is necessary for fertilization, since flagellated sperm cells must swim from the antheridium to the archegonium, where fertilization takes place.
- The sporophyte embryo develops protected within the archegonium.

E.   **The Coal Forests**

During the Carboniferous period, much of the land was covered in shallow seas and swamps.

- Organic rubble of the plants above accumulated as peat.
- When later covered by the sea and sediments, heat and pressure transformed the peat into coal.

VI.   Reproductive adaptations catalyzed the success of the seed plants

Three life cycle modifications contributed to seed plant success:

1. Gametophytes became reduced and were retained in the moist reproductive tissue of the sporophyte generation (not independent).
2. Pollination evolved, so plants were no longer tied to water for fertilization.
3. The evolution of the seed.
   - Zygote develops into an embryo packaged with a food supply within a protective seed coat.
   - Seeds replace spores as main means of dispersal.

VII.   Gymnosperms began to dominate landscapes as climates became drier at the end of the Paleozoic era

Gymnosperms appear in the fossil record much earlier than flowering plants, and they:

- Lack enclosed chambers in which seeds develop.
- Are grouped into four divisions: Cycadophpyta, Ginkgophyta, Gnetophyta and Coniferophyta.

A.   **Division Coniferophyta (Conifers)**

Division Coniferophyta is the largest division of gymnosperms:

- Most are evergreens and include pines, firs, spruces, larches, yews, junipers, cedars, cypresses, and redwoods.
- Includes some of the tallest (redwoods and some eucalyptus); largest (giant sequoias); and oldest (bristle cone pine) living organisms.
- Most lumber and paper pulp is from conifer wood.

Needle-shaped conifer leaves are adapted to dry conditions.

- Thick cuticle covers the leaf.
- Stomata are in pits, reducing water loss.
- Despite the shape, needles are megaphylls, as are leaves of all seed plants.

B. **The Life History of a Pine**

The life cycle of pine, a representative conifer, is characterized by the following:

- The multicellular sporophyte is the most conspicuous stage; the pine tree is a sporophyte, with its sporangia located on cones.
- The multicellular gametophyte generation is reduced and develops from haploid spores that are retained within sporangia.
  - ⇒ The male gametophyte is the pollen grain; note that there is *no antheridium*.
  - ⇒ The female gametophyte consists of multicellular nutritive tissue and an *archegonium* that develops within an *ovule*.

Conifer life cycles are *heterosporous*; male and female gametophytes develop from different types of spores produced by separate cones.

- Trees of most pine species bear both pollen cones and ovulate cones, which develop on different branches.
- Pollen cones have *microsporangia*; cells in these sporangia undergo meiosis producing haploid *microspores*, small spores that develop into pollen grains - the male gametophytes.
- Ovulate cones have *megasporangia*; cells in these sporangia undergo meiosis producing large *megaspores* that develop into the female gametophyte. Each ovule initially includes a sporangium (nucellus) enclosed in protective integuments with a single opening, the *micropyle*.

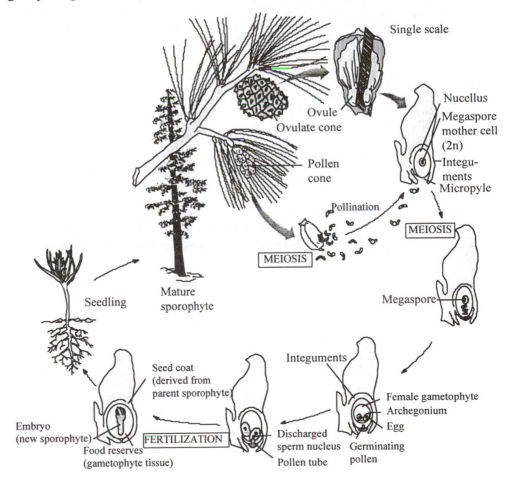

It takes nearly three years to complete the pine life cycle, which progresses through a complicated series of events to produce mature seeds.

- Windblown pollen falls onto the ovulate cone and is drawn into the ovule through the micropyle.

- The pollen grain germinates in the ovule, forming a pollen tube that begins to digest its way through the nucellus.

- A megaspore mother cell in the nucellus undergoes *meiosis* producing four haploid megaspores, one of which will survive; it divides repeatedly by *mitosis* producing the immature female gametophyte.

- Two or three archegonia, each with an egg, then develop within the multicellular gametophyte.

- More than a year after pollination, the eggs are ready to be fertilized; two sperm cells have developed and the pollen tube has grown through the nucellus to the female gametophyte.

- Fertilization occurs when one of the sperm nuclei unites with the egg nucleus. All eggs in an ovule may be fertilized, but usually only one zygote develops into an embryo.

- The pine embryo, or new sporophyte, has a rudimentary root and several embryonic leaves. It is embedded in the female gametophyte, which nourishes the embryo until it is capable of photosynthesis. The ovule has developed into a pine seed, which consists of an embryo (2n), its food source (n), and a surrounding seed coat (2n) derived from the parent tree.

- Scales of the ovulate cone separate, and the winged seeds are carried by the wind to new locations. Note, that with the seed plants, the seed has replaced the spore as the mode of dispersal.

- A seed that lands in a habitable place germinates, its embryo emerging as a pine seedling.

C.   The History of Gymnosperms

Gymnosperms descended from Devonian progymnosperms.

- Adaptive radiation during the Carboniferous and Permian periods led to today's divisions.

- During the Permian, Earth became warmer and drier; therefore, lycopods, horsetails and ferns (previously dominant) were largely replaced by conifers and their relatives, the cycads.

- This large change marks the end of the Paleozoic era and the beginning of the Mesozoic era.

VIII. The evolution of flowers and fruits contributed to the radiation of angiosperms

Flowering plants are the most widespread and diverse.

- There is only one division, Anthophyta, with two classes, Monocotyledones (monocots) and Dicotyledones (dicots).
- Most use insects and animals for transferring pollen, and, therefore, are less dependent on wind and have less random pollination.

Vascular tissue became more refined during angiosperm evolution.

- Conifers have *tracheids*, water-conducting cells that are:
  ⇒ An early type of xylem cell.
  ⇒ Elongate, tapered cells that function in <u>both</u> mechanical support and water movement up the plant.
- Most angiosperms also have *vessel elements* that are:
  ⇒ Shorter, wider cells than the more primitive tracheids.
  ⇒ Arranged end to end forming continuous tubes.
  ⇒ Compared to tracheids, vessel elements are more specialized for conducting water, but less specialized for support.
- Angiosperm xylem is reinforced by other cell types called *fibers*, which are:
  ⇒ Specialized for support with a thick lignified wall.
  ⇒ Evolved in conifers. (Note that conifer xylem contains both fibers and tracheids, but <u>not</u> vessel elements.)

A. The Flower

<u>Flower</u> = The reproductive structure of an angiosperm which is a compressed shoot with four whorls of modified leaves.

<u>Parts of the Flower</u>:

Sepals – Sterile, enclose the bud.

Petals – Sterile, aid in attracting pollinators.

Stamen – Produces the pollen.

Carpel – Evolved from a seed-bearing leaf that became rolled into a tube.

Stigma – Part of the carpel which is a sticky structure that receives the pollen.

Ovary – Part of the carpel that protects the *ovules*, which develop into seeds after fertilization.

Four Evolutionary Trends in Various Angiosperm Lineages:
1. The number of floral parts have become reduced.
2. Floral parts have become fused.
3. Symmetry has changed from radial to bilateral.
4. The ovary has dropped below the petals and sepals, where the ovules are better protected.

B. The Fruit

Fruit = A ripened ovary that protects dormant seeds and aids in their dispersal; some fruits (like apples) incorporate other floral parts along with the ovary.

Aggregate fruits = Several ovaries that are part of the same flower (e.g. raspberry).

Multiple fruit = One that develops from several separate flowers (e.g. pineapple).

Modifications of fruits that help disperse seeds include:
- Seeds are within fruits that act as kites or propellers to aid in wind dispersal.
- Fruits are modified as burrs that cling to animal fur.
- Fruits are edible and seeds pass through the digestive tract of herbivores unharmed, dispersing seeds miles away.

C. Life Cycle of an Angiosperm

Life cycles of angiosperms are heterosporous and the two types of sporangia are found in the flower:
- Microsporangia in anthers produce microspores that form male gametophytes.
- Megasporangia in ovules produce megaspores that develop into female gametophytes.

Immature male gametophytes:
- Are pollen grains, which develop within the anthers of stamens.
- Each pollen grain has two haploid nuclei that will participate in *double fertilization* characteristic of angiosperms.

Female gametophytes:
- Do *not* produce an archegonium.
- Are located within an ovule.
- Consist of only a few cells - an embryo sac with eight haploid nuclei in seven cells (a large central cell has two haploid nuclei).
- One of the cells is the egg.

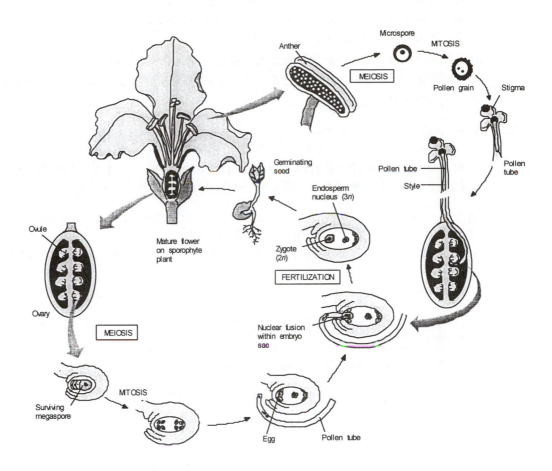

An outline of the angiosperm life cycle follows:

- Pollen from the anther lands on the sticky stigma at the carpel's tip; most flowers do not self-pollinate, but have mechanisms to ensure cross-pollination.
- The pollen grain germinates on the stigma by growing a pollen tube down the style of the carpel.
- When it reaches the ovary, the pollen tube grows through its micropyle and discharges two sperm cells into the embryo sac.
- Double fertilization occurs as one sperm nucleus unites with the egg to form a diploid zygote; the other sperm nucleus fuses with two nuclei in the embryo sac's central cell to form triploid ($3n$) endosperm.
- After double fertilization, the ovule matures into a seed.

The seed is a mature ovule, consisting of:

1. *Embryo*. The zygote develops into an embryo with a rudimentary root and one (in monocots) or two (in dicots) cotyledons or seed leaves.
2. *Endosperm*. The triploid nucleus in the embryo sac divides repeatedly forming triploid endosperm, rich in starch and other food reserves.
3. *Seed coat*. This is derived from the integuments (outer layers of the ovules).

Monocots and dicots use endosperm differently.
- Monocot seeds store most food in the endosperm;
- Dicots generally restock most of the nutrients in the developing cotyledons.

In a suitable environment the seed coat ruptures and the embryo emerges as a seedling, using the food stored in the endosperm and cotyledons.

D.   The Rise of Angiosperms

Angiosperms showed a relatively sudden appearance in the fossil record with no clear transitional links to ancestors.
- Earliest fossils are early Cretaceous (approximately 120 million years ago).
- By the end of the Cretaceous (65 million years ago), angiosperms became dominant, as they are today.

There are two theories about their sudden appearance:
1. Artifact of an imperfect fossil record: angiosperms originated where fossilization was unlikely.
2. *Punctuated equilibrium*: angiosperms did evolve and radiate relatively abruptly.

Angiosperms perhaps evolved from seed ferns, an extinct group of unspecialized gymnosperms.

E.   Relationships Between Angiosperms and Animals

Terrestrial plants and animals have *coevolved* – a consequence of their interdependence.

Coevolution = Reciprocal evolutionary responses among two or more interacting species; adaptive change in one species is in response to evolutionary change in the other species. For example,
- Coevolution between angiosperms and their pollinators led to diversity of flowers.
  ⇒ Some pollinators are specific for a particular flower. The pollinator has a monopoly on a food source and guarantees the flower's pollen will pollinate a flower of the same species.
  ⇒ Often, the relationship between angiosperms and their pollinators is not species specific; a pollinator may not depend exclusively on one flower species, or a flower species may not depend exclusively on one species of pollinator. However, flower color, fragrance and structure are usually adaptations for *types* of pollinators, such as various species of bees or hummingbirds.
- Edible fruits of angiosperms have coevolved with animals that can disperse seeds. Animals become attracted to ripening fruits as they:
  ⇒ Become softer, more fragrant and higher in sugar.
  ⇒ Change to a color that attracts birds and mammals, animals large enough to disperse the seeds.

F.   Angiosperms and Agriculture

Angiosperms provide nearly all our food:
- fruit
- vegetable crops
- grains, such as corn, rice, wheat

Flowering plants are also used for other purposes, such as:

- fiber
- medication
- perfume
- decoration

Through agriculture, humans have influenced plant evolution by artificially selecting for plants that improved the quantity and quality of foods and other crops.

- Many of our agricultural plants are so genetically removed from their origins that they probably could not survive in the wild.
- As a consequence, cultivated crops that require human intervention to water, fertilize, provide protection from insects and disease, and even to plant their seeds, are vulnerable to natural and human-caused diasters.

IX. Plant diversity is a nonrenewable resource

Plant diversity is a nonrenewable resource, and the irrevocable extinction of plant species is occurring at an unprecedented rate.

- The exploding human population demands space and natural resources.
- The toll of habitat destruction is greatest in the tropics because this is where:
  ⇒ Most species live.
  ⇒ More than half the human population lives and human population growth is fastest.
  ⇒ Most deforestation is caused by slash-and-burn clearing for agriculture.

As the forest disappears, so do thousands of plant and animal species.

- Habitat destruction also endangers animal species that depend on plants in the tropical rainforest.
- Habitat destruction by humans has not been limited to the tropics. Europeans eliminated most of their forests centuries ago, and in North America, destruction of habitat is endangering many species.

There are many reasons to value plant diversity and to find ways to protect it.

- Ecosystems are living treasures that can regenerate only slowly.
- Humans depend on plants for products such as medicines, food and building materials.
- We still know so little of the 250,000 known plant species. (Food agriculture is based on only about two dozen species.)

REFERENCES

Campbell, N. *Biology*. 4th ed. Menlo Park, California: Benjamin/Cummings, 1996.

Raven, P.H., R.F. Evert and S.E. Eichhorn. *Biology of Plants*. 4th ed. New York: Worth Publishers, Inc., 1986.

OUTLINE

I.   Structural and life history adaptations equip fungi for an absorptive mode of nutrition
   A.   Nutrition and Habitats
   B.   Structure
   C.   Growth and Reproduction

II.   The three major divisions of fungi differ in details of reproduction
   A.   Division Zygomycota (Zygote Fungi)
   B.   Division Ascomycota (Sac Fungi)
   C.   Division Basidiomycota (Club Fungi)

III.   Molds, yeasts, lichens, and mycorrhizae represent unique lifestyles that evolved independently in all fungal divisions
   A.   Molds
   B.   Yeasts
   C.   Lichens
   D.   Mycorrhizae

IV.   Fungi have a tremendous ecological impact
   A.   Fungi as Decomposers
   B.   Fungi as Spoilers
   C.   Pathogenic Fungi
   D.   Edible Fungi

V.   Fungi and animals probably evolved from a common protistan ancestor

OBJECTIVES

After reading this chapter and attending lecture, the student should be able to:

1. List characteristics that distinguish fungi from organisms in the other four kingdoms.
2. Explain how fungi acquire their nutrients.
3. Explain how non-motile fungi seek new food sources and how they disperse.
4. Describe the basic body plan of a fungus.
5. Distinguish between septate and aseptate (coenocytic) fungi.
6. Describe some advantages to the dikaryotic state.
7. Distinguish among the fungal divisions Zygomycota, Ascomycota, Basidiomycota and Deuteromycota and list some common examples of each.
8. Describe asexual and sexual reproduction in Zygomycota, Ascomycota and Basidiomycota, including a description of the sexual structure that characterizes each group.
9. Explain the difference between conidia and ascospores.
10. Explain why ascomycetes can be useful to geneticists studying genetic recombination.
11. Explain why the Deuteromycota are called imperfect fungi.
12. Describe the anatomy of lichens and explain how they reproduce.
13. Provide evidence for both sides of the debate whether symbiosis in lichens is parasitic or mutualistic.
14. Describe the ecological importance of lichens.
15. Explain why fungi are ecologically and commercially important.
16. Describe how the mutualistic relationship in mycorrhizae is beneficial to both the fungus and the plant, and explain its importance to natural ecosystems and agriculture.
17. Describe a scenario for fungal phylogeny and list two possible ancestors of Zygomycota.

KEY TERMS

hypha	dikaryons	ascospores	lichen
mycelium	Zygomycota	Basidiomycota	soredia
septa	sporangia	basidium	mycorrhizae
aseptate	Ascomycota	basidiocarp	
coenocytic	ascocarp	Deuteromycota	
haustoria	conidia	imperfect fungi	

LECTURE NOTES

I. **Structural and life history adaptations equip fungi for an absorptive mode of nutrition**

Fungi are eukaryotes, nearly all multicellular (although yeasts are unicellular). Their nutrition, structural organization, growth and reproduction distinguish them from organisms in the other four kingdoms.

A. Nutrition and Habitats

Fungi are heterotrophs that acquire nutrients by *absorption*.
- They secrete hydrolytic enzymes and acids to decompose complex molecules into simpler ones that can be absorbed.
- Are specialized into three main types:
    1. *Saprophytes*, which absorb nutrients from dead organic material (decomposers).
    2. *Parasitic fungi*, which absorb nutrients from body fluids of living hosts. Some are pathogenic.
    3. *Mutualistic fungi*, which absorb nutrients from a host but reciprocate to benefit the host.

Fungi exist in diverse habitats and form symbioses with many organisms. For example, fungi are found:
- commonly in terrestrial habitats.
- in aquatic habitats – both freshwater and marine.
- in symbiotic relationships with algae to form lichens.

B. Structure

The basic structural unit of a fungal vegetative body (*mycelium*) is the *hypha*. Except for yeasts, fungal bodies are diffuse, intertwining mats of *hyphae* that are organized around and within their food source. These hyphae:
- Are composed of tubular walls containing *chitin*, a strong, flexible nitrogen-containing polysaccharide similar to that found in arthropod exoskeletons.
- Provide enormous surface area for the absorptive mode of nutrition. Parasitic fungi have modified hyphae called *haustoria* – nutrient absorbing hyphae that penetrate host tissue, but remain outside the host cell membrane.

Fungal hyphae may be *aseptate* or *septate*.
- Hyphae of *aseptate fungi* lack cross-walls and are *coenocytic*, formed from repeated nuclear division without cytokinesis.
- Hyphae of *septate fungi* are divided into cells by crosswalls called *septa*. Pores in the septa allow organelles to move from cell to cell.

True fungi have no flagellated stages in their life cycle. This characteristic is partially why the Chytridiomycota and Oomycota have been moved to the Protista.

C. Growth and Reproduction

Mycelial growth is adapted to the absorptive node of nutrition.
- Mycelia grow in length, not girth, which maximizes the surface area for absorption.
- Mycelia grow rapidly, as much as a kilometer of hyphae each day. Fast growth can occur because cytoplasmic streaming carries molecules synthesized by the mycelium to the growing hyphal tips.
- Since fungi are nonmotile, they cannot search for food or mates. Instead, they grow in hyphal length to reach new food sources and territory.

Fungal chromosomes and nuclei are relatively small, and the nuclei divide differently from most other eukaryotes.

- During mitosis, the nuclear envelope remains intact from prophase to anaphase; the spindle is inside the nuclear envelope.
- After anaphase, the nuclear envelope pinches in two, and the spindle disappears.

Fungi reproduce by releasing spores that are:

- Usually unicellular, haploid, and of various shapes and sizes.
- Produced either sexually (by meiosis) or asexually (by mitosis). In favorable conditions, fungi generally produce enormous numbers of spores asexually. For many fungi (not all), sexual reproduction occurs only as a contingency for stressful environmental conditions.
- The agent of dispersal responsible for the wide geographic distribution of fungi. Carried by wind or water, spores germinate if they land in a moist place with an appropriate substratum.

Except for transient diploid stages in sexual life cycles, fungal hyphae and spores are haploid. Some mycelia may, however, be genetically heterogeneous resulting from fusion of hyphae with different nuclei.

- The different nuclei may stay in separate parts of the same mycelium.
- Alternatively, the different nuclei may mingle and even exchange genes in a process similar to crossing over.

The sexual cycle in fungi differs from other eukaryotic organisms in that *syngamy* occurs in two stages that are separated in time.

Syngamy = The sexual union of haploid cells from two individuals. In fungi, syngamy occurs in two stages:

1. *Plasmogamy*, the fusion of cytoplasm.
2. *Karyogamy*, the fusion of nuclei.

After plasmogamy, haploid nuclei from each parent pair up but do not fuse, forming a *dikaryon*.

- Nuclear pairs in dikaryons may exist and divide synchronously for months or years.
- The dikaryotic condition has some advantages of diploidy; one haploid genome may compensate for harmful mutations in the other nucleus.
- Eventually, the haploid nuclei fuse forming a diploid cell that immediately undergoes meiosis.

II.   The three major divisions of fungi differ in details of reproduction

Fungi in the three divisions differ in the:

- Structures involved in plasmogamy.
- Length of time spent as a dikaryon.
- Location of karyogamy. Note that the fungal divisions are named for the sexual structures in which karyogamy occurs.

A.   Division Zygomycota (Zygote Fungi)

Fungi in the division Zygomycota are characterized by the presence of dikaryotic *zygosporangia*, resistant structures formed during sexual reproduction.

- Zygomycetes are mostly terrestrial and live in soil or on decaying organic material.
- Some form *mycorrhizae*, mutualistic associations with plant roots.
- Zygomycete hyphae are coenocytic; septa are found only in reproductive cells.

Refer to Campbell, Figure 28.3 for the life cycle of the zygomycete, *Rhizopus stolonifer*, a common bread mold.

- The mycelium consists of horizontal hyphae that spread out and penetrate the food source.
- Under favorable environmental conditions, *Rhizopus* reproduces asexually:
  ⇒ Sporangia develop at the tips of upright hyphae.
  ⇒ Mitosis produces hundreds of haploid spores that are dispersed through the air.
  ⇒ If they land in a moist, favorable environment, spores germinate into new mycelia.
- In unfavorable conditions, *Rhizopus* begins its sexual cycle of reproduction:
  ⇒ Mycelia of opposite mating types (+ and −) form *gametangia* that contain several haploid nuclei walled off by the septum.
  ⇒ *Plasmogamy* of the + and − gametangia occurs, and the haploid nuclei pair up forming a dikaryotic *zygosporangium* that is metabolically inactive and resistant to desiccation and freezing.
  ⇒ When conditions become favorable, *karyogamy* occurs between paired nuclei; the resulting diploid nuclei immediately undergo meiosis producing genetically diverse haploid spores.
  ⇒ The zygosporangium germinates a sporangium that releases the genetically recombined haploid spores.
  ⇒ If they land in a moist, favorable environment, spores germinate into new mycelia.

Even though air currents are not a very precise way to disperse spores, *Rhizopus* releases so many that enough land in hospitable places. Some zygomycetes, however, can actually aim their spores.

- For example, *Pilobolus*, a fungus that decomposes animal dung, bends sporangium-bearing hyphae toward light, where grass is likely to be growing.
- The sporangium is shot out of the hypha, dispersing spores away from the dung and onto surrounding grass. If an herbivore eats the grass and consumes the spores, the asexual life cycle is completed when the animal disperses the spores in its feces.

B.   Division Ascomycota (Sac Fungi)

Ascomycetes include unicellular yeasts and complex multicellular cup fungi.

- Hyphae are septate.
- In asexual reproduction, the tips of specialized hyphae form *conidia*, which are chains of haploid, asexual spores that are usually wind dispersed.
- In sexual reproduction, haploid mycelia of opposite mating strains fuse. One acts as "female" and produces an *ascogonium* which receives haploid nuclei from the antheridium of the "male."

The ascogonium grows hyphae with dikaryotic cells. Syngamy is delayed.

↓

In terminal cells of dikaryotic hyphae – syngamy occurs.

↓

Meiosis forms 4 haploid nuclei which undergo mitotic division to yield eight haploid nuclei.

↓

The nuclei form walls and become ascospores within an *ascus*, the sac of sexually produced spores. Multiple asci may form an *ascocarp*.

Ascocarps = Fruiting structures of many asci packed together.

- The ascospores of each ascus are lined up in a row in the order in which they formed from a single zygote, allowing geneticists to study genetic recombination.
- Unicellular yeasts appear dissimilar, but produce the equivalent of an ascus during sexual reproduction and bud during asexual reproduction in a manner similar to the formation of conidia. Thus, they are classified as ascomycetes.
- Includes important decomposers and both mutualistic and parasitic symbionts.
- Many live symbiotically with algae as lichens.

C.   **Division Basidiomycota (Club Fungi)**

The division Basidiomycota, or club fungi, includes mushrooms, shelf fungi, puffballs, and stinkhorns. Basidiomycetes:

- Are named for a transient diploid stage called the *basidium*, a club-shaped spore-producing structure.
- Are important decomposers of wood and other plant material.   Saprobic basidiomycetes can decompose the complex polymer lignin, an abundant component of wood.
- Include mycorrhiza-forming mutualists and plant parasites. Many shelf fungi are tree parasites that function later as saprobes after the trees die.
- Include mushroom-forming fungi, only a few of which are strictly parasitic. About half are saprobic and the other half form mycorrhizae.
- Include the rusts and smuts, which are plant parasites.

Basidiomycete life cycles are characterized by a long-lived *dikaryotic* mycelium that reproduces sexually by producing fruiting bodies called *basidiocarps*. Refer to Campbell, Figure 28.7 for the life cycle of a mushroom-forming basidiomycete.

- Haploid basidiospores grow into short-lived haploid mycelia.   Under certain environmental conditions, *plasmogamy* occurs between two haploid mycelia of opposite mating types (+ and −).
- The resulting dikaryotic mycelium grows; depending upon the species, it may form mycorrhizae with trees. Certain environmental cues stimulate the mycelium to produce mushrooms (basidiocarps). A "fairy ring" is an expanding ring of living mycelium that produces mushrooms above it; it slowly increases in diameter, about 30 cm per year.
- The mushroom cap supports and protects a large surface area of gills; *karyogamy* in the terminal, dikaryotic cells lining the gills produces diploid *basidia*.
- Each basidium immediately undergoes meiosis producing four haploid basidiospores. When mature, these sexual spores drop from the cap and are dispersed by wind.

Asexual reproduction occurs less often than in ascomycetes, but also results in conidia formation.

III.   **Molds, yeasts, lichens, and mycorrhizae represent unique lifestyles that evolved independently in all fungal divisions**

A.   **Molds**

Mold = A rapidly growing, asexually reproducing fungus.

- May be saprobes or parasites on a great variety of substrates.
- Molds only include asexual stages; they may be zygomycetes, ascomycetes, basidiomycetes or fungi with no known sexual stage.

Since molds are classified by their sexual stages (zygosporangium, ascogonium or basidium), molds with no known sexual stage cannot be classified as zygomycetes, ascomycetes or basidiomycetes.

- Molds with no known sexual stages are classified as *Deuteromycota* or *imperfect fungi*.
- Imperfect fungi reproduce asexually by producing conidia on specialized hyphae called *conidiphores*. Their asexual formation of conidia suggests that many imperfect fungi are related to Ascomycetes.
- Deuteromycetes are sources of antibiotics. For example, *Penicillium* produces penicillin.
- Other commercial uses of imperfect fungi include flavoring for cheeses, such as blue cheese, Brie, Camembert and Roquefort; fermenting food products such as soybeans; and providing pharmaceuticals such as cyclosporine.
- Some deuteromycetes are predatory soil fungi that kill small animals such as soil nematodes.

B.   Yeasts

Yeasts are unicellular fungi that inhabit liquid or moist habitats; some can alternate between mycelium or yeast, depending on the amount of liquid in the environment.

Yeasts reproduce:
- Asexually by simple cell division or by budding off from a parent cell. Some are classified as Deuteromycota, if no sexual stages are known.
- Sexually by forming asci (Ascomycota) or basidia (Basidiomycota).

Though humans have used yeasts to raise bread and ferment alcoholic beverages for thousands of years, only recently have they been separated into pure culture for more controlled human use.
- *Saccharomyces cerevisiae* is the most important of all domesticated fungi. Highly active metabolically, this ascomycete is available as baker's and brewer's yeast.
- In an aerobic environment, baker's yeast respires, releasing small bubbles of carbon dioxide that leaven dough; cultured anaerobically, *Saccharomyces* ferments sugars to alcohol.
- Researchers use *Saccharomyces* to study eukaryotic molecular genetics because it is easy to culture and manipulate.

Some yeasts cause problems for humans.
- *Rhodotorula*, a pink yeast, grows on shower curtains and other moist surfaces.
- *Candida*, a normal inhabitant of moist human tissues, can become pathogenic when there is a change in pH or other environmental factor; or when someone's immune system is compromised.

C.   Lichens

Lichen = Highly integrated symbiotic association of algal cells (usually filamentous green algae or blue-green algae) with fungal hyphae (usually ascomycetes).

Though lichens vary in shape and physiology, some shared general features characterize the symbiotic relationship.

The alga, which is below the lichen's surface,
- Always provides the fungus with food.
- May fix nitrogen (i.e. Cyanobacteria).

The fungus provides a suitable environment for algal growth:
- The hyphal mass:
  ⇒ Absorbs needed minerals from airborne dust or rain.
  ⇒ Retains water and minerals.
  ⇒ Allows gas exchange.
  ⇒ Protects the algae.
- The fungus produces unique organic compounds with several functions.
  ⇒ Fungal pigments shade the algae from intense sunlight.
  ⇒ Toxic fungal compounds prevent lichens from being eaten by consumers.
  ⇒ Fungal acid secretion aids the uptake of minerals.

Most of the lichen's mass is hyphal tissue which gives the lichen its shape and structure. Named for their fungal component, lichens are informally categorized as:
- *Foliose* (leafy)
- *Fruticose* (shrubby)
- *Crustose* (crusty)

Lichen reproduction occurs as a combined unit or as independent reproduction of the symbionts.
- Many lichen fungi reproduce sexually by forming ascocarps or rarely, basidiocarps.
- Lichen algae reproduce independently by asexual cell division.
- Symbiotic units commonly reproduce asexually by:
  ⇒ Fragmentation of the parental lichen.
  ⇒ Formation of *soredia* - specialized reproductive structures that are small clusters of hyphae with embedded algae.

Though most evidence points to a mutualistic symbiosis, some debate that the relationship may actually be parasitic.
- The argument for mutualism is that fungi benefit the algae and that lichens can survive in habitats that are inhospitable to either organism alone.
- The argument for "controlled parasitism" is based on the fact that the fungus actually kills some algal cells, though not as fast as the algae replenishes itself.

Lichens are important pioneers, breaking down rock and allowing for colonization by other plants.

- Some can tolerate severe cold.
- Photosynthesis occurs when lichen water content is 65-90%.

Lichens are sensitive to air pollution due to their mode of mineral uptake.

D.   Mycorrhizae

Mycorrhizae are specific, mutualistic associations of plant roots and fungi.

- The fungi increase the absorptive surface of roots and exchanges soil minerals.
- Mycorrhizae are seen in 90% of trees (fungus is usually a basidiomycete) and the majority of small vascular plants.
- Are necessary for optimal plant growth. (See Campbell, Chapter 33)

IV.   Fungi have a tremendous ecological impact

A.   Fungi as Decomposers

Fungi and bacteria are the principal decomposers that reduce complex polysaccharides and proteins to simple organic compounds that plants can assimilate as raw material for photosynthesis.

B.   Fungi as Spoilers

Fungi decompose food, wood and even certain plastics.

C.   Pathogenic Fungi

Many fungi are pathogenic (e.g. athletes foot, ringworm and yeast infections).

Plants are particularly susceptible. For example, Dutch elm disease, caused by an ascomycete, drastically changed the landscape of northeastern U.S.

Ergots = Purple structure on rye caused by an ascomycete.

- Causes gangrene, hallucinations and burning sensations (St. Anthony's fire).
- Produces lysergic acid, from which LSD is made.

Toxins from fungi may be used in weak doses for medical purposes such as treating high blood pressure.

D. Edible Fungi

Fungi are consumed as food by a variety of animals, including humans.

- In the U.S., mushroom (basidiomycete) consumption is usually restricted to one species of *Agaricus*, which is cultivated commercially on compost in the dark.
- In many other countries, however, people eat a variety of cultivated and wild mushrooms.
- Truffles prized by gourmets are underground ascocarps of mycelia that are mycorrhizal on tree roots. The fruiting bodies (ascocarps) release strong odors that attract mammals and insects - consumers that excavate the truffles and disperse their spores.
- Since it is difficult for novices to distinguish between poisonous and edible mushrooms, only qualified experts at identification should collect wild mushrooms for eating.

V. Fungi and animals probably evolved from a common protistan ancestor

The Chytridiomycota and Fungi may share a protistan ancestor.

- Chytrids were placed in the Kingdom Protista because they form flagellated zoospores and gametes – a protistan characteristic.
- However, chytrids and fungi share many characteristics, such as:
  ⇒ An absorptive mode of nutrition.
  ⇒ Cell walls of chitin.
  ⇒ Most form hyphae.
  ⇒ Key enzymes and metabolic pathways that are not found in the other fungus-like protists (slime molds and water molds).
  ⇒ Similar sequences of proteins and nucleic acids.
- This evidence lends support for
  ⇒ Combining the chytrids with fungi as a monophyletic group.
  ⇒ The hypothesis that chytrids are the most primitive fungi, diverging earliest in fungal phylogeny.
  ⇒ The hypothesis that fungi evolved from protists with flagella, a feature retained by the chytrids.

There is compelling evidence that animals and fungi diverged from a common protistan ancestor.

- Animals probably evolved also from flagellated protists.
- Proteins and rRNA comparisons indicate that fungi and animals are more closely related to each other than to plants. Molecular systematists believe the most likely protistan ancestor common to fungi and animals was a choanoflagellate.

Perhaps the fungi are a consequence of adaptive radiation when life began to colonize land.

- The oldest undisputed fossils are 450-500 million years old.
- All major groups of fungi evolved by the end of the Carboniferous period (approximately 300 million years ago).
- Plants and fungi moved from water to land together. Fossils of the first vascular plants have mycorrhizae.

REFERENCES

Alexopoulos, C.J. and C.W. Mims. *Introductory Mycology*. 3rd ed. New York: John Wiley and Sons, Inc., 1979.

Campbell, N. *Biology*. 4th ed. Menlo Park, California: Benjamin/Cummings, 1996.

CHAPTER 29

INVERTEBRATES AND THE ORIGIN OF ANIMAL DIVERSITY

OUTLINE

I.   What is an animal?

II. Comparative anatomy and embryology provide clues to animal phylogeny: *an overview of animal diversity*
    A.   The Parazoa-Eumetazoa Split
    B.   The Radiata-Bilateria Split
    C.   The Acoelomate-Coelomate Split
    D.   The Protostome-Deuterostome Split

III. Sponges are sessile animals lacking true tissues

IV. Cnidarians and ctenophores are radiate, diploblastic animals with gastrovascular cavities
    A.   Phylum Cnidaria
    B.   Phylum Ctenophora

V.   Flatworms and other acoelomates are bilateral, triploblastic animals lacking body cavities
    A.   Phylum Platyhelminthes
    B.   Phylum Nemertea

VI. Rotifers, nematodes, and other pseudocoelomates have complete digestive tracts and blood vascular systems
    A.   Phylum Rotifera
    B.   Phylum Nematoda

VII. Mollusks and annelids are among the major variations on the protostome body plan
    A.   Phylum Mollusca
    B.   The Lophophorate Animals
    C.   Phylum Annelida

VIII.   The protostome phylum Arthropoda is the most successful group of animals ever to live
   A.   General Characteristics of Arthropods
   B.   Arthropod Phylogeny and Classification
   C.   Trilobites
   D.   Spiders and Other Chelicerates
   E.   Comparison of Chelicerates to Uniramians and Crustaceans
   F.   Insects and Other Uniramians
   G.   Crustaceans

IX.   The deuterostome lineage includes echinoderms and chordates
   A.   Phylum Echinodermata
   B.   Phylum Chordata

X.   The Cambrian explosion produced all the major animal body plans

OBJECTIVES

After reading this chapter and attending lecture, the student should be able to:
1.   List characteristics that distinguish animals from organisms in the other four kingdoms.
2.   Distinguish between radial and bilateral symmetry.
3.   Outline the major phylogenetic branches of the animal kingdom, which are based upon grade of organization; symmetry and embryonic germ layers; absence or presence of a body cavity; and protostome-deuterostome dichotomy.
4.   Distinguish among acoelomate, pseudocoelomate and coelomate.
5.   Distinguish between spiral and radial cleavage; determinant and indeterminate cleavage; schizocoelous and enterocoelous.
6.   Compare developmental differences between protostomes and deuterostomes including:
   a.   Plane of cleavage.
   b.   Determination.
   c.   Fate of the blastopore.
   d.   Coelom formation.
7.   From a diagram, identify the parts of a sponge and describe the function of each including: spongocoel, porocyte, epidermis, choanocyte, mesohyl, amoebocyte, osculum and spicule.
8.   List characteristics of the Phylum Cnidaria that distinguish it from the other animal phyla.
9.   Describe the two basic body plans in Cnidaria and their role in Cnidarian life cycles.
10.   List the three classes of Cnidaria and distinguish among them based upon life cycle and morphological characteristics.
11.   List characteristics of the Phylum Ctenophora that distinguish it from the other animal phyla.
12.   List characteristics that are shared by all bilaterally symmetrical animals.
13.   List characteristics of the Phylum Platyhelminthes that distinguish it from the other animal phyla.
14.   Distinguish among the four classes of Platyhelminthes and give examples of each.
15.   Describe the generalized life cycle of a trematode and give an example of one fluke that parasitizes humans.
16.   Describe the anatomy and generalized life cycle of a tapeworm.
17.   List distinguishing characteristics descriptive of the Phylum Nemertea.
18.   Explain why biologists believe proboscis worms evolved from flatworms.
19.   Describe features of digestive and circulatory systems that have evolved in the Nemertea and are not found in other acoelomate phyla.

20. Describe unique features of rotifers that distinguish them from other pseudocoelomates.
21. Define parthenogenesis and describe alternative forms of rotifer reproduction.
22. List characteristics of the phylum Nematoda that distinguish it from other pseudocoelomates.
23. Give examples of both parasitic and free-living species of nematodes.
24. List characteristics that distinguish the Phylum Mollusca from the other animal phyla.
25. Describe the basic body plan of a mollusk and explain how it has been modified in the Polyplacophora, Gastropoda, Bivalvia and Cephalopoda.
26. Distinguish among the following four Molluscan classes and give examples of each:
    a. Polyplacophora.
    b. Gastropoda.
    c. Bivalvia.
    d. Cephalopoda.
27. Explain why some zoologists believe the mollusks evolved from ancestral annelids while others propose that mollusks arose from flatworm-like ancestors.
28. List characteristics that distinguish the Phylum Annelida from the other animal phyla.
29. Explain how a fluid-filled septate coelom is used by annelids for burrowing.
30. Distinguish among the classes of annelids and give examples of each.
31. List characteristics of arthropods that distinguish them from the other animal phyla.
32. Describe advantages and disadvantages of an exoskeleton.
33. Distinguish between hemocoel and coelom.
34. Provide evidence for an evolutionary link between the Annelida and Arthropoda.
35. Describe major independent arthropod lines of evolution represented by the subphyla:
    a. Trilobitomorpha.
    b. Cheliceriformes.
    c. Crustacea.
    d. Uniramia.
36. Explain what arthropod structure was a preadaptation for living on land.
37. Distinguish among the following arthropod classes and give an example of each:
    a. Arachnida.
    b. Crustacea.
    c. Diplopoda.
    d. Chilopoda.
    e. Insecta.
38. Distinguish between incomplete metamorphosis and complete metamorphosis.
39. Define lophophore and list three lophophorate phyla.
40. Explain why lophophorates are difficult to assign as protostomes or deuterostomes.
41. List at least four characteristics shared by the deuterostome phyla that distinguish them from protostomes.
42. List characteristics of echinoderms that distinguish them from other animal phyla.
43. Describe the structures and function of a water vascular system including ring canal, radial canal, tube feet and ampulla.
44. Distinguish among the five classes of echinoderms and give examples of each.
45. Compare and contrast two hypotheses about animal origins from unicellular ancestors: syncytial hypothesis and colonial hypothesis.
46. Explain why it is difficult to resolve what the first animals looked like.
47. Describe two views about discontinuities between Ediacaran and Cambrian fauna.

KEY TERMS

ingestion
glycogen
detritus
cleavage
blastula
gastrulation
metamorphosis
Parazoa
Eumetazoa
Radiata
Bilateria
cephalization
diploblastic
triploblastic
ectoderm
endoderm
mesoderm
archenteron
acoelomate
pseudocoelomate
pseudocoelom
coelomate
coelom
protostome
deuterostome
spiral cleavage
radial cleavage
determinate cleavage
indeterminate cleavage
blastopore
schizocoelous
enterocoelous
Porifera
osculum
spongocoel
epidermis
porocyte

choanocyte
mesohyl
amoebocyte
spicule
spongin
Cnidaria
polyp
medusa
cnidocytes
gastrovascular cavity
Hydrozoa
Scyphozoa
Anthozoa
Ctenophora
Platyhelminthes
Turbellaria
flame cell
Trematoda
Monogenea
Cestoda
scolex
proglottid
Nemertea
proboscis
Rotifera
complete digestive
    system
parthenogenesis
Nematoda
Mollusca
foot
visceral mass
mantle
radula
Polyplacophora
Gastropoda
torsion

Bivalvia
mantle cavity
incurrent siphon
excurrent siphon
Cephalopoda
closed circulatory
    system
lophophore
Phoronida
Bryozoa
Brachiopoda
Annelida
pharynx
esophagus
crop
gizzard
metanephridia
cocoon
clitellum
fragmentation
Oligochaeta
casts
Polychaeta
parapodia
setae
Hirudinea
hirudin
Arthropoda
cuticle
chitin
hemolymph
trilobite
eurytherids
chelicerae
Arachnida
cephalothorax
pedipalps

book-lung
Uniramia
uniramous
Crustacea
biramous
mandibles
Diplopoda
Chilopoda
Insecta
head
thorax
abdomen
malpighian tubules
tracheal system
incomplete metamorphosis
complete metamorphosis
spermatophores
spermatheca
viviparous
decapods
carapace
isopods
copepods
barnacles
Echinodermata
water vascular system
Asteroidea
tube feet
ampulla
Ophiuroidea
Echinoidea
Crinoidea
Holothuroidea
Concentricycloidea
Chordata
Cambrian explosion
Ediacaran period

LECTURE NOTES

Over one million species of animals are living today; 95% of these are invertebrates.

- Grouped into about 35 phyla depending on the taxonomic view followed.
- Most are aquatic.
- The most familiar belong to the subphylum Vertebrata of the phylum Chordata. This is only about 5% of the total.

I.   What is an animal?

Although there is great animal diversity, most animals share the following features:

- Multicellular, eukaryotic organisms.
    - ⇒ Heterotrophy is by ingestion.

        Ingestion = Eating other organisms or decomposing organic matter (*detritus*). This mode of nutrition distinguishes animals from the plants and fungi.

- Carbohydrate reserves are stored as glycogen.
- No cell walls are present; animals do have intercellular junctions present: desmosomes, gap junctions, and tight junctions.
- Highly differentiated body cells which are organized into tissues, organs and organ systems for such specialized functions as digestion, internal transport, gas exchange, movement, coordination, excretion, and reproduction.
- Nervous tissue (impulse conduction) and muscle tissue (movement) are unique to animals.
- Reproduction is typically sexual with flagellated sperm fertilizing nonmotile eggs to form diploid zygotes. A diploid stage dominates the life cycle.
    - ⇒ The zygote undergoes a series of mitotic divisions known as *cleavage* which produces a *blastula* in most animals.
    - ⇒ *Gastrulation* occurs after the blastula has formed; during this process, the embryonic forms of adult body tissues are produced.
    - ⇒ Development in some animals is direct to maturation while the life cycles of others include larvae which undergo *metamorphosis* into a sexually mature adults.

        Larva = Free living sexually immature forms.

The seas contain the greatest diversity of animal phyla, although many groups live in fresh water and terrestrial habitats.

II. Comparative anatomy and embryology provide clues to animal phylogeny: *an overview of animal diversity*

Animals diversified so rapidly during the late Precambrian and early Cambrian periods that it is difficult to determine the exact sequence of branching from the fossil record.

- To reconstruct the evolutionary history of the animal phyla, zoologists use information from comparative anatomy, embryology of living animals, and molecular systematics.
- Most zoologists agree that the animal kingdom is monophyletic and the ancestral organism was probably a choanoflagellate.

A. The Parazoa-Eumetazoa Split

The Parazoa contains only the Porifera, which:

- Have unique development and simple anatomy that separates them from other animals.
- Lack true tissues.

The Eumetazoa include animals with true tissues, the presence of which has permitted the evolution of a more complex anatomy.

B. The Radiata-Bilateria Split

The division of eumetazoans into two branches is based partly on body symmetry.

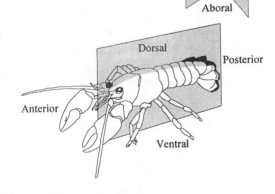

- The *Radiata* exhibit radial symmetry.
    - ⇒ These animals have an oral (top) and aboral (bottom) side, but no front, back, left, or right sides.

- The *Bilateria* exhibit bilateral symmetry.
    - ⇒ Bilaterally symmetrical animals have dorsal (top), ventral (bottom), anterior (head), posterior (tail), left and right body surfaces.
    - ⇒ These animals exhibit *cephalization* (an evolutionary trend toward concentration of sensory structures at the anterior end).

Care must be taken when assigning an animal to an evolutionary line as symmetry may change between the larval and adult forms. The phylum Echinodermata shows a secondary radial symmetry in adults, which evolved as an adaptation to their sedentary lifestyle. They are actually in the bilateria.

Examination of development and body plan can define the radiata-bilateria split better than symmetry.

- The early embryo of all eumetazoans undergoes *gastrulation*. Concentric germ layers develop which form the various tissues and organs as development continues.
  - ⇒ The Radiata develop only two germ layers (ectoderm and endoderm) and are termed *diploblastic* (e.g. phylum Cnidaria).
  - ⇒ The Bilateria develop three germ layers (ectoderm, endoderm, and mesoderm) and are termed *triploblastic*.

The Germ Layers of an Early Embryo:

Ectoderm

- Covers the surface of the embryo.
- Forms the animal's outer covering and the central nervous system in some phyla.

Endoderm

- Innermost germ layer which lines the *archenteron* (primitive gut).
- Forms the lining of the digestive tract, and outpocketings give rise to the liver and lungs of vertebrates.

Mesoderm

- Located between the ectoderm and endoderm in triploblastic animals.
- Forms the muscles and most organs located between the digestive tract and outer covering of the animal.

C. **The Acoelomate-Coelomate Split**

Triploblastic animals can also be grouped on the basis of whether a body cavity develops or not and, if so, how that cavity develops.

Animals in which no body cavity develops are termed *acoelomate*.

Acoelomate = An animal body plan characterized by no body cavity present between the digestive tract and the outer body wall. (See Campbell, Figure 29.4a)
- The area between the digestive tract and outer wall is filled with cells, producing a solid body. (e.g. Platyhelminthes)

Animals in which a body cavity develops may be termed *pseudocoelomate* or *coelomate*, depending on how the cavity develops.

Pseudocoelomate = Animal body plan characterized by a fluid-filled body cavity that separates the digestive tract and the outer body wall. (See Campbell, Figure 29.4b.)
- This cavity (the *pseudocoelom*) is not completely lined with tissue derived from mesoderm. (e.g. Nematoda)

Coelomate = Animal body plan characterized by a fluid-filled body cavity completely lined with tissue derived from mesoderm (the *coelom*) that separates the digestive tract from the outer body wall. (See Campbell, Figure 29.4c)

- Mesenteries connect the inner and outer mesoderm layers and suspend the internal organs in the coelom. (e.g. Annelida)

The functions of a fluid-filled body cavity are:

- The fluid cushions the organs thus preventing injury.
- Internal organs can grow and move independently of the outer body wall.
- Serves as a hydrostatic skeleton in soft bodied coelomates such as earthworms.

In addition to the presence of a body cavity, acoelomates differ from pseudocoelomates and eucoelomates by not having a blood vascular system.

D.   The Protostome-Deuterostome Split

Distinguished by differences in development, the coelomate phyla can be divided into two distinct evolutionary lines: *protostomes* and *deuterostomes*.

Developmental differences between protostomes and deuterostomes include: cleavage patterns, fate of the blastopore, and coelom formation.

1.   Cleavage Patterns

Most protostomes undergo *spiral cleavage* and *determinate cleavage* during their development.

Spiral cleavage = Cleavage in which the planes of cell division are diagonal to the vertical axis of the embryo.

Determinate cleavage = Cleavage in which the developmental fate of each embryonic cell is established very early. A cell isolated from the four-cell stage of an embryo will not develop fully.

Deuterostomes undergo *radial cleavage* and *indeterminate cleavage* during their development.

Radial cleavage = Cleavage during which the cleavage planes are either parallel or perpendicular to the vertical axis of the embryo.

Indeterminate cleavage = Cleavage in which each early embryonic cell retains the capacity to develop into a complete embryo if isolated from other cells. This type of cleavage in the human zygote results in identical twins.

2.   Blastopore Fate

Blastopore = The first opening of the archenteron which forms during gastrulation. (See Campbell, Figure 29.5b)

- In Protostomes, the blastopore forms the mouth.
- In Deuterostomes, the blastopore forms the anus.

3. **Coelom Formation**

   <u>Schizocoelous</u> = Descriptive term for coelom development during which, as the archenteron forms, the coelom begins as splits within the solid mesodermal mass; coelom formation found in Protostomes.

   <u>Enterocoelous</u> = Coelom development during which the mesoderm arises as lateral outpocketings of the archenteron with hollows that become the coelomic cavities; coelom formation found in Deuterostomes.

SUMMARY OF PROTOSTOME - DEUTEROSTOME SPLIT	
PROTOSTOMES	DEUTEROSTOMES
Spiral cleavage	Radial cleavage
Determinate cleavage	Indeterminate cleavage
Blastopore forms the mouth	Blastopore forms the anus
Schizocoelous coelom formation	Enterocoelous coelom formation

III.   Sponges are sessile animals lacking true tissues

The sponges, in the phylum Porifera, are the only members of the subkingdom Parazoa due to their unique development and simple anatomy.

- Approximately 9,000 species, mostly marine with only about 100 in fresh water.
- Lack true tissues and organs, and contain only two layers of loosely associated unspecialized cells.
- No nerves or muscles, but individual cells detect and react to environmental changes.
- Size ranges from 1 cm to 2 m.
- All are suspension-feeders (= filter-feeders).
- Possibly evolved from colonial choanoflagellates.

Parts of the Sponge:

<u>Spongocoel</u> = Central cavity of sponge.

<u>Osculum</u> = Larger excurrent opening of the spongocoel.

<u>Epidermis</u> = Single layer of flattened cells which forms outer surface of the sponge.

<u>Porocyte</u> = Cells which form pores; possess a hollow channel through the center which extends from the outer surface (incurrent pore) to spongocoel.

<u>Choanocyte</u> = Collar cell, majority of cells which line the spongocoel; possess a flagellum which is ringed by a collar of finger-like projections. Flagellar movement moves water and food particles which are trapped on the collar and later phagocytized.

<u>Mesohyl</u> = The gelatinous layer located between the two layers of the sponge body wall (epidermis and choanocytes).

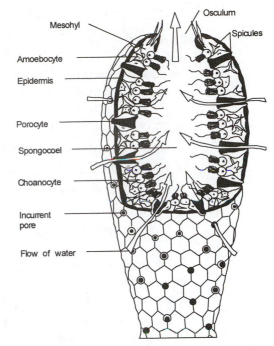

<u>Amoebocyte</u> = Wandering, pseudopod bearing cells in the mesohyl; function in food uptake from choanocytes, food digestion, nutrient distribution to other cells, formation of skeletal fibers, gamete formation.

<u>Spicule</u> = Sharp, calcium carbonate or silica structures in the mesohyl which form the skeletal fibers of many sponges.

<u>Spongin</u> = Flexible, proteinaceous skeletal fibers in the mesohyl of some sponges.

Sponges are hermaphroditic but usually cross-fertilize.

- Eggs and sperm form in the mesohyl from differentiated amoebocytes or choanocytes.
- Eggs remain in the mesohyl.
- Sperm are released into excurrent flow of the spongocoel and are then drawn in with incurrent flow of another sponge.
- Sperm penetrate into mesohyl and fertilize the eggs.
- The zygote develops into a flagellated larva which is released into the spongocoel and escapes with the excurrent water through the osculum.
- Surviving larvae settle on the substratum and develop. In most cases the larva turns inside-out during metamorphosis, moving the flagellated cells to the inside.

Sponges possess extensive regeneration abilities for repair and asexual reproduction.

IV. Cnidarians and ctenophores are radiate, diploblastic animals with gastrovascular cavities

A. Phylum Cnidaria

There are more than 10,000 species in the phylum Cnidaria, most of which are marine. The phylum contains hydras, jellyfish, sea anemones and coral animals.

Some characteristics of cnidarians include:

- Radial symmetry.
- Diploblastic.
- Simple, sac-like body.
- *Gastrovascular cavity* present: a central digestive cavity with only one opening (functions as mouth and anus).

There are two possible cnidarian body plans: sessile *polyp* and motile, floating *medusa*. Some species of cnidarians exist only as polyps, some only as medusae, and others are dimorphic (both polyp and medusa stages in their life cycles).

Polyp = Cylindrical form which adheres to the substratum by the aboral end of the body stalk and extends tentacles around the oral end to contact prey.

Medusa = Flattened, oral opening down, bell-shaped form. Moves freely in water by passive drifting and weak bell contractions. Tentacles dangle from the oral surface which points downward.

Cnidarians are carnivorous.

- Tentacles around the mouth/anus capture prey animals and push them through the mouth/anus into the gastrovascular cavity.
- Digestion begins in the gastrovascular cavity with the undigested remains being expelled through the mouth/anus.
- Tentacles are armed with stinging cells, called *cnidocytes* — after which the Cnidaria are named.

  Cnidocytes = Specialized cells of cnidarian epidermis that contain *nematocysts*, capsule-like structures with an eversible stinger; used in defense and capture of prey.

The simplest forms of muscles and nerves occur in the Phylum Cnidaria.

- Epidermal and gastrodermal cells have bundles of microfilaments arranged into contractile fibers.
  - ⇒ The gastrovascular cavity, when filled with water, acts as a hydrostatic skeleton against which the contractile fibers can work to change the animal's shape.
- A simple nerve net coordinates movement; no brain is present.
  - ⇒ The nerve net is associated with simple sensory receptors radially distributed on the body. This permits stimuli to be detected and responded to from all directions.

There are three major classes of cnidarians:

1. **Class Hydrozoa**

   Most alternate polyp and medusa forms in the life cycle although the polyp is the dominant stage. Some are colonial (e.g. *Obelia*, Campbell, Figure 29.11) while others are solitary (e.g. *Hydra*).

   Hydra is unique in that only the polyp stage is present.
   - They usually reproduce asexually by budding; however, in unfavorable conditions they reproduce sexually. In this case a resistant zygote is formed and remains dormant until environmental conditions improve.

2. **Class Scyphozoa**

   The planktonic medusa (jellyfish) is the most prominent stage of the life cycle.
   - Coastal species usually pass through a small polyp stage during the life cycle.
   - Open ocean species have eliminated the polyp entirely.

3. **Class Anthozoa**

   This class contains sea anemones and coral animals which only occur as polyps.

   Coral animals may be solitary or colonial and secrete external skeletons of calcium carbonate.
   - Each polyp generation builds on the skeletal remains of earlier generations. In this way, coral reefs are formed.
   - Coral is the rock-like external skeletons.

B. **Phylum Ctenophora**

This phylum contains the comb jellies. There are about 100 species, all of which are marine.

Some characteristics of Ctenophorans include:
- A resemblance to the medusa of Cnidarians in that the body of most is spherical or ovoid; a few are elongate and ribbon-like.
- Transparent body, 1-10 cm in diameter (spherical/ovoid forms) or up to 1 m long (ribbon-like forms).
- Possess eight rows of comblike plates composed of fused cilia which are used for locomotion.
- One pair of long retractable tentacles are present and function in capturing food; cnidocytes have been found in only one species.
- A sensory organ containing calcareous particles is present.
  ⇒ The particles settle to the low point of the organ which then acts as an orientation cue.
  ⇒ Nerves extending from the sensory organ to the combs of cilia coordinate movement.

V. Flatworms and other acoelomates are bilateral, triploblastic animals lacking body cavities

    A. Phylum Platyhelminthes

The phylum Platyhelminthes differs from the phylum Cnidaria in that they:
- Exhibit bilateral symmetry with moderate cephalization.
- Are triploblastic (develop from three-layered embryos: ectoderm, endoderm and mesoderm).
- Possess several distinct organs, organ systems and true muscles.

Although more advanced than cnidarians, two things point to the early evolution of platyhelminths in bilateria history.
- A gastrovascular cavity is present.
- They have an acoelomate body plan.

There are more than 20,000 species of Platyhelminthes which are divided into four classes:
- Class Turbellaria.
- Classes Trematoda and Monogenea.
- Class Cestoda.

    1. Class Turbellaria

Mostly free-living, marine species; a few species are found in freshwater and moist terrestrial habitats.

Planarians are familiar and common freshwater forms. (See Campbell, Figure 29.14)
- Carnivorous, feed on small animals and carrion.
- Lack specialized organs for gas exchange or circulation.
  - ⇒ Gas exchange is by diffusion (flattened body form places all cells close to water).
  - ⇒ Highly ramified gastrovascular cavity distributes food throughout the body.
- *Flame cell* excretory apparatus present which functions primarily to maintain osmotic balance of the animal.
  - ⇒ Nitrogenous waste (ammonia) diffuses directly from cells to the water.
- Move by using cilia on the ventral dermis to glide along a film of mucus. Muscular contractions produce undulations which allow some to swim.
- On the head are a pair of eyespots which detect light and a pair of lateral auricles that are olfactory sensors.
  - ⇒ Possess a rudimentary brain which is capable of simple learning.
- Reproduce either asexually or sexually.
  - ⇒ Asexual is by regeneration: mid-body constriction separates the parent into two halves, each of which regenerates the missing portion.
  - ⇒ Sexual is by cross-fertilization of these hermaphroditic forms.

2. **Classes Trematoda and Monogenea**

All members of these two classes are parasitic.

Flukes are members of the class Trematoda.

- Suckers are usually present for attaching to host internal organs.
- Primary organ system is the reproductive system; a majority are hermaphroditic.
- Life cycles include alternations of sexual and asexual stages with asexual development taking place in an intermediate host.
  - ⇒ Larvae produced by asexual development infect the final hosts where maturation and sexual reproduction occurs. (See Campbell, Figure 29.15)
- *Schistosoma* spp. (blood flukes) infect 200 million people worldwide.

Members of the class Monogenea are mostly external parasites of fish.

- Structures with large and small hooks are used for attaching to the host animal.
- All are hermaphroditic and reproduce sexually.

3. **Class Cestoda**

Adult tapeworms parasitize the digestive system of vertebrates.

- Possess a *scolex* (head) which may be armed with suckers and/or hooks that help maintain position by attaching to the intestinal lining.
- Posterior to the scolex is a long ribbon of units called *proglottids*.
  - ⇒ A proglottid is filled with reproductive organs.
- No digestive system is present.

The life cycle of a tapeworm includes an intermediate host.

- Mature proglottids filled with eggs are released from the posterior end of the worm and pass from the body with the feces.
- Eggs are eaten by an intermediate host and a larva develops, usually in muscle tissue.
  - ⇒ The final host becomes infected when it eats an intermediate host containing larvae.
- Humans can become infected with some species of tapeworms by eating poorly cooked beef or pork containing larvae.

B. **Phylum Nemertea**

There are about 900 species, most are marine with a few in fresh water and damp soil.

The phylum Nemertea contains the proboscis worms. (See Campbell, Figure 29.17)

- Size from 1 mm to more than 30 m.
- Some active swimmers, others burrow in sand.
- Possess a long, retractable hollow tube (proboscis) which is used to probe the environment, capture prey, and as defense against predators.
- Excretory, sensory and nervous systems are similar to planarians.
- Acoelomate.

There are some important differences between the Nemertea and Platyhelminthes:

- Nemertea possess a simple circulatory system which consists of vessels through which blood flows. Some species have red blood cells containing a form of hemoglobin which transports oxygen. No heart is present, but body muscle contractions move the blood through vessels.
- Nemertea possess a complete digestive system with a mouth and an anus.

The phylogenetic position of the Nemertea is uncertain.

- The body is structurally acoelomate, but the fluid-filled proboscis sac is considered a true coelom by some researchers.
- A simple blood vascular system and a complete digestive system are characteristics shared with more advanced phyla.

VI. Rotifers, nematodes, and other pseudocoelomates have complete digestive tracts and blood vascular systems

The pseudocoelomate body plan probably arose independently several times. Pseudocoelomates are thought to be more closely related to protostomes than deuterostomes, but more study of their true association is required.

A. Phylum Rotifera

There are approximately 1,800 species of rotifers. They are small, mainly freshwater organisms, although some are marine and others are found in damp soil.

- Size ranges from 0.05-2.0 mm
- Pseudocoelomate with the pseudocoelomic fluid serving as a hydrostatic skeleton and as a medium which transports nutrients and wastes when the body moves.
- Complete digestive system is present.
  - ⇒ Bear a crown of cilia that draws a vortex of water into the mouth.
  - ⇒ Posterior to the mouth, a jaw-like organ grinds the microscopic food organisms suspended in the water.

Reproduction in rotifers may be by *parthenogenesis* or sexual.

- Some species consist only of females with new females developing by parthenogenesis from unfertilized eggs.
- Other species produce two types of eggs, one that develops into females, the other into degenerate males.
  - ⇒ Males produce sperm that fertilize eggs which develop into resistant zygotes that survive desiccation.
  - ⇒ When conditions improve, the zygotes break dormancy and develop into a new female generation that reproduces by parthenogenesis until unfavorable conditions return.
- Rotifers have no regeneration or repair abilities.

Rotifers contain a certain and consistent number of cells as adults. The zygotes undergo a specific number of divisions and the adult contains a fixed number of cells.

B.   Phylum Nematoda

There are more than 80,000 species of roundworms ranging in size ranges from less than 1.0 mm to more than 1 m.

- Bodies are cylindrical with tapered ends.
- Very numerous in both species and individuals.
- Found in fresh water, marine, moist soil, tissues of plants, and tissues and body fluids of animals.
- A complete digestive tract is present and nutrients are transported through the body in the pseudocoelomic fluid.
- A tough transparent cuticle forms the outer body covering.
- Longitudinal muscles are present and provide for whip-like movements.
- Dioecious with females larger than males.
- Sexual reproduction only, with internal fertilization.
- Female may produce 100,000 or more resistant eggs per day.
- Like rotifers, nematodes have a fixed number of cells as adults.

Nematodes fill various roles in the community.

- Free-living forms are important in decomposition and nutrient cycling.
- Plant parasitic forms are important agricultural pests.
- Animal parasitic forms can be hazardous to health (*Trichinella spiralis* in humans via poorly cooked pork).

VII.   Mollusks and annelids are among the major variations on the protostome body plan

The protostome lineage of coelomate animals gave rise to many phyla. In many, the coelom functions as a hydrostatic skeleton.

A.   Phylum Mollusca

There are more than 50,000 species of snails, slugs, oysters, clams, octopuses, and squids.

- Mollusks are mainly marine, though some inhabit fresh water and many snails and slugs are terrestrial.

Mollusks are soft-bodied, but most are protected by a hard calcium carbonate shell.

- Squids and octopuses have reduced, internalized shells or no shell.

The molluscan body consists of three primary parts: muscular *foot* for locomotion, a *visceral mass* containing most of the internal organs, and a *mantle* which is a heavy fold of tissue that surrounds the visceral mass and secretes the shell.

A *radula* is present in many and functions as a rasping tongue to scrap food from surfaces.

Some species are monoecious while most are dioecious.
- Gonads are located in the visceral mass.

Some zoologists believe the mollusks evolved from annelid-like ancestors (although true segmentation is absent) while others believe that mollusks arose earlier in the protostome lineage before segmentation evolved.

1. **Class Polyplacophora**

    The Class Polyplacophora contains the marine species known as chitons.
    - They have an oval shape with the shell divided into eight dorsal plates. (See Campbell, Figure 29.21.)
    - Cling to rocks along the shore at low tide using the foot as a suction cup to grip the rock.
    - Move slowly on its broad, muscular foot when submerged.
    - A radula is used to cut and ingest ("graze") algae.

2. **Class Gastropoda**

    The Class Gastropoda contains the snails and slugs.
    - Largest molluscan class with more than 40,000 species.
    - Mostly marine but many species are freshwater or terrestrial.
    - *Torsion* during embryonic development is a distinctive characteristic:
        ⇒ Uneven growth in the visceral mass causes the visceral mass to rotate 180°, placing the anus above the head in adults.

- Body protected by a shell (absent in slugs and nudibranchs) which may be conical or flattened.
- Many species have distinct heads with eyes at the tips of tentacles.
- Movement results from a rippling motion along the elongated foot.
- Most gastropods are herbivorous, using the radula to graze on plant material; several groups are predatory and possess modified radulae.
- Most aquatic gastropods exchange gases via gills; terrestrial forms have lost the gills and utilize a vascularized lining of the mantle cavity for gas exchange.

3. **Class Bivalvia**

The Class Bivalvia contains the clams, oysters, mussels and scallops.

Possess a shell divided into two halves. (See Campbell, Figure 29.24.)

- The shell halves are hinged at the mid-dorsal line and are drawn together by two adductor muscles to protect the animal.
- Bivalves may extend the foot for motility or anchorage when the shell is open.
- The mantle cavity (between shells) contains gills which function in gas exchange and feeding.
- Most are suspension feeders that trap small food particles in the mucus coating of the gills and then use cilia to move the particles to the mouth.
  ⇒ Water enters the mantle cavity through an incurrent siphon, passes over the gills, and then exits through an excurrent siphon.
  ⇒ No radula or distinct head is present.

Bivalves lead sedentary lives. They use the foot as an anchor in sand or mud. Sessile mussels secrete threads that anchor them to rocks, docks or other hard surfaces.

Scallops can propel themselves along the sea floor by flapping their shells.

4. **Class Cephalopoda**

The Class Cephalopoda contains the squids and octopuses.

Cephalopods are agile carnivores.

- Use beak-like jaws to crush prey.
- The mouth is at the center of several long tentacles.

A mantle covers the visceral mass, but the shell is either reduced and internal (squids) or totally absent (octopuses).

- The chambered nautilus is the only shelled cephalopod alive today.
- Squids swim backwards in open water by drawing water into the mantle cavity, and then firing a jetstream of water through the excurrent siphon which points anteriorly.
  ⇒ Directional changes can be made by pointing the siphon in different directions.
  ⇒ Most squid are less than 75 cm long but the giant squid may reach 17 m and weigh 2 tons.

- Octopuses usually don't swim in open water, but move along the sea floor in search of food.

Cephalopods are the only mollusks with a *closed circulatory system* in which the blood is always contained in vessels.

Cephalopods have well developed nervous systems with complex brains capable of learning. They also have well developed sense organs.

The cephalopod ancestors were probably shelled, carnivorous forms - the *ammonites*.

- These cephalopods were the dominant invertebrate predators in the oceans until they became extinct at the end of the Cretaceous.

B. The Lophophorate Animals

The Lophophorate animals contain three phyla: Phoronida, Bryozoa and Brachiopoda.

- These three phyla are grouped together due to presence of a *lophophore*. (See Campbell, Figure 29.27)

Lophophore = Horseshoe-shaped or circular fold of the body wall bearing ciliated tentacles that surround the mouth at the anterior end of the animal.

- Cilia direct water toward the mouth between the tentacles which trap food particles for these suspension feeders.
- The presence of a lophophore in all three groups suggests a relationship among these phyla.

The three phyla also possess a U-shaped digestive tract (the anus lies outside of the tentacles) and have no distinct head – both adaptations for a sessile existence.

Lophophorates are difficult to assign as protostomes or deuterostomes.

- Their embryonic development more closely resembles deuterostomes; however, in the Phoronida, the blastopore develops into the adult mouth.
- Molecular systematics places the lophophorate phyla closer to the protostomes than the deuterostomes.

1. Phylum Phoronida

This phylum contains about 15 species of tube-dwelling marine worms.

- Length from 1 mm to 50 cm.
- Phoronids live buried in sand in chitinous tubes with the lophophore extended from the tube when feeding.

2. Phylum Bryozoa

This phylum contains the moss animals. There are about 5,000 species which are mostly marine and are widespread.

Bryozoans are small, colonial forms. (See Campbell, Figure 29.27a)

- In most, the colony is enclosed within a hard exoskeleton and the lophophores are extended through pores when feeding.
- Some are important reef builders.

3. Phylum Brachiopoda

The phylum Brachiopoda contains the lamp shells. There are approximately 330 extant species, all marine.

- More than 30,000 fossil species of the Paleozoic and Mesozoic have been identified.

The body of a brachiopod is enclosed by dorsal and ventral shell halves. (See Campbell, Figure 29.27b)

- Attach to the substratum by a stalk.
- Open the shell slightly to allow water to flow through the lophophore.

C. Phylum Annelida

The presence of a true coelom and segmentation are two important evolutionary advances present in the annelids.

- The coelom serves as a hydrostatic skeleton, permits development of complex organ systems, protects internal structures, and permits the internal organs to function separately from the body wall muscles.
- Segmentation also provided for the specialization of different body regions.

There are more than 15,000 species of annelids.

- They have segmented bodies and range in size from less than 1 mm to 3 m.
- There are marine, freshwater and terrestrial (in damp soil) annelids.

Annelids have a coelom partitioned by septa. The digestive tract, longitudinal blood vessels, and nerve cords penetrate the septa and extend the length of the animal. (See Campbell, Figure 29.28.)

The complete digestive system is divided into several parts, each specialized for a specific function in digestion:

pharynx $\longrightarrow$ esophagus $\longrightarrow$ crop $\longrightarrow$ gizzard $\longrightarrow$ intestine

Annelids have a closed circulatory system.

- Hemoglobin is present in blood cells.
- Dorsal and ventral longitudinal vessels are connected by segmental pairs of vessels.
- Five pairs of hearts circle the esophagus.
- Numerous tiny vessels in the skin permit gas exchange is across the body surface.

An excretory system of paired *metanephridia* is found in each segment; each metanephridium has a *nephrostome* (which removes wastes from the coelomic fluid and blood) and exits the body through an exterior pore.

The annelid nervous system is composed of a pair of cerebral ganglia lying above and anterior to the pharynx.

- A nerve ring around the pharynx connects these ganglia to a subpharyngeal ganglion, from which a pair of fused nerve cords run posteriorly.
- Along the ventral nerve cords are fused segmental ganglia.

Annelids are hermaphroditic but cross-fertilize during sexual reproduction.

- Two earthworms exchange sperm and store it temporarily.
- A special organ, the *clitellum*, secretes a mucous cocoon which slides along the worm, picking up its eggs and then the stored sperm.
- The cocoon slips off the worm into the soil and protects the embryos while they develop.
- Asexual reproduction occurs in some species by fragmentation followed by regeneration.

Movement involves coordinating longitudinal and circular muscles in each segment with the fluid-filled coelom functioning as a hydrostatic skeleton.

- Circular muscle contraction makes each segment thinner and longer; longitudinal muscle contraction makes the segment shorter and thicker.
- Waves of alternating contractions pass down the body.
- Most aquatic annelids are bottom-dwellers that burrow, although some swim in pursuit of food.

1. Class Oligochaeta

The Class Oligochaeta contains earthworms and a variety of aquatic species.

Earthworms ingest soil, extract nutrients in the digestive system and deposit undigested material (mixed with mucus from the digestive tract) as *casts* through the anus.

- Important to farmers as they till the soil and castings improve soil texture.
- Charles Darwin estimated that 1 acre of British farmland had about 50,000 earthworms that produced 18 tons of castings per year.

2. Class Polychaeta

The Class Polychaeta contains mostly marine species. (See Campbell, Figure 29.29a and b.)

A few drift and swim in the plankton, some crawl along the sea floor, and many live in tubes they construct by mixing sand and shell bits with mucus.

- Tube-dwellers include the fanworms that feed by trapping suspended food particles in their feathery filters which are extended from the tubes.

Each segment has a pair of *parapodia* which are highly vascularized paddle-like structures that function in gas exchange and locomotion.

- Traction for locomotion is proved by several chitinous *setae* present on each parapodium.

3. Class Hirudinea

The Class Hirudinea contains the leeches.

- A majority of species are freshwater but some are terrestrial in moist vegetation.
- Many are carnivorous and feed on small invertebrates while some attach temporarily to animals to feed on blood.
- Size ranges from 1 – 30 cm in length.

Some blood-feeding forms have a pair of blade-like jaws that slit the host's skin while others secrete enzymes that digest a hole in the skin.

- An anesthetic is secreted by the leech to prevent detection of the incision by the host.
- Leeches also secrete *hirudin* which prevents blood coagulation during feeding.
- Leeches may ingest up to 10 times their weight in blood at a single meal and may not feed again for several months.
- Leeches are currently used to treat bruised tissues and for stimulating circulation of blood to fingers and toes reattached after being severed in accidents.

VIII. The protostome phylum Arthropoda is the most successful group of animals ever to live

The Phylum Arthropoda is the largest phylum of animals with approximately 1,000,000 described species.

- Arthropods are the most successful phylum based on species diversity, distribution, and numbers of individuals.

A. General Characteristics of Arthropods

The success and great diversity of arthropods is related to their segmentation, jointed appendages, and hard exoskeleton.

The segmentation in this group is much more advanced than that found in annelids.

- In the arthropods, different segments of the body and their associated appendages have become specialized to perform specialized functions.
- Jointed appendages are modified for walking, feeding, sensory reception, copulation and defense.

The arthropod body is completely covered by the *cuticle*, an exoskeleton (external skeleton) constructed of layers of protein and chitin.

- The cuticle is thin and flexible in some locations (joints) and thick and hard in others.
- The exoskeleton provides protection and points of attachment for muscles that move the appendages.
- The exoskeleton is also relatively impermeable to water.
- The old exoskeleton must be shed for an arthropod to grow (molting) and a new one secreted.

Arthropods show extensive cephalization with many sensory structures clustered at the anterior end. Well developed sense organs including eyes, olfactory receptors, and tactile receptors are present.

An open circulatory system containing *hemolymph* is present.

- Hemolymph leaves the heart through short arteries and passes into the sinuses (open spaces) which surround the tissues and organs.
- The hemolymph reenters the heart through pores equipped with valves.
- The blood sinuses comprise the *hemocoel*. Though the hemocoel is the main body cavity, it is *not* part of the coelom.
- The true coelom is reduced in adult arthropods.

Gas exchange structures are varied and include:

- feathery *gills* in aquatic species.
- *tracheal systems* in insects.
- *book-lungs* in other terrestrial forms (e.g. spiders).

B.   Arthropod Phylogeny and Classification

Arthropods are segmented protostomes which probably evolved from annelids or a segmented protostome common ancestor.

- Early arthropods may have resembled onychophorans which have unjointed appendages.
   - ⇒ However, many fossils of jointed-legged animals resembling segmented worms support the evolutionary link between the Annelida and Arthropoda.
   - ⇒ Such comparisons also indicate that annelids and arthropods are *not* closely related.
- Parapodia may have been forerunners of appendages.
- Some systematists suggest that comparisons of ribosomal RNA and other macromolecules indicate that onychophorans are arthropods and not transitional forms.
- This evidence presents an alternative hypothesis that segmentation evolved independently in annelids and arthropods.
   - ⇒ Thus, the most recent common ancestor of these two phyla would have been an unsegmented protostome.

Although the origin of arthropods is unclear, most zoologists agree that four main evolutionary lines can be identified in the arthropods. Their divergence is represented by the subphyla: Trilobitomorpha (extinct trilobites), Cheliceriformes, Uniramia and Crustacea.

C. **Trilobites (Subphylum Trilobitomorpha)**

Early arthropods called *trilobites* were very numerous but became extinct approximately 250 million years ago. (See Campbell, Figure 29.32)

- Trilobites had extensive segmentation, but little appendage specialization.
- As evolution continued, the segments tended to fuse and appendages became specialized for a variety of functions.

D. **Spiders and Other Chelicerates (Subphylum Cheliceriformes)**

Other early arthropods included chelicerates such as the *eurypterids* (sea scorpions) which were predaceous and up to 3 m in length.

- The chelicerate body is divided into an anterior *cephalothorax* and a posterior *abdomen*.
- Their appendages were more specialized than those of trilobites, with the most anterior ones being either pincers or *fangs*.
- Chelicerates are named for their feeding appendages, the *chelicerae*.
  ⇒ Only one marine chelicerate remains, the horse-shoe crab.
  ⇒ Other chelicerates include terrestrial spiders, scorpions, ticks and mites of the Class Arachnida which lack antennae and have simple eyes.

Members of the Class Arachnida possess a cephalothorax with six pairs of appendages: chelicerae, *pedipalps* (used in sensing and feeding), and four pairs of walking legs. (See Campbell, Figure 29.34)

In spiders,

- Fang-like chelicerae, equipped with poison glands, are used to attack prey.
- Chelicerae and pedipalps masticate the prey while digestive juices are added to the tissues.
  ⇒ This softens the food and the spider sucks up the liquid.
- Gas exchange is by book-lungs (stacked plates in an internal chamber).
  ⇒ The structure of the book-lung provides an extensive surface area for exchange.

Spiders weave silken webs to capture prey.

- The proteinaceous silk is produced as a liquid by abdominal glands and spun into fibers by *spinnerets*. The fibers harden on contact with air.
- Web production is apparently an inherited complex behavior.
- Silk fibers are also used for escape, egg covers, and wrapped around food presented to females during courtship.

E.   **Comparison of Chelicerates to Uniramians and Crustaceans**

Another major line of arthropod evolution gave rise to the subphyla Uniramia and Crustacea. These groups are characterized by:

- Jaw-like mandibles instead of claw-like chelicerae.
- One or two pairs of sensory antennae and a pair of compound eyes.

The subphylum Uniramia includes millipedes, centipedes and insects:

- All possess one pair of antennae and unbranched (uniramous) appendages.
- Are believed to have evolved on land, aided by the presence of the exoskeleton first evolved in marine forms.

The subphylum Crustacea includes the crustaceans, which:

- Have two pairs of antennae and branched (biramous) appendages.
- Are primarily aquatic and believed to have evolved in the ocean.

Chelicerates and uniramians moved onto land during the Devonian period.

- They were able to move into terrestrial habitats partly due to the presence of the exoskeleton.
  - ⇒ The main functions of the exoskeleton in the seas, where it evolved, were protection and muscle attachment.
  - ⇒ In terrestrial habitats, the exoskeleton provided support and prevented desiccation.
- Fossilized burrows of millipede-like arthropods are the oldest (about 450 million years) evidence of terrestrial animals.
  - ⇒ The presence of a protective cuticle reduces water loss.
  - ⇒ A firm exoskeleton also provided support without the buoyancy of water.

F.   **Insects and Other Uniramians**

The subphylum Uniramia includes three classes: Diplopoda, Chilopoda, and Insecta.

The class Diplopoda includes the millipedes.

- Worm-like with a large number of walking legs (2 pair per segment).
- Eat decaying leaves and other plant matter.
- Probably among the earliest land animals.

The class Chilopoda includes the centipedes.

- They are carnivorous.
- One pair of antennae and three pairs of appendages modified as mouthparts (including mandibles) are located on the head.
- Each trunk segment has one pair of walking legs.
- Poison claws on the most anterior trunk segment are used to paralyze prey and for defense.

The class Insecta has a greater species diversity than all other forms of life combined.

- There are about 26 orders of insects. (See Campbell, Table 29.1)
- They inhabit terrestrial and freshwater environments, but only a few marine forms exist.

Entomology = The study of insects.

The oldest insect fossils are from the Devonian period (about 400 million years ago), and an increase in insect diversity can be attributed to:

- The evolution of flight during the Carboniferous and Permian.
- The evolution of specialized mouth parts for feeding on gymnosperms and other Carboniferous plants.
  - ⇒ The fossil record holds examples of a diverse array of specialized mouth parts.

A second major radiation of insects which occurred during the Cretaceous period was once thought to have paralleled radiation of flowering plants.

- Current research indicates the major diversification of insects preceded angiosperm radiation during the Cretaceous period (65 million years ago).
- If this is true, insect diversity played a major role in angiosperm radiation, the reverse of the original hypothesis.

Flight is the key to the success of insects, enabling them to escape predators, find food and mates, and disperse more easily than non-flying forms.

- One or two pairs of wings emerge from the dorsal side of the thorax in most species.
- Wings are extensions of the cuticle and not modified appendages.
- Wings may have first evolved to help absorb heat, then developed further for flight.
- Other views suggest wings may have initially served for gliding, as gills in aquatic forms, or even as structures for swimming.

Dragonflies were among the first to fly and have two coordinated pairs of wings. Modifications are found in groups which evolved later.

- Bees and wasps hook their wings together (act as one pair).
- Butterflies have overlapping anterior and posterior wings.
- Beetles have anterior wings modified to cover and protect the posterior (flying) wings.

There are three main regions to the anatomy of an insect: *head, thorax* and *abdomen.* (See Campbell, Figure 29.38)

- Head segments are fused while segmentation is apparent on the thorax and abdomen.
- The head has: one pair of antennae, one pair of compound eyes, and several pairs of appendages modified as feeding mouthparts (for chewing or lapping, piercing, sucking – depending on the insect).
- The thorax has three pairs of walking legs and the wings.

Insects have several complex internal organ systems.

- Complete digestive system with specialized regions.
- Open circulatory system with hemolymph.
- Excretory organs are the *Malpighian tubules* which are outpocketings of the gut.
- Gas exchange is by a *tracheal system* which opens to the outside via spiracles that can open or close to regulate air and limit water loss.
- Nervous system is composed of a pair of ventral nerve cords (with several segmental ganglia) which meet in the head where the anterior ganglia are fused into a dorsal brain close to the sense organs.
  ⇒ Insects show complex behavior which is apparently inherited (e.g. social behavior of bees and ants).

Many insects undergo metamorphosis during their development.

Incomplete metamorphosis = A type of development during which young resemble adults but are smaller and have different body proportions.

- For example, in grasshoppers a series of molts occur with each stage looking more like an adult until full size is reached.

Complete metamorphosis = A type of development characterized by larval stages (e.g. maggot, grub, caterpillar) which are very different in appearance from adults.

- Larva eat and grow before becoming adults.
- Adults find mates and reproduce with the females laying eggs on the appropriate food source for the larval forms.

Insects are dioecious and usually reproduce sexually with internal fertilization.

- In most, sperm are deposited directly into the female's vagina during copulation. Some males produce spermatophores which are picked up by the female.
- Inside the female, sperm are stored in the *spermatheca*.
- Most insects produce eggs although some flies are viviparous.
- Many insects mate only once in a lifetime with stored sperm capable of fertilizing many batches of eggs.

Insects impact terrestrial organisms in a number of ways. They:

- Compete for food.
- Serve as disease vectors.
- Pollinate many crops and orchards.

G. **Crustaceans**

There are more than 40,000 species of crustaceans in marine and fresh waters.

The crustaceans have extensive specialization of their appendages.
- Two pairs of antennae, three or more pairs of mouthparts including mandibles, walking legs on the thorax, appendages are present on the abdomen.
- Lost appendages can be regenerated.

Characteristics of their physiology include:
- Gas exchange may take place across thin areas of the cuticle (small forms) or by gills (large forms).
- An open circulatory system is present with hemolymph.
- Nitrogenous wastes are excreted by diffusion across thin areas of the cuticle.
- Salt balance of the hemolymph is regulated by a pair of specialized antennal or maxillary glands.

Most are dioecious and some males (e.g. lobsters) have a specialized pair of appendages to transfer sperm to the female's reproductive pore during copulation.
- Most aquatic crustaceans have at least one swimming larval stage.

The *decapods* are relatively large crustaceans that have a *carapace* (calcium carbonate hardened exoskeleton over the cephalothorax). Examples include:
- Freshwater crayfish.
- Marine lobsters, crabs and shrimp.
- Tropical land crabs.

The *isopods* are mostly small marine crustaceans but include terrestrial sow bugs and pill bugs.
- Terrestrial forms live in moist soil and damp areas.

Copepods are numerous small marine and freshwater planktonic crustaceans.
- The larvae of larger crustaceans may also be planktonic.

Barnacles are sessile crustaceans with parts of their cuticle hardened into shells by calcium carbonate.
- Barnacles feed by directing suspended particles toward the mouth with specialized appendages.

IX.   The deuterostome lineage includes echinodermats and chordates

The deuterostomes, while a very diverse group, share characteristics which indicate their association: radial cleavage, enterocoelous coelom formation, and the blastopore forms the anus.

A.   Phylum Echinodermata

Most echinoderms are sessile or sedentary marine forms with radial symmetry as adults. (See Campbell, Figure 29.41)

- Internal and external parts radiate from the center, often as five spokes.
- A thin skin covers a hard calcareous plate-like exoskeleton.
- Most have bumps and spines which serve various functions.

A unique feature of Echinoderms is the *water vascular system*.

Water vascular system = A network of hydraulic canals which branch into extensions called *tube feet* that function for locomotion, feeding and gas exchange.

Echinoderms are dioecious with sexual reproduction and external fertilization.

- Bilaterally symmetrical larvae metamorphose into radial adults.
- Early embryonic development exhibits the characteristics of deuterostomes.

There are about 7,000 species of echinoderms, all of which are marine. The six recognized classes are:

1.   Class Asteroidea

This class includes the sea stars which have five or more arms extending from a central disc. (See Campbell, Figure 29.41)

- *Tube feet* on the undersurface of the arms are extended by fluid forced into them by contraction of their *ampulla*.
- Suction cups at the end of each tube foot attach to the substratum and muscles in the tube foot wall contract and shorten the foot.

Coordination of extension, attaching, contraction and release allow slow movement and attachment to prey.

- Prey are obtained by attaching tube feet to the shells of clams and oysters; the arms of the sea star wrap around the prey and hold tightly using the tube feet.
- The muscles of the mollusk fatigue and the shell is pulled open.
- The sea stars evert their stomachs between the shell halves and secrete digestive juices onto the soft tissues of the mollusk.

Sea stars have a strong ability to regenerate, although a single arm cannot regenerate an entire body. Fishermen chopping up sea stars may actually increase their numbers.

2. Class Ophiuroidea

This class contains the brittle stars which differ from sea stars in that they have:

- Smaller central discs than sea stars.
- Longer, more flexible arms than sea stars.
- No suckers on their tube feet.
- Locomotion is by serpentine lashing of flexible arms.
- Varying feeding mechanisms.

3. Class Echinoidea

The Class Echinoidea contains the sea urchins and sand dollars.

- Sea urchins are spherical in shape while sand dollars are flattened in the oral-aboral axis.

Echinoideans lack arms but have:

- Five rows of tube feet present that provide slow movement.
- Spines are pivoted to aid in locomotion.

Echinoideans have a complex jaw-like structure present around the mouth which is used for feeding on seaweeds and other food.

4. Class Crinoidea

The Class Crinoidea contains the sea lilies.

- Most sea lilies are sessile, living attached to substratum by stalks.
- Motile sea lilies use their arms for a crawling form of locomotion as well as for feeding.
- Arms circle the mouth (which points upward) and are used in suspension feeding.

Crinoids have exhibited a very conservative evolution.

- Extant forms are very similar to fossilized forms from Ordovician period (500 million years ago).

5. Class Holothuroidea

The Class Holothuroidea contains the sea cucumbers which have little resemblance to other echinoderms.

- The lack spines.
- The hard endoskeleton is reduced.
- The body is elongated in the oral-aboral axis.

Species in the Holothuroidea do possess five rows of tube feet, a part of the unique water vascular system.

- Some tube feet around the mouth have developed into feeding tentacles.

6. **Class Concentricycloidea**

   This class contains the sea daisies which are small (less than 1 cm), disc-shaped marine animals.

   - Live in deep water.
   - Do not possess arms.
   - Tube feet are located around the disc margin.
   - Possess a rudimentary digestive system or an absorptive velum on the oral surface.
   - Water vascular system consists of two concentric ring canals.

B. **Phylum Chordata**

   The Chordata diverged from a common deuterostome ancestor with echinoderms at least 500 million years ago.
   - The two phyla are grouped together due to similarities in early embryonic development.
   - This phylum contains three subphyla: Urochordata, Cephalochordata and Vertebrata.

X. **The Cambrian explosion produced all the major animal body plans**

The animal kingdom probably originated from colonial protists related to choanoflagellates. (See Campbell, Figure 29.43)

The diversification that produced many phyla occurred in a relatively short time on the geological scale. This evolutionary episode is call the *Cambrian explosion*.

The Cambrian explosion encompassed a 5 to 10 million year time span at the beginning of the Cambrian period (545 million years ago).
- All the major animal body plans seen today evolved during this time.
- New taxa appeared later but were variations on the basic plans already evolved. For example, mammals evolved about 220 million years ago but are only a variation of the chordate body plan which evolved during the Cambrian explosion.

A much less diverse fauna preceded the Cambrian explosion.
- This Precambrian fauna dated back to the Ediacaran period (700 million years ago).
  - ⇒ This period is named for the Ediacara Hills of Australia where Precambrian animal fossils were first discovered.
  - ⇒ Fossils similar in age to these have since been discovered on other continents.
- Most Ediacaran fossils appear to be cnidarians although bilaterial animals are also indicated by fossilized burrows probably left by worms.

The diversity of Cambrian animals is represented in three fossil beds:
- The Burgess Shale in British Columbia is the best known.
- A fossil bed in Greenland and one in the Yunnan region of China predate the Burgess Shale by 10 million years.
  - ⇒ These beds indicate animal diversification occurred within 5 million years of the start of the Cambrian.

Two contrasting interpretations of Burgess Shale fossils have been proposed:
- The Cambrian explosion resulted in a large number of phyla which included the current phyla, many of which are now extinct.
  - ⇒ During the mass extinction at the end of the Cambrian, only the base stock of 35 or so extant phyla survived.
- The diversity of the Cambrian fossils represents ancient variations within the taxonomic boundaries of extant phyla.
  - ⇒ As these fossils undergo continued study, many are classified into extant phyla. Thus, the number of exclusively Cambrian fossils is decreasing.

Several hypotheses about external factors have been proposed as explanations for the Cambrian explosion and the lack of subsequent major diversification.
- The Cambrian explosion was an adaptive radiation resulting from the origin of the first animals.
  - ⇒ These early animals diversified as they adapted to the various, previously unoccupied, ecological niches.
- Predator-prey relationships emerged and triggered diverse evolutionary adaptations.
  - ⇒ Various kinds of shells and different forms of locomotion evolved as defense mechanisms against predation.
  - ⇒ Predators also evolved new mechanisms to capture prey.
- Major environmental change provided an opportunity for diversification during the Cambrian explosion.
  - ⇒ The accumulation of atmospheric oxygen may have finally reached a concentration to support the more active metabolism needed for feeding and other activities by mobile animals.

Other hypotheses for the Cambrian explosion have emphasized internal changes in the organisms.
- The origin of mesoderm may have stimulated diversification of the body plan.
  - ⇒ This third tissue layer permits development of more complex anatomical structure.
- Variation in genes that control pattern formation during animal development may have played a role in diversification.
  - ⇒ Some of the genes that determine features such as segmentation and placement of appendages and other structures are common to diverse animal phyla.
  - ⇒ Variation in expression of these genes during development results in morphological differences that distinguish the phyla.

⇒ This same kind of variation in expression may have resulted in the relatively rapid origin of diverse animal types during the Cambrian explosion.

⇒ The phyla, once developed, may have become locked into developmental patterns that permitted subtle variation to allow speciation and the origin of lower taxa, but prevented large scale morphological evolution resulting in new phyla.

The hypotheses presented for external and internal factors are not mutually exclusive. A combination of factors may have combined to produce the Cambrian explosion.

REFERENCES

Barnes, Robert D. *Invertebrate Zoology*. 5th ed. Philadelphia, Pennsylvania: Saunders College/Holt, Rinehart and Winston, 1987.

Campbell, N. *Biology*. 4th ed. Menlo Park, California: Benjamin/Cummings, 1996.

Hickman, Cleveland P., Jr., Larry S. Roberts, and Allan Larson. *Integrated Principles of Zoology*. 9th ed. St. Louis, Missouri: Mosby Publishing, 1993.

CHAPTER 30

THE VERTEBRATE GENEALOGY

OUTLINE

I. Vertebrates belong to the phylum Chordata
    A. Chordate Characteristics

II. Invertebrate chordates provide clues to the origin of vertebrates
    A. Subphylum Urochordata
    B. Subphylum Cephalochordata
    C. Relationship of Invertebrate Chordates to the Vertebrates

III. The evolution of vertebrate characteristics is associated with increased size and activity

IV. Vertebrate diversity and phylogeny: *an overview*

V. Agnathans are jawless vertebrates

VI. Placoderms were armored fishes with jaws and paired fins

VII. Sharks and their relatives have adaptations for powerful swimming

VIII. Bony fishes are the most abundant and diverse vertebrates

IX. Amphibians are the oldest class of tetrapods
    A. Early Amphibians
    B. Modern Amphibians

X. Evolution of the amniotic egg expanded the success of vertebrates on land

XI. A reptilian heritage is evident in all amniotes
    A. Reptilian Characteristics
    B. The Age of Reptiles
    C. Reptiles of Today

XII. Birds began as flying reptiles
    A. Characteristics of Birds
    B. The Origin of Birds
    C. Modern Birds

XIII. Mammals diversified extensively in the wake of the Cretaceous extinctions
    A. Mammalian Characteristics
    B. The Evolution of Mammals
    C. Monotremes
    D. Marsupials
    E. Placental Mammals

XIV. Primate evolution provides a context for understanding human origins
    A. Evolutionary Trends in Primates
    B. Modern Primates

XV. Humanity is one very young twig on the vertebrate tree
    A. Some Common Misconceptions
    B. Early Anthropoids
    C. Australopithecines: The First Humans
    D. *Homo habilis*
    E. *Homo erectus* and Descendants
    F. The Emergence of *Homo sapiens*: Out of Africa...But When? *Science as a Process*
    G. Cultural Evolution: A New Force in the History of Life

OBJECTIVES

After reading this chapter and attending lecture, the student should be able to:
1. Describe the four unique characteristics of chordates.
2. Distinguish between the three subphyla of the Phylum Chordata and give examples of each.
3. Describe the specialized characteristics found in the Subphylum Vertebrata and explain how each is beneficial to survival.
4. Compare and contrast members of the Classes Agnatha, Placadermi and Chondrichthyes.
5. Explain how members of the Class Osteichthyes have becomes so diversified.
6. Summarize the evidence supporting the fact that amphibians evolved from crossopterygians.
7. Distinguish between the three orders of living amphibians.
8. List the distinguishing characteristics of members of the Class Reptilia and explain any special adaptations to the terrestrial environment.
9. Explain how environmental changes during the Cretaceous Period may have affected the dinosaurs.
10. List the distinguishing characteristics of members of the Class Aves and explain any special adaptations for flight.
11. Summarize the evidence supporting the fact that birds evolved from reptilian ancestors.
12. Explain why mammals underwent an adaptive radiation during the Cenozoic.
13. Distinguish between monotreme, marsupial and placental mammals.

14. Explain how convergent evolution produced marsupial and placental ecological counterparts on different continents.
15. Compare and contrast the four main evolutionary lines of placental mammals.
16. Describe the characteristics found in early primates which indicate an arboreal existence.
17. Appraise the significance of the three most prominent misconceptions about human evolution.
18. Diagram an evolutionary tree for humans.
19. Explain how humans have influenced the extinction rates of other organisms.

KEY TERMS

Chordata
notochord
nerve cord
pharyngeal slits
postanal tail
Urochordata
tunicin
atrium
atriopore
Cephalochordata
filter feeders
somites
paedogenesis
cephalization
axial skeleton
cranium
vertebral column
neural crest
amniotic egg
amniote
Agnatha
ostracoderms
hagfishes
lampreys
Placodermi

acanthodians
Chondrichthyes
cartilage
pectoral fin
pelvic fin
spiral valve
lateral line system
claspers
oviparous
ovoviviparous
viviparous
cloaca
dorsoventrally flattened
Osteichthyes
operculum
fusiform body shape
Actinopterygii
Sarcopterygii
rhipidistians
coelocanth
Amphibia
Eusthenopteron
Urodela
Anura
Apoda

extraembryonic
   membranes
amnion
Reptilia
keratin
ectothermic
synapsids
sauropods
therapsid
anapsids
diapsids
endothermy
Cretaceous Crisis
Chelonia
Squamata
Crocodilia
Aves
Archaeopteryx
   lithographica
ratites
carinates
Mammalia
placenta
monotremes
marsupials

marsupium
convergent evolution
plate tectonics
continental drift
Pangaea
adaptive radiation
placental mammals
Purgatorius unio
brachiate
prosimian
arboreal
paleoanthropology
mosaic evolution
Aegyptopithecus
Australopithecus africanus
"Lucy"
Australopithecus afarensis
Australopithecus ramidus
Homo habilis
Homo erectus
Homo sapiens
Neanderthal
Multiregional model
Monogenesis model

LECTURE NOTES

I. **Vertebrates belong to the phylum Chordata**

The phylum Chordata includes three subphyla: Urochordata, Cephalochordata and Vertebrata.

A. **Chordate Characteristics**

Chordates are deuterostomes with four unique characteristics which appear at some time during the animal's life. These characteristics are: the notochord; a dorsal, hollow nerve cord; pharyngeal slits; and a muscular postanal tail.

1. **Notochord**

Notochord = A longitudinal, flexible rod located between the gut and nerve cord.
- Present in all chordate embryos.
- Composed of large, fluid-filled cells encased in a stiff, fibrous tissue.
- Extends through most of the length of the animal as a simple skeleton.

In some invertebrate chordates and primitive vertebrates it persists to support the adult.

In most vertebrates, a more complex, jointed skeleton develops and the notochord is retained in adults as the gelatinous material of the discs between the vertebrae.

2. **Dorsal, Hollow Nerve Cord**

Develops in the embryo from a plate of dorsal ectoderm that rolls into a tube located dorsal to the notochord.
- Unique to chordates.
  - ⇒ Other animal phyla have solid, usually ventral nerve cords.
- The brain and spinal cord (central nervous system) develops from this nerve cord.

3. **Pharyngeal Slits**

Chordates have a complete digestive system (mouth and anus). The pharynx is the region just posterior to the mouth and it opens to the outside through several pairs of slits.
- The presence of these pharyngeal slits permits water entering the mouth to exit without passing through the entire digestive system.
- These pharyngeal gill slits function for suspension-feeding in invertebrate chordates.
- They have become modified for gas exchange and other functions during the evolution of vertebrates.

4. **Muscular Postanal Tail**

   A tail extending beyond the anus, it is found in most chordates and contains skeletal elements and muscles.

   - Provides much of the propulsive force in many aquatic species.

   The digestive tract in most nonchordates extends nearly the whole length of the body.

II. Invertebrate chordates provide clues to the origin of vertebrates

A. Subphylum Urochordata

Species in the subphylum Urochordata are called tunicates.

- Entire animal is cloaked in a tunic made of a celluloselike carbohydrate.
- Most are sessile marine animals which adhere to rocks, docks and boats.
- Some species are planktonic, while others are colonial.

The tunicates are filter feeders. (See Campbell, Figure 30.2b)

- Seawater enters through an incurrent siphon, passes through the slits of the pharynx into a chamber called the *atrium*, and exits via an excurrent siphon, the *atriopore*.
- Food filtered from the water by a mucous net of the pharynx is moved by cilia into the intestine.
- The anus empties into the excurrent siphon.
- When disturbed, tunicates eject a jet of water through the excurrent siphon, so they are commonly called sea squirts.

Adult tunicates bear little resemblance to other chordates.

- They lack a notochord, a nerve cord and tail.
- They possess only pharyngeal slits.

Larval tunicates are free swimmers and possess all four chordate characteristics.

- Larva attach by the head on a surface and undergo metamorphosis to adult form.

B. Subphylum Cephalochordata

Animals in the subphylum Cephalochordata are known as *lancelets* due to their bladelike shape. (See Campbell, Figure 30.3) Chordate characteristics are prominent and persist in the adult. These include:

- Notochord
- Dorsal nerve cord
- Numerous gill slits
- Postanal tail

Cephalochordates are marine filter feeders.

- They burrow tail first into the sand with only the anterior exposed.
- Water is drawn into mouth by ciliary action and food is trapped on a mucous net secreted across the pharyngeal slits.
- Water exits through the slits and trapped food passes down the digestive tube.

Cephalochordates are feeble swimmers with fishlike motions.

- Frequently move to new locations
- Muscle segments are serially arranged in chevronlike rows, and coordinated contraction flexes the notochord from side to side in a sinusoidal pattern.
- Muscle segments develop from blocks of mesoderm called *somites* that are arranged along each side of the notochord in the embryo.
- The serial musculature is evidence of chordate segmentation which developed independently of that in annelids and arthropods.

C. Relationship of Invertebrate Chordates to the Vertebrates

Vertebrates first appear in the fossil records in Cambrian rocks.

- Fossilized invertebrates (about 550 million years old) resembling cephalochordates are found in Burgess Shale of British Columbia.
- This is about 50 million years older than the oldest known vertebrates.

Most zoologists feel the vertebrate ancestors possessed all four chordate characteristics and were suspension-feeders.

- They may have resembled lancelets but were less specialized.
- Information provided by molecular systematics supports the idea that cephalochordates are the closest relatives of vertebrates.

Cephalochordates and vertebrates may have evolved from a sessile ancestor by *paedogenesis*.

Paedogenesis = Precocious attainment of sexual maturity in a larva.

- Cephalochordates more closely resemble urochordate larvae than adult urochordates.
- Changes in the developmental control genes can alter the timing of developmental events (i.e. gonad maturation).
- Zoologists postulate that some early urochordatelike larval forms became sexually mature and reproduced before undergoing metamorphosis.
- If reproducing larvae were successful, natural selection may have reinforced the absence of metamorphosis and a vertebrae life cycle may have evolved.

The divergence of cephalochordates from vertebrates occurred about 500 million years ago.

- The differences between the two groups can be viewed as vertebrate adaptations to larger size and a more mobile life style.

III. The evolution of vertebrate characteristics is associated with increased size and activity

Vertebrates have retained the chordate features while adding other specializations. These synapomorphies distinguish the vertebrates from urochordates and cephalochordates.

The unique vertebrate structures probably evolved in association with increased size and more active foraging for food. The unique vertebrate adaptations include:

- Vertebrates show a much greater degree of *cephalization* than cephalochordates.
  ⇒ The brain and sense organs are located at the anterior end which is the portion of the body which is in contact with the most environmental stimuli.
- A skeleton including a *cranium* and *vertebral column* is the main axis of the body, replacing the notochord as the basic skeleton.
  ⇒ The cranium protects the brain.
  ⇒ The vertebral column provides support and a strong, jointed anchor that provides leverage to the segmental swimming muscles.
  ⇒ The axial skeleton made larger size and stronger, faster movement possible.
  ⇒ Most vertebrates also have ribs (anchor muscles and protect internal organs) and an appendicular skeleton supporting two pairs of appendages.
- The vertebrate skeleton may be composed of bone, cartilage, or a combination of the two.
  ⇒ A majority of the skeleton is a non-living matrix which is secreted and maintained by living skeletal cells.
  ⇒ The living vertebrate endoskeleton can grow with the animal unlike the exoskeleton of arthropods.
- The *neural crest*, a group of embryonic cells found only in vertebrates, contributes to the formation of certain skeletal components and many other structures distinguishing vertebrates from other chordates.
  ⇒ The dorsal, hollow nerve cord develops from an infolding of the edges of an ectodermal plate on the surface of the embryo.
  ⇒ The neural crest forms near the dorsal margins of the tube resulting from this infolding.
  ⇒ Cells from the neural crest then migrate to various specific areas of the embryo and help form a variety of structures including some of the bones and cartilage of the head.

Some anatomical adaptations also support the greater metabolic demands of increased activity.
- The generation of ATP by cellular respiration, to replace the energy used by vertebrates in obtaining food or escaping predators, consumes oxygen.
  ⇒ The respiratory and circulatory systems of vertebrates show adaptations which support the mitochondria of muscles and other active tissues.
- Vertebrates have a closed circulatory system composed of a ventral chambered (2 to 4) heart, arteries, capillaries and veins.
  ⇒ The heart pumps the blood through the system.
  ⇒ The blood becomes oxygenated as it passes through the capillaries of the gills or lungs.

- The more active the life style, the larger the amounts of organic molecules necessary to produce energy.
    - ⇒ Vertebrates have several adaptations for feeding, digestion and nutrient absorption.
    - ⇒ For example, muscles in the walls of the digestive tract move food from organ to organ along the tract.

IV.   Vertebrate diversity and phylogeny: *an overview*

Seven extant classes comprise the subphylum Vertebrata: Agnatha, Chondrichthyes, Osteichthyes, Amphibia, Reptilia, Aves, and Mammalia. The first three of these classes are fishes, while the remaining four are the tetrapod vertebrates.

<u>Tetrapod</u> = An animal possessing two pairs of limbs that support it on land.

In addition to being tetrapods, the reptiles, birds and mammals have other adaptations for a terrestrial lifestyle which are not found in the amphibians.

- The *amniotic egg* (a shelled, water resistant egg) allows completion of the life cycle on land.

- Most mammals do not lay eggs but retain other features of the amniotic condition, consequently, they are also considered *amniotes* along with the birds and reptiles.

It is recommended that figure 30.6 be reviewed to establish the relationships among the classes of the subphylum Vertebrata.

V.   Agnathans are jawless vertebrates

The class Agnatha contains the jawless fishes and the extinct, heavily armored ostracoderms.

The oldest fossilized vertebrates are classified as agnathans. Some are found in Cambrian strata but most date to the Ordovician and Silurian (400 to 500 million years ago).

- Early agnathans were small, less than 50 cm in length.
- They were jawless with oval or slitlike mouths, most lacked paired fins and were bottom-dwellers.
- Some were active and had paired fins.
- Were probably bottom- or suspension-feeders that trapped organic debris in their gill slits.
- Ostracoderms and most other agnathans declined and disappeared during the Devonian.

Extant forms include about 60 species of lampreys and hagfishes which lack paired appendages and external armor.

Lampreys are eel-shaped and feed by clamping their round mouths onto live fish.
- Once attached, they use a rasping tongue to penetrate the skin and feed on the prey's blood.
- Sea lampreys spend their larval development in freshwater streams and migrate to the sea or lakes as they mature.
- Larva are suspension-feeders that resemble lancelets.
- Some lamprey species feed only as larvae. Once they mature and reproduce, they die within a few days.

Hagfishes superficially resemble lampreys.
- They are scavengers without rasping mouthparts.
- Some species will feed on sick or dead fish while others feed on marine worms.
- Lack a larval stage and are entirely marine.

VI. Placoderms were armored fishes with jaws and paired fins

The class Placodermi is a group of armored fishes which replaced agnathans during the late Silurian and early Devonian.
- Most were less than 1 m in length, but some were up to 10 m long.
- Differed from agnathans in that they possessed paired fins and hinged jaws.
- Paired fins enhanced swimming ability and hinged jaws allowed more varied feeding habits including predation.

Hinged jaws evolved as modifications of the skeletal rods which previously supported the anterior pharyngeal (gill) slits.
- Remaining gill slits retained function as major gas exchange sites.
- Hinged jaws of vertebrates work in an up and down direction - those in arthropods work from side to side.

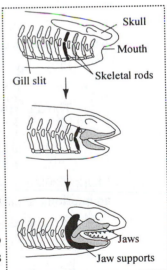

Placoderms and another group of jawed fishes, the acanthodians, radiated during the Devonian period (the Age of Fishes) and many new forms evolved in fresh and salt waters.
- Placoderms and acanthodians disappeared by the start of the Carboniferous period (350 million years ago).
- Ancestors of the placoderms and acanthodians also gave rise to early sharks (class Chondrichthyes) and bony fishes (class Osteichthyes).

VII. Sharks and their relatives have adaptations for powerful swimming

The class Chondrichthyes contains about 750 extant species of cartilaginous fishes (e.g. sharks, skates, rays).

Species in the class Chondrichthyes have flexible skeletons composed of cartilage, well-developed jaws and paired fins.
- The ancestors of members of this class had bony skeletons.
- The characteristic cartilaginous skeleton is thus a derived characteristic, having evolved secondarily.
- The developmental sequence in cartilaginous fishes differs from other vertebrates in that the initial (first) cartilaginous skeleton does not become ossified.
- The cartilaginous skeleton is more elastic and lighter than a bone skeleton.

Sharks have streamlined bodies and are swift swimmers.
- The tail provides propulsion.
- The dorsal fins serve as stabilizers.
- Pectoral and pelvic fins produce lift.
- Some buoyancy is provided by large amounts of oil stored in liver, but most swim continuously to remain in the water column.
  - ⇒ Continual swimming also produces water flow through mouth and over gills for gas exchange.
  - ⇒ Some sharks are known to rest on the sea floor and in caves; water is continuously pumped over the gills while resting.

Most sharks are carnivorous, although the largest (whale shark) is a suspension feeder.
- Prey may be swallowed whole or pieces may be torn from large prey.
- Teeth evolved as modified scales.
- The digestive tract is proportionately shorter than in other vertebrates.
  - ⇒ A spiral valve, which increases surface area and slows food movement, is present in the intestine.

Sharks possess sharp vision (can not distinguish color), and olfactory senses and other adaptations to their life style.
- Electric sensory regions that detect muscle contractions of prey are located on the head.
- A *lateral line system* is present along the flanks.
  - ⇒ It is composed of rows of microscopic organs sensitive to water pressure changes and detects vibrations.
- A pair of auditory organs also detect sound waves passing through the water.

Sharks reproduce sexually with internal fertilization.

- A pair of claspers on the pelvic fins of males transfers sperm into the female reproductive tract.
- Some species are *oviparous*, some are *ovoviviparous*, and a few are *viviparous*.
- A *cloaca* (common chamber for reproductive, digestive and excretory systems) is present.

Rays are adapted to a bottom-dwelling life style.

- They have dorsoventrally flattened bodies.
- Their jaws are used to crush mollusks and crustaceans.
- Enlarged pectoral fins provide propulsion for swimming.
- The tail in many species is whiplike and, in some, bears venomous barbs.

VIII. Bony fishes are the most abundant and diverse vertebrates

The class Osteichthyes contains the bony fishes which are represented by more than 30,000 extant species.

- Abundant in marine and fresh waters.
- Range from 1 cm to 6 m in length.
- Skeleton is bony, reinforced with a matrix of calcium phosphate.
- Skin is covered with flattened bony scales.
- Skin glands produce mucus that reduces drag when swimming.
- A lateral line system is present as a row of tiny pits in the skin on both sides of the body.

Gas exchange occurs by drawing water over the four or five pairs of gills located in chambers covered by an *operculum*.

- Water is drawn into the mouth, through the pharynx and out between the gills by movement of the operculum and contraction of muscles within the gill chambers.
  - ⇒ Allows bony fishes to breath while stationary.

A swim bladder located dorsal to the digestive system provides buoyancy.

- Transfer of gases between blood and swim bladder varies bladder inflation and adjusts the density of the fish.

Bony fishes are very maneuverable swimmers. The flexible fins provide better steering and propulsion than the stiff fins of sharks.

- The fastest bony fish can swim to 80 km per hour in short bursts.
- A fusiform body shape is common to all fast fishes and aquatic mammals.
  - ⇒ This body shape reduces drag produced by the density of water (convergent evolution).

Most bony fish are oviparous and utilize external fertilization.

- Some are ovoviviparous or viviparous and utilize internal fertilization.
- Some display complex mating behavior.

The cartilaginous and bony fishes diversified during the Devonian and Carboniferous periods.

- Sharks arose in the sea, bony fishes probably originated in fresh water.
- The swim bladder was modified from lungs of ancestral fishes which supplemented the gills for gas exchange in stagnant waters.

The two extant subclasses of bony fishes had diverged by the end of the Devonian: the Actinopterygii (ray-finned fishes) and the Sarcopterygii (fleshy-finned fishes).

The subclass *Actinopterygii* includes fish with fins supported mainly by flexible rays.

- These are the most familiar fishes.
- They spread from fresh water to the seas and many returned to fresh water during evolution of the taxon.
- Some bony fish (e.g. salmon and sea-run trout) reproduce in fresh water and mature in the sea.

The subclass *Sarcopterygii* includes three orders of fleshy-finned fishes that evolved in fresh water: lungfishes, coelocanths and rhipidistans.

- The ancestors of these fishes continued to use their lungs to aid the gills in gas exchange.

Coelocanths and rhipidistians are referred to as *lobe-finned* fishes.

- Their fins were fleshy, muscular and supported by extensions of the bony skeleton.
- Many were large, bottom-dwelling forms that used their paired fins to walk on the substratum.
- The only extant species, the coelocanth, is marine and lungless. It belongs to a lineage that became marine at some point in its evolution.
- All rhipidistians are extinct.

Three genera of *lungfishes* exist in the Southern Hemisphere.

- They live in stagnant ponds and swamps where they surface to gulp air into lungs connected to the pharynx.
- When ponds dry, lungfishes burrow in the mud and aestivate.

Lobe-finned fishes of Devonian were numerous and important in vertebrate genealogy because they gave rise to amphibians. (See Campbell, Figure 30.13)

IX. Amphibians are the oldest class of tetrapods

The first vertebrates to move onto land were members of the class Amphibia.

- Today there are about 4,000 extant species of frogs, salamanders and caecilians.

A. Early Amphibians

Early amphibians evolved from lobe-finned fishes that adapted to environmental variations (drought and flooding) of the Devonian.

- The skeletal structure of the lobed fins suggest they could have assisted in movement on land.
- Fossil lobe-fins, such as the rhipidistian *Eusthenopteron*, exhibited many anatomical similarities to early amphibians. (See Campbell, Figure, 30.14a)

Oldest amphibian fossils are from the late Devonian (350 million years ago).

- Early amphibians were predators that ate insects and other invertebrates that had moved previously onto land.
- Amphibians were the only vertebrates on land in the late Devonian and early Carboniferous.

Radiation of forms occurred during the early Carboniferous period.

- Some forms reached 4 m in length and some resembled reptiles.
- Amphibians began to decline during the late Carboniferous. At the beginning of the Triassic period (230 million years ago), most of the survivors resembled modern species.

B. Modern Amphibians

There are three extant orders of amphibians: Urodela (salamanders), Anura (frogs and toads) and Apoda (caecilians).

The Order Urodela (salamanders) contains about 400 species.

- Some are aquatic and some are terrestrial.
- Terrestrial forms walk with a side-to-side bending of the body. Aquatic forms swim sinusoidally or walk along the bottom of streams or ponds.

The Order Anura (frogs and toads) contains about 3,500 species which are better adapted to the terrestrial habitat than urodeles.

- Enlarged hindlegs provide better movement (hopping) than in urodeles.
- They capture prey by flicking the sticky tongue which is attached anteriorly.
- Predator avoidance aided by camouflage color patterns and distasteful or poisonous mucus secreted by skin glands.
  - ⇒ Bright coloration is common in poisonous species.

The Order Apoda (caecilians) contains about 150 species.

- They are legless and almost blind.
- Most species burrow in moist tropical soils; a few species inhabit freshwater ponds and streams.

The Frog Life Cycle:

The frog exhibits a metamorphosis from the larval to adult stage. (See Campbell, Figure 30.16)

- The tadpole (larval stage) is usually aquatic, herbivorous and possesses internal gills, a lateral line system and a long, finned tail.
- The tadpole lacks legs and swims by undulating the tail.
- During metamorphosis, legs develop and the gills and lateral line system disappear.
- A young frog is tetrapod, has air-breathing lungs, a pair of external eardrums and a digestive system that can digest animal protein.
- The adult is usually terrestrial or semiaquatic and a predator.

Many amphibians, including some frogs, do not have a tadpole stage.

- Some species in each order are strictly aquatic while others are strictly terrestrial.
- Urodeles and apodans have larva that more closely resemble adults and both larva and adults are carnivorous.
- Paedogenesis is common in some groups of urodeles.

Most amphibians maintain close ties with water and are most abundant in damp habitats.

- Terrestrial forms in arid habitats spend much of their time in burrows where humidity is high.
- Gas exchange is primarily cutaneous and terrestrial forms must keep the skin moist.
    - ⇒ Lungs can aid in gas exchange although most are small and inefficient. Some forms lack lungs.
    - ⇒ Many species also exchange gases across moist surfaces of the mouth.

Amphibians are dioecious.

- The reproduce sexually usually with external fertilization in water (e.g. ponds, streams, temporary pools).
- In frogs, the male grasps the female and sperm are released as the female sheds her eggs. (See Campbell, Figure 30.16a)
- Eggs are unshelled and produced in large numbers by most species.

Some species exhibit parental behavior and produce small numbers of eggs.

- Males or females (species dependent) incubate eggs on their back, in the mouth or in the stomach.
- Some tropical species lay eggs in a moist foamy nest that prevents drying.
- Some species are ovoviviparous and a few are viviparous with the eggs developing in the female's reproductive tract.

Amphibians exhibit complex and diverse social behavior especially during breeding season (e.g. vocalization by male anurans, migrations, navigation or chemical signaling).

X.   **Evolution of the amniotic egg expanded the success of vertebrates on land**

Many specialized adaptations for living in a terrestrial habitat were necessary for reptiles to evolve from their amphibian ancestor.

- The amniotic egg was important as its development broke the last ties with the aquatic environment by allowing life cycles to be completed on land.
  - ⇒ Amphibians must deposit their shell-less eggs in water.
  - ⇒ The shell of the amniotic egg helps prevent desiccation.
- The *extraembryonic membranes* within the egg develop from tissue layers that grow out from the embryo.
  - ⇒ These specialized membranes function in gas exchange, transfer of stored nutrients to the embryo, and waste storage.
  - ⇒ One of these membranes, the *amnion*, encloses a compartment filled with amniotic fluid that bathes the embryo and provides a cushion against shocks.

Birds and mammals evolved from reptiles.

- Birds produce shelled amniotic eggs.
- Placental mammals do not produce a shelled egg, the developing embryo is nourished by the mother after it becomes implanted in the uterine wall.
  - ⇒ The extraembryonic membranes are retained and serve important functions in embryo development.

XI.   **A reptilian heritage is evident in all amniotes**

The class Reptilia is a diverse group with about 7,000 extant species and a wide array of extinct forms.

- Reptiles possess several adaptations to live on land not found in amphibians.

A.   **Reptilian Characteristics**

Adaptations for a terrestrial existence include:

- Scales contain the protein *keratin* which helps prevent dehydration.
- Gas exchange via lungs although many turtles also use moist cloacal surfaces.
- Dioecious with sexual reproduction and internal fertilization.
  - ⇒ Most are oviparous and produce an *amniotic egg*.
  - ⇒ Some species of snakes and lizards are viviparous with the young obtaining nutrients from the mother across a "placenta" which forms from the extraembryonic membranes.

Reptiles are *ectotherms* and use behavioral adaptations to regulate their body temperature.

Ectotherm = An animal that uses behavioral adaptations to absorb solar energy and regulate its body temperature.

- Due to ectothermy, reptiles can survive on a much lower caloric intake than mammals of comparable size.

Reptiles were abundant and diverse in the Mesozoic era.

B. The Age of Reptiles

1. Origin and Early Evolutionary Radiation of Reptiles

The oldest reptilian fossils are found in upper Carboniferous rock (300 million years old).

- Ancestors were probably Devonian amphibians.
- Two waves of adaptive radiation resulted in reptiles being the dominant terrestrial vertebrates for 200 million years.

The first reptilian radiation was in the early Permian period and gave rise to two main evolutionary branches: the *synapsids* and *sauropsids*.

- Synapsids were terrestrial predators and gave rise to the *therapsid* lineage which were mammal-like reptiles.

  ⇒ Therapsids were large, dog-sized predators from which mammals are believed to have evolved.

- Sauropsids gave rise to the modern amniote groups other than mammals. They split into two lineages early in their history: the *anapsids* and the *diapsids*.

  ⇒ The anapsids are presently represented only by the turtles.

  ⇒ The extant diapsids are represented by the lizards, snakes, and crocodilians; dinosaurs and some other extinct groups were also diapsids.

2. Dinosaurs and Their Relatives

The second reptilian radiation began in the late Triassic (about 220 million years ago) and several lineages evolved during this event.

- Two groups are most important: the dinosaurs and pterosaurs (flying reptiles).
- Dinosaurs varied in body shape, size and habitat.

  ⇒ Some fossilized forms measure 45 meters in length.

- Pterosaurs had wings formed from skin stretched from the body wall, along the forelimb to the tip of an elongate finger and supported by stiff fibers.

Evidence indicates that dinosaurs were agile, fast moving and social, some may have even exhibited parental care of the young.

There is also some anatomical evidence supporting the hypothesis that dinosaurs were *endothermic*.

<u>Endothermy</u> = The ability to keep the body warm through an animal's own metabolism.

- Skeptics of this hypothesis feel that the Mesozoic climate was warm and consistent, and that basking may have been sufficient for maintaining body temperature.
- Low surface-to-volume ratios of large forms reduced fluctuations of body temperature vs air temperature; thus, dinosaurs may not have been endothermic.

3. The Cretaceous Crisis

In the Cretaceous (last period of the Mesozoic), the climate became cooler and more variable, and mass extinctions occurred.

- Twenty-five percent of the families of marine invertebrates disappeared.
- Nearly all dinosaurs disappeared by the end of the Mesozoic (65 million years ago) over a period of only five to ten million years.
- The few reptilian groups to survive gave rise to extant forms.

C. Reptiles of Today

The largest and most diverse extant orders are the: Chelonia (turtles), Squamata (lizards and snakes), and Crocodilia (alligators and crocodiles).

Turtles of the order *Chelonia* evolved from anapsids during the Mesozoic.
- They show little change from the earliest forms.
- They are protected from predators by a hard shell.
- All turtles, even aquatic species, lay their eggs on land.

Lizards and snakes are classified in the order *Squamata*.

Lizards are the most numerous and diverse group of extant reptiles.
- They evolved from the diapsid lineage.
- Most are small.
- Many nest in crevices and decrease activity during cold periods.

Snakes probably descended from burrowing lizards.
- They are limbless and most live above ground.
- Vestigial pelvic and limb bones present in primitive snakes (boas) are evidence of a limbed ancestor.

Snakes are carnivorous and have a number of adaptations for hunting prey.
- They have acute chemical sensors.
- They are sensitive to ground vibrations (although lacking eardrums).
- Pit vipers have sensitive heat-detecting organs between their eyes and nostrils.
- Flicking tongue helps transmit odors toward olfactory organs on roof of mouth.
- Poisonous snakes inject a toxin through a pair of sharp, hollow teeth and loosely articulated jaws allow them to swallow large prey.

Crocodiles and alligators of the order Crocodilia are among the largest living reptiles. (See Campbell, Figure 30.21d)

- Crocodilians also evolved from the diapsid lineage.
- The spend most of their time in the water, breathing air through upturned nostrils.
- They are confined to warm regions of Africa, China, Indonesia, India, Australia, South America and the southeastern United States.
- They are the living reptiles most closely related to dinosaurs.

XII.   Birds began as flying reptiles

The class Aves contains the birds which evolved from reptiles during the Mesozoic era.

- They possess distinct reptilian characteristics such as the amniotic egg and scales on the legs.

A.   Characteristics of Birds

Each part of the bird's anatomy is modified in some way that enhances flight.

- The bones have a honeycombed internal structure that provides strength while reducing weight.
- Some organ systems are reduced (only one ovary in females).
- Birds have no teeth (reduces weight) and food is ground in the gizzard.
  ⇒ The beak is made of keratin and evolution has produced many shapes in relation to the bird's diet.

Flying requires much energy production from an active metabolism.

- Birds are endothermic with insulation provided by feathers and a fat layer.
- Birds have efficient circulatory system with a four-chambered heart that segregates oxygenated blood from unoxygenated blood.
- They have efficient lungs with tubes connecting to elastic air sacs that help dissipate heat and reduce the body density.

Birds also have a very well developed nervous system.

- Acute vision and well-developed visual and coordinating areas of the brain aid in flying.
- They show complex behavior especially during breeding season when elaborate courtship rituals are performed.

Birds are dioecious with sexual reproduction and internal fertilization.

- Sperm are transferred from the cloaca of the male to the cloaca of the female (males of most species lack a penis) during copulation.
- Eggs are laid and must be kept warm through brooding by the female, male or both depending on the species.

Wings are airfoils, formed by the shape and arrangement of the feathers, that illustrate the same aerodynamic principles as airplane wings.

- Power is supplied to the wings by contraction of the large pectoral (breast) muscles which are anchored to a keel on the sternum (breastbone).
- Some birds have wings adapted for soaring (hawks) while others must beat their wings continuously to stay aloft (hummingbirds).

Feathers are made of keratin and are extremely light and strong. (See Campbell, Figure 30.23)

- Feathers evolved from the scales of reptiles and may have first functioned as insulation.
- Feathers also function to control air movements around the wing.

Radical alteration of body form was necessary for evolution of flight, but flight provides many benefits.

- Allows aerial reconnaissance that enhances hunting and scavenging.
- Birds can exploit flying insects as an abundant, highly nutritious food resource.
- Flight provides an escape mechanism from land-bound predators.
- Flight also allows migration to utilize different food resources and seasonal breeding areas.

B. The Origin of Birds

Birds shared a common ancestor with *Archaeopteryx lithographica*.

- Fossils of *Archaeopteryx* have been recovered from limestone dated to the Jurassic period (150 million years ago).
- *Archaeopteryx* had clawed forelimbs, teeth, a long tail containing vertebrae and feathers.
- *Archaeopteryx* is not considered the ancestor to modern birds, but a side branch of the avian lineage.
  ⇒ The skeleton indicates a weak flyer which may have been a tree-dwelling glider.

C. Modern Birds

There are about 8,600 extant species in 28 orders. Most birds can fly but several are flightless (ostrich, kiwi, and emu).

- Flightless birds are called *ratites* because the breastbone lacks a keel and large breast muscles used for flying are absent.
- Flying birds are referred to as *carinates* due to the presence of a sternal keel (carina) that supports the large breast muscles used in flying.
- Carinate birds exhibit a variety of feather colors, beak and foot shape, behavior and flight ability.
  ⇒ Penguins are also carinate birds which do not fly but use powerful breast muscles in swimming.
- Almost 60% of extant species belong to one order of carinate birds (the *Passeriformes*, or *perching birds*) which includes the jays, swallows, sparrows, warblers and many others.

XIII. Mammals diversified extensively in the wake of the Cretaceous extinctions

There are about 4,500 species of extant mammals.
- Extinction of the dinosaurs opened new adaptive zones at the end of the Mesozoic era.
- Mammals underwent a massive radiation to fill these vacant zones.

A. Mammalian Characteristics

Species in the class Mammalia have the following characteristics:
- Possess hair which is composed of keratin but is not believed to have evolved from reptilian scales.
  ⇒ The hair provides insulation.
- Mammals are endothermic with an active metabolism.
  ⇒ An efficient respiratory system which utilizes a diaphragm for ventilation supports the metabolism.
  ⇒ A four-chambered heart segregates oxygenated from unoxygenated blood.
- Mammary glands are present and produce milk to nourish the young.
- Teeth are differentiated into various sizes and shapes which are adapted to chewing many types of food.
  ⇒ The jaw apparatus of the ancestral reptiles was also modified during evolution with two of the jaw bones becoming incorporated into the middle ear.

Mammals are dioecious with sexual reproduction and internal fertilization.
- Most are viviparous with the developing embryo receiving nutrients from the female across the *placenta*.
- A few are oviparous.

Mammals have large brains in comparison to other vertebrate groups and are capable of learning.
- Parental care of long duration helps young learn from the parents.

B. Evolution of Mammals

Mammals evolved from therapsid ancestors (part of the synapsid branch) during the Triassic period.
- The oldest fossil mammals are dated to 220 million years ago.
- Early mammals coexisted with dinosaurs throughout the Mesozoic era.
- Most Mesozoic mammals were small, probably insectivorous and nocturnal.

Mammals continued to diversify during the Cenozoic.

- During the Cretaceous period (last of the Mesozoic), mass extinctions and mass radiations transformed the flora and fauna of Earth.
    ⇒ Angiosperms replaced gymnosperms as the dominant plants in many areas.
- By the beginning of the Cenozoic, mammals were undergoing an adaptive radiation, and their diversity is represented today by three major groups: monotremes, marsupials and placental mammals.

C.   Monotremes

The Monotremes include the platypuses and echidnas which are characterized by:

- Oviparity.
- A reptilian-like egg with large amounts of yolk that nourishes the developing embryos.
- Hair.
- Milk production from specialized glands on the belly of females.
    ⇒ After hatching, young suck milk from the fur of the mother who lacks nipples.

The mixture of ancestral reptilian and derived mammalian traits suggests that monotremes descended from an early branch of the mammalian lineage.

- Extant monotremes are found in Australia and New Guinea.

D.   Marsupials

The Marsupials include opossums, kangaroos, koalas, and other mammals that complete their development in a *marsupium* (maternal pouch).

Marsupial eggs contain a moderate amount of yolk that nourishes the embryo during early development in the mother's reproductive tract.

- Young are born in an early stage of development and are small (about the size of a honeybee in kangaroos).
- The hindlegs are simple buds, but the forelimbs are strong enough for the young to climb from the female reproductive tract exit to the marsupium.
- In the marsupium, the young attaches to a teat and completes its development while nursing.

Convergent evolution in Australian marsupials has produced a diversity of forms which resemble placental counterparts in all ecological roles.

- Opossums are the only extant marsupials outside of the Australian region.
- South America had an extensive marsupial fauna during the Tertiary period as seen in the fossil record.

Plate tectonics and continental drift provide a mechanism which explains the distribution of fossil and modern marsupials.

- Fossil evidence indicates marsupials probably originated in what is now North America and spread southward while the land masses were joined.
- The breakup of Pangaea produced two island continents: South America and Australia.
  - ⇒ With isolation, their marsupial faunas diversified away from the placental mammals that began adaptive radiation on the northern continents.
- Australia has remained isolated from other continents for about 65 million years, thus isolating its developing fauna.
- When North and South America joined at the isthmus of Panama, extensive migrations took place over the land bridge in both directions.
  - ⇒ The most important migrations occurred about 12 million years ago and again about 3 million years ago.

E. Placental Mammals

In placental mammals, embryonic development is completed within the uterus where the embryo is joined to the mother by the *placenta*. (See Campbell, Figure 30.27c)

Adaptive radiation during the late Cretaceous and early Tertiary periods (about 70 – 45 million years ago) produced the orders of extant placental mammals. (See Campbell, Table 30.2)

- Fossil evidence indicates that placentals and marsupials diverged from a common ancestor about 80 – 100 million years ago; thus, they are more closely related than either is to the monotremes.

Most mammalogists favor a genealogy that recognizes at least four main evolutionary lines of placental mammals.

- One lineage consists of the orders Chiroptera (bats) and Insectivora (shrews) which resemble early mammals.
  - ⇒ The modified forelimbs which serve a wings in bats probably evolved from insectivores that fed on flying insects.
  - ⇒ Some bats feed on fruits while others bite mammals and lap the blood.
  - ⇒ Most bats are nocturnal.
- A second lineage consists of medium-sized herbivores that underwent a massive adaptive radiation during the Tertiary period.
  - ⇒ This led to such modern orders as the Lagomorpha (rabbits), Perissodactyla (odd-toed ungulates), Artiodactyla (even-toed ungulates), Sirenia (sea cows), Proboscidea (elephants) and Cetacea (whales, porpoises).
- The third evolutionary lineage produced the Order Carnivora which probably first appeared during the Cenozoic.
  - ⇒ Included in this order are the cats, dogs, raccoons, skunks, and pinnipeds (seals, sea lions, walruses).
  - ⇒ Seals and their relatives evolved from middle Cenozoic carnivores that became adapted for swimming.

- The fourth lineage had the greatest adaptive radiation and produced the primate-rodent complex.
  - ⇒ Includes the orders Rodentia (rats, squirrels, beavers) and Primates (monkeys, apes, humans).

XIV. Primate evolution provides a context for understanding human origins

A. Evolutionary Trends in Primates

The first primates were small arboreal mammals.
- Dental structure suggests they descended from insectivores in the late Cretaceous.
- *Purgatorius unio*, found in Montana, is considered to be the oldest primate.

Primates have been present for 65 million years (end of Mesozoic era) and are defined by characteristics shaped by natural selection for living in trees. These characteristics include:
- Limber shoulder joints which make it possible to *brachiate* (swing from one hold to the next).
- Dexterous hands for hanging on branches and manipulating food.
- Sensitive fingers with nails, not claws.
- Eyes are close together on the front of the face, giving overlapping fields of vision for enhanced depth perception (necessary for brachiating).
- Excellent eye-hand coordination.
- Parental care with usually single births and long nurturing of offspring.

B. Modern Primates

Modern primates are divided into two suborders: Prosimii (premonkeys) and Anthropoidea (monkeys, apes, humans).
- Prosimians (lemurs, lorises, pottos, tarsiers) probably resemble early arboreal primates.

There is a question as to which early prosimian lineage is ancestral to the anthropoids.
- Two groups of prosimian fossils are recognized by paleontologists.
  - ⇒ One ancestral to the tarsiers, the other to lemurs, lorises, and pottos.
- The divergence of these two groups occurred at least 50 million years ago and it has been debated as to which of these two groups was also ancestral to the anthropoids.
- Recently discovered fossils raise another possibility.
  - ⇒ Fossils found in Asia and Africa which date at least 50 million years ago appear to be more similar to anthropoids than to either groups of prosimian fossils.
  - ⇒ These fossils indicate an early divergence of prosimians into three lineages with the third being ancestral to the anthropoids.

Fossils of monkeylike primates indicate anthropoids were established in Africa and Asia by 40 million years ago in Africa or Asia (South America and Africa had already separated).

- Ancestors of New World monkeys may have reached South America by rafting from Africa or migration southward from North America.

New World monkeys and Old World monkeys have evolved along separate pathways for many millions of years. (See Campbell, Figure 30.30)

- All New World monkeys are arboreal.
- Old World monkeys include arboreal and ground-dwelling forms.
- Most monkeys, both New and Old World, are diurnal and usually live in social bands.

There are also four genera of apes included in the anthropoid suborder: *Hylobates* (gibbons), *Pongo* (orangutans), *Gorilla* (gorillas) and *Pan* (chimpanzees). (See Campbell, Figure 30.31)

- Apes are confined to the tropical regions of the Old World.
- They are larger than monkeys (except the gibbons) with relatively long legs, short arms and no tails.
- Only gibbons and orangutans are primarily arboreal although all are capable of brachiation.
- Social organization varies with the gorillas and chimpanzees being highly social.
- Apes have larger brains than monkeys and thus exhibit more adaptable behavior.

XV. Humanity is one very young twig on the vertebrate tree

Paleoanthropology concentrates on the small span of geological time during which humans and chimpanzees diverged from a common ancestor.

Paleoanthropology = The study of human origins and evolution.

Competition between researchers has often clouded the field of paleoanthropology.

- Researchers gave new names to fossil forms which were actually the same species recovered by others. (This practice ended about 20 years ago.)
- Theories were proposed on insufficient evidence; often a few teeth or a jawbone fragment.
- Such actions resulted in many persistent misconceptions about human evolution even though fossil discoveries have disproved many of the myths.

A. Some Common Misconceptions

Our ancestors were chimpanzees or other modern apes.

- Humans and chimpanzees represent two divergent branches of the anthropoid lineage which evolved from a common, less specialized ancestor.

Human evolution represents a ladder with a series of steps leading directly from an ancestral anthropoid to <u>Homo</u> <u>sapiens</u>.

- This progression is usually shown as a line of fossil hominids becoming progressively more modern.
- Human evolution included *many* branches which led to dead ends with several different human species coexisting at times (See Campbell, Figure 30.32).
- If punctuated equilibrium applies to humans, most evolutionary change occurred with the appearance of new hominid species, not phyletic (anagenic) change in an unbranched lineage.

Various human characteristics like upright posture and an enlarged brain evolved in unison.

- *Mosaic evolution* occurred with different features evolving at different times.
  ⇒ Some ancestral forms walked upright but had small brains.

Present understanding of our ancestry remains unclear even after dismissing many of these myths.

B. Early Anthropoids

The oldest known fossils of apes are of *Aegyptopithecus*, the "dawn ape," which was a cat-sized tree-dweller about 35 million years ago.

About 25 million years ago (during the Miocene epoch) descendants of the first apes diversified and spread to Eurasia.

About 20 million years ago, the Indian plate collided with Asia and the Himalayan range formed.

- The climate became drier and the African and Asian forests contracted.
  ⇒ This isolated these regions of anthropoid evolution from each other.

Most anthropologists believe that humans and apes diverged from a common African anthropoid ancestor 6 – 8 million years ago.

- Evidence from the fossil record and DNA comparisons between humans and chimpanzees supports this conclusion.

C. Australopithecines: The First Humans

Australopithecus africanus was discovered by Raymond Dart in 1924.

- Additional fossils proved that *Australopithecus* was a hominid that walked fully erect and had humanlike teeth and hands.
- The brain was about one-third the size of modern humans.
- Various species of *Australopithecus* began appearing about 4.4 million years ago and existed for over 3.0 million years.

"Lucy", an *Australopithecus* skeleton, was discovered in 1974 in the Afar region of Ethiopia by paleoanthropoligists.

- Lucy is 3.18 million years old.
- The skeleton was 40% complete and small, about one meter tall with a head the size of a softball.
  - ⇒ The structure indicates an upright posture.
- It was different enough to be placed in a different species, *Australopithecus afarensis*.
- Similar fossils have been discovered which indicate the species existed for about one million years.

The history of hominids is now believed to extend back to 4.4 million years.

- Bone fragments from a hominid were recovered in 1994 in an area 75 km from the site of Lucy's discovery.
  - ⇒ These fragments were so different from *A. afarensis* that a new hominid species was named, *Australopithecus ramidus*.
- *A. ramidus*, which represents the oldest known hominid, exhibited some interesting characteristics.
  - ⇒ Skull fragments indicated the head balanced on top of the spinal column – evidence of the early evolution of upright posture.
  - ⇒ The skeletons of forest-dwelling animals were found among the bones – this challenges the view that bipedalism evolved when humans began living on the savanna.

The discovery of *A. ramidus* and *A. afarensis* has raised several questions.

- Is *A. ramidus* the ancestor of *A. afarensis* or an extinct evolutionary branch?
- Is *A. afarensis* ancestral to other hominids or did it share a common ancestor with *Homo*?
  - ⇒ *A. afarensis* underwent little change during its one million year span.
  - ⇒ Several new hominid species resulted from an adaptive radiation which began about 3 million years ago.
- The new species included *A. africanus*, several heavy-boned species of *Australopithecus*, and *Homo habilis* (appeared about 2.5 million years ago).

While the phylogeny of early hominids is uncertain, one fact is clear: hominids walked upright for two million years without a substantial increase in brain size.

- This posture may have freed the hands for other things such as gathering food or caring for infants.

D. *Homo habilis*

Enlargement of the human brain is first evident in fossils dating to about 2.5 million years ago.

- Skulls with brain capacities of about 650 cubic centimeters have been found compared with the 500 cc capacity of *A. africanus*.
- Simple stone tools have been found at times with the larger-brained fossils.
- Most paleoanthropologists believe these advances warrant placing the larger-brained fossils in the genus *Homo* and naming them *Homo habilis*.
- It is clear from the fossil record that after walking upright for more than two million years, hominids began to use their brains and hands to fashion tools.

Homo habilis and other new hominids were a part of a larger speciation event among African mammals.

- About 2.5 million years ago, Africa's climate began to become drier and savannas started to replace forests.
- The fauna began to adapt to these new conditions.
- *Homo habilis* coexisted with the smaller-brained *Australopithecus* for nearly one million years.
- One hypothesis is that *Australopithecus africanus* (and other australopithecines) and *Homo habilis* were two distinct lines of hominids.
    - ⇒ *Australopithecus africanus* was an evolutionary dead end with *Homo habilis* leading first to *Homo erectus* which later gave rise to *Homo sapiens*.

E. *Homo erectus* **and Descendants**

Homo erectus was the first hominid to migrate out of Africa into Europe and Asia.

- Fossils known as Java Man and Beijing Man are examples if *H. erectus*.

Homo erectus lived from about 1.8 million years ago until 300,000 years ago.

- Fossils found in Africa cover the entire span of *H. erectus*' existence.
- These populations existed during the same period as *H. erectus* populations on other continents.
- The spread to new continents may have resulted from a gradual range expansion associated with a shift in diet to include a larger portion of meat.
    - ⇒ In general, carnivores need a larger range than herbivores.

Homo erectus was taller and had a larger brain than *H. habilis*.

- The *H. erectus* brain capacity increased to as large as 1200 cc during the 1.5 million years of its existence.
    - ⇒ This overlaps the normal range of modern humans.

The intelligence that evolved in *H. habilis* allowed early humans to survive in the colder climates to which they migrated.

- *Homo erectus* lived in huts or caves, built fires, wore clothes of skins, and designed more refined stone tools than *H. habilis*.

- *Homo erectus* was poorly equipped in a physical sense to live outside of the tropics but made up for the deficiencies with intelligence and social cooperation.

Some descendants of *H. erectus* developed larger brain capacities and exhibited regional diversity in populations.

- The Neanderthals are the best known descendants of *H. erectus*.

- Neanderthals lived in Europe, the Middle East, and Asia from 130,000 to 35,000 years ago.

- They had heavier brows, less pronounced chins, and slightly larger brain capacities than modern man.

- They were skilled tool makers who participated in burials and other rituals requiring abstract thought.

Many paleoanthropologists group the African post-*Homo erectus* fossils with Neanderthals and other descendants from Asia and Australasia. They believe these fossils represent the earliest forms of *Homo sapiens*.

- Some post-*H. erectus* fossils date to 300,000 years ago.

F.   The Emergence of *Homo sapiens*: Out of Africa...But When? *Science as a Process*

The debate over the origin of modern humans continues unabated with two widely divergent models currently being discussed. These are the *Multiregional Model* and the *Monogenesis Model*.

The *Multiregional Model* proposes: 1) Neanderthals and other post-*Homo erectus* hominids were ancestors to modern humans; and 2) modern humans evolved along the same lines in different parts of the world.

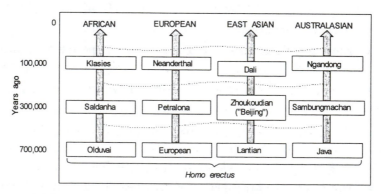

- If this model is correct, the geographic diversity of humans originated between one and two million years ago when *Homo erectus* spread from Africa to other continents.

- Supporters feel that interbreeding among neighboring populations provided opportunities for gene flow over the entire range and resulted in the genetic similarity of modern humans.

During the 1980's, some paleoanthropologists who interpreted the fossil record in a different way began to develop an alternative to the Multiregional Model. This alternative became known as the "Out of Africa" or *Monogenesis Model*.

The *Monogenesis Model* proposes: 1) *Homo erectus* was the ancestor to modern humans who evolved in Africa; and 2) modern humans dispersed from Africa, displacing the Neanderthals and other post-*H. erectus* hominids.

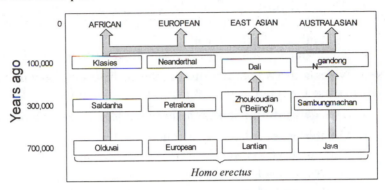

- If this model is correct, the diversity of modern humans has developed from geographic diversification within the last 100,000 years.
- To supporters of this model, an exclusively African genesis for modern humans is strongly indicated by the fact that the complete transition from archaic *Homo sapiens* to modern humans is found only in African fossils.
- The focus of their interpretations is on the relationship between Neanderthals and modern humans in Europe and the Middle East.
- The oldest fossils of modern *Homo sapiens* are about 100,000 years old. These were found in Africa and similar fossils have been recovered from caves in Israel.
- The fossils from Israel were found in caves near other caves containing Neanderthal-like fossils which date from 120,000 to 60,000 years ago – overlapping *H. sapiens* by about 40,000 years.
- Supporters of the Monogenesis Model interpret this information to mean that no interbreeding occurred during the time of coexistence since the two types of hominids persisted as distinct forms.
- This interpretation means the Neanderthals were not ancestors of modern humans since they coexisted and were probably evolutionary dead ends along with other dispersed post-*H. erectus* hominids.

Modern molecular techniques are being used to examine the question of human origins.
- In the late 1980's, a group of geneticists compared the mitochondrial DNA (mtDNA) from a multiethnic group of more than 100 people from four different continents.
- The premise for this analysis is that the greater the differences in the mtDNA of two people, the longer the time of divergence from a common source.
- Analysis of the mtDNA comparisons resulted in the tracing of the source of all human mtDNA back to Africa with the divergence from that common source beginning about 200,000 years ago.
- These results appeared to support the Monogenesis Model in that the divergence from the common source was too late to represent dispersal of *Homo erectus*, but supported a later dispersal of modern humans.

Several researchers have challenged the interpretation of the mtDNA study, especially the methods used to construct evolutionary trees from this type of data and the reliability of mtDNA as a biological clock.

- This criticism has encouraged advocates of the Multiregional Model to argue that the fossil evidence supports the multiregional evolution of humans more strongly than African monogenesis.

- These scientists also consider certain fossils from different geographical regions to be links between that region's archaic *Homo sapiens* and the modern humans currently on that continent.

New evidence from comparisons of nuclear DNA and new methods for using the mtDNA data to trace the relationships of populations are stimulating further debate.

- This new information strengthens the Monogenesis Model.

- Greater genetic diversity is found in African populations south of the Sahara than in other parts of the world.

  ⇒ If modern humans evolved in southern Africa, these populations would have the longest history of genetic diversification.

  ⇒ If populations of humans in other parts of the world resulted from migrations out of Africa, the smaller genetic diversity could be the result of the founder effect and genetic drift.

Debate continues about whether the Multiregional Model or the Monogenesis Model is more accurate.

- New evidence obtained by attempts to extract DNA from fossilized Neanderthal and other archaic *Homo sapiens* skulls may shed new light on this question.

G.   Cultural Evolution: A New Force in the History of Life

Erect stance was a very radical anatomical change in our evolution and required major changes in the foot, pelvis and vertebral column.

Enlargement of the brain was a secondary alteration made possible by prolonging the growth period of the skull and its contents.

The brains of nonprimate mammalian fetuses grow rapidly, but growth slows and stops soon after birth.

- The brains of primates continue to grow after birth and the period is longer for a human than other primates.

- Parental care is lengthened due to this extended development and this contributes to the child's learning.

Learning from the experiences of earlier generations is the basis of culture (transmission of accumulated knowledge over generations); the transmission is by written and spoken language.

A cultural evolution is a continuum, but three stages are recognized:

- Nomads of the African grasslands made tools, organized communal activities and divided labors about 2 million years ago.
- The development of agriculture in Africa, Eurasia, and the Americas about 10,000 to 15,000 years ago encouraged permanent settlements.
- The Industrial Revolution began in the eighteenth century.
  - ⇒ Since then, new technology and the human population have escalated exponentially.

No significant biological change in humans has occurred from the beginning to now.

Evolution of the human brain may have been anatomically simpler than acquiring an upright stance, but the consequences of cerebral growth have been enormous.

- Cultural evolution resulted in *Homo sapiens* becoming a species that could change the environment to meet its needs and not have to adapt to an environment through natural selection.
  - ⇒ Humans are the most numerous and widespread of large animals.
- Cultural evolution outpaces biological evolution and we may be changing the world faster than many species can adapt.
  - ⇒ The rate of extinctions this century is 50 times greater than the average for the past 100,000 years.
  - ⇒ The overwhelming rate of extinction is due primarily to habitat destruction and chemical pollution, both functions of human cultural changes and overpopulation.
  - ⇒ Global temperature increase and alteration of world climates are a result of escalating fossil fuel consumption.
  - ⇒ Destruction of tropical rain forests, which play a role in maintenance of atmospheric gas balance and moderating global weather, is startling.

The effect of *Homo sapiens* is the latest and may be the most devastating crisis in the history of life.

REFERENCES

Campbell, N. *Biology*. 4th ed. Menlo Park, California: Benjamin/Cummings, 1996.

Coppens, Y. "The East Side Story: The Origin of Humankind." Scientific American, May, 1994.

Dean, D., and E. Delson. "Homo at the Gates of Europe." Nature, February 9, 1995.

CHAPTER 31

PLANT STRUCTURE AND GROWTH

OUTLINE

I.   Plant biology reflects the major themes in the study of life

II. A plant's root and shoot systems are evolutionary adaptations to living on land
    A.   The Root System
    B.   The Shoot System

III. The many types of plant cells are organized into three major tissue systems
    A.   Types of Plant Cells
    B.   The Three Tissue Systems of a Plant

IV. Meristems generate cells for new organs throughout the lifetime of a plant: *an overview of plant growth*

V.   Apical meristems extend roots and shoots: *a closer look at primary growth*
    A.   Primary Growth of Roots
    B.   Primary Growth of Shoots

VI. Lateral meristems add girth to stems and roots: *a closer look at secondary growth*
    A.   Secondary Growth of Stems
    B.   Secondary Growth of Roots

OBJECTIVES

After reading this chapter and attending lecture, the student should be able to:
1. List the characteristics of an angiosperm.
2. Explain the differences between monocots and dicots.
3. Describe the importance of root systems and shoot systems to plants and explain how they work together.
4. Explain how taproot systems and fibrous root systems differ.
5. Explain the differences between stolons and rhizomes.

6. Describe how plant cells grow.
7. Distinguish between parenchyma and collenchyma cells with regards to structure and function.
8. Describe the differences in structure and function of the two types of sclerenchyma cells.
9. Explain the importance of tracheids and vessel elements to plants.
10. Distinguish between water-conducting cells and sieve-tube members with regards to structure and function.
11. Explain the differences between simple tissues and complex tissues.
12. Explain the importance of a cuticle on the aerial parts of a plant and its absence on roots.
13. Describe the functions of the dermal tissue system, vascular tissue system and ground tissue system.
14. Distinguish among annual, biennial and perennial plants.
15. Explain the importance of the zones of cell division, cell elongation, and cell differentiation in primary growth of roots.
16. Explain the importance of the endodermis to a plant.
17. Describe the importance of an apical meristem to the primary growth of shoots.
18. Distinguish between the arrangement of vascular tissues in roots and shoots.
19. Describe how "wood" forms due to secondary growth of stems.
20. Using a diagram, describe the basic structure of a root, a stem, and a leaf.

KEY TERMS

monocot	parenchyma cell	annual	pericycle
dicot	collenchyma cell	biennial	vascular bundle
xylem	sclerenchyma cell	perennial	stomata
phloem	sclerids	apical meristem	guard cells
taproot system	tracheids	lateral meristem	transpiration
fibrous root system	vessel elements	primary growth	mesophyll
root hairs	sieve-tube members	rootcap	leaf trace
adventitious roots	sieve plates	zone of cell division	secondary growth
stem	companion cells	zone of cell elongation	vascular cambium
node	simple tissue	zone of maturation	secondary xylem
internode	complex tissue	protoderm	secondary phloem
axillary bud	dermal tissue system	procambium	rays
terminal bud	cuticle	ground meristem	annual growth ring
apical dominance	vascular tissue system	stele	cork cambium
stolon	ground tissue system	pith	suberin
rhizome	indeterminate growth	cortex	phelloderm
bulb	meristem	endodermis	periderm
petiole	determinate growth	lateral roots	lenticel

LECTURE NOTES

Plants form the foundation on which most terrestrial ecosystems are built.

- As the primary producers in most systems (through the process of photosynthesis), plants serve as the first link in the food chain which affects all species of animals in the system.

Plants were first studied by early humans who had to distinguish between edible and poisonous plants.

- These early humans later began to use plant products to make useful tools and other items.
- Modern plant biology continues to center on how to use plants and plant products to benefit humans.

I. Plant biology reflects the major themes in the study of life

New methods and the discovery of unique, interesting experimental organisms has expanded the knowledge base in many areas of plant biology.

- Studying such organisms (e.g. *Arabidopsis*) helps relate processes which occur at the molecular and cellular level to what is observed at the whole plant level.

Correlating the structure and functions of plants is a second focus of plant biology.

- Structure and function of plants are the result of interactions with the environment over both evolutionary and short-term time scales.
  - ⇒ On the evolutionary scale, plants adapted to the problems of a terrestrial existence as they moved from water to land (i.e. woody tissues provide support).
  - ⇒ On the short-term scale, environmental stimuli cause individual plants to show structural and physiological responses (i.e. stomata close during the hottest part of the day, thus conserving water).

While plants and animals faced many of the same problems (support, internal transport, etc.) they solved such problems in different ways.

- The two groups evolved independently from unicellular ancestors which had different modes of nutrition; this alone established a different evolutionary direction.

II. A plant's root and shoot systems are evolutionary adaptations to living on land

Plant biologists study two levels of plant architecture: morphology and anatomy.

Plant morphology = The study of the external structure of plants (e.g. arrangement of the parts of a flower).

Plant anatomy = The study of the internal structure of plants (e.g. arrangement of the cells and tissues in a leaf).

Angiosperms (flowering plants) are the most diverse and widespread of the plants with about 275,000 extant species.

- Characterized by flowers and fruits which are believed to be adaptations for reproduction and seed dispersal.
- Taxonomists divide angiosperms into two taxonomic classes: *monocots* and *dicots* which possess either one or two seed leaves, respectively, in combination with other characteristics. (See Campbell, Figure 31.3)

A plant can be divided into two basic systems, a subterranean *root system* and an aerial *shoot system* (stems, leaves, flowers).

- This two system arrangement reflects the evolutionary history of plants as terrestrial organisms. (See Campbell, Figure 31.4)
- Unlike the algal ancestors which were completely surrounded by nutrient rich water, terrestrial plants face a divided habitat:
  ⇒ Air is the source of CO_2 for photosynthesis and sunlight cannot penetrate into the soil.
  ⇒ Soil provides water and dissolved minerals to the plant.
- Each system depends on the other for survival of the whole plant.
  ⇒ Roots depend on shoots for sugar and other organic nutrients.
  ⇒ Shoots depend on roots for minerals, water and support.
- Materials are transported through the plant by vascular tissues.
  ⇒ *Xylem* conveys water and dissolved minerals to the shoots.
  ⇒ *Phloem* conveys food from shoots to roots and other nonphotosynthetic parts, and from storage roots to actively growing shoots.

A.   The Root System

Root structure is well adapted to:

- Anchor plants.
- Absorb and conduct water and nutrients.
- Store food.

There are two major types of root systems:

The *taproot system* is seen in many dicots.

- One large, vertical root (the *taproot*) produces many smaller secondary roots.
- Provides firm anchorage.
- Some taproots such as carrots, turnips, and sweet potatoes, are modified to store a large amount of reserve food.

A *fibrous root system* is found primarily in monocots (palms, bamboo, grasses).

- A mat of threadlike roots spreads out below the soil surface.
- Provides extensive exposure to soil water and minerals.
- Roots are concentrated in the upper few centimeters of soil, preventing soil erosion.

Absorption of water is greatly enhanced by *root hairs*, which increase the surface area of the root.

- Root hairs are normally most numerous near the root tips.
- Water and mineral absorption are also enhanced by *mycorrhizae*, symbioses between roots and fungi.

Adventitious roots = Roots arising above ground from stems or leaves.
- Form in addition to the normal root system.
- Some, such as prop roots of corn, help support the plant stem.

B.   The Shoot System

Shoot systems are comprised of vegetative shoots and floral shoots.
- Vegetative shoots consist of a *stem* and attached *leaves*; may be the main shoot or a vegetative branch.
- Floral shoots terminate in flowers.

1.   Stems

Stem morphology includes: (See Campbell, Figure 31.4)

Nodes = The points where leaves are attached to stems.

Internodes = The stem segments between the nodes.

Axillary bud = An embryonic side shoot found in the angle formed by each leaf and the stem; usually dormant.

Terminal bud = The bud on a shoot tip; it usually has developing leaves and a compact series of nodes and internodes.

Growth of a shoot is usually concentrated at the *apex* of the shoot where the terminal bud is located.

- The presence of a terminal bud inhibits development of axillary buds, a condition called *apical dominance*.
- Apical dominance appears to be an evolutionary adaptation to increase exposure of plant parts to light by concentrating resources on increasing plant height.

Axillary buds begin to grow under certain conditions and after damage or removal of the terminal bud.

- Some may develop into floral shoots.
- Others develop into vegetative shoots with terminal buds, leaves, and axillary buds.
    ⇒ This development results in branching which also increases exposure of plant parts to light.

Some plants have *modified stems* which are often mistaken for roots. There are several types of modified stems, each of which performs a specific function. (See Campbell, Figure 31.6)

- *Stolons* are horizontal stems growing along the surface of the ground (e.g. strawberry plant runners).
- *Rhizomes* are horizontal stems growing underground (e.g. irises).
    ⇒ Some end in enlarged tubers where food is stored (e.g. potatoes).
- *Bulbs* are vertical, underground shoots with leaf bases modified for food storage (e.g. onions).

2. Leaves

Leaves are the main photosynthetic organs of a plant.

- A leaf usually exists in the shape of a flattened *blade* which is joined to the node of a stem by a *petiole*.
    ⇒ Most monocots lack petioles; instead, the leaf base forms a sheath surrounding the stem.
- Monocot leaves have parallel major veins running the length of the blade.
- Dicot leaves have a multi-branched network of major veins. Can be *palmate* or *pinnate*.
- All leaves have numerous minor cross-veins.

Plant taxonomists use a variety of leaf characteristics to classify plants. (See Campbell, Figure 31.7)

Classification of leaf arrangement on a stem. The arrangement may be:

- *Opposite* – 2 leaves at each node 180° apart.
- *Alternate* – each node has one leaf and leaves at adjacent nodes point in opposite directions.
- *Whorled* – a node has 3 or more leaves attached.

Leaves may also be classified as:

- *Simple leaf* – one undivided blade.
- *Compound leaf* – divided into several leaflets.

Classification by leaf shape. Leaves can be:

- Lanceolate.
- Oval.
- Cordate (heart-shaped).
- Triangular.

Classification by leaf margin, which can be:

- Entire (smooth).
- Undulate.
- Serrate.
- Lobed.

Some plants have leaves that have become adapted for functions other than photosynthesis. (See Campbell, Figure 31.8)

- *Tendrils* are modified leaflets that cling to supports.
- *Spines* of cacti function in protection.
- Many succulents have leaves modified for storing water.
- Some plants have brightly colored leaves that help attract pollinators to the flower.

III. The many types of plant cells are organized into three major tissue systems

A. Types of Plant Cells

Each type of plant cell has structural adaptations that make it possible to perform that cell's function. Some are coupled with specific characteristics of the *protoplast*. (See Campbell, Figure 31.10)

<u>Protoplast</u> = Contents of a plant cell exclusive of the cell wall.

1. Parenchyma Cells

Parenchyma cells are the least specialized plant cells.

- Primary walls are thin and flexible.
- Most lack secondary walls.
- The protoplast usually has a large central vacuole.
- Function in synthesizing and storing organic products.
    ⇒ Photosynthesis occurs in the chloroplasts of mesophyll cells.
- Some in stems and roots have colorless plastids that store starch.
- Most mature cells do not divide, but retain the ability to divide and **differentiate** into other cell types under special conditions (e.g. repair and replacement after injury).

2. Collenchyma Cells

Collenchyma cells usually lack secondary walls.

- The primary cell wall is thicker than in parenchyma cells but is of an uneven thickness.
- They are usually grouped in strands or cylinders to support young parts of plants without restraining growth.
- They are living cells which elongate as the stems and leaves they support grow.

3. **Sclerenchyma Cells**

Sclerenchyma cells function in support.

- They have very rigid, thick secondary walls strengthened by *lignin*.
- Many lack protoplasts at functional maturity, so they can not elongate and may be dead, functioning only as support.
- There are two forms: *fibers* (long, slender, tapered cells occurring in bundles) and *sclerids* (shorter, irregularly-shaped cells).

4. **Tracheids and Vessel Elements: Water-Conducting Cells**

Xylem consists of two cell types, both with secondary walls and both dead at functional maturity.

- Before the protoplast dies, secondary walls are deposited in spiral or ring patterns (which allows them to stretch) in parts of the plant that are still growing.

Tracheids are long, thing tapered cells having lignin-hardened secondary walls with *pits* (thinner regions where only primary walls are present). (See Campbell, Figure 31.11a)

- Water flows from cell to cell through pits.
- Also function in support.

Vessel elements are wider, shorter, thinner-walled, and less tapered than tracheids. (See Campbell, Figure 31.11b and c)

- Vessel elements are aligned end to end.
- The end walls are perforated, permitting the free flow of water through long chains of vessel elements called xylem vessels.

5. **Sieve-Tube Members: Food-Conducting Cells**

Sieve-tube members are chains of phloem cells that transport sucrose, other organic compounds, and some minerals.

- The cells are alive at functional maturity.
- Protoplasts lack a nucleus, ribosomes and a distinct vacuole.

In angiosperms, the end walls of sieve-tube members have pores and are called *sieve plates*.

- The pores probably facilitate the movement of fluid between cells.
- At least one *companion cell* is connected to each sieve-tube member by many plasmodesmata; the companion's nucleus and ribosomes may also serve the sieve-tube member which lacks these organelles.
  - ⇒ Companion cells also help load sugar produced in the mesophyll into sieve-tubes of leaves of some plants.

NOTE: It would be useful to emphasize how close inspection of cell structure often reveals its function, and vice versa.

B.   **The Three Tissue Systems of a Plant**

Plant cells are arranged into three tissue systems: *dermal, vascular,* and *ground* tissue systems.

Each of the three systems is continuous throughout the plant, although, specific characteristics and spatial relationships vary in different plant organs. (See Campbell, Figure 31.12)

1.   **Dermal Tissue System**

Dermal tissue system (or epidermis) = Single layer of tightly packed cells covering and protecting the young parts of the plant.
- Functions in protection and has special characteristics consistent with the function of the organ it covers.
    ⇒ Root hairs specialized for water and mineral absorption are extensions of epidermal cells near root tips.
    ⇒ The waxy *cuticle* that helps the plant retain water is secreted by epidermal cells of leaves and most stems.

2.   **Vascular Tissue System**

Vascular tissue system = The xylem and phloem that functions in transport and support; is continuous throughout the plant.

3.   **Ground Tissue System**

Ground tissue system = Predominantly parenchyma, with some collenchyma and sclerenchyma, that fills the space between dermal and vascular tissue systems; functions in photosynthesis, storage and support.

IV.   **Meristems generate cells for new organs throughout the lifetime of a plant:** *an overview of plant growth*

Plant growth begins with germination of the seed and continues for the lifespan of the plants.

Indeterminate growth = Continued growth as long as the plant lives.
- In contrast, most animals cease growing after reaching a certain size (determinate growth).
- Certain plant organs, such as flower parts, show determinate growth.

Plants do not live indefinitely, they do have finite life spans.
- Most have genetically determined lifespans.
- Some are environmentally determined.
- *Annuals* complete their life cycles in one year or growing season.
- *Biennials* typically have life spans of two years.
- *Perennials*, such as trees and some grasses, live many years.

Indeterminate growth is made possible by *meristems* (perpetually embryonic tissues).

- Meristematic cells are unspecialized and divide to generate new cells near the growing point.
- New meristematic cells formed by division may remain in the region and produce new cells (*initials*) or be displaced and become incorporated and specialized into tissues (*derivatives*).
- *Apical meristems*, located in root tips and shoot buds, supply cells for plants to grow in length.
    - ⇒ *Primary growth* (elongation) is initiated by apical meristems and forms primary tissues organized into the 3 tissue systems.
- *Secondary growth* (increased girth) is the thickening of roots and shoots which occurs in woody plants due to development of *lateral meristems*.

    <u>Lateral meristems</u> = Cylinders of dividing cells extending along the lengths of roots and shoots.

    - ⇒ Cell division in the lateral meristems produces secondary dermal tissues which are thicker and tougher than the epidermis it replaces.
    - ⇒ Also adds new layers of vascular tissues.

V.   **Apical meristems extend roots and shoots:** *a closer look at primary growth*

The *primary plant body* with its three tissue systems is produced by primary growth.

- The youngest portions of woody plants and herbaceous plants are examples of primary plant bodies.
- Apical meristems are responsible for the primary growth of roots and shoots.

A.   **Primary Growth of Roots**

Root growth is concentrated near its tip and results in roots extending through the soil.

- The root tip is covered by a *root cap*, which protects the meristem and secretes a polysaccharide coating that lubricates the soil ahead of the growing root.
- The root tip contains 3 zones of cells in successive stages of primary growth. Although described separately, these zones blend into a continuum. (See Campbell, Figure 31.14)

1. <u>Zone of Cell Division</u>
    - Located near the tip of the root; includes the apical meristem and its derivatives, the *primary meristems*.
    - The apical meristem is centrally located; it produces the primary meristems and replacement cells of the root cap.
    - A *quiescent center* is located near the center of the apical meristem. It is composed of resistant, slowly dividing cells which may serve as reserve replacement cells in case of damage to the meristem.
    - The primary meristems form as three concentric cylinders of cells (the *protoderm, procambium, ground meristem*) which will produce the tissue systems of the roots.

2. <u>Zone of Cell Elongation</u>
   - In this region, cells elongate to at least 10 times their original length.
   - The elongation of cells in this region pushes the root tip through the soil.
   - Continued growth is sustained by the meristem's constant addition of new cells to the youngest end of the elongation zone.

3. <u>Zone of Maturation</u>
   - Farthest from the root tip.
   - Region where the new cells become specialized in structure and function and where the three tissue systems complete their differentiation.

The apical meristem produces three primary meristems, which in turn give rise to the three *primary tissues of roots*:

1. The *protoderm* is the outermost primary meristem.
   - Gives rise to the epidermis over which water and minerals must cross.
   - Root hairs are epidermal extensions which increase surface area, thus enhancing uptake.

2. The *procambium* forms a *stele* (central cylinder) where xylem and phloem develop.
   - In dicots, xylem radiates from the stele's center in 2 or more spokes, with phloem in between the spokes.
   - In monocots, a central *pith* (core of parenchyma cells) is ringed by vascular tissue in an alternating pattern of xylem and phloem.

3. The *ground meristem* is located between the protoderm and procambium; it gives rise to the *ground tissue system*. The ground tissue:
   - Is mostly parenchyma and fills the *cortex* (root area between the stele and epidermis).
   - Stores food; the cell membranes are active in mineral uptake.
   - Has *endodermis*, the single-cell thick, innermost layer of the cortex that forms the boundary between the cortex and the stele.
     ⇒ It selectively regulates passage of substances from soil to the vascular tissue of the stele.

Lateral roots may sprout from the outermost layer of the stele of a root.
- The *pericycle*, just inside the endodermis, is a layer of cells that may become meristematic and divide to form the lateral root.
- A lateral root forms as a clump of cells in the pericycle, then elongates and pushes through the cortex until it emerges from the primary root.
- The lateral root maintains its vascular connection to the stele of the main root.

B. **Primary Growth of Shoots**

Shoot apical meristem is a dome-shaped mass of dividing cells at the tip of the terminal bud.

- Forms the primary meristems that differentiate into the 3 tissue systems.
- On the flanks of the apical meristem dome are *leaf primordia* which form leaves.
- Meristematic cells left by the apical meristem at the base of the leaf primordia develop into axillary buds.

Most shoot elongation actually occurs due to growth of slightly older internodes below the shoot apex.

- Growth is a result of both cell division and elongation within the internode.
- *Intercalary meristems* are present at the base of each internode in grasses and some other plants.
  ⇒ These tissues permit prolonged internode elongation along the length of the shoot.

Axillary buds may form branches later in the life of the plant.

- Branches originate at the surface of the shoot and are connected to the vascular system which lies near the surface.
- This is a direct contrast to development of lateral roots which form deep in the pericycle of the root.

1. **Primary Tissues of Stems**

The vascular tissue of the stem is organized into strands of *vascular bundles* that run the length of the stem.

- Converge at the transition zone (shoot → root) to join the root stele.
- Each bundle is surrounded by ground tissue, including pith and cortex.

In dicots, bundles are arranged in a ring with pith inside and cortex outside.

- Xylem faces the pith, phloem faces the cortex.
- Pith and cortex are connected by *pith rays*, thin layers of ground tissue.

In monocots, vascular bundles are scattered throughout the ground tissue of the stem.

- The stem ground tissue is mostly parenchyma.
- Ground tissue of stems is mostly parenchyma, strengthened in many plants by collenchyma located beneath the epidermis.

The protoderm of the terminal bud gives rise to the epidermal portion of the dermal tissue system.

2. **Tissue Organization of Leaves**

Leaves are cloaked by an *epidermis* of tightly interlocked cells.

- It protects against physical damage and pathogens.
- A waxy cuticle prevents water loss.

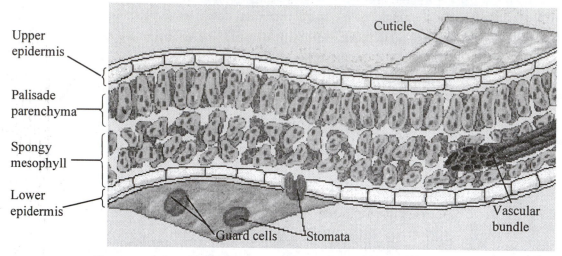

- *Stomata* are pores flanked by *guard cells* which regulate gas exchange with the surrounding air and photosynthetic cells inside the leaf.
- Stomata also allow *transpiration* (water loss from plant by evaporation).
- Stomata are usually more numerous on the bottom surface of the leaf. This location minimizes water loss.

The ground tissue of a leaf is *mesophyll*.

- Consists mainly of parenchyma cells equipped with chloroplasts for photosynthesis.
- Dicots usually have two distinct regions of mesophyll:

  Palisade parenchyma = One or more layers of columnar cells of the upper half of a leaf.

  Spongy mesophyll = Irregularly-shaped cells surrounded by air spaces through which oxygen and carbon dioxide circulate. Located in the lower half of the leaf.

The leaf vascular tissue is continuous with that of the stem through *leaf traces* which are branches from the stem vascular bundles.

- Leaf traces continue in the petiole as a vein, which branches repeatedly throughout the mesophyll of the leaf blade.
- This arrangement brings the photosynthetic tissue of the leaf into close contact with xylem and pholem.
- Functions also as a skeleton to support the shape of the leaf.

3. Modular Shoot Construction and Phase Changes During Development

Modular construction of a Shoot:

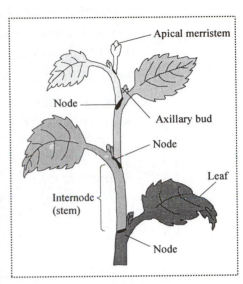

- Shoots are constructed of a series of modules produced by the serial development of nodes and internodes within the shoot apex.
- Each module consists of a stem, one or more leaves, and an axillary bud associated with each leaf.
- Elongation of the internode provides for primary growth of the plant.

Modules found in the body of a plant represent various ages of tissue. The age of a module is proportional to its distance from the apical meristem.

- The developmental phase of each module changes over time.
- There is a gradual change from a juvenile vegetative state to a mature vegetative state.
  ⇒ Leaf morphology usually changes during this transition.
- In some plants, the mature shoot apex may undergo a second phase change to a flower-producing (reproductive) state.
  ⇒ Primary growth of these shoot tips is terminated since the apical meristem is consumed in production of the flower.

VI. Lateral meristems add girth to stems and roots: *a closer look at secondary growth*

A *secondary plant body* results from secondary growth.

- The secondary plant body is comprised of the secondary tissues produced during growth in diameter.
- Secondary growth results from two lateral meristems: the *vascular cambium* and the *cork cambium*.
  ⇒ Vascular cambium produces secondary xylem and pholem.
  ⇒ Cork cambium produces a tough, thick covering for roots and stems that replaces the epidermis.
- Secondary growth occurs in all gymnosperms and most dicot angiosperms.
  ⇒ It is rare in monocots.

A. **Secondary Growth of Stems**

Vascular cambium forms when meristemic parenchyma cells develop between the primary xylem and primary phloem of each vascular bundle and in the rays of ground tissue between the bundles.

Fasicular cambium = Cambium within the vascular bundle.

Interfasicular cambium = Cambium in the rays between vascular bundles.

- Together, meristematic bands in the fasicular and interfasicular regions form a continuous cylinder of dividing cells around the xylem and pith of the stem.
- *Ray initials* (meristematic cells of the interfasicular cambium) produce radial files of parenchyma cells (= *xylem and phloem rays*) which permit lateral transport of water and nutrients as well as storage of starch and other reserves.
- *Fusiform initials* (cells of fasicular cambium) produce new vascular tissues; secondary xylem to the inside of the vascular cambium and secondary phloem to the outside.

Accumulated layers of secondary xylem produces *wood* which consists mostly of tracheids, vessel elements and fiber.

- The hardness and strength of wood results from these cells which, while dead at maturity, have thick, lignified walls.
- Forms annual growth rings due to yearly activity: cambium dormancy, spring wood production and summer wood production.

The secondary phloem does not accumulate extensively. The secondary phloem, and all tissues external to it, develop into bark which eventually sloughs off the tree trunk.

Cork cambium is a cylinder of meristematic tissue that forms protective layers of the secondary plant body.

- First forms in the outer cortex of the stem.
- *Phelloderm* (parenchyma cells) forms to the inside of cork cambium initials as they divide.
- *Cork cells* form to the outside of cork cambium initials as they divide.
- As cork cells mature, they deposit *suberin* (a waxy material) in their walls and die.
- These dead cork tissues protect the stem from damage and pathogens while reducing water loss.
- The combination of cork cambium, layers of cork, and phelloderm form the *periderm* (the protective coat of the secondary plant body that replaces primary epidermis).
- The term bark refers to all the tissues external to the vascular cambium (phloem, phelloderm, cork cambium and cork).

Cork cambium is a cylinder of fixed size and does not grow in diameter.

- As continued secondary growth splits this, it is replaced by new cork cambium formed deeper in the cortex.
- When no cortex is left, it develops from parenchyma cells in the secondary phloem (only the youngest secondary phloem, internal to cork cambium, functions in sugar transport).
- *Lenticels* are present in the bark.

  Lenticel= Spongy region in the bark which permit gas exchange by living cells within the trunk.

B.   Secondary Growth of Roots

Vascular cambium and cork cambium also function in secondary growth of roots.

- The vascular cambium produces secondary xylem to its inside and secondary phloem to its outside.
  - ⇒ It is first located between xylem and phloem of the stele.
  - ⇒ The cortex and epidermis split and are shed as the stele grows in diameter.
- Cork cambium forms from the pericycle of the stele and produces the *periderm*, which becomes secondary dermal tissue.
  - ⇒ Periderm is impermeable to water, consequently, the roots with secondary growth function to anchor the plant and transport water and solutes between the younger roots and the shoot system.

Older roots become woody, and annual rings appear in the secondary xylem.

- Old roots and old stems may look very similar.

REFERENCES

Campbell, N. *Biology*. 4th ed. Menlo Park, California: Benjamin/Cummings, 1996.

Raven, P.H., R.F. Evert and S.E. Eichhorn. *Biology of Plants*. 5th ed. New York: Worth Publishers, Inc., 1992.

OUTLINE

I. The traffic of water and solutes occurs on cellular, organ, and whole-plant levels: *an overview of transport in plants*
    A. Transport at the Cellular Level
    B. Short Distance (Lateral) Transport at the Level of Tissues and Organs
    C. Long Distance Transport at the Whole-Plant Level

II. Roots absorb water and minerals from soil

III. The ascent of xylem sap depends mainly on transpiration and the physical properties of water
    A. Pushing Xylem Sap: Root Pressure
    B. Pulling Xylem Sap: The Transpiration-Cohesion-Tension Mechanism
    C. Review: The Long-Distance Transport of Water in Plants by Solar-Powered Bulk Flow

IV. Guard cells mediate the transpiration-photosynthesis compromise
    A. The Photosynthesis-Transpiration Compromise
    B. How Stomata Open and Close
    C. Evolutionary Adaptations that Reduce Transpiration

V. A bulk-flow mechanism translocates phloem sap from sugar sources to sugar sinks
    A. Source-to-Sink Transport
    B. Phloem Loading and Unloading
    C. Pressure Flow (Bulk Flow) of Phloem Sap

OBJECTIVES

After reading this chapter and attending lecture, the student should be able to:
1. List three levels in which transport in plants occurs.
2. Trace the path of water and minerals from outside the root to the shoot system.
3. Provide experimental evidence that links plant cellular respiration to mineral accumulation.

4. Explain how a proton pump may affect mineral transport in plants.
5. Describe the symplast and apoplast routes for the transit of water and minerals across the root cortex from the epidermis to the stele.
6. Explain the function of the Casparian strip.
7. Explain how solutes are transferred between the symplast and apoplast.
8. Define water potential.
9. Explain how solute concentration and pressure affects water potential.
10. Predict the direction of net water movement based upon differences in water potential between a plant cell and a hypoosmotic environment, a hyperosmotic environment or an isosmotic environment.
11. Explain how root pressure is created by some plants and how it causes guttation.
12. According to the transpiration-cohesion-adhesion theory, describe how xylem sap can be pulled upward in xylem vessels.
13. Explain why a water potential gradient is required for the passive flow of water through a plant, from soil.
14. Compare the transpiration-to-photosynthesis ratio between C_3 and C_4 plants.
15. Describe both the disadvantages and benefits of transpiration.
16. Explain how guard cells control the stomatal aperture and how this, in turn, can affect photosynthetic rate and transpiration.
17. Explain how K^+ fluxes across the guard cell membrane affects guard cell function.
18. List three cues that contribute to stomatal opening at dawn.
19. Describe environmental stresses that can cause stomata to close during the daytime.
20. Explain how xerophytes can be adapted to arid climates.
21. Explain how crassulacean acid metabolism allows CAM plants to reduce the transpiration rate.
22. Describe source-to-sink transport in phloem and explain what determines the direction of phloem sap flow.
23. Compare the process of phloem loading between plants such as corn and squash.
24. Give one explanation for how a proton pump can allow for selective accumulation of sucrose in the symplast.
25. Explain what causes phloem sap to flow from source to sink and describe how a scientist can study pressure-flow in phloem.

KEY TERMS

cortex	endodermis	cavitation	xerophytes
symplast	transpiration	cohesion	CAM (crassulacean
apoplast	water potential	adhesion	acid metabolism)
tracheids	bar	stomata	translocation
xylem vessels	megapascal	guard cells	source
Casparian strip	plasmolysis	turgor pressure	sink
suberin	guttation	circadian rhythm	tonoplast
transfer cells	transpirational pull		

LECTURE NOTES

Land plants require a transport system, because unlike their aquatic ancestors, photosynthetic plant organs have no direct access to water and minerals.

I. **The traffic of water and solutes occurs on cellular, organ, and whole-plant levels:** *an overview of transport*

Three levels of transport occur in plants:

1. Uptake of water and solutes by individual cells.
2. Short-distance cell-to-cell transport at the level of tissues and organs.
3. Long-distance transport of sap in xylem and phloem at the whole-plant level.

A. **Transport at the Cellular Level**

1. **Review of Active and Passive Transport of Solutes**

The plasma membrane's selective permeability controls the movement of solutes between a plant cell and the extracellular fluids. Solutes may move by *passive* or *active transport*.

Passive transport occurs when a solute molecule diffuses across a membrane down a concentration gradient.

- No direct expenditure of energy by the cell.
- Transport proteins embedded in the membrane may increase the speed at which solutes cross.

Transport proteins may facilitate diffusion by serving as carrier proteins or forming selective channels.

- Carrier proteins bind selectively to a solute molecule on one side of the membrane, undergo a conformational change, and release the solute molecule on the opposite side of the membrane.
- Selective channels are simply passageways by which selective molecules may enter and leave a cell; some gated selective channels are stimulated to open or close by environmental conditions.

Active transport occurs when a solute molecule is moved across a membrane against the concentration gradient.

- Energy requiring process.
- The *proton pump* is an active transporter important to plants.

A proton pump hydrolyzes ATP and uses the energy to pump hydrogen ions (H^+) out of the cell.

- Produces a proton gradient with a higher concentration outside of the cell.
- Produces a membrane potential since the inside of the plant cell is negative in relation to the outside.

- This membrane potential and the stored energy of the proton gradient are used by the plant to transport many different molecules.

- Potassium ions (K^+) are pulled into the cells because of the electrochemical gradient.

- Nitrate (NO_3^-) enters plant cells against the electrochemical gradient by *cotransporting* with hydrogen ions.

2. Water Potential and Osmosis

Osmosis results in the net uptake or loss of water by the cell and depends on which component, the cell or extracellular fluids, has the highest *water potential*.

Water potential (Ψ) = The free energy of water that is a consequence of solute concentration and applied pressure; physical property predicting the direction water will flow.

- Water will always move across the membrane from the solution with the higher water potential to the one with lower water potential.

- Water potential is measured in units of pressure called *megapascals* (MPa); one MPa is equal to ten atmospheres of pressure.

- Pure water in an open container has a water potential of zero megapascals ($\Psi = 0$ MPa).

- Addition of solutes to water lowers the Ψ into the negative range.

- Increased pressure raises the Ψ into the positive range.

- A negative pressure, or *tension*, may also move water across a membrane; this *bulk flow* (movement of water due to pressure differences) is usually faster than the movement caused by different solute concentrations.

NOTE: You may wish to call attention to Campbell, Figure 32.3.

The effects of pressure and solute concentration on water potential are represented by:

$$\Psi = \Psi_P - \Psi_S$$

- Ψ_P = pressure potential.

- Ψ_S = solute potential or osmotic potential.

- A 0.1M solution has a pressure potential of 0.23 MPa; in an open container, physical pressure is zero; the water potential of this 0.1M solution would be -0.23 ($\Psi = 0 - 0.23 = -0.23$). Water would enter this solution due only to osmotic pressure.

- The addition of pressure to the solution could counter the affects of osmotic pressure by stopping net water movement (if $P = 0.23$) or by forcing water from the solution back into the pure water (if $P \geq 0.3$).

- Similar changes would result if a negative pressure were applied to the pure water side of the membrane.

Plant cells will gain or lose water to intercellular fluids depending upon their water potential.

- A flaccid cell (P = 0) placed in a hyperosmotic solution will lose water by osmosis; the cell will plasmolyze (protoplast pulls away from the cell wall) in response.
- A flaccid cell placed in a hypoosmotic solution will gain water by osmosis; the cell will swell and a *turgor pressure* develops; when pressure from the cell wall is equal to the osmotic pressure, an equilibrium is reached and no net water movement occurs ($P = \pi$; $\Psi = 0$).

3. The Role of the Tonoplast

Tonoplast = Membrane surrounding the large central vacuole found in plant cells; is important in regulating intracellular conditions.

- Contains integral transport proteins that control the movement of solutes between the cytosol and the vacuole.
- Has a membrane potential. Proton pumps in the tonoplast help the plasma membrane maintain a low H^+ concentration in the cytosol by moving H^+ *into* the vacuole.
- Several solutes are transported between the cytosol and vacuole due to this membrane potential and proton gradient.

B. Short-Distance (Lateral) Transport at the Level of Tissues and Organs

Lateral transport is usually along the radial axis of plant organs, and can occur by three routes in plant tissues and organs: (See Campbell, Figure 32.5)

1. *Across the plasma membranes and cell walls.* Solutes move from one cell to the next by repeatedly crossing plasma membranes and cell walls.
2. *The symplast route.* A symplast is the continuum of cytoplasm within a plant tissue formed by the plasmodesmata which pass through pores in the cell walls. Once water or a solute enters a cell by crossing a plasma membrane, the molecules can enter other cells by traveling through the plasmodesmata.
3. *The apoplast route.* An apoplast is the continuum between plant cells which is formed by the continuous matrix of cell walls. Water and solute molecules can move from one area of a root or other organ via the apoplast without entering a cell.

Water and solute molecules can move laterally in a plant organ by any one of these routes or by a combination through switching from one to another.

C. Long-Distance Transport at the Whole-Plant Level

This type of transport is usually along the vertical axis of the plant (up and down) from the roots to the leaves and vice versa.

Vascular tissues are involved in this type of transport as diffusion would be too slow.

- Bulk flow (movement due to pressure differences) moves water and solutes through xylem vessels and sieve tubes.

- Transpiration reduces pressure in the leaf xylem; this creates a tension which pulls sap up through the xylem from the roots.
- Hydrostatic pressure develops at one end of the sieve tubes in the phloem; this forces the sap to the other end of the tube.

II.   Roots absorb water and minerals from soil

Water and minerals enter plants through the following transport pathway:

soil ———▶ epidermis ———▶ root cortex ———▶ xylem. (See Campbell, Figure 32.6)

Soil ———▶ Epidermis:

- Most absorption occurs near root tips where the epidermis is permeable to water.
- Root hairs, extensions of epidermal cells, increase the surface area available for absorption.

Epidermis ———▶ Root cortex:

Lateral transport of minerals and water through the root is usually by a combination of apoplastic and symplastic routes.

- The apoplastic route exposes parenchymal, cortex cells to soil solution.
    - ⇒ Soil solution, containing soil particles, water and dissolved minerals, flows into hydrophilic walls of epidermal cells and passes freely along the apoplast into root cortex.
    - ⇒ Compared to the epidermis, the apoplastic route exposes greater membrane surface area for water and mineral uptake into cytoplasm.
- The symplastic route makes selective mineral absorption possible.
    - ⇒ As soil solution moves along cell walls, some water and solutes cross the plasma membrane of epidermal and cortex cells.
    - ⇒ Cells cannot absorb a sufficient supply of mineral ions by diffusion alone – the soil solution is too dilute. Active transport permits root cells to accumulate essential minerals in very high concentrations.
    - ⇒ For example, transport proteins of the plasma membrane and tonoplast actively transport K^+ into root cells as Na^+ is pumped out.

Root Cortex ———▶ Xylem:

- Only minerals using the symplastic route may move directly into the vascular tissues. They have been previously selected by a membrane.
- Minerals and water passing through apoplasts are blocked at the endodermis by a *Casparian strip* (a ring of suberin around each cell in the endodermis) and must enter an endodermal cell.

Water and mineral entrance into the stele must be through the cells of the endodermis.

- Casparian strips ensure that all substances entering the stele pass through at least one membrane, allowing only selected ions to pass into the stele. Also prevents stele contents from leaking back into the apoplast and out into the soil.
- Water and minerals enter the stele via symplast, but tracheids and xylem vessels are part of the apoplast.
- Endodermal and parenchymal cells selectively discharge minerals into the apoplast so they may enter the xylem. This action probably involves diffusion and active transport.
- Those minerals and water which move into the apoplast are free to enter the tracheids and xylem vessels.

III. The ascent of xylem sap depends mainly on transpiration and the physical properties of water

The shoot depends upon an efficient delivery of its water supply.

- Xylem sap flows upward at 15m per hour or faster.
- Xylem vessels are close to each leaf cell, because veins branch throughout the leaves.

Water transported up from roots must replace that lost by *transpiration*.

- Transpiration is the evaporation of water from the aerial parts of a plant.
- The upward flow of xylem sap also provides nutrients (minerals) to the shoot system.

A. Pushing Xylem Sap: Root Pressure

When transpiration is low, active transport of ions into the xylem decreases the stele's water potential and causes water flow into the stele. This osmotic water uptake increases pressure which forces fluid up the xylem (= *root pressure*).

- Root pressure causes *guttation* (exudation of water droplets at leaf margins).
- The water droplets escape through specialized structures called *hydrathodes* which relieve the pressure caused by more water entering the leaves than is lost by transpiration.

Root pressure is not the major mechanism driving the ascent of xylem sap.

- Cannot keep pace with transpiration.
- Can only force water up a few meters.

B. Pulling Xylem Sap: The Transpiration-Cohesion-Tension Mechanism

Transpiration pulls xylem sap upward, and cohesion of water transmits the upward-pull along the entire length of xylem.

Transpirational Pull:

- Transpirational pull depends upon the creation of negative pressure.
- Gaseous water in damp intercellular leaf spaces diffuses into the drier atmosphere through stomata.

- The lost water vapor is replaced by evaporation from mesophyll cells bordering the airspaces.
- The remaining water film, adhering to the hydrophilic cell walls, retreats into the cell wall pores.
- Cohesion in this surface film of water resists an increase in the surface area of the film – a surface tension effect.
- The water film forms a meniscus due to the negative pressure caused by the adhesion and cohesion.
- This negative pressure pulls water from the xylem, through the mesophyll, toward the surface film on cells bordering the stomata.
- Water moves through symplasts and apoplasts to a region of low water potential.
- Mechanism results in water from the xylem replacing water transpired through the stomata.

Cohesion and *Adhesion* of Water:
- The transpirational pull on the xylem sap is transmitted to the soil solution. Cohesion of water due to H bonds allows for the pulling of water from the top of the plant without breaking the "chain."
- The adhesion of water (by H bonds) to the hydrophilic walls of xylem cells also helps pull against gravity.
- The small diameter of vessels and tracheids is important to the adhesion effect.
- The upward pull of sap causes tension (negative pressure) in xylem, which decreases water potential and allows passive flow of water from soil into stele.

Cavitation = Formation of a water vapor pocket in xylem.
- Breaks the chain of water molecules and the pull is stopped.
- Vessels cannot function again unless refilled with water by root pressure. (This can only occur in small plants.)
- Pits between adjacent xylem vessels allow for detours around a cavitated area.
- Secondary growth also adds new xylem vessels each year.

C. **Review: The Long-Distance Transport of Water in Plants by Solar-Powered Bulk Flow**

Bulk flow is the movement of fluid due to pressure differences at opposite ends of a conduit.

The ascent of xylem sap is ultimately solar-powered, by causing evaporative water loss and, thus, negative pressure.
- Xylem vessels or chains of tracheids serve as the conduits in plants.
- Transpirational pull lowers the pressure at the upper (leaf) end of the conduit.
- Osmotic movement of water from cell to cell in roots and leaves are due to small gradients in water potential caused by both solute and pressure gradients.
- In contrast, bulk flow through the xylem vessels depends only on pressure.

IV.   Guard cells mediate the transpiration-photosynthesis compromise

Transpiration results in a tremendous water loss from the plant. This water is replaced by the upward movement of water through the xylem, and guard cells surrounding stomata balance the requirements for photosynthesis with the need to conserve water.

A.   The Photosynthesis-Transpiration Compromise

Large surface areas along a leaf's airspaces are needed for CO_2 intake for photosynthesis, but also results in greater surface area for evaporative water loss.
- Internal surface area of a leaf may be 10×–30× the external surface area.
- Stomata are more concentrated on the bottom of leaves, away from the sun; this reduces evaporative loss.
- The waxy leaf cuticle prevents water loss from the rest of the leaf surface.

Transpiration:photosynthesis ratios measure efficiency of water use. This ratio is g H_2O lost/g CO_2 assimilated into organic material.
- Ratio of 600:1 is common in C_3 plants; 300:1 in C_4 plants.
- C_4 plants can assimilate CO_2 at greater rates than C_3 plants.

Benefits of transpiration:
- Assists in mineral transfer from roots to shoots.
- Evaporative cooling reduces risk of leaf temperatures becoming too high for enzymes to function.

If transpiration exceeds delivery of water by xylem, plants wilt.
- Plants can adjust to reduce risk of wilting.
- Regulating the size of stomatal openings also reduces transpiration.

B.   How Stomata Open and Close

Guard cells = Cells that flank stomata and control stomatal diameter by changing shape.
- When turgid, guard cells "buckle" due to radially-arranged microfibrils and stomata open; when flaccid, guard cells sag and stomatal openings close.

The change in turgor pressure that regulates stomatal opening results from reversible uptake and loss of K^+ by guard cells. (See Campbell, Figure 32.8)
- Uptake of K^+ decreases guard cell water potential so H_2O is taken up, cells become turgid, and stoma opens. The tonoplast plays a role as most of the K^+ and water are stored in the vacuole.
- The increase in positive charge is countered by the uptake of chloride (Cl^-), export of H^+ ions released from organic acids, and the negative charges acquired by organic acids as they lose their protons.
- Closing of the stomata results when K^+ exits the guard cells and creates an osmotic loss of water.

Evidence from studies using patch clamping techniques indicates that K^+ fluxes across the guard cell membrane are likely coupled to membrane potentials created by proton pumps.

- Stomatal opening correlates with active transport of H^+ out of the guard cell.
- The resulting membrane potential drives K^+ into the cell through specific membrane channels.

By integrating internal and external environmental cues, guard cells open and close, balancing the requirements for photosynthesis with the need to conserve water from transpirational loss.

- Stomata open at dawn in response to three cues:
  - $\Rightarrow$ Light induces guard cells to take up K^+ by:
    - $\Diamond$ Activating a blue-light receptor which stimulates proton pumps in the plasma membrane.
    - $\Diamond$ Driving photosynthesis in guard cell chloroplasts, making ATP available for the ATP-driven proton pumps.
  - $\Rightarrow$ Decrease of CO_2 in leaf air spaces due to photosynthesis in the mesophyll.
  - $\Rightarrow$ An internal clock of guard cells will make them open even if kept in dark (a *circadian rhythm* approximates a 24 hour cycle).
- Guard cells may close stomata during the daytime if:
  - $\Rightarrow$ There is a water deficiency resulting in flaccid guard cells.
  - $\Rightarrow$ Mesophyll production of abscisic acid (a hormone) in response to water deficiency signals guard cells to close.
  - $\Rightarrow$ High temperature increases CO_2 in leaf air spaces due to increased respiration, closing guard cells.

C. Evolutionary Adaptations that Reduce Transpiration

Xerophytes, plants adapted to arid climates, have some of the following evolutionary adaptations that reduce transpiration:

- Small, thick leaves (reduced surface area:volume, so less H_2O loss).
- A thick cuticle.
- Stomata are in depressions on the underside of leaves to protect from water loss due to drying winds.
- Some shed leaves in the driest time of the year.
- Cacti and others store water in stems during the wet season.

Plants of family Crassulaceae use CAM (crassulacean acid metabolism) to assimilate CO_2.

- At night, mesophyll cells assimilate CO_2 into organic acids.
- During the day, the acids are broken down, releasing CO_2 which is used to synthesize sugars by the conventional C_3 pathway.
- Thus, stomata can close during the day, the plant conserves water, and there is still an ample supply of CO_2 for photosynthesis.

V.   A bulk-flow mechanism translocates phloem sap from sugar sources to sugar sinks

Translocation = The transport of the products of photosynthesis by phloem to the rest of the plant.

- In angiosperms, *sieve-tube members* are the specialized cells of phloem that function in translocation.
    ⇒ Sieve-tube members are arranged end-to-end forming long *sieve tubes*.
    ⇒ Porous cross walls called *sieve plates* are in between the members and allow phloem to move freely along the sieve tubes.
- Phloem sap contains primarily sucrose, but also minerals, amino acids and hormones.

A.   Source-to-Sink Transport

Phloem sap movement is not unidirectional; it moves through the sieve tubes from a source (production area) to a sink (use or storage area).

Source = Organ where sugar is produced by photosynthesis or by the breakdown of starch (usually leaves).

Sink = Organ that consumes or stores sugar (growing parts of plants, fruits, non-green stems and trunks, and others).

Sugar flows from source to sink.

- Source and sink depend on season. A tuber is the sink when stockpiling in the summer, but is the source in the spring.
- Minerals may also be transported to sinks.
- The sink is usually supplied by the nearest source.
- Direction of flow within a phloem element can change, depending on locations of the source and sink.

B.   Phloem Loading and Unloading

Sugar produced at a source must be loaded into sieve-tube members before it can be translocated to a sink.

- In some plant species, the sugar may move through the symplast from mesophyll cells to the sieve members.
- In other species, the sugar uses a combination of symplastic and apoplastic routes. (See Campbell, Figure 32.9)
- Some plants have *transfer cells*. These are modified companion cells which have numerous ingrowths of their walls. These structures increase the cells' surface area and enhances solute transfer between apoplast and symplast.
- In plants such as corn, active transport accumulates sucrose in sieve-tube members to two to three the concentration in mesophyll cells.
    ⇒ Proton pumps power this transport by creating a H^+ gradient.
    ⇒ A membrane protein uses the potential energy stored in the gradient to drive the cotransport of sucrose by coupling sugar transport to the diffusion of H^+ back into the cell.

Sucrose is unloaded at the sink end of sieve tubes.

- In some plants, sucrose is unloaded from the phloem by active transport.
- In other species diffusion moves the sucrose from the phloem into the cells of the sink.
- Both symplastic and apoplastic routes may be involved.

C.   Pressure Flow (Bulk Flow) of Phloem Sap

Phloem sap flows up to 1 meter per hour, too fast for just diffusion or cytoplasmic streaming.

The flow is by a bulk flow (pressure-flow) mechanism; buildup of pressure at the source and release of pressure at the sink causes source-to-sink flow. (See Campbell, Figure 32.10)

- At the source end, phloem loading causes high solute concentrations.
  - ⇒ Water potential decreases, so water flows into tubes creating hydrostatic pressure.
  - ⇒ Hydrostatic pressure is greatest at the source end of the tube.
- At the sink end, the water potential is lower outside the tube due to the unloading of sugar; osmotic loss of water releases hydrostatic pressure.
- Xylem vessels recycle water from the sink to the source.

Aphids have been used to study flow in phloem. (See Campbell, Figure 32.11)

- The aphid stylet punctures phloem and the aphid is "force-fed" by the pressure.
- The aphid is severed from its stylet and flow is measured through the stylet.
- The flow exerts greater pressure and has a higher sugar concentration the closer to the source.

REFERENCES

Campbell, N. *Biology*. 4th ed. Menlo Park, California: Benjamin/Cummings, 1996.

Raven, P.H., R.F. Evert and S.E. Eichhorn. *Biology of Plants*. 5th ed. New York: Worth Publishers, Inc., 1992.

CHAPTER 33

PLANT NUTRITION

OUTLINE

I.   Plants require at least seventeen essential nutrients
    A.   The Chemical Composition of Plants
    B.   Essential Nutrients

II.   The symptoms of mineral deficiency depend on the function and mobility of the element

III.   Soil characteristics are key environmental factors in terrestrial ecosystems
    A.   Texture and Composition of Soils
    B.   The Availability of Soil Water and Minerals

IV.   Soil conservation is one step toward sustainable agriculture

V.   The metabolism of soil bacteria makes nitrogen available to plants
    A.   Nitrogen Fixation
    B.   Symbiotic Nitrogen Fixation

VI.   Improving the protein yield of crops is a major goal of agricultural research

VII.   Predation and symbiosis are evolutionary adaptations that enhance plant nutrition
    A.   Parasitic Plants
    B.   Carnivorous Plants
    C.   Mycorrhizae

OBJECTIVES

After reading this chapter and attending lecture, the student should be able to:
1.   Describe the chemical composition of plants including:
    a. Percent of wet weight as water.
    b. Percent of dry weight as organic substances.
    c. Percent of dry weight as inorganic minerals.
2.   Explain how hydroponic culture is used to determine which minerals are essential nutrients.

3. Distinguish between macronutrient and micronutrient.
4. Recall the nine macronutrients required by plants and describe their importance in normal plant structure and metabolism.
5. List seven micronutrients required by plants and explain why plants need only minute quantities of these elements.
6. Explain how a nutrient's role and mobility determine the symptoms of a mineral deficiency.
7. Explain how soil is formed.
8. Explain what determines the texture of topsoil and list the type of soil particles from coarsest to smallest.
9. Describe the composition of loams and explain why they are the most fertile soils.
10. Explain how humus contributes to the texture and composition of soil.
11. Explain why plants cannot extract all of the water in soil.
12. Explain how the presence of clay in soil helps prevent the leaching of mineral cations.
13. Define cation exchange, explain why it is necessary for plant nutrition, and describe how plants can stimulate the process.
14. Explain why soil management is necessary in agricultural systems but not in natural ecosystems such as forests and grasslands.
15. List the three mineral elements that are most commonly deficient in farm soils.
16. Describe the environmental consequence of overusing commercial fertilizers.
17. Explain how soil pH determines the effectiveness of fertilizers and a plant's ability to absorb specific mineral nutrients.
18. Describe problems resulting from farm irrigation in arid regions and list several current approaches to solving these problems.
19. Describe precautions that can reduce wind and water erosion.
20. Define nitrogen fixation and write the overall equation representing the conversion of gaseous nitrogen to ammonia.
21. Distinguish between nitrogen-fixing bacteria and nitrifying bacteria.
22. Recall the forms of nitrogen that plants can absorb and describe how they are used by plants.
23. Beginning with free-living rhizobial bacteria, describe the development of a root nodule.
24. Explain why the symbiosis between a legume and its nitrogen-fixing bacteria is considered to be mutualistic.
25. Recall two functions of leghemoglobin and explain why its synthesis is evidence for coevolution.
26. Describe the basis for crop rotation.
27. Describe agricultural research methods used to improve the quality and quantity of proteins in plant crops.
28. Describe modifications for nutrition that have evolved among plants including parasitic plants, carnivorous plants and mycorrhizae.

KEY TERMS

macronutrients	humus	nitrogen fixation	leghemoglobin
micronutrients	cation exchange	nitrifying bacteria	haustoria
horizons	chemical fertilizers	nodules	epiphytes
topsoil	drip irrigation	*Rhizobium*	mycorrhizae
loams			

LECTURE NOTES

Plants and other photosynthetic autotrophs play a critical role in the energy flow and chemical cycling of ecosystems by transforming:

- Light energy into chemical bond energy.
- Inorganic compounds into organic compounds.

To accomplish these tasks, plants require:

- Sunlight as the energy source for photosynthesis.
- Inorganic, raw materials, such as:
  ⇒ carbon dioxide
  ⇒ water
  ⇒ variety of inorganic, mineral ions in the soil

To acquire the resources of light and inorganic nutrients, plants have highly ramified root and shoot systems with large surface areas for exchange with their environment – soil and air.

I. Plants require at least seventeen essential nutrients

A. The Chemical Composition of Plants

Early ideas about plant nutrition:

- In the fourth century B.C., Aristotle thought that soil was the substance for plant growth and that leaves only provided shade for fruit.
- In the 1600's, Jean-Baptiste von Helmont concluded plants grow mainly from the water it absorbs. To discover if plants grew by absorbing soil, he:
  ⇒ planted a seed in a measured amount of soil;
  ⇒ weighed the grown plant;
  ⇒ measured how much soil was left. The soil did not decrease proportionately.
- In the 1700's, Stephen Hales postulated plants were nourished mostly by air.

Plants grow mainly by accumulating water in their cells' central vacuoles.

- Water is a nutrient since it supplies most of the hydrogen and some oxygen incorporated into organic compounds by photosynthesis.
- 80–85% of an herbaceous plant is water.
- More than 90% of the water absorbed is lost by transpiration.
- Retained water functions as a solvent, allows cell elongation, and keeps cells turgid.

By weight, the bulk of a plant's organic material is from assimilated CO_2. The composition of the dry weight of plants is:

- 95% organic substances, mostly carbohydrates.
- 5% minerals, to some extent determined by soil composition.

B.   **Essential Nutrients**

An *essential nutrient* is one that is required for a plant to grow from a seed and complete its life cycle. (See Campbell, Table 33.1)

- Determined by hydroponic culture, in which roots are bathed in an aerated aqueous solution of known mineral content. If a mineral is omitted and plant growth is abnormal, it is essential.

Macronutrients = Elements required by plants in large amounts. (See Campbell, Table 33.1)

Micronutrients = Elements required by plants in small amounts.
- These function primarily as cofactors of enzymatic reactions.
- Optimal concentrations vary for different plant species.

II.   **The symptoms of a mineral deficiency depend on the function and mobility of the element**

Symptoms of mineral deficiencies depend on:
1.   The role of the nutrient in the plant.
2.   Its mobility within the plant.
   - Deficiencies of nutrients mobile in the plant appear in older organs first since some are preferentially shunted to growing parts.
   - Deficiencies of immobile nutrients affect young parts of plant first because older tissues may have adequate reserves.

Deficiencies of N, K, and P are the most common.
- Shortages of micronutrients are less common and often localized.
- Overdoses of some micronutrients can be toxic.

III.   **Soil characteristics are key environmental factors in terrestrial ecosystems**

Plants growing in an area are adapted to the texture and chemical composition of the soil.

A.   **Texture and Composition of Soils**

Soil is produced by the weathering of solid rock. Living organisms may accelerate the process once they become established.

Horizons = Distinct soil layers.

Topsoil = Mixture of decomposed rock of varying texture, living organisms, and humus. The texture of a topsoil depends on the particle size.

- Coarse sand has diameter of 0.2 – 2 mm.
- Sand is 20 – 200 mm.
- Silt is 2 – 20 mm.
- Clay is less than 2 mm.

Loams = Mixture of sand, silt and clay; are the most fertile soils.

- Fine particles retain water and minerals.
- Coarse particles provide air spaces with oxygen for cellular respiration.

Soil contains many bacteria, fungi, algae, protists, insects, earthworms, nematodes and plant roots. These affect the soil composition. For example, earthworms aerate soil; bacteria alter soil mineral composition.

Humus = Decomposing organic material.

- Prevents clay from packing together.
- Builds a crumbly soil that retains water but is still porous for good root aeration.
- Acts as a reservoir of mineral nutrients.

Soil composition determines which plants may grow in it, and plant growth, in turn, affects soil characteristics.

B.   The Availability of Soil Water and Minerals

Some water is bound so tightly to hydrophilic soil that it cannot be extracted by plants.

Water bound less tightly is generally available to the plant as a soil solution containing minerals. This solution is absorbed into the root hairs and passes via the apoplast to the endodermis. (See Campbell, Figure 33.5)

Positively charged minerals (K^+, Ca^+, Mg^+) adhere by electrical attraction to negatively charged clay particles.

- Clay provides much surface area for binding.
- Prevents leaching of mineral nutrients.

Cation exchange = H ions in soil displace positively charged mineral ions from clay, making them available to plants.

- Stimulated by roots which release acids to add H^+ to the soil solution.

Negatively charged minerals (NO_3^-, $H_2PO_4^-$, SO_4^-) are not tightly bound to soil particles.

- Tend to leach away more quickly.

Water and minerals are depleted from soil surrounding roots, but roots grow to bring the plant into contact with a new supply of nutrients.

IV. **Soil conservation is one step toward sustainable agriculture**

Good soil management is necessary to maintain soil fertility which may have taken centuries to develop through decomposition and accumulation of organic matter.

- Agriculture is unnatural and depletes the mineral content of the soil, making soil less fertile. Crops also use more water than natural vegetation.
- Three important aspects of soil management are:
  1. Fertilizers
  2. Irrigation
  3. Erosion prevention

1. **Fertilizers**

   Fertilizers may be mined, chemically produced or organic.

   - Usually enriched in N, P, K.
   - Marked with 3 numbers corresponding to % nitrogen (as ammonium or nitrate), % phosphorus (as phosphoric acid), and % potassium (as potash).

   Organic fertilizers are manure, fishmeal, and compost.

   - Release minerals more gradually than chemical fertilizers, thus, soil retains minerals longer. Excess minerals from chemical fertilizers may be leached from soil and may pollute streams and lakes.

   To properly use fertilizers, one must consider the soil pH.

   - Acidity affects cation exchange and the chemical form of the minerals.
   - A change in soil pH may make one essential element more available while causing another to adhere so tightly to soil particles that it is unavailable.

2. **Irrigation**

   Availability of water limits plant growth in many environments.

   Problems of irrigation arid land:

   - Huge drain of water resources.
   - Can gradually make soil salty and infertile.

   Solutions to these problems:

   - Use of *drip irrigation*. Perforated pipes slowly drip water close to plant roots, which reduces water evaporation and drainage.
   - Development of plant varieties that require less water or can tolerate more salinity.

3. **Erosion**

   Wind and water erode away much of the topsoil each year. Measures to prevent these losses include:

   - Rows of trees to divide fields act as windbreaks.
   - Terracing hillsides helps prevent water erosion.
   - Planting alfalfa and wheat provide good ground cover and protection.

   Proper management makes soil a renewable resource which will remain fertile for many generations.

V. **The metabolism of soil bacteria makes nitrogen available to plants**

Plants require nitrogen to produce proteins, nucleic acids and other organic molecules.

- Plants can not use nitrogen in gaseous form (N_2).
- To be assimilated by plants, nitrogen must be in the form of ammonium (NH_4^+) or nitrate (NO_3^-).

There is a complex cycling of nitrogen in ecosystems:

- Over the short term, the main source of nitrogenous minerals is the decomposition of humus by microbes.

  ⇒ For example, *ammonifying bacteria*.

  ⇒ Nitrogen in organic compounds is repackaged into inorganic compounds that can be absorbed as minerals by roots.

- Nitrogen is lost from this local cycle when soil *denitrifying bacteria* convert NO_3^- to N_2, which diffuses form the soil to the atmosphere.

- *Nitrogen fixing bacteria* restock nitrogenous minerals in the soil by converting N_2 to NH_3 (ammonia), a metabolic process called *nitrogen fixation*.

A. **Nitrogen Fixation**

Nitrogen fixation = The process of converting atmospheric nitrogen (gaseous state) to nitrogenous compounds that can be directly used by plants (nitrate or ammonia). (See Campbell, Figure 33.8)The process is catalyzed by the enzyme *nitrogenase*:

$$N_2 + 8e^- + 8H^+ + 16ATP \xrightarrow{\text{nitrogenase}} 2NH_3 + H_2 + 16ADP + 1 P_i$$

- Some soil bacteria possess nitrogenase.
- Very energy consuming process, costing the bacteria at least 12 ATPs for each ammonia molecule synthesized.
- In the soil, ammonia is converted to the ammonium ion which plants can absorb:

$$NH_3 + H^+ \longrightarrow NH_4^+$$

- Plants acquire most of their nitrogen in the form of nitrate (NO_3^-), which is produced in soil by *nitrifying bacteria* that oxidize ammonium.
- Nitrogen absorbed by the plant is incorporated into organic compounds.
- Most plant species export nitrogen from the roots to the shoots (through xylem) in the form of organic compounds and amino acids.

B.   Symbiotic Nitrogen Fixation

Legumes (e.g. peas, beans, soybeans, peanuts, alfalfa, clover) have a built-in source of fixed nitrogen because they possess root *nodules*.

Nodules = Root swellings composed of plant cells that contain nitrogen-fixing bacteria of the genus *Rhizobium*. Each species of legume is associated with a particular species of *Rhizobium*.
Nodules form as follows: (See Campbell, Figure 33.9)

- Roots secrete chemicals that attract nearby bacteria.
- Attracted bacteria emit chemicals which stimulate root hairs to elongate and curl to prepare for bacterial infection.
- Bacteria enter the root through an "infection thread" that carries them to the root cortex.
- Bacteria become enclosed in vesicles and assume a form called *bacteroids*.
- Bacteroids produce a chemical that induces the host's cells to divide and form a nodule.
- Nodules continue to grow and a connection with the xylem and phloem develops.

This association is mutualistic; the bacteria supplies fixed nitrogen, and the plant supplies carbohydrates and other organic compounds.

Leghemoglobin = An iron-containing protein that binds oxygen.

- The plant and the bacteria each make a part of the molecule.
- Releases oxygen for the intense respiration needed to produce ATP for nitrogen fixation.
- Keeps the free oxygen concentration low in root nodules so that the oxygen cannot inhibit the function of nitrogenase.

Most of the ammonium produced is used by the nodules to make amino acids for export to the shoots and leaves.

The basis for crop rotation is that, under favorable conditions, root nodules fix more nitrogen than the legume uses. The excess is secreted as ammonium into the soil.

- One year a nonlegume crop is planted, and the next year a legume is planted to restore the fixed nitrogen content of the soil.
- Legumes may be plowed under to further increase the fixed nitrogen content of the soil.

Some non-legumes host nitrogen-fixing symbionts.

- Alders and tropical grasses may host nitrogen-fixing actinomycetes.
- Rice farms culture a fern (*Azolla*), containing symbiotic nitrogen-fixing cyanobacteria, with the rice.

VI.   Improving the protein yield of crops is a major goal of agricultural research

A majority of the world's population depends mainly on plants for protein. Some food plants have a low protein content while others may be deficient in certain amino acids.

Ways in which to increase the protein content of plants are:

1.   Plant breeding to create new varieties enriched in protein.

   - Unfortunately these "super" varieties require large quantities of nitrogen in the form of commercial fertilizer too expensive for many countries to afford.

2.   Improving the productivity of symbiotic nitrogen fixation.

   - Mutant strains of *Rhizobium* have been isolated that continue to produce nitrogenase even after fixed nitrogen accumulates, thus releasing the excess into the soil.

   - *Rhizobium* varieties may be selected that fix N_2 at a lower cost in photosynthetic energy, yielding a higher total food content in the plant.

3.   Genetic engineering.

   - May create varieties of *Rhizobium* that can infect nonlegumes.

   - May transfer the genes required for nitrogen-fixation directly into plant genomes, using bacterial plasmids as vectors. (See Campbell, Chapter 19)

VII.   Predation and symbiosis are evolutionary adaptations that enhance plant nutrition

Some plant adaptations enhance nutrition through interactions with other organisms.

A.   Parasitic Plants

Some parasitic plants:

   - Are photosynthetic, and only supplement nutrition by using *haustoria* (<u>not</u> homologous to those of parasitic fungi) to obtain xylem sap from its host plant (e.g. mistletoe).

   - Have lost photosynthesis entirely, drawing all nutrients from the host plant (e.g. dodder).

Epiphytes are plants that:

   - Grow on the surface of other plants, anchored by roots, but are *not parasitic*.
   - Nourish themselves from the water and minerals absorbed from rain.
   - Examples include Spanish moss and staghorn ferns.

B.   Carnivorous Plants

Carnivorous plants:

   - Live in habitats with poor (usually nitrogen deficient) soil conditions.
   - Are photosynthetic, but obtain some nitrogen and minerals by killing and digesting insects.
   - Most insect traps evolved by modifications of leaves and usually are equipped with glands that secrete digestive juices.

C.   Mycorrhizae

<u>Mycorrhizae</u> = Symbiotic associations (mutualistic) between plant roots and fungi. The fungus either forms a sheath around the root or penetrates root tissue.
- Helps the plant absorb water.
- Absorbs minerals and may secrete acid that increases mineral solubility and converts minerals to forms easily used by the plant.
- May help protect the plant against certain soil pathogens.
- The plant nourishes the fungus with photosynthetic products.

Almost all plants are capable of forming mycorrhizae if exposed to the proper species of fungi. Plants grow more vigorously when mycorrhizae are present.

Mycorrhizae may have permitted early plants to colonize land.
- Fossils indicate the earliest land plants possessed mycorrhizae.
- This mutualistic association may have allowed the early plants to obtain enough nutrients to survive colonization.

REFERENCES

Campbell, N. *Biology*. 4th ed. Menlo Park, California: Benjamin/Cummings, 1996.

Raven, P.H., R.F. Evert and S.E. Eichhorn. *Biology of Plants*. 5th ed. New York: Worth Publishers, Inc., 1992.

Reganold, J.P., R.I. Papendick and J.F. Parr. "Sustainable Agriculture." *Scientific American*, June 1990.

CHAPTER 34

PLANT REPRODUCTION AND DEVELOPMENT

OUTLINE

I. Sporophyte and gametophyte generations alternate in the life cycles of plants: *an overview*
   A.   More About Flowers

II. Male and female gametophytes develop within anthers and ovaries, respectively
   A.   Pollen Development
   B.   Ovule Development

III. Pollination brings female and male gametophytes together
   A.   Pollination
   B.   The Molecular Basis of Self-Incompatibility
   C.   Double Fertilization

IV. The ovule develops into a seed containing a sporophyte embryo and a supply of nutrients
   A.   Endosperm Development
   B.   Embryo Development (Embryogenesis)
   C.   Structure of the Mature Seed

V. The ovary develops into a fruit adapted for seed dispersal

VI. Evolutionary adaptations in the process of germination increase the probability that seedlings will survive
   A.   Seed Dormancy
   B.   From Seed to Seedling

VII. Many plants can clone themselves by asexual reproduction

VIII. Vegetative reproduction of plants is common in agriculture
   A.   Clones from Cuttings
   B.   Test-Tube Cloning and Related Techniques
   C.   Benefits and Risks of Monoculture

OBJECTIVES

After reading this chapter and attending lecture, the student should be able to:
1. Outline the angiosperm life cycle.
2. List the four floral parts in their order from outside to inside of the flower.
3. From a diagram of an idealized flower, correctly label the following structures and describe their function:
     a. Sepals
     b. Petals
     c. Stamen: filament and anther
     d. Carpel: style, ovary, ovule and stigma
4. Distinguish between complete and incomplete flowers.
5. Distinguish between a perfect and imperfect flower.
6. Distinguish between monoecious and dioecious.
7. Explain by which generation, structure and process spores are produced.
8. Explain by which generation, structures and process gametes are produced.
9. Explain why it is technically incorrect to refer to stamens and carpels as male and female sex organs.
10. Describe the formation of a pollen grain in angiosperms.
11. With reference to the developing pollen grain, distinguish among generative nucleus, tube nucleus and sperm nucleus.
12. Describe the development of an embryo sac and explain what happens to each of its cells.
13. Distinguish between pollination and fertilization.
14. Describe how pollen can be transferred between flowers.
15. Describe mechanisms that prevent self-pollination and explain how this contributes to genetic variation.
16. Outline the process of double fertilization and describe the function of endosperm.
17. Describe the development of a plant embryo from the first mitotic division to an embryonic plant with rudimentary organs.

18. From a diagram, identify the following structures of a seed and recall a function for each:
    a. Seed coat
    b. Embryo
    c. Hypocotyl
    d. Radicle
    e. Epicotyl
    f. Plumule
    g. Endosperm
    h. Cotyledons
    i. Shoot apex
19. Explain how a monocot and dicot seed differ.
20. Describe several functions of fruit and explain how fruits form.
21. Distinguish among simple, aggregate and multiple fruits and give examples of each.
22. Explain how seed dormancy can be advantageous to a plant and describe some conditions for breaking dormancy.
23. Using a cereal as an example, explain how a seed mobilizes its food reserves and describe the function of aleurone, α-amylase and gibberellic acid.
24. Describe variations in the process of germination including the fate of the radicle, shoot tip, hypocotyl, epicotyl and cotyledons.
25. Distinguish between sexual reproduction and vegetative reproduction.
26. Describe natural mechanisms of vegetative reproduction in plants including fragmentation and apomixes.
27. Describe various methods horticulturists use to vegetatively propagate plants from cuttings.
28. Explain how the technique of plant tissue culture can be used to clone and genetically engineer plants.
29. Describe the process of protoplast fusion and its potential agricultural impact.
30. Define monoculture and list its benefits and risks.
31. Compare sexual and asexual reproduction in plants and explain their adaptive roles in plant populations.

KEY TERMS

flower
complete flower
sepal
petal
stamen
carpel
incomplete flower
perfect flower
imperfect flower
monoecious
dioecious
alternation of
    generations
sporophyte
sporangia
gametophyte
pollen grain
microspore
generative nucleus

tube nucleus
ovules
megasporocyte
megaspore
embryo sac
synergids
micropyle
pollination
double fertilization
endosperm
seed
cotyledon
suspensor
proembryo
torpedo stage
primary meristem
protoderm
ground meristem
procambium

hypocotyl
radicle
epicotyl
plumule
coleorhiza
coleoptile
fruit
simple fruit
aggregate fruit
multiple fruit
inflorescence
fruit set
climacteric stage
seed dormancy
seed germination
imbibition
gibberellic acid (GA)
aleurone
fragmentation

apomixis
callus
stock
scion
explants
protoplast fusion
monoculture
"gene banks"
growth
morphogenesis
cellular
    differentiation
preprophase band
pattern formation
positional
    information
clonal analysis
organ-identity genes

LECTURE NOTES

Modifications in reproduction were key adaptations enabling plants to spread into a variety of terrestrial habitats.

- Water has been replaced by wind and animals as a means for spreading gametes.
- Embryos are protected in seeds.
- Vegetative reproduction is an asexual mechanism for propagation in many environments.

I. **Sporophyte and gametophyte generations alternate in the life cycles of plants:** *an overview*

The angiosperm (flowering plant) life cycle includes *alternation of generations* during which multicellular haploid gametophyte generations alternate with diploid sporophyte generations. (See Campbell, Figure 34.1)

- The *sporophyte* is the recognizable "plant" most familiar to us. It produces haploid spores by meiosis in *sporangia*.
- Spores will undergo mitotic division and develop into a multicellular male or female *gametophyte*.
- Gametophytes produce gametes (sperm and egg) by mitosis. The gametes fuse to form a zygote which develops into a multicellular sporophyte.
- The sporophyte is dominant in the angiosperm life cycle with the gametophyte stages being reduced and totally dependent on the sporophyte.

Flowers are the reproductive structure of angiosperm sporophytes.

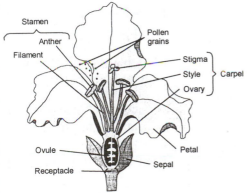

- Evolved from compressed shoots with four whorls of modified leaves separated by very short internodes.
- The four sets of modified leaves are the: sepals, petals, stamens, and carpels.
- Stamens and carpels contain the sporangia and are the reproductive parts of the flower.
  - ⇒ Female gametophytes develop in carpel sporangia as *embryo sacs* which contain the eggs. This occurs inside the *ovules* which are at the base of the carpel and surrounded by ovaries.
  - ⇒ Male gametophytes develop in the stamen sporangia as *pollen grains*. These form at the stamen tips within chambers of the anthers.

Pollination occurs when wind- or animal-born pollen released from anthers lands on the stigma at the tip of a carpel.

- A pollen tube grows from the pollen grain, down the carpel, into the embryo sac.
- Sperm are discharged resulting in fertilization of the eggs.
- The zygote will develop into an embryo; as the embryo grows, the ovule surrounding it develops into a *seed*.
- While seed formation is taking place, the entire ovary is developing into a *fruit* which will contain one or more seeds.

Seeds are dispersed from the source plant when fruits are moved about by the wind or animals.

- Seeds deposited in soil of the proper conditions (moisture, nutrients) will *germinate*.
- The embryo starts growing and develops into a new sporophyte.
- After flowers are produced by the sporophyte, a new generation of gametophytes develop and the life cycle continues.

A.   More About Flowers

Several variations on the basic flower structure have evolved during the angiosperm evolutionary history.

Complete flower = A flower with sepals, petals, stamens and carpels.

Incomplete flower = A flower that is missing one or more of the parts listed for a complete flower (e.g. most grasses do not have petals on their flowers).

Perfect flower = A flower having both stamens and carpels (may be incomplete by lacking either sepals or petals).

Imperfect flower = A flower that is either *staminate* (having stamens but no carpels) or *carpellate* (having carpels but no stamens) - a unisex flower.

Monoecious = Plants having both staminate flowers and carpellate flowers on the same individual plant.

Dioecious = Plants having staminate flowers and carpellate flowers on separate individual plants of the species.

See Campbell, Figure 34.3, for a review of additional differences in flower structure.

II. **Male and female gametophytes develop within anthers and ovaries, respectively**

    A. **Pollen Development**

Pollen grain = The immature male gametophyte that develops within the anthers of stamens in an angiosperm.

- Extremely durable; their tough coats are resistant to biodegradation.
- Fossilized pollen has provided many important evolutionary clues.
- Formation of a pollen grain is as follows: (See Campbell, Figure 34.4a)

Within the sporangial chamber of an anther, diploid microsporocytes undergo meiosis to form four haploid microspores.

↓

The haploid microspore nucleus undergoes mitotic division to give rise to a *generative cell* and a *tube cell*.

↓

The wall of the microspore then thickens and becomes sculptured into a species-specific pattern.

↓

These two cells and the thickened wall are the pollen grain, an immature male gametophyte.

    B. **Ovule Development**

Ovule = Structure which forms within the chambers of the plant ovary and contains the female sporangium.

The female gametophyte is the *embryo sac*, and it develops as follows:

A megasporocyte in the sporangium of each ovule grows and goes through meiosis to form 4 haploid *megaspores* (only one usually survives).

↓

The remaining megaspore grows and its nucleus undergoes 3 mitotic divisions, forming 1 large cell with 8 haploid nuclei.

↓

Membranes partition this into a multicellular *embryo sac*.

Within the embryo sac:
- The egg cell is located at one end and is flanked by two other cells (*synergids*).
- At the opposite end are 3 antipodal cells.
- The other 2 nuclei (*polar nuclei*) share the cytoplasm of the large central cell.
- At the end containing the egg is the *micropyle* (an opening through the integuments surrounding the embryo sac).

III. Pollination brings female and male gametophytes together

A. Pollination

Pollination = The placement of pollen onto the stigma of a carpel.
- Some plants use wind to disperse pollen.
- Other plants interact with animals that transfer pollen directly between flowers.
- Some plants self-pollinate, but most cross-pollinate.

Most monoecious angiosperms have mechanisms to prevent self-pollination. These mechanisms thus contribute to genetic variation in the species by ensuring sperm and eggs are from different plants.
- The stamens and carpels mature at different times in some species.
- Structural arrangement of the flower in many species pollinated by animals reduces the chance that pollinators will transfer pollen from anthers to the stigma of the same flower.
- Other species are *self-incompatible*. If a pollen grain lands on the stigma of the same flower, a biochemical block prevents the pollen grain from developing and fertilizing the egg.

B. The Molecular Basis of Self-Incompatibility

Self-Incompatibility = The rejection of pollen from the same, or closely related, plant by the stigma.

Self-incompatibility is a single-gene-based mechanism.
- The S-locus is responsible for this reaction.
- Many alleles for the S-locus are found in a plant population's gene pool.
- A pollen grain that lands on a stigma with matching alleles at the S-locus is self-incompatible.
    ⇒ The pollen grain will either not adhere strongly to the stigma or the pollen tube will not develop and invade the ovary.
    ⇒ This prevents self-fertilization and fertilization between plants with a common S-locus (usually closely related plants).

The S-locus consists of multiple transcription units (codes for several proteins) that are so tightly linked that they are inherited as if they were a single gene.

- One protein produced by a unit of the S-locus is a ribonuclease (RNA digesting enzyme).
  - ⇒ If the pollen and stigma have matching S-alleles, the stigma releases a ribonuclease that destroys RNA in the pollen tube, thus preventing growth of the pollen tube.
- Current research indicates recognition of "self" pollen and the stigma's response are based in other proteins produced by the S-locus.
  - ⇒ A second protein produced by the S-locus is secreted into the cell walls of the stigma cells; this protein may function as a specific receptor for molecules secreted by pollen of matching S-type.
  - ⇒ A third enzyme produced by the S-locus is a kinase (regulatory protein).
- The current hypothesis is that the receptor molecule signals the presence of molecules from "self" pollen.
  - ⇒ The kinase relays this signal across the plasma membrane to the cytoplasm of the stigma cell.
  - ⇒ The stigma cell then triggers a rejection response to the pollen.

Studies on self-incompatibility may lead to benefits for agricultural production.

- Many important agricultural plants are self-compatible.
  - ⇒ Different varieties of these crop plants are hybridized to combine the best traits of the varieties and prevent loss of vigor from excessive inbreeding.
  - ⇒ To maximize the numbers of hybrids, plant breeders prevent self-fertilization by laboriously removing the anthers from parent plants that provide the seeds.
- If the molecular mechanism responsible for self-incompatibility can be imposed on normally self-compatible crop species, production of hybrids would be simplified.

C. Double Fertilization

When a compatible pollen grain (different S-locus alleles) lands on a stigma of an angiosperm, *double fertilization* occurs. (See Campbell, Figure 34.5)

Double fertilization = The union of two sperm cells with two cells of the embryo sac.

- After adhering to a stigma, the pollen grain germinates and extends a pollen tube between the cells of the style toward the ovary.
- The generative cell divides (mitosis) to form two sperm. (A pollen grain with a tube enclosing two sperm = mature male gametophyte.)
- Directed by a chemical attractant (usually calcium), the tip of the pollen tube enters through the micropyle and discharges its two sperm nuclei into the embryo sac.
- One sperm unites with the egg to form the zygote.
- The other sperm combines with the two polar nuclei to form a 3N nucleus in the large central cell of the embryo sac.
  - ⇒ This central cell will give rise to the *endosperm* which is a food storing tissue.

After double fertilization is completed, each ovule will develop into a seed and the ovary will develop into a fruit surrounding the seed(s).

IV.   **The ovule develops into a seed containing a sporophyte embryo and a supply of nutrients**

    A.   **Endosperm Development**

        Endosperm development begins before embryo development.
- The triploid nucleus divides to form a milky, multinucleate "supercell" after double fertilization.
- This endosperm undergoes *cytokinesis* to form membranes and cell walls between the nuclei, thus, becoming multicellular.
  - ⇒ Endosperm is rich in nutrients, which it provides to the developing embryo.
  - ⇒ The endosperm in most monocots stocks nutrients which are available to the seedling after germination.
  - ⇒ In many dicots, food reserves of the endosperm are exported to the cotyledons, thus mature seeds have no endosperm.

    B.   **Embryo Development (Embryogenesis)**

        During embryogenesis: (See Campbell, Figure 34.6)
- The zygote's first mitotic division is transverse, creating a larger basal cell and a smaller terminal cell.
- The basal cell divides transversely to form the *suspensor*, which anchors the embryo and transfers nutrients to it from the parent plant.
- The terminal cell divides several time to form a spherical *proembryo* attached to the suspensor.
- Cotyledons appear as bumps on the proembryo and the embryo elongates.
  - ⇒ The apical meristem of the embryonic shoot is located between the cotyledons.
- Where the suspensor (the opposite end of the axis) attaches is the apex of the embryonic root with its meristem.
  - ⇒ The basal cell gives rise to part of the root meristem in some species.
- After germination, the apical meristems at the root and shoot tips will sustain primary growth.
  - ⇒ The embryo also contains *protoderm*, *ground meristem* and *procambium*.
- Two features of plant form are established during embryogenesis.
  - ⇒ The root-shoot axis with meristems at opposite ends.
  - ⇒ A radial pattern of protoderm, ground meristem, and procambrium ready to produce the dermal, ground, and vascular tissue systems.

C. Structure of the Mature Seed

In mature seeds, the embryo is quiescent until germination.
- The seed dehydrates until its water content is only 5-15% by weight.
- The embryo is surrounded by endosperm, enlarged cotyledons, or both.
- The seed coat is formed from the integuments of the ovule.

The arrangement within the seed of a dicot is shown in Figure 34.7a.
- Below the cotyledon attachment point, the embryonic axis is termed the *hypocotyl*, which terminates in the *radicle*, or embryonic root.
- Above the cotyledons, the embryonic axis is termed the *epicotyl*, which terminates in the *plumule* (shoot tip with a pair of tiny leaves).
- Fleshy cotyledons are present in some dicots before germination due to their absorption of nutrients from the endosperm.
- In other dicots, thin cotyledons are found and nutrient absorption and transfer occurs only after germination.

A monocot seed has a single cotyledon called the *scutellum*.
- The scutellum has a large surface area and absorbs nutrients from the endosperm during germination.
- The embryo is enclosed in a sheath comprised of the *coleorhiza* (covers the root) and the *coleoptile* (covers the shoot).

V. The ovary develops into a fruit adapted for seed dispersal

A fruit develops from the ovary of the flower while seeds are developing from the ovules.
- A fruit protects the seeds and aids in their dispersal by wind or animals.
- In some angiosperms, other floral parts also contribute to formation of what we call fruit:
  ⇒ The core of an apple is the true fruit.
  ⇒ The fleshy part of the apple is mainly derived from the fusion of flower parts located at the base of the flower.

A true fruit is a ripened ovary.
- Pollination triggers hormonal changes that cause the ovary to grow.
- The wall of the ovary thickens to become the *pericarp*.
- Transformation of a flower into a fruit parallels seed development and is called *fruit set*.
- In most plants, fruit does not develop without fertilization of ovules. (In *parthenocarpic* plants, fruit does develop without fertilization.)

Depending upon their origin, fruits can be classified as:

1. Simple fruits.
   - Fruit derived from a single ovary. For example, cherry (fleshy) or soybean (dry).
2. Aggregate fruits.
   - Fruit derived from a single flower with several separate carpels. For example, a strawberry.
3. Multiple fruits.
   - Fruit derived from an *inflorescence* or separate tightly clustered flowers. For example, pineapple.

Fruits ripen about the time seeds are becoming fully developed.
- In dry fruits, such as soybean pods, the fruit tissues age and the fruit (pod) opens and releases the seeds.
- Fleshy fruits ripen through a series of steps guided by hormonal interactions.
  ⇒ The fruit becomes softer as a result of enzymes digesting the cell wall components.
  ⇒ Colors usually change and the fruit becomes sweeter as organic acids or starch are converted to sugar.
  ⇒ These changes produce an edible fruit which entices animals to feed, thus dispersing the seeds.

VI. Evolutionary adaptations in the process of germination increase the probability that seedlings will survive

Seed germination represents the continuation of growth and development which was interrupted when the embryo became quiescent at seed maturation.
- Some seeds germinate as soon as they reach a suitable environment.
- Other seeds require a specific environmental cue before they will break dormancy.

A. Seed Dormancy

The evolution of the seed was an important adaptation by plants to living in terrestrial habitats.
- The environmental conditions in terrestrial habitats fluctuate more often than conditions in aquatic habitats.

Seed dormancy prevents germination when conditions for seedling growth are unfavorable.
- It increases the chance that germination will occur at a time and place most advantageous to the success of the seedling.

Conditions for breaking dormancy vary depending on the type of environment the plant inhabits.

- Seeds of desert plants may not germinate unless there has been heavy rainfall (not after a light shower).
- In chaparral regions where brushfires are common, seeds may not germinate unless exposed to intense heat, after a fire has cleared away older, competing vegetation.
- Other seeds may require exposure to cold, sunlight or passage through an animal's digestive system before germination will occur.

A dormant seed may remain viable for a few days to a few decades (most are viable for at least a year or two). This provides a pool of ungerminated seeds in the soil which is one reason vegetation appears so rapidly after environmental disruptions.

B.   From Seed to Seedling

The first step in seed germination in many plants is *imbibition* (absorption of water).

- Hydration causes the seed to swell and rupture the seed coat.
- Hydration also triggers metabolic changes in the embryo that cause it to resume growth.
- Storage materials of the endosperm or cotyledons are digested by enzymes and the nutrients transferred to the growing regions of the embryo.
  - ⇒ For example, the embryo of a cereal grain releases a hormone (a gibberellin) as a messenger to the *aleurone* (outer layer of endosperm) to initiate production of α-amylase and other enzymes that digest starch stored in the endosperm.
- The *radicle* (embryonic root) then emerges from the seed.

The next step in the change from a seed to a seedling is the shoot tip breaking through the soil surface.

- In many dicots, a hook forms in the *hypocotyl*.
  - ⇒ Growth pushes the hypocotyl above ground.
- Light stimulates the hypocotyl to straighten, raising the cotyledons and epicotyl.
- The epicotyl then spreads the first leaves which become green and begin photosynthesis.

Germination may follow different methods depending on the plant species.

- In peas, a hook forms in the *epicotyl* and the shoot tip is lifted by elongation of the epicotyl and straightening of the hook.
  - ⇒ The cotyledons remain in the ground.
- In monocots, the coleoptile pushes through the soil and the shoot tip grows up through the tunnel of the tubular coleoptile.

Only a small fraction of the seedlings will survive to the adult plant stage.

- Large numbers of seeds and fruits are produced to compensate for this loss.
- This utilizes a large proportion of the plant's available energy.

VII. Many plants can clone themselves by asexual reproduction

<u>Asexual reproduction</u> (or <u>vegetative reproduction</u>) = The production of offspring from a single parent; occurs without genetic recombination, resulting in a clone.
- Meristematic tissues composed of dividing, undifferentiated cells can sustain or renew growth indefinitely.
- Parenchyma cells can also divide and differentiate into various types of specialized cells.

There are two major natural mechanisms of vegetative reproduction:

<u>Fragmentation</u> = Separation of a parent plant into parts that reform whole plants.
- The most common form of vegetative reproduction.
- Some species of dicots exhibit a variation of fragmentation during which the parental root system develops adventitious shoots that become separate shoot systems.

<u>Apomixis</u> = The production of seeds without meiosis and fertilization.
- A diploid cell in the ovule gives rise to an embryo.
- The ovules mature into seeds which are dispersed.
- An example would be a dandelion.

VIII. Vegetative reproduction of plants is common in agriculture

There are several methods of vegetative propagation in agriculture:
- Most are based on the ability of plants to form adventitious roots or shoots.
- The objective being to improve crops, orchards, and ornamental plants.

A. Cones from Cuttings

Clones may be obtained from either shoot or stem cuttings (plant fragments).
- At the cut end of the shoot, a mass of dividing, undifferentiated cells form (called a *callus*).
- Adventitious roots then form from the callus.
- Cuttings may come from stems, leaves (African violets), or specialized storage stems (potatoes).

It is possible to combine the best qualities of different varieties or species by grafting a twig of one plant onto a closely related species or different variety of the same species.
- The plant providing the root system is the *stock*.
- The twig grafted onto the stock is the *scion*.
- The quality of a fruit is usually determined by the scion, although sometimes the stock can alter the characteristics of the shoot system that develops from the scion.

B. Test-Tube Cloning and Related Techniques

Test-tube cloning makes it possible to culture small *explants* (pieces of parental tissue) or single parenchyma cells on an artificial medium containing nutrients and hormones. (See Campbell, Figure 34.12)

- The cultured cells divide to form an undifferentiated callus.
- The callus sprouts fully differentiated roots and shoots when the hormone balance of the culture media is manipulated.
- A single plant can be cloned into thousands by subdividing calluses as they grow.

Cultured explants may also be used to produce "artificial seeds."

- Composed of somatic embryos, nutrients, and small amounts of fertilizer in an artificial seed coat made of polysaccharide gel.
- The somatic embryos are derived asexually from somatic cells.
- Has the advantage that all plants growing from the same set will mature at the same time.

Tissue culture is often used to regenerate genetically engineered plants.

- Foreign genes are typically introduced into small pieces of plant tissue or into single plant cells.
- The use of test-tube culture techniques permits the regeneration of genetically altered plants from a single plant cell that received foreign DNA.
  - ⇒ The protein quality of sunflower seeds has been improved in transgenic plants which received a gene for bean protein as cultured cells.

Protoplast fusion, coupled with tissue culture methods, can produce new plant varieties that can be cloned.

- Protoplasts are plant cells which have had their cell walls removed.
- Protoplasts may be fused to form hybrid protoplasts.
- Protoplast can be screened for mutations that will improve the agricultural value of the plant.
- Protoplasts regenerate cell walls and become hybrid plantlets.

C. Benefits and Risks of Monoculture

Monoculture = The cultivation of large areas of land with a single plant variety.

Genetic variability in many crops has been purposefully reduced by plant breeders who have elected self-pollinating varieties or used vegetative reproduction to clone exceptional plants.

- Benefits of such genetic unity are: plant growth is uniform; fruits ripen in unison; crop yields are dependable.
- A great disadvantage is that little genetic variability means little adaptability. One disease could destroy a whole plant variety.
- "Gene banks", where seeds of many plant varieties are stored, are maintained to retain diverse varieties of crop plants.

IX. **Sexual and asexual reproduction are complementary in the life histories of many plants:** *a review*

Both sexual and asexual reproduction have had featured roles in the adaptation of plant populations to their environments.

Benefits of sexual reproduction:

- Generates variation, an asset when the environment (biotic and abiotic) changes.
- Production of the seed, which can disperse to new locations and wait until hostile environments become favorable.

Benefits of asexual reproduction:

- In a stable environment, plants can clone many copies of themselves in a short period.
- Progeny are mature fragments of the parent plant, and not as fragile as seedlings produced by sexual reproduction.

X. **Growth, morphogenesis, and differentiation produce the plant body:** *an overview of developmental mechanisms in plants*

Regardless of whether a plant is sexually produced or results from vegetative reproduction, the initial individual will go through a series of changes that will produce a whole plant.

Development = The sum of all changes that progressively elaborate an organism's body.

- These changes will include a number of mechanisms that shape the leaves, roots, and other organs into functional structures.

The change from a fertilized egg to a plant involves *growth, morphogenesis*, and *cellular differentiation*.

Growth is an irreversible increase in size resulting from cell division and cell enlargement.

- The zygote divides mitotically to produce a multicellular embryo in the seed.
- Mitosis resumes in the root and shoot apical meristems after germination.
- Enlargement of the newly produced cells results in most of the actual size increase.

Morphogenesis is the development of body shape and organization.

- Begins in the early divisions of the embryo to produce the cotyledons and rudimentary roots and shoots.
- Continues to shape the root and shoot systems as the plant grows.
  ⇒ The meristems, which remain embryonic, continue growth and morphogenesis throughout the life of the plant.

Cellular differentiation is the divergence in structure and function of cells as they become specialized during the plant's development.

- Every organ of the plant body has a diversity of cells within its total structure.
- Each cell of each organ is fixed in a certain location and performs a specific function (e.g. guard cells, xylem).

XI. **The cytoskeleton guides the geometry of cell division and expansion:** *a closer look*

Plant shape depends on the spatial orientations of cell divisions and cell expansions.
- Plant cells cannot move about as individuals within a developing organ due to their cell walls being cemented to those of neighboring cells.
- Since movement is eliminated, when the cell elongates, its growth is perpendicular to the plane of division.

A. **Orienting the Plane of Cell Division**

During late interphase (G_2), the cytoskeleton of the cell becomes rearranged with the microtubules of the cortex becoming concentrated into the *preprophase band*.
- The microtubules of the preprophase band disperse leaving behind an array of actin microfilaments.
- These microfilaments hold the nucleus in a fixed orientation until the spindle forms and then direct movement of the vesicles that produce the cell plate.
- The walls that develop at the end of cell division form along the plane established by the preprophase band.

B. **Orienting the Direction of Cell Expansion**

Plant cells expand (elongate) when the cell wall yields to the turgor pressure of the cell.
- Cross-links between cellulose microfibrils in the cell wall are weakened by an acid secreted by the cell.
- The loosened wall permits uptake of water by the hypertonic cell; water uptake causes the cell to expand.
- Growth continues until the cross links become re-established firmly enough to offset the turgor pressure.
- About 90% of the cell's expansion is due to water uptake, although some cytoplasm is also produced by the cell.
- Most of the water entering the cell is stored in the large central vacuole which forms due to coalescence of small vacuoles as the cell grows.

Plant cells show very little increase in width as they elongate.
- Cellulose microfibrils in the innermost cell wall layers stretch very little; consequently, the cell expands in the direction perpendicular to the orientation of the microfibrils.
- Alignment of microfibrils in the wall mirrors the microtubule orientation found in the cortex. This is believed to result from microtubular control of the flow of cellulose-producing enzymes in a specific direction along the membrane.

XII. Cellular differentiation depends on the control of gene expression: *a closer look*

The progressive development of specialized structures and functions in plant cells reflects the different types of proteins synthesized by different types of cells.

Xylem cells function in both transport within the plant and structural support.
- Cell walls are hardened by lignin which is produced by enzymes made by the cell.
- The final stage of differentiation includes the production of hydrolytic enzymes which destroy the protoplast.
  ⇒ This leaves only the cell walls intact and permits the movement of xylem sap through the cells.

Guard cells regulate the size of the stomatal opening.
- Must have flexible walls, thus the enzymes that produce lignin are not produced.
- The protoplast remains intact and regulates ion exchange necessary to increase and decrease turgor.

All cells in a plant possess a common genome. This has been proven by cloning whole plants from single somatic cells.
- All the genes necessary are present since these cells dedifferentiate in tissue cultures and then redifferentiate to produce the diversity of cells found in the plant.
- This ability indicates that cellular differentiation is controlled by gene expression leading to the production of specific proteins.
- Different cell types (like xylem and guard cells) selectively express certain genes at different times during their differentiation; this results in the different developmental pathways that gives rise to the diverse cell types.

XIII. Mechanisms of pattern formation determine the location and tissue organization of plant organs: *a closer look*

The organization in a plant can be seen in the characteristic pattern of cells in each tissue, the pattern of tissues in each organ, and the spatial organization of the organs on the plant.

Pattern formation = The development of specific structures in specific locations.

A. Positional Information

Pattern formation depends on *positional information*.

Positional information = Signals indicating a cell's location relative to other cells in an embryonic mass.
- Genes respond to these signals and their response affects the localized rates and planes of cell division and expansion.
- This signal detection continues in each cell as the organs develop and cells respond by differentiating into particular cell types.

Several hypotheses have been proposed as to how embryonic cells detect their positions. One hypothesis about positional information transmission is that it relies on gradients of proteins called *chemical signals*.

- A chemical signal might diffuse from a shoot's apical meristem and the decreasing gradient farther from the source would indicate to cells their relative position from the tip.
- A second chemical signal released from the outermost lateral cells would diffuse inward indicating the radial position to each cell.
- Each cell could thus determine its relative longitudinal and radial position from the gradients of these two chemical signals.

B.   Clonal Analysis of the Shoot Apex

Positional information is the basis for the processes involved in plant development: growth, morphogenesis, and differentiation. Plant developmental biologists have developed the technique of *clonal analysis* to study the relationships of these processes.

- Clonal analysis involves mapping the cell lineages derived from each cell of the apical meristem, noting their position as the plant organs develop.
- Mapping is possible due to induced somatic mutations in each cell which can be used to distinguish that cell and its derived cells from neighboring cells.

This technique has been used to determine that the developmental fates of cells in the apex are somewhat predictable.

- For example, almost all cells developing from the outermost meristematic cells become part of the dermal tissues of leaves and stems.

It is not currently possible to predict what meristematic cells will develop into specific tissues and organs.

- The outermost cells usually divide on a plane perpendicular to the shoot surfaces, thus adding cells to the surface layer.
- Random changes can result in one of these cells dividing on a plane parallel to the surface; this indicates meristematic cells are not dedicated early in their development to forming specific tissues and organs.
- Consequently, it is the cell's final position in a developing organ which determines what type of cell it becomes, not the particular cell lineage to which it belongs.

C.   The Genetic Basis of Pattern Formation in Flower Development

The shoot tip in flowering plants shifts from indeterminate growth to determinate growth when the flower is produced.

- The meristem is consumed during the formation of primordia for sepals, petals, stamens, and carpels.

Positional information commits each primordium to develop into an organ of specific structure and function.

- Some organ-identity genes that function in development of the floral pattern and are regulated by positional information have been identified.
- Mutations in these organ-identity genes can cause abnormal floral patterns.
    ⇒ For example, an extra whorl of sepals may develop instead of petals.
    ⇒ Such abnormal patterns indicate wild-type alleles are responsible for normal floral pattern development.

Arabidopsis thaliana is the experimental organism many plant biologists are now using to study plant development.

- It is a small plant with a relatively rapid life cycle.
- It also has a small genome that simplifies the search for specific genes.
- Several organ-identity genes affecting floral pattern development have been identified and a few have been cloned.
- Similar organ-identity genes have been found in a distantly related plant, the snapdragon.
    ⇒ This finding suggests a conservative evolution of the genes controlling basic angiosperm body plan development.

One hypothesis about how positional information influences a particular floral-organ primordium is based on the overall genetic basis of pattern formation.

- Organ-identity genes code for transcription factors that are regulatory proteins which help control expression of other genes.
- This control involves binding of the transcription factor to specific sites on the DNA, which affects transcription.
- It is believed that positional information determines which organ-identity gene is expressed, and the resulting transcription factor induces expression of these genes controlling development of specific organs.

REFERENCES

Campbell, N. Biology. 4th ed. Menlo Park, California: Benjamin/Cummings, 1996.

Raven, P.H., R.F. Evert and S.E. Eichhorn. Biology of Plants. 5th ed. New York: Worth Publishers, Inc., 1992.

OUTLINE

I. Research on how plants grow toward light led to the discovery of plant hormones: *science as a process*

II. Plant hormones help coordinate growth, development and responses to environmental stimuli
   A. Auxin
   B. Cytokinins
   C. Gibberellins
   D. Abscisic Acid
   E. Ethylene

III. Tropisms orient the growth of plant organs toward or away from stimuli

IV. Turgor movements are relatively rapid, reversible plant responses
   A. Rapid Leaf Movements
   B. Sleep Movements

V. Biological clocks control circadian rhythms in plants and other eukaryotes

VI. Photoperiodism synchronizes many plant responses to changes of season
   A. Photoperiodic Control of Flowering

VII. Phytochrome functions as a photoreceptor in many plant responses to light and photoperiod
   A. Ecological Significance of Phytochrome as a Photoreceptor
   B. Interaction of Phytochrome and the Biological Clock in Photoperiodism

VIII.   Control systems enable plants to cope with environmental stress
    A.   Responses to Water Deficit
    B.   Responses to Oxygen Deprivation
    C.   Responses to Salt Stress
    D.   Responses to Heat Stress
    E.   Responses to Cold Stress
    F.   Responses to Herbivores
    G.   Defense Against Pathogens: Systemic Acquired Resistance

IX.   Signal-transduction pathways mediate the responses of plant cells to environmental and hormonal stimuli

OBJECTIVES

After reading this chapter and attending lecture, the student should be able to:

1. For each of the following scientists, describe their hypothesis, experiments and conclusions about the mechanism of phototropism:

    a.   Charles Darwin   c.   Peter Boysen Jensen
    b.   Francis Darwin   d.   F.W. Went

2. List five classes of plant hormones, describe their major functions and recall where they are produced in the plant.
3. Explain how a hormone may cause its effect on plant growth and development.
4. Describe a possible mechanism for polar transport of auxin.
5. According to the acid-growth hypothesis, explain how auxin can initiate cell elongation.
6. Explain why 2,4-D is widely used as a weed killer.
7. Explain how the ratio of cytokinin to auxin affects cell division and cell differentiation.
8. Define apical dominance and describe the check-and-balance control of lateral branching by auxins and cytokinins.
9. List several factors besides auxin from the terminal bud that may control apical dominance.
10. Describe how stem elongation and fruit growth depend upon a synergism between auxin and gibberellins.
11. Explain the probable mechanism by which gibberellins trigger seed germination.
12. Describe how abscisic acid (ABA) helps prepare a plant for winter.
13. Explain the antagonistic relationship between ABA and gibberellins, and how it is possible for growing buds to have a higher concentration of ABA than dormant buds.
14. Give an example of how ABA can act as a "stress hormone".
15. Describe the role of ethylene in plant senescence, fruit ripening and leaf abscission.
16. List two environmental stimuli for leaf abscission.
17. Define tropism and list three stimuli that induce tropisms and a consequent change of body shape.
18. Explain how light causes a phototropic response.
19. Describe how plants apparently tell up from down, and explain why roots display positive gravitropism and shoots exhibit negative gravitropism.
20. Distinguish between thigmotropism and thigmomorphogenesis.
21. Describe how motor organs within pulvini can cause rapid leaf movements and sleep movements.
22. Provide a plausible explanation for how a stimulus that causes rapid leaf movement can be transmitted through the plant.

23. Define circadian rhythm and explain what happens when an organism is artificially maintained in a constant environment.
24. List some common factors that entrain biological clocks.
25. Define photoperiodism.
26. Distinguish among short-day plants, long-day plants and day-neutral plants; give common examples of each and explain how they depend upon critical night length.
27. Provide evidence for the existence of a florigen.
28. Explain how the interconversion of phytochrome can act as a switching mechanism to help plants detect sunlight and trigger many plant responses to light.
29. Using photoperiodism as an example, explain how an integrated control system can regulate a plant process such as flowering.
30. Describe the components of a signal-transduction pathway.

KEY TERMS

hormone
phototropism
Charles Darwin
Francis Darwin
Peter Boysen-Jensen
F.W. Went
auxin
indoleacetic acid
   (IAA)
polar transport
acid-growth
   hypothesis

cytokinins
apical dominance
gibberellins
Gibberella
bolting
abscisic acid
stress hormone
ethylene
leaf abscission
tropisms
phototropism

gravitropism
statoliths
thigmotropism
thigmomorphogenesis
turgor movements
rapid leaf movements
pulvini
sleep movements
circadian rhythm
free-running periods
photoperiod

short-day plants
long-day plants
day-neutral plants
vernalization
phytochrome
chromophore
signal-transduction
   pathway
target cells
second messenger
calmodulin

LECTURE NOTES

Control systems in plants are adaptations that evolved over time in response to interactions with their environment. Plants respond to environmental stimuli by:

- Sending signals between different parts of the plant.
- Tracking the time of day and the time of year.
- Sensing and responding to gravity and direction of light, etc.
- Adjusting their growth pattern and development.

I. **Research on how plants grow toward light led to the discovery of plant hormones:** *science as a process*

Hormone = A compound produced by one part of an organism that is transported to other parts where it triggers a response in target cells and tissues.

Phototropism = Growth toward or away from light.

- Growth of a shoot toward light is *positive* phototropism; growth away from the light is *negative* phototropism.
- Results form differential growth of cells on opposite sides of a shoot or, in the case of a grass seedling, coleoptile.
- Cells on the darker side elongate faster than those on the light side.

Experiments on phototropism led to the discovery of a plant hormone.

1. Charles and Francis Darwin removed the tip of the coleoptile from a grass seedling (or covered it with an opaque cap) and it failed to grow toward light. They concluded that:
   - The coleoptile tip was responsible for sensing light.
   - Since the curvature occurs some distance below the tip, the tip sends a signal to the elongating region.

2. Peter Boysen-Jensen separated the tip from the remainder of the coleoptile by a block of gelatin, preventing cellular contact but allowing chemical diffusion.
   - Seedlings behaved normally.
   - If an impenetrable barrier was substituted, no phototropic response occurred.
   - These experiments demonstrated that the signal was a mobile substance.

3. F.W. Went removed the coleoptile tip, placed it on an agar block, and then put the agar (without the tip) on decapitated coleoptiles kept in the dark.
   - A block centered on the coleoptile caused the stem to grow straight up.
   - If the block was placed off-center, the plant curved away from the side with the block.
   - Went concluded the agar block contained a chemical that diffused into it from the coleoptile tip, and that this chemical stimulated growth.
   - Went called this chemical an *auxin*.

4. Kenneth Thimann later purified and characterized auxin.

II. **Plant hormones help coordinate growth, development, and responses to environmental stimuli**

Plant hormones may function in one of many ways:

- Control plant growth and development by affecting division, elongation, and differentiation of cells.
    - $\Rightarrow$ Effects depend on site of action, stage of plant growth and hormone concentration.
- The hormonal signal is amplified, perhaps by affecting gene expression, enzyme activity, or membrane properties.
- Reaction to hormones depends on hormonal balance (relative concentration of one hormone compared with others).

Five classes of plant hormones have been identified. (See Campbell, Table 35.1)

1. Auxin (such as IAA)
2. Cytokinins (such as zeatin)
3. Gibberellins (such as GA_3)
4. Abscisic acid
5. Ethylene

A. **Auxin**

Auxin = A hormone that promotes elongation of young developing shoots or coleoptiles.

- The natural auxin found in plants is a compound named indoleacetic acid (IAA).

The *apical meristem* is a major site of auxin production.

- Stimulates cell growth only at concentrations between 10^{-8} to 10^{-3} M.
- Moves from the apex down to the zone of cell elongation at a rate of about 10 mm per hour.
    - $\Rightarrow$ This is faster than would be found in diffusion but much slower than in phloem translocation.

Polar transport of auxin is unidirectional and requires metabolic energy. (See Campbell, Figure 35.4)

- IAA is actively transported down a stem by auxin carriers located on the basal ends of cells (carriers are absent on the apical ends).
- Energy for active transport is provided by chemiosmosis.
- Movement of auxin is aided by the differences in pH between the acidic cell wall and the neutral cytoplasm.
    - $\Rightarrow$ ATP-driven pumps maintain a proton gradient across the plasma membrane.
    - $\Rightarrow$ As auxin passes through the acidic cell wall, it picks up a proton to become electrically neutral, which allows it to pass through the plasma membrane.
    - $\Rightarrow$ Auxin is ionized in the neutral intracellular environment which temporarily traps it within the cell since the plasma membrane is less permeable to ions.
    - $\Rightarrow$ Auxin can only exit the cell by the basal end, where specific carrier proteins are built into the membrane. The proton gradient contributes to auxin efflux by favoring the transport of anions out of the cell.

The *acid-growth hypothesis* states that cell elongation is due to stimulation of a proton pump which acidifies the cell wall.

- Acidification causes the crosslinks between the cellulose myofibrils of the cell walls to break.
- This loosens the wall, allowing water uptake which results in elongation of the cell.

In addition to stimulating cell elongation, auxin also:
- Affects secondary growth by inducing vascular cambium cell division and differentiation of secondary xylem.
- Promotes formation of adventitious roots.
- Promotes fruit growth in many plants.

Auxins are used as herbicides. 2,4-D is a synthetic auxin which affects dicots selectively, allowing removal of broadleaf weeds from a lawn or grain field.

B. **Cytokinins**

Cytokinins = Modified forms of adenine that stimulate cytokinesis.
- Move from the roots to target tissues by moving up in the xylem sap.
- Stimulate RNA and protein synthesis. The new proteins produced by stimulation of RNA appear to be involved in cell division.

Cytokinins function in several areas of plant growth:
- Cell division and differentiation.
- Apical dominance.
- Anti-aging hormones.

Cytokinins, in conjunction with auxin, control cell division and differentiation.
- Stem parenchyma cells cultured without cytokinins grow very large and do not divide.
- Cytokinins added alone have no effect on cells grown in tissue culture.
    - ⇒ Equal concentrations of cytokinins and auxins stimulate cells to grow and divide, but they remain an undifferentiated callus.
    - ⇒ More cytokinin than auxin causes shoot buds to develop from the callus.
    - ⇒ More auxin than cytokinin causes roots to form.

Cytokinins and auxin contribute to apical dominance through an antagonistic mechanism.
- Auxin from the terminal bud restrains axillary bud growth causing the shoot to lengthen.
- Cytokinins (from the roots) stimulate axillary bud growth.
- Auxin cannot suppress axillary bud growth once it has begun.
- Lower buds thus grow before higher ones since they are closer to the cytokinin source than the auxin source.
- Auxin stimulates lateral root formation while cytokinins restrain it.
- This stimulation-inhibition action may help balance plant growth since an increase in the root system would signal the plant to produce more shoots.

Cytokinins can retard aging of some plant organs, perhaps by inhibiting protein breakdown, stimulating RNA and protein synthesis, and mobilizing nutrients.
- May slow leaf deterioration on plants since detached leaves dipped in a cytokinin solution stay green longer.

C. Gibberellins

More than 70 different gibberellins, many naturally occurring, have been identified.

Gibberellins are produced primarily in roots and young leaves. They:
- Stimulate growth in leaves and stems but show little effect on roots.
- Stimulate cell division and elongation in stems, possibly in conjunction with auxin.
- Gibberellins cause *bolting* (rapid growth of floral stems, which elevates flowers).

Fruit development is controlled by both gibberellins and auxin.
- In some plants, both must be present for fruit set.
- Commercial application of gibberellins is the spraying of Thompson seedless grapes. The hormones cause the grapes to grow larger and farther apart after treatment.

The release of gibberellins signals seeds to break dormancy and germinate.
- A high concentration of gibberellins is found in many seeds, especially in the embryo.
- Imbibed water appears to stimulate gibberellin release.
- Environmental cues may also cause gibberellin release in seeds which require special conditions to germinate.
- In cereal grains, gibberellins stimulate germination and support growth by stimulating synthesis of mRNA which codes for α-amylase. The α-amylase then digests the stored nutrients, making them available to the embryo and seedling.

In breaking both seed dormancy and apical bud dormancy, gibberellins act antagonistically with *abscisic acid*, which inhibits plant growth.

D. Abscisic Acid (ABA)

Abscisic acid is produced in the terminal bud and helps prepare plants for winter by suspending both primary and secondary growth.
- Directs leaf primordia to develop scales that protect dormant buds.
- Inhibits cell division in vascular cambium.

The onset of seed dormancy is another time it is advantageous to suspend growth.
- In most cases, the ratio of ABA:gibberellins determines whether seeds remain dormant or germinate.
- In other plants, seeds germinate when ABA is washed out of the seeds (desert plants) or degraded by some other stimulus such as sunlight.

ABA also acts as a stress hormone, closing stomata in times of water-stress thus reducing transpirational water loss.

E.   Ethylene

<u>Ethylene</u> = A gaseous hormone that diffuses through air spaces between plant cells.
- Ethylene can also move in the cytosol, traveling from cell to cell in the phloem or symplast.
- High auxin concentrations induce release of ethylene, which acts as a growth inhibitor.

Senescence (aging) is a natural process in plants that may occur at the cellular, organ, or whole plant level. Ethylene probably plays an important role at each level.
- Examples include: xylem vessel elements and cork cells which die before becoming fully functional; leaf fall in the autumn; withering of flowers; death of annuals after flowering.
- The best studied forms of senescence are fruit ripening and leaf abscission.

During fruit ripening, ethylene triggers senescence, and then the aging cells release more ethylene.
- The breakdown of cell walls and loss of chlorophyll are considered aging processes.
- The signal to ripen spreads from fruit to fruit since ethylene is a gas.

Leaf abscission is an adaptation that prevents deciduous trees from desiccating during winter when roots cannot absorb water from the frozen ground.
- Before abscission, the leaf's essential elements are shunted to storage tissues in the stem from which they are recycled to new leaves in the spring.
- Environmental stimuli are shortening days and cooler temperatures.

When a leaf falls, the breakpoint is an abscission layer near the petiole base.
- Weak area since the small parenchyma cells have very thin walls and there are no fiber cells around the vascular tissue.
- Mechanics of abscission are controlled by a change in the balance of ethylene and auxin.
  - ⇒ Auxin decrease makes cells in the abscission layer more sensitive to ethylene. Cells then produce more ethylene which inhibits auxin production.
  - ⇒ Ethylene induces synthesis of enzymes that digest the polysaccharides in the cell walls, further weakening the abscission layer.
- Wind and weight cause the leaf to fall by causing a separation in the abscission layer.
- Even before the leaf falls, a layer of cork forms a protective scar on the twig's side of the abscission layer. The cork prevents pathogens from entering the plant.

III. **Tropisms orient the growth of plant organs toward or away from stimuli**

<u>Tropisms</u> = Growth responses that result in curvatures of whole plant organs toward or away from stimuli.

- Mechanism is a differential rate of cell elongation on opposite sides of the organ.

Three primary stimuli which result in tropisms are: light (photo), gravity (gravit), and touch (thigmo).

1. *Phototropism* is a response to light.
2. *Gravitropism* is a response to gravity.
3. *Thigmotropism* is a response to touch.

A. **Phototropism**

<u>Phototropism</u> = Growth either toward or away from light.

- It is generally accepted that cells on the darker side of a grass coleoptile elongate faster than cells on the bright side due to asymmetric distribution of auxins moving down from the shoot tip.
- For organs, other than grass coleoptiles, the mechanism may be different.
    ⇒ No evidence exists that unilateral light causes an asymmetric distribution of auxins in the stems of many dicots.
    ⇒ There is evidence that other substances which act as growth inhibitors do have an asymmetric distribution toward the lighted side of the stem.
- Regardless of the mechanism, the shoot tip is the site of the photoreception that triggers the growth response.
    ⇒ A photoreceptor sensitive to blue light is present in the shoot tip; this receptor is believed to be a yellow pigment related to riboflavin.
    ⇒ The same receptor may be involved in other plant responses to light.

B. **Gravitropism**

<u>Gravitropism</u> = Orientation of a plant in response to gravity.

- Roots display *positive gravitropism* (curve downward).
- Shoots display *negative gravitropism* (bend upward).

The possible mechanism of gravitropism in roots:

- Specialized plastids containing dense starch grains (*statoliths*) aggregate in the low points of plant cells.
- In roots, statoliths occur in certain root cap cells.
    ⇒ Aggregating statoliths trigger calcium redistribution which results in lateral transport of auxin in the root.
    ⇒ Calcium and auxin accumulate on the lower side of the elongation zone.
    ⇒ Roots curve down because, at high concentrations, auxin *inhibits* root cell elongation, so cells on the upper side elongate faster than those on the lower side.

Researchers are challenging the falling statolith hypothesis for positive gravitropism in root growth.

- Insufficient energy is released by starch grains settling to the bottom of cells to account for gravitational detection.
- Many plants lacking starch grains distinguish up from down.
- Studies on *Chara*, a green alga closely related to plants, indicate the settling of the entire protoplasm provides a cell with its up-down orientation.
    ⇒ The protoplast is attached to the inside of the cell wall by proteins.
    ⇒ When the protoplast settles, the protein tethers at the top of the cell are stretched and those at the bottom are compressed.
    ⇒ The sense of up and down is related to this stretching and compressing of the proteins.
    ⇒ Experiments where *Chara* was placed in a solution more dense than the protoplast resulted in the protoplast floating upward and an upside down growth pattern.
    ⇒ Whether this mechanism is at work in true plants is currently under investigation.

C. Thigmotropism

Thigmotropism = Directional growth in response to touch.
- Contact of tendrils stimulates a coiling response caused by differential growth of cells on opposite sides of the tendril.

Thigmomorphogenesis = Developmental response to mechanical perturbation.
- Usually results from increased ethylene production in response to chronic mechanical stimulation.
- Stem lengthening is decreased while stem thickening increases.

IV. Turgor movements are relatively rapid, reversible plant responses

Turgor movements = Reversible movements caused by changes in turgor pressure of specialized cells in response to stimuli.

A. Rapid Leaf Movements

Rapid leaf movements occur in plants such as *Mimosa*.
- When the compound leaf is touched, it collapses and folds together.
- Results from rapid a loss of turgor within *pulvini* (special motor organs located in leaf joints).
- Motor cells lose potassium, which causes water loss by osmosis.
- Turgor pressure is regained and natural leaf form restored in about 10 minutes.

Rapid leaf movements travel from the leaf that was stimulated to adjacent leaves along the stem.

- May be a response to reduce water loss or protect against herbivores.
- The stimulus and response travel wavelike through the plant at 1 cm/sec.
- This transmission is correlated with *action potentials* (electrical impulses) resembling those in animals but thousands of times slower.
- Action potentials may be widely used as a form of internal communication since they have been found in many algae and plants.

B.   Sleep Movements

Sleep movements = Lowering of leaves to a vertical position in evening and raising of leaves to a horizontal position in morning.

- Occurs in many legumes.
- Due to daily changes in turgor pressure of motor cells of pulvini.
- Cells on one side of the pulvinus are turgid while those on the other side are flaccid.
- Migration of potassium ions from one side of the pulvinus to the other is the osmotic agent leading to reversible uptake and loss of water by motor cells.

V.   Biological clocks control circadian rhythms in plants and other eukaryotes

Biological clocks (internal oscillators that keep accurate time) are common in all eukaryotes and control many rhythmic phenomena.

- Many human features (blood pressure, temperature, metabolic rate, etc.) fluctuate with the time of the day.
- Certain fungi produce spores for only certain hours during the day.
- Plants display sleep movements and a rhythmic pattern of opening and closing stomata.

Circadian rhythm = A physiological cycle with a frequency of about 24 hours.

- Persists even when an organism is sheltered from environmental cues.
- The oscillator is probably endogenous and is set to a 24-hour period by daily signals from the environment.
- When the organism is sheltered from environmental cues, rhythm may deviate from 24 hours (called *free-running periods*) and can vary from 21 to 27 hours.

Deviation of a free-running period from 24 hours does not indicate erratic drift of a biological clock, just absence of a synchronizing cue.

- Most biological clocks are cued to the light-dark cycle resulting from the Earth's rotation.
  - ⇒ The clock may take days to reset once the cues change.
  - ⇒ Jet lag is a human condition resulting from a lack of synchronization of the internal clock to the time zone.

The nature of the internal oscillator is still unknown.
- Most researchers believe biological clocks are present at the cellular level; either in membranes or in the machinery of protein synthesis.
    - ⇒ Researchers found that the sleep movements of legumes are correlated with the rhythmic opening and closing of potassium channels in motor cell membranes.

VI. Photoperiodism synchronizes many plant responses to changes of season

Photoperiodism = A physiological response to day length.
- Seasonal events (seed germination, flowering) are important in plant life cycles.
- Plants detect the time of year by the *photoperiod* (relative lengths of night and day).

A. Photoperiodic Control of Flowering

W.W. Garner and H.A. Allard (1920) postulated that the amount of day length controls flowering. Based on their studies, they classified plants into three categories:
- *Short-day plants* require a light period *shorter* than a critical length and generally flower in late summer, fall and winter.
- *Long-day plants* flower only when the light period is longer than a certain number of hours, generally in late spring and summer.
- *Day-neutral plants* are unaffected by photoperiod and flower when they reach a certain stage of maturity.

It was discovered in the 1940's that a *critical night length*, not day length, actually controls flowering and other responses to photoperiod.
- If the daytime period is broken by a brief exposure to darkness, there is no effect on flowering.
- If the nighttime period is interrupted by short exposure to light, photoperiodic responses are disrupted and the plants do not flower.
- Therefore, short-day plants flower if night is longer than a critical length and long-day plants need a night shorter than a critical length.

Some plants flower after a single exposure to the proper photoperiod.
- Some require several successive days of the proper photoperiod to bloom.
- Others respond to photoperiod only if they have been previously exposed to another stimulus. For example, *vernalization* is a requirement for pretreatment with cold before flowering.

There is evidence that a *flowering hormone* is present in plants since leaves detect the photoperiod while buds produce flowers.
- Only requires one leaf for a plant to detect photoperiod and floral buds develop.
- If all leaves are removed, no photoperiod detection occurs.
- Most plant physiologists believe an as yet unidentified hormone is produced in the leaves and moves to the buds.
    - ⇒ The hormone (or mixture of hormones) appears to be the same in both long-day and short-day plants.

VII.   **Phytochrome functions as a photoreceptor in many plant responses to light and photoperiod**

A pigment named *phytochrome* helps plants measure the length of darkness in a photoperiod.

<u>Phytochrome</u> = A protein containing a *chromophore* (light-absorbing component) responsible for a plant's response to photoperiod.

- It was discovered during studies on how different colors of light affect responses to photoperiod.
- Red light (λ of 660 nm) is most effective in interrupting night length.
  - ⇒ Brief exposure of short-day plants to red light prevents flowering even if the plant is kept at critical night-length conditions.
  - ⇒ A long-day plant is induced to flower by a brief exposure to red light even if kept at a night length exceeding the critical number of hours.
- If a flash of red light (R flash) is followed by a flash of far-red (FR) light (λ of 730 nm), the plant perceives no interruption of night length.
- Only the wavelength of the last flash affects the plant's measurement of night length, regardless of the number of alternating flashes.

Phytochrome alternates between two photoreversible forms: P_r (red absorbing) and P_{fr} (far-red absorbing). The $P_r \leftrightarrows P_{fr}$ interconversion is a switching mechanism controlling various plant events.

A.   **Ecological Significance of Phytochrome as a Photoreceptor**

Phytochrome functions as a photodetector that tells the plant if light is present.

- Plants synthesize phytochrome as P_r and, if kept in dark, it remains as P_r, but if the phytochrome is illuminated, some P_r is converted to P_{fr}.
- P_{fr} triggers many plant responses to light (e.g. seed germination).
- A shift in the $P_r \leftrightarrows P_{fr}$ equilibrium indicates the relative amounts of red and far-red light present in the sunlight.
- Shifts in the $P_r \leftrightarrows P_{fr}$ ratio may cause changes (e.g. increased growth) which would adjust a plant's growth and development in response to some environmental changes.

Photoreception by phytochrome has a large effect on the whole plant even though very little of the pigment is present in plant cells.

- This fact implies the photoconversion from P_r to P_{fr} produces a signal that is amplified in some way.
- The amplification may be by either an alteration of membrane permeability and/or by affecting gene expression.
  - ⇒ Photoconversion of phytochrome triggers the potassium fluxes in cells of the pulvini that produces the sleep movements in legumes.
  - ⇒ Light induces the synthesis of starch digesting α-amylase required for seed germination in some species.

Complementing phytochrome's effect, other photoreceptors help coordinate a plant's growth and development with its environment.

B. **Interaction of Phytochrome and the Biological Clock in Photoperiodism**

P_{fr} gradually reverts to P_r.

- This occurs every day after sunset.
- The pigment is synthesized as P_r and degradative enzymes destroy more P_{fr} than P_r.
- At sunrise, the P_{fr} level increases due to photoconversion of P_r.

Plants do not use the disappearance of P_{fr} to measure night length since:

- The conversion is complete within a few hours after sunset.
- Temperature affects the conversion rate, thus, it would not be reliable.

Night length is measured by the biological clock, not by phytochrome.

- Perhaps phytochrome synchronizes the clock to the environment.
- The clock measures night length very accurately (some short-day plants will not flower if night is even 1 minute shorter than the critical length).

VIII. Control systems enable plants to cope with environmental stress

A plant must adjust to environmental fluctuations every day of its life. Severe fluctuations may put plants under *stress*.

> Stress = An environmental condition that can have an adverse effect on a plant's growth, reproduction, and survival.

Some plants have evolutionary adaptations that enable them to live in environments that are stressful to other plants. For example,

- *Halophytes* have special anatomical and physiological adaptations which permit them to grow best in salty soils.

⇒ Salt glands on the leaves eliminate excess salt from the plants – thus the saline environment is not an environmental stress.

A. **Responses to Water Deficit**

Plants have control systems in both the leaves and roots that help them cope with water deficits.

Most control systems in leaves help the plant conserve water by reducing transpirational water loss.

- Guard cells lose turgor and the stomata close when a leaf faces a water deficit.
- Mesophyll cells in the leaf are also stimulated to increase synthesis and release abscisic acid which acts on guard cell membranes to help keep the stomata closed.
- Growth of young leaves is inhibited by a water deficit since cell expansion is a turgor-dependent process.

⇒ this reduces transpiration by slowing the increase in leaf surface area.

- The leaves of many grasses and other plants wilt; they roll into a shape that reduces the surface area exposed to the sun, thus reducing transpiration.

Roots respond to water deficits by reducing growth.
- Drying of the soil from the surface down inhibits the growth of shallow roots.
  - ⇒ The cells can not retain the turgor necessary for elongation.
- Deeper roots surrounded by moist soil continue to grow.
  - ⇒ This maximizes root exposure to soil moisture.

B. Responses to Oxygen Deprivation

Waterlogged soil lacks the air spaces that provide oxygen for cellular respiration in the roots.
- Some plants form air tubes that extend from submerged roots to the surface, thus oxygen can reach the roots.
- Mangroves are structurally adapted to their coastal marsh environments in that their submerged roots are continuous with aereal roots that provide access to oxygen.

C. Responses to Salt Stress

Excess salts (sodium chloride or others) in the soil may:
- Lower the water potential of the soil solution causing a water deficit even though sufficient water is present.
  - ⇒ A water potential in the soil that is more negative than that of the root tissue will cause roots to *lose* water instead of absorb it.
- Have a toxic effect on the plant at relatively high concentrations.
  - ⇒ The uptake of most harmful ions is impeded by the selectively permeable membranes of root cells.
  - ⇒ This causes a problem with acquiring water from solute rich soils.
- Many plants produce *compatible solutes* in response to moderately saline soils.

  Compatible solute = An organic compound that keeps the water potential of cells more negative than the soil solution without admitting toxic quantities of salt.

D. Responses to Heat Stress

Transpiration is one mechanism that helps plants respond to excessive heat and prevent the denaturing of enzymes and damage to metabolism.
- The evaporative cooling associated with transpiration keeps the temperature of the leaf 3° to 10°C lower than ambient temperature.
- Cooling via transpiration will continue while stomata remain open; however, if a water deficit occurs, the stomata close and the cooling function is lost in order to conserve water.

Most plants will begin producing *heat-shock proteins* when exposed to excessive temperatures (40°C or above for temperate zone plants).

- This is a back-up system to transpiration.
- Some heat-shock proteins are identical to chaperone proteins found in unstressed cells.
  - ⇒ Chaperone proteins serve as temporary supports which help other proteins fold into their functional conformations.
  - ⇒ Heat-shock proteins may help enzymes and other proteins maintain their conformation, thus preventing denaturation.

E.   **Responses to Cold Stress**

Chilling of a plant (reduction of ambient temperature to a non-freezing level) causes a change in the *fluidity* of cell membranes.

Fluidity = The lateral drifting of proteins and lipids in the plane of the membrane; a result of the fluid mosaic structure of membranes.

- At a critical point, lipids become locked into crystalline structures causing a loss of fluidity.
- Solute transport and membrane protein function are adversely affected by the loss of fluidity.

Plants respond to the cold stress of chilling by altering the lipid composition of their membranes.

- The proportion of saturated fatty acids in the membrane is increased.
  - ⇒ The shape of the fatty acids reduces crystal formation, thus maintaining fluidity at lower temperatures.
- This modification works best for gradual temperature changes as it takes several hours to days to occur.

Subfreezing temperatures are the most severe form of cold stress because ice crystals begin to form in the plant.

- Less threat to plant survival occurs if the ice crystals form only in the cell walls and intracellular spaces.
- When ice crystals form in the protoplasts, the cell usually dies.
  - ⇒ The ice crystals perforate the membranes and organelles.
- Wood plants which are native to regions where cold winters occur have adaptations to cope with the stress of freezing.
  - ⇒ The solute composition of live cells is changed in a way that prevents ice crystal formation even when the cytosol is supercooled.
  - ⇒ Effective even when ice crystals form in the cell walls.

F.   Responses to Herbivores

Plants counter excessive grazing by herbivores with both physical and chemical defense measures.

- Physical defenses include structures such as thorns and spines.
- Chemical defenses take the form of distasteful or toxic compounds such as *canavanine*.
    ⇒ Similar to arginine, canavanine is an unusual amino acid produced by some plants.
    ⇒ When ingested by insects, it is incorporated in place of arginine in the insect's proteins.
    ⇒ Incorporation of canavanine disrupts protein conformation and the insect dies.

G.   Defense Against Pathogens: Systemic Acquired Resistance

The epidermis of the primary plant body and the periderm of the secondary plant body serve as surface barriers against pathogenic microorganisms.

- Pathogens can cross this barrier and invade the plant through injuries or natural epidermal openings such as stomata.

When a pathogen invades the plant, a variety of compounds known as *phytoalexins* are produced as a defense measure.

Phytoalexin = An antibiotic that destroys or inhibits the growth of microorganisms.

Plants also exhibit a *systemic acquired resistance* (SAR) response to help protect uninfected tissue from a pathogen that might spread from its point of invasion.

- When a pathogen invades a plant, molecules from the pathogen and compounds from the injured plant tissue serve as "alarm substances".
- The alarm substances cause rapid responses at the infection which slows the spread of the pathogen to other parts of the plant.
    ⇒ One such response is the cross-linking of molecules in cell walls which form a local barrier.
- The alarm substances also stimulate cells in infected areas to synthesize and release *salicylic acid*, a hormone that is transported throughout the plant. (A form of salicylic acid is the active ingredient in aspirin.)
- The hormone stimulates phytoalexin production and other defense measures in cells throughout the plant, even those far from the infection.

IX. Signal-transduction pathways mediate the responses of plant cells to environmental and hormonal stimuli

Plant cell responses to hormones and environmental stimuli are mediated by intracellular signals (*signal-transduction pathways*).

<u>Signal-transduction pathway</u> = A mechanism linking a mechanical or chemical stimulus to a cellular response.

- Three steps are involved in each pathway: *reception, transduction* and *induction*. (See Campbell, Figure 35.19)

Reception is the detection of a hormone or environmental stimulus by the cell.

- May take various forms depending on the stimulus. For example,
  ⇒ Absorption of a particular wavelength of light by a pigment within a cell.
  ⇒ The binding of a hormone to a specific protein receptor in the cell or on its membrane.
- Reception of a hormone only occurs in *target cells* for that hormone.
  ⇒ Target cells possess the specific protein receptor to which the hormone must bind; other cells do not possess the receptor.

Transduction in the pathway results in an amplification of the stimulus and its conversion into a chemical form which can activate the cell's responses.

- The hormone (*first messenger*) binds to a specific receptor and the hormone-receptor combination stimulates the *second messenger* (a substance that increases in concentration within a cell stimulated by the first messenger).
- The receptor may be bound to the cell membrane and its activation results in a chemical change to the cell.
  ⇒ Calcium ions appear to be important second messengers in many plant responses. Calcium ion concentration increases in the cell and the ions bind to the protein *calmodulin*.
  ⇒ The calmodulin-calcium complex then activates other target molecules within the cell.
- Amplification of the signal results from a single first messenger molecule binding to its receptor giving rise to many second messengers which activate an even larger number of proteins and other molecules.
- A second part of transduction that is important is the specificity of the responses.
  ⇒ Two cell types may both have receptors for a hormone but respond differently because each contains different target proteins for the second messenger.

Induction is the pathway step in which the amplified signal induces the cell's specific response to the stimulus.

- Some responses occur rapidly. For example,
  ⇒ ABA stimulation of stomatal closing.
  ⇒ Auxin-induced acidification of cell walls during cell elongation.
- Other responses take longer, especially if they require changes in gene expression (thigmomorphogenesis).

REFERENCES

Campbell, N. *Biology*. 4th ed. Menlo Park, California: Benjamin/Cummings, 1996.

Raven, P.H., R.F. Evert and S.E. Eichhorn. *Biology of Plants*. 5th ed. New York: Worth Publishers, Inc., 1992.

CHAPTER 36

AN INTRODUCTION TO ANIMAL STRUCTURE AND FUNCTION

OUTLINE

I. The functions of animal tissues and organs are correlated with their structures
    A. Animal Tissues
    B. Organs and Organ Systems

II. Bioenergetics is fundamental to all animal functions
    A. Metabolic Rate
    B. Body Size and Metabolic Rate

III. An animal's size and shape affect its interactions with the external environment
    A. Body Size, Proportions, and Posture
    B. Body Plans and Exchange with the Environment

IV. Homeostatic mechanisms regulate an animal's internal environment

OBJECTIVES

After reading this chapter and attending lecture, the student should be able to:
1. Define tissue and explain where it falls in the hierarchy of structural organization.
2. From micrographs or diagrams, correctly identify the following animal tissues, explain how their structure relates to function and give examples of each.
    a. Epithelial tissue
        1. Cuboidal   2. Columnar   3. Squamous
    b. Connective Tissue
        1. Adipose   2. Cartilage   3. Bone
    c. Muscle.
        1. Skeletal (striated)   2. Cardiac   3. Visceral (smooth).
    d. Nervous
3. Describe several body shapes that maximize external surface area in contact with the environment.
4. Explain how animals with complex internal organization and relatively small surface area to volume ratio can have adequate surface area for materials exchange with the environment.

5. Describe the location and function of interstitial fluid.
6. Define homeostasis.
7. Distinguish between negative and positive feedback.

KEY TERMS

epithelial

connective

muscle

nervous

simple epithelia

stratified epithelia

pseudostratified

cuboidal

columnar

squamous

loose connective
   tissue

collagenous fibers

elastic fibers

reticular fibers

fibroblasts

macrophages

adipose tissue

fibrous connective
   tissue

tendons

ligaments

chondrin

chondrocytes

osteocytes

Haversian systems

blood

leukocytes

erythrocytes

platelets

skeletal muscle

cardiac muscle

visceral muscle

nervous tissue

neuron

dendrites

axons

organs

mesenteries

organ systems

thoracic cavity

abdominal cavity

interstitial fluid

homeostasis

negative feedback

positive feedback

LECTURE NOTES

There are several unifying themes in the study of animal anatomy and physiology.

- There is a correlation between form and function; functions are properties that emerge from the specific shape and order of body parts.
- A comparative approach allows us to see how species of diverse evolutionary history and varying complexity solve problems common to all.
- Animals, as all living organisms, have the capacity to respond and adjust to environmental change in two temporal scales:
  ⇒ Over the long term by adaptation due to natural selection.
  ⇒ Over the short term by physiological responses.

The purpose of this introductory chapter is to:

- Illustrate the hierarchy of structural order characterizing animals.
- Emphasize the importance of energetics in animal life (how animals obtain, process, and use chemical energy).
- See how animal body forms affect their interactions with the environment.

I. **The functions of animal tissues and organs are correlated with their structures**

There is a structural hierarchy of life:
- Atoms → molecules → supramolecular structures → cell.
- The cell is the lowest level of organization that can live as an organism.

The hierarchy of multicellular organisms is: cell → tissues → organs → organ systems.

A. **Animal Tissues**

<u>Tissues</u> = Groups of cells with common structure and function.
- Cells may be held together by a sticky coating or woven together in a fabric of extracellular fibers.

There are four main categories of tissues: *epithelial* tissue, *connective* tissue, *muscle* tissue and *nervous* tissue.

1. **Epithelial Tissue**

Formed from sheets of tightly packed cells, epithelial tissue covers the outside of the body and lines organs and body cavities. Characteristics of epithelium include:
- Cells are closely joined and are riveted by tight junctions in some. (See Campbell, Chapter 7)
- It functions as barriers against mechanical injury, invading microbes, and fluid loss.
- Its free surface is exposed to air or fluid. Cells at the base are attached to a *basement membrane* which is a dense layer of extracellular material.

Epithelial tissue cells are categorized by the number of layers and shape of the free surface cells:
- *Simple epithelium* is one layer of cells.
- *Stratified epithelium* has multiple tiers of cells.
- *Pseudostratified epithelium* is one layer of cells that appear to be multiple since the cells vary in length.
- Cell shapes are: *cuboidal* (like dice), *columnar* (bricks on end), or *squamous* (like flat floor tile).
- A tissue may be described by a combination of terms such as stratified squamous epithelium.

Some epithelia are specialized for absorption or secretion of chemical solutions in addition to their protective role.
- Some epithelia are ciliated (e.g. the lining of the respiratory system).
- The *mucous membranes* lining the oral cavity and nasal passageways secrete mucus which moistens and lubricates the surfaces.
- The structure fits function. For example, simple squamous epithelium is leaky and is specialized for exchange of materials by diffusion. It is found in blood vessel linings and air sacs in the lungs.

2. **Connective Tissue**

Connective is characterized by a sparse cell population scattered through an extensive extracellular *matrix*.

*   Functions to bind and support other tissues.
*   Matrix is a web of fibers embedded in a homogenous ground substance.
*   Major types of connective tissue include:
    ⇒ loose connective tissue
    ⇒ adipose tissue
    ⇒ fibrous connective tissue
    ⇒ cartilage
    ⇒ bone
    ⇒ blood.

Loose connective tissue holds organs in place and attaches epithelia to underlying tissues.

*   Consists of a loose weave of 3 types of proteinaceous fibers:
    ⇒ *Collagenous fibers* are bundles of fibers containing 3 collagen molecules each. Great tensile strength; resist stretching.
    ⇒ *Elastic fibers* are long threads of the protein elastin. Elastic properties lend tissue a resilience to quickly return to the original shape.
    ⇒ *Reticular fibers* are branched and form a tightly woven fabric joining connective tissue to adjacent tissues.
*   Consists of two types of cells;
    ⇒ *Fibroblasts* secrete the proteins of the extracellular fibers.
    ⇒ *Macrophages* are phagocytic amoeboid cells that function in immune defense of the body.

Adipose tissue:

*   Is loose connective tissue specialized to store fat in adipose cells distributed throughout its matrix.
*   Insulates the body and stores fuel molecules.
*   Each adipose cell has one large fat droplet which can vary in size as fats are stored or utilized.

Fibrous connective tissue:

*   Is dense due to the arrangement of a large number of collagenous fibers in parallel bundles, which imparts great tensile strength.
*   Found in tendons (attach muscles to bones) and ligaments (attach bones together at joints).

Cartilage:

- Is composed of collagenous fibers embedded in *chondroitin sulfate*, a protein-carbohydrate ground substance.
- Cartilage cells, or *chondrocytes*:
  - ⇒ Secrete both collagen and chondroitin sulfate, which make cartilage both strong and flexible.
  - ⇒ Are confined to *lacunae*, scattered spaces within the ground substance.
- Comprises the skeleton of all vertebrate embryos.
  - ⇒ Some vertebrates (e.g. sharks) retain the cartilaginous skeleton in adults.
  - ⇒ Most vertebrates eventually replace most of the cartilage with *bone*. Cartilage is retained in areas such as the nose, ears, trachea, intervertebral discs and ends of some bones.

Bone is a mineralized connective tissue.

- *Osteoblasts*, bone-forming cells, deposit a matrix of collagen and calcium phosphate which hardens into the mineral hydroxyapatite. The combination of collagen and hydroxyapatite makes the bone harder than cartilage, but not brittle.
- Bone consists of repeating *Haversian systems* (concentric layers or *lamellae* deposited around a central canal containing blood vessels and nerves).
- *Osteocytes* are located in spaces called *lacunae* surrounded by a hard matrix and are connected to each other by cell extensions called *canaliculi*.
- In long bones, only the outer area is hard and compact; the inner area is filled with spongy bone tissue called *marrow*.

Blood is a connective tissue composed of:

- Liquid extracellular matrix of plasma, which contains water, salts and proteins.
- Cellular component which contains:
  - ⇒ *Leukocytes*, white blood cells that function in immune defense.
  - ⇒ *Erythrocytes*, red blood cells that transport oxygen.
  - ⇒ *Platelets*, cell fragments that function in blood clotting.
- Blood cells are made in red marrow near the ends of long bones.

3. **Muscle Tissue**

Muscle tissue consists of long, excitable cells capable of contraction.

- In the muscle cell cytoplasm are parallel bundles of *microfilaments* made of the contractile proteins, actin and myosin.
- Muscle is the most abundant tissue in most animals.

There are three types of vertebrate muscle tissue:

- *Skeletal muscle* is responsible for voluntary movements.
    - ⇒ Attached to bones by tendons.
    - ⇒ Microfilaments are aligned to form a banded *striated* appearance.
- *Cardiac muscle* forms the contractile wall of the heart.
    - ⇒ Cells are striated and *branched*.
    - ⇒ Ends of cells are joined by *intercalated disks*, which relay the contractile impulse from cell to cell.
- *Visceral muscle* is smooth (unstriated) tissue in walls of internal organs.
    - ⇒ Spindle-shaped cells contract slowly, but can retain contracted condition longer than skeletal muscle.
    - ⇒ Responsible for involuntary movements (e.g. churning of the stomach).

4. **Nervous Tissue**

Nervous tissue senses stimuli and transmits signals from one part of the animal to another.

Neuron = Nerve cell specialized to conduct an impulse or bioelectric signal. Consists of:

- *Cell body*.
- *Dendrites*, extensions that conduct impulses to the cell body.
- *Axons*, extensions that transmit impulses away from the cell body. (See Campbell, Chapter 44)

B. **Organs and Organ Systems**

Tissues are organized into *organs* in all but the simplest animals.

- May be layered, such as the dermis of a human.
- Many organs are suspended by sheets of connective tissue called *mesenteries*.
- Organs may be organized into *organ systems*.

Organ systems = Several organs with separate functions that act in a coordinated manner (e.g. digestive, circulatory and respiratory systems).

- Systems are interdependent: an organism is a living whole greater than the sum of its parts.

II. Bioenergetics is fundamental to all animal functions

Animals, as living organisms, exchange energy with the environment. Since they are heterotrophic, animals acquire energy from organic molecules synthesized by other organisms.

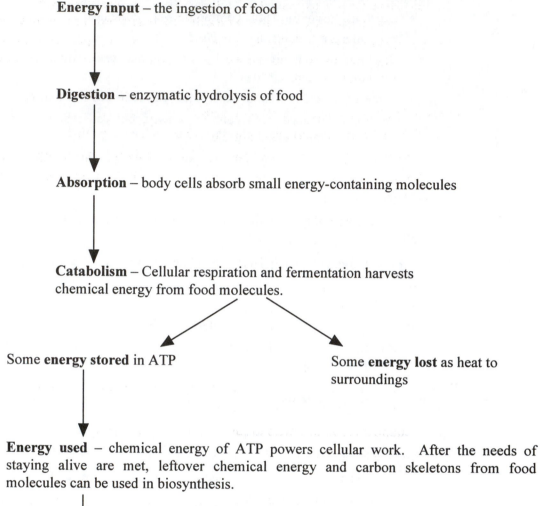

Energy input – the ingestion of food

Digestion – enzymatic hydrolysis of food

Absorption – body cells absorb small energy-containing molecules

Catabolism – Cellular respiration and fermentation harvests chemical energy from food molecules.

Some **energy stored** in ATP

Some **energy lost** as heat to surroundings

Energy used – chemical energy of ATP powers cellular work. After the needs of staying alive are met, leftover chemical energy and carbon skeletons from food molecules can be used in biosynthesis.

Energy lost – cellular work generates heat, which is lost to the surroundings.

A. Metabolic Rate

Bioenergetics, the study of the dynamic balance between energy intake and loss in an organism, gives clues to how an animal adapts to its environment. By measuring the rate of energy use, physiologists can determine.

- How much food energy an animal needs just to stay alive.
- The energy costs for specific activities such as walking or running.

Metabolic rate = total amount of energy an animal uses per unit of time; usually measured in kilocalories (kcal or CAL = 1000 calories). Can be determined by:
- Measuring the amount of oxygen used for an animal's cellular respiration.
- Measuring an animal's heat loss per unit of time.
  - ⇒ Heat loss, a byproduct of cellular work, is measured with a *calorimeter* - a closed, insulated chamber with a device that records heat production.
  - ⇒ Calorimeters are effectively used with small animals that have high metabolic rates, but are less precise with small animals that have low metabolic rates and with large animals.

Every animal has a range of metabolic rates.
- Minimal rates support basic life functions, such as breathing.
- Maximal rates occur during peak activity, such as all-out running.
- Between these extremes, metabolic rates can be influenced by many factors, such as:
  - ⇒ age, sex, and size
  - ⇒ body temperature
  - ⇒ environmental temperature
  - ⇒ food quality and quantity
  - ⇒ activity level
  - ⇒ amount of available oxygen
  - ⇒ hormonal balance
  - ⇒ time of day.

Endotherms = Animals that generate their own body heat metabolically.
- Examples include birds and mammals.
- Require more kilocalories to sustain minimal life functions that ectotherms.
- Many are also *homeothermic*, that is their body temperature must be maintained within narrow limits.

Basal Metabolic Rate (BMR) = An endothermic animal's metabolic rate measured under resting, fasting and stress-free conditions.
- Average human BMR is 1600-1800 kcal/day for adult males; 1300-1500 kcal/day for adult females.

Ectotherms = Animals that acquire most of their body heat from the environment.
- Include most fish, amphibians, reptiles, and invertebrates.
- Are energetically different from endotherms; body temperature and metabolic rate changes with environmental temperature.
- Because it is influenced by temperature, an ectotherm's minimal metabolic rate (SMR) must be determined at a specific temperature.

Standard Metabolic Rate (SMR) = An ectotherm's metabolic rate measured under controlled temperature and under resting, fasting and stress-free conditions.

B.   Body size and Metabolic Rate

There is an inverse relationship between metabolic rate and size.
- Smaller animals consume more calories per gram than larger animals
- Correlated with a higher metabolic rate and need for faster O_2 delivery to the tissues, small animals also have higher:
  ⇒ breathing rates
  ⇒ blood volume
  ⇒ heart rate
- This inverse relationship between metabolic rate and body size holds true for both endotherms and ectotherms, and is not simply a function of surface area to volume ratio.

III.   An animal's size and shape affect its interactions with the external environment

An animal's body plan results from a developmental pattern programmed by its genome – a product of millions of years of evolution due to natural selection.

A.   Body Size, Proportions, and Posture

Body proportions and size-weight relationships change in animal bodies as they become larger.
- Body design must accommodate the greater demand for support that comes with increasing size. (The strain on body supports depends on an animal's weight, which increases as the cube of its height or other linear dimension.)
- In mammals and birds, the most important design feature in supporting body weight is posture – leg position relative to the main body – rather than leg bone size. For example:
  ⇒ The legs of an elephant are in a more upright position than those of small mammals.
  ⇒ Large mammals run with legs nearly extended, which reduces strain; whereas, small mammals run with legs bent and crouch when standing.
- Bioenergetics also plays an important role in load-bearing, since crouched posture is partly a function of muscle contraction, powered by chemical energy.

B.   Body Plans and Exchange with the Environment

Animal cells must have enough surface area in contact with an aqueous medium to allow adequate environmental exchange of dissolved oxygen, nutrients and wastes. This requirement imposes constraints on animal size and shape.
- Unicellular organism, such as protozoans, must have sufficient surface area of plasma membrane to service the entire volume of cytoplasm and are thus limited in size. Recall that:
  ⇒ The upper limits of cells size are imposed by the surface area to volume ratio.
  ⇒ As cell size increases, volume increases proportionately more than surface area.

- Some multicellular animals have a body plan that places all cells in direct contact with their aqueous environments. Two such body plans include:
  ⇒ Two-layered sac – a body wall only two cell layers thick. For example, the body cavity of *Hydra* opens to the exterior, so both outer and inner layers of cells are bathed in water.
  ⇒ Flat-shaped body with maximum surface area exposed to the aqueous environment. For example, tapeworms are thin and flat, so most cells are bathed in the intestinal fluid of the worm's vertebrate host.
- Most complex animals have a smaller surface area to volume ratio and thus lack adequate exchange area on the outer surface.
  ⇒ Instead, highly folded, moist *internal* surfaces exchange materials with the environment.
  ⇒ The circulatory system shuttles materials between these specialized exchange surfaces.

Though logistical problems exist with environmental exchange, there are some distinct advantages to a compact body form.

- Environmental exchange surfaces are internal and protected from desiccation, so the animal can live on land.
- Cells are bathed with internal body fluid, so the animal can control the quality of the cells' immediate environment.

IV. Homeostatic mechanisms regulate an animal's internal environment

Interstitial fluid = The internal environment of vertebrates, composed of fluid between the cells.
- Fills spaces between cells.
- Exchanges nutrients and wastes with blood carried in capillaries.

Homeostasis = Dynamic state of equilibrium in which internal conditions remain relatively stable; steady state.
- French physiologist Claude Bernard first described the "constant internal milieu" in animals; he recognized many animals can maintain constant conditions in their internal environment – even when the external environment changes.
- Maintained by control systems which include three components:
  ⇒ *receptor* – detects internal change.
  ⇒ *control center* – processes information from the receptor and directs the effector to respond.
  ⇒ *effector* – provides the response.
- As a control system operates, the effector's response feeds back and influences the magnitude of the stimulus by either depressing it (*negative feedback*) or enhancing it (*positive feedback*).

<u>Negative Feedback</u> = Homeostatic mechanism that stops or reduces the intensity of the original stimulus and consequently causes a change in a variable that is opposite in direction to the initial change.

- Most common homeostatic mechanism in animals.
- There is a lag time between sensation and response, so the variable drifts slightly above and below the *set point*.
- A nonbiological example is the thermostatic control of room temperature. (See Campbell, Figure 36.11.) Human examples include hormonal control of blood glucose levels and the regulation of body temperature by the hypothalamus.
  - ⇒ If the human hypothalamus detects a high blood temperature, it sends nerve impulses to sweat glands, which increase sweat output and cause evaporative cooling.
  - ⇒ When the body temperature returns to normal, no additional signals are sent.

<u>Set Point</u> = A variable's range of values that must be maintained to preserve homeostasis.

<u>Positive Feedback</u> = Homeostatic mechanism that enhances the initial change in a variable.

- Rarer than negative feedback and usually controls only episodic events.
- Examples include blood clotting and the heightening of labor contractions during childbirth.
  - ⇒ During childbirth, the baby's head against the uterine opening stimulates contractions which cause greater pressure of the head against the uterine opening.
  - ⇒ The greater pressure, in turn, enhances uterine contractions further.

REFERENCES

Campbell, N. *Biology*. 4th ed. Menlo Park, California: Benjamin/Cummings, 1996.

Marieb, E.N. *Human Anatomy and Physiology*. 3rd ed. Redwood City, California: Benjamin/Cummings, 1995.

Schmidt-Nielsen, Knut. *Animal Physiology: Adaptation and Environment*. 3rd ed. New York: Cambridge University Press, 1990.

CHAPTER 37

ANIMAL NUTRITION

OUTLINE

I. Diets and feeding mechanisms vary extensively among animals

II. Ingestion, digestion, absorption, and elimination are the four main stages of food processing

III. Digestion occurs in food vacuoles, gastrovascular cavities, and alimentary canals
   A. Intracellular Digestion in Food Vacuoles
   B. Digestion in Gastrovascular Cavities
   C. Digestion in Alimentary Canals

IV. A tour of the mammalian digestive system
   A. The Oral Cavity
   B. The Pharynx
   C. The Esophagus
   D. The Stomach
   E. The Small Intestine
   F. The Large Intestine

V. Vertebrate digestive systems exhibit many evolutionary adaptations associated with diet

VI. An adequate diet provides fuel, carbon skeletons for biosynthesis, and essential nutrients
   A. Food as Fuel
   B. Food for Biosynthesis
   C. Essential Nutrients

OBJECTIVES

After reading this chapter and attending lecture, the student should be able to:

1. Distinguish among herbivores, carnivores and omnivores.
2. Describe the following feeding mechanisms and give examples of animals that use each:
    a. Filter-feeding.
    b. Substrate-feeding.
    c. Deposit-feeding.
    d. Fluid-feeding.
3. Define digestion and describe why it is a necessary process.
4. Explain how anhydro bonds are formed and describe the role of hydrolysis in digestion.
5. Distinguish between intracellular and extracellular digestion.
6. Explain why intracellular digestion must be sequestered in a food vacuole and give examples of organisms which digest their food in vacuoles.
7. Define gastrovascular cavity and explain why extracellular digestive cavities are advantageous.
8. Using *Hydra* as an example, describe how a gastrovascular cavity functions in both digestion and distribution of nutrients.
9. List major animal phyla which use gastrovascular cavities for digestion.
10. Describe some distinct advantages that complete digestive tracts have over gastrovascular cavities and list the major animal phyla with alimentary tracts.
11. Define peristalsis and describe its role in the digestive tract.
12. Describe how salivation is controlled and list the functions of saliva.
13. Describe the role of salivary amylase in digestion.
14. Describe the sequence of events which occur as a result of the swallowing reflex.
15. Describe the function of the esophagus and explain how peristalsis in the esophagus is controlled.
16. Describe the role of the cardiac and pyloric sphincters.
17. List the three types of secretory cells found in stomach epithelium and what substances they secrete.
18. Recall the normal pH of the stomach and explain the function of stomach acid.
19. Describe the function of pepsin.
20. Explain why the stomach normally does not digest itself.
21. Explain how pepsin and acid secretion are regulated and describe the roles of the hormones gastrin and enterogastrone.
22. Describe the cause of ulcers and explain why they are frequently found in the duodenum.
23. Explain how chyme is moved through the small intestine.
24. Describe the sequence of events which occur in response to acid chyme entering the duodenum and include the roles of:
    a. Secretin
    b. Bicarbonate
    c. Cholecystokinin (CCK)
    d. Gall bladder.
    e. Bile.
    f. Pancreatic enzymes.
    g. Enterogastrone.
25. Describe how pancreatic zymogens for proteolytic enzymes are activated in the duodenum and include the role of the intestinal enzyme enterokinase.
26. Describe enzymatic digestion of carbohydrates, proteins, lipids and nucleic acids including the reactants and products for each enzymatic reaction and whether they occur in the:
    a. Oral cavity.
    b. Stomach.
    c. Lumen of small intestine.
    d. Brush border of small intestine.
27. Explain the function of bile; describe where it is produced and stored; and describe its composition.
28. State whether the lumen of the digestive tract is technically inside or outside the body.
29. Explain where most nutrient absorption occurs.

30. Explain why the many folds, villi and microvilli are important in the small intestine.
31. Describe how specific nutrients are absorbed across the intestinal epithelium and across the capillary or lacteal wall and indicate whether the transport is with or against the concentration gradient.
32. Explain what happens to glycerol and fatty acids after they are absorbed into the intestinal epithelium and describe the fate of chylomicrons and lipoproteins.
33. Explain the function of the hepatic portal vein.
34. Explain where in the digestive tract that most reabsorption of water occurs.
35. Describe the composition of feces and explain what the main source of vitamin K is for humans.
36. Give examples of vertebrates with the following digestive adaptations and explain how these adaptations are related to diet:
    a. Variation in dentition.
    b. Variation in length of the digestive tract.
    c. Fermentation chambers.
37. Explain why animals need a nutritionally adequate diet.
38. Describe the effects of undernourishment or starvation.
39. List some of the risks of obesity.
40. Distinguish between malnourished and undernourished.
41. List four classes of essential nutrients and describe what happens if they are deficient in the diet.
42. List and distinguish between water-soluble and fat-soluble vitamins and explain how they are used by the body.
43. Describe the dietary sources, major body functions and effects of deficiency for the following required minerals in the human diet: calcium, phosphorus, sulfur, potassium, chlorine, sodium, magnesium and iron.

KEY TERMS

heterotroph	salivary amylase	lipase	microvilli
herbivores	bolus	pancreatic amylase	lacteal
carnivores	pharynx	maltose	chylomicrons
omnivores	epiglottis	maltase	lipoproteins
filter-feeders	glottis	disaccharidases	hepatic portal vein
substrate-feeders	esophagus	trypsin	large intestine
deposit-feeders	stomach	chymotrypsin	cecum
fluid-feeders	rugae	carboxypeptidase	appendix
hydrolysis	lumen	aminopeptidase	*Escherichia coli*
intracellular digestion	gastric juice	dipeptidase	rectum
extracellular digestion	pepsin	enteropeptidase	essential nutrients
gastrovascular cavity	pepsinogen	nucleases	undernourished
alimentary canal	zymogens	enterogastrone	malnourished
peristalsis	gastrin	peristalsis	complete proteins
salivary gland	pyloric sphincter	emulsification	incomplete proteins
pancreas	acid chyme	lipase	kwashiorkor
liver	duodenum	jejunum	beriberi
gall bladder	secretin	ileum	vitamins
oral cavity	cholecystokinin (CCK)	villi	minerals

LECTURE NOTES

Animals are heterotrophic and, like all heterotrophs, must rely on organic compounds in their food to supply energy and the raw materials for growth and repair.

- A variety of nutritional adaptations have resulted from natural selection during the evolutionary development of animals.

I.   **Diets and feeding mechanisms vary extensively among animals**

Animals usually ingest other organisms.

- The food organism may be either dead or alive.
- The food organism may be ingested whole or in pieces.
- Some parasitic animals (i.e. tapeworms) are exceptions.

Animals are categorized based on the kinds of food they usually eat and their adaptations for obtaining and processing food items.

- *Herbivores* eat autotrophic organisms (plants, algae, and autotrophic bacteria).
- *Carnivores* eat other animals.
- *Omnivores* eat other animals and autotrophs.

The varied diets exhibited by animals are accompanied by a variety of mechanisms used to obtain food.

- *Suspension-feeders* sift small food particles from the water.
    ⇒ Many are aquatic animals such as clams and oysters (trap food on gills) and baleen whales (strain food from water forced through the screen-like plates on their jaws).
- *Substrate-feeders* live on or in their food source and eat their way through the food.
    ⇒ Leaf miners (larvae of various insects) tunnel through the interior of leaves.
- *Deposit-feeders* are a type of substrate-feeder that ingests partially decayed organic materials along with their substrate.
    ⇒ Earthworms ingest soil and their digestive systems extract the organic materials.
- *Fluid-feeders* suck nutrient-rich fluids from a living host.
    ⇒ Aphids ingest the phloem sap from plants; leeches and mosquitoes suck blood from animals; hummingbirds and bees ingest nectar from flowers.
- *Bulk-feeders* eat relatively large pieces of food.
    ⇒ Most animals; they possess various adaptations to kill prey or tear off pieces of meat or vegetation.

II.   Ingestion, digestion, absorption, and elimination are the four main stages of food processing

Food processing involves four stages: *ingestion*, *digestion*, *absorption*, and *elimination*.

Ingestion is the first stage and is the act of eating.

Digestion is the second stage and is the process of breaking down food into small molecules the body can absorb.

- Organic food material is composed of macromolecules (proteins, fats, carbohydrates) that are too large to cross the membranes and enter an animal's cells.
- Digestion enzymatically cleaves these macromolecules into component monomers that can be used by the animal (polysaccharides and disaccharides to simple sugars; proteins to amino acids; fats to glycerol and fatty acids).
- Digestion uses *enzymatic hydrolysis* to break bonds in macromolecules.
  ⇒ Hydrolytic enzymes catalyze the digestion of each class of macromolecule by adding water.
  ⇒ The chemical digestion is usually preceded by mechanical fragmentation (e.g. chewing) that increases the surface area exposed to digestive juices.
  ⇒ Occurs in a specialized compartment where the enzymes are contained so they don't damage the animal's own cells.

Absorption is the third stage of the process and involves the uptake of the small molecules resulting from digestion.

Elimination is the fourth, and last, stage where undigested material passes out of the digestive compartment.

III.   Digestion occurs in food vacuoles, gastrovascular cavities, and alimentary canals

A.   Intracellular Digestion in Food Vacuoles

Food vacuoles are the simplest digestive compartments.

- They are organelles where a single cell digests its food without hydrolytic enzymes mixing with the cell's cytoplasm.
- Protozoa have food vacuoles that form around food by endocytosis.
- Hydrolytic enzymes are secreted into the food vacuole and digestion occurs; this is referred to as *intracellular digestion*. (See Campbell, Figure 37.5)
- Sponges differ from other animals in that all digestion is by the intracellular mechanism. (See Campbell, Figure 37.6)

Extracellular digestion occurs within compartments that are continuous via passages with the outside of the body.

- At least some hydrolysis occurs in most animals by this mechanism.

B.   **Digestion in Gastrovascular Cavities**

Most animals that have simple body plans possess a *gastrovascular cavity*.

Gastrovascular cavity = Digestive sac with a single opening.

- Function in both digestion and nutrient distribution.

Digestion in *Hydra* involves both extracellular and intracellular digestion (See Campbell, Figure 37.7a)

- Hydra is a carnivore which captures prey.
- Food items are immobilized by stings from nematocysts on the tentacles.
- Tentacles then force prey through the mouth into the gastrovascular cavity.
- Specialized gastrodermal cells secrete digestive enzymes that fragment the soft tissues of the prey into tiny pieces.
- Some gastrodermal cells also possess flagella whose movement prevents settling of food particles and distributes them through the cavity.
- The small pieces are phagocytized by nutritive gastrodermal cells and surrounded by food vacuoles.
- Hydrolysis is completed by intracellular digestion.
- Undigested materials are expelled from the gastrovascular cavity through the single opening.

The combination of extracellular and intracellular digestion that occurs in most animals permits these organisms to feed on larger prey items.

- Phagocytosis is limited to microscopic food.
- Extracellular digestion begins the digestive process by breaking-down large food items into smaller particles.

Extracellular digestion may even begin outside of the animal's body as in planarians.

- Planarians possess a gastrovascular cavity that is extensively branched.
- Associated with the cavity is a muscular, tubular *pharynx* that can be everted ventrally through the mouth.
- The pharynx penetrates the prey (small invertebrates and dead organic matter) and releases digestive juices.
- The partially digested food is sucked into the gastrovascular cavity where digestion continues.
- Digestion is completed by intracellular digestion in cells that phagocytize the tiny particles.

C.   Digestion in Alimentary Canals

Animals with body plans more complex than cnidarians and platyhelminths have *alimentary canals*.

> Alimentary canal = Digestive tube running between two openings: the *mouth* (where food is initially ingested) and the *anus* (where undigested wastes are egested). (See Campbell, Figure 37.8)

- Since food moves in one direction along the tube, the tube can be organized into specialized regions that carry out digestion and absorption of nutrients in a stepwise fashion.

- The unidirectional passage of food and the specialization of function for different regions makes the alimentary canal more efficient.

IV.   A tour of the mammalian digestive system

The digestive system in mammals includes the alimentary canal and accessory glands that secrete digestive juices into the canal through ducts.

- The digestive tract has a four-layered wall: the lumen is lined by a mucous membrane (*mucosa*), then a connective tissue layer followed by a layer of smooth muscle, and outermost is a connective tissue layer attached to the membrane of the body cavity.

- *Peristalsis* (rhythmic smooth muscle contractions) pushes food along the tract.

- *Sphincters* (modifications of the muscle layer into ringlike valves) occur at some junctions between compartments and regulate passage of materials through the system.

- The accessory glands are: three pairs of *salivary glands*, the *pancreas*, the *liver*, and the *gall bladder*.

Refer to Figure 37.9 for orientation of the specialized compartments in the human system.

A.   The Oral Cavity

Physical and chemical digestion begin in the *oral cavity*.

- Chewing breaks down large pieces of food into smaller pieces.

- This makes food easier to swallow and increases the surface area available for enzyme action.

The presence of food in the oral cavity stimulates the salivary glands to secrete *saliva* into the oral cavity.

- Saliva contains *mucin* (protects the mouth from abrasion and lubricates food), buffers that neutralize acids, antibacterial agents, and *salivary amylase* – an enzyme that hydrolyzes starch and glycogen to the disaccharide maltose or small polysaccharides.

The tongue manipulates food during chewing and forms it into a *bolus* which is swallowed.

- The tongue also tastes the food and pushes the bolus to the pharynx.

B.   The Pharynx

The pharynx is an intersection for both the digestive and respiratory systems.

- The movement of swallowing moves the *epiglottis* to block the entrance of the windpipe (the *glottis*).
- This directs food through the *pharynx* and into the *esophagus*. (See Campbell, Figure 37.10)

C.   The Esophagus

The esophagus is a muscular tube which conducts food from the pharynx to the stomach.

- Peristalsis moves the bolus along the esophagus to the stomach.
- The initial entrance of the bolus into the esophagus is voluntary (swallowing); once in, the peristalsis results from involuntary contraction of the smooth muscles.
- Salivary amylase remains active as the bolus moves through the esophagus.

D.   The Stomach

The stomach is a large, saclike structure located just below the diaphragm on the left side of the abdominal cavity. It functions in:

- Food storage.
  - ⇒ The stomach has an elastic wall with *rugae* – folds that can expand to accommodate up to 2 liters of food.
  - ⇒ Storage capacity permits periodic feeding (meals).
- Churning.
  - ⇒ The stomach has longitudinal, vertical and diagonal muscles that contract the stomach in churning movements that mix the food.
  - ⇒ Stomach contents are mixed about every 20 minutes.
  - ⇒ Churning, mixing and addition of stomach acid convert food to a nutrient broth called *acid chyme*.
    - ◊ The passage of acid chyme into the small intestine is regulated by the *pyloric sphincter* at the bottom of the stomach.
    - ◊ The pyloric sphincter relaxes at intervals and permits small quantities of chyme to pass.
- Secretion. Gastric secretion is controlled by nerve impulses and the hormone *gastrin*. The stomach epithelium contains three types of secretory cells.
  - ⇒ *Mucous cells* secrete:
    - ◊ *mucin* – a thin mucus that protects the stomach lining from being digested.
    - ◊ gastrin – a hormone produced by the stomach. Gastrin is released into the bloodstream and its action is to stimulate further secretion of gastric juice (HCl and pepsin).
  - ⇒ *Chief cells* secrete:
    - ◊ *pepsinogen*, an inactive protease or zymogen that is the precursor to pepsin.
      <u>Zymogen</u> = An inactive form of a protein-digesting enzyme.

⇒ *Parietal cells* secrete:

◊ HCl

- Protein digestion. Both components of *gastric juice*, HCl and pepsin, are involved with protein digestion:

⇒ HCl – provides acidity (pH 1 - 4) which:

◊ kills bacteria.

◊ denatures protein.

◊ starts the conversion of pepsinogen to pepsin. Newly formed pepsin can also catalyze this reaction:

Pepsinogen $\xrightarrow{\text{HCl}}$ Pepsin

⇒ Pepsin – splits peptide bonds next to some amino acids.

◊ Does not hydrolyze protein completely.

◊ Is an endopeptidase that splits peptide bonds located within the polypeptide chain.

E.   The Small Intestine

NOTE: A discussion of hormonal control of digestion includes several examples of positive and negative feedback. This is a good place to reinforce the concepts learned in Chapter 36 and to make a transition to the topic of the endocrine system (Chapter 41).

The human small intestine is about 6 m in length and is the site of most enzymatic hydrolysis of food and absorption of nutrients.

- Remember, only limited digestion of carbohydrates occurs in the oral cavity and esophagus (by salivary amylase) and of proteins in the stomach (by pepsin).

The *pancreas*, *liver*, *gall bladder* and small intestine all contribute to the digestion that occurs in the small intestine. Their products are released into the *duodenum*, the first 25 cm of the small intestine.

- The *pancreas* produces:

⇒ hydrolytic enzymes that break down all major classes of macromolecules – carbohydrates, lipids, proteins and nucleic acids.

⇒ bicarbonate buffer that helps neutralize the acid chyme coming from the stomach.

- The *liver* performs many functions including the production of bile, which:

⇒ is stored in the *gall bladder*.

⇒ does not contain digestive enzymes.

⇒ contains bile salts which emulsify fat.

⇒ contains pigments that are byproducts of destroyed red blood cells.

Hormonal control of digestion involves the following regulatory hormones:

- *Gastrin* – Released from the stomach in response to presence of food; stimulates the stomach to release gastric juice (HCl and pepsin).
- *Secretin* – Released from the duodenum in response to acid chyme entering from the stomach; signals the pancreas to release bicarbonate buffer to neutralize acid chyme.
- *Cholecystokinin* (*CCK*) – Released from the duodenum in response to chyme entering from the stomach; signals the gall bladder to release bile and the pancreas to release pancreatic enzymes into the duodenum.
- *Enterogastrone* – Released from the duodenum in response to the presence of fat in the chyme; inhibits peristalsis in the stomach and slows digestion.

Carbohydrate digestion:

- Begins with the action of salivary amylase in the mouth.
- Begins again in the duodenum where *pancreatic amylases* hydrolyze starch and glycogen into disaccharides.
- *Disaccharidases* attached to the surface of the duodenal epithelium hydrolyze disaccharides into monosaccharides.

  ⇒ Each disaccharide has its own disaccharidase. For example, maltose is hydrolyzed by *maltase*, sucrose by *sucrase*, lactose by *lactase*, etc.

  ⇒ Since the disaccharidases are on the surface of the epithelium, the final breakdown of carbohydrates occurs where the sugars will be absorbed.

Protein digestion involves the efforts of teams of enzymes:

- Pepsin, an endopeptidase, begins protein digestion in the stomach.
- The pancreas secretes proteases in the form of zymogens that will be activated only in the lumen of the duodenum by the intestinal enzyme, *enteropeptidase*.

  ⇒ Enteropeptidase converts trypsinogen to trypsin

$$\text{Trypsinogen} \xrightarrow{\text{enteropeptidase}} \text{Trypsin}$$

  ⇒ Trypsin catalyzes the conversion of more trypsinogen to trypsin.
  ⇒ Trypsin catalyzes the conversion of the other zymogens.

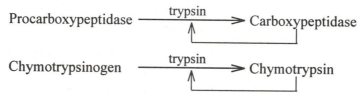

$$\text{Procarboxypeptidase} \xrightarrow{\text{trypsin}} \text{Carboxypeptidase}$$

$$\text{Chymotrypsinogen} \xrightarrow{\text{trypsin}} \text{Chymotrypsin}$$

- Trypsin and chymotrypsin (endopeptidases) digest large polypeptides into shorter chains by breaking internal peptide bonds adjacent to certain amino acids.
- Carboxypeptidase (exopeptidase) splits amino acids, one at a time, off the end of a polypeptide that has a free carboxyl group.
- The lining of the small intestine also secretes protein-digesting enzymes, *aminopeptidase* and *dipeptidases*.

  ⇒ Aminopeptidase begins at the end of a polypeptide that has a free amino group and splits off one amino acid at a time.

⇒ Dipeptidases attached to the intestinal lining split small polypeptides.

- Since the protein-digesting enzymes from the pancreas and small intestine break bonds in specific areas of the polypeptide, protein digestion to amino acids is a combined effort from all of these enzymes.

Nucleic acid digestion also involves teams of enzymes.

- *Nucleases* hydrolyze DNA and RNA into nucleotides.
- Other hydrolytic enzymes (nucleotidases and nucleosidases) break nucleotides into nucleosides and nitrogenous bases, sugars and phosphates.

Fat digestion occurs only in the small intestine, so most fat in food is undigested when it reaches the duodenum. If fat is present in chyme,

- The duodenum secretes *enterogastrone*, a hormone that inhibits peristalsis in the stomach, slowing the entry of food into the duodenum.
- In response to CCK, the gall bladder releases bile into the upper duodenum. Bile salts emulsify the insoluble fat by coating fat droplets and thus preventing them from coalescing.
- *Emulsification* produces many small fat droplets that collectively have a large surface area exposed for digestion.
- Pancreatic *lipase*, secreted into the duodenum, hydrolyzes fats into the building blocks, glycerol and fatty acids.

Macromoleucles are completely hydrolyzed as peristalsis moves the digestive juice-chyme mixture through the duodenum. See Table 37.1 for a summary of digestion.

- The remaining areas of the small intestine, the *jejunum* and *ileum*, are specialized for absorption of nutrients.

Nutrients resulting from digestion must cross the digestive tract lining to enter the body. While a small number of nutrients are absorbed by the stomach and large intestine, most absorption occurs in the small intestine.

- Large folds in the walls are covered with projections called *villi*, which in turn have many microscopic *microvilli*; this results in a surface area for absorption of about 300 m^2. (See Campbell, Figure 37.14)

  ⇒ This *brush border* (microvillar surface) is exposed to the lumen of the intestine.
- Penetrating the hollow core of each villus are capillaries and a tiny lymph vessel called a *lacteal*.
- Nutrients are absorbed by diffusion or active transport across the 2 cell-thick epithelium and into the capillaries or lacteals.

  ⇒ Amino acids and sugars enter the capillaries and are transported by the blood.

  ⇒ Absorbed glycerol and fatty acids are recombined in epithelial cells to form fats; most are coated with proteins to form *chylomicrons* which enter the lacteals.
- Capillaries and veins draining nutrients away from the villi converge into the *hepatic portal vein*, which leads directly to the liver.

  ⇒ Here various organic molecules are used, stored, or converted to different forms.

  ⇒ Blood flows at a rate of about 1 L per minute through the hepatic portal vessel.

F.   The Large Intestine

The large intestine, or *colon*, connects to the small intestine at a T-shaped junction containing a sphincter; the blind end of the T is called the *cecum*.

- The *appendix* is a fingerlike extension of the cecum and is composed of lymphoid tissue.
- The *colon* is about 1.5 m long and is in the shape of an inverted "U". Its major function is water reabsorption.
- *Feces* (wastes of the digestive tract) are moved through the colon by peristalsis.
  - ⇒ Intestinal bacteria live on organic material in the feces and some produce vitamin K which is absorbed by the host.
  - ⇒ An abundance of salts is excreted by the colon lining.
  - ⇒ Feces are stored in the *rectum* and pass through the two sphincters (one involuntary, one voluntary) to the *anus* for elimination.

V.   Vertebrate digestive systems exhibit many evolutionary adaptations associated with diet

The digestive systems of vertebrates are variations on a common plan, and many adaptations associated with diet are found.

Variation in the dentition (assortment of teeth) of mammals reflects the animal's diet. See Campbell, Figure 37.15.

- Carnivores (i.e. dogs and cats) generally have pointed *canines* and *incisors*.
  - ⇒ These teeth are used to kill prey and rip away pieces of flesh.
  - ⇒ *Premolars* and *molars* are jagged and used to crush and shred the food.
- Herbivores (i.e. cows) have teeth with broad, ridged surfaces that are used for grinding vegetation.
  - ⇒ Incisors and canines are modified for biting off pieces of vegetation.
- Omnivores (i.e. humans) have relatively unspecialized dentition.
  - ⇒ They are adapted to eat both vegetation and meat and teeth similar to those of both herbivores and carnivores are found.

Nonmammalian vertebrates typically have less specialized dentition, although there are exceptions.

- Poisonous snakes have fangs which are teeth modified to inject venom into prey.
  - ⇒ Some fangs are hollow, others are grooved.
  - ⇒ Snakes can also swallow very large prey items due to the loosely hinged lower jaw-skull articulation.

A correlation is also found between length of the vertebrate digestive system and diet.

- Herbivores and omnivores have longer alimentary canals than carnivores relative to size.
- The cell walls in vegetation make it more difficult to digest than meat, and nutrients are less concentrated.
- The longer tract allows for more time for digestion and provides a greater surface area for absorption.
- The *functional* length of an alimentary canal may be increased by increased folding as with the spiral valve structure in sharks.

Special fermentation chambers are present in the alimentary canals of many herbivores.

- Symbiotic bacteria and protozoa present in these chambers produce cellulase which can digest the cellulose. (Animals do not produce cellulase.)
- The microorganisms digest cellulose to simple sugars and convert the sugars to nutrients essential to the animal.
- The microorganisms may be housed in the cecum (i.e. horses), cecum and colon (i.e. rabbit), or the much more elaborate structure found in *ruminants*.

VI. An adequate diet provides carbon skeletons for biosynthesis and essential nutrients

A nutritionally adequate diet provides an animal with: fuel (chemical energy) for cellular respiration, raw organic materials for biosynthesis, and *essential nutrients* which must be obtained in prefabricated form.

A. Food as Fuel

Chemical energy is obtained from the oxidation of complex organic molecules.

- Monomers from any of the complex organic molecules can be used to produce energy, although those from carbohydrates and fats are used first.
- Oxidation of a gram of fat liberates 9.5 kcal, twice that of a gram of carbohydrate or protein.

The basal energy requirements of an animal must be met to sustain their metabolic functions.

- When an animal takes in more calories than are consumed, the liver and muscles store the excess in the form of glycogen; further excess is stored in adipose tissue in the form of fat.
- When the diet is deficient in calories, glycogen stored in the liver and muscles is utilized first and fat is then withdrawn from adipose tissues.
- An *undernourished* person or animal is one whose diet is deficient in calories.
  - ⇒ If starvation persists, the body begins to breakdown its own proteins as a source of energy.
  - ⇒ The breakdown of the body's own proteins can cause muscles to atrophy and can result in the consumption of the brain's proteins.

Obesity (overnourishment) is a greater problem in the United States and other developed countries than undernourishment.

- It increases the risk of heart attack, diabetes, and other disorders.

B. Food for Biosynthesis

Heterotrophs cannot use inorganic materials to make organic molecules; they must obtain organic precursors for these molecules from the food they ingest.

Given a source of carbon and nitrogen, heterotrophs can fabricate a great variety of organic molecules by using enzymes to rearrange the molecular skeletons of precursors acquired from food.

- A single type of amino acid can supply the nitrogen necessary to build other amino acids.
- Fats can be synthesized from carbohydrates.
- The liver is responsible for most of the conversion of nutrients from one type of organic molecule to another.

C. Essential Nutrients

An animal's diet must include *essential nutrients* in addition to providing fuel and carbon skeletons.

Essential nutrients = Chemicals an animal requires but cannot synthesize.

- Vary from species to species.
- An animal is *malnourished* if its diet is missing one or more essential nutrients.
- Includes essential amino acids, essential fatty acids, vitamins, and minerals.

Essential amino acids are those that must be obtained in the diet in a prefabricated form.

- Most animals can synthesize about half of the 20 kinds of amino acids needed to make proteins.
- Human adults can produce 12, leaving 8 as essential in the diet. (Human infants can only produce 11.)
- *Protein deficiency* results when the diet lacks one or more essential amino acids.
  ⇒ The syndrome known as *kwashiorkor* is a form of protein deficiency in some parts of Africa.
- The human body cannot store essential amino acids, thus a deficiency retards protein synthesis.
  ⇒ This is most frequent in individuals, who for economic or other reasons, have unbalanced diets.

Essential fatty acids are those unsaturated fatty acids that cannot be produced by the body.

- An example in humans is linoleic acid which is required to produce some of the phospholipids necessary for membranes.
- Fatty acid deficiencies are rare as most diets include sufficient quantities.

Vitamins are organic molecules required in the diet in much smaller quantities (0.01 – 100 mg/day) than essential amino acids or fatty acids.

- Many serve a catalytic function as coenzymes or parts of coenzymes.
- Vitamin deficiencies can cause very severe effects as shown in Table 37.2.
- Water-soluble vitamins are not stock-piled in the body tissues; amounts ingested in excess of body needs are excreted in the urine.
- Fat-soluble vitamins (vitamins A, D, E and K) can be held in the body; excess amounts are stored in body fat and may accumulate over time to toxic levels.
- If the body of an animal can synthesize a certain compound, it is not a vitamin.
  - ⇒ A compound such as ascorbic acid is a vitamin for humans (vitamin C) and must be included in our diets; it is not a vitamin in rabbits where the normal intestinal bacteria produce all that is needed.

Minerals are inorganic nutrients required in the diet in small quantities ranging from 1 mg to 2500 mg per day, depending on the mineral.

- Some minerals serve structural and maintenance roles in the body (calcium, phosphorous) while others serve as part of enzymes (copper) or other molecules (iron).
- Refer to Table 37.3 for the mineral requirements of humans.

REFERENCES

Campbell, N. *Biology.* 4th ed. Menlo Park, California: Benjamin/Cummings, 1996.

Marieb, Elaine N. *Human Anatomy and Physiology.* 3rd ed. Redwood City, California: Benjamin/Cummings, 1995.

CHAPTER 38

CIRCULATION AND GAS EXCHANGE

OUTLINE

VIII. Gas exchange supplies oxygen for cellular respiration and disposes of carbon dioxide: *an overview*
    A. Gas Exchange and the Respiratory Medium
    B. Respiratory Surfaces: Form and Function

IX. Gills are respiratory adaptations of most aquatic animals

X. Tracheae are respiratory adaptations of insects

XI. Lungs are the respiratory adaptation of most terrestrial vertebrates
    A. Form and Function of Mammalian Respiratory Systems
    B. Ventilating the Lungs
    C. The Control of Breathing
    D. Loading and Unloading of Oxygen and Carbon Dioxide
    E. Respiratory Pigments and Oxygen Transport
    F. Carbon Dioxide Transport
    G. Adaptations of Diving Mammals

OBJECTIVES

After reading this chapter and attending lecture, the student should be able to:

1. List major animal phyla with gastrovascular cavities and explain why they do not need a circulatory system.
2. Distinguish between open and closed circulatory systems.
3. Using an arthropod as an example, describe the circulation of hemolymph.
4. Explain how hemolymph differs from blood.
5. Using an earthworm as an example, describe circulation of blood and explain how it exchanges materials with interstitial fluid.
6. List the components of a vertebrate cardiovascular system.
7. Distinguish between an artery and a vein.
8. Using diagrams, compare and contrast the circulatory schemes of birds, amphibians and mammals.
9. Distinguish between pulmonary and systemic circuits and explain the function of each.
10. Explain the advantage of double circulation over a single circuit.
11. Trace a drop of blood through the human heart, listing the structures it passes through en route.
12. List the four heart valves, describe their location and explain their function.
13. Distinguish between systole and diastole.
14. Describe the events of the cardiac cycle and explain what causes the first and second heart sounds.
15. Define heart murmur and explain its cause.
16. Define pulse and describe the relationship between size and pulse rate among different mammals.
17. Define cardiac output and explain how it is affected by a change in heart rate or stroke volume.
18. Define myogenic and describe some unique properties of cardiac muscle which allow it to contract in a coordinated manner.
19. Define pacemaker and describe the location of two patches of nodal tissue in the human heart.
20. Describe the origin and pathway of the action potential (cardiac impulse) in the normal human heart.
21. Explain why it is important that the cardiac impulse be delayed at the AV node and describe the function and importance of the Purkinje fibers.

22. Explain how the pace of the SA node can be modulated by sympathetic and parasympathetic nerves, changes in temperature, physical conditioning and exercise.
23. Compare the structures of arteries and veins and explain how differences in their structures are related to differences in their functions.
24. Describe how capillary structure differs from other vessels and explain how this structure relates to its function.
25. Recall the law of continuity and explain why blood flow through capillaries is substantially slower than it is through arteries and veins.
26. Define blood pressure and describe how it is measured.
27. Explain how peripheral resistance and cardiac output affect blood pressure.
28. Explain how blood returns to the heart, even though it must travel from the lower extremities against gravity.
29. Define microcirculation and explain how blood flow through capillary beds is regulated.
30. Explain how osmotic pressure and hydrostatic pressure regulate the exchange of fluid and solutes across capillaries.
31. Describe the composition of lymph and explain how the lymphatic system helps the normal functioning of the circulatory system.
32. Explain why protein deficiency can cause edema.
33. Explain how the lymphatic system helps defend the body against infection.
34. Explain why vertebrate blood is classified as connective tissue.
35. List the components of blood and describe a function for each.
36. Outline the formation of erythrocytes from stem cells to destruction by phagocytic cells.
37. Outline the sequence of events that occur during blood clotting and explain what prevents spontaneous clotting in the absence of injury.
38. Explain how atherosclerosis affects the arteries.
39. Distinguish between thrombus and embolus; atherosclerosis and arteriosclerosis; low-density lipoproteins (LDLs) and high-density lipoproteins (HDLs).
40. List the factors that have been correlated with an increased risk of cardiovascular disease.
41. Describe the general requirements for a respiratory surface and list the variety of respiratory organs adapted for this purpose.
42. Describe respiratory adaptations of aquatic animals.
43. Describe countercurrent exchange and explain why it is more efficient than concurrent flow of water and blood.
44. Describe the advantages and disadvantages of air as a respiratory medium and explain how insect tracheal systems are adapted for efficient gas exchange in a terrestrial environment.
45. For the human respiratory system, describe the movement of air through air passageways to the alveolus, listing the structures it must pass through on the journey.
46. Define negative pressure breathing and explain how respiratory movements in humans ventilate the lungs.
47. Define the following lung volumes and give a normal range of capacities for the human male:
    a. Tidal volume
    b. Vital capacity
    c. Residual volume
48. Explain how breathing is controlled.
49. List three barriers oxygen must cross from the alveolus into the capillaries and explain the advantage of having millions of alveoli in the lungs.
50. Describe how oxygen moves from the alveolus into the capillary and explain why a pressure gradient is necessary.

51. Distinguish between hemocyanin and hemoglobin.
52. Describe the structure of hemoglobin, explain the result of cooperative binding and state how many oxygen molecules a saturated hemoglobin molecule can carry.
53. Draw the Hb-oxygen dissociation curve, explain the significance of its shape and explain how the affinity of hemoglobin for oxygen changes with oxygen concentration.
54. Describe the Bohr effect and explain how the oxygen dissociation curve shifts with changes in carbon dioxide concentration and changes in pH.
55. Explain the advantage of the Bohr shift.
56. Describe how carbon dioxide is picked up at the tissues and deposited in the lungs, describe the role of carbonic anhydrase and state the form most of the carbon dioxide is in as it is transported.
57. Explain how hemoglobin acts as a buffer.
58. Describe respiratory adaptations of diving mammals including the role of myoglobin.

KEY TERMS

open circulatory system
sinuses
closed circulatory systems
heart
atria
ventricles
arteries
capillaries
venules
veins
pulmonary circuit
systemic circuit
double circulation
heart cycle
systole
diastole
atrioventricular valves
semilunar valves
heart murmur
heart rate
cardiac output
stroke volume
sinoatrial (SA) node

atrioventricular (AV) node
blood pressure
hydrostatic force
peripheral resistance
microcirculation
precapillary sphincters
capillary bed
lymphatic system
lymph
lymph nodes
lymph capillaries
electrolytes
plasma proteins
lipid escorts
immunoglobins
fibrinogens
erythrocytes
hemoglobin
erythropoietin
leukocytes
stem cells
basophils
eosinophils
neutrophils
lymphocytes

monocytes
platelets
clotting factors
fibrinogen
fibrin
hemophilia
thrombus
embolus
arrhythmia
artherosclerosis
plaques
low-density lipo-proteins (LDLs)
high-density lipo-proteins (HDLs)
arteriosclerosis
angina pectoris
hypertension
respiratory medium
respiratory surface
gills
lungs
tracheae
ventilation
countercurrent exchange

spiracles
pharynx
larynx
trachea
bronchi
bronchioles
alveoli
negative-pressure breathing
diaphragm
tidal volume
residual volume
vital capacity
parabronchi
breathing centers
medulla oblongata
pons
partial pressure
respiratory pigments
hemocyanin
hemoglobin
cooperativity
dissociation curve
Bohr shift
carbonic anhydrase

LECTURE NOTES

The exchange of materials (whether nutrients, gases, or waste products) between an organism and its environment must take place across a moist cell membrane.

- The molecules must be dissolved in water in order to diffuse or be transported across the membrane.
- In protozoans, the entire external surface may be used for this exchange.
- The simple, multicellular animals (sponges, cnidarians) have body structures such that each cell is exposed to the surrounding waters.

Three-dimensional animals face the problem that some of their cells are isolated from the surrounding environment.

- These animals have specialized organs, where exchange with the environment occurs, coupled with special systems for the internal transport through body fluids to the cells.
- The association of specialized organs with an internal transport system not only reduces the distance over which molecules must diffuse to enter or leave a cell, but permits regulation of internal fluid composition.

I.   **Transport systems functionally connect body cells with the organs of exchange:** *an overview*

Diffusion is too slow of a process to transport chemicals through the body of an animal.

- Time of diffusion is proportional to the square of the distance the chemical must travel.
- If a glucose molecule takes 1 second to diffuse 1 μm, it will take 100 seconds to diffuse 1 mm.

The presence of a circulatory system reduces the distance a substance must diffuse to enter or leave a cell.

- The distance is reduced because the circulatory system connects the aqueous environment of the cell with organs specialized for exchange.
- Molecules will diffuse or be transported into the blood which *carries* these molecules to the cells or to the exchange surface depending on whether the molecules are used (nutrients) or produced (wastes) by the cells.
  - ⇒ Oxygen diffuses from air in the lungs across the thin lung epithelium into the blood.
  - ⇒ This oxygenated blood is carried by the circulatory system to all parts of the body.
  - ⇒ As the blood passes through capillaries in the tissues, oxygen diffuses from the blood into the cells across the cells' plasma membranes.
  - ⇒ Carbon dioxide is produced by the cells and moves in the opposite direction through the same system.

The circulatory system does more than move gases, it is a critical component to maintaining homeostasis of the body.

- The chemical and physical properties of the immediate surroundings of the cells can be controlled using the carrying capacity of the blood.
  - ⇒ The blood passes from the cells through organs (liver, kidneys) which regulate the nutrient and waste content of the blood.

II. **Most invertebrates have a gastrovascular cavity or a circulatory system for internal transport**

A. **Gastrovascular Cavities**

The cnidarian body plan does not require a specialized internal transport system. The body wall is only two cells thick and encloses a central gastrovascular cavity.

- The water inside the gastrovascular cavity is continuous with the surrounding water.
- The gastrovascular cavity functions in digestion and distribution of nutrients.
- Gastrodermal cells lining the cavity have direct access to the nutrients produced by digestion, however, the structure of the body is such that nutrients only have a short distance to diffuse to the outer cell layer.

Planarians and other flatworms also have a gastrovascular cavity.

- The highly ramified structure of the cavity and the flattened body shape ensure all cells are exposed to cavity contents.

A gastrovascular cavity cannot perform the necessary internal transport required by more complex animals, especially if they are terrestrial.

- These animals have some type of *circulatory system*.

B. **Open and Closed Circulatory Systems**

Open and *closed* circulatory systems are alternative solutions to moving materials efficiently through the bodies of animals.

- They usually function in combination with other organ systems which increases the efficiency of the overall body.

Insects, other arthropods, and mollusks have an *open circulatory system*.

Open circulatory system = Circulatory system in which hemolymph bathes internal organs directly while moving through sinuses. (See Campbell, Figure 38.2a)

- *Hemolymph* is a body fluid which acts as both blood and interstitial fluid.
- Circulation results from contractions of the dorsal vessel (heart) and body movements.
- Relaxation of the "heart" draws blood through the ostia (pores) into the vessel.
- Chemical exchange between the hemolymph and cells occurs in the *sinuses* which are an interconnected system of spaces surrounding the organs.

A *closed circulatory system* is found in annelids, squids, octopuses, and vertebrates.

Closed circulatory system = Circulatory system in which blood is confined to vessels and a distinct interstitial fluid is present.

- The heart (or hearts) pumps blood into large vessels.
- The major vessels branch into smaller vessels which supply blood to organs. (See Campbell, Figure 38.2b)
- Blood exchanges materials with the interstitial fluids which come into direct contact with the cells.

III.   Diverse adaptations of a cardiovascular system have evolved in vertebrates

A *cardiovascular system* consists of a *heart, blood vessels* and *blood*.

- The heart has of one or two *atria*, chambers that receive blood, and one or two *ventricles*, chambers that pump blood out.

- *Arteries* carry blood away from the heart to organs where they branch into smaller *arterioles* that give rise to microscopic *capillaries* (the site of chemical exchange between blood and interstitial fluid).

   ⇒ Capillaries have thin, porous walls and are usually arranged into networks called capillary beds that infiltrate each tissue.

- Capillaries rejoin to form *venules* which converge to form the *veins* that return blood to the heart.

An examination of vertebrate circulatory systems shows various adaptations have evolved within this taxon.

Fish have a 2-chambered heart with one atrium and one ventricle. (See Campbell, Figure 38.3a)

- Blood pumped from the ventricle goes to the gills; here oxygen diffuses into the gill capillaries and CO_2 diffuses out; gill capillaries converge into arteries that carry blood to capillary beds in other organs; blood from the organs travel through veins to the atrium of the heart, then into the ventricle.

- Blood flows through two capillary beds during each complete circuit: one in the gills and a second in the organ systems (systemic capillaries).

   ⇒ As blood flows through a capillary bed, blood pressure drops substantially.

- Blood flow to the tissues and back to the heart is aided by swimming motions.

Amphibians have a three-chambered heart with two atria and one ventricle. (See Campbell, Figure 38.3b)

- Blood flows through a *pulmonary circuit* (to the lungs and skin) and a *systemic circuit* (to all other organs) in a scheme called *double circulation*.

- Blood flow pattern: ventricle → lungs and skin to become oxygenated → left atrium → ventricle → all other organs → right atrium.

   ⇒ The second passage through the ventricle ensures sufficient blood pressure for the systemic circulation.

- There is some mixing of oxygen-rich and oxygen-poor blood in the single ventricle although a ridge present in the ventricle diverts most of the oxygenated blood to the systemic circuit and most of the deoxygenated blood to the pulmonary circuit.

Most reptiles (excluding crocodilians) have a three chambered heart although the ventricle is partially divided.

- This reduces mixing of oxygenated and deoxygenated blood.

Birds and mammals have a four-chambered heart with two atria and two ventricles. (Please note that crocodiles also have a four-chambered heart.)

- Double circulation is similar to that of amphibians except that oxygenated and deoxygenated blood does not mix due to the presence of two separate ventricles.
- The complete separation of oxygenated and deoxygenated blood increases the efficiency of oxygen delivery to the cells.

IV. Rhythmic pumping of the mammalian heart drives blood through pulmonary and systemic circuits

A. The Heart: General Form and Function

The human heart is:

- Located beneath the sternum.
- Cone-shaped and about the size of a clenched fist.
- Surrounded by a sac with a two-layered wall.
- Comprised mostly of cardiac muscle tissue.
- The two atria have thin walls and function as collection chambers for blood returning to the heart.
- The ventricles have thick, powerful walls that pump blood to the organs.

The heart chambers alternately contract and relax in a rhythmic cycle.

- A complete sequence of contraction and relaxation is the *cardiac cycle*.
- During *systole*, heart muscle contracts and the chambers pump blood.
- During *diastole*, the heart muscles relax and the chambers fill with blood.

There are four valves in the heart which prevent backflow of blood during systole: (See Campbell, Figure 38.4b)

- The valves consist of flaps of connective tissue.
- *Atrioventricular valves* are found between each atrium and ventricle and keep blood from flowing back into the atria during ventricular contraction.
- *Semilunar valves* are located where the aorta leaves the left ventricle and where the pulmonary artery leaves the right ventricle; these prevent blood from flowing back into ventricles when they relax.
- A *heart murmur* is a defect in one or more of the valves that allows backflow of blood. Serious defects are usually corrected by surgical replacement of the valve.

Heart rate = The number of heartbeats per minute.

- Usually measured by taking the pulse.
- In humans, the average resting heart rate of a young adult is 60 beats per minute; rates vary depending on the individual's activity level.
- It is not uncommon for heart rates to change during the day.
- There is an inverse relationship between animal body size and pulse; elephants have a rate of 25 beats per minute while some shrews have 600 beats per minute.

<u>Cardiac output</u> = The volume of blood per minute that the left ventricle pumps into the systemic circuit; depends on heart rate and stroke volume.

- *Stroke volume* is the amount of blood pumped by the left ventricle each time it contracts. The average human stroke volume is about 75 ml per beat.
- The cardiac output of an average human is 5.25 L per minute.

B.   Control of the Heart

Cardiac muscle cells are *myogenic* (self-excitable) and can contract without input from the nervous system. The tempo of contraction is controlled by the *sinoatrial (SA) node*, a specialized region of the heart, sometimes called the pacemaker.

- Located in the right atrium wall near the entrance of the superior vena cava.
- The SA node is composed of specialized muscle tissue which has characteristics of both muscle and nerve tissue.
- Contraction of the SA node initiates a wave of excitation that spreads rapidly from the node and causes the two atria to contract in unison.
- This wave of contraction will pass down the atria until it reaches the *atrioventricular (AV) node*; a second mass of specialized muscle tissue located near the base of the wall separating the atria.
- The impulse is delayed at the AV node for 0.1 second to ensure the atria are completely empty before the ventricles contract.
- The impulse is then carried by a mass of specialized muscle fibers to the tip of the ventricles; the impulse then travels through the Purkinje fibers upward through the ventricular walls.
- The impulses produce electrical currents as they pass through the cardiac muscle.
- An *electrocardiogram* (EKG) is the detection of these currents which are conducted through the body fluids to the body's surface.

Although the SA node controls the rate of heartbeat, it is influenced by several factors:

- Two antagonistic sets of nerves influence the heart rate; one speeds up contractions in the SA node, the other slows contractions.
- Hormones influence the SA node with some like epinephrine increasing the rate.
- Other factors, including body temperature changes and exercise, also have an influence.
  - ⇒ Exercise creates a greater demand for oxygen in the muscles and an increase in heart rate is an adaptation to meet this demand.

C.   Blood Vessel Structure

The walls of arteries and veins have 3 layers:

- An outer layer of connective tissue with elastic fibers that permits stretching and recoil of the vessel.
- A middle layer of smooth muscle and elastic fibers.
- An inner endothelium of simple squamous epithelium. (See Campbell, Figure 38.7)

- The middle and outer layers of arteries are thicker than those in veins. Capillaries are comprised only of the endothelial lining which permits the exchange of chemicals with the interstitial fluids.

D.   Blood Flow Velocity

There is a great difference in the speed at which blood flows through the various parts of the circulatory system. Blood travels about 30 cm/sec in the aorta and about 0.026 cm/sec in capillaries.

- The velocity decreases in accordance with the *law of continuity* which states that a fluid will flow faster through narrow portions of a pipe than wider portions if the volume of flow remains constant.

- An artery gives rise to so many arterioles and then capillaries that the *total* diameter of vessels is much greater in capillary beds than in the artery, thus blood flows more slowly in capillaries.

- Resistance to blood flow is greater in the smaller vessels since the blood contacts more epithelial surface area.

- Blood flows faster as it enters the venules and veins since the cross-sectional area is decreased.

E.   Blood Pressure

Fluids are driven through pipes by *hydrostatic pressure* which is the force exerted by fluids against the surfaces they contact.

<u>Blood pressure</u> = The hydrostatic force that blood exerts against a vessel wall.

- Pressure is greater in arteries than in veins and greatest during ventricular systole.
  - ⇒ This is the main force propelling blood from the heart through the vessels.
- *Peripheral resistance* results from impedance by arterioles; blood enters the arteries faster than it can leave.
  - ⇒ Thus, there is a pressure even during diastole, driving blood into capillaries continuously.
- Determined by cardiac output and degree of peripheral resistance.
  - ⇒ Stress may trigger neural and hormonal responses that cause the smooth muscles of vessel walls to contract, constricting blood vessels and increasing resistance.
- In veins pressure is near zero; blood returns to the heart by the action of skeletal muscles around the veins.
  - ⇒ Veins have valves that allow blood to flow only towards the heart.
- Breathing also helps return blood to the heart since the pressure change in the thoracic cavity during inhalation cause the vena cavae and large veins near the heart to expand and fill.

F.   Blood Flow Through Capillary Beds

All tissues and organs receive a sufficient supply of blood even though only 5-10% of the capillaries are carrying blood at any one time; the supply is adequate due to the vast number of capillaries present in each tissue.

- Capillaries in the brain, heart, kidneys, and liver usually carry a full load of blood.

- Two mechanisms regulate the distribution of blood in capillary beds. Both are controlled by nerve signals and hormones.

- One mechanism involves the contraction and relaxation of the smooth muscle layer in the walls of arterioles.

    ⇒ Contraction of the muscle layer constricts the arteriole and reduces the blood flow from it into the capillary bed.

    ⇒ Relaxation of the muscle layer dilates the arteriole and increases the blood flow into the capillary bed.

- The second mechanism involves the contraction and relaxation of *precapillary sphincters*, rings of smooth muscle located at the entrance to capillary beds.

    ⇒ Contraction of these sphincters reduces blood flow into the capillary bed; relaxation increases blood flow into the capillary bed.

The diversion of blood from one area of the body to another changes the blood supply to capillary beds.

- After ingesting food, the digestive tract receives a large supply of blood due to dilation of arterioles and the opening of precapillary sphincters associated with the system.

- During exercise, blood is diverted to the skeletal muscles.

G. Capillary Exchange

The exchange of materials between the blood and interstitial fluids that are in direct contact with the cells occurs across the thin walls of capillaries.

- The capillary wall is a single, "leaky" layer of flattened endothelial cells that overlap at their edges.

- Materials may cross these cells in vesicles (via endocytosis and exocytosis), by diffusion through the cell, or by bulk flow between the cells due to hydrostatic pressure.

Direction of fluid movement at any point along a capillary depends on the relative forces of hydrostatic pressure and osmotic pressure. (See Campbell, Figure 38.11)

- Fluid flows out of a capillary at the upstream end near an arteriole and into a capillary at the downstream end near a venule.

- About 85% of the fluid which leaves the blood at the arteriole end of the capillary, re-enters from the interstitial fluid at the venous end.

    ⇒ The remaining 15% of the fluid is eventually returned by the lymphatic system.

V.   The lymphatic system returns fluid to the blood and aids in body defense

Capillary walls leak fluid and some blood proteins which return to the blood via the *lymphatic system*. This is the 15% (4 L/day) of fluid that does not re-enter the capillaries.

- The fluid, *lymph*, is similar in composition to interstitial fluid.

- The lymph enters the system by diffusing into lymph capillaries which intermingle with the blood capillaries.
- The lymphatic system drains into the circulatory system at two locations near the shoulders.
- Lymph vessels have valves that prevent backflow and depend mainly on movement of skeletal muscles to squeeze the fluid along.
- Rhythmic contractions of vessel walls also help draw fluid into the lymphatic capillaries.
- *Lymph nodes* are specialized swellings along the system that filter the lymph and attack viruses and bacteria.
  ⇒ This defense is conducted by specialized white blood cells inhabiting the lymph nodes.
  ⇒ The nodes become swollen and tender during an infection due to rapid multiplication of the white blood cells present.
- Lymph capillaries penetrate small intestine villi and absorb fats, thus transporting them from the digestive system to the circulatory system.

VI. Blood is a connective tissue with cells suspended in plasma

Vertebrate blood is connective tissue with several cell types suspended in a liquid matrix called *plasma*.

- The average human has 4 to 6 liters of whole blood (plasma + cellular elements).
- Cellular elements are the cells and cell fragments of the blood and represent about 45% of the blood volume.

A. Plasma

Water accounts for 90% of plasma which also contains electrolytes and plasma proteins.

Electrolytes = Inorganic salts in the form of dissolved ions that help maintain osmotic balance of the blood; some also help buffer the blood.

- Electrolyte balance in the blood is maintained by the kidneys.

Plasma proteins help buffer blood, help maintain the osmotic balance between blood and interstitial fluids, and contribute to its viscosity.

- Some escort lipids through blood, some are immunoglobins, some (*fibrinogens*) are clotting factors.
- Serum is blood plasma that has had the clotting factors removed.

Plasma also contains substances in transit through the body such as nutrients, metabolic wastes, respiratory gases, and hormones.

B. Cellular Elements

Erythrocytes = Red blood cells; biconcave discs that function in transport of oxygen.

- Each cubic millimeter of human blood contains 5-6 million erythrocytes.
- Lack nuclei and mitochondria; generate ATP exclusively by anaerobic metabolism.

- Contain *hemoglobin*, an iron-containing protein that reversibly binds oxygen. About 250 million molecules of hemoglobin are in each erythrocyte.
- Erythrocyte production is controlled by a negative-feedback mechanism; if tissues are not receiving enough oxygen, the kidneys convert a plasma protein to the hormone *erythropoietin*, which stimulates production of erythrocytes in the bone marrow.

Leukocytes = White blood cells that function in defense and immunity.

- There are 5 types of leukocytes: basophils, eosinophils, neutrophils, lymphocytes and monocytes.
- There are usually 5,000-10,000 leukocytes per cubic millimeter of blood although this number increases during an infection.
- Actually spend most of their time outside the circulatory system in the interstitial fluid and lymphatic system; large numbers are found in the lymph nodes.
- Lymphocytes become specialized during an infection and produce the body's immune response.

Platelets = Fragments of cells 2 to 3 μm in diameter.

- Originate as pinched-off cytoplasmic fragments of large cells in the bone marrow.
- Lack nuclei.
- Function in blood clotting.

C.   The formation of Blood Cells and Platelets

The cellular elements of the blood must be replaced as they wear out.

- The average erythrocyte circulates in the blood for 3 – 4 months before being destroyed by phagocytic cells in the liver and spleen.
- The components are usually recycled with new molecules being constructed from the macromolecule components of the old cells.

The *multipotent stem cells* give rise to all three of the cellular elements.

- These are found in the red marrow of bones, especially the ribs, vertebrae, breastbone, and pelvis.
- Form in the early embryo and are renewed by mitosis.
- Produce a number of new blood cells equivalent to the number of dying cells.

Using DNA technology to correct genetic defects in multipotent stem cells may provide a treatment for such diseases as leukemia and sickle-cell anemia.

D.   Blood Clotting

A clot forms when platelets clump together to form a temporary plug and release clotting factors (some are also released from damaged cells) that initiate a complex reaction resulting in conversion of inactive *fibrinogen* to active *fibrin*. (See Campbell, Figure 38.14)

- Fibrin aggregates into threads that form the clot.
- Anticlotting factors normally prevent spontaneous clotting in the absence of injury.

VII. **Cardiovascular diseases are the leading cause of death in the United States and many other developed nations**

Diseases of the heart and blood vessels are referred to as *cardiovascular diseases*.

- Account for more than 50% of all deaths in the United States.
- May culminate in a *heart attack* or *stroke*.

  Heart attack = Death of the cardiac muscle resulting from prolonged blockage of one or more coronary arteries.

  Stroke = Death of nervous tissue in the brain often resulting from blockage of arteries in the brain.

A *thrombus* is often associated with a heart attack or stroke.

Thrombus = A blood clot that blocks a key blood vessel.

- If it blocks the coronary arteries, a heart attack occurs.
- An *embolus* is a moving clot.
- If the thrombus or embolus blocks an artery in the brain, a stroke results.

Arteries may gradually become impaired by *atherosclerosis*.

Atherosclerosis = Chronic cardiovascular disease characterized by plaques that develop on the inner walls of arteries and narrow the bore of the vessels.

- Decreases blood flow through the vessels.
- Increase the risk of clot formation and heart attack.

Arteriosclerosis = Degenerative condition of arteries (form of atherosclerosis) in which plaques become hardened by calcium deposits.

Angina pectoris = Chest pains that occur when the heart receives insufficient oxygen due to the build up of plaques in the arteries.

- An indicator that the heart is not receiving sufficient oxygen.

Hypertension = High blood pressure; promotes atherosclerosis.

- Increases risk of heart attack and stroke.

Smoking, lack of exercise and a diet rich in animal fats correlate with increased risk of cardiovascular disease.

Abnormally high concentrations of *LDL's* (low density lipoproteins) in the blood correlate with atherosclerosis; *HDL's* (high density lipoproteins) actually *reduce* deposition of cholesterol in arterial plaques.

- Exercise tends to increase HDL concentration.
- Smoking increases the LDL to HDL ratio.

VIII. Gas exchange supplies oxygen for cellular respiration and disposes of carbon dioxide: *an overview*

Circulatory systems transport oxygen and carbon dioxide between respiratory organs and other parts of the body.

A. Gas Exchange and the Respiratory Medium

Gas exchange = The movement of O_2 and CO_2 between the animal and its environment.

- Supports cellular respiration by supplying O_2 and removing CO_2.

The *respiratory medium* (source of oxygen) for terrestrial animals is the air, while it is water for aquatic animals.

- Air is 21% oxygen while the amount of dissolved oxygen in water varies due to temperature, solute concentrations, and other factors.

Respiratory surface = Portion of the animal surface where gas exchange with the respiratory medium occurs. Oxygen diffuses in; carbon dioxide diffuses out.

- O_2 and CO_2 can only diffuse through membranes if they are first dissolved in the water that coats the respiratory surface.
- This surface must be large enough to provide O_2 and expel CO_2 for the entire body.

B. Respiratory Surfaces: Form and Function

A variety of respiratory surfaces have evolved that are adaptive for organism size and environment.

- Protozoa and other unicellular organisms exchange gases over their entire surface area.
- Animals such as sponges, cnidarians and flatworms have body structures such that the plasma membrane of each body cell contacts the outside environment and can function in exchange.
- In animals with a more three-dimensional body plan, most of the body cells are isolated from the environment and the respiratory surfaces are generally thin, moist epithelium with a rich blood supply.
  ⇒ Usually only a single cell layer separates the respiratory medium from the blood or capillaries.

Animals that are relatively small or have a shape (long, thin) that results in a high surface to volume ratio may use their outer skin as a respiratory organ.

- Earthworms have moist skin which overlays a dense network of capillaries.
- Gases diffuse across the entire surface (O_2 in; CO_2 out) and into the circulatory system.
- Earthworms and other animals (some amphibians) that use their skin for gas exchange must live in water or damp places in order to keep the exchange surface moist.

Most other animals lack sufficient body surface area to exchange gases for the entire body. These animals possess a region of the body surface that is extensively branched or folded, thus providing a large enough respiratory surface area for gas exchange.

- In most aquatic animals, external *gills* are present and are in direct contact with the water.
- Terrestrial animals have internal respiratory surfaces that open to the atmosphere through narrow tubes.
    ⇒ Air lacks the supportive density and moisturizing qualities of water.
    ⇒ Lungs and insect tracheae are two variations.

IX. Gills are respiratory adaptations of most aquatic animals

Gills are outfoldings of the body surface specialized for gas exchange. (See Campbell, Figure 38.17)

- In some invertebrates (i.e. echinoderms), gills have simple shapes and are distributed over the entire body.
- In other invertebrates (i.e. annelids), gills may be flap-like and extend from each body segment or be clustered at one end and be long and feathery.
- Other animals (mollusks, fishes) have gills that are localized on a body region where the surface is finely subdivided to provide a large amount of surface area.
- The gills must be efficient since although water keeps the respiratory surface wet, it has a lower oxygen concentration than air.

Ventilation = Any method of increasing the flow of the respiratory medium over the respiratory surface; brings in a fresh supply of O_2 and removes CO_2.

Due to the density and low oxygen concentration of water, fish must expend a large amount of energy to ventilate water.

- Fish have a unique arrangement of blood vessels in their gills, which maximizes O_2 uptake from H_2O.
- Called *countercurrent exchange*, blood flows opposite to the direction in which water passes over gills, maintaining a constant concentration gradient for oxygen between the blood and the water passing over the gill surface. (See Campbell, Figures 38.18 and 35.19)

X. Tracheae are respiratory adaptations of insects

Air has several advantages over water as a respiratory medium:

- A higher oxygen concentration.
- Oxygen and carbon dioxide diffuse faster through air than water.
- Respiratory surfaces do not have to be ventilated as thoroughly.

The major disadvantage is that respiratory surfaces are continually desiccated.

- The evolution of respiratory surfaces within the body (tracheae, lungs) helped solve this problem.

Tracheae are tiny air tubes that branch throughout the insect body; air enters the system through pores called *spiracles* and diffuses through the small branches which extend to the surface of nearly every cell.

- Some small insects rely on diffusion alone to move O_2 into the system and CO_2 out; others use rhythmic body movements for ventilation.
- Cells are exposed directly to the respiratory medium so insects do not use their circulatory systems to transport O_2 and CO_2.
    - $\Rightarrow$ This is a major reason why the open circulatory system works so well in insects.

XI.   Lungs are the respiratory adaptation of most terrestrial vertebrates

Lungs are highly vascularized invaginations of the body surface that are restricted to one location.

- The circulatory system must transport oxygen from the lungs to the rest of the body.
- Land snails use an internal mantle as a lung.
- Spiders possess booklungs.
- Various degrees of lung development are found in terrestrial vertebrates: frogs have simple balloonlike lungs with limited surface area; mammals have highly subdivided lungs with a large surface area.

A.   Form and Function of Mammalian Respiratory Systems

Located in the thoracic cavity, mammalian lungs are enclosed in a sac consisting of two layers held together by the surface tension of fluid between the layers.

- Air entering the *nostrils* is filtered by hairs, warmed and moistened.
- The air then travels through the *pharynx*, then through the *glottis*, and into *larynx* (which possesses vocal cords and functions as a voice box).
- The flow then enters the cartilage-lined *trachea* that forks into two *bronchi* which further branch into finer bronchioles that dead-end in *alveoli*.
- Alveoli are lined with a thin layer of epithelium which serves as the respiratory surface.
- Oxygen dissolves in the moist film covering the epithelium and diffuses across to the capillaries covering each alveolus; carbon dioxide moves in the opposite direction by diffusion.

B.   Ventilating the Lungs

Vertebrates ventilate lungs by *breathing* (alternate inhalation and exhalation of air).

- Maintains a maximum O_2 concentration and minimum CO_2 concentration in the alveoli.

Frogs ventilate the lungs by *positive pressure breathing*. Air is pushed down the windpipe into the lungs.

- Air is pulled into the mouth by lowering the floor region; this enlarges the oral cavity.
- The nostrils and mouth are closed and the floor of the mouth is raised, forcing air down the trachea.
- Air is exhaled by elastic recoil of lungs and by compression of the lungs by the muscular body wall.

Mammals ventilate their lungs by *negative pressure breathing*. During *inhalation*, air is pulled into the lungs by the negative pressure created as the thoracic cavity enlarges by two possible mechanisms:

- When a mammal is at rest, most of the shallow inhalation results from contraction of the *diaphragm*.
  - ⇒ The diaphragm is a dome-shaped, thin sheet of muscle that forms the bottom wall of the thoracic cavity. When it contracts, it pushes downward towards the abdomen, enlarging the thoracic cavity.
  - ⇒ This lowers the air pressure in the lungs below atmospheric pressure and causes inhalation.
- Action of the rib muscles in increasing lung volume is important during vigorous exercise.
  - ⇒ Contraction of the rib muscles pulls the ribs upwards, which expands the rib cage.
  - ⇒ As the thoracic cavity enlarges, the lungs also expand, since the surface tension of the fluid between the layers of the lung sac causes the lungs to follow.
  - ⇒ Lung volume increases, resulting in negative pressure within the alveoli, causing air to rush in.

Exhalation occurs when the diaphragm and/or the rib muscles relax, decreasing the volume of the thoracic cavity.

The amount of air inhaled and exhaled depends upon size, activity level and state of health.

- *Tidal volume* is the volume of air an animal inhales and exhales with each breath during normal quiet breathing. Averages about 500 ml in humans.
- *Vital capacity* is the maximum air volume that can be inhaled and exhaled during forced breathing. Averages 3400 – 4800 ml in college-age females and males, respectively.
- *Residual volume* is the amount of air that remains in the lungs even after forced exhalation.

Birds have a more complex process for ventilation:

- Besides lungs, birds have 8 or 9 air sacs in their abdomen, neck and wings that serve to trim the density of the body and act as sinks for the heat dissipation by metabolism of flight muscles.
- The air sacs do not exchange gases, they serve as bellows to keep the air moving.

- The lungs also have tiny channels called *parabronchi*; air flows through the entire system, lungs and air sacs, in only one direction regardless of whether it is inhaling or exhaling.
- The continuous flow of air through the parabronchi provides a constant supply of oxygen to the blood whether the bird is inhaling or exhaling.

C.   The Control of Breathing

Breathing is an automatic action, we inhale when nerves in the *breathing control centers* of the medulla oblongata and pons send impulses to the rib muscles or diaphragm, stimulating the muscles to contract.

- This occurs about 10 – 14 times per minute and the degree of lung expansion is controlled by a negative-feedback mechanism involving stretch receptors in the lungs.
- The medulla's control center also monitors blood and cerebrospinal fluid pH, which drops as blood CO_2 concentrations increase. When it senses a drop in pH (increased CO_2 level), the tempo and depth of breathing are increased and the excess CO_2 is removed in exhaled air.
- O_2 concentration in the blood only effects the breathing control centers when it becomes severely low.
  ⇒ O_2 sensors in the aorta and carotids send signals to the breathing control centers and the centers respond by increasing the breathing rate.

The breathing control centers thus respond to a variety of neural and chemical signals.

- The response is an adjustment in the rate and depth of breathing to meet the demands of the body.
- The control is effective since it is coordinated with control of the circulatory system.

D.   Loading and Unloading of Oxygen and Carbon Dioxide

Whether a gas enters or leaves the blood depends on the concentration gradient of gases which is measured as differences in *partial pressure* (e.g. P_{O2} = partial pressure of oxygen). Oxygen comprises about 21% of the atmosphere and carbon dioxide about 0.03%.

- The partial pressure of a gas is the proportion of the total atmospheric pressure (760 mm) contributed by the gas; P_{O2} = 160 mm Hg (0.21×760) and P_{CO2} = 0.23 mm Hg.
- Gases always diffuse from areas of high partial pressure to those of low partial pressure.
- Blood arriving at the lungs from the systemic circulation has a lower P_{O2} and a higher P_{CO2} than air in the alveoli.
  ⇒ Thus, the blood exchanges gases with air in the alveoli and the P_{O2} of the blood increases while the P_{CO2} decreases.
- In systemic capillaries gradients of partial pressure favor diffusion of oxygen out of the blood and diffusion of CO_2 into it, since cellular respiration rapidly depletes interstitial fluid of O_2 and adds CO_2. (See Campbell, Figure 38.25)

E.   Respiratory Pigments and Oxygen Transport

Oxygen is carried by respiratory pigments in the blood of most animals since oxygen is not very soluble in water.

- The pigments are proteins which contain metal atoms; the metal atoms are responsible for the color of the protein.

In arthropods and mollusks, *hemocyanin* is the O_2 carrying pigment.

- The oxygen-binding component is copper and results in a blue color.
- It is dissolved directly in plasma, not confined to cells.

Hemoglobin is the oxygen-transporting pigment in almost all vertebrates.

- Consists of 4 subunits, each containing a heme group that bonds O_2; an iron atom is at the center of each heme group and actually binds to oxygen.
- The binding of oxygen to hemoglobin is reversible. (Binding occurs in the lungs; release occurs in the tissues.)
- Binding of O_2 to one subunit induces a shape change that increases the affinity of the other 3 subunits for oxygen (*cooperativity*).
- The unloading of oxygen from one heme group results in a conformational change that stimulates unloading from the other three.
- The cooperative nature of this mechanism is evident in the dissociation curve depicted in Figure 38.26.
- *Bohr shift* is the lowering of hemoglobins' affinity for oxygen upon a drop in pH. (See Campbell, Figure 38.26b) This occurs in active tissues due to the entrance of CO_2 into the blood.

F.   Carbon Dioxide Transport

Hemoglobin not only transports oxygen but also helps the blood transport carbon dioxide and assists in buffering the blood against harmful pH changes.

Carbon dioxide is transported by the blood in three forms:

- Dissolved CO_2 in the plasma (7%).
- Bound to the amino groups of hemoglobin (23%).
- As bicarbonate ions in the blood (70%).

Carbon dioxide from cells diffuses into the blood plasma and then into erythrocytes; in the erythrocytes, *carbonic anhydrase* catalyzes a reversible reaction wherein CO_2 is converted into bicarbonate.

- The CO_2 reacts with water to form carbonic acid.
- The carbonic acid quickly dissociates to bicarbonate and hydrogen ions.
- Bicarbonate then diffuses out of the erythrocyte and into the blood plasma.
- The hydrogen ions attach to hemoglobin and other proteins which results in only a slight change in the pH.
- The process is reversed in the lungs.

G.   Adaptations of Diving Mammals

Diving mammals such as seals, dolphins, and whales have special adaptations which allow them to make long underwater dives.

Weddell seals, which make dives to 200-500 m depths for 20 minutes or more, have been extensively studied.

- These mammals can store large amounts of oxygen, about twice as much per kilogram body weight as humans.
  - ⇒ 70% of the oxygen load is found in the blood (51% in humans) and 5% in the lungs (36% in humans).
  - ⇒ They also possess a higher myoglobin (oxygen storing pigment) concentration in their muscles.
- The high oxygen load in the blood may be due to the higher blood volume in these seals (twice the amount of blood per kilogram of body weight as humans).
- The seals also have a very large spleen which can store about 24 L of blood.
  - ⇒ The spleen probably contracts after a dive and forces more erythrocytes loaded with oxygen into the blood.
- The high concentration of myoglobin permits these seals to store about 25% of their oxygen in muscles (compared to 13% in humans).

These adaptations provide diving mammals with a large oxygen reservoir at the beginning of a dive, but they also have several adaptations to conserve oxygen during the dive.

- A diving reflex slows the pulse, and oxygen consumption declines as the cardiac output slows.
- Most of the blood is routed to the brain, spinal cord, eyes, adrenal glands, and placenta due to regulatory mechanisms that reduce blood flow to the muscles.
- The reduced blood flow to the muscles results in a depletion of the myoglobin oxygen reserves after about 20 minutes, the muscles then shift to anaerobic production of ATP.

The special adaptations found in diving mammals emphasizes that animals can respond to environmental pressures in the short term through physiological adaptation as well as in the long term by natural selection.

REFERENCES

Campbell, N. *Biology*. 4th ed. Menlo Park, California: Benjamin/Cummings, 1996.

Marieb, E.N.   *Human Anatomy and Physiology*.   3rd ed.   Redwood City, California: Benjamin/Cummings, 1995.

OUTLINE

I. Nonspecific mechanisms provide general barriers to infection
    A. The Skin and Mucous Membranes
    B. Phagocytic White Cells and Natural Killer Cells
    C. Antimicrobial Proteins
    D. The Inflammatory Response

II. The immune system defends the body against specific invaders: *an overview*
    A. Key Features of the Immune System
    B. Active Versus Passive Acquired Immunity
    C. Humoral immunity and Cell-Mediated Immunity
    D. Cells of the Immune System

III. Clonal selection of lymphocytes is the cellular basis for immunological specificity and diversity

IV. Memory cells function in secondary immune responses

V. Molecular markers on cell surfaces function in self/nonself recognition

VI. In the humoral response, B cells defend against pathogens in body fluids by generating specific antibodies
    A. The Activation of B Cells
    B. T-Dependent and T-Independent Antigens
    C. The Molecular Basis of Antigen-Antibody Specificity
    D. How Antibodies Work
    E. Monoclonal Antibody Technology

VII. In the cell-mediated response, T cells defend against intracellular pathogens
    A. The Activation of T Cells
    B. How Cytotoxic T Cells Work

VIII. Complement proteins participate in both nonspecific and specific defenses

IX.   The immune system's capacity to distinguish self from nonself is critical in blood transfusion and transplantation
    A.   Blood Groups
    B.   Tissue Grafts and Organ Transplants

X.   Abnormal immune function leads to disease states
    A.   Autoimmune Diseases
    B.   Allergy
    C.   Immunodeficiency
    D.   Acquired Immunodeficiency Syndrome (AIDS)

XI.   Invertebrates exhibit a rudimentary immune system

OBJECTIVES

After reading this chapter and attending lecture, the student should be able to:

1.   Explain what is meant by nonspecific defense and list the nonspecific lines of defense in the vertebrate body.
2.   Explain how the physical barrier of skin is reinforced by chemical defenses.
3.   Define phagocytosis and list two types of phagocytic cells derived from white blood cells.
4.   Explain how the function of natural killer cells differs from the function of phagocytes.
5.   Describe the inflammatory response including how it is triggered.
6.   Explain how the inflammatory response prevents the spread of infection to surrounding tissue.
7.   List several chemical signals that initiate and mediate the inflammatory response.
8.   Describe several systemic reactions to infection and explain how they contribute to defense.
9.   Describe a plausible mechanism for how interferons can fight viral infections and might act against cancer.
10.   Explain how complement proteins may be activated and how they function in cooperation with other defense mechanisms.
11.   Explain how the immune response differs from nonspecific defenses.
12.   Distinguish between active and passive immunity.
13.   Explain how humoral immunity and cell-mediated immunity differ in their defensive activities.
14.   Outline the development of B and T lymphocytes from stem cells in red bone marrow.
15.   Describe where T and B cells migrate and explain what happens when they are activated by antigens.
16.   Characterize antigen molecules, in general, and explain how a single antigen molecule may stimulate the immune system to produce several different antibodies.
17.   Describe the mechanism of clonal selection.
18.   Distinguish between primary and secondary immune response.
19.   Describe the cellular basis for immunological memory.
20.   Describe the cellular basis for self-tolerance.
21.   Explain how the humoral response is provoked.
22.   Explain how B cells are activated.

23. Diagram and label the structure of an antibody and explain how this structure allows antibodies to perform the functions of:
    a. Recognizing and binding to antigens.
    b. Assisting in destruction and elimination of antigens.
24. Distinguish between variable (V) regions and constant (C) regions of an antibody molecule.
25. Compare and contrast the structure and function of an enzyme's active site and an antibody's antigen-binding site.
26. List and distinguish among the five major classes of antibodies in mammals.
27. Describe the following effector mechanisms of humoral immunity triggered by the formation of antigen-antibody complexes:
    a. Neutralization
    c. Precipitation
    b. Agglutination
    d. Activation of complement system
28. Explain how monoclonal antibodies are produced and give examples of current and potential medical uses.
29. Explain how T-cell receptors recognize "self" and how macrophages, B cells and some T cells recognize one another in interactions.
30. Describe an antigen-presenting cell (APC).
31. Design a flow chart describing the sequence of events which follows the interaction between antigen presenting macrophages and helper T cells, including both cell-mediated and humoral immunity.
32. Define cytokine and distinguish between interleukin I and interleukin II.
33. Distinguish between T-independent antigens and T-dependent antigens.
34. Describe how cytotoxic T cells recognize and kill their targets.
35. Explain how the function of cytotoxic T cells differs from that of complement and natural killer cells.
36. Describe the function of suppressor T cells.
37. Distinguish between complement's classical and alternative activation pathways.
38. Describe the process of opsonization.
39. For ABO blood groups, list all possible combinations for donor and recipient in blood transfusions; indicate which combinations would cause an immune response in the recipient, and state which blood type is the universal donor.
40. Explain how the immune response to Rh factor differs from the response to A and B blood antigens.
41. Describe the potential problem of Rh incompatibility between a mother and her unborn fetus and explain what precautionary measures may be taken.
42. Explain why, other than with identical twins, it is virtually impossible for two people to have identical MHC markers.
43. Describe the rejection process of transplanted tissue in terms of normal cell-mediated immune response and describe how the immune system can be suppressed in transplant patients.
44. List some known autoimmune disorders and describe possible mechanisms of autoimmunity.
45. Explain why immunodeficient individuals are more susceptible to cancer than normal individuals.
46. Describe an allergic reaction including the role of IgE, mast cells and histamine.
47. Explain what causes anaphylactic shock and how it can be treated.
48. Recall the infectious agent that causes AIDS and explain how it weakens the immune system.
49. Explain how AIDS is transmitted and why it is difficult to produce vaccines to protect uninfected individuals.
50. Describe what it means to be HIV-positive.
51. Explain how general health and mental well being might affect the immune system.

KEY TERMS

nonspecific defense
  mechanism
inflammatory
  response
lysozyme
phagocytes
phagocytosis
natural killer cells
monocytes
macrophages
neutrophils
eosinophils
pus
histamine
systemic reaction
pyrogenes
interferon
anti-viral proteins
complement system
opsonization
antigen
vaccination
active immunity
passive immunity
humoral immunity
immune system
immunological
  specificity
immunological
  diversity

immunological memory
self/nonself recognition
cell-mediated immunity
lymphocytes
B cells
T cells
effector cells
cytokine
antigenic
  determinants
clonal selection
clone
primary immune
  response
memory cells
secondary immune
  response
humoral response
plasma cells
antibody molecule
disulfide bridge
constant regions
humoral effector
  mechanism
self-tolerance
antigen-antibody
  complexes
monoclonal antibodies
hybridoma
myeloma

helper T cells
cytokine
interleukin-1
interleukin-2
T dependent antigens
cytotoxic T cell
T-independent
  antigens
membrane attack
  complex
classical pathway
alternative pathway
perforin
suppressor T cells
T cell receptors
antigen presenting
  cells (APCs)
major histocompatibility
  complex (MHC)
immunoglobins (Ig)
light (L) chains
heavy (H) chains
antigen binding site
antigen receptor
epitope
IgM
IgG
IgA
IgD
IgE

agglutination
neutralization
precipitation
activation of
  complement
Rh factor
ABO blood group
cyclosporine
autoimmune
systemic lupus
  erythematosus
allergies
allergens
degranulation
anaphylactic shock
mast cells
immunodeficiency
severe combined
  immunodeficiency
acquired immune
  deficiency
  syndrome (AIDS)
Kaposi's sarcoma
human
  immunodeficiency
  virus (HIV)
HIV positive
immune adherence

LECTURE NOTES

The vertebrate body possesses two mechanisms which protect it from potentially dangerous viruses, bacteria, other pathogens, and abnormal cells which could develop into cancer.

- One of these mechanisms is *nonspecific*, that is, it does not distinguish between infective agents.
- The second mechanism is *specific* in that it responds in a very specific manner (production of antibodies) to the particular type of infective agent.

I.   Nonspecific mechanisms provide general barriers to infection

Nonspecific defense mechanisms help prevent entry and spread of invading microbes in an animal's body.

- An invading microbe must cross the external barrier formed by the skin and mucous membranes.
- If the external barrier is penetrated, the microbe encounters a second line of defense: interacting mechanisms of phagocytic white blood cells, antimicrobial proteins, and the inflammatory response.

A.   The Skin and Mucous Membranes

The skin and mucous membranes act as physical barriers preventing entry of pathogens, and as chemical barriers of anti-pathogen secretions.

- In humans, oil and sweat gland secretions acidify the skin (pH 3 – 5) which discourages microbial growth.
- The normal bacterial flora of the skin (adapted to the acidity) may release acids and other metabolic wastes to further inhibit pathogen growth.
- Saliva, tears and mucous secretions also wash away potential invading microbes in addition to containing antimicrobial proteins.
- An enzyme (*lysozyme*) in perspiration, tears, and saliva attacks the cell walls of many bacteria and destroys other microbes entering the respiratory system and eyes.
- In the respiratory tract, nostril hairs filter inhaled particles and mucus traps microorganisms that are then swept out of the upper respiratory tract by cilia, thus preventing their entrance into the lungs.
- In the digestive tract, stomach acid kills many bacteria that enter with foods or those trapped in swallowed mucus from the upper respiratory system.

B.   Phagocytic White Cells and Natural Killer Cells

Microbes that penetrate the skin or mucous membranes encounter amoeboid white blood cells capable of phagocytosis or cell lysis.

Neutrophils are cells that become phagocytic in infected tissue.

- Comprise 60% – 70% of total white cells.
- Attracted by chemical signals, they enter infected tissues by amoeboid movement; only live a few days as they destroy themselves when destroying pathogens.

Monocytes comprise only about 5% of the total white blood cells. They mature, circulate for a few hours, then migrate to the tissues where they enlarge and become *macrophages*.

Macrophages are large amoeboid cells that use pseudopodia to phagocytize microbes that are destroyed by digestive enzymes and reactive forms of oxygen within the cell.

- Most wander through interstitial fluid phagocytosing bacteria, viruses and cell debris.
- Some reside permanently in organs and connective tissues. They are fixed in place, but are located where they will have contact with infectious agents circulating in the blood and lymph.

- Fixed macrophages are especially numerous in the lymph nodes and spleen.

Eosinophils represent about 1.5% of the total white cell count but have limited phagocytic activity.

- Contain destructive enzymes in cytoplasmic granules which are discharged against the outer covering of the invading pathogen.

- Main contribution is defense against larger invaders such as parasitic worms.

Natural killer cells destroy the body's own infected cells, especially those harboring viruses.

- Also assault aberrant cells that could form tumors.

- Are not phagocytic, but attack the membrane, causing cell lysis.

C.   Antimicrobial Proteins

A number of proteins function in nonspecific defense by either directly attacking microorganisms or impeding their reproduction.

The two most important nonspecific protein groups are *complement proteins* and the *interferons*.

The *complement system* is a group of at least 20 proteins which interact with other defense mechanisms.

- These proteins interact in a series of steps which results in lysis of the invading microbes.

- Some components of the system function in chemotaxis as attractants to stimulate phagocyte movement into the infected site.

The *interferons* are substances produced by virus-infected cells which help other cells resist infection by the virus.

- First discovered in 1957, three major types have been identified: *alpha*, *beta*, and *gamma*.

- They are secreted by infected cells as a nonspecific defense earlier than specific antibodies appear.

- Cannot save the infected cell, but their diffusion to neighboring cells stimulates production of proteins in those cells that inhibit viral replication.

- Not a virus-specific defense; interferon produced to infection by one strain of virus produces resistance in cells to other unrelated viruses.

- Most effective against short-term infections (colds and influenza).

- Interferon-gamma also activates phagocytes which enhances their ability to ingest and kill microorganisms.

- Interferons are now being mass produced using recombinant DNA technology and are being tested as treatments for viral infections and cancer.

D.   The Inflammatory Response

A localized *inflammatory response* occurs when there is damage to a tissue due to physical injury or entry of microorganisms.

- Vasodilation of small vessels near the injury increases the blood supply to the area which produces the characteristic redness.
- The dilated vessels become more permeable allowing fluids to move into surrounding tissues resulting in localized edema.

Chemical signals are important in initiating an inflammatory response. (See Campbell, Figure 39.5)

- *Histamine* is released from injured circulating *basophils* and *mast cells* in the connective tissue.
  - ⇒ Released histamine causes localized vasodilation and the capillaries in the area become leakier.
- *Prostaglandins* are also released from white blood cells and damaged tissues.
  - ⇒ These and other substances promote increased blood flow to the injured area.
- Increased blood flow to the site of injury delivers clotting elements which help block the spread of pathogenic microbes and begin the repair process.

Migration of phagocytic cells into the injured area is also a result of increased blood flow and increased leakage from the capillaries.

- Phagocytes are attracted to the damaged tissues by several chemical mediators including some complement system proteins.
- Neutrophils arrive first, followed closely by monocytes which develop into macrophages.
- The neutrophils eliminate microorganisms and then die.
- Macrophages destroy pathogens and clean up the remains of damaged tissue cells and dead neutrophils.
- Dead cells and fluid leaked from the capillaries may accumulate as pus in the area before it is absorbed by the body.

More widespread (systemic) inflammatory responses may also occur in cases of severe infections (meningitis, appendicitis).

- The bone marrow may be stimulated to release more neutrophils by molecules emitted by injured cells.
- There may also be a severalfold increase in the number of leukocytes within a few hours of response onset.
- A fever may develop in response to toxins produced by pathogens or due to *pyrogens* released by leukocytes.
- While a high fever is dangerous, moderate fevers inhibit the growth of some microorganisms.
- Moderate fevers may facilitate phagocytosis and speed up tissue repairs.

II.   The immune system defends the body against specific invaders: *an overview*

A.   Key Features of the Immune System

The *immune system* is the body's third line of defense and is very specific in its response.

- Distinguished from nonspecific defenses by: specificity, diversity, self/nonself recognition, and memory.

Specificity refers to this system's ability to recognize and eliminate particular microorganisms and foreign molecules.

   Antigen = A foreign substance that elicits an immune response.

   Antibody = An antigen-binding immunoglobulin (protein), produced by B cells, that functions as the effector in an immune response.

- Antigens may be molecules exhibited on the surface of, produced by, or released from bacteria, viruses, fungi, protozoans, parasitic worms, pollen, insect venom, transplanted organs, or worn-out cells.
- Each antigen has a unique molecular shape and stimulates production of an antibody that defends specifically against that particular antigen.
- The immune response is thus very specific and distinguishes between even closely related invaders.

Diversity refers to the immune system's ability to respond to numerous kinds of invaders which are recognized by their antigenic markers.

- Based on the wide variety of lymphocyte populations in the immune system.
- Each population of antibody-producing lymphocytes is stimulated by a specific antigen; the stimulated lymphocytes synthesize and secrete the appropriate antibody.

Memory refers to the immune system's ability to recognize previously encountered antigens and to react faster and more effectively to subsequent exposures.

- This *acquired immunity* has long been recognized as a resistance to some infections encountered earlier in life (e.g. chicken pox).

Self/nonself recognition is the ability of the immune system to distinguish between the body's own molecules and foreign molecules (antigens).

- Failure of this system leads to autoimmune disorders which destroy the body's own tissues.

B.   Active Versus Passive Acquired Immunity

Active immunity is the immunity conferred by recovery from an infectious disease.

- Depends on response by the person's own immune system.
- May be acquired *naturally* from an infection to the body or *artificially* by vaccination.
- Vaccines may be inactivated bacterial toxins, killed microorganisms, or weakened living microorganisms.
    ⇒ In all cases the organisms can no longer cause the disease but can act as antigens and stimulate an immune response.

- A person vaccinated against an infectious agent will show the same rapid, memory-based immunological response upon encountering the pathogen as someone who has had the disease.

Passive immunity is immunity which has been transferred from one individual to another by the transfer of antibodies.

- Natural occurrence when antibodies cross the placenta from a pregnant woman's system to her fetus.
- Passive immunity provides temporary protection to newborns whose immune systems are not fully operational at birth.
- Some antibodies are transferred to nursing infants though the milk.
- Persists for only a few weeks or months after which the infant's own system defends its body.
- May also be transferred artificially from an animal or human already immune to the disease.
  ⇒ Rabies is treated by injecting antibodies from people vaccinated against rabies; produces an immediate immunity important to quickly progressing infections.
  ⇒ Artificial passive immunity is of short duration but permits the body's own immune system to begin to produce antibodies against the rabies virus.

C. Humoral Immunity and Cell–Mediated Immunity

The body will mount either a *humoral* response or a *cell-mediated* response depending on the antigen which stimulates the system.

Humoral immunity produces antibodies in response to toxins, free bacteria, and viruses present in the body fluids.

- Antibodies to these types of antigens are synthesized by certain lymphocytes and then secreted as soluble proteins which circulate through the body in blood plasma and lymph.

Cell-mediated immunity is the response to intracellular bacteria and viruses, fungi, protozoans, worms, transplanted tissues, and cancer cells.

- Depends on the direct action of certain types of lymphocytes rather than antibodies.

D. Cells of the Immune System

Lymphocytes are responsible for both humoral and cell-mediate immunity; the different responses are due to the two main classes of lymphocytes in the body: *B cells* and *T cells*.

- Early B and T cells (as well as other lymphocytes) develop from multipotent stem cells in the bone marrow and are very much alike, they only differentiate after reaching their site of maturation.
- B cells (B lymphocytes) are responsible for the humoral immune response.
  ⇒ They form in the bone marrow and remain there to complete their maturation.
- T cells (T lymphocytes) are responsible for the cell-mediated immune response.
  ⇒ They also form in the bone marrow, then migrate to the thymus gland to mature.

Mature B cells and T cells are concentrated in the lymph nodes, spleen, and other lymphatic organs.

- These positions place lymphocytes where they are most likely to contact antigens.
- *Antigen receptors* are present on the membranes of both B cells and T cells.
- The antigen receptors on a B cell are membrane-bound antibody molecules which will recognize specific antigens.
- The T cell antigen receptors are proteins (not antibodies) embedded in the membrane which recognize specific antigens.

Effector cells are the cells which actually defend the body during an immune response.

- Effector cells are populations of cells resulting from division of lymphocytes which were activated by the binding of antigens to their antigen receptors.
- Activated B cells give rise to effector cells called *plasma cells* which secrete antibodies (humoral response) that eliminate the activating antigen.
- Activated T cells may produce two types of effector cells: *cytotoxic T cells* which destroy infected cells and cancer cells; and *helper T cells* which secrete *cytokines*.

Cytokine = Molecules secreted by one cell as a regulator of neighboring cells.

- Cytokines help regulate both B and T cells and thus are involved in both the humoral and cell-mediated responses.

III. Clonal selection of lymphocytes is the cellular basis for immunological specificity and diversity

The ability of the immune system to respond to the wide variety of antigens which enter the body is based in the enormous diversity of antigen-specific lymphocytes present in the system.

- Each lymphocyte will recognize and respond to only one antigen.
- This specificity is determined during embryonic development before any antigens are encountered, and is the consequence of the antigen receptor on the lymphocyte's surface.
- When an antigen enters the body and binds to receptors on the specific lymphocytes, those lymphocytes are activated and begin to divide.
  ⇒ The divisions produce a large number of identical effector cells (*clones*) which bind to the antigen that stimulated the response.
  ⇒ If, for example, a B cell is activated, it will proliferate to produce a large number of plasma cells that will each secrete an antibody which functions as an antigen receptor for the specific antigen that activated the original B cell.
- Each antigen thus activates only a small number of the diverse group of lymphocytes. The activated cells proliferate to produce a clone of millions of effector cells which are specific for the original antigen (clonal selection).

Clonal selection = Antigenic-specific selection of a lymphocyte that activates it to produce clones of effector cells dedicated to eliminating the antigen that provoked the initial immune response.

IV.   Memory cells function in secondary immune responses

The *primary immune response* is the proliferation of lymphocytes to form clones of effector cells specific to an antigen during the body's *first* exposure to the antigen.

- There is a 5 to 10 day lag period between exposure and maximum production of effector cells.
- The lymphocytes selected by the antigen are differentiating into effector T cells and plasma cells during the lag period.

A *secondary immune response* occurs when the body is exposed to a previously encountered antigen.

- The response is faster (3 to 5 days) and more prolonged than a primary response.
- The antibodies produced are also more effective at binding the antigen.

This ability to recognize a previously encountered antigen is known as *immunological memory*.

- Based on *memory cells* which are produced during clonal selection for effectors in a primary immune response.
- Memory cells are not active during the primary response and survive in the system for long periods. (Effector cells produced in the primary response are active and, thus, short-lived.)
- When the same antigen that caused a primary immune response again enters the body, the memory cells are activated and rapidly proliferate to form a new clone of effector cells and memory cells.
- These new clones of effector and memory cells are the secondary immune response.

V.   Molecular markers on cell surfaces function in self/nonself recognition

Antigen receptors on the surfaces of lymphocytes are responsible for detecting foreign molecules that enter the body. There are no lymphocytes reactive against the body's own molecules under normal conditions.

Self-tolerance = The lack of a destructive immune response to the body's own cells.

- Develops before birth when T and B lymphocytes begin to mature in the thymus and bone marrow of the embryo.
- Any lymphocytes with receptors for molecules present in the body at that time are destroyed; consequently, the body contains no lymphocytes with antigen receptors for its own molecules, only for foreign molecules.

The *major histocompatibility complex* (*MHC* or *HLA* in humans) is a group of glycoproteins embedded in the plasma membranes of cells.

- Important "self-markers" coded for by a family of genes.
- There are at least 20 MHC genes and at least 100 alleles for each gene.
- The probability that two individuals will have matching MHC sets is virtually zero unless they are identical twins.

- There are two main classes of MHC molecules in the body:
  - ⇒ *Class I MHC* molecules are located on all nucleated cells of the body.
  - ⇒ *Class II MHC* molecules are found only on specialized cells such as macrophages, B cells, and activated T cells.
- Class II MHC molecules are important in interactions between cells of the immune system.

VI. In the humoral response, B cells defend against pathogens in body fluids by generating specific antibodies

The humoral response occurs when an antigen binds to B cell receptors which are specific for the antigen epitopes.

- The B cells differentiate into a clone of plasma cells which begin to secrete antibodies.
- These antibodies are most effective against pathogens circulating in the blood or lymph.
- Memory cells are also produced and form the basis for secondary immune responses.

A. The Activation of B Cells

The selective activation of a B cell by an antigen results in the formation of a clone of plasma cells and memory cells. This is often a two-step process.

- One step is the binding of the antigen to specific antigen-receptors on the surface of the B cell.

The other step in B cell activation involves macrophages and helper T cells; this step ends with the production of plasma cells. (See Campbell, Figure 39.9)

- After macrophages phagocytize pathogens, pieces of the partially digested antigen molecules are bound to class II MHC molecules which are moved to and presented on the surface of the macrophage.
- These presented antigen molecules in complexes with class II MHC proteins result in the macrophage functioning as an *antigen-presenting cell*.
- A helper T cell with a receptor specific for the presented antigen binds to the self/nonself class II MHC protein-antigen complex.
- The T cell is activated by contact with the macrophage and proliferates to form a clone of helper T cells specific for the presented antigen.
- These helper T cells secrete cytokines which stimulate B cells that have encountered the same antigen; recognition of these B cells also involves a class II MHC protein-antigen complex to which the receptor on the T cell binds.
- The T cell contact activates these B cells to form a clone of plasma cells.
- Each plasma cell (= effector cell) then secretes antibodies specific for the antigen.

Both macrophages and B cells act as antigen-presenting cells in their interactions with helper T cells, but there is one major difference:

- Each macrophage can display a number of different antigens depending on the type of pathogen phagocytized.

- B cells are specific and can bind to and display only one type of antigen.
- Macrophages, which are nonspecific, can thus enhance specific defense by selectively activating helper T cells which in turn activate B cells specific for the antigen.
- Helper T cells are also antigen-specific and are activated only by macrophages presenting the proper class II MHC protein-antigen complex.

B.   **T-Dependent and T-Independent Antigens**

Antigens may be either *T-dependent* or *T-independent*.

T-dependent antigens = Antigens that evoke the cooperative response involving macrophages, helper T cells, and B cells.

- These antigens cannot stimulate antibody production without T cell involvement.
- Most antigens are T-dependent.
- Memory cells are produced in T-dependent responses.

T-independent antigens = Antigens that trigger humoral immune responses without macrophage or T cell involvement.

- Usually long chains of repeating units such as polysaccharides or protein subunits often found in bacterial capsules and flagella.
- B cells are stimulated directly by the antigen which probably binds simultaneously to several antigen receptors on the B cell surface.
- The antibody production (humoral response) is usually much weaker to that of T-dependent antigens.
- No memory cells are generated in T-independent responses.

Whether activated by T-dependent or T-independent antigens, a B cell gives rise to a clone of plasma cells.

- Each of these effector cells secretes up to 2,000 antibodies per second into the body fluids for its 4 to 5 day lifespan.
- The specific antibodies help eliminate the foreign invader from the body.

C.   **The Molecular Basis of Antigen-Antibody Specificity**

Antigens are usually proteins or large polysaccharides that make up a portion of the outer covering of pathogens or transplanted cells.

- May be components of the coats of viruses, capsules and cell walls of bacteria, or surface molecules of other cell types.
- Molecules on the cell surface of transplanted tissues and organs or blood cells from other individuals are also recognized as foreign.
- Antibodies recognize a localized region on the surface of an antigen (the *epitope*), not the entire antigen molecule. (See Campbell, Figure 39.10)

Epitope = On an antigen's surface, a localized region that is chemically recognized by antibodies; also called an *antigenic determinant*.

- Several types of antibodies from several different B cells may be produced to a single bacterial cell since it may have different antigens on different areas and each bacterial antigen may possess more than one recognizable epitope.

Antibodies comprise a specific class of proteins called *immunoglobulins* (Igs). (See Campbell, Figure 39.11a)

- The structure of the immunoglobulin is associated with its function.
- Antibodies are Y-shaped molecules comprised of four polypeptide chains: two identical light chains and two identical heavy chains.
- All four chains have *constant* (C) regions that vary little in amino acid sequence among antibodies that perform a particular type of defense.
- At the tips of the Y are found *variable* (V) regions in all four chains; the amino acid sequences in the variable region show extensive variation from antibody to antibody.
- The variable regions function as *antigen-binding sites* and their amino acid sequences result in specific shapes that fit and bind to specific antigen epitopes.
- The antigen-binding site is responsible for the antibody's ability to identify its specific antigen epitope and the stem (constant) regions are responsible for the mechanism by which the antibody inactivates or destroys the antigenic invader.

There are five types of constant regions, each of which characterizes one of the five major classes of mammalian immunoglobins. (See Table 39.1 for a summary.)

- *IgM.* Consists of five Y-shaped monomers arranged in a pentamer structure. Circulating antibodies which appear in response to an initial exposure to an antigen.
- *IgG.* A Y-shaped monomer. Most abundant circulating antibody; readily crosses blood vessels and enters tissue fluids; protects against bacteria, viruses, and toxins circulating in blood and lymph; triggers complement system action.
- *IgA.* A dimer consisting of two Y-shaped monomers. Produced primarily by cells abundant in mucous membranes; prevents attachment of bacteria and viruses to epithelial surfaces; also found in saliva, tears, perspiration, and colostrum.
- *IgD.* A Y-shaped monomer. Found primarily on external membranes of B cells; probably functions as an antigen-receptor which initiates differentiation of B cells.
- *IgE.* A Y-shaped monomer. Stem regions attach to receptors on mast cells and basophils; stimulates these cells to release histamine and other chemicals that cause allergic reactions when triggered by an antigen.

D. How Antibodies Work

Antibodies do not directly destroy an antigenic pathogen. The antibody binds to the antigen to form an antigen-antibody complex which tags the invader for destruction by one of several effector mechanisms. (See Campbell, Figure 39.12)

- *Neutralization* is the simplest mechanism.
  - ⇒ The antibody blocks viral attachment sites or coats a bacterial toxin, making them ineffective. Phagocytic cells eventually destroy the complex.

- *Agglutination* is another mechanism.

  ⇒ Each antibody has two or more antigen-binding sites and can cross-link adjacent antigens. The cross-linking can result in clumps of a bacteria being held together by the antibodies, making it easier for phagocytes to engulf the mass.

- *Precipitation* is similar to agglutination but involves the cross-linking of soluble antigen molecules instead of cells.

  ⇒ These immobile precipitates are easily engulfed by phagocytes.

- *Activation* of the complement system is another mechanism.

  ⇒ Antibodies combine with complement proteins; this combination activates the complement proteins which produce lesions in the foreign cell's membrane that result in cell lysis.

E.   Monoclonal Antibody Technology

A new method of obtaining antibodies which did not depend on the *polyclonal antibodies* isolated from the blood of immunized animals was developed in 1975.

- This new method produced *monoclonal antibodies*.

Monoclonal antibodies = Defensive proteins produced by cells descended from a single cell; all antibodies produced by these cells are identical.

This technology permits the production of large quantities of antibodies quickly and at relatively little expense. These monoclonal antibodies:

- Can be used in diagnostic labs to detect pathogenic microbes in clinical samples.

- Form the basis for over-the-counter pregnancy tests.

- Are used as therapeutic agents.

- Show promise in treating cancer when combined with a toxin that would destroy the cancer cell.

Monoclonal antibodies are produced by *hybridoma* cells.

- Hybridoma cells are hybrid cells resulting from the fusion of certain cancer cells (*myelomas*) with normal antibody-producing plasma cells.

- The cancer cell is used since it can be cultured indefinitely (other cells can only be cultured for a few generations).

- Plasma cells expressing the desired antibody are mixed with myeloma cells and some of the two cell types fuse.

- The hybridoma cells are isolated and cultured.

  ⇒ These cells exhibit the key qualities of the two cell types.

  ⇒ They will produce a single type of antibody and can be cultured indefinitely to manufacture that antibody on a large scale.

VII. In the cell-mediated response, T cells defend against intracellular pathogens

The humoral immune response is that portion of the body's defenses that identifies and destroys extracellular pathogens.

The *cell-mediated immune response* is the defense mechanism that combats pathogens that have already entered cells.

- The key components of cell-mediated immunity are helper T cells (T_H) and cytotoxic T cells (T_C).
- These lymphocytes complete their maturation in the thymus and migrate to lymphoid organs such as the spleen and lymph nodes.

A. The Activation of T Cells

T cells respond only to antigenic epitopes displayed on the surfaces of the body's own cells.

- T cells cannot detect free antigens in the body fluids.
- Specific T-cell receptors embedded in the T cell plasma membrane recognize the bound antigens.
- The receptor of a helper T cell (T_H) recognizes the molecular combination of an antigen fragment with a *class II MHC*.
- The receptor of a cytotoxic T cell (T_C) recognizes the combination of an antigen fragment with a *class I MHC* molecule.
- The antigen is nestled within the MHC protein in both cases.
- While an MHC molecule can associate with a variety of antigens, each combination is a unique complex recognized by specific T cells.

The presence of a T cell surface molecule called *CD4* enhances the interaction between T_H cells and *antigen-presenting cells (APC)*.

- CD4 is present on most T_H cells and has an affinity for a part of the class II MHC molecule.
- The CD4-class II MHC interaction helps keep the T_H cell and APC engaged while antigen specific contact is occurring. (See Campbell, Figure 39.13)
- T_C cells carry a surface molecule called CD8 which has an affinity for class I MHC molecules.

The MHC-antigen complex displayed on an infected body cell stimulates T cells with the proper receptor to multiply and form clones of activated T_H and T_C cells which recognized the pathogen.

- T_H cells stimulate B cells to secrete antibodies against T-dependent antigens in a humoral response.
- T_H cells also activate other types of T cells to mount cell-mediated responses to antigens.

Helper T cells (T_H) are able to stimulate other lymphocytes by receiving and sending cytokines.

- When a T_H cell binds to an antigen-presenting macrophage, the macrophage releases *interleukin-1* (a cytokine).

- The presence of interleukin-1 stimulates the T_H cell to release its own cytokine, *interleukin-2*.

- In a positive feedback mechanism, interleukin-2 stimulates T_H cells to grow and divide more rapidly; this results in the production of more T_H cells and an increased supply of interleukin-2.

- The humoral response against the antigen is enhanced because interleukin-2 and other cytokines secreted by the increasing numbers of T_H cells activate B cells.

- The increased levels of cytokines also increase the cell-mediated response by stimulating another class of T lymphocytes to differentiate into *cytotoxic T cells* (effector cells).

B. How Cytotoxic T Cells Work

Cytotoxic T cells (T_C) are the cells which actually destroy infected host cells.

- Host cells infected by viruses and other pathogens display antigens complexed with *class I MHC* molecules on their surfaces.

- T_C cells have specific receptors which recognize and bind to antigen-class I MHC markers. (Note that this differs from T_H cells which bind to antigen-class II MHC complexes.)

- The T_C receptor can bind to any cell in the body displaying the antigen-class I MHC marker since class I MHC is present on all nucleated cells.

- When a T_C cell binds to an infected cell, it releases *perforin* which is a protein that forms a lesion in the infected cell's membrane.

- Cytoplasm escapes through the lesion and eventually cell lysis occurs.

- Destruction of the host cell not only removes the site where pathogens can reproduce, but also exposes the pathogens to circulating antibodies from the humoral response.

- T_C cells continue to live after destroying the infected cell and may kill many others displaying the same antigen-class I MHC marker.

Cytotoxic T cells also function to destroy cancer cells which develop periodically in the body.

- Cancer cells possess distinctive markers not found on normal cells.

- T_C cells recognize these markers as nonself and attach and lyse the cancer cells.

- Cancers develop primarily in individuals with defective or declining immune systems.

A third type of T lymphocyte, *suppressor T cells* (T_S), has been found in the body.

- Probably function to suppress the immune system when an antigen is no longer present.

- Action is not well understood and some immunologists feel T_S cells are actually a form of T_H cells.

VII.   Complement proteins participate in both nonspecific and specific defenses

The 20 or so complement proteins circulate in the blood in inactive forms.

- These proteins become activated in a series fashion with each activating the next in the series.

The *classical pathway* describes complement's activation in the specific defense mechanism. (See Campbell, Figure 39.16)

- Initiated when antibodies bind to a specific pathogen which targets the cell for destruction.
- A complement protein attaches to, and bridges the gap between, two adjacent antibody molecules.
- The antibody-complement association activates complement proteins to form, in a step-by-step sequence, a *membrane attack complex*.
- The membrane attack complex lyses the pathogen's membrane producing a lesion.
- Lysis of the pathogenic cell then occurs.

The *alternative pathway* is how complement is activated in nonspecific defense mechanisms.

- Does not require cooperation with antibodies.
- Complement proteins are activated by substances found in many pathogens (yeasts, viruses, virus-infected cells, protozoans) to form a membrane attack complex.
- The complex lyses the pathogen without the aid of antibodies.
- Complement proteins also contribute to inflammation by binding to histamine-containing cells; this association triggers the release of histamine from those cells.
- Several complement proteins also attract phagocytic cells to infected sites.

Complement and phagocytes also work together in two ways to destroy pathogens.

- *Opsonization* is a cooperative mechanism in which complement proteins attach to a foreign cell and stimulate phagocytes to engulf the cell.
- In *immune adherence*, complement proteins and antibodies coat a microbe which causes it to adhere to blood vessel walls and other surfaces; this makes the cell easy prey for circulating phagocytes.

IX.   The immune system's capacity to distinguish self from nonself is critical in blood transfusion and transplantation

The body's immune system distinguishes between self (the body's own cells) and nonself (foreign cells).

- Nonself includes not only the pathogens discussed earlier, but also cells from other individuals of the same species.

A.   **Blood Groups**

The human ABO blood groups provide a good example for nonself recognition. The antigen present on the surface of the erythrocytes is *not* antigenic to that person but may be recognized as foreign if placed in the body of another individual.

- Individuals of blood type A have the A antigen and make anti-B antibodies.
- Individuals of blood type B have the B antigen and make anti-A antibodies.
- Individuals of blood type AB have the A and B antigen and make no antibodies.
- Individuals of blood type O have neither the A nor B antigen and make anti-A and anti-B antibodies.

Blood group antibodies can cause blood of a different antigenic type to agglutinate, a life-threatening reaction.

- Type O individuals are universal donors since their blood has neither antigen.
- Type AB individuals are universal recipients since they produce neither antibody A or antibody B.
- The blood group antibodies are present in the body before a transfusion occurs as they form in response to the bodies normal bacterial flora and cross-react with blood group antigens.
  ⇒ Usually IgM class antibodies do not cross the placenta, thus they present no harm to a developing fetus with a blood type different from the mother.

The *Rh factor* is another blood group antigen. Rh factor causes problems when a mother is Rh negative and her fetus is Rh positive (inherited from the father).

- When small amounts of fetal blood cross the placenta and come into contact with the mother's lymphocytes, the mother develops antibodies against the Rh factor.
- Usually only a problem in the second child since the response will be quick due to sensitization and formation of memory cells during the first baby's gestation.
  ⇒ Unlike blood group antibodies, Rh antibodies are IgG class which can cross the placenta.
  ⇒ The mother's antibodies cross the placenta and destroy the red blood cells of the Rh positive fetus.
- Can be prevented by injection of anti-Rh antibodies which destroy Rh-positive red cells before the mother develops immunological memory.

B.   **Tissue Grafts and Organ Transplants**

The MHC is a biochemical fingerprint unique to each individual.

- Remember there are at least 20 MHC genes and about 50 alleles for each gene.
- Complicates tissue grafts and organ transplants since foreign MHC molecules are antigens and cause cytotoxic T cells to mount a cell-mediated response.
- Cyclosporine and FK506 suppress cell-mediated immunity without crippling humoral immunity, thus increasing the chance of successful grafts and transplants.

Note the reactions of the immune system to transfusions, tissue grafts, and organ transplants are normal reactions of a healthy immune system, not disorders of the system.

X. **Abnormal immune function leads to disease states**

    A. **Autoimmune Diseases**

        <u>Autoimmune Disease</u> = The immune system reacts against self.

- Some cases involve immune reactions against components of the body's own cells which are released by the normal breakdown of skin and other tissues; especially nucleic acids in lupus erythematous.

- Rheumatoid arthritis is an autoimmune disease in which inflammation damages cartilage and bones in joints.

- Destruction of insulin-producing pancreas cells by an autoimmune reaction appears to cause insulin-dependent diabetes.

- Antibodies produced to repeated streptococcal infections may react with heart tissues and cause valve damage in some people.

- Other autoimmune diseases are Grave's disease and rheumatic fever.

    B. **Allergy**

        <u>Allergy</u> = A hypersensitivity of the body's defense system to an environmental antigen called an *allergen*.

- Some believe these reactions to be evolutionary remnants to infection by parasitic worms due to similarities in the responses.

        IgE class antibodies are commonly involved in allergic reactions; these antibodies recognize pollen as allergens. (See Campbell, Table 39.1)

- IgE antibodies attach by their tails to noncirculating mast cells found in connective tissues.

- When a pollen grain bridges the gap between two adjacent IgE monomers, the mast cell responds with a reaction called *degranulation*.

- Degranulation involves the release of histamine and other inflammatory agents.

- Histamine causes dilation and increased permeability of small blood vessels which results in the common symptoms of an allergy.

- Antihistamines are drugs used to treat allergies since they interfere with the action of histamine.

        Anaphylactic shock is a life-threatening reaction to injected or ingested antigens; it is the most serious type of acute allergic response.

- Occurs when mast cell degranulation causes a sudden dilation of peripheral blood vessels and a drastic drop in blood pressure.

- Death may occur in a few minutes.

- This hypersensitivity may be associated with foods (peanuts, fish) or insect venoms (wasp or bee stings).

- Epinephrine may be injected to counteract the allergic response.

C.   Immunodeficiency

Immunodeficiency refers to a condition where an individual is inherently deficient in either humoral or cell-mediated immune defenses.

- *Severe combined immunodeficiency (SCID)* is a congenital disorder in which both the humoral and cell-mediated immune defenses fail to function.
- Gene therapy has had some success in the treatment of a type of SCID where there is a deficiency of the enzyme adenosine deaminase.

Not all cases of immunodeficiency are inborn conditions.

- Some cancers, like Hodgkin's disease, damage the lymphatic system and make the individual susceptible to infection.
- Some viral infections cause depression of the immune system (i.e. AIDS).
- Physical and emotional stress may compromise the system; adrenal hormones secreted by stressed individuals affect the number of leukocytes and may suppress the system in other ways.

Evidence suggests direct links between the nervous system and the immune system.

- There is a network of nerve fibers which penetrates into lymphoid tissues including the thymus.
- Lymphocytes have also been found to possess surface receptors for chemical signals secreted by nerve cells.

D.   Acquired Immunodeficiency Syndrome (AIDS)

Acquired immunodeficiency syndrome is a severe immune system disorder caused by infection with the *human immunodeficiency virus (HIV)*.

- Characterized by a reduction of T cells and the appearance of characteristic secondary infections.
- Mortality rate approaches 100%.
- HIV probably evolved from another virus in central Africa and may have gone unrecognized for many years.
- HIV infects cells, including T_H cells, which carry the CD4 receptor on their surface.
- A glycoprotein on the HIV envelope binds specifically to the CD4 receptor.
  - ⇒ After attaching, the HIV enters the cell and begins to replicate.
  - ⇒ Newly formed viruses bud continuously from the host cell, circulate, and infect other cells.
- Infected cells may be killed quickly by the virus or by the immune response; they may also live for an extended time.
- The immune system is devastated by HIV since the virus targets T_H cells which play a central role in both the humoral and cell-mediated responses.
  - ⇒ Macrophages and a few subclasses of B cells carrying the CD4 receptor can also be infected.
- The HIV may also remain as a provirus in the infected cell genome for many years before becoming active.

HIV is not eliminated from the body by antibodies for several reasons:

- The latent provirus is invisible to the immune system.
- The virus undergoes rapid mutational changes in antigens during replication which eventually overwhelms the immune system.
- The population of helper T-cells eventually declines to the point where cell-mediated immunity collapses.
- Secondary infections characteristic of HIV infection develop (*Pneumocystis pneumonia* and Kaposi's sarcoma).

AIDS is the late stage of HIV infection and is defined by a reduced T cell population and the appearance of secondary infections.

- Takes an average of about ten years to reach this stage of infection.
- During most of this time, only moderate symptoms are shown.
- Progression of infection is more rapid in infants who were infected *in utero*.
- Individuals exposed to HIV have circulating antibodies that can be detected; displaying these antibodies designates and individual as *HIV-positive*.

HIV is only transmitted through the transfer of body fluids, blood or semen, containing infected cells.

- Most commonly transmitted in the U.S. and Europe through unprotected sex between male homosexuals and unsterilized needles in intravenous drug users.
- In Africa and Asia, transmission through unprotected heterosexual sex is rapidly increasing; especially in areas with a high incidence of other sexually transmitted diseases.
- Transmission to nursing infants through breast milk has been reported.
- Transmission through blood transfusions has also been reported, but the incidence has declined greatly with implementation of screening procedures.

AIDS is currently considered an incurable disease.

- AZT, ddC, and ddI are antiviral drugs used to extend the lives of infected individuals.
  ⇒ They do not eliminate the virus but inhibit the viral enzyme reverse transcriptase.
- Other drugs are used to fight opportunistic infections common in AIDS patients.
- The best way to prevent additional infections is to educate people on how the disease is transmitted and how to protect themselves.

XI. Invertebrates exhibit a rudimentary immune system

How invertebrates react against pathogens that enter their bodies is poorly understood, although it is known that they have a well developed ability to distinguish self from nonself.

- Experiments have shown that if the cells from two sponges of the same species are mixed, the cells from each individual will aggregate in separate groups, excluding cells from the other individual.
- *Coelomocytes*, amoeboid cells that destroy foreign materials, have been found in many invertebrates.

A memory response has also been identified in earthworms.

- A body wall graft from one worm to another will survive for about eight months before rejection if the worms are from the same population.
- A graft involving worms from different populations is rejected in two weeks.
- A second graft from the same donor to the same recipient is rejected in less than one week due to coelomocyte activity.

REFERENCES

Beardsley, T. "Better Than a Cure." *Scientific American*, 272(1). 1995.

Campbell, N. *Biology*. 4th ed. Menlo Park, California: Benjamin/Cummings, 1996.

Coleman, R.M., M.F. Lombard, and R.E. Sicard. *Fundamental Immunology*. 2nd ed. Dubuque, IA: Wm. C. Brown, Publishers, 1992.

Gallo, R.C. "The AIDS Virus." *Scientific American*. 256(1). 1987.

Laurence, J. "The Immune System in AIDS." *Scientific American*. 253(6). 1985.

Marrack, P. and J. Kappler. "The T Cell and Its Receptor." *Scientific American*. 254(2). 1986.

Newman, J. "How Breast Milk Protects Newborns." *Scientific American*, 273(6). 1995.

Tonegawa, S. "The Molecules of the Immune System." *Scientific American*. 253(4). 1985.

CHAPTER 40

CONTROLLING THE
INTERNAL ENVIRONMENT

OUTLINE

IX. An animal's nitrogenous wastes are correlated with its phylogeny and habitat
   A. Ammonia
   B. Urea
   C. Uric Acid

X. Thermoregulation maintains body temperature within a range conducive to metabolism
   A. Heat Transfer Between Organisms and Their Surroundings

XI. Ectotherms derive body heat mainly from their surroundings and endotherms derive it mainly from metabolism

XII. Thermoregulation involves physiological and behavioral adjustments

XIII. Comparative physiology reveals diverse mechanisms of thermoregulation
   A. Invertebrates
   B. Amphibians and Reptiles
   C. Fishes
   D. Mammals and Birds
   E. Thermoregulation in Humans
   F. Torpor: Conserving Energy During Environmental Extremes
   G. Temperature Range Adjustments

XIV. Regulatory systems interact in the maintenance of homeostasis
   A. Roles of the Liver in Homeostasis

OBJECTIVES

After reading this chapter and attending lecture, the student should be able to:
1. Distinguish between osmoregulators and osmoconformers.
2. Discuss the problems that marine, freshwater and terrestrial organisms face in maintaining homeostasis and explain what osmoregulatory adaptations serve as solutions to these problems.
3. Explain the role of transport epithelia in osmoregulation.
4. Describe how a flame-bulb (protonephridial) excretory system functions.
5. Explain how the metanephridial excretory tubule of annelids functions, and describe any structural advances over a protonephridial system.
6. Explain how the Malpighian tubule excretory system contributed to the success of insects in the terrestrial environment.
7. Using a diagram, identify and give the function of each structure in the mammalian excretory system.
8. Using a diagram, identify and give the function of each part of the nephron.
9. Describe and show the relationship among the processes of filtration, secretion and reabsorption.
10. Explain the significance of the fact that juxtamedullary nephrons are only found in birds and mammals.
11. Explain how the loop of Henle enhances water conservation by the kidney.
12. Describe the mechanisms involved in the hormonal regulation of the kidney.

13. Describe structural and physiological adaptations in the kidneys of non-mammalian species, which allow them to osmoregulate in different environments.
14. Explain the correlation between the type of nitrogenous waste produced (ammonia, urea, or uric acid) by an organism and its habitat.
15. Describe the adaptive advantages of endothermy.
16. Discuss the four general categories of physiological and behavioral adjustments used by land mammals to maintain relatively constant body temperatures.
17. Distinguish between the two thermoregulatory centers of the hypothalamus.
18. Describe the thermoregulatory adaptations found in animals other than terrestrial mammals.
19. Describe several mechanisms for physiological acclimation to new temperature ranges.
20. Distinguish between hibernation and aestivation.

KEY TERMS

osmoregulation	urine	peritubular	radiation
excretion	ureter	capillaries	evaporation
thermoregulation	urethra	vasa recta	ectotherm
osmolarity	urinary bladder	podocytes	endotherm
stenohaline	filtrate	filtrate	nonshivering
euryhaline	Bowman's capsule	antidiuretic hormone	thermogenesis
anhydrobiosis	glomerulus	osmoreceptor cells	brown fat
transport epithelia	proximal tubule	juxtaglomerular	vasodilation
flame cell	loop of Henle	apparatus	vasoconstriction
protonephridium	distal tubule	renin	counter-current heat
metanephridium	collecting duct	angiotensin	exchanger
nephrostome	cortex	angiotensin II	rete mirabile
nephridiopore	cortical nephrons	aldosterone	acclimation
Malpighian tubules	juxtamedullary	atrial natriuretic	torpor
nephron	nephrons	protein	hibernation
kidney	medulla	conduction	aestivation
renal artery	afferent arteriole	convection	diurnation
renal vein	efferent arteriole		

LECTURE NOTES

Most animals can survive environmental fluctuations more extreme than any of their individual cells could tolerate. This is possible because mechanisms of *homeostasis* maintain internal environments within ranges tolerable to body cells.

- Are long-term adaptations that evolved in populations facing environmental problems.
- Include cellular mechanisms and short-term physiological adjustments.

I.   Homeostatic mechanisms protect an animal's internal environment from harmful fluctuations: *an overview*

The majority of cells in most animals (all but sponges and cnidarians) are not exposed to the external environment, but are bathed by an extracellular fluid.

- Animals with open circulatory systems have an extracellular compartment containing *hemolymph* which bathes the cells.
- Animals with closed circulatory systems have two extracellular compartments – *interstitial fluid* and *blood plasma.*
- By moderating changes in the extracellular fluid, an animal's homeostatic mechanisms prevent environmental fluctuations from having a harmful impact.

II.   Cells require a balance between water uptake and loss

Water may enter the body of a terrestrial animal through food, drinking, and oxidative metabolism; water exits the body via evaporation and excretion. Aquatic animals are not affected by evaporation, but face the problem of osmosis where water may enter (freshwater) or leave (marine) the body.

- Even animals with specialized body coverings that retard water gain or loss have some unprotected structures exposed to the environment for gas exchange (lungs, gills).

Animals cells cannot survive a net gain (swell and burst) or loss (shrivel and die) of water.

- Solutes in the blood help ensure the proper water balance in the cell.

Osmosis = Diffusion of water across a selectively permeable membrane.

⇒ Occurs when two solutions separated by a membrane differ in *osmolarity* (total solute concentration).

⇒ If a selectively permeable membrane separates two solutions of differing osmolarities, water flows from the *hypotonic* solution to the *hypertonic* solution.

Hypertonic Solution = When comparing two solutions, the solution with a greater solute concentration; net water movement occurs into the solution.

Hypotonic Solution = When comparing two solutions, the solution with a lower solute concentration; net water movement occurs out of the solution.

Isotonic Solution = When comparing two solutions, a solution with a solute concentration equal to that of the other solution; no net water movement occurs between the solutions.

A.   Osmoconformers and Osmoregulators

An animal may be an *osmoconformer* or *osmoregulator* depending on how they balance water loss with water gain.

Osmoconformers = Animals that do not actively adjust their internal osmolarity.

- Many salt water animals.
- Body fluids are isotonic with surroundings.

Osmoregulators = Animals that regulate internal osmolarity by discharging excess water or taking in additional water.

- Many saltwater animals, all freshwater animals, and terrestrial animals.
- Net movement of water in or out, requires a concentration gradient – the maintenance of which requires energy.
- Osmoregulation permits animals to live in a variety of habitats, but the tradeoff is that it requires an energy expenditure by the animal.

A large change in external osmolarity is fatal to most animals, although some can survive radical fluctuations.

Stenohaline animal = Animal that cannot survive a wide fluctuation in external osmolarity.

Euryhaline animal = Animal that can survive wide fluctuations in external osmolarity; may be:

- Osmoconformers.
- Osmoregulators which minimize the osmotic shock by a variety of mechanisms.

B.   Maintaining Water Balance in Different Environments

Marine animals live in a saline environment consisting of about 96.5% water and 3.5% dissolved substances.

- The dissolved substances are collectively referred to as salts and the total amount of these dissolved substances in the water is *salinity*.

Most marine invertebrates are osmoconformers.

- Body fluids are isotonic to the environment.
- Body fluid composition usually differs from the external medium due to internal regulation of specific ions.

Some vertebrates of the Class Agnatha (hagfishes) are also osmoconformers. Most cartilaginous fishes, including sharks, maintain internal salt concentrations lower than sea water by pumping salt out through rectal glands and through the kidneys, yet their osmolarity is slightly hypertonic to seawater.

- Sharks retain urea as a dissolved solute in the body fluids.
- Sharks also produce and retain trimethylamine oxide (TMAO), which protects their proteins from denaturation by urea.
- Retention of these organic solutes (urea, TMAO) in the body fluids actually makes them slightly hypertonic to seawater.
- Do not drink water, but balance osmotic uptake of water by excreting urine.

Marine bony fishes are hypotonic to seawater.

- Compensate for osmotic water loss by drinking large amounts of seawater and pumping excess salt out with their gill epithelium.
- Excrete only a small amount of urine.

Freshwater animals are hypertonic to their environment and constantly take in water by osmosis.

- Freshwater protozoa compensate with *contractile vacuoles* that pump out excess water.
- Many freshwater animals, including fish, compensate by excreting large amounts of very dilute urine.
  - ⇒ Since salts are lost in this process, salt is replenished either by eating substances with a higher salt content or, in the case of some fish, by active uptake of sodium and chloride ions from the surrounding water by gill epithelium.
  - ⇒ Anadromous fishes such as salmon are *euryhaline* and migrate between seawater and fresh water.
    - ♦ While in the ocean, they osmoregulate like other marine fishes.
    - ♦ When in fresh water, they alter their osmoregulation to that of freshwater fishes.

Temporary waters present a special problem to animals that live in such environments.

- *Anhydrobiosis* is an adaptation found in a small number of aquatic invertebrates which permits them to survive when their habitat dries up.
  - ⇒ Best exemplified by the tardigrades. (See Campbell, Figure 40.3)
  - ⇒ Hydrated animals are about 1 mm long and are about 85% water.
  - ⇒ As the water around the animal disappears, water is lost from the tissues.
  - ⇒ Tardigrades can survive many years in this state and will rehydrate and become active when water returns.
- Dehydrated and frozen animals face the problem of keeping their cell membranes intact.
  - ⇒ Researchers have found that dehydrated anhydrobiotic animals contain large amounts of the disaccharide trehalose along with other sugars.
  - ⇒ Trehalose appears to replace the water associated with membranes and proteins.
  - ⇒ Trehalose is also found in insects that survive freezing.

Terrestrial animals live in a dehydrating environment and cannot survive desiccation.

- Humans die if 12% of their body water is lost.

Osmoregulatory mechanisms in terrestrial animals include protective outer layers, drinking and eating moist foods, behavioral adaptations and excretory organ adaptations that conserve water.

- Arthropods have waxy cuticles, land snails possess shells, and vertebrates are covered by a multi-layer skin comprised of dead, keratinized cells.
- Drinking and eating moist foods replaces much of the water lost during gas exchange.
- Some desert animals are nocturnal; being active only at night reduces dehydration and some like the kangaroo rat produce large amounts of metabolic water.
- The excretory organs of terrestrial animals are adapted to conserve water while eliminating wastes.

III.   Osmoregulation depends on transport epithelia

Osmoregulators utilize *transport epithelia* to regulate the movement of solutes between their internal fluids and the external environment. (See Campbell, Figure 40.5)

- Usually a single sheet of cell, joined by impermeable tight junctions, facing the external environment.
- May be a channel that leads to the exterior through an opening on the body surface.
- The molecular composition of the epithelium's plasma membrane determines the specific osmoregulatory functions. (Remember that gill epithelium pumps salt out of marine fishes and pumps salts into freshwater fishes.)
- The transport epithelium in the nasal glands of marine birds are very efficient at eliminating the excess salts obtained from drinking seawater.
- May also function in excretion of nitrogenous wastes in some animals.

IV.   Tubular systems function in osmoregulation and excretion in many invertebrates

A.   Protonephridia: The Flame-Bulb System of Flatworms

Flatworms, which have neither circulatory systems nor coeloms, have a simple tubular excretory system called a *protonephridium*. (See Campbell, Figure 40.6)

Protonephridium = A network of closed tubules lacking internal openings that branch throughout the body; the smallest branches are capped by a cellular *flame bulb*.

- Interstitial fluid passes through a flame bulb and is propelled by a tuft of cilia (in the flame bulb) along the branched system of tubules.
- In planaria, this fluid drains into excretory ducts that empty out of the body through numerous *nephridiopores*.
- Transport epithelium lining the tubules function in osmoregulation by absorbing salts before the fluid exits the body.
- Some parasitic flatworms are isotonic to their hosts and this closed system is used mainly to excrete nitrogenous wastes.

Protonephridia are also found in other invertebrates and lancets.

B.   Metanephridia of Earthworms

Each segment of most annelids, including earthworms, contains a pair of *metanephridia*, excretory tubules that have internal openings to collect body fluids. (See Campbell, Figure 40.7)

- Coelomic fluid enters the funnel-shaped *nephrostome* which is surrounded with cilia.
- The fluid passes through the metanephridium and empties into a storage bladder that empties outside the body through the *nephridiopore*.
- The nephrostome collects coelomic fluid from the body segment just anterior.

- A network of capillaries envelops each metanephridium.
    - ⇒ These capillaries reabsorb essential salts pumped out of the collecting tubules by transport epithelium bordering the lumen.
- Excretion of hypotonic, dilute urine offsets the continual osmosis of water from damp soil across the skin.

C. **Malpighian Tubules of Insects**

Malpighian tubules = Excretory organs of insects and other terrestrial arthropods that remove nitrogenous wastes from the hemolymph and function in osmoregulation. (See Campbell, Figure 40.8)

- Are outpocketings of the gut that open into the digestive tract at the midgut-hindgut juncture.
    - ⇒ The tubules dead-end at the tips away from the gut and are bathed in the hemolymph.
- Transport epithelium lining each tubule moves solutes (salts and nitrogenous wastes) from the hemolymph into the tubule's lumen.
- Accumulates nitrogenous wastes from the hemolymph and water follows by osmosis.
- The fluid in the tubule then passes through the hindgut to the rectum.
- Salts and water are reabsorbed across the epithelium of the rectum and dry nitrogenous wastes are excreted with feces.

V. **The kidneys of most vertebrates are compact organs with many excretory tubules**

The invertebrate chordate ancestors of vertebrates probably possessed segmentally arranged excretory structures arranged throughout the body.

- Extant hagfishes have segmentally arranged excretory tubules associated with their kidneys.

The kidneys of vertebrates (other than hagfishes) are compact organs and contain large numbers of non-segmentally arranged tubules.

- Kidney structure also includes a dense capillary network intimately associated with the tubules.
- The tubules function in both excretion and osmoregulation in vertebrates that osmoregulate.

The vertebrate excretory system is comprised of the kidneys, blood vessels serving the kidneys, and the structures that carry urine from the kidneys out of the body.

- Variations of the basic system are found among the vertebrate classes.

A. The Mammalian Excretory System

The human kidneys are a pair of bean-shaped organs about 10 cm long. (See Campbell, Figure 40.9).

- Blood enters each kidney via the *renal artery* and exits via the *renal vein*.
- About 20% of the blood pumped by each heartbeat passes through the kidneys.

Urine exits each kidney through a *ureter* and both ureters drain into a common *urinary bladder*.

Urine leaves the body from the urinary bladder through the *urethra*.

- Sphincter muscles near the junction of the urethra and bladder control urination.

B. The Nephron and Associated Structures

The two distinct regions of the kidney are the outer *renal cortex* and inner *renal medulla*.

- Each region contains many microscopic *nephrons* and *collecting ducts*.
- Associated with each excretory tubule is a network of capillaries.

<u>Nephron</u> = Functional unit of the kidney, consisting of a single long tubule and its associated capillaries.

- The blind end of the renal tubule that receives filtrate from the blood forms a cup-shaped *Bowman's capsule* which embraces a ball of capillaries, the *glomerulus*. (See Campbell, Figure 40.9d)
- Water, salts, urea and other small molecules are separated from the blood passing through the glomerulus by blood pressure.
    - ⇒ This *filtrate* enters the lumen of the Bowman's capsule.
- Filtrate then passes through the *proximal tubule*, the *loop of Henle* (a long hairpin turn with a descending limb and an ascending limb) and the *distal tubule*, which empties into a *collecting duct*.
- The collecting duct receives filtrate from many nephrons.
- Filtrate, now called *urine*, flows from the collecting ducts into the *renal pelvis*. Urine then drains from the chamberlike pelvis into the ureter.

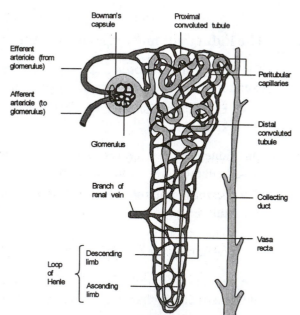

Two types of nephrons are found in mammals and birds: *cortical nephrons* and *juxtamedullary nephrons*.

> Cortical nephrons = Nephrons that have reduced loops of Henle and are confined to the renal cortex. 80% of the nephrons in humans are cortical nephrons.

> Juxtamedullary nephrons = Nephrons that have long loops that extend into the renal medulla. 20% of the nephrons are juxtamedullary nephrons.

- The nephrons in other vertebrates lack loops of Henle.

The nephron and collecting duct are lined by transport epithelium that processes the filtrate into urine.

- About 1100 – 1200 L of blood flow through the human kidneys each day.
- The nephrons process 180 L of filtrate per day, and the transport epithelium processes this filtrate to form the approximately 1.5 L urine excreted daily.
- The rest of the filtrate is reabsorbed into the blood.

The Bowman's capsules and the proximal and distal convoluted tubules are located in the cortex.

- The loops of Henle and collecting tubules extend into the medulla.

Each nephron is closely associated with blood vessels:

- *Afferent arteriole* is a branch of the renal artery that divides to form the capillaries of the glomerulus.
- *Efferent arteriole* forms from the converging capillaries as they leave the glomerulus. This subdivides to form the *peritubular capillaries* which intermingle with the proximal and distal tubules.
- *Vasa recta* is the capillary system branching downward from the peritubular capillaries that serves the loop of Henle.

Materials are exchanged between capillaries and nephrons through interstitial fluids.

VI. The kidney's transport epithelia regulate the composition of blood

The composition of blood is regulated by transport epithelia of the nephrons and collecting ducts through three processes: *filtration*, *secretion*, and *reabsorption*.

A. Production of Urine from a Blood Filtrate

1. Filtration of the Blood.

Blood pressure forces fluid from the glomerulus across the Bowman's capsule epithelium into the lumen of the nephron tubule.

- Porous capillaries and *podocytes* (specialized cells of the capsule) nonselectively filter out blood cells and large molecules; any molecule small enough to be forced through the capillary wall enters the nephron tubule.
- *Filtrate* at this point contains a mixture of glucose, salts, vitamins, nitrogenous wastes, and small molecules in concentrations similar to that in blood plasma.

2. **Secretion.**

Filtrate is joined by substances transported across the tubule epithelium from the surrounding interstitial fluid as it moves through the nephron tubule.

- Adds plasma solutes to the filtrate.
- The proximal and distal tubules are the most common sites of secretion.
- A very selective process involving both passive and active transport.
  ⇒ For example, controlled secretion of H^+ ions helps maintain constant body fluid pH.

3. **Reabsorption.**

Reabsorption is the selective transport of filtrate substances across the excretory tubule epithelium from the filtrate back to the interstitial fluid.

- Reclaims small molecules essential to the body.
- Occurs in the proximal tubule, distal tubule, loop of Henle, and collecting duct.
- Nearly all sugar, vitamins, organic nutrients and, in mammals and birds, water are reabsorbed.

The composition of the filtrate is modified by selective secretion and reabsorption.

- The concentration of beneficial substances in the filtrate is reduced as they are returned to the body.
- The concentration of wastes and nonuseful substances is increased and excreted from the body.

The kidneys are central to the process of homeostasis as they clear metabolic wastes from the blood and respond to body fluid imbalances by selectively secreting ions.

B. **Transport Properties of the Nephron and Collecting Duct**

Reclamation of small molecules and water from the filtrate as it flows through the nephron and collecting duct converts the filtrate into urine.

1. The *proximal tubule* alters the volume and composition of filtrate by reabsorption and secretion.

   - In this area ammonia, drugs and poisons processed in the liver are secreted to join the filtrate.
   - Helps maintain a constant body fluid pH by controlled secretion of H^+ and reabsorption of bicarbonate.
   - Nutrients such as glucose and amino acids are reabsorbed (by active transport) from the filtrate and returned to the interstitial fluid from which they enter the blood.

   The reabsorption of NaCl and water is also an important function of the proximal tubule.

   - Salt diffuses into the transport epithelium cells; the membranes facing the interstitial fluid then actively pump Na^+ out of the cells, which is balanced by passive transport of Cl^-. Water follows passively by osmosis.
   - Cells facing the interstitial fluid (outside the tubule) have a small surface area to minimize leakage of salt and water back into the filtrate.

2. In the *descending limb of the loop of Henle*, the transport epithelium is freely permeable to water but not to salt and other small solutes.

    - Filtrate moving down the tubule from the cortex to the medulla continues to lose water by osmosis since the interstitial fluid in this region increases in osmolarity.
    - The NaCl concentration of the filtrate increases due to the water loss.

3. In the *ascending limb of the loop of Henle*, the transport epithelium is very permeable to salt, but not to water.

    - As the filtrate ascends through the thin segment near the loop tip, NaCl diffuses out and contributes to the high osmolarity of interstitial fluids of the medulla.
    - In the thick segment leading to the distal convoluted tubule, NaCl is actively transported out into the interstitial fluid.
    - The filtrate becomes more dilute due to the removal of salts without loss of water.

4. The *distal tubule* is an important site of selective secretion and absorption.

    - It regulates K^+ and NaCl concentrations of body fluids by regulating K^+ secretion into the filtrate and NaCl reabsorption from the filtrate.
    - This region also contributes to pH regulation by controlled secretion of H^+ and reabsorption of bicarbonate.

5. The *collecting duct* carries filtrate back through the medulla into the renal pelvis.

    - The transport epithelium here is permeable to water but not to salt.
      $\Rightarrow$ The filtrate loses water by osmosis to the hypertonic fluid outside the duct which results in a concentration of urea in the filtrate.
    - The lower portion of the duct is permeable to urea, some of which diffuses out.
      $\Rightarrow$ This contributes to the high osmolarity of the interstitial fluid of the renal medulla, which enables the kidney to conserve water by excreting a hypertonic urine.

VII. The water-conserving ability of the mammalian kidney is a key terrestrial adaptation

Cooperative action between the loop of Henle and the collecting duct maintain the osmolarity gradient in the tissues of the kidney.

- The two solutes responsible for the gradient are NaCl (deposited by the loop of Henle) and urea which leaks across the epithelium of the collecting ducts.
- The urine formed is up to four times as concentrated as the blood (about 1200 mosm/L).

A. Conservation of Water by Two Solute Gradients

The juxtamedullary nephron, with its urine-concentrating features, is a key adaptation to terrestrial life that enables mammals to excrete nitrogenous waste without squandering water. Filtrate passing from the Bowman's Capsule to the proximal tubule has about the same osmolarity as blood (300 mosm/L). (See Campbell, Figure 40.12)

- A large amount of water and salt is reabsorbed as filtrate passes through the proximal tubule which is located in the renal cortex.
  - ⇒ The osmolarity remains about the same during the decrease in volume.
- Water moves out of the filtrate by osmosis as it flows from the cortex into the medulla through the descending loop of Henle.
  - ⇒ The osmolarity steadily increases due to loss of water until it peaks at the apex.
- As filtrate moves up the ascending loop of Henle back to the cortex, salt leaves the filtrate first by passive then by active transport.
  - ⇒ Osmolarity decreases to about 100 mosm/L at this point.
- Osmolarity changes very little as filtrate flows through the distal tubule as it is hypotonic to the interstitial fluids of the cortex.
- After entering the collecting duct, the filtrate passes back through the medulla.
  - ⇒ The filtrate loses water which increases osmolarity.
  - ⇒ Some urea also leaks out of the lower portion of the collecting duct.
- The passage of the filtrate back through the hypertonic medulla causes a gradual increase in osmolarity to about 1200 mosm/L; the remaining molecules are excreted in a minimal amount of water as urine which passes to the renal pelvis to the ureter.

NOTE: The loss of salt *from* the filtrate passing through the ascending limb of the loop of Henle *to* the interstitial fluid of the medulla contributes to the high osmolarity of the medulla. This, in turn, helps conserve water.

The *vasa recta* (capillary network of the renal medulla) does not dissipate the crucial osmolarity gradient in the kidney.

- As the descending vessel conveys blood toward the inner medulla, water is lost from the blood and NaCl diffuses into the blood.
- These fluxes are reversed as blood flows back toward the cortex in the ascending vessel.
- This countercurrent system allows the vasa recta to supply the tissues with necessary substances without interfering with the osmolarity gradient.

Urine, at its most concentrated, is isotonic to the interstitial fluid of the inner medulla, but is hypertonic to body fluids elsewhere.

B. Regulation of Kidney Function by Feedback Circuits

The kidney is a versatile osmoregulatory organ subject to a combination of nervous and hormonal controls.

- It excretes hyper- or hypotonic urine as necessary.

Three mechanisms regulate the kidney's ability to change the osmolarity, salt concentration, volume, and blood pressure: 1) *antidiuretic hormone*; 2) *juxtaglomerular apparatus*; and 3) *atrial natriuretic factor*.

Antidiuretic hormone (*ADH*) enhances fluid retention by increasing the water permeability of epithelium of the distal tubules and the collecting duct. (See Campbell, Figure 40.13)

- ADH is produced in the hypothalamus and stored and released from the pituitary gland.
- ADH release is triggered when osmoreceptor cells in the hypothalamus detect increased blood osmolarity due to an excessive loss of water from the body.
- Increased water reabsorption reduces blood osmolarity and reduces stimulation of the osmoreceptor cells.
  - ⇒ Results in less ADH being secreted.
  - ⇒ Ingestion of water returns blood osmolarity to normal.
- When a large volume of water has been ingested, little ADH is released and the kidneys produce a dilute urine since little water is absorbed.
- Alcohol can inhibit ADH release, causing dehydration.

The *juxtaglomerular apparatus* (*JGA*) is a specialized tissue near the afferent arterioles which carry blood to the glomeruli. It responds to a decrease in blood pressure or blood volume by releasing the enzyme *renin* into the blood.

- Renin converts inactive *angiotensin* to active *angiotensin II* which functions as a hormone.
- Angiotensin II directly increases blood pressure by:
  - ⇒ Causing arteriole constriction.
  - ⇒ Stimulating the proximal tubules to absorb more NaCl and water. Reduction in the amounts of salt and water excreted raises blood volume and pressure.
- Angiotensin II acts indirectly by:
  - ⇒ Signaling adrenal glands to release *aldosterone*, which stimulates Na^+ reabsorption by the distal tubules (water follows by osmosis).
  - ⇒ Increases blood pressure and blood volume.
- The increased blood pressure and blood volume suppresses further release of renin.
- The *renin-angiotensin-aldosterone system* (RAAS) is part of a complex feedback circuit that functions in homeostasis.

ADH and the RAAS cooperate in homeostasis.

- ADH is released in response to increased blood osmolarity, but this does not compensate for excessive loss of salts and body fluids if blood osmolarity does not change.
- RAAS responds to a decrease in blood volume caused by fluid loss.
- ADH alone would lower blood Na^+ concentration by increasing water reabsorption, but the RAAS maintains balance by stimulating Na^+ reabsorption.

The hormone *atrial natriuretic factor* (*ANF*) opposes the RAAS.

- Released by the heart's atrial walls in response to increased blood volume and pressure.
- Inhibits release of renin from the JGA, inhibits NaCl absorption by the collecting ducts, and reduces aldosterone release from the adrenal glands. This decreases volume and lowers blood pressure.

VIII. Diverse adaptations of the vertebrate kidney have evolved in different habitats

Nephrons vary in structure and physiology and help different vertebrates to osmoregulate in their various habitats.

- Desert mammals have very long loops of Henle to maintain steep osmotic gradients that conserve water by allowing urine to become very concentrated.
- Mammals living in aquatic environments (e.g. beavers) have nephrons with very short loops of Henle resulting in production of dilute urine.
- Birds have shorter loops of Henle and produce a more dilute urine than mammals.
- Reptiles have only cortical nephrons and produce isotonic urine, but the epithelium of their cloaca conserves fluid by reabsorbing water from urine and feces.

  ⇒ Also most excrete nitrogenous wastes as uric acid which conserves water.

- Freshwater fish (hypertonic to surroundings) nephrons use cilia to sweep the large volume of very dilute urine from the body.

  ⇒ Salts are conserved by efficient ion reabsorption from filtrate.

- Amphibians excrete dilute urine and accumulate certain salts from the water by active transport across the skin. On land, body fluid is conserved by water reabsorption across urinary bladder epithelium.
- Bony marine fishes (hypotonic to surroundings) excrete very little concentrated urine (many lack glomeruli and capsules).

  ⇒ The kidneys function mainly to rid the body of divalent ions such as Ca^{2+}, Mg^{2+} and SO_4^{2-} taken in by drinking seawater.

  ⇒ Monovalent ions like Na^+ and Cl^-, and the majority of nitrogenous waste (in the form of ammonium) is excreted mainly by the gills.

IX. An animal's nitrogenous wastes are correlated with its phylogeny and habitat

The metabolism of proteins and nucleic acids produces ammonia, a small and very toxic waste product. Some animals excrete the ammonia directly, while other first convert it to urea or uric acid which are less toxic.

A.   Ammonia

Ammonia is excreted directly by most aquatic animals.
- Easily permeates membranes since molecules are very water soluble.
- In soft-bodied invertebrates, ammonia just diffuses out across the body surface.
- In fishes, ammonium ions are excreted as ammonium ions across gill epithelium.
  ⇒ The gill epithelium of freshwater fish exchange Na^+ from the water for NH_4^+, thus maintaining a higher Na^+ concentration in the blood.

B.   Urea

Ammonia excretion is unsuitable for animals in a terrestrial habitat.
- It requires large amounts of water and is so toxic it must be eliminated quickly.

Urea is the nitrogenous waste excreted by mammals and most adult amphibians.
- Can be more concentrated in the body since it is much less toxic than ammonia; reduces water loss for terrestrial animals.
- Produced in liver by a metabolic cycle combining ammonia with CO_2. It is transported to kidneys via the circulatory system.
- Some urea is retained in the kidney where it contributes to osmoregulation by maintaining the osmolarity gradient in the medulla.
- Sharks also produce and retain urea in the blood as an osmoregulatory agent.
- Amphibians that undergo metamorphosis and move to land as adults switch from excreting ammonia to excreting urea.

C.   Uric Acid

Uric acid is the primary form of nitrogenous waste excreted by land snails, insects, birds and many reptiles.
- Much less soluble in water than ammonia or urea and can be excreted as a precipitate after reabsorption of nearly all the water from the urine.
- Eliminated in a pastelike form through the cloaca (mixed with feces) in birds and reptiles.

The mode of reproduction is an important factor in determining whether uric acid or urea excretion evolved in a particular group.
- If an embryo released ammonia or urea within a shelled egg, the soluble waste would accumulate to toxic concentrations: uric acid precipitates out of solution to be stored as a solid.

The animal's habitat, along with the phylogenetic position, influences the type of nitrogenous waste produced.
- Terrestrial reptiles excrete mostly uric acid; crocodiles excrete ammonia and uric acid; aquatic turtles excrete urea and ammonia.

Some animals can modify their nitrogenous wastes when the temperature or water availability changes.

X.   Thermoregulation maintains body temperature within a range conducive to metabolism

Metabolism and membrane properties are very sensitive to changes in an animal's internal temperature.

- Each animal lives in, and is adapted to, an optimal temperature range in which it can maintain a constant internal temperature when external temperatures fluctuate.
- Maintaining the body temperature within a range that permits cells to function efficiently is known as *thermoregulation*.

A.   Heat Transfer Between Organisms and Their Surroundings

An organism exchanges heat with its environment by four physical processes:

1. *Conduction* is the direct transfer of thermal motion (heat) between molecules of the environment and a body surface.

    - Heat is always conducted from a body of higher temperature to one of lower temperature.
    - Water is 50-100 times more effective than air in conducting heat.
    - For example, on a hot day, an animal in cold water cools more rapidly than one on land.

2. *Convection* is the transfer of heat by the movement of air or liquid past a body surface.

    - For example, breezes contribute to heat loss from an animal with dry skin.

3. *Radiation* is the emission of electromagnetic waves produced by all objects warmer than absolute zero.

    - Can transfer heat between objects not in direct contact.
    - For example, an animal can be warmed by the heat radiating from sun.

4. *Evaporation* is the loss of heat from a liquid's surface that is losing some molecules as gas.

    - Cooling increases greatly by production of sweat.
    - Can only occur if surrounding air is not saturated with water molecules.
    - Along with convection, is most variable cause of heat loss.

XI.   Ectotherms derive body heat mainly from their surroundings and endotherms derive it mainly from metabolism

Animals may be classified as either *ectotherms* or *endotherms* depending on their major source of body heat.

<u>Ectotherm</u> = An animal that warms its body mainly by absorbing heat from surroundings.

- Most invertebrates, fishes, reptiles and amphibians.
- May derive a small amount of body heat from metabolism.

<u>Endotherm</u> = An animal that derives most or all of its body heat from its own metabolism.
- Mammals, birds, some fishes, and numerous insects.
- Many maintain a consistent internal temperature even as the environmental temperature fluctuates.

The main source of body heat distinguishes endotherms from ectotherms, not the body temperature.
- Distinction is not absolute.
- Most endothermic insects and some fishes are partial endotherms.
    ⇒ These organisms retain metabolic heat to warm only certain body parts (i.e. locomotor muscles).
- Some birds and mammals acquire additional body heat by basking in the sun.

A terrestrial life style presents certain problems which have been solved by the evolution of endothermy.
- Environmental temperatures fluctuate more in terrestrial habitats than in aquatic habitats.
    ⇒ Endothermic vertebrates are usually warmer than their surroundings, but also have mechanisms to cool their bodies.
- Maintaining a warm body temperature requires an active metabolism which contributes to a high level of cellular respiration
    ⇒ This permits endotherms to be more physically active for a longer period of time in comparison to most ectotherms.

Endothermy requires more energy than ectothermy.
- At 20°C, a human has a resting metabolic rate of 1300 to 1800 kcal/day while an alligator (ectotherm) of similar weight has a resting metabolic rate of about 60 kcal/day.
- Endotherms usually consume more food than ectotherms of similar size to offset the energy requirements.

The bioenergetic connections between body temperature, an active metabolism, and mobility were important to the evolution of endothermy.
- The evolution of endothermy and a high metabolic rate were accompanied by an increase in efficiency in circulatory and respiratory systems as seen in the birds and mammals.
- Terrestrial ectotherms exhibit their own adaptations for adjusting to temperature changes in the terrestrial environment.

XII. Thermoregulation involves physiological and behavioral adjustments

Endotherms *and* ectotherms use a combination of strategies to thermoregulate. For example:

1. *Adjusting the rate of heat exchange between the animal and its surroundings.*

    Heat loss is reduced by the presence of hair, feathers, and fat just below the skin.

    Adaptations found in the circulatory system also help regulate heat exchange.

    - The amount of blood flowing to the skin can be change by many endotherms and some ectotherms.
    - *Vasodilation* increases the blood flow to the skin due to an increase in the diameter of blood vessels near the body surface.
        - ⇒ Nerve signals cause muscles in the walls of the blood vessels to relax which permits more blood to flow through the vessel.
        - ⇒ As blood flow increases, more heat is transferred to the environment by conduction, convection, and radiation.
    - *Vasoconstriction* reduces blood flow to the skin due to a decrease in the diameter of blood vessels near the body surface.
        - ⇒ Reduced blood flow decreases the amount of heat transferred to the environment.

    Heat exchange with the environment is also altered by a *countercurrent heat exchanger*. (See Campbell, Figure 40.16)

    - This is a special arrangement of arteries and veins found in the extremities of many endothermic animals.
    - Arteries carrying warm blood from the body to the legs of a bird or flipper of a dolphin are in close contact with veins carrying blood from the appendage into the body.
    - This vessel arrangement enhances heat transfer from arteries to veins along the entire length of the vessel.
        - ⇒ This is possible since venous blood returning from the tip of the appendage is always cooler than the arterial blood.
    - The venous blood entering the body has been warmed to almost core temperature by the exchange.
    - In some species, an alternative set of vessels which by-pass the exchanger is present.
        - ⇒ The rate of heat loss is controlled by the relative amount of blood that enters the appendage by the two paths.

2. *Cooling by evaporative heat loss.*

    Water is lost by terrestrial endotherms and ectotherms through breathing and across the skin.

    - In low humidity, water evaporates and heat is lost by evaporative cooling.
    - Panting increases evaporation from the respiratory system.
    - Sweating in mammals increases evaporative cooling across the skin.

3. *Behavioral responses.*

   Relocating allows animals to increase or decrease heat loss from the body.
   - In winter, many animals bask in the sun or on warm rocks.
   - In summer, many animals burrow or move to damp areas.
   - Some animals migrate to more suitable climates.

4. *Changing the rate of metabolic heat production.*

   This is found only in birds and mammals.
   - Increased skeletal muscle activity and non-shivering thermogenesis can greatly increase the amount of metabolic heat produced.

XIII. Comparative physiology reveals diverse mechanisms of thermoregulation

A. Invertebrates

Most invertebrates have little control over body temperature, but some do adjust temperature by behavioral or physiological mechanisms.
- The desert locust orients the body to maximize heat absorption from the sun.
- Some large flying insects (e.g. bees) can generate internal heat by contracting all flight muscles in synchrony (functionally analogous to shivering).
  ⇒ Little wing movement occurs but large amounts of heat are produced.
  ⇒ Allows activity even on cold days and at night.
- Endothermic insects such as honeybees, bumblebees, and noctuid moths have a countercurrent heat exchanger that maintains a high temperature in the thorax where flight muscles are located.

Honeybees also use social organization to regulate temperature.
- Increase movements and huddle to retain heat in cold weather.
  ⇒ Maintain constant temperature by changing the density of huddling; heat is distributed by the movement of individuals from the core to the margins of the huddle.
- In warm weather, they cool hives by transporting water to it and fanning with their wings to promote evaporation and convection.

B. Amphibians and Reptiles

Amphibians produce little heat and most lose heat rapidly by evaporative cooling from their body surface, making thermoregulation difficult.
- The optimal temperature range varies greatly depending on the species.
- Seek cooler or warmer microenvironments as necessary (behavioral adaptation).
- Some can vary the amount of mucus they secrete to regulate evaporative cooling.

Reptiles are generally ectotherms that warm themselves mainly by behavioral adaptations.
- Orient the body toward the heat sources to increase uptake and maximize the body surface exposed to, the heat source.
- Along with orientations and body surface exposure, they regulate body temperature by seeking warm places or other favorable microclimates in the environment.

Some reptiles have physiological adaptations that help regulate heat loss.
- When swimming in cold water, the Galapagos iguana utilizes superficial vasoconstriction to reduce heat loss.
- Female pythons that are incubating eggs increase their metabolic rate by shivering.
  ⇒ Maintains their body temperature 5° – 7°C above ambient temperature.

C. Fishes

Fish are generally ectotherms although some are endothermic.
- The body temperature of most is within 1° – 2°C of the water temperature.
- Metabolic heat produced by the swimming muscles is lost to the water as blood passes through the gills.
  ⇒ The dorsal aorta carries blood directly from the gills to the tissues, cooling the body core.

Endothermic fishes have an adaptation to reduce heat loss. (See Campbell, Figure 40.19)
- Includes large active species such as the bluefin tuna, swordfish, and great white shark.
  ⇒ Large arteries carry most of the blood from the gills to tissues just under the skin.
  ⇒ Branches carry blood to the deep muscles where smaller vessels are arranged into a countercurrent heat exchanger.
- The swimming muscles produce enough metabolic heat to raise temperatures at the body core.
- Adaptations in the circulatory system help retain the heat.
- Swimming muscles remain several degrees warmer than tissues closer to the animal's surface.
  ⇒ Endothermy thus enhances the activity level of these fishes.

D. Mammals and Birds

These organisms maintain high body temperatures within a narrow range.
- 36° – 38°C for most mammals.
- 40° – 42°C for most birds.

Birds and mammals regulate the rate of metabolic heat production and balance it with the rate of heat loss or gain from their surroundings.

The rate of heat production may be increased by:

- The increased contraction of muscles.
  - ⇒ Moving or shivering produce metabolic heat.
- The action of hormones that increase the metabolic rate and the production of heat instead of ATP.
  - ⇒ *Nonshivering thermogenesis* is the hormonal triggering of heat production.
  - ⇒ Found in numerous mammals and a few birds.
  - ⇒ Occurs throughout the body and some mammals have *brown fat* between the shoulders and in the neck that is specialized for rapid heat production.

Birds and mammals also use other mechanisms to regulate environmental heat exchange.

- Vasodilation and vasoconstriction permit the maintenance of a proper core temperature even when the extremities are cooler.
- Fur and feathers trap a layer of insulating air next to the body.
  - ⇒ Most land mammals and birds raise their hair or feathers in response to cold.
  - ⇒ This traps a thicker layer of air next to the body, thus providing more insulation.
- A layer of fat just below the skin provides insulation.

The insulating power of hair is greatly reduced in water, and marine mammals show various adaptations to reduce heat loss. For example, whales and seals live in water colder than their body temperature.

- They are insulated by a very thick layer of fat called *blubber* under the skin.
  - ⇒ Helps them maintain a body temperature of 36° – 38°C even in Arctic and Antarctic waters.
- Countercurrent heat exchangers reduce heat loss in the extremities where no blubber is found.

Thermoregulation by endotherms also involves cooling the body. Many birds and mammals have behavioral and physiological adaptations to cool the body.

- When whales and other marine mammals move into warm waters, excess heat is eliminated by vasodilation.
  - ⇒ A large number of blood vessels in the outer skin dilate which permits increased blood flow (and heat loss by conduction) to occur.
- Terrestrial birds and mammals depend on evaporative cooling.
  - ⇒ Panting increases heat loss.
  - ⇒ The fluttering of vascularized pouches in the floor of the mouth in some birds increases evaporative cooling.
- Many terrestrial mammals have sweat glands.
- Some mammals spread saliva (some kangaroos and rodents) or a combination of saliva and urine (some bats) over the body to increase evaporative heat loss.

E.   Thermoregulation in Humans

Thermoregulation in humans involves a complex homeostatic system and feedback mechanisms. (See Campbell, Figure 40.21)

- The *hypothalamus* contains nerve cells which control thermoregulation and many other aspects of homeostasis.
- The hypothalamus responds to changes in body temperature which are above or below the normal range. (36.1° – 37.8°C).

The body's temperature is monitored by nerve cells in the skin, hypothalamus, and other parts of the body.

- Some function as warm receptors that signal the hypothalamus when the skin or blood temperature increases.
- Others function as cold receptors that signal the hypothalamus when there is a decrease in temperature of the skin or blood.

When the body's temperature increases above the normal range, *cooling mechanisms* are activated.

- Vasodilation of skin vessels occurs and capillaries fill with blood; the heat radiates from the skin's surface.
- Sweat glands are activated which increases evaporative cooling.
- These changes result in a decrease in body temperature to within the normal range.
- The cooling mechanisms are "turned off" by the hypothalamus when normal body temperature is achieved.

When the body's temperature decreases below the normal range due to a cold environment, *warming mechanisms* are activated.

- Vasoconstriction of skin vessels occurs.
  - ⇒ A smaller volume of blood passes to the skin, thus reducing heat loss from the body surface.
  - ⇒ The warm blood is diverted from the skin to deeper tissues.
- Skeletal muscles are activated and shivering occurs which generates metabolic heat.
- Nonshivering thermogenesis increases heat production.
- Changes resulting in an increase in body temperature to within the normal range.
- The warming mechanisms are "turned off" by the hypothalamus when normal body temperature is achieved.

F.   Torpor: Conserving Energy During Environmental Extremes

When food supply is low and/or environmental temperatures are extreme, many endotherms will enter a state of *torpor*.

<u>Torpor</u> = Alternative physiological state in which metabolism decreases and the heart and respiratory systems slow down.

- This is a mechanism to conserve energy.

Torpor may be a long-term state (hibernation, aestivation) or short-term as in the daily period of torpor seen in many small mammals and birds.

Hibernation and aestivation are often triggered by seasonal changes in daylength.
- In *hibernation*, the body temperature is lowered; this allows an animal to survive long periods of cold and diminished food supplies.
  ⇒ Some animals will begin to eat huge quantities of food as the amount of daylight decreases.
- *Aestivation* is characterized by slow metabolism and inactivity.
  ⇒ Allows the animal to survive long periods of high temperature and diminished water supply.
  ⇒ Also known as *summer torpor*.

Daily periods of torpor appear to be adapted to feeding patterns.
- They allow many relatively small endotherms to survive on stored energy during hours when they cannot feed.
  ⇒ These animals have a very high metabolic rate and rate of energy consumption when active.
- Most bats and shrews feed at night and enter torpor during daylight hours.
- Chickadees and hummingbirds feed during the day and undergo torpor on cold nights.
  ⇒ The body temperature of chickadees in the cold, northern forests may drop 10°C during winter nights.

An animal's biological clock appears to control its daily cycle and torpor.
- For example, shrews continue to undergo a daily torpor even when food is continually available.
- Human sleep periods and the corresponding drop in body temperature may be a remnant of a daily torpor in ancestral mammals.

G.   **Temperature Range Adjustments**

A physiological response called *acclimatization* allows many animals to adjust to a new range of environmental temperature.
- The adjustment takes place over a period of days or weeks.
- The process is important to animals that must adjust to seasonal changes in temperature.

Acclimatization of an animal to a new external temperature range may involve several physiological changes.
- May involve changes in the cooling and warming mechanisms which maintain the internal temperature.
- May involve adjustments at the cellular level.
  ⇒ Cells may increase production of certain enzymes to compensate for the lowered activity of each molecule at lower temperatures.

⇒ Cells may produce enzyme variants having the same function but different temperature optima.

- Membranes may remain fluid by changing the proportions of saturated and unsaturated lipid in their composition.

Cells can also make rapid adjustments to temperature changes.

- Severe stress due to a large temperature increase (or other stress) stimulate the accumulation of *stress-induced proteins*.
  ⇒ Found in animal cells, yeast, and bacteria.
  ⇒ Help prevent denaturation of other proteins by high temperatures.
  ⇒ Help prevent cell death and may help maintain homeostasis while the organisms is adjusting to the external environment.

XIV. Regulatory systems interact in the maintenance of homeostasis

Maintaining homeostasis in the internal environment involves a number of regulatory systems within an animal's body. For example:

- Regulation of body temperature involves mechanisms that control osmolarity, metabolic rate, blood pressure, tissue oxygenation, and body weight.

While the regulatory systems normally work in concert to maintain homeostasis, under extreme physiological stress, demands of one regulatory system may conflict with those of other systems.

- Water conservation takes precedence over evaporative heat loss in very warm, dry climates.
- Many desert animals must tolerate occasional hyperthermia (abnormally high body temperature) in order to maintain body water.

A. Roles of the Liver in Homeostasis

The functions of the vertebrate liver are critical to maintaining homeostasis and involve interactions with most of the body's organ systems. (See Campbell, Figure 40.1)

- Excretion by the kidneys is supported by the liver which synthesizes ammonia, urea, or uric acid from the nitrogen of amino acids.
  ⇒ The liver also detoxifies many chemical poisons.
- The bioenergetics of the body are influenced by many of the liver's activities.
  ⇒ Liver cells take up glucose from the blood and store excess amounts as glycogen.
  ⇒ When glucose is needed by the body, the liver converts glycogen back to glucose and releases it into the blood.
  ⇒ Blood glucose levels are closely controlled by feedback circuits which regulate the homeostatic mechanisms.
- Thus, while the liver performs the essential functions, feedback circuits are in turn controlled by the nervous and endocrine systems.

REFERENCES

Campbell, N. *Biology*. 4th ed. Menlo Park, California: Benjamin/Cummings, 1996.

Eckert, Roger, and David Randall. *Animal Physiology: Mechanisms and Adaptations*. 2nd ed. San Francisco: W.H. Freeman and Company, 1983.

Marieb, E.N. *Human Anatomy and Physiology*. 3rd ed. Redwood City, California: Benjamin/Cummings, 1995.

Nybakken, James W. *Marine Biology: An Ecological Approach*. New York: Harper & Row Publishers, Inc., 1982.

CHAPTER 41

CHEMICAL SIGNALS
IN ANIMALS

OUTLINE

I. A variety of chemical signals coordinate body functions: *an overview*
    A.   Hormones
    B.   Pheromones
    C.   Local Regulators

II. Hormone binding to specific receptors triggers signaling mechanisms at the cellular level
    A.   Steroid Hormones and Gene Expression
    B.   Peptide Hormones and Signal Transduction

III. Many chemical signals are relayed and amplified by second messengers and protein kinases
    A.   Cyclic AMP
    B.   Signal Amplification and Specificity
    C.   Cytoplasmic Ca^{2+} and Signal Transduction

IV. Invertebrate control systems often integrate endocrine and nervous system functions

V. The hypothalamus and pituitary integrate many functions of the vertebrate endocrine system
    A.   Structural and Functional Relationships of the Hypothalamus and the Pituitary Gland
    B.   Posterior Pituitary Hormones
    C.   Anterior Pituitary Hormones
    D.   Hypothalamic Hormones

VI. The vertebrate endocrine system coordinates homeostasis and regulates growth, development, and reproduction
    A.   The Thyroid Gland
    B.   The Parathyroid Glands
    C.   The Pancreas
    D.   The Adrenal Glands
    E.   The Gonads
    F.   Other Endocrine Organs

VII. The endocrine system and the nervous system are structurally, chemically, and functionally related

OBJECTIVES

After reading this chapter and attending lecture, the student should be able to:

1. Compare the response times of the two major systems of internal communication: the nervous system and the endocrine system.
2. On the basis of structure and function, distinguish among types of chemical messengers.
3. Distinguish between endocrine and exocrine glands.
4. Describe the relationships among endocrine system components: hormones, endocrine glands, target cells and target cell receptors.
5. List the three general classes of hormones and give examples of each.
6. Explain how pheromone function differs from hormone function.
7. Provide indirect evidence that humans may communicate with pheromones.
8. State which of the two classes of hormones is lipid soluble and explain how this property affects hormone function.
9. Describe the mechanism of steroid hormone action and explain the location and role of steroid hormone receptors.
10. Using epinephrine as an example, explain how nonsteroid hormones act via second messengers, including the roles of adenylate cyclase, cyclic AMP, G proteins and cAMP-dependent protein kinase.
11. Explain how adenylate cyclase activity is controlled.
12. Using epinephrine as an example, explain how an enzyme cascade amplifies response to a hormone.
13. Explain how to account for specificity in target cell response to hormonal signals.
14. Outline the sequence of events which occurs as inositol triphosphate mediates between a hormonal signal and an increase in cytoplasmic calcium ions.
15. Compare and contrast the two general modes of hormone action.
16. Describe hormonal regulation of insect development including the roles of ecdysone, brain hormone and juvenile hormone.
17. Describe the location of the hypothalamus and explain how its hormone-releasing cells differ from both endocrine gland secretory cells and other neurons.
18. Describe the location of the pituitary and explain the functions of the posterior and anterior lobes.
19. List the posterior pituitary hormones and describe their effects on target organs.
20. Using antidiuretic hormone as an example, explain how a hormone contributes to homeostasis and how negative feedback can control hormone levels.
21. Define tropic hormone and describe the functions of tropic hormones produced by the anterior pituitary.
22. Explain how the anterior pituitary is controlled.
23. List hormones of the thyroid gland and explain their role in development and metabolism.
24. Diagram the negative feedback loop which regulates the secretion of thyroid hormones.
25. State the location of the parathyroid glands and describe hormonal control of calcium homeostasis.
26. Distinguish between α and β cells in the pancreas and explain how their antagonistic hormones (insulin and glucagon) regulate carbohydrate metabolism.
27. List hormones of the adrenal medulla, describe their function and explain how their secretion is controlled.
28. List hormones of the adrenal cortex, describe their function and explain how their secretion is controlled.
29. Describe both the short term and long term endocrine responses to stress.
30. Identify male and female gonads and list the three categories of gonadal steroids.

31. Define gonadotropin and explain how estrogen and androgen synthesis is controlled.
32. Describe the location of the pineal and thymus glands, list their hormone products and describe their functions.
33. Explain how the endocrine and nervous systems are structurally, chemically and functionally related.

KEY TERMS

hormone
steroid
tyrosine
peptide hormone
signal molecules
target cells
endocrine gland
exocrine gland
pheromone
neurotransmitter
histamine
interleukin
nerve growth
    factors (NGF)
epidermal growth
    factors (EGF)
prostaglandins (PGs)
second messengers
cytoplasmic
    intermediates
cyclic AMP (cAMP)
inositol triphosphate
melanocyte-stimulating
    hormone (MSH)
Earl W. Sutherland
G protein

adenylate cyclase
phosphodiesterase
hormone-receptor
    complex
enzyme cascade
protein kinase
inositol triphosphate
calmodulin
ecdysone
juvenile hormone (JH)
hypothalamus
neurosecretory cells
releasing factors
posterior pituitary
    (neurohypophysis)
antidiuretic
hormone (ADH)
oxytocin
anterior pituitary
    (adenohypophysis)
growth hormone (GH)
insulinlike growth
    factors (IGFs)
prolactin (PRL)
follicle-stimulating
    hormone (FSH)

luteinizing
    hormone (LH)
thyroid-stimulating
    hormone (TSH)
pro-opiomelonocortin
adrenocorticotropin
    (ACTH)
melanocyte-stimulating
    hormone (MSH)
endorphins
hypothalamic
    releasing factors
median eminence
thyroid gland
triiodothyrone
thyrodoxine
hyperthyroidism
hypothyroidism
calcitonin
parathyroid
    hormone (PTH)
insulin
β islet cells
proinsulin
α islet cells

glucagon
adrenal medulla
epinephrine
norepinephrine
catecholamines
acetylcholine
adrenal cortex
corticosteroids
glucocorticoids
cortisol
mineralocorticoids
aldosterone
estrogen
androgen
testosterone
progestins
gonadotropins
gonadotropin-
    releasing
    hormone (GnRH)
gastrin
secretin
erythropoietin
melatonin
thymosin

LECTURE NOTES

The activities of the various specialized parts of an animal are coordinated by the two major systems of internal communication: the *nervous system* and the *endocrine system*.

- The nervous system is involved with high-speed messages.
- The endocrine system is slower and involves the production, release, and movement of chemical messages.

I. **A variety of chemical signals coordinate body functions:** *an overview*

All animals exhibit some coordination by chemical signals:

- Hormones produced by the endocrine system convey information between organs of the body.
- Pheromones are chemical signals used to communicate between different individuals.
- Other messengers, such as neurotransmitters, act between cells on a localized scale.

A. **Hormones**

The endocrine system has 3 key components: *hormones*, *endocrine glands*, and *target cells* with their molecular receptors for hormones.

Hormone = Secreted by endocrine or neurosecretory cells, a chemical signal that travels in body fluids to target cells where it elicits a specific response.

Neurosecretory cell = Neuron that receives signals from other nerve cells and responds by releasing hormones into body fluids or into a storage organ from which they are later released.

Target cell = A cell equipped to respond to a given hormone.

Endocrine glands = Ductless glands that secrete hormones into the body fluids for distribution through the body.

- Composed of the specialized cells which secrete hormones.
- May comprise only a portion of organs that produce other materials (e.g. pancreas).

Exocrine glands = Glands that produce a variety of substances (e.g. sweat, mucus, digestive enzymes) and convey their products by means of ducts.

- Many organs perform both exocrine and endocrine functions.
- Exocrine glands are *not* a part of the endocrine system.

Hormones are grouped into two general classes according to similarities in chemical structure: *steroid hormones* and *hormones derived from amino acids*.

- Steroid hormones are lipid molecules fashioned from cholesterol.
  - ⇒ Includes the sex hormones.
- Hormones derived from amino acids may be small or large. Includes
  - ⇒ Amine hormones (modified versions of single amino acids).
  - ⇒ Peptide hormones (short chains of amino acids).
  - ⇒ Protein hormones.
  - ⇒ Glycoproteins.

A hormone's target cell possesses a *hormone receptor* to which the hormone molecule must bind.

- The hormone receptor is a protein within the cell or on the plasma membrane of the target cell.
- The hormone and receptor are shape specific and must bind together to elicit a response in the cell to the hormone.

The actions of antagonistic hormones are important in maintaining homeostasis.

- Feedback mechanisms are usually involved in adjusting hormone actions.

B. Pheromones

Pheromones = Chemical signals that function *between* animals of the same species.

- Classified according to function (e.g. mate attractant, territorial marker, alarm substance).
- Are small, volatile, easily dispersed molecules that are active in minute amounts.
- Have an important role in social behavior.

C. Local Regulators

Local regulators are chemical messengers that affect target cells adjacent to or near their point of secretion. Includes *synaptic signaling* and *paracrine signaling*.

- *Synaptic signaling* is direct and involves the release of a neuron's transmitter into a synapse.
- *Paracrine signaling* is less direct but very localized.
  - ⇒ Regulatory substances are released into interstitial fluids but affect only nearby target cells (i.e. histamine and interleukins).

Local regulators may be gases produced by cells, growth factors, or *prostaglandins*.

1. *Gases produced by cells.* Many types of cells produce gases which can act as neurotransmitters and paracrine signals.
   - Nitric oxide (NO) released by endothelial cells of blood vessels makes the adjacent smooth muscle cells relax, dilating the vessel.
   - Nitric oxide released by white blood cells kills certain cancer cells and bacteria in the body fluids.

2. *Growth Factors*

    Growth factors = Peptides and proteins that regulate the behavior of cells in growing and developing tissues.

    - Must be present in the extracellular environment for certain cell types to grow and develop normally.
    - Studied mainly in cultures of mammalian cells but also regulate development within the animal body.
    - Binding of the growth factor to a specific receptor protein on the outer membrane surface triggers the target cell's response – a change in gene expression in the target-cell nucleus.

3. Prostaglandins

    Prostaglandins (PGs) = Modified fatty acids released into interstitial fluid that function as local regulators.

    - Often derived from lipids of the plasma membrane.
    - Very subtle differences in their molecular structure profoundly affect how these signals affect target cells (e.g. antagonistic actions of PGE and PGF).
    - PGs secreted by the placenta help induce labor during childbirth by causing chemical changes in the nearby uterine muscles.
    - Other PGs help defend the body by inducing fever and inflammation.

II. Hormone binding to specific receptors triggers signaling mechanisms at the cellular level

Hormones can act in very low concentration and can affect different target cells in an animal or different species of animals very differently.

- A hormone-induced change in a target organ cell results from one of two possible *signal transduction pathways*.
- Signal transduction pathways include a series of steps between hormone-receptor binding and target-cell response.

A. Steroid Hormones and Gene Expression

Steroid hormones enter the target cell by passing through the target cell membrane and binding to a receptor protein in the cytoplasm. (See Campbell, Figure 41.3)

- The *hormone-receptor complex* then has the proper configuration to bind to specific regulatory sites located along the genome.
- Binding of the hormone-receptor complex to regulatory sites may induce or suppress expression of specific genes.
- When expression is stimulated, mRNA molecules are produced, exit to the cytoplasm, and direct the synthesis of new proteins.
- Two cell types can respond differently to a single hormone since the hormone-receptor complex probably binds with regulatory sites controlling different genes in the two kinds of cells.

Steroid hormones primarily affect the synthesis of proteins.

B.   Peptide Hormones and Signal Transduction

Peptide hormones and most hormones derived from amino acids are unable to pass through the target cell plasma membrane, they thus require a different mechanism of action.

- These hormones attach to specific receptor proteins embedded in the target cell's plasma membrane.
- The receptors are components of a signal-transduction pathway that converts the extracellular signal of the hormone to intracellular signals that alter the target-cell's behavior.

Peptide hormones usually affect the activity of enzymes and other proteins already present in the cell.

III.   Many chemical signals are relayed and amplified by second messengers and protein kinases

Earl W. Sutherland and his colleagues discovered the signal-transduction pathway used by nonsteroid hormones.

- They discovered that a second messenger transmits the signal from the plasma membrane to the metabolic machinery in the cytoplasm.

A.   Cyclic AMP

Cyclic AMP (cyclic adenosine monophosphate) was discovered as the second messenger in Sutherland's studies of how epinephrine stimulates glycogen hydrolysis in liver and muscle cells. (See Campbell, Figure 41.4)

- Binding of a hormone with its receptor activates a *G protein*, which binds with GTP (guanosine triphosphate), an energy carrying molecule.
- Activated G protein binds to *adenylyl cyclase*, an effector enzyme embedded in the plasma membrane.
- Adenylyl cyclase catalyzes the conversion of ATP to cyclic AMP (cAMP).
- cAMP relays the signal from the membrane to the metabolic machinery of the cytoplasm.
- A second type of G protein *inhibits* adenylyl cyclase, when activated by an inhibitory hormone-receptor complex. This reduces cAMP concentration in the cytoplasm.
- Regulation of adenylyl cyclase by the opposing effects of two G proteins enables a cell to respond to slight changes in the proportions of antagonistic hormones.

B.   Signal Amplification and Specificity

Many metabolic responses to hormones involve an *enzyme cascade*, where each step activates an enzyme that in turn activates the next enzyme in the series (Campbell, Figure 41.6). For example, a cell's response to epinephrine:

- cAMP activates a *cAMP dependent protein kinase* (an enzyme that catalyzes the transfer of a phosphate group from ATP to proteins = *phosphorylation*), which in turn can activate phosphorylase kinase.
- Phosphorylase kinase adds a phosphate group to glycogen phosphorylase which actually hydrolyzes glycogen.
- The cascade thus amplifies response to the hormone since the number of activated products increases with each step.

Although a single basic mechanism is used to mediate responses of diverse cells to many different hormones, response is still specific. Specificity arises since:

- Particular cells can only interact with certain hormones — those for which they have receptors.
- cAMP-dependent protein kinases differ in structure and function from tissue to tissue.
- Post-protein kinase activation steps in the cascade differ between cell types; thus, individual cell types produce a unique set of proteins in response to cAMP-dependent protein kinase.

In addition to working as a second messenger, cAMP has been found to inhibit or stimulate key enzyme components of the signal-transduction pathway of many growth factors.

C.   Cytoplasmic Ca^{2+} and Signal Transduction

Many chemical messengers in animals (neurotransmitters, growth factors, some hormones) stimulate responses in their target cells by increasing the calcium ion concentration of the cytoplasm.

- The second messenger in this pathway is *inositol triphosphate*.

Inositol triphosphate (IP_3) acts as follows: (See Campbell, Figure 41.7)

- The hormone (first messenger) binds to its receptor, activating a G protein (different from those that function in the cAMP system) that in turn stimulates an enzyme in the plasma membrane called phospholipase C.
- Phospholipase C cleaves a membrane phospholipid into IP_3 and diacylglycerol; both may function as second messengers.
- Diacylglycerol stimulates another membrane enzyme, protein kinase C, which triggers various target cell responses by phosphorylating specific proteins.
- IP_3 released from the plasma membrane stimulates movement of Ca^{2+} from the ER into the cytoplasm.
- The Ca^{2+} regulates cellular protein activity either by itself or by first binding to *calmodulin* which binds to other proteins.
- The signal is thus transmitted to various enzymes and other proteins by the increased concentration of Ca^{2+} in the cytoplasm.
- The target cell then produces a specific response to the original hormonal signal.

IV.   **Invertebrate control systems often integrate endocrine and nervous system functions**

Invertebrates possess a diversity of hormones which function in homeostasis, reproduction, development and behavior.

In many cases, the hormone may stimulate one activity while inhibiting another.
- In Hydra, one hormone stimulates growth and budding while inhibiting sexual reproduction.
- In the sea slug, a peptide hormone secreted by specialized neurons stimulates egg laying while inhibiting feeding and locomotion.

All arthropods have extensive endocrine systems that regulate growth and reproduction, water balance, pigment movement in the integument and eyes, and metabolism.

Hormones may act together such as in the case of insect development and molting.
- *Ecdysone* is secreted by a pair of *prothoracic glands*; it triggers molting and favors development of adult characteristics and metamorphosis.
    - ⇒ Ecdysone stimulates transcription of specific genes, the same mechanism of action as vertebrate steroid hormones.
- *Brain hormone* promotes development by stimulating production of ecdysone.
- Brain hormone and ecdysone are balanced by *juvenile hormone* (JH), which actively promotes retention of larval characteristics; JH is secreted by the *corpora allata*.
- When JH levels decrease, ecdysone-induced molting produces a pupa; in the pupa, adult anatomy replaces larval anatomy during metamorphosis.

V.   **The hypothalamus and pituitary integrate many functions of the vertebrate endocrine system**

Vertebrates possess a number of tissues and organs that secrete hormones. (See Campbell, Figure 41.9 and Table 41.1)
- Some hormones affect most tissues of the body (e.g. sex hormones).
- Some hormones affect only one or a few tissues.
- Some affect other endocrine glands; these are called *tropic hormones*.

A.   **Structural and Functional Relationships of the Hypothalamus and the Pituitary Gland**

The *hypothalamus* is a region of the lower brain that receives information from nerves throughout the body and brain and initiates endocrine signals appropriate to the environmental conditions.
- Contains two sets of *neurosecretory cells* whose secretions are stored in or regulate activity of the *pituitary gland*.

The *pituitary gland* is an extension of the brain located at the base of the hypothalamus. It consists of two lobes and has numerous endocrine functions. (See Campbell, Figure 41.10)

1. The *neurohypophysis*, or posterior pituitary, stores and secretes two peptide hormones which are made by the hypothalamus: *oxytocin* and *antidiuretic hormone*.

   - These hormones act directly on muscles of the uterus and on the kidneys.

2. The *adenohypophysis*, or anterior pituitary, is composed of non-nervous tissues and synthesizes its own hormones.

   - Controlled by two kinds of hormones secreted by neurosecretory cells in the hypothalamus: *releasing hormones* and *inhibiting hormones*.
   - Releasing hormones stimulate the anterior pituitary to secrete its hormones.
   - Inhibiting hormones stop the anterior pituitary from secreting its hormones.

B. **Posterior Pituitary Hormones**

The hypothalamic peptide hormones oxytocin and antidiuretic hormone are stored in and released from the posterior pituitary.

- They are synthesized in the hypothalamus and secreted into the posterior pituitary.

Oxytocin induces uterine muscle contraction and causes the mammary glands to eject milk during nursing.

Antidiuretic hormone (ADH) is part of the feedback mechanism that helps regulate blood osmolarity.

- It acts on the kidneys to increase water retention which results in a decrease in urine volume.
- *Osmoreceptors* (specialized nerve cells) in the hypothalamus monitor blood osmolarity.
- When plasma osmolarity increases, the osmoreceptors shrink slightly (lose water by osmosis) and transmit nerve impulses to certain hypothalamic neurosecretory cells.
- These neurosecretory cells respond by releasing ADH into the general circulation from their tips in the posterior pituitary.
- The target cells for ADH are cells lining the collecting ducts of nephrons in the kidneys.
- ADH binds to receptors on the target cells and activates a signal-transduction pathway that increases the water permeability of the collecting duct.
- Water retention is increased as water exits the collecting ducts and enters nearby capillaries.
- The osmoreceptors also stimulate a thirst drive.
  ⇒ Drinking water reduces blood osmolarity to the set point.
- As more dilute blood (lower osmolarity) arrives at the brain, the hypothalamus responds by reducing ADH secretion and lowering thirst sensations.
  ⇒ This prevents over compensation by stopping hormone secretion and quenching thirst.

Note that this negative feedback scheme includes a hormonal action and a behavioral response (drinking).

C.   Anterior Pituitary Hormones

The *adenohypophysis* produces many different hormones and is regulated by releasing factors from the hypothalamus.
- Four are tropic hormones that stimulate other endocrine glands to synthesize and release their hormones.

Growth hormone (GH) is a protein hormone which affects a wide variety of tissues.
- It promotes growth of some tissues directly *and* promotes growth of others indirectly by stimulating the production of growth factors by other tissues.
- For example, GH stimulates the liver to secrete *insulin-like growth factors* (IGFs) which stimulate bone and cartilage growth.

Prolactin (PRL) is a protein hormone similar in structure to GH although their physiological roles are very different.
- PRL produces a diversity of effects in different vertebrates:
  ⇒ It stimulates mammary gland development and milk synthesis in mammals.
  ⇒ Regulates fat metabolism and reproduction in birds.
  ⇒ Delays metamorphosis and may function as a larval growth hormone in amphibians.
  ⇒ Regulates salt and water balance in freshwater fish.

Follicle-stimulating hormone (FSH) is a tropic hormone which affects the gonads (= gonadotropin).
- In males, it is necessary for spermatogenesis.
- In females, it stimulates ovarian follicle growth.

Luteinizing hormone (LH) is another gonadotropin.
- It stimulates ovulation and corpus luteum formation in females and spermatogenesis in males.

Thyroid-stimulating hormone is a tropic hormone which stimulates the thyroid gland to produce and secrete its own hormones.

The remaining hormones from the anterior pituitary are formed by the cleaving of a single large protein, *pro-opiomelanocortin*, into short fragments.
- At least three of these fragments become active peptide hormones:
  ⇒ *Adrenocorticotropin (ACTH)* stimulates the adrenal cortex to produce and secrete its steroid hormones.
  ⇒ *Melanocyte-stimulating hormone (MSH)* regulates the activity of pigment-containing skin cells in some vertebrates.
  ⇒ *Endorphins* inhibit pain perception.

D.   Hypothalamic Hormones

Regulation of the anterior pituitary by the hypothalamus involves releasing hormones produced by neurosecretory cells in the hypothalamus into capillaries.

- The releasing hormones are secreted into capillaries that drain into short portal vessels that subdivide into a second capillary bed within the anterior pituitary.

- These hormones then travel through the second capillary network within the anterior pituitary, where they stimulate or inhibit release of specific hormones by pituitary cells.

- Every hormone produced by the anterior pituitary is controlled by at least one releasing hormone; some have both a releasing and an inhibiting hormone.

VI.   The vertebrate endocrine system coordinates homeostasis and regulates growth, development, and reproduction

There are several endocrine glands that play major roles in the control of homeostasis, growth, development, and reproductive functions of vertebrates.

A.   The Thyroid Gland

The *thyroid gland* consists of two lobes located on the ventral surface of the trachea in mammals, and on the two sides of the pharynx in other vertebrates.

- Produces two hormones, T_3 (triiodothyronine) and T_4 (thyroxine or tetraiodothyronine) derived from the amino acid tyrosine.

    $\Rightarrow$ These differ in structure by only one iodine atom.

- Both have the same effects on their target although T_3 is usually more reactive than T_4 in mammals.

The thyroid gland plays a major role in vertebrate development and maturation.

- Thyroid hormones control metamorphosis in amphibians.

- Normal function of bone-forming cells and the branching of nerve cells during embryonic brain development of nonhuman animals also requires the presence of the thyroid hormones.

The thyroid gland is critical for maintaining homeostasis in mammals.

- Help maintain normal blood pressure, heart rate, muscle tone, digestion, and reproductive functions.

- T_3 and T_4 tend to increase the rate of oxygen consumption and cellular metabolism.

- Serious metabolic disorders can result from a deficiency or excess of thyroid hormones.

    $\Rightarrow$ *Hyperthyroidism*, excessive secretion of thyroid hormones, causes high body temperature, sweating, weight loss, irritability and high blood pressure.

    $\Rightarrow$ *Hypothyroidism*, low secretion of thyroid hormones, can cause cretinism in infants and weight gain, lethargy and cold-intolerance in adults.

    $\Rightarrow$ *Goiter* (enlarged thyroid) is caused by a dietary iodine deficiency and thyroid hormones cannot be synthesized.

Thyroid hormone secretion is regulated by the hypothalamus and pituitary through a negative feedback system.

- The hypothalamus secretes TRH (TSH-releasing hormone) which stimulates TSH (thyroid-stimulating hormone) secretion by the anterior pituitary.
- When TSH binds to receptors in the thyroid, cAMP is generated and triggers release of T_3 and T_4.
- High levels of T_3, T_4, and TSH inhibits TRH secretion.

The thyroid gland in mammals also produces and secretes *calcitonin* which is a peptide hormone that lowers blood calcium levels.

B. The Parathyroid Glands

Four parathyroid glands are embedded in the surface of the thyroid and function in the homeostasis of calcium ions.

- They secrete *PTH* (*parathyroid hormone*) which raises blood Ca^{2+} levels (antagonistic to calcitonin) and needs vitamin D to function.
- PTH stimulates Ca^{2+} reabsorption in the kidney and induces osteoclasts to decompose bone and release Ca^{2+} into the blood.

PTH and calcitonin (which have opposite affects) work in an antagonistic manner and their balance in the body maintains the proper blood calcium levels.

C. The Pancreas

The pancreas is composed primarily of exocrine tissue which produces digestive enzymes; however, scattered among the exocrine tissue are clusters of endocrine cells called the *islets of Langerhans.*

- Each islet is composed of *alpha* (α) *cells*, which secrete the peptide hormone *glucagon*, and *beta* (β) *cells* which secrete the hormone *insulin*.

Glucagon and insulin work together in an antagonistic manner to regulate the concentration of glucose in the blood. (See Campbell, Figure 41.13)

- Blood glucose levels must remain near 90 mg/mL in humans for proper body function.
- At glucose levels above the set point, insulin is secreted and lowers blood glucose concentration by stimulating body cells to take up glucose from the blood.
    - $\Rightarrow$ It also slows glycogen breakdown in the liver and inhibits the conversion of amino acids and fatty acids to sugar.
- When blood glucose levels drop below the set point, glucagon is secreted and increases blood glucose concentrations by stimulating the liver to increase the hydrolysis of glycogen, convert amino acids and fatty acids to glucose, and slowly release glucose into the blood.

Glucose homeostasis is critical due to its function as the major fuel for cellular respiration and a key source of carbon for the synthesis of other organic compounds.

Serious conditions can result when glucose homeostasis is unbalanced. *Diabetes mellitus* is caused by a deficiency of insulin or a loss of response to insulin in target tissues. Diabetes occurs in two forms:

- *Type I diabetes mellitus* is an autoimmune disorder that appears in juveniles and is insulin dependent.
  - ⇒ Results from an immune system attack on the cells of the pancreas.
  - ⇒ Usually occurs suddenly during childhood and destroys the ability of the pancreas to produce insulin.
  - ⇒ Treated by insulin injections several times each day.
- *Type II diabetes mellitus* occurs most frequently in adults over 40.
  - ⇒ May be due to an insulin deficiency but more commonly results from reduced responsiveness in target cells because of changes in insulin receptors.
  - ⇒ Is non-insulin-dependent and can be treated with exercise and dietary controls.

Both types of diabetes mellitus will result in high blood sugar concentrations if untreated.

- Kidneys excrete glucose, resulting in higher concentrations in the urine.
- More water is excreted due to the high concentration of glucose (results in the symptoms of copious urine production accompanied by thirst).
- Fat must serve as the major fuel source for cellular respiration since glucose does not enter the cells.
- In severe cases, acidic metabolites formed during fat metabolism may lower the blood pH to a critical level.

D.   The Adrenal Glands

Adrenal glands are located adjacent to kidneys.

- In mammals, each gland has an outer *adrenal cortex* and inner *adrenal medulla* which are composed of different cell types, have different functions, and are of different embryonic origin.
- Different arrangements of these same tissues are found in other vertebrates.

The adrenal medulla synthesizes and secretes *catecholamines* (*epinephrine* and *norepinephrine*).

- Catecholamines are secreted in times of stress when nerve cells excited by stressful stimuli release the neurotransmitter acetylcholine in the medulla. Acetylcholine combines with cell receptors, stimulating release of epinephrine.
  - ⇒ Norepinephrine is released independently of epinephrine.
- Epinephrine and norepinephrine released into the blood results in rapid and dramatic effects on several targets:
  - ⇒ Glucose is mobilized in skeletal muscle and liver cells.
  - ⇒ Fatty acid release from fat cells is stimulated (may serve as extra energy sources).

⇒ The rate and stroke volume of the heartbeat is increased.

⇒ Blood is shunted away from the skin, gut, and kidneys to the heart, brain, and skeletal muscles by stimulation of smooth muscle contraction in some blood vessels and relaxation in others.

⇒ Oxygen delivery to the body cells is increased by dilation of bronchioles in the lungs.

The adrenal cortex synthesizes and secretes *corticosteroids*. In humans the two primary types are: *glucocorticoids* (cortisol) and *mineralocorticoids* (aldosterone).

- Stressful stimuli cause the hypothalamus to secrete a releasing hormone that stimulates release of ACTH from the anterior pituitary.

  ⇒ ACTH stimulates release of corticosteroids from the adrenal cortex.

- *Glucocorticoids* promote glucose synthesis from noncarbohydrate substances such as proteins.

  ⇒ Skeletal muscle proteins are broken down and the carbon skeletons transported to the liver and kidneys.

  ⇒ The liver and kidneys convert the carbon to glucose which is released into the blood to increase the fuel supply.

  ⇒ Also have immunosuppressive effects and are used to treat inflammation.

- *Mineralocorticoids* affect salt and water balance.

  ⇒ Aldosterone stimulates kidney cells to reabsorb sodium ions and water from the filtrate.

  ⇒ This raises blood volume and blood pressure.

- Aldosterone from the RAAS (renin-angiotensin-aldosterone system), ADH, and ANF (atrial natriuretic factor) from the heart form a regulatory complex which influences the kidneys ability to maintain the blood's ion and water concentrations.

  ⇒ The RAAS regulates aldosterone secretion in response to changes in plasma ion concentrations.

  ⇒ Severe stress also stimulates aldosterone secretion by causing the hypothalamus to secrete releasing hormones that increase ACTH secretion from the anterior pituitary that increases aldosterone secretion by the adrenal cortex.

Current evidence indicates glucocorticoids and mineralocorticoids are important to maintaining body homeostasis during extended periods of stress.

A third group of corticosteroids are sex hormones.

- Include androgens (male hormones) similar to testosterone and small amounts of estrogens and progesterone (female hormones).
- The roles of adrenal sex hormones are not well understood.

E.   The Gonads

The testes of males and ovaries of females produce steroid hormones that affect growth and development as well as regulate reproductive cycles and behaviors.

The gonads of both males and females produce all three categories of gonadal steroids (androgens, estrogens, and progestins) although the proportions differ.

Androgens generally stimulate the development and maintenance of the male reproductive system.
- Produced in greater quantities in males than females.
- Primary androgen is *testosterone*.
- Androgens produced during early embryonic development determine whether the fetus will be male or female.
- High androgen concentrations at puberty stimulate development of male secondary sex characteristics.

Estrogens perform the same functions in females as androgens do in males.
- Estradiol is the primary estrogen produced.
- Maintain the female system and stimulate development of female secondary sex characteristics.

Progestins are primarily involved with preparing and maintaining the uterus for reproduction in mammals.
- Include progesterone.

The gonadotropins from the anterior pituitary (FSH and LH) control the synthesis of both androgens and estrogens.
- FSH and LH are in turn controlled by gonadotropin-releasing hormone (GnRH) from the hypothalamus.

F.   Other Endocrine Organs

Many organs with primarily nonendocrine functions also secrete hormones.
- The digestive tract secretes at least eight hormones (e.g. gastrin, secretin).
- Kidneys secrete *erythropoietin* which stimulates red blood cell production.
- Two other important endocrine glands are the *pineal gland* and the *thymus*.

The *pineal gland* is a small mass of tissue near the center of the mammalian brain although it is closer to the surface in some other vertebrates.
- The pineal contains light sensitive cells or has nervous connections with the eyes.
- It secretes *melatonin*, which is a modified amino acid.
- Melatonin regulates functions related to light and to seasons marked by changes in day length, such as biological rhythms associated with reproduction.
- Melatonin is secreted only at night; larger amounts are thus secreted during the winter than the summer.
    ⇒ The role of melatonin in regulating biological rhythms is not yet understood.

The *thymus* is located under the breast-bone in humans.
- It is large in children and reduced in adults.
- It secretes several messengers including *thymosin* which stimulates T lymphocyte development and differentiation after they leave the thymus.

VII. The endocrine system and the nervous system are structurally, chemically, and functionally related

The endocrine system and nervous system interact and are often inseparable. Relationships between the two systems may be: *structural*, *chemical*, or *functional*.

Structural relationships result from the fact that many endocrine glands are either made of nerve tissue or evolved from nervous tissue.
- The hypothalamus and posterior pituitary are nervous tissues that secrete hormones into the blood.
- The adrenal medulla evolved from the same cells that produce certain ganglia of the nervous system.

Chemical relationships center around those vertebrate hormones that are used as signals by both systems.
- Norepinephrine functions as both an adrenal hormone and as a neurotransmitter.

Functional relationships relate to the coordination of physiological processes by both nervous and hormonal components arranged in series.
- Each system can affect the output of the other.

REFERENCES

Campbell, N. *Biology*. 4th ed. Menlo Park, California: Benjamin/Cummings, 1996.

Marieb, E.N. *Human Anatomy and Physiology*. 3rd ed. Redwood City, California: Benjamin/Cummings, 1995.

CHAPTER **42**

ANIMAL REPRODUCTION

OUTLINE

OBJECTIVES

After reading this chapter and attending lecture, the student should be able to:

1. Distinguish between asexual and sexual reproduction.
2. List and describe four forms of asexual reproduction.
3. Explain how asexual reproduction may be advantageous for a population of organisms living in a stable, favorable environment.
4. Explain the advantages of sexual reproduction.
5. Explain the importance of reproductive cycles.
6. Distinguish among parthenogenesis, hermaphroditism, and sequential hermaphroditism.
7. Describe three mechanisms which increase the probability of successful fertilization that are found in organisms which use external fertilization.
8. List and describe the various methods of parental care by animals.
9. Using a diagram, identify and give the function of each part of the reproductive systems of an insect and a platyhelminth.
10. Using a diagram, identify and give the function of each component of the reproductive system of the human male.
11. Using a diagram, identify and give the function of each component of the human female reproductive system.
12. Discuss the hormonal control of reproduction in male mammals.
13. Explain the differences between menstrual cycles and estrous cycles.
14. Discuss the hormonal control of reproduction in female mammals.
15. Explain how the menstrual cycle and ovarian cycle are synchronized in female mammals.
16. Describe spermatogenesis.
17. Describe oogenesis.
18. Compare and contrast oogenesis and spermatogenesis.
19. Describe the hormonal changes which occur at puberty in humans.
20. Describe the four phases of the sexual response cycle.
21. Describe the changes which occur in the developing embryo and the mother during each trimester of a human pregnancy.
22. Describe the hormonal control of a pregnancy in a human female.
23. Explain the possible mechanisms which prevent the mother's immune system from rejecting the developing embryo.
24. List the various methods of contraception and explain how they work.
25. Explain how technological advancements are used to study human reproductive problems.

KEY TERMS

asexual reproduction
clone
budding
fragmentation
gemmules
regeneration
sexual reproduction
gametes
zygote
ovum
spermatozoon
parthenogenesis
hermaphroditism
sequential
   hermaphroditism
protogynous
protandrous
external fertilization
internal fertilization
gonads
spermatheca
cloaca
uterus
penes
scrotum
penis
glans penis
prepuce

testes
seminiferous tubules
interstitial cells
epididymis
vas deferens
ejaculatory duct
urethra
seminal vesicles
prostaglandin
semen
prostate gland
bulbourethral glands
baculum
ovaries
follicles
ovulation
progesterone
estrogen
oviduct
uterus
endometrium
cervix
vagina
hymen
vestibule
labia minora
labia majora
clitoris

mammary glands
alveoli
primordia
androgens
testosterone
GnRH
LH
FSH
estrous cycle
estrus
menstrual cycle
flow phase
proliferation phase
secretory phase
ovarian cycle
follicular phase
luteal phase
corpus luteum
gametogenesis
spermatogenesis
acrosome
oogenesis
puberty
vasocongestion
myotonia
excitement phase
plateau phase
orgasm

resolution phase
pregnancy (gestation)
conception
cleavage
blastocyst
placenta
organogenesis
fetus
embryo
human chorionic
   gonadotropin (HCG)
labor
birth (parturition)
lactation
contraception
barrier methods
rhythm method
intrauterine
   device (IUD)
chemical
   contraception
tubal ligation
vasectomy
ultrasound imaging
amniocentesis
chorionic villi
   sampling
in vitro fertilization

LECTURE NOTES

An individual organism exists for only a length of time, its lifespan.

- For a species to remain viable, its members must reproduce.
- Only through reproduction can extinction be avoided.

I. Animals produce offspring by asexual or sexual reproduction

Two principal modes of reproduction are found in animals: *asexual reproduction* and *sexual reproduction*.

- Asexual reproduction is the production of offspring whose genes come from one parent without the fusion of egg and sperm.
    - ⇒ Relies completely on mitotic cell division.
- Sexual reproduction is the production of offspring by the fusion of *gametes* (egg and sperm) to form a diploid *zygote*.
    - ⇒ Usually involves two parents, each contributes genes to the offspring.
    - ⇒ The gametes are formed by meiosis.

A. Asexual Reproduction

There are various types of asexual reproduction found in invertebrates: fission, budding, the release of specialized cells and regeneration, and fragmentation.

- *Fission* involves the separation of a parent into two or more individuals of approximately equal size.
- *Budding* occurs when a new individual splits off from an existing one.
    - ⇒ In cnidarians, new individuals grow out from the parental body.
    - ⇒ Offspring may detach from the parent or remain joined to form extensive colonies.
- Some invertebrates release specialized groups of cells that grow into new adults.
    - ⇒ Freshwater sponges produce *gemmules* which form from aggregates of several types of cells surrounded by a protective coat.
- *Fragmentation* is the breaking of the body into several pieces, each of which develops into a complete adult.
    - ⇒ Must be accompanied by regeneration – the regrowth of lost body parts.
    - ⇒ Found in sponges, cnidarians, polychaete annelids, and tunicates.
    - ⇒ Regeneration allows many animals to replace lost parts following an injury. (When the arm of a sea star is removed, it will grow a new arm).

Several advantages are associated with asexual reproduction:

- Sessile animals that cannot seek mates or more active animals in conditions of low population density can reproduce without finding a mate.
- Allows production of many offspring in a short time.
- Most advantageous in stable, favorable environments because it perpetuates successful genotypes.

B. Sexual Reproduction

In sexual reproduction, two individuals produce offspring having a combination of genes inherited from both parents.

- The female gamete (*ovum*) is usually larger and nonmotile.
- The male gamete (*spermatozoon*) is usually a smaller, flagellated cell (except in arthropods which produce nonmotile sperm).
- Increases genetic variability, it is thus advantageous in a fluctuating environment.

C. Reproductive Cycles and Patterns

Most animals have reproductive cycles, often related to changing seasons.

- Allows conservation of resources; so animals reproduce when energy is above that needed for maintenance.
- Allows reproduction when environmental conditions favor offspring survival.
- Cycles are controlled by a combination of hormonal and environmental cues.

Animals employ various patterns of reproduction and may use either sexual or asexual reproduction exclusively or alternate between the two.

- In animals that employ both mechanisms, asexual reproduction often occurs under favorable conditions and sexual reproduction during times of environmental stress.

Environmental conditions will stimulate *parthenogenesis* in aphids, rotifers, some freshwater crustaceans, and other animals.

Parthenogenesis = Development of an egg without fertilization.

- *Daphnia*, a freshwater crustacean, will switch from sexual to asexual reproduction depending on the season and environmental conditions.
    - ⇒ Each female can produce two types of eggs, one type is fertilized, the other develops by parthenogenesis.
    - ⇒ Adults developing by parthenogenesis are often haploid and their cells do not undergo meiosis when forming new eggs.
- Plays a role in the social organization of some insects; male honey bees (drones) are produced parthenogenetically, females develop from fertilized eggs.
- Some fish, amphibians and lizards reproduce via a more complex form of parthenogenesis which involves a doubling of chromosomes after meiosis to create diploid "zygotes".

Hermaphroditism is a solution found in many animals which may have difficulty finding a member of the opposite sex.

- Each individual has both functional male and female reproductive parts.
- Solves the problem of finding a mate of the opposite sex for some sessile, burrowing, and parasitic animals.
- Some self-fertilize, but most mate with another; each donates and receives sperm, thus potentially producing twice as many offspring from one mating.

Sequential hermaphroditism occurs in some species.
- An individual reverses its sex during its lifetime.
- Some species are *protogynous* (female first); others are *protandrous* (male first).
- Reversal often associated with age and size.
    ⇒ In protogynous species such as the blue-head wrasse, the largest (=oldest) fish in the harem becomes a male and defends the harem.
    ⇒ Some oysters are protandrous, the largest individuals become females and can produce more egg cells.

II. In sexual reproduction, gametes unite in the external environment or within the female

Two major mechanisms of *fertilization* (the union of sperm and egg) have evolved in animals: *external fertilization* and *internal fertilization*. Each has specific environmental and behavioral requirements.

In *external fertilization*, eggs are shed by a female and fertilized by a male in the environment.
- Occurs almost exclusively in moist habitats where development can occur without desiccation or heat stress.
- Some aquatic invertebrates release their eggs and sperm into the surrounding water with no contact occurring between the parents.
- Environmental cues and pheromones trigger release of mature gametes in close proximity which increases the probability of successful fertilization.
- In vertebrates exhibiting external fertilization (fishes and amphibians), courtship behavior increases the probability of successful fertilization and permits mate selection.

Internal fertilization occurs when sperm are deposited in or near the female reproductive tract and fertilization occurs within the female's body.
- Usually requires more sophisticated reproductive systems and cooperative mating behaviors.
- Copulatory organs for sperm delivery and receptacles for sperm storage and transport must be present.
- Mating behaviors must include specific reproductive signals for copulation to occur.

A. Protection of the Embryo

Once fertilization occurs, the zygote undergoes embryonic development. The degree to which these developing embryos are protected, varies with the mechanism of fertilization and the organisms involved.

Organisms using external fertilization do not have protective coverings for the embryos.

- Fish and amphibian eggs are covered with a gelatinous coat which permits the free exchange of gases and water.
- The water or moist habitat into which the gametes are released prevent desiccation and temperature stress.
- Very large numbers of zygotes are usually formed although only a small proportion survive to complete development.

Many types of protection are found in those animals which use internal fertilization.

- Eggs resistant to harsh environments are produced by many reptiles, all birds, and monotremes. The amniotic egg of these animals has a protective shell of protein (reptiles and monotremes) or calcium (birds) which makes them resistant to water loss and physical damage.
- Mammals (other than monotremes) do not produce a shelled egg, but possess reproductive tracts that permit embryonic development within the female parent.
  ⇒ In marsupials, the embryo develops for a short time in the uterus, then moves to the mother's pouch to complete development.
  ⇒ Placental mammals retain the embryo in the uterus until development is completed.
- Parental protection is also important to survival of the developing embryo.
  ⇒ While some organisms using external fertilization show forms of parental care (nesting fish will guard the eggs against predators), parental care is most highly developed in animals using internal fertilization.
- Internal fertilization usually produces fewer zygotes than external, but survival through development is much greater due to the protection and care of the developing offspring.

III. Diverse reproductive systems have evolved in the animal kingdom

The complexity of the reproductive system is *not* necessarily related to the phylogenetic position of the animal.

- Only requires that the animal possess the capability to produce and deliver gametes to the gametes of the opposite sex.

A. Invertebrate Reproductive Systems

Diverse systems have evolved among invertebrates; they range from simple systems without distinct gonads to very complex systems.

Most polychaetes have separate sexes without distinct gonads.

- Gametes develop from undifferentiated cells lining the coelom and fill the coelom as they mature.
- In some species they are released through excretory openings; in others the parental body splits open to release the eggs to the environment as the parent dies.

Insects have separate sexes and complex reproductive structures. (See Campbell, Figure 42.6)

- In males, sperm develop in a pair of testes, pass through a coiled duct to and are stored in seminal vesicles.

  ⇒ At copulation, the sperm are ejaculated into the female's system.

- In females, eggs develop in a pair of ovaries and pass through ducts to the vagina where they are fertilized.

  ⇒ May have a *spermatheca*, a blind-ended sac in which sperm may be stored for a year or more.

Flatworms are hermaphroditic with very complex reproductive systems. (See Campbell, Figure 42.7)

- The female system includes an ovary, oviduct, seminal receptacle, yolk, yolk glands, and uterus.

- The male system includes testes, vas efferentia, sperm duct, seminal vesicle, and a copulatory apparatus.

  ⇒ Copulation may occur in a variety of ways in different groups.

B. Vertebrate Reproductive Systems

Vertebrate reproductive systems are all similar, but some important differences are found:

- Most mammals have separate openings for digestive, excretory, and reproductive tracts while many nonmammalian vertebrates have only a common opening, the *cloaca*.

- Some mammals, birds and snakes have a uterus with only one branch, while most other vertebrates have a uterus that is partly or completely divided into two chambers.

- Nonmammalian vertebrates do not have well-developed *penises* and use other mechanisms to transfer spermatozoa.

IV. Human reproduction involves intricate anatomy and complex behavior

A. Reproductive Anatomy of the Human Male

The *human male reproductive system* includes the external *genitalia* and the internal reproductive organs.

- The external genitalia includes the *scrotum* and *penis*.

- The internal reproductive organs consist of the gonads (*testes*), accessory glands, and associated ducts.

The male genitalia (*scrotum* and *penis*) aid the reproductive process in different ways.

- Testes develop in the abdomen and descend into the scrotum just before birth. This is important since sperm cannot develop at normal body temperature.

- By having the testes hang outside the abdominal cavity in the scrotum, the temperature is 2°C lower and sperm production can occur.

- The penis serves as the male copulatory organ. The ejaculatory duct joins the urethra (from the excretory system) which opens at the tip of the penis.
- The movement of semen through the urethra during copulation results in the sperm being deposited directly in the female system.

Internal male reproductive organs are the gonads and associated ducts: (See Campbell, Figure 42.8)

- The *testes* are comprised of highly coiled tubes surrounded by layers of connective tissue.
    ⇒ The tubules, or *seminiferous tubules*, are where sperm form.
    ⇒ *Interstitial cells* are scattered between the tubules and produce testosterone and other androgens.
- Sperm pass from the seminiferous tubules into the tubules of the *epididymis*.
    ⇒ These coiled tubules are where sperm are stored and mature (gain motility and fertilizing power).
- At *ejaculation*, sperm are forced through the *vas deferens* which is a muscular duct running from epididymis to the *ejaculatory duct*.
    ⇒ The ejaculatory duct forms by the joining of the two vas deferens ducts (one from each testis) with the duct from the seminal vesicles.
    ⇒ The ejaculatory duct and opens into the urethra.
- The *urethra* is the tube that runs through the penis and drains both the excretory and reproductive systems.

There are three sets of accessory glands associated with the male system. These glands add their secretions to the semen.

- A pair of *seminal vesicles* is located below and behind the bladder and empty into the ejaculatory duct.
    ⇒ They secrete a fluid containing mucus, amino acids (causes semen to coagulate after deposited in female), fructose (provides energy for sperm) and prostaglandins (stimulate female uterine contractions to help move semen to the uterus).
    ⇒ Seminal vesicle secretions make up about 60% of the total semen volume.
- The *prostate gland* is a large gland that surrounds the upper portion of, and empties directly into, the urethra.
    ⇒ It secretes a thin, milky alkaline fluid that contains several enzymes.
    ⇒ Prostatic fluid balances the acidity of residual urine in the male system and the acidity of the vagina and helps activate sperm.
- The *bulbourethral glands* are a pair of small glands below the prostate that empty into the urethra at the base of the penis.
    ⇒ Secrete a viscous fluid before sperm ejaculation.
    ⇒ The fluid's function is unclear but it may neutralized any acidic urine remaining in the urethra.

The human penis is composed of 3 cylinders of spongy erectile tissue that fill with blood during sexual arousal.

- Rodents, raccoons, walruses and some other mammals also possess a *baculum* (a bone that helps stiffen the penis).
- The head of the penis, the *glans penis*, is covered by *prepuce*, or foreskin.

B. Reproductive Anatomy of the Human Female

The *human female reproductive system* is more complicated than that of the male; it possesses structures not only for the production of female gametes, but also to house the embryo and fetus.

- The internal reproductive organs are the gonads (*ovaries*) and associated ducts and chambers which are involved with gamete movement and embryo development.
- The external genitalia include the *clitoris* and the two sets of *labia* that surround the clitoris and vaginal opening.

The *ovaries* are located in the abdominal cavity and enclosed in a tough protective capsule.

- A mesentery flanks and attaches each ovary to the uterus.
- Each ovary contains many *follicles* (one egg cell surrounded by follicle cells, which nourish and protect the developing egg).
  ⇒ All are formed at birth.
- Follicle cells also produce estrogens.
- Starting at puberty, and continuing to menopause, one follicle matures and releases its egg cell during each menstrual cycle.

During *ovulation*, the egg is expelled from the follicle. The remaining follicular tissue forms the *corpus luteum*, which secretes progesterone (maintains the uterine lining) and additional estrogen. (See Campbell, Figure 42.10)

- If the egg is not fertilized, the corpus luteum degenerates.
- The egg cell is expelled into the abdominal cavity near the opening of the *oviduct*.
- Cilia lining the oviduct draw in the egg cell and convey it to the *uterus*.
- The *uterus* (or womb) is a thick muscular organ shaped like an inverted pear and approximately 7 cm long and 4-5 cm wide.
- The inner uterine lining, the *endometrium*, is richly supplied with blood vessels.

The remaining female reproductive structures are:

Cervix = The neck of the uterus which opens into the vagina.

Vagina = Thin walled chamber that is the repository for semen during copulation and forms the birth canal.

- The *hymen*, a vascularized membrane, usually covers the vaginal opening from birth until ruptured by vigorous physical activity or sexual intercourse.

Vestibule = Chamberlike area formed by the two pairs of skin folds covering the vaginal orifice and urethral opening.

Labia minora = The slender skin folds bordering the vestibule.

Labia majora = A pair of thick, fatty ridges enclosing and protecting the labia minora and vestibule.

Clitoris = Bulb of erectile tissue at the front edge of the vestibule which is covered by a prepuce (small hood of skin).

Bartholin's glands = Small glands located near the vaginal opening that secrete mucus into the vestibule during sexual arousal.

- The mucus facilitates intercourse by lubricating the vagina.

Mammary glands are important to mammalian reproduction, although not actually a part of the reproductive system.

- Consist of small sacs of epithelial tissue that secrete milk.
- The milk drains into a series of ducts that open at the nipple.
- In a nonlactating female mammal, the mammary glands are composed primarily of adipose tissue.

C. **Human Sexual Response**

Human sexuality includes a diversity of stimuli and responses.

- Although variable, human sexual behavior is based on a common physiological pattern, the *sexual response cycle*.
- The sexual responses of males and females show similarities and differences.

Physiological reactions which predominate in both sexes can be divided into two types: *vasocongestion* and *myotonic*.

- Vasocongestion is the filling of a tissue with blood (i.e. penis, clitoris) due to an increased in blood flow through the arteries of that tissue.
- Myotonia is increased muscle tension and both skeletal and smooth muscles may show sustained or rhythmic contractions.

There are four phases in the sexual response cycle in human males and females.

- In the *excitement* phase, the vagina and penis are prepared for *coitus* (sexual intercourse).
  ⇒ Vasocongestion of the penis and clitoris occurs along with enlargement of testes, labia and breasts. Vaginal lubrication and myotonia also occur.
- The *plateau* phase continues the responses of the excitement phase.
  ⇒ The vagina forms a depression to receive the sperm; the outer third becomes vasocongested, the inner third slightly expands and the uterus elevates.
  ⇒ In both sexes, the heart rate rises and breathing rates increase in response to stimulation of the autonomic nervous system.
- *Orgasm* is the third and shortest phase, it is characterized by rhythmic, involuntary contractions in the reproductive systems of both sexes.
  ⇒ Two stages occur in males:
    ◊ *Emission* is the forcing of semen into the urethra due to contraction of the glands and ducts of the reproductive system.
    ◊ *Expulsion* (ejaculation) expels the semen due to contraction of the urethra.

⇒ In females, the uterus and outer vagina contract but the inner two-thirds of the vagina do not.

- The *resolution* phase reverses the responses of earlier phases and completes the cycle.
  ⇒ Vasocongested organs return to normal size and color.
  ⇒ Muscles relax.

V. Spermatogenesis and oogenesis both involve meiosis but differ in three significant ways

Spermatogenesis is the production of mature sperm cells in adult males. (See Campbell, Figure 42.12)

- A continuous process in adult males and may result in 250 to 400 million sperm cells per ejaculate.
- Occurs in seminiferous tubules of the testes.
- Begins with the differentiation of primordial germ cells into spermatogonia in the embryonic testes. Both of these types of cells are diploid.
- The spermatogonia are located near the outer wall of the seminiferous tubules. They increase in numbers through repeated mitoses throughout development and early life.
- When the male matures, spermatogonia begin to differentiate into primary spermatocytes which are diploid.
- Primary spermatocytes will pass through several stages before giving rise to mature spermatozoa.
  ⇒ Each primary spermatocyte undergoes meiosis I to produce two haploid secondary spermatocytes.
  ⇒ Each secondary spermatocyte undergoes meiosis II to form two spermatids.
  ⇒ Thus, each primary spermatocyte forms four haploid spermatids through meiotic division.
- Each spermatid becomes associated with a large Sertoli cell from which it receives nutrients.
- All four spermatids from each primary spermatocyte differentiate into mature spermatozoa.

In a spermatozoon, the thick head containing the haploid nucleus is tipped with the *acrosome*, which contains enzymes to aid in egg penetration. (See Campbell, Figure 42.11)

- Behind the head, the sperm cell contains many mitochondria that provide ATP for movement of the flagellum.

Oogenesis is the development of ova (mature, unfertilized egg cells). (See Campbell, Figure 42.13)

- Begins in the embryo when primordial germ cells undergo mitotic divisions to produce diploid oogonia.
- Each oogonium will develop into a primary oocyte by the time of birth of the female, resulting in all potential ova being present in the ovaries at birth.
- Between birth and puberty, primary oocytes enlarge and their surrounding follicles grow.
  ⇒ They replicate their DNA and enter prophase I and remain there until activated by hormones.
- After puberty, during each ovarian cycle, FSH stimulates a follicle to enlarge and the primary oocyte within completes meiosis I to produce a haploid secondary oocyte and the first polar body.
  ⇒ Meiosis then stops again.
- LH triggers ovulation and the secondary oocyte is released from the follicle.
- If a sperm cell penetrates the secondary oocyte's membrane, meiosis II will occur and the second polar body will separate from the ovum, this completes oogenesis.

The three important differences between spermatogenesis and oogenesis are:

1. In spermatogenesis, all four products of meiosis I and II become mature spermatozoa; in oogenesis, the unequal cytokinesis which occurs during meiosis I and II results in most of the cytoplasm being in one daughter cell which will form the single ovum, the other cells (polar bodies) will degenerate.

2. Spermatogenesis is a continuous process throughout the reproductive life of the male; all potential ova that can be produced via oogenesis will be present as primary oocytes in the ovaries at the time of the female's birth.

3. Spermatogenesis occurs as an uninterrupted sequence; in oogenesis, long "resting" periods occur between the formation of the initial steps and final production of the ovum.

VI. A complex interplay of hormones regulates reproduction

A. The Male Pattern

In males, androgens are directly responsible for formation of primary sex characteristics (reproductive organs) and secondary sex characteristics (deepening of voice, hair growth pattern, muscle growth).

- Androgens are steroid hormones produced primarily by the interstitial tissues of the testes.
- Testosterone is the most important androgen produced.
- Androgens are also potent determinants of sexual and aggressive behaviors.
- GnRH from the hypothalamus stimulates the pituitary to release LH (stimulates androgen production) and FSH (acts on seminiferous tubules to increase sperm production).

B.   The Female Pattern

Hormonal control in females is more complicated and reflects the cyclic nature of female reproduction. Female mammals display two different types of cycles: *estrous cycles* and *menstrual cycles*.

Estrous cycles occur in non-primate mammals.

- Ovulation occurs after the endometrium thickens and vascularizes.
- If pregnancy does not occur, the endometrium is reabsorbed by the uterus.
- Involves pronounced behavioral changes; seasonal and climatic changes effect the estrous cycle more than the menstrual cycle.
- *Estrus* is the period of sexual activity surrounding ovulation and is the only time most mammals will copulate. The length and frequency varies widely among species.

Menstrual cycles occur in humans and many other primates.

- Ovulation occurs after the endometrium thickens and vascularizes (as in the estrous cycle).
- If pregnancy does not occur, the endometrium is shed from the uterus through the cervix and vagina during menstruation.

The term *menstrual cycle* refers to the changes that occur in the uterus during the reproductive cycle.

- In human females, the cycle varies from one woman to another.
- Usually ranges from 20 – 40 days with an average of 28 days.
- Some women have very regular cycles; others vary from one cycle to the next.

There are three phases of the menstrual cycle. (See Campbell, Figure 42.15)

- The *menstrual flow phase* is that time during which most of the endometrium is being lost from the uterus (= menstruation).
    ⇒ Persists only a few days.
    ⇒ The first day of this phase is usually designated day 1 of the cycle.
- The *proliferative phase* lasts for one to two weeks and involves the regeneration and thickening of the endometrium.
    ⇒ The endometrium is thin at the beginning since most of its structure was lost during the preceding phase.
- The *secretory phase* lasts about two weeks and is a time when the endometrium continues to develop.
    ⇒ The endometrium continues to thicken, becomes more vascularized, and develops glands which secrete a glycogen rich fluid.
    ⇒ If an embryo does not implant in the uterine lining by the end of this phase, a new menstrual flow phase begins.

An *ovarian cycle* parallels the menstrual cycle.

- The *follicular phase* begins this cycle and is a time during which several follicles in the ovaries begin to grow.
  - ⇒ The egg cells within the follicles enlarge and the follicle cell coat becomes multilayered.
  - ⇒ Only one of the growing follicles will continue to mature while the others degenerate.
  - ⇒ A fluid-filled cavity develops in the maturing follicle and grows large enough to form a bulge on the surface of the ovary.
- The *ovulatory phase* consists of that time when ovulation occurs.
  - ⇒ The follicle and adjacent wall of the ovary rupture, releasing the egg cell.
- The *luteal phase* begins after ovulation.
  - ⇒ Follicular tissue remaining in the ovary after ovulation forms a *corpus luteum*.
  - ⇒ The corpus luteum is endocrine tissue that secretes female hormones.

Five hormones work together (by positive and negative feedback mechanisms) to coordinate the menstrual and ovarian cycles. This coordination synchronizes follicle growth and ovulation with preparation of the uterine lining for embryo implantation. (See Campbell, Figure 42.15)

- During the follicular phase of the ovarian cycle, *GnRH* secreted by the hypothalamus stimulates the anterior pituitary to secrete small quantities of *FSH* and *LH*.
- FSH stimulates the immature follicles in the ovary to grow and these follicle cells secrete small amounts of *estrogen*.
- As the follicle continues to grow, the amount of estrogen secreted increases. The mature follicle secretes a large amount of estrogen.
- The high concentration of estrogen stimulates the hypothalamus to increase secretion of GnRH which results in a sudden increase in FSH and LH secretion.
  - ⇒ LH increases more than FSH because high estrogen levels increases the sensitivity of LH-releasing mechanisms (in the pituitary) to GnRH.
  - ⇒ The follicles have LH receptors and can respond directly to the hormone.
- The sudden surge in LH concentration stimulates final maturation of the follicle and ovulation (after about 24 hours). The high concentration of LH stimulates the ruptured follicular tissue to transform into the corpus luteum.
- The presence of LH during the luteal phase of the ovarian cycle stimulates the corpus luteum to continue to secrete estrogen, but to also secrete increasing amounts of *progesterone*.
- Increasing concentrations of estrogen and progesterone inhibits GnRH secretion by the hypothalamus resulting in a decrease in FSH and LH.
- As the LH concentration declines, the corpus luteum begins to atrophy; this results in a sudden drop in estrogen and progesterone concentrations.
- Decreasing levels of estrogen and progesterone removes the inhibition exerted on the hypothalamus, which begins to secrete small amounts of GnRH that stimulates the anterior pituitary to secrete low levels of FSH and LH.
- A new follicular phase begins at this point.

Coordination of the menstrual cycle with the ovarian cycle depends primarily on the levels of estrogen and progesterone.

- Increasing amounts of estrogen secreted by growing follicles during the follicular phase stimulate the endometrium lining the uterus to thicken in preparation for the embryo.

  ⇒ Coordination of the follicular phase of ovarian cycle with proliferative phase of the menstrual cycle.

- After ovulation, the estrogen and progesterone secreted by the corpus luteum stimulate continued development and maintenance of the endometrium.

  ⇒ Arteries supplying blood to the endometrium enlarge and the endometrial glands that supply the nutritional fluid to the early embryo grow and mature.

  ⇒ Coordination of the luteal phase of the ovarian cycle with the secretory phase of the menstrual cycle.

- Decreasing concentrations of estrogen and progesterone due to the disintegration of the corpus luteum reduce blood flow to the endometrium.

  ⇒ The endometrium breaks down and passes out of the uterus as the menstrual flow.

- A new menstrual cycle begins with a new ovarian cycle.

- Estrogens are also responsible for development of the female secondary sex characteristics.

C.   Hormones and Sexual Maturation

Mammals can not reproduce until they reach sexual maturity.

- In humans, the onset of reproductive ability is called *puberty*.
- Usually occurs in humans between the ages of 8 and 14.
- The hypothalamus secretes increasing amounts of GnRH, leading to higher FSH and LH levels. These gonadotropins trigger maturation of the reproductive system and development of secondary sex characteristics.
- The indication of puberty is the ejaculation of viable sperm by males and the first menstruation by females.

VII.   Embryonic and fetal development occur during pregnancy in humans and other placental mammals

A.   From Conception to Birth

In placental mammals, *pregnancy (gestation)* is the condition of carrying one or more developing embryos in the uterus.

- It is preceded by *conception* (fertilization of an egg by a spermatozoon) and ends with birth of the offspring.
- Human pregnancy averages 266 days from conception.
- Duration in other species correlates with body size and extent of development of young at birth.

Human gestation is divided into three *trimesters*, each of about 3 months duration.

The *first trimester* is when the most radical changes occur for both the mother and baby.

- Fertilization occurs in the oviduct and *cleavage* (cell division) begins in about 24 hours.
- As cleavage continues, the zygote develops into a ball of cells passing down the oviduct to the uterus.
- The embryo reaches the uterus in 3-4 days and develops into a hollow ball of cells called a *blastocyst*.
  ⇒ This stage develops about one week after fertilization.
- The blastocyst will implant into the endometrium in the next 5 days.
- During implantation, the blastocyst bores into the endometrium which grows over the blastocyst.
  ⇒ For the first 2-4 weeks of development, nutrients are obtained directly from the endometrium.
- Embryonic tissues begin to mingle with the endometrium to form the *placenta*, which functions in respiratory gas exchange, nutrient transfer, and waste removal for the embryo.
  ⇒ Blood from the embryo passes through the umbilical arteries to the placenta and returns through the umbilical vein.
- This is also the main period of *organogenesis* (development of organs).
  ⇒ After eight weeks, the embryo develops into a *fetus* stage and possesses all organs of the adult in rudimentary form.
  ⇒ The fetus is about 5 cm in length by the end of the first trimester.
- During this trimester, the embryo secretes hormones that signal its presence and controls the mother's reproductive system.
- *Human chorionic gonadotropin* (HCG) is an embryonic hormone that maintains progesterone and estrogen secretion by the corpus luteum to prevent menstruation which would end the pregnancy.
- High progesterone levels also stimulate formation of a protective mucous plug in the cervix, growth of the maternal part of the placenta, uterus enlargement, and cessation of ovulation and menstrual cycling.

In the *second trimester*, rapid growth occurs and the fetus is very active.

- The fetus grows rapidly to about 30 cm in length.
- The mother may feel movement during the early part of this trimester.
- HCG declines, the corpus luteum degenerates, hormone levels stabilize and the placenta secretes its own progesterone to maintain the pregnancy.
- The uterus grows sufficiently for the pregnancy to become obvious.

In the *third trimester*, growth is rapid and fetal activity decreases.

- The fetus grows to about 50 cm in length and 3 – 3.5 kg in weight.
- The maternal abdominal organs become compressed and displaced.
- Labor is induced and regulated by interplay among estrogen, oxytocin, and prostaglandins.
  - ⇒ High estrogen levels during the last weeks of pregnancy trigger the formation of oxytocin receptors on the uterus.
  - ⇒ Oxytocin (from the fetus and maternal posterior pituitary) stimulate the smooth muscles of the uterus to contract.
  - ⇒ Oxytocin also stimulates prostaglandin secretion by the placenta (the prostaglandins enhance the muscle contractions).
  - ⇒ The physical and emotional stresses caused by the uterine muscle contractions stimulate secretion of additional oxytocin and prostaglandins.

Parturition (birth) occurs through a series of strong, rhythmic contractions of the uterus usually called *labor*.

- The first stage of labor involves the opening and thinning of the cervix until it is completely dilated.
- The second stage is the expulsion of the baby from the uterus.
  - ⇒ During this time there are continuous strong uterine contractions.
- The last stage is expulsion of the placenta from the uterus.

Mammals are unique among animals due to the *lactation* component of their postnatal care.

- Decreasing levels of progesterone after birth remove the inhibition from the anterior pituitary which allows *prolactin* secretion.
- Prolactin stimulates milk production after 2 - 3 days delay.
- Oxytocin controls the release of milk from the mammary glands.

B. Reproductive Immunology

One function of the immune system is to protect the body from bodies identified as foreign (nonself).

- Although the embryo possesses many chemical markers that are "foreign" to the mother, an immune response is not mounted against the embryo.
  - ⇒ Foreign markers result from the half of the genome inherited from the father.
- Reasons why the mother does not reject the developing embryo are only partly known but include the presence of the trophoblast and a number of hypotheses.

The *trophoblast* is a protective layer that forms a physical barrier preventing the embryo from actually contacting maternal tissues.

- Develops from embryonic cells of the blastocyst and penetrates the endometrium.
- May not be detected as foreign due to production of a chemical signal that stimulates production of specialized type of white blood cells in the uterus.

⇒ These special white blood cells act to suppress other white blood cells by secreting a chemical that blocks the interleukin-2 action required for normal immune responses.

- Some researchers believe this interference with localized immune responses may occur only after the trophoblast has been identified as foreign by uterine white blood cells which have begun an immune response.

⇒ No suppressor cells are produced if the paternal cellular markers are very similar to the maternal markers.

Many women who have multiple miscarriages may be rejecting the embryo as foreign tissue.

- Some researchers have speculated that in some cases, spontaneous abortions are due to a failure to suppress the immune response in the uterus.
- The hypothesis is that the mother's immune response is too weak to trigger suppression and that the continued immune response is sufficient to reject the embryo.

VIII. Contraception prevents pregnancy

There are three major ways to achieve contraception: prevent fertilization by keeping sperm and eggs apart; prevent implantation of the embryo; and prevent the release of mature eggs and sperm from the gonads. (See Campbell, Figure 42.20)

Preventing the egg and sperm from meeting in the female reproductive tract will not permit fertilization.

- The *rhythm method* involves refraining from intercourse when conception is most likely (failure rate = 10-20%).

⇒ Temporary abstinence during the days before and after ovulation is necessary since the egg can survive for 24 – 48 hours in the oviduct and sperm up to 72 hours.

⇒ The time of ovulation may be hard to predict even in women with regular menstrual cycles.

- Condoms (male and female), diaphragms, cervical caps, and contraceptive sponges are physical barriers fitting over the penis (condom) or into the vagina.

⇒ Failure rate < 10%; most effective when used in conjunction with spermicidal foam or jelly.

- Withdrawal is unreliable since sperm may be present in secretions preceding ejaculation.

Preventing implantation of a blastocyst in the uterus can be accomplished by using an *intrauterine device* (*IUD*).

- IUDs are small plastic devices fitted into the uterine cavity.
- Probably works by irritating the endometrium but precise mechanism is unknown.
- May have serious side effects such as persistent bleeding, uterine infections and other complications.

Chemical contraception prevents the release of mature gametes from the gonads.

- The commonly used birth control pills are combination of synthetic estrogen and progestin (similar to progesterone).
- These hormones stop release of GnRH, LH and FSH through a negative feedback mechanism.
  - ⇒ By blocking LH release, progestin prevents ovulation.
  - ⇒ Estrogen inhibits FSH so no follicles develop.
  - ⇒ Failure rates < 1%.
- The *minipill* contains only progestin and may be in the form of an oral pill or an implant.
  - ⇒ Prevents sperm from entering the uterus by altering the woman's cervical mucus.

Sterilization by *tubal ligation* (cutting the oviduct in women) or *vasectomy* (cutting the vas deferens of men) is nearly 100% effective, safe, but difficult to reverse.

If contraceptive methods are unsuccessful, a pregnancy will occur. The pregnancy may be terminated in several ways.

- A *miscarriage* (spontaneous abortion) occurs in as many as one-third of all pregnancies.
  - ⇒ May occur so early that the woman isn't aware of the pregnancy.
- Physician assisted surgical abortions are chosen by about 1.5 million women annually in the United States.
- A drug induced abortion using RU-486 is a nonsurgical method which is effective in the first few weeks of the pregnancy.
  - ⇒ RU-486 is a progesterone analog which blocks progesterone receptors in the uterus.

IX. New technologies offer help for reproductive problems

Technological advances have made it possible to detect problems in the developing fetus.

Ultrasound imaging is a noninvasive procedure in which a scanner emits high-frequency sound waves that are reflected back by tissues of varying densities to form an image of the fetus.

Amniocentesis is an invasive procedure in which a needle is inserted into the amnion and a sample of fluid is withdrawn.

- The fluid contains fetal cells that are cultured and analyzed for chromosomal defects and genetic disorders.
- Often takes weeks to obtain the results.
- 1% risk of spontaneous abortion following procedure.

Chorionic villus sampling is another invasive procedure where a small portion of the chorion (a fetal part of the placenta) is removed, cultured and analyzed for genetic and metabolic disorders.

- 5%-20% risk of spontaneous abortion following procedure.
- Can be performed earlier in pregnancy and results are obtained in days.

Scientific discovery has solved some fertility problems. *In vitro fertilization* can permit women with blocked oviducts to become pregnant.

- Ova are surgically removed from hormonally stimulated follicles, fertilized in petri dishes and the embryo placed in the uterus for implantation.
- Difficult and costly procedure with a success rate of about 1 in every 6 attempts.
- Embryos may be frozen for use after unsuccessful attempts.

REFERENCES

Campbell, N. *Biology*. 4th ed. Menlo Park, California: Benjamin/Cummings, 1996.

Marieb, E.N. *Human Anatomy and Physiology*. 3rd ed. Redwood City, California: Benjamin/Cummings, 1995.

CHAPTER 43

ANIMAL DEVELOPMENT

OUTLINE

I. From egg to organism, an animal's form develops gradually: *the concept of epigenesis*

II. Embryonic development involves cell division, differentiation, and morphogenesis

III. Fertilization activates the egg and brings together the nuclei of sperm and egg
   A. The Acrosomal Reaction
   B. The Cortical Reaction
   C. Activation of the Egg
   D. Fertilization in Mammals

IV. Cleavage partitions the zygote into many smaller cells

V. Gastrulation rearranges the blastula to form a three-layered embryo with a primitive gut

VI. Organogenesis forms the organs of the animal body from the three embryonic germ layers

VII. Amniote embryos develop in a fluid-filled sac within a shell or uterus
   A. Avian Development
   B. Mammalian Development

VIII. The developmental fate of cells depends on cytoplasmic environment, location, and cell-cell interactions
   A. Polarity and the Basic Body Plan
   B. Fate Maps and the Analysis of Cell Lineages
   C. Determination and Restriction of Cellular Potency
   D. Morphogenetic Movements
   E. Induction
   F. Differentiation
   G. Pattern Formation

IX. Pattern formation in *Drosophilia* is controlled by a hierarchy of gene activations
   A. Egg-Polarity Genes
   B. Segmentation Genes
   C. Homeotic Genes

X. Comparisons of genes that control development reveal homology in animals as diverse as flies and mammals

OBJECTIVES

After reading this chapter and attending lecture, the student should be able to:

1. List the two functions of fertilization.
2. Describe the acrosomal reaction and explain how it ensures the gametes are conspecific.
3. Describe the cortical reaction.
4. Explain how the acrosomal and cortical reactions function sequentially to prevent polyspermy.
5. Describe the changes that occur in an activated egg and explain the importance of cytoplasmic materials to egg activation.
6. Explain the importance of embryo polarity during cleavage.
7. Describe the process of gastrulation and explain its importance.
8. List adult structures derived from each of the primary tissue layers.
9. Using diagrams, identify the various stages of embryonic development of an amphibian.
10. Distinguish between meroblastic cleavage and holoblastic cleavage.
11. List and explain the functions of the extraembryonic membranes in bird and reptile eggs.
12. Compare and contrast development in birds and mammals.
13. Explain the relationships among polarity, cytoplasmic determinants and development.
14. Describe how cell extension, contraction and adhesion are involved in shaping the embryo.
15. Explain how interactions among the three primary tissue layers influence organogenesis.
16. Explain the relationship between cytoplasmic cues and cell determination.
17. Describe the importance of cell location and orientation along the three body axes with respect to polarity in the embryo, morphogenetic movements, and pattern formation.
18. Explain how positional cues influence pattern formation.

KEY TERMS

differentiation	animal pole	neural plate	cell lineage analysis
morphogenesis	gray crescent	neural tube	apoptosis
fertilization	morula	somites	totipotent
zygote	blastocoel	neural crest	determination
acrosome	blastula	blastodisc	laminin
acrosomal reaction	meroblastic cleavage	primitive streak	fibronectins
acrosomal vesicle	holoblastic cleavage	extraembryonic	cell adhesion
acrosomal process	syncytium	membranes	molecules (CAMs)
vitelline layer	gastrulation	yolk sac	induction
fast block to polyspermy	gastrula	amnion	organizer
cortical reaction	invagination	chorion	pattern formation
cortical granules	archenteron	allantois	positional information
fertilization membrane	blastopore	blastocyst	egg-polarity genes
slow block to polyspermy	ectoderm	epiblast	maternal-effect genes
capacitation	mesoderm	hypoblast	morphogen
zona pellucida	endoderm	inner cell mass	segmentation genes
cleavage	dorsal lip	trophoblast	gap genes
blastomeres	involution	cytoplasmic determinants	homeotic genes
vegetal pole	yolk plug	fate map	homeobox
radial cleavage	organogenesis	W. Vogt	homeodomain
spiral cleavage	notochord		

LECTURE NOTES

Animals develop throughout their lifetime. Development begins with the changes that form a complete animal from the zygote and continue as progressive changes in form and function.

I. **From egg to organism, an animal's form develops gradually:** *the concept of epigenesis*

Two early views of how animals developed from an egg competed for supporters until modern techniques were developed. One view was termed *preformation* and the other *epigenesis*.

Preformation suggested that the embryo contained all of its descendants as a series of successively smaller embryos within embryos.

- Based on this idea, by dissecting an egg an individual would be able to find smaller and smaller individuals as the dissection continued.
- Was a popular idea as recently as the eighteenth century.

Epigenesis proposed that the form of an embryo gradually emerged from a formless egg.

- Originally proposed by Aristotle.
- Gained support in the nineteenth century as improved microscopy permitted biologists to view embryos as they gradually developed.

Modern biology has found that an organism's development is mostly determined by the zygote's genome and the organization of the egg cell's cytoplasm.

- The heterogenous distribution of messenger RNA, proteins, and other components in the unfertilized egg greatly impacts the development of the embryo in most animals.
- Division of the zygote after fertilization partitions the heterogenous cytoplasm in such a way that nuclei of different embryonic cells are exposed to different cytoplasmic environments.
- These different cytoplasmic environments result in the expression of different genes in different cells.
- This leads to an emergence of inherited traits that is ordered in space and time by mechanisms controlling gene expression.

II. Embryonic development involves cell division, differentiation, and morphogenesis

Embryonic development includes a succession of mitotic divisions which result in the single cell zygote becoming a complete organism consisting of trillions of cells.

- In order for an organism to develop instead of a ball of identical cells, other processes must be involved.
- The three component processes involved in embryonic development are: *cell division*, *differentiation* and *morphogenesis*.

Cell division results in an increase in the number of cells.

- All of the cells result from mitotic divisions beginning with the zygote.

Differentiation is the development of specialized cells that are organized into tissues and organs.

- The changes that result in two cells with the same genes becoming two different types of cells depends upon which genes are expressed.

Morphogenesis includes the physical processes that give shape to an animal's body and organs.

- This involves the movement and arrangement of cells and tissues in the early embryo and produces the characteristic three-dimensional form of the organism.
- Cell division, differentiation and the programmed death of certain cells also are a part of morphogenesis.

Precise regulation of the three developmental processes is necessary for proper development. A breakdown in this regulation can result in developmental disorders. For example,

- A cleft palate results from an improper morphogenesis in which the upper palate fails to close.
- Webbed fingers and toes result from the failure of normal cell death during morphogenesis.

III.   Fertilization activates the egg and brings together the nuclei of sperm and egg

Fertilization is important because:
- It forms a diploid zygote from the haploid sets of chromosomes from two individuals.
- It triggers onset of embryonic development.

A.   The Acrosomal Reaction

The acrosomal reaction is the discharge of hydrolytic enzymes from a vesicle in the acrosome of a sperm cell (based on studies with sea urchins):
- Upon contacting the egg's jelly coat, the acrosomal vesicle in the head of the sperm releases hydrolytic enzymes via exocytosis.
- These enzymes enable an *acrosomal process* to elongate and penetrate the jelly coat. (See Campbell, Figure 43.2)
- A protein coating the tip of the process attaches to specific receptors on the egg's *vitelline layer* (just external to the plasma membrane).

   ⇒ This provides species specificity for fertilization.

- Enzymes of the acrosomal process probably digest vitelline layer materials allowing the tip of the process to contact the egg's plasma membrane.
- The sperm and egg's plasma membranes fuse, allowing the sperm nucleus to enter the egg and causing a depolarization of the plasma membrane that prevents other sperm cells from also uniting with the egg (*fast block to polyspermy*).

B.   The Cortical Reaction

The fusion of the egg and sperm membranes stimulates a series of changes in the egg's cortex known as a *cortical reaction*. (See Campbell, Figure 43.2, step 6)
- A signal transduction pathway involving a G protein is activated by egg-sperm fusion.
- The G-protein stimulates the release of calcium (Ca^{2+}) from the egg cell's endoplasmic reticulum.
- Ca^{2+} acts as a second messenger and results in a change in the egg's *cortical granules* (special vesicles).
- The increase in Ca^{2+} causes the cortical granules to fuse with the plasma membrane and release their contents into the perivitelline space outside the plasma membrane.

   ⇒ The contents of the cortical granules include enzymes that separate the vitelline layer from the plasma membrane.

   ⇒ Mucopolysaccharides produce an osmotic gradient which stimulates osmosis into the perivitelline space causing the area to swell.

   ⇒ The swelling elevates the vitelline layer and other granule enzymes harden it to form the *fertilization* membrane.

- The fertilization membrane prevents entry of additional sperm.
- The fast block to polyspermy no longer functions and is replaced by a *slow block to polyspermy* consisting of the fertilization membrane and other changes to the egg's surface.

C. Activation of the Egg

The sharp rise in cytoplasmic Ca^{2+} concentration also incites metabolic changes that *activates* the egg cell.

- Cellular respiration and protein synthesis rates increase.
- Cytoplasmic pH changes from slightly acidic to slightly alkaline due to H^+ extrusion.
- Activation can be artificially induced by injection of Ca^{2+}.
- The sperm nucleus within the egg swells and merges with the egg nucleus to form the zygote (actual fertilization).
- DNA replication begins and the first division occurs in about 90 minutes.

D. Fertilization in Mammals

The events of internal fertilization in mammals are similar to those of external fertilization discussed for sea urchins but include some important differences.

Capacitation (enhanced sperm function) results from secretion in the female's reproductive tract.

- Alters certain molecules on the surface of sperm cells and increase sperm motility.

The capacitated sperm cell must reach the *zona pellucida* for the process to continue.

- The secondary oocyte (egg) released at ovulation is surrounded by follicle cells released at the same time.
    - ⇒ The sperm cell must migrate through this layer of cells.
- The zona pellucida (extracellular matrix of the egg) is a three dimensional network of cross-linked filaments formed by three different glycoproteins.
    - ⇒ The glycoprotein ZP3 acts as a sperm receptor by binding to a complementary molecule on the surface of the sperm head.
- The binding of ZP3 and its complementary molecule stimulates an acrosomal reaction similar to that described earlier.
    - ⇒ Protein-digesting and other enzymes from the acrosome allow the sperm cell to penetrate the zona pellucida and reach the plasma membrane of the egg.
    - ⇒ The acrosomal reaction also exposes a sperm membrane protein that binds and fuses with the egg membrane.

Binding of the sperm to the egg depolarizes the egg's plasma membrane (a fast block to polyspermy).

- A cortical reaction occurs, however, cortical granule contents stimulate a hardening of the zona pellucida (slow block to polyspermy) instead of raising a fertilization membrane.
- Microvilli from the egg pull the whole sperm cell into the egg cell.
    - ⇒ The sperm cell's flagellar basal body divides and forms the zygote's centrioles.
    - ⇒ Nuclear envelopes disperse and the chromosomes from the gametes share a common spindle apparatus for the first mitotic division of the zygote.
    - ⇒ The chromosomes from the two parents form a common nucleus (offspring's genome) as diploid nuclei form in the daughter cells after the first division.

III. Cleavage partitions the zygote into many smaller cells

The basic body plan of an animal is established in the early stages of postfertilization development by *cleavage*, *gastrulation* and *organogenesis*. The following information is based upon studies of the sea urchin, frogs and *Drosophila*.

Cleavage is a succession of rapid mitotic cell divisions following fertilization that produces a multicellular embryo, the *blastula*.

- During cleavage, the cells undergo the S and M phases of the cell cycle but the G_1 and G_2 phases are virtually skipped.
- Very little gene transcription occurs during cleavage and the embryo does not grow.
- The cytoplasm of the zygote is simply divided into many smaller cells called *blastomeres*, each of which has a nucleus.
- The heterogenous nature of the zygote's cytoplasm results in blastomeres with differing cytoplasmic components.

A definite polarity is shown by the eggs of most animals and the planes of division during cleavage follow a specific pattern relative to the poles of the zygote.

- The polarity results from concentration gradients in the egg of such cellular components as mRNA, proteins, and *yolk* (stored nutrients).
- The yolk gradient is a key factor in determining polarity and influencing the cleavage pattern in frogs and other animals.
- The *vegetal pole* of the egg has the highest concentration of yolk.
- The *animal pole*, opposite the vegetal pole, has the lowest concentration of yolk and is the site where polar bodies are budded from the cell.
- The animal pole also marks the area where the most anterior part of the embryo will form in most animals.

The zygote is composed of two hemispheres named for the respective poles: vegetal and animal.

- The hemispheres in the egg of many frogs have different coloration due to the heterogeneous distribution of cytoplasmic substances.
  ⇒ The animal hemisphere has a gray hue due to the presence of melanin granules in the outer cytoplasm.
  ⇒ The vegetal hemisphere has a light yellow hue due to the yellow yolk.
- The cytoplasm in amphibian eggs is rearranged at fertilization.
  ⇒ The outer cytoplasm rotates toward the point of sperm entry.
  ⇒ This rotation may be due to a reorganization of the cytoskeleton around the centriole introduced by the sperm.
  ⇒ A narrow *gray crescent* is seen due to the exposure of a light-colored region of cytoplasm by the rotation.
  ⇒ The gray crescent is located near the equator of the egg on the side opposite of sperm entry.
  ⇒ The presence of the gray crescent is an early marker of the egg's polarity.

Cleavage in the animal hemisphere of a frog's zygote is more rapid than in the vegetal hemisphere.

- Large amounts of yolk impedes cell division.
- This discrepancy in the rate of cleavage divisions results in a frog embryo with different size cells.
    - ⇒ Animal hemisphere cells are smaller than those in the vegetal pole.
- In sea urchins and many other animals, the blastomeres are about equal in size due to small amounts of yolk.
    - ⇒ Animal-vegetal pole axes are present but are due to concentration gradients of cytoplasmic components other than yolk.
    - ⇒ The absence of yolk permits cleavage divisions to occur at about the same rate.

The first two cleavage divisions in sea urchins and frogs are vertical and divide the embryo into four cells that extend from the animal pole to the vegetal pole. (See Campbell, Figure 43.7)

- The third cleavage plane is horizontal and produces an eight cell embryo with two tiers (animal and vegetal) of four cells each.
- In deuterostomes which have radial cleavage, the upper tier of cells is aligned directly over the lower tier.
- In protostomes which have spiral cleavage, the upper tier of cells align with the grooves between cells of the lower tier.

A continuation of cleavage produces a solid ball of cells called a *morula*.

- The *blastocoel*, a fluid-filled cavity, develops within the morula as cleavage continues which changes the embryo from the solid morula to a hollow ball of cells, the *blastula*.
- In sea urchins, the blastocoel is centrally located in the blastula due to equal cell divisions.
- Unequal cell divisions in the frog embryo produces a blastocoel in the animal hemisphere.

The amount of yolk present in an egg greatly effects the cleavage.

- In eggs with little yolk (sea urchins) or moderate amounts of yolk (frogs), a complete division of the egg occurs and is called *holoblastic cleavage*.
- In eggs which contain large amounts of yolk (birds, reptiles), cleavage is incomplete and confined to a small disc of yolk-free cytoplasm at the animal pole of the egg; this is termed *meroblastic cleavage*.

A unique type of meroblastic cleavage occurs in the yolk-rich eggs of *Drosophila* and other insects.

- The zygote's nucleus is located within a mass of yolk.
- Cleavage begins with the nucleus undergoing a series of mitotic division *without* cytoplasmic divisions.
- The several hundred nuclei produced by these division migrate to the outer margin of the egg.
- Cytoskeletal proteins surround each nucleus and separate them from each other without membranes.
    - ⇒ At this point the embryo is a *syncytium*, a cell containing many nuclei within a common cytoplasm.

- Several more mitotic divisions occur before plasma membranes form around the nuclei.
- The formation of the plasma membranes forms a blastula consisting of about 6000 cells in a single layer surrounding a yolk mass.

V. Gastrulation rearranges the blastula to form a three-layered embryo with a primitive gut

Gastrulation involves an extensive rearrangement of cells which transforms the blastula, a hollow ball of cells, into a three-layered embryo called the *gastrula*.

- A set of common cellular changes is involved in all animals: changes in cell motility; changes in cell shape; and changes in cellular adhesion to other cells and to molecules of the extracellular matrix.
  ⇒ Some cells at or near the surface move to a more interior location.
- Specific details of processes may differ from one animal group to the next.

The three layers produced by gastrulation are embryonic tissues called embryonic germ layers. These three cell layers will eventually develop into all parts of the adult animal.

- The *ectoderm* is the outermost layer of the gastrula.
  ⇒ The nervous system and outer layer of skin in adult animals develop from ectoderm.
- The *endoderm* lines the archenteron.
  ⇒ The lining of the digestive tract and associated organs (i.e. liver, pancreas) develop from endoderm.
- The *mesoderm* partly fills the space between the ectoderm and endoderm.
  ⇒ The kidneys, heart, muscles, inner layer of the skin, and most other organs develop from mesoderm.

Gastrulation in *sea urchins* begins at the vegetal pole. (See Campbell, Figure 43.9)

- The sea urchin blastula consists of a single layer of cells.
- Vegetal pole cells form a flattened plate that buckles inward (= *invagination*).
- Cells near the plate detach and enter the blastocoel as migratory *mesenchyme* cells.
- The invaginated plate undergoes rearrangement to form a deep, narrow pouch, the *archenteron* or primitive gut.
  ⇒ The archenteron opens to the surface through the *blastopore* which will become the anus.
- A second opening forms at the other end of the archenteron and the digestive tube is developing.
- At this point, gastrulation has produced an embryo with a primitive gut and three germ layers (ectoderm, endoderm, mesoderm).
- The sea urchin embryo will develop into a ciliated larva that feeds on bacteria and unicellular algae while drifting near the ocean's surface as plankton.

Gastrulation during *frog development* also results in an embryo with the three embryonic germ layers and an archenteron that opens through a blastopore.

- The mechanics of gastrulation are more complicated than in the sea urchin because of the large, yolk-laden cells in the vegetal hemisphere and the presence of more than one cell layer in the blastula wall. (See Campbell, Figure 43.10)
- A small crease forms on one side of the blastula where the blastopore will eventually form.
  ⇒ Invagination is due to a cluster of cells burrowing inward.
- The invagination produces an external tuck that becomes the *dorsal lip* (upper edge) of the blastopore.
  ⇒ The dorsal lip forms where the gray crescent was located in the zygote.
- *Involution* then occurs. This is a process in which cells on the surface of the embryo roll over the dorsal lip and move into the embryo's interior away from the blastopore.
  ⇒ These cells continue to migrate along the dorsal wall of the blastocoel.
- As involution continues, migrating internal cells become organized into layered mesoderm and endoderm.
  ⇒ The archenteron forms within the endoderm.
- The cell movements of gastrulation produce a three-layered embryo.
- As gastrulation continues, more and more lateral cells involute with the dorsal cells resulting in a wider blastopore lip which eventually forms a complete circle.
- The circular blastopore lip surrounds a group of large, food-laden cells from the vegetal pole called the *yolk plug*.
- Gastrulation is complete with the formation of the three germ layers.
- Ectoderm is formed from those cells remaining on the surface (other than the yolk plug).
  ⇒ Surrounds the layers of mesoderm and endoderm.

VI.   Organogenesis forms the organs of the animal body from the three embryonic layers

The three germ layers that develop during gastrulation will give rise to rudimentary organs through the process of *organogenesis*.

- The first evidence of organ development is morphogenetic changes (folds, splits, condensation of cells) that occur in the layered embryonic tissues.

The neural tube and notochord are the first organs to develop in frogs and other chordates.

- The dorsal mesoderm above the archenteron condenses to form the *notochord* in chordates.
- Ectoderm above the rudimentary notochord thickens to form a *neural plate* that sinks below the embryo's surface and rolls itself into a *neural tube*, which will become the brain and spinal cord. (See Campbell, Figure 43.11)
- The notochord elongates and stretches the embryo lengthwise; it functions as the core around which mesoderm cells that form the vertebrae gather.
- Strips of mesoderm lateral to the notochord condense into blocks of mesodermal cells called *somites* from which will develop the vertebrae and muscles associated with the axial skeleton.

- As organogenesis continues, other organs and tissues develop from the embryonic germ layers.
  - ⇒ Ectoderm also gives rise to epidermis, epidermal glands, inner ear and eye lens.
  - ⇒ Mesoderm also gives rise to the notochord, coelom lining, muscles, skeleton, gonads, kidneys and most of the circulatory system.
  - ⇒ Endoderm forms the digestive tract linings, liver, pancreas and lungs.
- The *neural crest* forms from ectodermal cells which develop along the border where the neural tube breaks off from the ectoderm.
  - ⇒ These cells migrate to other parts of the body and form pigment cells in the skin, some bones and muscles of the skull, teeth, adrenal medulla, and parts of the peripheral nervous system.

The end result of embryonic development in the frog is an aquatic, herbivorous tadpole.

- Metamorphosis will later change this larval stage into a terrestrial, carnivorous adult.

VII. Amniote embryos develop in a fluid-filled sac within a shell or uterus

All vertebrate embryos require an aqueous environment for development.

- Fish and amphibians lay their eggs in water.
- Terrestrial animals live in dry environments and have evolved two solutions to this problem: the *shelled egg* in birds and reptiles and the *uterus* in placental mammals.
- The embryos of reptiles, birds, and mammals develop in a fluid-filled sac, the *amnion*.
- These three classes of animals are referred to as *amniotes* due to the presence of the amnion around the embryo.
- Important differences between cleavage and gastrulation occur in the development of birds and mammals, as well as differences between these amniotes and nonamniotes such as the frog previously discussed.

A. Avian Development

The larger yellow "yolk" of a bird egg is actually the ovum containing a large food reserve properly called yolk.

- Surrounding this large cell is a protein-rich solution (the egg white) that provides additional nutrients during development.
- After fertilization, meroblastic cleavage will be restricted to a small disc of yolk-free cytoplasm at the animal pole.
- Cell division partitions the yolk-free cytoplasm into a cap of cells called the *blastodisc* which rests on large undivided yolk mass.
- The blastomeres sort into an upper layer (epiblast) and lower layer (hypoblast) with a cavity (blastocoel) forming between them.
- Although different in appearance, this blastula is equivalent to the hollow ball stage in the frog.

Gastrulation begins with the movement of some epiblast cells toward the midline of the blastodisc. These cells then detach and move inward toward the yolk.

- The medial and inward movement of these cells produce a groove called the *primitive streak*, in the upper cell layer along the anterior-posterior axis. (See Campbell, Figure 43.12)

- Some cells in the epiblast migrate through the primitive streak and into the blastocoel to form mesoderm; others invade the hypoplast and contribute to endoderm.

- The ectoderm forms from cells remaining in the epiblast.

- Hypoblast cells appear to direct formation of the primitive streak and are necessary for development even though they contribute no cells to the embryo.

   ⇒ Hypoblast cells segregate from the endoderm and form portions of a sac surrounding the yolk and a stalk which connects the embryo and yolk.

- Embryonic disc borders fold downward, come together below the embryo, and form a three-layered embryo attached to the yolk below by a stalk at midbody.

Subsequent organogenesis occurs as in the frog, except that primary germ layers also form four extraembryonic membranes: the *yolk sac, amnion, chorion* and *allantois*. (See Campbell, Figure 43.14)

B.   Mammalian Development

Fertilization occurs in the oviducts of most mammals and the early development occurs while the embryo travels down the oviduct to the uterus.

- The egg of placental mammals stores little nutrients and the zygote displays *holoblastic cleavage*.

- Gastrulation and organogenesis follow a similar pattern to that in birds.

- Cleavage is relatively slow in mammals and the zygote has no apparent polarity.

- Cleavage planes appear randomly oriented and blastomeres are of equal size.

The development of a human embryo can represent mammalian development.

- Cleavage is relatively slow with the first, second and third divisions being completed at 36, 60, and 72 hours respectively.

- At 7 days post-fertilization the embryo consists of about 100 cells arranged around a central cavity forming the *blastocyst*. (See Campbell, Figure 43.15)

- The *inner cell mass* protrudes into one end of the cavity and will develop into the embryo and some of its extraembryonic membranes.

- The *trophoblast* is the outer epithelium surrounding the cavity which will, along with mesodermal tissue, form the fetal part of the placenta.

- During implantation, the inner cell mass forms a flat disc with an epiblast and hypoblast similar to those in birds; the embryo develops from epiblast cells and the yolk sac from hypoblast cells.

The blastocyst stage reaches the uterus and begins to implant.

- The trophoblast layer:
    ⇒ Secretes enzymes that enable blastocyst implantation in the uterus.
    ⇒ Thickens and extends fingerlike projections into the endometrium.
- Gastrulation occurs by inward movement of mesoderm and endoderm through a primitive streak.

Four extraembryonic membranes homologous to those in birds and reptiles form in mammals.

- The chorion forms from the trophoblast and surrounds the embryo and all extraembryonic membranes.
- The amnion forms as a dome above the epiblast and encloses the embryo in a fluid-filled cavity.
- The yolk sac encloses a fluid-filled cavity but no yolk; its membrane is the site of early blood cell formation.
    ⇒ These cells later migrate into the embryo.
- The allantois develops from an outpocketing of the rudimentary gut and is incorporated into the umbilical cord where it forms blood vessels that transport oxygen and nutrients from the placenta to the embryo and waste products from the embryo to the placenta.

Organogenesis begins with formation of the neural tube, notochord, and somites.

- Rudiments of all major organs have developed from the three germ layers by the end of the first trimester in humans. (See Campbell, Table 43.1)

V. The developmental fate of cells depends on cytoplasmic environment, location, and cell-cell interactions

Early in the embryonic development of an animal, a sequence of changes takes place that establishes the basic body plan of that animal.

- These include not only morphogenetic changes which result in characteristic shapes, but also differentiation of many kinds of cells in specific locations.

Two general principles appear to integrate the genetic and cellular mechanisms underlying embryonic development.

1. *The heterogeneous organization of cytoplasm in the unfertilized egg leads to regional differences in the early embryos of many animal species.*
    - Different blastomeres receive different substances (mRNA, proteins, etc.) during cleavage due to the partitioning of the heterogenous cytoplasm of the ovum.
    - Gene expression in, and the developmental fate of, cells in the early embryo are influenced by these local differences in cytoplasmic composition.

2. *Cell-cell interactions compound differences initiated by cytoplasmic organization and by the relocation of cells during morphogenesis.*

- One cell of an embryo may transmit chemical signals to a neighboring cell.
  - ⇒ The chemical signals induce cytoplasmic changes affecting gene expression in that cell.
- Membrane interaction between cells that are in contact may also induce cytoplasmic changes.
  - ⇒ These can affect gene expression.
- Differences between early blastomeres due to cytoplasmic heterogeneity determine the basic features of an embryo.
- Interactions among embryonic cells induce development of many specialized cell types.
- As cells from different regions of the embryo come together due to morphogenetic movements, cell interactions can cause changes in gene expression among neighboring cells.

A. Polarity and the Basic Body Plan

Bilaterally symmetrical animals have an anterior-posterior axis, a dorsal-ventral axis, and left and right sides.

- The first step in morphogenesis is establishment of this body plan which is prerequisite to tissue and organ development.
- In humans and other mammals, the basic polarities do not appear to be established until after cleavage.
- In most animals polarity of the embryo is established in the unfertilized egg or during early cleavage.

In frogs and many other animals, all three axes of the embryo are defined before cleavage begins.

- The animal-vegetal axis of the egg is due to concentrations gradients of cytoplasmic contents and marks the anterior-posterior axis.
- The dorsal-ventral and left-right axes are marked by the gray crescent which forms opposite of the sperm entry point.
- The embryo continues development along all three axes.

B. Fate Maps and the Analysis of Cell Lineages

It is often possible to develop a *fate map* for embryos whose axes are defined early in development.

- The fate map shows which parts of the embryo are derived from each region of the zygote or blastula.
- First accomplished by W. Vogt in the 1920s, using vital dyes to color different regions of the amphibian blastula surface and subsequently sectioning the embryo to see where each color turned up.

Cell lineage analysis is a more detailed form of fate mapping which is possible due to the development of new techniques.

- Modern techniques permit the marking of an individual blastomere during cleavage and following the marker as it is distributed to the mitotic descendants of that cell.
- Using such techniques, it has been possible to map the developmental fate of every cell in the nematode *Caenorhabditis elegans*. (See Campbell, Figure 43.16)

Events crucial to the normal development and growth of *C. elegans* and other animals have been determined through cell lineage analyses.

- One such event is *apoptosis* which is programmed cell death.
- Normal development of *C. elegans* includes 131 episodes of timed cell death.
    ⇒ The genome of *C. elegans* includes two "suicide" genes that code for proteins involved in apoptosis.
    ⇒ Cell death results from toxic proteins produced by the cell or from metabolites that are changed into toxins by proteins.
- These genes are expressed in cells destined to die at precisely the same points in development in each worm's cell lineage.
    ⇒ The cells shrink, their nuclei condense, and they are quickly engulfed and digested by neighboring cells.
- A third gene in *C. elegans* codes for a protein called *ced-9* which suppresses the two suicide genes in all surviving cells.

Normal development and growth in embryos and adults results from the timely activation of an inherent cell-death mechanism.

- Apoptosis is essential in vertebrates for normal development of the nervous system, sculpting of body parts, and normal operations of the immune system.

C. Determination and Restriction of Cellular Potency

The developmental fate of different regions of the embryos of some animals are fixed very early in development by *cytoplasmic determinants*.

- These are substances localized within specific regions of the egg's cytoplasm which leads them to be included in specific blastomeres.
- May control gene expression.

Two blastomeres with equal developmental potential can develop from a zygote with asymmetrically distributed cytoplasmic determinants.

- Requires that the axis of the first cleavage equally distribute these substances.
- In amphibians, a normal first cleavage divides the gray crescent between the two blastomeres.
    ⇒ If the two blastomeres are separated, each will develop into a normal tadpole.
    ⇒ The cells are *totipotent* in that each cell retains the zygote's potential to form all parts of the animal.
    ⇒ Experimental manipulation of the first cleavage plane so that it bypasses the gray crescent produces one cell that will develop into a normal tadpole and one that fails to develop properly. (See Campbell, Figure 43.18)

Only the zygote is totipotent in many species while in others there is a progressive restriction in potency of the cells.

- Only the zygote is totipotent in those where the first cleavage plane divides the cytoplasmic determinants in a way that each blastomere will give rise to only certain parts of the embryo.
- In mammals, cells of the embryo remain totipotent until they become arranged into the trophoblast and inner cell mass of the blastocyst.
- The cells of the early gastrula of some species can still give rise to more than one type of cell even though they have lost their totipotency.
- By the late gastrula stage, the developmental fate of all cells is fixed.

Determination is the progressive restriction of a cell's developmental potential.

- A determined cell is one whose developmental fate can not be changed by moving it to a different location in the embryo.
- Daughter cells receive a developmental commitment from the original cell.
- Involves the cytoplasmic environment's control of the genome of the cell.
- The partitioning of heterogeneous cytoplasm of an egg during cleavage exposes the nuclei of the cells to cytoplasmic determinants that will affect which genes are expressed as the cells begin to differentiate.

D. Morphogenetic Movements

The changes in cell shape and the cell migrations during cleavage, gastrulation and organogenesis are *morphogenetic movements*. These various morphogenetic movements help shape an embryo.

- Cell extension, contraction and adhesion are involved in these movements.
- Changes in shape usually involve reorganization of the cytoskeleton.
- Amoeboid movement based on extension and contraction is an important factor, especially during gastrulation.
- Morphogenetic movements are partially guided by the extracellular matrix.
  ⇒ The matrix contains adhesive substances and fibers that may direct migrating cells. (*Laminin* and *fibronectins* are extracellular glycoproteins that help cells adhere to their substratum as they migrate.)
- Nonmigrant cells along the pathways may promote or inhibit migration depending on the substances they secrete.
- Migrating cells receive direction cues, by way of surface receptor proteins, as they migrate along specific pathways.
  ⇒ Signals from the receptors direct the cytoskeletal elements to propel the cell in the proper direction.
- Other substances in the extracellular matrix direct cells by preventing migration along certain paths.
- *Cell adhesion molecules* (*CAMs*) are substances on a cell surface that contribute to the selective association of certain cells with each other.
  ⇒ CAMs vary either in amount or chemical identity from one type of cell to another which helps regulate morphogenetic movements and tissue building.

E.   Induction

Morphogenetic movements are important to cellular determination and differentiation since cells move to areas with differing physical and chemical environments.

- Interactions among cells and cell layers are crucial during and after gastrulation.

Induction = The ability of one cell group to influence the development of another.

- The rudimentary notochord induces the dorsal ectoderm of a gastrula to form the neural plate.

    ⇒ Transplanting the chordamesoderm, which forms the notochord, to an abnormal site in the embryo will cause the neural plate to develop in an abnormal location.

- The dorsal lip of the blastopore acts as a primary *organizer*, setting up the interaction between chordamesoderm and the overlying ectoderm.

- Certain peptide growth factors have been discovered that apparently serve as a chemical signal causing cells migrating inward over the dorsal lip to form chordamesoderm.

- Many inductions appear to involve a sequence of inductive steps as in the development of an amphibian's eye. (See Campbell, Figure 43.21)

F.   Differentiation

Cells become specialized during *cellular differentiation* which occurs as tissues and organs form in the embryo.

- First appears as alterations in cellular structure.

- Molecular changes first appear with production of *tissue-specific proteins*, proteins found only in a specific type of cell.

Cells continue to make certain proteins as they become specialized during differentiation.

- In vertebrates, developing lens cells produce large quantities of crystallins, proteins that will aggregate to form the transparent lens fibers.

- Immature retinal cells induce immature lens cells to begin crystallin production.

- Chemical signals from the retinal cells stimulate immature lens cells to activate their crystallin genes which transcribe specific mRNA molecules coding for crystallin.

- Those mRNA molecules accumulate in the immature lens cells' cytoplasm.

- Production of crystallin begins when lens cells mature and the cells flatten and elongate to accommodate the crystallin fibers.

Genomic equivalence refers to the presence of all of an organism's genes in all of its cells.

- The cells of an organism differ in structure and function because they express different portions of a common genome.

- Transplantation experiments where nuclei from differentiated and undifferentiated cells were placed in enucleated eggs have shown that normal development occurs in eggs receiving nuclei from undifferentiated cells while abnormal development occurs in 98% of eggs receiving nuclei from differentiated cells. (See Campbell, Figure 43.22)

- These results have led to two conclusions: 1) nuclei do change in some way as they prepare for differentiation; 2) the change is not always irreversible.

G. **Pattern Formation**

Pattern formation is the development of an animal's spatial organization with organs and tissues in their characteristic places in the three dimensions of the animal. The emergence of a body form with specialized organs and tissues all in their characteristic places.

- Occurs in addition to the determination and differentiation of cells.

Pattern formation is controlled by *positional information* which is a set of molecular cues that indicate a cell's location relative to other cells in an embryonic structure.

- Vertebrate limbs all develop from undifferentiated limb buds. (See Campbell, Figure 43.23)
- A specific pattern of tissues (i.e. bone, muscle) emerges as the limb develops.
- Every component has a precise location and orientation relative to three axes.
    ⇒ The proximal-distal axis (from the limb base to tips of the digits).
    ⇒ The anterior-posterior axis (front edge to rear edge of the limb).
    ⇒ The dorsal-ventral axis (top to bottom of the limb).
- For proper development to occur, embryonic cells in the limb bud must receive positional information indicating location along all three axes.

Experiments on the development of chick limbs have provided information about positional information.

- A region of mesoderm located beneath an *apical epidermal ridge* at the wing bud tip is believed to assign embryonic tissue position along the proximal-distal axis. Transplantation experiments reversing the position of this tissue have produced limbs with reversed orientation.
- A *zone of polarized activity* (*ZPA*) has been found to be the point of reference for the anterior-posterior axis.
    ⇒ The ZPA is located where the posterior part of the bud is attached to the body.
- Cells nearest the ZPA give rise to posterior structures, while cells farthest from the ZPA give rise to anterior structures. (See Campbell, Figure 43.24)

Such experiments have led to the conclusion that pattern formation requires cells to receive and interpret environmental cues that vary between locations.

- Certain polypeptides are believed to be the cues which function as positional signals for vertebrate limb development.
- Regional gradients of these polypeptides along the three orientation axes would provide a cell positional information needed to determine its position in a three-dimensional organ.
- The regional variation in production of these polypeptides result from differential gene expression in different locations of the embryo.

IX. Pattern formation in *Drosophila* is controlled by a hierarchy of gene activations

Drosophila has proven to be a valuable experimental animal for the study of pattern formation.

- Genes controlling the overall body plan are the basis of pattern formation.
- Fruitflies (and other arthropods) possess an ordered series of segments giving them a modular arrangement.
- The modular body plan emerges in a stepwise fashion with very similar embryonic segments become anatomically distinct as development progresses.
- The gross form of the animal (anterior-posterior) is set first by positional information.
- Positional information then works on an increasingly refined scale to direct further development.
  - ⇒ The origin of a specific number of segments with the correct orientation relative to body axes is cued.
  - ⇒ Each segment then develops its specific anatomical features.
- The positional information results from regional differences in the type or concentration of regulatory proteins that control gene expression.
- The process begins before the egg is fertilized.

A. Egg-Polarity Genes

Egg-polarity genes are maternal-effect genes that set the polar axes of the future embryo.

- One group of these genes determines the anterior-posterior axis, another the dorsal-ventral axis.
- Other *maternal-effect genes* (phenotypes produced are a product of the female's genes regardless of the males contributions) have been identified.

The axes of a future embryo are defined by egg-polarity genes before the egg leaves the female's ovary. (See Campbell, Figure 43.25)

- Follicle cells and nurse cells (both are types of ovarian cells) nourish and support the unfertilized egg.
  - ⇒ Also produce mRNA molecules and proteins that regulate early development.
- For example, the *bicoid* gene is an egg-polarity gene active in nurse cells at the end of the egg that will become the anterior of the embryo.
  - ⇒ Transcription of the gene results in mRNA (for the bicoid protein) which is secreted by the nurse cell into the egg.
  - ⇒ This bicoid mRNA is not translated until the egg is fertilized and the zygote laid.
  - ⇒ Translation produces the bicoid protein which diffuses through the cytoplasm of the syncytial embryo.
  - ⇒ When the blastoderm forms at cytokinesis, cells at the "anterior pole" of the embryo will have the highest concentration of bicoid protein while those at the "posterior pole" will have the lowest concentration.
  - ⇒ The bicoid protein binds to DNA and regulates expression of certain genes in the blastoderm cells.
  - ⇒ The hierarchy of pattern formation results from some of the genes activated by bicoid (and other) proteins which code for proteins that establish basic segmentation patterns.

Bicoid protein is a *morphogen*.

<u>Morphogen</u> = A substance that provides positional information in the form of a gradation in its concentration along an embryonic axis.

- Many morphogens result from differential activity of egg polarity genes in follicles around the *Drosophila* egg.
- Working together, the morphogens determine the three body axes.

B.   Segmentation Genes

Morphogen (including bicoid protein) gradients resulting from spatial differences in the activity of egg-polarity genes, cause regional differences in expression of *segmentation genes*.

- Segmentation genes regulate the origin of segmentation once the embryo's anterior-posterior and dorsal-ventral axes are defined.
- Positional information to establish the animal's body plan is provided by the sequential activation of three sets of segmentation genes.
  - ⇒ *Gap genes* are activated first and their products help establish the basic subdivision along the anterior-posterior axis.
  - ⇒ *Pair-rule genes* are activated next and their products define the modular pattern in terms of pairs of segments.
  - ⇒ *Segment-polarity genes* are the last activated and define the anterior-posterior axis of each segment.
- The products of each type of gene activates the next in the series and the result is the establishment of the boundaries and axis of each segment in the body.

C.   Homeotic Genes

In *Drosophila*, once segmentation genes have established the basic body plan, *homeotic genes* set the anatomical identity of the segment.

- Homeotic genes specify the type of appendages and other structures (antennae, wings) specific for that segment.
- The activation of particular homeotic genes in each segment is a response to the positional imprints left by earlier activations of segmental genes.

The products of homeotic genes are regulatory proteins that bind to DNA.

- These proteins affect the selective gene expression for continuing development of the segmental body plan of the fly embryo.

X. **Comparisons of genes that control development reveal homology in animals as diverse as flies and mammals**

Research on the basic processes of epigenesis has shown a very conservative evolution in the genetic mechanisms regulating pattern formation.

- All of the homeotic genes in *Drosophila* include a common sequence of 180 nucleotides.
- Such a specific DNA sequence within a homeotic gene is termed a *homeobox*.
- Homeoboxes have also been found in genes that regulate development.
- Very similar or identical homeobox sequences have been found in a number of other eukaryotes (i.e. *Hydra*, frogs, mammals).
- Homeotic genes are a good example of homology in genes that control development.

Many protein-coding genes associated with development also contain homeobox nucleotide sequences.

- The homeobox region specifies a 60-amino-acid-long *homeodomain* that is part of the protein produced when the gene is translated.
- Homeodomains are the portions of the regulatory proteins that bind to DNA.
  - ⇒ Variability in other domains of regulatory proteins are responsible for specific functions such as binding to targets and activation of certain genes.
- To this point, all homeodomain-containing proteins examined have been activating or repressing transcription factors.
- Pattern formation in the embryo is controlled by the selective expression, over time and space, of regulatory genes.

Homeoboxes appear to be derived from a nucleotide sequence that arose very early in the history of life and has undergone a very conservative evolution.

- Many animals have homeobox-containing genes organized in the same way as those in *Drosophila*.
- Eukaryotic homeodomains show some similarity to DNA-binding domains of prokaryotic regulatory proteins.

REFERENCES

Browder, L.W. *Developmental Biology*. 2nd ed. Philadelphia: Saunders College/Holt, Rinehart and Winston, 1984.

Campbell, N. *Biology*. 4th ed. Menlo Park, California: Benjamin/Cummings, 1996.

Fjose, A. "Spatial Expression of Homeotic Genes in *Drosophila*." *Bioscience*, September 1986.

Gehring, Walter J. "The Molecular Basis of Development." *Scientific American*, October 1985.

Gilbert, S.F. *Developmental Biology*. 2nd ed. Sunderland, MA: Sinauer Associates, 1988.

Walbot, V., and N. Holder. *Developmental Biology*. New York: Random House, Inc., 1987.

CHAPTER 44

NERVOUS SYSTEMS

OUTLINE

I. Nervous systems perform the three overlapping functions of sensory input, integration, and motor output: *an overview*

II. The nervous system is composed of neurons and supporting cells
   A. Neurons
   B. Supporting Cells

III. Impulses are action potentials, electrical signals propagated along neuronal membranes
   A. The Origin of Electrical Membrane Potential
   B. Membrane Potential Changes and the Action Potential
   C. Propagation of the Action Potential
   D. Action Potential Transmission Speed

IV. Chemical or electrical communication between cells occurs at synapses
   A. Electrical Synapses
   B. Chemical Synapses
   C. Summation: Neural Integration at the Cellular Level
   D. Neurotransmitters and Receptors
   E. Gaseous Signals of the Nervous System
   F. Neural Circuits and Clusters

V. Invertebrate nervous systems are highly diverse

VI. The vertebrate nervous system is a hierarchy of structural and functional complexity
   A. The Peripheral Nervous System
   B. The Central Nervous System
   C. Evolution of the Vertebrate Brain

VII. The human brain is a major research frontier
   A. Anatomy of the Brain
   B. Integration and Higher Brain Functions

OBJECTIVES

After reading this chapter and attending lecture, the student should be able to:

1. Compare the two coordinating systems in animals.
2. Describe the three major functions of the nervous system.
3. List and describe the three major parts of a neuron and explain the function of each.
4. Explain how neurons can be classified by function.
5. Describe the function and location of each type of supporting cell.
6. Explain what a resting potential is, and list four factors that contribute to the maintenance of the resting potential.
7. Define equilibrium potential, and explain why the K^+ equilibrium potential is more negative than the resting potential.
8. Define graded potential, and explain how it is different from a resting potential or action potential.
9. Describe the characteristics of an action potential, and explain the role membrane permeability changes and ion gates play in the generation of an action potential.
10. Explain how the action potential is propagated along a neuron.
11. Describe two ways to increase the effectiveness of nerve transmission.
12. Describe synaptic transmission across an electrical synapse and a chemical synapse.
13. Describe the role of cholinesterase and explain what would happen if acetylcholine were not destroyed.
14. List some other possible neurotransmitters.
15. Define neuromodulator and describe how it may affect nerve transmission.
16. Explain how excitatory postsynaptic potentials (EPSP) and inhibitory postsynaptic potentials (IPSP) affect the postsynaptic membrane potential.
17. Explain how a neuron integrates incoming information, including a description of summation.
18. List three criteria for a compound to be considered a neurotransmitter.
19. List two classes of neuropeptides and explain how they illustrate overlap between endocrine and nervous control.
20. Describe two mechanisms by which a neurotransmitter affects the postsynaptic cell.
21. Diagram or describe the three major patterns of neural circuits.
22. Compare and contrast the nervous systems of the following invertebrates and explain how variation in design and complexity correlate with phylogeny, natural history and habitat.
    a. *Hydra.*
    b. Jellyfish, ctenophores and echinoderms.
    c. Flatworms.
    d. Annelids and arthropods.
    e. Mollusks.
23. Outline the divisions of the vertebrate nervous system.
24. Distinguish between sensory (afferent) nerves and motor (efferent) nerves.
25. Define reflex and describe the pathway of a simple spinal reflex.
26. Distinguish between the functions of the autonomic nervous system and the somatic nervous system.
27. List the major components of the central nervous system.
28. Distinguish between white matter and gray matter.
29. Describe three major trends in the evolution of the vertebrate brain.
30. From a diagram, identify and describe the functions of the major structures of the human brain:
    a. Medulla oblongata.
    b. Pons.
    c. Cerebellum.
    d. Superior and inferior colliculi.
    e. Telencephalon.
    f. Diencephalon.
    g. Thalamus.
    h. Hypothalamus
    i. Cerebral cortex
    j. Corpus Callosum

31. Explain how electrical activity of the brain can be measured and distinguish among alpha, beta, theta and delta waves.
32. Describe the sleep-wakefulness cycle, the associated EEG changes, and the parts of the brain that control sleep and arousal.
33. Define lateralization and describe the role of the corpus callosum.
34. Describe the positions and functions of Wernicke's area and Broca's area.
35. Distinguish between short-term and long-term memory.
36. Using a flowchart, outline a possible memory pathway in the brain.

KEY TERMS

supporting cells
neurons
cell body
dendrites
axon
Schwann cells
myelin sheath
axon hillock
telodendria
synaptic knobs
synapse
effector cells
central nervous system
sensory neurons
motor neurons
peripheral nervous
 system
interneurons
glial cells
astrocytes
blood-brain barrier
oligodendrocytes
multiple sclerosis
polarization
membrane potential
resting potential
depolarization
hyperpolarization
graded potentials
action potential
voltage-sensitive
 gates
threshold potential
Na$^+$ gates
refractory period
all-or-none event

repolarization
saltatory conduction
node of Ranvier
electrical synapse
chemical synapse
gap junctions
synaptic cleft
synaptic vesicles
neurotransmitter
 molecules
presynaptic
 membrane
postsynaptic
 membrane
excitatory postsynaptic
 potential (EPSP)
inhibitory postsynaptic
 potential (IPSP)
summation
temporal summation
spatial summation
acetylcholine
biogenic amines
epinephrine
norepinephrine
dopamine
tyrosine
serotonin
tryptophan
gamma
 aminobutyric
 acid (GABA)
neuropeptides
endorphins
enkephalins
analgesics

convergent circuits
divergent circuits
reverberating
 circuits
ganglia
nucleus
nerve net
cephalization
central nervous
 system
brain
spinal cord
peripheral nervous
 system
sensory (afferent)
 division
somatic sensory
 neurons
visceral sensory
 neurons
motor (efferent)
 division
somatic system
autonomic system
sympathetic division
parasympathetic
 division
reflex
meninges
white matter
gray matter
cerebrospinal fluid
ventricles
central canal of the
 spinal cord
rhombencephalon

mesencephalon
prosencephalon
brainstem
hindbrain
medulla oblongata
pons
cerebellum
midbrain
superior colliculi
inferior colliculi
reticular formation
forebrain
diencephalon
thalamus
hypothalamus
telencephalon
cerebral cortex
corpus callosum
basal ganglia
sleep
arousal
alpha waves
beta waves
delta waves
REM sleep
electroencephalogram
 (EEG)
limbic system
lateralization
Wernicke's area
Broca's area
aphasia
short-term memory
long-term memory
hippocampus
amygdala

LECTURE NOTES

The *endocrine* and *nervous* systems of animals often cooperate and interact to maintain homeostasis and control behavior. Though structurally and functionally linked, these two systems play different roles.

	Nervous System	**Endocrine System**
Complexity	More structurally complex; can integrate vast amounts of information & stimulate a wide range of responses.	Less structurally complex.
Structure	System of *neurons* that branch throughout the body.	*Endocrine glands* secrete *hormones* into the bloodstream where they are carried to the *target organ.*
Communication	Neurons conduct electrical signals directly to and from specific targets; allows fine pin-point control.	As chemical messengers, hormones circulate throughout the body in the bloodstream; exposes most body cells to the hormone & only target cells with receptors respond.
Response Time	Fast transmission of nerve impulses up to 100m/sec.	May take minutes, hours or days - time for hormones to be produced, carried by blood to target organ, & for response to occur.

I.   **Nervous systems perform the three overlapping functions of sensory input, integration, and motor output:** *an overview*

The nervous system has three overlapping functions:

1. *Sensory input* is the conduction of signals from sensory receptors to integration centers of the nervous system.

2. *Integration* is a process by which information from sensory receptors is interpreted and associated with appropriate responses of the body.

3. *Motor output* is the conduction of signals from the processing center to *effector cells* (muscle cells, gland cells) that actually carry out the body's response to stimuli.

These functions involve both parts of the nervous system:

1. *Central nervous system* (*CNS*): comprised of the brain and spinal cord; responsible for integration of sensory input and associating stimuli with appropriate motor output.

2. *Peripheral nervous system* (*PNS*): consists of the network of nerves extending into different parts of the body that carry sensory input to the CNS and motor output away from the CNS.

II.   The nervous system is composed of neurons and supporting cells

The nervous system includes two main types of cells: *neurons* which actually conduct messages; and *supporting cells* which provide structural reinforcement as well as protect, insulate and assist the neuron.

A.   Neurons

Neurons = Cells specialized for transmitting chemical and electrical signals from one location in the body to another.

- Have a large *cell body*.
    ⇒ Contains most of the cytoplasm, the nucleus, and other organelles.   (See Campbell, Figure 44.2)
    ⇒ The cell bodies of most neurons are located in the CNS, although certain types of neurons have their cell bodies located in *ganglia* outside of the CNS.
- Have *two* types of fiberlike extensions (*processes*) that increase the distance over which the cells can conduct messages:
    ⇒ *Dendrites* convey signals to the cell body; are short, numerous and extensively branched to increase surface area where the cell is most likely to be stimulated.
    ⇒ *Axons* conduct impulses away from the cell body; are long, single processes.
        ◊ Vertebrate axons in PNS are wrapped in concentric layers of *Schwann cells* which form an insulating *myelin sheath*.
        ◊ Axons extend from the *axon hillock* (where impulses are generated) to many branches called *telodendria*, which are tipped with *synaptic terminals* that release *neurotransmitters*.

Synapse = Gap between a synaptic terminal and a target cell – either dendrites of another neuron or an effector cell.

Neurotransmitters = Chemicals that cross the *synapse* to relay the impulse.

There are three major classes of neurons: (See Campbell, Figure 44.3)

1. *Sensory neurons* convey information about the external and internal environments from sensory receptors to the central nervous system. Most sensory neurons synapse with *interneurons*.

2. *Interneurons* integrate sensory input and motor output; located within the CNS, they synapse only with other neurons..

3. *Motor neurons* convey impulses from the CNS to effector cells.

B.   Supporting Cells

Supporting cells structurally reinforce, protect, insulate and generally assist neurons.
- Do not conduct impulses.
- Outnumber neurons 10- to 50-fold.

Glial cells = Supporting cells of the CNS; several types are present.
- *Astrocytes* encircle capillaries in the brain.
  ⇒ Contribute to the blood-brain barrier which restricts passage of most substances into the CNS.
  ⇒ Probably communicate with one another and with neurons via chemical signals.
- *Oligodendrocytes* form myelin sheaths that insulate nerve processes.

Schwann cells are the supporting cells in the PNS. They form the insulating myelin sheath around axons.

Myelination of neurons in a developing nervous system occurs when Schwann cells or oligodendrocytes grow around an axon so their plasma membranes form concentric layers.
- Provides electrical insulation since membranes are mostly lipid which is a poor current conductor.
- Insulating myelin sheath increases the speed of nerve impulse propagation.
- In multiple sclerosis, myelin sheaths deteriorate causing a disruption of nerve impulse transmission and consequent loss of coordination.

III.   Impulses are action potentials, electrical signals propagated along neuronal membranes

Signal transmission along the length of a neuron depends on voltages created by ionic fluxes *across* neuron plasma membranes.

A.   The Origin of Electrical Membrane Potential

All cells have an electrical *membrane potential* or voltage across their plasma membranes.
- Ranges from -50 to -100 mV in animal cells.
- The charge outside the cell is designated as zero, so the minus sign indicates that the cytoplasm inside is negatively charged compared to the extracellular fluid.
- Is about -70mV in a resting neuron.

The membrane potential arises from two things:

1. *Differences in the ionic composition of the intracellular and extracellular fluids.* For example, the approximate concentrations in millimoles per liter (mM) are listed below for ions in mammalian cells:

Ion	Inside Cell	Outside Cell
$[Na^+]$	15 mM	150mM
$[K^+]$	150mM	5mM
$[Cl^-]$	10mM	120mM
$[A-]$	100mM	---------------

   Note: [A-] symbolizes all anions within the cell including negatively charged proteins, amino acids, sulfate and phosphate.

   - Principal cation inside the cell is K^+, while the principal cation outside the cell is Na^+.
   - Principal anions inside the cell are proteins, amino acids, sulfate, phosphate, and other negatively charged ions (A-); principal anion outside the cell is Cl^-.
   - Because internal anions (A-) are primarily large organic molecules, they cannot cross the membrane and remain in the cell as a reservoir of internal negative charge.

2. *Selective permeability of the plasma membrane.*
   - As charged molecules, ions cannot readily diffuse through the hydrophobic core of the plasma membrane's phospholipid bilayer.
   - Ions can only cross membranes by carrier-mediated transport or by passing through *ion channels.*

Ion channel = Integral transmembrane protein that allows a specific ion to cross the membrane.

- May be passive and open all the time; or may be *gated,* requiring a stimulus to change into an open conformation.
- Are selective for a specific ion, such as Na^+, K^+ and Cl^-.
- Membrane permeability to each ion is a function of the type and number of ion channels. For example, membranes are usually more permeable to K^+ than to Na^+, suggesting that there are more potassium channels than sodium channels.

How does the cell create and maintain the membrane potential?. (See Campbell, Figure 44.5)

- K^+ diffuses out of the cell down its concentration gradient, since the K^+ concentration is greater inside cell and the membrane has a high permeability to potassium.
- K^+ diffusion out of the cell transfers positive charges from inside of the cell to outside.

- The cell's interior becomes progressively more negative as K^+ leaves because the molecules of the impermeant anion pool (A^-) are too large to cross the membrane.
- As the electrical gradient increases, the negatively charged interior attracts K^+ back into the cell.
- If K^+ was the only ion to cross the membrane, an *equilibrium potential* for potassium ions (= -85 mV) would be reached; this is the potential at which no net movement of K^+ would occur since the electrical gradient attraction of K^+ would balance the K^+ loss due to the concentration gradient.
- However, K^+ is not the only ion to cross the membrane; although the membrane is less permeable to Na^+ than K^+, some Na^+ diffuses into the cell down *both* its concentration gradient <u>and</u> the electrical gradient.
- The Na^+ trickle into the cell transfers positive charge to the inside resulting in a slightly more positive charge (-70 mV) than if the membrane were permeable only to K^+ (-85 mV).
- If left unchecked, the Na^+ trickle into the cell would cause a progressive increase in Na^+ concentration and a decrease in K^+ concentration (the electrical gradient would be weakened by the positive sodium ions and the potassium ions would diffuse down its concentration gradient).
- This shift in ionic gradients is prevented by sodium-potassium pumps, special transmembrane proteins which use energy from ATP to:
  - $\Rightarrow$ Pump sodium back out of the cell against concentration and electrical gradients.
  - $\Rightarrow$ Pump potassium *into* the cell, restoring its concentration gradient.

B. Membrane Potential Changes and the Action Potential

All cells in the body exhibit the properties of membrane potential described above. However, only neurons and muscle cells can *change* their membrane potentials in response to stimuli.

- Such cells are called *excitable cells*.
- The membrane potential of an excitable cell at rest (unexcited state) is called a *resting potential*.

The presence of *gated ion channels* in neurons permits these cells to change the plasma membrane's permeability and alter the membrane potential in response to stimuli received by the cell.

- Sensory neurons are stimulated by receptors that are triggered by environmental stimuli.
- Interneurons normally receive stimuli produced by activation of other neurons.

The effect on the neuron depends on the type of gated ion channel the stimulus opens.

- Stimuli that open potassium channels *hyperpolarize* the neuron.. K^+ effluxes from the cell; this increases the electrical gradient making the interior of the cell more negative.
- Stimuli that open sodium channels *depolarize* the neuron. Na^+ influxes into the cell; this reduces the electrical gradient and membrane potential because the inside of the cell becomes more positive.

Voltage changes caused by stimulation are called *graded potentials* because the magnitude of the change depends on the strength of the stimulus. (See Campbell, Figure 44.6)

- Each excitable cell has a *threshold* to which depolarizing stimuli are graded. This threshold potential is usually slightly more positive (−50 to −55 mV) than the resting potential.
- If depolarization reaches the threshold, the cell responds differently by triggering an *action potential*.
- Hyperpolarizing stimuli do not produce action potentials since they cause the potential to become more negative; actually reduces the probability an action potential will occur by making it more difficult for depolarizing stimuli to reach the threshold.

An *action potential* is the rapid change in the membrane potential of an excitable cell, caused by stimulus-triggered selective opening and closing of *voltage-gated ion channels*.

- Voltage-gated ion channels open and close in response to changes in membrane potential.
- Voltage-gated sodium channels have two gates: the activation gate opens *rapidly* at depolarization, the inactivation gate closes slowly at depolarization.
- The voltage-gated potassium channel has one gate that opens *slowly* in response to depolarization.

An action potential has four phases. (See Campbell, Figure 44.7)

- *Resting state*, no channels are open.
- *Large depolarizing phase* during which the membrane briefly reverses polarity (cell interior becomes positive to the exterior). The Na^+ activation gates open allowing an influx of Na^+, while potassium gates remain closed.
- The *steep repolaring phase* follows quickly and returns the membrane potential to its resting level; inactivation gates close the sodium channels and the potassium channels open.
- The *undershoot phase* is a time when the membrane potential is temporarily more negative than the resting state (hyperpolarized); sodium channels remain closed but potassium channels remain open since the inactivation gates have not had time to respond to repolarization of the membrane.

A *refractory period* occurs during the undershoot phase; during this period, the neuron is insensitive to depolarizing stimuli. The refractory period limits the maximum rate at which action potentials can be stimulated in a neuron.

Action potentials are *all-or-none* events and their amplitudes are not affected by stimulus intensity. The nervous system distinguishes between strong and weak stimuli based on the frequency of action potentials generated.

- Strong stimuli produce action potentials more rapidly than weak stimuli.
- Maximum frequency is limited by the refractory period of the neuron.

C.   **Propagation of the Action Potential**

A neuron is stimulated at its dendrites or cell body and the action potential travels along the axon to the other end of the neuron.

- Action potentials in the axon are usually generated at the axon hillock.
- Strong depolarization in one area results in depolarization above the threshold in neighboring areas.
- The action potential does not travel down the axon, but is regenerated at each position along the membrane.

The signal travels in a perpendicular direction along the axon regenerating the action potential. (See Campbell, Figure 44.8)

- Na^+ influx in the area of the action potential results in depolarization of the membrane just ahead of the impulse, surpassing the threshold.
- The voltage-sensitive channels in the new location will go through the same sequence previously described regenerating the action potential.
- Subsequent portions of the axons are depolarized in the same manner.
- The action potential moves in only one direction (down the axon) since each action potential is followed by a refractory period when sodium channel inactivation gates are closed and no action potential can be generated.

D.   **Action Potential Transmission Speed**

Factors affecting the speed of action potential propagation:

1. The larger the diameter of the axon, the faster the rate of transmission since resistance to the flow of electrical current is inversely proportional to the cross-sectional area of the "wire" conducting the current.
2. Saltatory conduction.

Saltatory conduction = The action potential "jumps" from one *node of Ranvier* to the next, skipping the myelinated regions of membrane. (See Campbell, Figure 44.9)

- *Nodes of Ranvier* are gaps in the myelin sheath between successive glial cells.
- Voltage-sensitive ion channels are concentrated in node regions of the axon.
- Extracellular fluid only contacts the axon membranes at the nodes; restricts the area for ion exchange to these regions.
- Results in faster transmission of the nerve impulse.

IV.   **Chemical or electrical communication between cells occurs at synapses**

Synapse = Tiny gap between a synaptic terminal of an axon and a signal-receiving portion of another neuron or effector cell.

- Also found between sensory receptors and sensory neurons, and between motor neurons and muscle cells.
- *Presynaptic cell* is the transmitting cell; *postsynaptic cell* is the receiving cell.
- There are two types of synapses: electrical and chemical.

A.   **Electrical Synapses**

Electrical synapses allow action potentials to spread directly from pre- to postsynaptic cells via gap junctions (intercellular channels).

- Allows impulses to travel from one cell to the next without delay or loss of signal strength.
- Much less common than chemical synapses.
- Examples are the giant neuron processes of crustaceans.

B.   **Chemical Synapses**

At a *chemical synapse* a synaptic cleft separates the pre- and postsynaptic cells so they are not electrically coupled. (See Campbell, Figure 44.10)

- Within the cytoplasm of the synaptic terminal of a presynaptic cell are numerous *synaptic vesicles* containing thousands of neurotransmitter molecules.
- An action potential arriving at the synaptic terminal depolarizes the *presynaptic membrane* causing Ca^{2+} to rush through voltage-sensitive channels.
- The sudden rise in Ca^{2+} concentration stimulates synaptic vesicles to fuse with the presynaptic membrane and release neurotransmitter into the synaptic cleft by exocytosis.
- The neurotransmitter diffuses to the *postsynaptic membrane* where it binds to specific receptors, causing ion gates to open.
- Depending on the type of receptors and the ion gates they control, the neurotransmitter may either excite the membrane by depolarization or inhibit the postsynaptic cell by hyperpolarization.
- The neurotransmitter molecules are quickly degraded by enzymes and the components recycled to the presynaptic cell.

Chemical synapses allow transmission of nerve impulses in only one direction.

- Synaptic vesicles and their neurotransmitter molecules are found only in the synaptic terminals at the tip of axons.
- Receptors for neurotransmitters are located only on postsynaptic membranes.

C.   **Summation: Neural Integration at the Cellular Level**

One neuron may receive information from thousands of synapses. Some synapses are excitatory, others are inhibitory.

- *Excitatory postsynaptic potentials* (*EPSP*) occur when excitatory synapses release a neurotransmitter that opens gated channels allowing Na^+ to enter the cell and K^+ to leave (depolarization).
- *Inhibitory postsynaptic potentials* (*IPSP*) occur when neurotransmitters released from inhibitory synapses bind to receptors which open ion gates that make the membrane more permeable to K^+ (which leaves the cell) and/or to Cl^- (which enters the cell) causing hyperpolarization.

EPSPs and IPSPs are graded potentials; they vary in magnitude with the number of neurotransmitter molecules binding to postsynaptic receptors.

- Change in voltage lasts only a few milliseconds since neurotransmitters are inactivated by enzymes soon after release.
- The electrical impact on the postsynaptic cell also decreases with distance from the synapse.

A single EPSP is rarely strong enough to trigger an action potential, although an additive effect (*summation*) from several terminals or repeated firing of terminals can change membrane potential.

- *Temporal summation* is when chemical transmissions from one or more synaptic terminals occur so close in time that each affects the membrane while it is partially depolarized and before it has returned to resting potential.
- *Spatial summation* is when several different synaptic terminals, usually from different presynaptic neurons, stimulate the postsynaptic cell at the same time and have an additive effect on membrane potential.
- EPSPs and IPSPs can summate, each countering the effects of the other.

At any instant, the axon hillock's membrane potential is an average of the summated depolarization due to all EPSPs and the summated hyperpolarization due to all IPSPs.

- An action potential is generated when the EPSP summation exceeds the IPSP summation to the point where the membrane potential of the axon hillock reaches threshold voltage.

D. Neurotransmitters and Receptors

Dozens of different molecules are known to be neurotransmitters and many others are suspected to function as such.

Criteria for neurotransmitters are:

- Must be present in and discharged from synaptic vesicles in the presynaptic cell when stimulated and affect the postsynaptic cell's membrane potential.
- Must cause an IPSP or EPSP when experimentally injected into the synapse.
- Must be rapidly removed from the synapse by an enzyme or uptake by a cell permitting the postsynaptic membrane to return to resting potential.

Types of neurotransmitters:

1. *Acetylcholine* may be excitatory or inhibitory depending on the receptor; functions in the vertebrate neuromuscular junction (between a motor neuron and muscle cell) and in the central nervous system.
   - The most common neurotransmitter in both vertebrates and invertebrates.

2. *Biogenic amines* are derived from amino acids.
   - Epinephrine, norepinephrine, dopamine are produced from tyrosine; serotonin is synthesized from tryptophan.
   - Commonly function in the central nervous system; imbalances in dopamine and serotonin are associated with mental illness. Norepinephrine also functions in the peripheral nervous system.

3. Amino acids *glycine*, *glutamate*, *aspartate*, and *gamma aminobutyric acid* (*GABA*) function as neurotransmitters in the central nervous system.
   - GABA is the most abundant inhibitory transmitter in the brain.

4. *Neuropeptides* are short chains of amino acids.
   - Substance P is an excitatory signal that mediates pain perception.
   - Endorphins (or enkephalins) function as natural analgesics in the brain.

Neurotransmitters may affect the postsynaptic cell by:
- Altering the permeability of the postsynaptic membrane to specific ions (e.g. acetylcholine, amino acid transmitters).
- Affecting postsynaptic cell metabolism (e.g. biogenic amines, neuropeptides) by triggering a signal transduction pathway.

E.   Gaseous signals of the Nervous System

Some neurons of the vertebrate PNS and CNS release gas molecules, such as nitric oxide (NO) and carbon monoxide (CO), as local regulators. For example,
- During sexual arousal, neurons release NO into erectile tissue of the penis, which causes blood vessels to dilate and fill with blood, producing an erection.
- Acetylcholine released by neurons into blood vessel walls stimulates their endothelial cells to produce and release NO. In response, neighboring smooth muscle cells relax, dilating the vessels.
- Similarly, nitroglycerin is effective in treating angina because it is converted to NO, which dilates the heart's blood vessels.

Cells do not store gaseous messengers, so they must be produced on demand.
- Within a few seconds, they diffuse into target cells, produce a change, and are broken down.
- NO often works by a signal-transduction pathway; it stimulates a membrane-bound enzyme to synthesize a second messenger that directly affects cellular metabolism.

F.   Neural Circuits and Clusters

Neurons are arranged in groups referred to as circuits:

Convergent circuits = Neural circuit in which information from several presynaptic neurons come together at a single postsynaptic neuron; permits integration of information from several sources.

Divergent circuits = Neural circuit in which information from a single neuron spreads out to several postsynaptic neurons; permits transmission of information from a single source to several parts of the brain.

Reverberating circuits = Circular circuits in which the signal returns to its source. Believed to play a role in memory storage.

Nerve cell bodies in the PNS are arranged into functional clusters called *ganglia*.
- A *nucleus* is a similar functional cluster within the brain.
- Ganglia and nuclei allow coordination of activities by only part of the nervous system.

V. Invertebrate nervous systems are highly diverse

There is great diversity in invertebrate nervous system organization. (See Campbell, Figure 44.13)

The *Hydra*, a cnidarian, has a *nerve net* — a loosely organized system of nerves with no central control.

- Impulses are conducted in both directions causing movements of the entire body.
- Some Cnidaria, Ctenophora and echinoderms have modified nerve nets with rudimentary centralization.

Cephalization = Evolutionary trend for concentration of sensory and feeding organs on the anterior end of a moving animal; gave rise to the first brains.

- Found in bilaterally symmetrical animals.

Most bilaterally symmetrical animals also have a peripheral nervous system and a central nervous system.

- Flatworms have a simple "brain" containing many large interneurons that coordinate most nervous functions. Two or more nerve trunks travel posteriorly in a ladderlike system with transverse nerves connecting the main trunks.
- Annelids and arthropods have a well-defined ventral nerve cord and a prominent brain. Often contain ganglia in each body segment to coordinate actions of that segment.
- Cephalopods have the most sophisticated invertebrate nervous system containing a large brain and giant axons.

Nervous system complexity often correlates with phylogeny, habitat and natural history. For example, sessile animals such as clams show little or no cephalization.

VI. The vertebrate nervous system is a hierarchy of structural and functional complexity

Because vertebrate nervous systems are so complex, it is useful to group them into functional components: the *peripheral nervous system* and the *central nervous system*.

A. The Peripheral Nervous System

The *peripheral nervous system* consists of:

- *Sensory (afferent) nervous system* which brings information from sensory receptors to the CNS.
- *Motor (efferent) nervous system* which carries signals from the CNS to effector cells.

The peripheral nervous system of humans consists of 12 pairs of cranial nerves and 31 pairs of spinal nerves.

- Cranial nerves originate from the brain and innervate organs of the head and upper body; most contain both sensory and motor neurons, although some are sensory only (e.g. optic nerve).
- Spinal nerves innervate the entire body and contain both sensory and motor neurons.

The two basic functions of a nervous system are to:
- Control responses to external environment.
- Maintain homeostasis by coordinating internal organ functions.

The sensory nervous system contributes to both functions by carrying stimuli from the external environment and monitoring the status of the internal environment.

The motor nervous system has two separate divisions associated with these functions.

1. The *somatic nervous system*'s neurons carry signals to skeletal muscles in response to external stimuli; includes *reflexes* (automatic responses to stimuli) and is often considered "voluntary" since it is subject to conscious control.

2. The *autonomic nervous system* controls primarily "involuntary," automatic, visceral functions of smooth and cardiac muscles and organs of the gastrointestinal, excretory, cardiovascular and endocrine systems.
   - Divided into a *parasympathetic* division that enhances activities that gain and conserve energy, and a usually antagonistic *sympathetic* division that increases energy expenditures.

B.   The Central Nervous System (CNS)

The CNS bridges the sensory and motor functions of the peripheral nervous system.
- Consists of the *spinal cord*, which is located inside the vertebral column and *receives* information from skin and muscles and *sends* out motor commands for movement; and the *brain*, which carries out complex integration for homeostasis, perception, movement, intellect and emotions.
- Covered with *meninges*, three protective layers of connective tissue.
- In the brain, white (*myelinated*) matter is in the inner and gray matter is in the outer regions. This orientation is reversed in the spinal cord.
- *Cerebrospinal fluid* fills the ventricles in the brain and the central canal of the spinal cord; it functions in circulation of hormones, nutrients and white blood cells and in absorption of shock.

The spinal cord integrates simple responses to certain stimuli (*reflexes*) and carries information to and from the brain.
- The patellar (knee-jerk) reflex is one of the simplest and involves only two neurons. A stretch receptor in the quadriceps muscle is stimulated by stretching of the patellar tendon; this activates a sensory neuron that carries the information to the spinal cord where it synapses with a motor neuron; if an action potential is generated in the motor neuron, it travels back to the quadriceps which contracts and causes the forward knee jerk. (See Campbell, Figure 44.16)
- Larger-scale, more complex responses result when branches of a reflex pathway carry signals to other parts of the spinal cord or to the brain.

C.   Evolution of the Vertebrate Brain

The vertebrate brain has shown an evolutionary trend toward greater complexity which has resulted in more complex behavioral patterns.

All vertebrates posses a *rhombencephalon* (hindbrain), *mesencephalon* (midbrain) and *prosencephalon* (forebrain). (See Campbell, Figure 44.17)
- More complex brains have further subdivisions.

Trends in the evolution of the vertebrate brain are: (See Campbell, Figure 44.17)
- Relative brain size increases in certain evolutionary lineages.
- Increased compartmentalization of function with certain areas of the brain assuming specific responsibilities.
- Increasing complexity and sophistication of the forebrain; increased complexity of behaviors parallels an increase in growth of the cerebrum. ∘

VII.   The human brain is a major research frontier

The human brain weights about 1.35 kg and is one of the largest organs in the body.
- It develops at the anterior end of the spinal cord from three primary bulges that later differentiate into distinct structures with specific functions.

The human hindbrain consists of three parts: (See Campbell, Figure 44.18)
- The *medulla oblongata* and *pons* control visceral functions like breathing, heart and blood vessel activity, swallowing, vomiting and digestion; also coordinates large-scale body movements like walking.
- The *cerebellum* functions in balance and coordination of movement.

The human midbrain together with the hindbrain forms the *brainstem*.
- The *superior* and *inferior colliculi* are areas of the midbrain that function in the visual and auditory systems.
- The *reticular formation* regulates states of arousal.

The *forebrain* contains sensory and motor pathways and integrating centers involved with pattern and image formation, and associative functions, such as memory, learning and emotions. The forebrain has two major divisions:
- Lower *diencephalon* that contains two integrating centers, the *thalamus* and the *hypothalamus*.
- Upper *telencephalon* that consists of the *cerebrum* – complex integrating center in the CNS.

The *thalamus*, a prominent integrating center in the diencephalon, relays sensory information to the cerebrum.

- Contains many different nuclei, each one dedicated to one type of sensory information.
- Sorts incoming sensory information and sends it to appropriate higher brain centers for further interpretation and integration.
- Receives input from the cerebrum and from parts of the brain that regulate emotion and arousal.

The *hypothalumus* is one of the most important regulators of homeostasis.

- Is the source of releasing hormones of the anterior pituitary and two posterior pituitary hormones.
- Contains the body's thermostat and centers for regulating hunger and thirst.
- Plays a role in sexual response and mating behavior, the fight-or-flight response, and pleasure.
- Contains the *suprachiasmiatic nucleus* which uses visual information to synchronize certain bodily functions with the natural cycles of day length and darkness. This biological clock maintains daily biorhythms such as when:
    ⇒ sleep occurs.
    ⇒ blood pressure is highest.
    ⇒ sex drive peaks.

The *cerebrum* is divided into the right and left *cerebral hemispheres*. Each hemisphere consists of:

- Outer covering of gray matter, the *cerebral cortex*
- Internal white matter
- Cluster of nuclei deep within the white matter, the *basal ganglia* which:
    ⇒ Are centers for motor coordination, relaying impulses from other motor systems.
    ⇒ Send motor impulses to the muscles.
    ⇒ If damaged, passivity and immobility result, because they no longer allow motor impulses to be sent to the muscles. Degeneration of cells entering the basal ganglia occurs in Parkinson's disease.

The largest, most complex part of the human brain is the *cerebral cortex* which is:

- Highly folded with a surface area of about 0.5 m^2.
- Bilaterally symmetrical with two *hemispheres* connected by a thick band of fibers (white matter) known as the *corpus callosum*. Each hemisphere is divided into four lobes; some functional areas within each lobe have been identified.

Two functional cortical areas, *motor cortex* and *somatosensory cortex*, form the boundary between the *frontal lobe* and the *parietal lobe*.

- In response to sensory stimuli, the motor cortex sends appropriate commands to skeletal muscles.
- The somatosensory cortex receives and partially integrates signals from the body's touch, pain, pressure and temperature receptors

The proportion of somatosensory or motor cortex devoted to a particular body region depends upon how important sensory or motor information is for that part.

- For example, more brain surface is committed to sensory and motor communication with the hands than with the entire torso.
- Impulses transmitted from receptors to specific areas of somatosensory cortex enable us to associate pain, touch, pressure, heat or cold with specific parts of the body receiving those stimuli.

A complicated interchange of signals among receiving centers and association centers produces our sensory perceptions.

- The special senses – vision, hearing, smell and taste – are integrated by cortical regions other than the somatosensory cortex.
- Each of these functional regions, as well as, the somatosensory cortex, cooperate with an adjacent association area.

A.   Integration and Higher Brain Functions

1.   Arousal and Sleep

Sleep and *arousal* are controlled by several centers in the cerebrum and brainstem; the most important is the *reticular formation*, which:

- Is a group of 90 separate brain nuclei (brain ganglia) which extend from the medulla to the thalamus; almost all neuron processes which reach the cerebral cortex pass through the reticular formation.
- Serves as a filter that selects what sensory information will reach the cortex.

Electrical potential between areas of the cortex can be measured and recorded in a graph called an *electroencephalogram* or *EEG*. Different states of arousal produce different patterns of electrical activity or *brain waves*, which can be classified into four types:

- *Alpha waves* – slow synchronous waves; produced in the relaxed closed-eye state of wakefulness.
- *Beta waves* – faster and less synchronous than alpha waves; produced during mental alertness such as occurs during problem solving.
- *Theta waves* – more irregular than beta waves; predominate in the early stages of sleep.
- *Delta waves* – slow, high amplitude, highly synchronized waves; occur during deep sleep

Sleep is a dynamic process during which a sleeping person alternates between two types of sleep: *nonrapid eye movement sleep (NREM)* and *rapid eye movement sleep (REM)*.

- *NREM sleep.* In the early stages, the brain produces slow, regular theta waves; during deeper sleep, it produces high amplitude delta waves.
- *REM sleep.* Characterized by rapid eye movement and desynchronized EEG similar to that of wakefulness. Most dreaming occurs during REM sleep.

2. **Right Brain/Left Brain**

   Lateralization (right brain/left brain) refers to the fact that the association areas of the cerebral cortex are not bilaterally symmetrical; each side of the brain controls different functions.
   - The left hemisphere controls speech, language, and calculation.
   - The right hemisphere controls artistic ability and spatial perception.
   - The *corpus callosum* transfers information between the left and right hemispheres. Severing the corpus callosum will not alter perception, but will dissociate sensory input from spoken response.

3. **Language and Speech**

   Language and *speech* are controlled by two areas on the left hemisphere of the cerebral cortex.
   - *Wernicke's area* stores information required for speech content (arrangement of learned words in grammatical order).
   - *Broca's area* contains necessary information for speech production.
   - Information for speech delivery is formulated in the Wernicke's area and verbalization is coordinated by the Broca's area which programs the motor cortex to move the tongue, lips, and speech muscles.
   - Damage to either area results in some form of *aphasia*, the inability to speak coherently.

4. **Emotions**

   Emotions depend on interactions between the cerebral cortex and the *limbic system*, a group of giant nuclei and interconnecting axon tracts in the forebrain.
   - Includes parts of the thalamus, hypothalamus, and inner portions of the cerebral cortex, including two nuclei called the *amygdala* and *hippocampus*.
   - Cerebral components of the limbic system are linked to the *prefrontal cortex*, which is involved in complex learning, reasoning, and personality, so there is a close relationship between emotion and thought.
   - Frontal lobotomy, used to treat severe mental illness, is the surgical destruction of the limbic cortex or its connection with the prefrontal cortex.

5. **Memory**

   Memory is the ability to store and retrieve information related to previous experiences.
   - Memory occurs in two stages: short-term memory and long-term memory.
   - *Short-term memory* reflects immediate sensory perceptions of an object or idea and occurs before the image is stored.
   - *Long-term memory* is stored information that can be recalled at a later time.
   - Transfer of information from short-term to long-term memory is enhanced by rehearsal, favorable emotional state, and association of new information with previously learned and stored information.

Fact memory differs from skill memory.

- Fact memory involves conscious and specific retrieval of data from long-term memory.
- Skill memory usually involves motor activities learned by repetition which are recalled without consciously remembering specific details.

There is no highly localized memory trace in the nervous system; instead, memories are stored in certain association areas of the cortex.

Fact memory involves a pathway in which sensory information is transmitted from the cerebral cortex to the *hippocampus* and *amygdala* which are two parts of the limbic system.

- The *amygdala* may filter memory, labeling information to be saved by tying it to an event or emotion.
- In the hippocampus, certain synapses functionally change and show an enhanced response by a postsynaptic cell. Such a change is called *long-term potentiation (LTP)*:
  ⇒ Results from brief, repeated action potentials that strongly depolarize the postsynaptic membrane, so an action potential from the presynaptic cell has a much greater effect at the synapse than before.
  ⇒ Lasts for hours, days or weeks and may occur when a memory is stored or learning takes place.
  ⇒ Mechanism involves presynaptic release of glutamate – an excitatory neurotransmitter.
    ◊ Glutamate binds with postsynaptic receptors and opens gated channels highly permeable to calcium ions.
    ◊ Ca^{2+} influx triggers intracellular changes that induce LTP.
  ⇒ Postsynaptic neurons may also change to enhance LTP.
    ◊ In a positive feedback loop, the affected postsynaptic cell may signal the presynaptic cell to release more glutamate, enhancing LTP.
    ◊ A likely messenger for this backward signaling is the local mediator nitric oxide.

REFERENCES

Campbell, N. *Biology*. 4th ed. Menlo Park, California: Benjamin/Cummings, 1996.

Marieb, E.N. *Human Anatomy and Physiology*. 3rd ed. Redwood City, California: Benjamin/Cummings, 1995.

CHAPTER 45

SENSORY AND MOTOR MECHANISMS

OUTLINE

I. Sensory receptors detect changes in the external and internal environment: *an overview*
   A. Sensation and Perception
   B. The General Function of Sensory Receptors
   C. Types of Receptors

II. Photoreceptors contain light-absorbing visual pigments: *a closer look*
   A. Invertebrates
   B. Structure and Function of the Vertebrate Eye
   C. Signal Transduction in the Eye
   D. Visual Integration

III. Hearing and balance are related in most animals: *a closer look*
   A. The Human Ear
   B. Balance and Equilibrium in Humans
   C. Hearing and Equilibrium in Other Vertebrates
   D. Sensory Organs for Hearing and Equilibrium in Invertebrates

IV. The interacting senses of taste and smell enable animals to detect many different chemicals: *a closer look*

V. Movement is a hallmark of animals

VI. Skeletons support and protect the animal body and are essential to movement
   A. Hydrostatic Skeletons
   B. Exoskeletons
   C. Endoskeletons

VII.   Muscles move skeletal parts by contracting
    A.   Structure and Function of Vertebrate Skeletal Muscle
    B.   The Molecular Mechanism of Muscle Contraction
    C.   The Control of Muscle Contraction
    D.   Graded Contractions of Whole Muscles
    E.   Fast and Slow Muscle Fibers
    F.   Other Types of Muscles

OBJECTIVES

After reading this chapter and attending lecture, the student should be able to:
1.   Differentiate between sensation and perception.
2.   Give the general function of a receptor cell and explain the five processes involved in this function.
3.   Explain the difference between exteroreceptors and interoreceptors.
4.   List and describe the energy stimulus of the five types of receptors.
5.   Using a cross-sectional diagram of human skin, identify the various receptors present and explain the importance of having near-surface and deep-layer receptors for such stimuli as pressure.
6.   Compare and contrast the structure and processing of light in eye cups of *Planaria*, compound eyes of insects, and single-lens eyes of mollusks.
7.   Using a diagram of the vertebrate eye, identify and give the function of each structure.
8.   Describe how the rod cells and cone cells found in the vertebrate eye function.
9.   Explain how retinal signals following a horizontal pathway can enhance visual integration.
10.   Using a diagram of the human ear, identify and give the function of each structure.
11.   Explain how the mammalian ear functions as a hearing organ.
12.   Explain how the mammalian ear functions to maintain body balance and equilibrium.
13.   Compare the hearing and equilibrium systems found in non-mammalian vertebrates.
14.   Describe the structure and function of statocysts.
15.   Explain how the chemoreceptors involved with taste and smell function.
16.   List any advantages or disadvantages associated with moving through:
    a.   An aquatic environment.
    b.   A terrestrial environment.
17.   Give the three functions of a skeleton.
18.   Describe how hydrostatic skeletons function and explain why they are not found in large, terrestrial organisms.
19.   Explain how the structure of the arthropod exoskeleton provides both strength and flexibility.
20.   Explain the adaptive advantage of having different types of joints in different locations in the vertebrate skeleton.
21.   Explain how the skeleton combines with an antagonistic muscle arrangement to provide a mechanism for movement.
22.   Using a diagram, identify the basic components of skeletal muscle.
23.   Explain how muscles contract according to the sliding filament model of contraction.
24.   Describe the processes involved in excitation-contraction coupling.
25.   List and explain the two mechanisms responsible for graded contraction of muscles.
26.   Explain the adaptive advantage of possessing both slow and fast muscle fibers.
27.   Distinguish among skeletal muscle, cardiac muscle and smooth muscle.

KEY TERMS

sensation
perception
sensory receptor
reception
transduction
amplification
transmission
receptor potential
integration
sensory adaptation
exteroreceptors
interoreceptors
mechanoreceptors
Pacinian corpuscles
Meissner's
  corpuscles
Merkel's discs
muscle spindle
hair cell
chemoreceptor
gustatory receptor
olfactory receptor
electromagnetic
  receptor
photoreceptor
thermoreceptor
Ruffini's end organs
end-bulbs of Krause
pain receptor
nociceptor
eye cup

compound eyes
ommatidia
single-lens eye
sclera
choroid
cornea
iris
pupil
retina
lens
ciliary body
aqueous humor
vitreous humor
accommodation
rod cells
cone cells
fovea
retinal
opsin
rhodopsin
photopsins
bipolar cells
ganglion cells
horizontal cells
amacrine cells
lateral inhibition
optic chiasma
lateral geniculate
  nuclei
primary visual
  cortex

pinna
auditory canal
tympanic membrane
middle ear
malleus
incus
stapes
inner ear
oval window
auditory tube
cochlea
organ of Corti
perilymph
endolymph
round window
utricle
saccule
semicircular canals
otoliths
lateral line system
neuromast
Weberian apparatus
statocysts
setae
taste buds
hydrostatic skeleton
exoskeleton
cuticle
chitin
endoskeleton
skeletal muscle

myofibrils
myofilaments
thin filaments
thick filaments
sarcomere
Z lines
I band
A band
H zone
sliding filament
  model of muscle
  contraction
cross-bridges
tropomyosin
troponin complex
sarcoplasmic
  reticulum
transverse tubules
phosphagens
creatine phosphate
motor unit
multiple motor unit
  summation
wave summation
muscle twitch
tetanus
myoglobin
cardiac muscle
intercalated discs
smooth muscles

LECTURE NOTES

I. Sensory receptors detect changes in the external and internal environment: *an overview*

An animal's interaction with its environment depends on the processing of sensory information and the generation of motor output.

- Sensory receptors receive information from the environment.
- Motor effectors carry out the movement in response to the sensory information.

A. Sensation and Perception

<u>Sensations</u> = Action potentials traveling along sensory neurons that are interpreted by the brain as perceptions; what one perceives depends on what part of the brain receives the impulse.

B. The General Function of Sensory Receptors

<u>Sensory receptors</u> = Structures that transmit information about changes in an animal's internal and external environment.

- Are usually modified neurons occurring singly or within groups in sensory organs.
- Specialized to respond to specific stimuli and convert stimuli energy into electrochemical energy of action potentials.
- All receptor cells have the same five functions: reception, transduction, amplification, transmission, and integration.

<u>Reception</u> = Ability of a cell to absorb the energy of a stimulus.

- Each receptor has a specific region which absorbs a particular type of energy.

<u>Transduction</u> = The conversion of stimulus energy into electrochemical activity of nerve impulses (action potentials).

- Stimulus energy may change membrane permeability of the receptor cell, open or close ion channel gates, or increase ion flow by stretching the receptor cell membrane.

Amplification of stimulus energy which is too weak to be carried into the nervous system often occurs.

- May take place in accessory structures or be a part of the transduction process.

Initial responses to stimuli during transmission are graded changes in membrane potential (*receptor potential*); the stimulus modulates action potential *frequency*.

Receptor signals are *integrated* through summation of graded potentials.

- *Sensory adaptation* is a decrease in sensitivity during continued stimulation; a type of integration which results in selective information being sent to the CNS.
- The threshold for firing in receptor cells varies with conditions resulting in a change in receptor sensitivity.
- Sensory information integration occurs at all levels in the nervous system.

C. **Types of Receptors**

Exteroreceptors monitor the external environment while *interoreceptors* monitor changes in the body's internal environment. All receptors can be grouped into 5 types depending on the type of energy to which they respond:

1. *Mechanoreceptors* are stimulated by their physical deformation caused by pressure, stretch, motion, sound, etc. (See Campbell, Figure 45.1)
   - Bending of the plasma membrane increases its permeability to Na^+ and K^+ resulting in a receptor potential.
   - In human skin, *Pacinian corpuscles* deep in the skin respond to strong pressure, while *Meissner's corpuscles* and *Merkel's discs*, closer to the surface, detect light touch.
   - Muscle spindles are stretch receptors (a type of interoreceptor) that monitor the length of skeletal muscles, as in the reflex arc.
   - *Hair cells* detect motion.

2. *Chemoreceptors* include general receptors that sense total solute concentration (e.g. osmoregulators of the mammalian brain), receptors that respond to individual molecules, and those that respond to categories of related chemicals (e.g. *gustatory* and *olfactory receptors*).

3. *Electromagnetic receptors* respond to electromagnetic radiation such as light (*photoreceptors*) and magnetic fields (*magnetoreceptors*).

4. *Thermoreceptors* respond to heat or cold and help regulate body temperature.
   - In human skin, *Ruffini's end organs* may be heat receptors, and *end-bulbs of Krause* may be cold receptors.
   - The interothermoreceptors in the hypothalamus function as the primary temperature control of the mammalian body.

5. *Nociceptors* = A class of naked dendrites that function as *pain receptors*.
   - Different groups respond to excess heat, pressure, or specific chemicals released from damaged or inflamed tissue.
   - Prostaglandins increase pain by lowering receptor thresholds; aspirin reduces pain by inhibiting prostaglandin synthesis.

II. **Photoreceptors contain light-absorbing visual pigments:** *a closer look*

A. **Invertebrates**

The *eye cup* of planarians is a simple light receptor that responds to light intensity and direction without forming an image. (See Campbell, Figure 45.4)
   - An opening on one side of the cup permits light to enter; the opening to one cup faces left and slightly forward, the other cup opens right and slightly forward.
   - Light enters the opening and stimulates photoreceptors that contain light-absorbing pigments.
   - Planaria move away from light sources to avoid predators.

- The proper direction is determined by the brain, which compares the rate of nerve impulses coming from the two cups; the animal turns until the impulses from each cup are equal and minimal.

Two types of image-forming eyes evolved in invertebrates:

1. A *compound eye* contains thousands of light detectors called *ommatidia* each with its own cornea and lens. (See Campbell, Figure 45.5)
   - Found in insects, crustaceans, and some polychaete worms.
   - Results in a mosaic image.
   - More acute at detecting movement partly due to rapid recovery of photoreceptors.
   - Superimposition eyes have ommatidia with lens that work like prisms and parabolic mirrors, focusing light entering several ommatidia onto photoreceptor (increases sensitivity to light).

2. In a *single-lens eye*, one lens focuses light onto the *retina*, a bilayer of photosensitive receptor cells.
   - Found in some jellyfish, polychaetes, spiders and many mollusks.

B. Structure and Function of the Vertebrate Eye

The parts of the vertebrate eye are structurally and functionally diverse. (See Campbell, Figure 45.6)

- The vertebrate eye consists of a tough outer layer of connective tissue, the *sclera*, and a thin inner pigmented layer, the *choroid*.
- The *cornea* is located in front and is a transparent area of the sclera; it allows light to enter the eye and acts as a fixed lens.
- The anterior choroid forms the *iris*, which regulates the amount of light entering the *pupil*. The iris is pigmented and gives the eye color; the pupil is the hole in the center of the iris.
- The *retina* is the innermost layer of the eyeball; it contains photoreceptor cells which transmit signals from the optic disc, where the optic nerve attaches to the eye.
- The lens and ciliary body divide the eye into two chambers: a small chamber between the lens and cornea; and a large chamber within the eyeball.
- The *ciliary body* produces *aqueous humor* that fills the cavity between the lens and cornea.
- *Vitreous humor* fills the cavity behind the lens and comprises most of the eye's volume.
- Both aqueous humor and vitreous humor help to focus light onto the retina.

The *lens* is a transparent, protein disc that focuses an image onto the retina by changing shape (*accommodation*). (See Campbell, Figure 45.7)

- Is nearly spherical when focusing on near objects and is flatter when focusing at a distance.
- Controlled by the ciliary muscle.

C. **Signal Transduction in the Eye**

Cells in the retina transduce stimuli (caused by the lens focusing a light image onto the retina) into action potentials.

The photoreceptors of the eye are rod cells and cone cells. (See Campbell, Figure 45.8)

- Their relative numbers in the retina are partly correlated with whether an animal is diurnal or nocturnal.
- The light absorbing molecule is *retinal*, synthesized from vitamin A, which is bonded to a characteristic kind of protein, *opsin*.

Rod cells are sensitive to light but do not distinguish colors.

- Found in greatest density at lateral regions of the retina; completely absent from the *fovea* (visual field center).
- Contains *rhodopsin*, which is inactivated by light.

Cone cells are responsible for daytime color vision.

- Consists of 3 subclasses, red cones, green cones and blue cones, each with its own type of opsin associated with retinal for form visual pigments (*photopsins*).
- Most dense at the fovea.

Retinal responds to light by changing shape; this triggers a chain of metabolic events that hyperpolarizes the photoreceptor cell membrane; thus, a decrease in chemical signal to the cells with which photoreceptors synapse serves as the message.

- Rod and cone cell axons synapse with neurons called *bipolar cells*, which in turn synapse with *ganglion cells*.
- *Horizontal cells* and *amacrine cells*, neurons in the retina, help integrate the information, after which ganglion cell axons convey action potentials along the optic nerve to the brain.

D. **Visual Integration**

Integration of visual information begins at the retina.

Rod and cone cell signals may follow vertical or lateral pathways.

- Vertical pathways involve information passing directly from receptor cells to bipolar cells to ganglion cells.
- Lateral pathways involve: horizontal cells carrying signals from one rod or cone to other receptor cells and several bipolar cells; amacrine cells spread the signals from one bipolar cell to several ganglion cells.

Lateral inhibition = Horizontal cells, stimulated by rod or cone cells, stimulate nearby receptors and inhibit more distant receptors and non-illuminated bipolar cells, thus sharpening image edges and enhancing contrast. (See Campbell, Figure 45.10)

- Occurs at all levels of visual processing.

Optic nerves from each eye meet at the *optic chiasma*. (See Campbell, Figure 45.11)

- Has nerve tracts arranged so that what is viewed in the left field of view is transmitted to the right side of the brain, and vice versa.

- Ganglion axons usually continue through the *lateral geniculate nuclei* of the thalamus, and these neurons continue back to the *primary visual cortex* in the occipital lobe of the cerebrum.

- Additional neurons carry information to more sophisticated visual processing centers in the cortex.

III.   Hearing and balance are related in most animals: *a closer look*

Hearing and balance are related in most animals and involve mechanoreceptors.

A.   The Human Ear

Sound waves are collected by the outer ear (the external *pinna* and the *auditory canal*) and are channeled to the *tympanic membrane* of the *middle ear*. (See Campbell, Figure 45.12)

The sound waves cause the tympanic membrane to vibrate at the same frequency; the tympanic membrane transmits the waves to three small bones — the *malleus, incus* and *stapes* — which amplify and transmit the mechanical movements of the membrane to the *oval window*, a membrane of the cochlea surface.

- Oval window vibrations produce pressure waves in the fluid (endolymph) in the coiled cochlea of the inner ear.

The pressure waves vibrate the *basilar membrane* (forms the floor of the cochlear duct) and the attached *organ of Corti*, which contains receptor *hair cells*.

- The bending of the hair cells against the *tectorial membrane* depolarizes the hair cells and causes them to release neurotransmitter that triggers an action potential in a sensory neuron which then carries sensations to the brain through the auditory nerve.

Volume is determined by the amplitude of the sound wave; pitch is a function of sound wave frequency.

- The greater the amplitude of a sound, the more vigorous the vibrations; this results in more bending of the hair cells and more action potentials.

- Different sound frequencies affect different areas of the basilar membrane, thus some receptors send more action potentials than others.

B.   Balance and Equilibrium in Humans

The sense of balance is centered in the inner ear; behind the oval window is a vestibule that contains two chambers, the *utricle* and the *saccule*; the utricle opens into three *semicircular canals*.

- Hair cells in the utricle and saccule respond to changes in head position with respect to gravity and movement in one direction.

- Hair cells are arranged in clusters with their hairs projecting into a gelatinous material containing numerous *otoliths* (small calcium carbonate particles).

- The otoliths are heavier than endolymph in the saccule and utricle; gravity pulls them down on the hairs of receptor cells, thus causing a constant series of action potentials indicating position of the head.
- Semicircular canals detect rotation of the head due to endolymph movement against the hair cells.

C. Hearing and Equilibrium in Other Vertebrates

Fishes and aquatic amphibians have a *lateral line system* running along both sides of the body. (See Campbell, Figure 45.15)

- Mechanoreceptors called *neuromasts* contain hair cell clusters whose hairs are embedded in a gelatinous cap, the *cupula*.
- Water enters the system through numerous pores on the animal's surface and flows along the tube past the neuromasts.
- Pressure of moving water bends the cupula causing an action potential in the hair cells.
- Provides information about the body's movement direction and velocity of water currents, and movements or vibrations caused by predators and prey.

The inner ear of a fish has no eardrum, does not open to the outside of the body and has no cochlea; a saccule, utricle, and semicircular canals are present.

- Sound waves are conducted through the skeleton of the head to the inner ear, this sets otoliths in motion, stimulating the hair cells.
- Some fish have a *Weberian apparatus*, a series of 3 bones which conducts vibrations from the swim bladder to the inner ear.
- Fish can hear higher frequencies due to their inner ears.

In terrestrial amphibians, reptiles and birds, sound is conducted from the tympanic membrane to the inner ear by a single bone, the stapes.

D. Sensory Organs for Hearing and Equilibrium in Invertebrates

The body hairs of many insects vibrate in response to sound waves of specific frequencies.

- Fine hairs on the antennae of male mosquitoes detect the hum produced by a female's wings.
- Vibrating body hairs of some caterpillars detect predatory wasps.

Many insects also have "ears," commonly located on their legs, consisting of a tympanic membrane stretched over an internal air chamber containing receptor cells that send nerve impulses to the brain. (See Campbell, Figure 45.16)

Most invertebrates have mechanoreceptors called *statocysts* that function in their sense of equilibrium. (See Campbell, Figure 45.17)

- Gravity causes statoliths (dense granules) to settle to the low point in a chamber, stimulating hair cells in that location.
- Statocysts are located along the bell fringe of many jellyfish and at the antennule bases in lobsters and crayfish.

IV.   The interacting senses of taste and smell enable animals to detect many different chemicals: *a closer look*

Gustation (taste) and olfaction (smell) depend on chemoreceptors that detect specific chemicals in the environment.

Insects have taste receptors within sensory hairs called *setae* on the feet and mouthparts; olfactory setae are usually located on antennae.

- Several chemoreceptor cells, each responding to a particular chemical, are located on each tasting hair; integrating impulses from the different receptors permits distinguishing many tastes.

In humans and other mammals, receptor cells for taste are organized into *taste buds* scattered in several areas of the tongue and mouth.

- Sweet, sour, salty and bitter are detected in distinct regions of the tongue.
- These tastes are associated with specific molecular shapes and charges that bind to separate receptor molecules.

In humans, olfactory receptor cells line the upper nasal cavity and send impulses along their axons directly to the olfactory bulb of the brain. (See Campbell, Figure 45.20)

- Receptive ends of the cells contain cilia that extend into the coating layer of mucus lining the nasal cavity.
- Specific receptors respond to certain odorous molecules by depolarizing.
- Respond to airborne chemicals that diffuse into the region.
- Taste and olfaction have different receptors but interact.

V.   Movement is a hallmark of animals

Different modes of transportation (running, flying, swimming) have evolved along with adaptations of animals to overcome the difficulties associated with each type of locomotion.

- Swimming animals must overcome resistance; thus, many are fusiform in body shape.
- A walking or running land animal must support itself and move against gravity, and also overcome inertia with each step by accelerating a leg from a standing start; strong skeletal support and muscles evolved.
- Flying animals do not use a skeleton for support during motion, and must almost completely overcome gravity to become airborne; wings must provide enough lift to overcome gravity.

At the cellular level, all movements are based on the contractile systems of microfilaments and microtubules.

VI.   Skeletons support and protect the animal body and are essential to movement

Skeletons function in support, protection and movement.
- Help maintain shape of aquatic animals.
- Hard skeletons protect soft body tissues.
- Skeletons provide a firm attachment against which muscles can work during movement.

A.   Hydrostatic Skeletons

Hydrostatic skeletons consist of fluid held under pressure in a closed body compartment.
- Found in most cnidarians, flatworms, nematodes and annelids.
- Control form and movement by using muscles to change the shape of fluid-filled compartments.
- Provide no protection and could not support a large land animal.

B.   Exoskeletons

Exoskeleton = Hard encasement deposited on the surface of an animal.
- Mollusks are enclosed in calcareous shells secreted by the mantle.
- Arthropods are enclosed in a *cuticle* secreted by the epidermis that contains *chitin* and also contains quinones that add extra strength in areas where protection is most important.
- Exoskeletons must be shed (molted) as the animals grow.

C.   Endoskeletons

Endoskeleton = Hard supporting elements buried within the soft tissues of an animal.
- Sponges possess hard spicules of inorganic material or softer protein fibers.
- Echinoderms have ossicles composed of magnesium carbonate and calcium carbonate forming hard plates beneath the skin.
- Chordates have cartilage and/or bone skeletons divided into several areas.
- The vertebrate frame is divided into an axial skeleton (the skull, vertebral column, and rib cage) and an appendicular skeleton (limb bones, pectoral and pelvic girdles).
- Bones of the vertebrate act in support and as levers when their attached muscles contract.

VII.   Muscles move skeletal parts by contracting

Animal movement is based on contraction of muscles working against some kind of skeleton.
- Muscles always contract actively; they can extend only passively.
- Ability to move a body part in opposite directions requires that muscles be attached to the skeleton in antagonistic pairs.

A. Structure and Function of Vertebrate Skeletal Muscle

Skeletal muscle = Bundle of long fibers running the length of the muscle; attached to and responsible for movement of bones.

Each fiber is a single cell with many nuclei; consists of bundles of smaller *myofibrils* arranged longitudinally. Two kinds of myofilaments are found in each myofibril.

- *Thin filaments* consist of two strands of actin and one strand of regulatory protein coiled together.
- *Thick filaments* are staggered arrays of myosin molecules.

Sarcomere = Unit of organization of skeletal muscle. (See Campbell, Figure 45.25)
- *Z lines* are the borders of the sarcomere; aligned in adjacent myofibrils.
- *I bands* are areas near the edge of the sarcomere containing only thin filaments.
- *A bands* are regions where thick and thin filaments overlap and correspond to the length of the thick filaments.
- *H zones* are areas in the center of the A bands containing only thick filaments.

B. The Molecular Mechanism of Muscle Contraction

Muscle contraction reduces the length of each sarcomere.

Sliding filament model = Mechanism of muscle contraction. (See Campbell, Figure 45.26)
- Thin filaments ratchet across thick filaments to pull the Z lines together and shorten the sarcomere; the myofilaments themselves do not contract.
- Myosin molecules on thick filaments attach to actin on the thin filament to form a crossbridge. It then bends inward, pulling the thin filament toward the center of the sarcomere, breaks the crossbridge, and forms a new crossbridge further down.
- Energy for crossbridge formation comes from the hydrolysis of ATP by the head region of myosin. (See Campbell, Figure 45.27)

Skeletal muscles contract when stimulated by motor neurons.

In a muscle at rest, myosin-binding sites on the actin are blocked by the regulatory protein strand (*tropomyosin*) in the thin filament and by the *troponin complex* located at each binding site.

- An action potential in the motor neuron innervating the muscle cell causes the axon to release acetylcholine.
- Infoldings in the muscle cell plasma membrane (*transverse tubules*) carry the resulting graded depolarization deep into the muscle cell.
- The *sarcoplasmic reticulum* membrane becomes depolarized and releases its store of Ca^{2+}.
- The calcium ions bind to troponin, causing the thin filament to change shape and expose the myosin-binding sites; the muscle can then contract.
- The contraction is terminated as calcium is pumped out of the cytoplasm by the sarcoplasmic reticulum; as the calcium concentration falls, the tropomyosin-troponin complex again blocks the myosin-binding sites.
- *Phosphagens* (*creatine phosphate* in vertebrates) provide most of the necessary energy.

C. Graded Contractions of Whole Muscles

Graded contractions of skeletal muscles are due to summation of multiple motor unit activity (*recruitment*) and wave summation.

- Motor neurons usually deliver their stimuli rapidly, resulting in smooth contraction typical of *tetanus* (sustained contraction) rather than the jerky actions of muscle twitches.
- A motor unit consists of a single motor neuron and all the muscle fibers it controls; all fibers in the motor unit contract as a group when the motor neuron fires.
- As more motor neurons are recruited by the brain, tension in the muscle progressively increases.

D. Fast and Slow Muscle Fibers

Duration of muscle contraction is controlled by how long the calcium concentration in the cytoplasm remains elevated.

Slow fibers have longer-lasting twitches because they have less sarcoplasmic reticulum; thus, Ca^{2+} remains in the cytoplasm longer.

- Have many mitochondria, a rich blood supply, and the oxygen-storing protein *myoglobin*.
- Used to maintain posture since they can sustain long contractions.

Fast fibers have short duration twitches and are used in fast muscles for rapid, powerful contractions.

- Some are able to sustain long periods of repeated contractions without fatiguing.

E. Other Types of Muscles

Vertebrate *cardiac muscle* is found only in the heart.

- Is striated.
- Muscle cells are branched, and the junction between cells contain *intercalated discs* that electrically couple all heart muscle cells, allowing coordinated action.
- Cells can also generate their own action potentials.
- Rhythmic depolarizations due to pacemaker channels in the plasma membrane trigger action potentials which last up to 20 times longer than those for skeletal muscle and have long refractory periods.

Smooth muscles lack striations and contain less myosin; the myosin is not associated with specific actin strands.

- Generates less tension but can contract over a greater range of lengths.
- Does not have a transverse tubule system or a well developed sarcoplasmic reticulum; calcium ions must enter the cytoplasm through the plasma membrane during an action potential.
- Contractions are relatively slow but there is greater range of control.
- Found mainly in the walls of blood vessels and digestive tract organs.

Invertebrates have muscles similar to the skeletal and smooth muscles of vertebrates, but with some interesting adaptations.

- Arthropod skeletal muscles are very similar to vertebrate skeletal muscle.
- Insect wings actually beat faster than action potentials arrive from the CNS since the flight muscles are capable of independent, rhythmic contractions.
- The thick filaments of muscles in clams which hold the shell closed contain *paramyosin*; this unique protein allows the muscles to stay in a fixed state of contraction for up to a month.

REFERENCES

Campbell, N. *Biology*. 4th ed. Menlo Park, California: Benjamin/Cummings, 1996.

Marieb, E.N. *Human Anatomy and Physiology*. 3rd ed. Redwood City, California: Benjamin/Cummings, 1995.

CHAPTER 46

AN INTRODUCTION TO ECOLOGY: DISTRIBUTION AND ADAPTATIONS OF ORGANISMS

OUTLINE

I. Ecology is the scientific study of the interactions between organisms and their environments

II. Basic ecology provides a scientific context for evaluating environmental issues

III. Ecological research ranges from the adaptations of organisms to the dynamics of ecosystems
   A. The Questions of Ecology
   B. Ecology as an Experimental Science
   C. Ecology and Evolution

IV. Climate and the other abiotic factors are important determinants of the biosphere's distribution of organisms
   A. Important Abiotic Factors
   B. Climate and the Distribution of Biomes
   C. Global Climate Patterns
   D. Local and Seasonal Effects on Climate

V. The costs and benefits of homeostasis affect an organism's responses to environmental variation
   A. Regulators and Conformers
   B. The Principle of Allocation

VI. The response mechanisms of organisms are related to environmental grain and the time scale of environmental variation
   A. Environmental Grain
   B. Behavioral Responses
   C. Physiological Responses
   D. Morphological Responses
   E. Adaptation Over Evolutionary Time

VII. The geographical distribution of terrestrial biomes is based mainly on regional variations in climate
    A. Tropical Forest
    B. Savanna
    C. Desert
    D. Chaparral
    E. Temperate Grassland
    F. Temperate Deciduous Forest
    G. Taiga (Coniferous Forest)
    H. Tundra

VIII. Aquatic ecosystems, consisting of freshwater and marine biomes, occupy the largest part of the biosphere
    A. Ponds and Lakes
    B. Streams and Rivers
    C. Wetlands
    D. Estuaries
    E. Marine Zones: An Introduction
    F. The Intertidal Zones
    G. Coral Reefs
    H. The Oceanic Pelagic Biome
    I. Benthos

OBJECTIVES

After reading this chapter and attending lecture, the student should be able to:
1. Explain why the field of ecology is a multidisciplinary science.
2. Distinguish among physiology, ecology, community ecology and ecosystem ecology.
3. Describe the relationship between ecology and evolution.
4. Explain the importance of temperature, water, light, soil and wind to living organisms.
5. Explain the "Principle of Allocation."
6. Describe how environmental changes may produce behavioral, physiological, morphological or adaptive responses in organisms.
7. Explain the concept of environmental grain and under what situation(s) a single environment may be both coarse-grained and fine-grained.
8. Describe the characteristics of the major biomes: tropical forest, savanna, desert, chaparral, temperate grassland, temperate forest, taiga, tundra.
9. Compare and contrast the types of freshwater communities.
10. Using a diagram, identify the various zones found in the marine environment.

KEY TERMS

ecology	biosphere	tropical dry forest	aphotic zone
abiotic factor	rainshadow	tropical deciduous forest	benthic
biotic factor	principle of allocation	savanna	intertidal zone
organismal ecology	regulator	chaparral	neritic zone
population ecology	conformer	taiga	oceanic zone
community ecology	acclimation	tundra	pelagic zone
ecosystem	succession	permafrost	benthic zone
biome	climograph	photic zone	abyssal zone
ecosystem ecology	forb	wetlands	

LECTURE NOTES

Ecology is the scientific study of the interactions of organisms and their environments.
- A complex and critical area of biology.
- There are three key words in the definition: *scientific*, *environment*, and *interactions*.

I. **Ecology is the scientific study of the interactions between organisms and their environments**

The *scientific* nature of ecology involves using observations and experiments to test hypothetical explanations of ecological phenomena.
- A multidisciplinary field examining questions from all areas of biology as well as many physical sciences.

The *environment* of an organism includes both *biotic* and *abiotic* factors.
- Biotic factors include *all* other organisms that are a part of any individual organism's environment.
- Abiotic factors are the temperature, light, water, nutrients, etc. to which an organism is exposed.

The *interactions* between organisms and their environments involve how the environment affects an organism *and* how an organism can change the environment.
- Cyanobacteria began to use sunlight for photosynthesis about three billion years ago.
  ⇒ Oxygen, a by-product of photosynthesis, accumulated and resulted in an aerobic environment.
- The shading of the forest floor by trees sometimes makes the floor unsuitable (due to reduced light) for their offspring to grow.

II. Basic ecology provides a scientific context for evaluating environmental issues

Although distinct, basic ecology and environmental issues have many connections.

- To properly address environmental problems, it is necessary to understand the relationships between organisms and their environments.
- Current environmental awareness began with Rachel Carson's *Silent Spring* which exposed the fact that widespread use of pesticides often affects nontarget organisms.
- The realization that Earth is a finite resource and that pollution, overpopulation, and habitat destruction are threatening that resource is a concern of most people.

III. Ecological research ranges from the adaptations of organisms to the dynamics of ecosystems

A. The Questions of Ecology

Ecology can be divided into four comprehensive levels of inquiry:

- *Organismal ecology* studies the behavioral, physiological, and morphological ways individuals meet abiotic environmental challenges.
    - ⇒ The distribution of organisms is limited by their tolerance of abiotic conditions.
- *Population ecology* studies groups of individuals of the same species in a particular geographic area.
    - ⇒ Questions in population ecology concern factors that affect population size and composition.
- *Community ecology* studies all organisms that inhabit a particular area.
    - ⇒ Questions concern predation, competition and other interactions that affect community structure and organization.
- *Ecosystem ecology* studies all abiotic factors as well as communities of organisms in an area.
    - ⇒ Questions concern energy flow and chemical cycling among the abiotic and biotic components.

B. Ecology as an Experimental Science

Ecology has a long history as a descriptive science but is young as an experimental science.

- Very difficult to conduct experiments and control variables.
- Some test hypotheses in lab experiments and by manipulating populations and communities in field experiments.
- Some devise mathematical models which include important variables and hypothetical relationships; usually studied with the aid of a computer.

C. **Ecology and Evolution**

Short term interactions of organisms with their environments could have long term effects through natural selection. *Ecological time* translates into effects over *evolutionary time*.

- For example, predator-prey interactions may affect gene pools where individuals with protective coloration would become more prevalent.
- The current distribution and abundance of organisms are products of long-term evolutionary changes and on going interactions.

IV. **Climate and other abiotic factors are important determinants of the biosphere's distribution of organisms**

The *biosphere* is that portion of Earth inhabited by life and represents the sum of all communities and ecosystems.

- It is a thin layer consisting of seas, lakes, rivers, streams, the land to a soil depth of a few meters, and the atmosphere to an altitude of a few kilometers.

Biomes are the major types of communities and ecosystems typical of broad geographic areas.

- The different biomes form global patterns which reflect regional differences in climate and other abiotic factors.
- Abiotic factors such as temperature, humidity, salinity, and light influence the distribution of organisms.
- The patchiness of the global biosphere illustrates how different physical environments produce a mosaic of habitats.

A. **Important Abiotic Factors**

Some of the important abiotic factors that affect distribution of species include: temperature, water, sunlight, wind, rocks and soil, and periodic disturbances.

1. *Environmental temperature* affects biological processes and the ability of most organisms to regulate their body temperature.
    - Temperature greatly affects metabolism: few organisms have active metabolisms at temperatures close to 0°C, and temperatures above 45°C denature most essential enzymes.
    - The actual body temperature of ectotherms is affected by heat exchange with the environment.
        ⇒ Most animals maintain a body temperature only a few degrees above or below ambient temperature.
    - Even endotherms function best within the environmental temperature range to which they are adapted.

2. *Water* is essential and adaptations for water balance and conservation help determine a species' habitat range.
    - Marine and freshwater animals face the problems of regulating intracellular osmolarity; terrestrial animals face the problem of desiccation.

3. *Sunlight* provides the energy that drives nearly all ecosystems although only photosynthetic organisms use it directly as an energy source.
   - Light is not the most important limiting factor in terrestrial environments but may play a role as in the reduction of competition due to shading in a forest.
   - In aquatic environments, the distribution of photosynthetic organisms is limited by the intensity and quality of light.
     ⇒ Water selectively reflects and absorbs certain wavelengths; therefore, most photosynthesis occurs near the water surface.
   - The physiology, development, and behavior of many animals and plants are often sensitive to photoperiod.

4. *Wind* amplifies the effects of temperature by increasing heat loss by evaporation and convection.
   - Wind also increases the evaporation rate of animals and transpiration rate of plants, resulting in more rapid water loss.
   - Mechanical pressure of wind can effect plant morphology (for example, inhibiting growth of limbs on windward side of trees).

5. *Rocks and soil.* The physical structure, pH, and mineral composition of soil limit distribution of plants and hence animals that feed on those plants.
   - The composition of the substrate in a stream or river greatly influences the water chemistry, which in turn influences the plants and animals.
   - The type of substrate influences what animals can attach or burrow in intertidal zones.

6. *Periodic disturbances* such as fire, hurricanes, typhoons, and volcanic eruptions can devastate biological communities.
   - After which the area is recolonized by organisms or repopulated by survivors.
     ⇒ May go through a succession of changes.
   - Those disturbances that are infrequent (volcanic eruptions) do not illicit adaptations. Adaptations do evolve to periodically recurring disturbances such as fires.

B. Climate and the Distribution of Organisms

Climate is the prevailing weather conditions at a locality.
- The major components of climate are temperature, water, light, and wind.
- Climate has a major impact on the distribution of organisms.

A *climatograph* plots temperature and rainfall in a particular region.
- Usually in terms of annual means.
- Must be careful to distinguish a *correlation* between climate variables and biomes from *causation*.
  ⇒ Only a detailed study of a species' tolerances to water and temperature ranges could establish the direct effects of these variables.
- The fact that biomes overlap on climatographs indicate that factors other than mean temperature and rainfall play roles in determining biome location (i.e. soil composition).

C. Global Climate Patterns

Solar energy input and the Earth's movement in space determine the planet's global climate patterns.

- About 50% of the solar energy that reaches the atmosphere's upper layers is absorbed before it reaches the surface.

- Ultraviolet and certain other wavelengths are more readily absorbed by oxygen and ozone than other wavelengths.

- Some of the solar energy that reaches the Earth's surface is reflected back into the atmosphere; large amounts are absorbed by land, water, and organisms.

- The atmosphere, land, and water are heated when they absorb solar energy; this heating establishes the temperature variations, air movement cycles, and evaporation of water responsible for the latitudinal variations in climate.

Latitudinal variation in the intensity of sunlight results from the Earth's spherical shape; seasonal variation in solar radiation in the Northern and Southern Hemispheres are due to the Earth's tilt of 23.5° relative to its plane of orbit. (See Campbell, Figures 46.4 and 46.5)

- Only the *tropics* (23.5°N to 23.5°S) receive sunlight from directly overhead year round; the tilt causes solar radiation to change daily as the Earth rotates around the sun.

- The tropics receive the greatest annual input of solar radiation and show the least seasonal variation; only small variations in daylength and temperature occur.

- Seasonal variation in light and temperature increases steadily toward the poles; polar regions have long, cold winters with periods of continual darkness and short summers with periods of continual light.

A global circulation of air which creates precipitation and winds results from the intense solar radiation near the equator. (See Campbell, Figure 46.6)

- Evaporation of surface water due to high tropical temperatures causes warm, wet air masses to rise near the equator; this rising air creates an area of light, shifting winds (doldrums) along the equator.

- As these warm air masses rise, they expand and cool; cool air can hold less water vapor so the rising air masses drop large amounts of rain in the tropics.

- The cool, dry air masses flow toward the poles at high altitudes; they continue to cool as they move farther from the equator.

- Air mass density increases as they become cooler; they begin to descend toward the surface as cool, dry air masses at about 30° latitude.

- The air masses absorb water as they descend, thus creating arid climates around 30°N and 30°S.

- Some of the descending air masses flows toward the poles at low altitudes, the rest flows toward the equator.

- The air masses flowing toward the poles are warmed and rise again at about 60°N and 60°S; as these masses of air begin to cool with increased altitude, the water vapor is lost as precipitation.

- As air from this third cell reaches the higher altitude and cools, it flows toward the poles where it descends and flows back toward the equator.

The Earth's predictable wind patterns are established by air flowing in the circulation cells. The rotation of the Earth deflects the winds from a vertical path since land near the equator is moving faster than that at the poles.

- Tropical and subtropical tradewinds blow from east to west.
- In temperate zones, the predominating winds flow from west to east.

D. Local and Seasonal Effects on Climate

Although global climate patterns explain the geographic distribution of major biomes, local variations due to bodies of water and topographical features create a regional patchiness in climatic conditions.

- The local variation in climate and soil have a major influence on less widely distributed communities and individual species.

Proximity of large bodies of water affect local climates.

- Ocean currents are generated by the Earth's rotation and may heat or cool (depending on whether they are tropical or polar currents) air masses passing over them toward land; evaporation is also greater over the ocean than over land.
- Coastal areas are moister than inland areas at the same latitude; the California current flows from North to South along the west coast and helps form the cool, moist climate in this area.
- During warm summer days, air over land heats faster than that over the ocean or large inland lakes; this warmer air rises, drawing a cool breeze from the water across the land.

Topographical variations, such as mountains, also exert an influence on solar radiation, local temperature and rainfall.

- In the Northern Hemisphere, south-facing slopes receive more sunlight and are therefore warmer and drier than north-facing slopes.
  - ⇒ The vegetation differs with south-facing slopes being covered by shrubby, drought resistant plants while the north-facing slopes have forests.
- Air temperature declines 6° C for each 1000m increase in elevation.
  - ⇒ This parallels the decline in temperature associated with increasing latitude.
  - ⇒ For this reason, mountain communities are similar to those at lower elevations farther from the equator.
- Deserts commonly occur on leeward sides of mountains.
  - ⇒ This results from the movement of air masses over the mountain which produces a *rainshadow*.
  - ⇒ When warm, moist air moves over a mountain, its altitude is increased and it cools; cooling causes the air to release its moisture as rain on the windward side.
  - ⇒ The cooler, drier air then flows down the leeward side of the mountain; the cool dry air is warmed and absorbs moisture, producing the rainshadow.

The Earth's orbit around the sun causes seasonal changes in local conditions.

- The changing angle of the sun causes slight shifts in the wet and dry air masses on either side of the equator.
  - ⇒ These shifts result in the wet and dry seasons at 20° latitude where tropical deciduous forests grow.
- Seasonal changes in wind patterns produce variations in ocean currents, sometimes causing *upwellings* that bring nutrient rich water from the deep ocean layer to the surface.
- Seasonal temperature changes also cause the temperature profiles that develop during the summer in temperate zone ponds and lakes.
  - ⇒ These profiles reverse in the autumn and spring resulting in biannual mixing that brings nutrient-rich water from the bottom to the top. (See Campbell, Figure 46.8)

Climate also varies on a smaller scale, the *microclimate*. Microclimate refers to small areas within a habitat that may have very different conditions than the overall area (e.g. under a rock, a forest floor).

- Cleared areas in a forest generally show greater temperature extremes than the shaded forest floor due to greater solar radiation and wind currents.
- Low lying areas are usually moister than high ground and support different forms of vegetation.
- The area under a large stone or log is protected from extremes of temperature and moisture; a large number of small organisms usually live in such sheltered areas.

V. The costs and benefits of homeostasis affect an organism's responses to environmental variation

Organisms live in an integrated environment which includes various combinations of abiotic factors. The success of an organism reflects overall tolerance to the complete set of variables it encounters.

- Some polar organisms tolerate air temperatures of -70°C, while desert organisms survive at 45°C or higher.
- Aquatic organisms are found in waters with near zero salinity, while some tolerate salinities several times that of sea water.
- No organism can survive the full range of environmental conditions present on Earth; each population or species is distributed, at least in part, based on its tolerance for a specific subset of environmental conditions.
- Various anatomical structures and physiological mechanisms have evolved as adaptations to environmental constraints.

Organisms survive and reproduce in areas where environmental conditions to which they are adapted are found.

- The ability to tolerate one factor may be dependent on another factor; for example, many aquatic ectotherms can tolerate reduced oxygen at low temperatures, but not at high temperatures which cause higher metabolic rates.

A. Regulators and Conformers

Organisms faced with a fluctuation in an environmental variable may maintain the homeostasis of their bodies through behavioral and physiological mechanisms (*regulators*) or by allowing their internal conditions to vary with external conditions (*conformers*).

- Some species are conformers under certain conditions and become regulators under others. (See Campbell, Figure 46.9)

Energy expenditure by the animal is necessary if behavioral or physiological mechanisms are used to maintain homeostasis.

- For organisms to survive and reproduce, the energy "cost" of regulation cannot exceed the benefits of homeostasis.

- Since few organisms are perfect regulators or perfect conformers, a majority of organisms represent a group of evolved strategies which permits them to live in their specific environment.

B. The Principle of Allocation

The *Principle of Allocation* is an important concept for assessing the responses of organisms to a complex environment.

- This principle holds that each organism has a limited amount of energy that can be allocated for obtaining nutrients, escaping predators, coping with environmental fluctuations, growth, and reproduction.

- Energy expended for one function reduces the amount of energy available for other functions.

- If an organism expends a large amount of energy to maintain homeostasis, less is available for growth, reproduction, and other functions.

The distribution of organisms and the homeostatic mechanisms they possess establish different priorities for energy allocation.

- Conformers living in a stable environment may have more energy available for growth and reproduction; however, their geographic distribution is restricted due to intolerance to environmental change.

- Regulators that allocate a large amount of energy to survive environmental changes have less available for other functions so they grow and reproduce less efficiently; however, they can survive and reproduce over a wider range because they can cope with changing environmental conditions.

VI. **The response mechanisms of organisms are related to environmental grain and the time scale of environmental variation**

Since environments vary over space and time, the impact of the variations on a particular species depends on the scale of the variation in relation to the species' overall life history.

A. **Environmental Grain**

The concept of *environmental grain* is used by ecologists to define an organism's use of spatial variation in the environment.

- A *coarse-grained* environment is one in which environmental patches are so large in relation to the size and activity of the organism that an individual organism can choose among patches.
- A *fine-grained* environment is one in which environmental patches are small relative to the size and activities of an organism, and the organism may not behave as though patches exist.
- An environment may be course-grained to one organism and fine-grained to another.
    ⇒ A field of wild flowers is coarse-grained for small herbivorous insects since some plants may be preferential reproductive sites or food sources.
    ⇒ The same field would be fine-grained to a large herbivorous mammal that feeds indiscriminately on all the plants.

Temporal variation in an environment may also be coarse-grained or fine-grained depending on the periodicity of the variation relative to the organism's lifespan.

- Daily environmental variations are usually fine-grained for most organisms.
- Seasonal and long-term shifts in climate are usually coarse-grained to all organisms.

Organisms exhibit a variety of adaptations as responses to spatial and temporal variations in the environments. Some of these adaptations have short activation times while others take much longer.

- Behavioral adaptations are instantaneous and easily reversed.
- Physiological adaptations are activated and reversed on a time scale from seconds to weeks, depending on the specific response.
- Morphological adaptations may occur during an individual's lifespan or over generations.
- Adaptive genetic changes in populations take several generations to appear.
- The appropriate response to environmental change depends to a large degree on the duration of the change.

B. **Behavioral Responses**

Behavioral responses to unfavorable environmental changes include an animal's movement to a new, more favorable location and modification of the immediate environment by cooperative social behavior. For example,

- Fish descend into deeper, cooler waters of a lake during summer when the upper layer becomes too warm.

- Desert animals return to burrows or move into the shade during the day when heat is most intense.
- Migratory birds migrate to warmer climates to overwinter.
- Honeybees seal the hive during cold periods to conserve heat and cool the hive on hot days by the collective beating of their wings.
- Small mammals may huddle in burrows in cold weather which minimizes the total surface area exposed to cold air, thus reducing heat loss.

C. Physiological Responses

Physiological responses to environmental change are generally slower than behavioral responses although some may occur very rapidly.

- A slow change occurs when a human moves to an area of less oxygen such as at higher altitudes.
    - ⇒ After several days to a few weeks the person responds with an increased number of red blood cells being produced.
- An example of a faster change would be when blood vessels in the skin constrict within seconds to reduce loss of body heat when the skin is exposed to very cold air.

Physiological adaptation is centered around regulation and homeostasis.

- Both regulators and conformers function most efficiently under certain environmental conditions which are optimal for the organism.
- Efficiency declines both above and below optimal values. (See Campbell, Figure 46.10)

Physiological responses to changing environments can shift tolerance limits of organisms (*acclimation*).

- Acclimation is a gradual process and is related to the range of environmental conditions experienced under natural conditions.

D. Morphological Responses

Organisms can react to environmental change with responses that alter body form or internal anatomy.

- May be forms of acclimation since they are reversible.
    - ⇒ Increase in coat fur or feather density in winter.
    - ⇒ Change in coat color between winter and summer.
- Other morphological changes are irreversible over an individuals lifespan since the environmental variation may have affected growth and differentiation patterns.
    - ⇒ Plants are more morphologically plastic than animals.
    - ⇒ The arrowleaf plant lacks a waxy leaf cuticle when growing in water with submerged leaves (allows absorption of materials directly from the water).
    - ⇒ Arrowleaf plants growing on land have a thick cuticle (reduces water loss) and an extensive root system.

E. Adaptation over Evolutionary Time

It is important to remember that behavioral, physiological, and morphological responses to environmental change have evolved over evolutionary time to their current levels through natural selection.

- What appear to be short-term adjustments are actually evolutionary adaptations to maintain homeostasis.

- Natural selection also places constraints on the distribution of populations by adapting them to localized environments.

- Organisms adapted to one type of environment may not survive if dispersed to a foreign environment or may become extinct if the local environment changes to beyond their tolerance limits.

- The existence of a species in a particular location depends on the species reaching that location and being able to survive and reproduce after getting there.

VII. The geographical distribution of terrestrial biomes is based mainly on regional variations in climate

Biomes are communities and ecosystems typical of broad geographic regions.

- Ecologists recognize and organize biomes in different ways since there are no strict guidelines for defining biomes.

- Distributional patterns of organisms within the biosphere are the major focus of ecological research.

- The climate and other abiotic factors are important in determining why a particular biome is found in an area.

- There are latitudinal patterns of biome distribution over the Earth's surface due to the latitudinal patterns in climate.

Terrestrial biomes are often named for the predominant vegetation, but each is also characterized by microorganisms, fungi, and animals adapted to that particular environment. (See Campbell, Figure 46.11)

- Biomes grade into each other without sharp boundaries and may form *ecotones*, transitional areas between two communities.

- Species composition may vary from one location to another within a biome.

- May be patchy, with several communities represented in one biome.

  ⇒ Usually recognized on the basis of communities that develop through succession.

A. Tropical Forest

Tropical forest is found near the equator (within 23.5° latitude) where temperatures vary little from about 23°C and the length of daylight varies only slightly from 12 hours.

Rainfall is variable and the amount determines what vegetation is found in an area.

- In lowlands with prolonged dry seasons and scarce rain, *tropical dry forests* occur; these are a mixture of thorny trees and shrubs, and succulents.

- In regions with distinct wet and dry seasons, *tropical deciduous forests* occur. Trees releaf following heavy rains and drop leaves during the dry season.

Near the equator where rainfall is abundant (> 250 cm/year) and the dry season lasts less than a few months is *tropical rain forest*. (See Campbell, Figure 46.13)

- Has the greatest diversity, harboring more plant and animal species than any other biome.
- The vegetation is divided into five general layers: trees emerging above the canopy; the *canopy* which is the topmost continuous layer of foliage; the low tree stratum; the shrub understory; and the ground layer of herbaceous plants and ferns.
- A horizontal mosaic of changing vegetation is also present.
- The great plant diversity supports a great animal diversity.
- Widespread individuals of plant species depend on animals for pollination and dispersal of fruits and seeds.
- Competition for light is a strong selective force in plant communities.
- Soils are typically poor due to rapid nutrient recycling.

Destruction of tropical rain forests, occurring rapidly due to human intervention, may cause large-scale changes in world climate as well as destruction of many species.

B. Savanna

Savanna is grassland with scattered individual trees. (See Campbell, Figure 46.14)

- Cover areas of central South America, central and southern Africa, and parts of Australia.
- The three distinct seasons occur in sequence: cool-dry; hot-dry; warm-wet.
- Soils are generally: poor in nutrients; porous with a thin humus layer; have rapid water drainage.
- Usually have a simple physical structure but are often rich in species numbers.
- Frequent fires inhibit invasion by trees to maintain the grasses (wind pollinated) and forbs (often insect pollinated).
- The dominant herbivores are insects (ants, termites) although they are less obvious than other animals.
- Large herbivores (zebras, giraffes, kangaroos) and burrowing animals are commonly most active in the rainy season and many are nocturnal.
  - ⇒ During the dry season, many small animals are dormant or survive on seeds and dead plant parts while large animals migrate.

Savannas are examples of ecotones, where forest and grassland biomes integrate.

- The climatic conditions and community features are intermediate between grassland and forest.

Human impact on savannas has existed for centuries although the modern impact has been severe and detrimental.

- Population pressure, intensified grazing, and firewood collection have destroyed much of the vegetation in the Sahel region of Africa.
- Loss of vegetation has reduced water vapor recirculation to the atmosphere which has devastating climatic consequences for the region.

C. **Desert**

Desert is characterized by low and unpredictable precipitation (< 30 cm/year), not by temperature: both cold and hot deserts exist.

*   Occur in two distinct belts on Earth: between 15° and 35° latitude in both hemispheres.
    ⇒ A result of global air circulation with the descending air absorbing the available moisture.
*   Hot deserts occur in S.W. United States, W. South America, North Africa, the Middle East, and Central Australia.
*   Cold deserts occur in E. Argentina, central Asia and west of the Rocky Mountains.

The density of vegetation cycles, with the growth and reproductive cycles being keyed to precipitation.

*   No perennial vegetation is found in the driest deserts due to a lack of rainfall which is usually <2 cm/year.
*   Dominant vegetation in less arid deserts includes scattered shrubs, cacti and other succulents that store water.
    ⇒ Many bloom abundantly after a rainfall.

Reptiles and seed eaters such as ants, birds and rodents are common. Many live in burrows, are nocturnal or have other adaptations for conserving water.

*   Light coloration reflects sunlight.
*   Some mice never drink and derive all of their water from the metabolic breakdown of the seeds they eat.
*   The reproductive cycle of some (i.e. spadefoot toads) is completed quickly in temporary pools.

D. **Chaparral**

Chaparrals are regions of dense, spiny shrubs with tough evergreen leaves. They are found along coasts where cool ocean currents circulate offshore resulting in mild, rainy winters and long, hot, dry summers.

*   Are found between 30° and 40° latitude.
*   Occur in Mediterranean and coastlines of California, Chile, S.W. Africa, and S.W. Australia. (See Campbell, Figure 46.16)
*   Shrubby vegetation dominates since: low nutrient soils, aridity, short growing seasons, and frequent fires prevent trees from becoming established.
*   Vegetation from these areas are similar in appearance but taxonomically unrelated.

Chaparral is maintained by periodic fires.

*   Many shrubs have root systems and seeds adapted for fire; root crowns may be fire resistant and resprout quickly, others have seeds that only germinate after a fire.
*   Other plants are clonal and use asexual reproduction.

Browsers such as deer, fruit-eating birds, seed-eating ants and rodents, snakes and lizards are common animals of these biomes.

E.   Temperate Grasslands

Temperate grasslands have some characteristics of tropical savannas but occur in regions with relatively cold winters.

- The veldts of southern Africa, the pampas of Uruguay and Argentina, the steppes of the former Soviet Union and the plains and prairies of the U.S. are examples. (See Campbell, Figure 46.17)
- Occasional fires and seasonal drought prevent encroachment of trees into grasslands.
- Soil tends to be deep and rich with a large amount of decaying plant matter (mulch) being deposited each year.
    ⇒ Mulch decreases runoff and erosion, stabilizes soil temperature, and improves conditions for seed germination.
    ⇒ Too much mulch may suppress grass growth and allow invasion by wood plants.
- Grazing and burning reduce mulch accumulation.
- Large grazing mammals and large carnivores are common, and burrowing rodents and other small mammals are present in dense populations.
- Underground invertebrates graze on the below-ground parts of the plants.
- Few natural grasslands remain due to their conversion to farm and pasture lands.

F.   Temperate Deciduous Forest

Temperate forests grow throughout midlatitude regions with sufficient moisture to support growth of large, broad-leaved deciduous trees. (See Campbell, Figure 46.18)

- Occur in Eastern U.S., Middle Europe and E. Asia.
- Temperatures range from very cold in winter to very hot in summer (-30°C to 30°C), with a 5 to 6 month growing season.
- Precipitation is fairly high and evenly distributed throughout the year.
- Soil is rich in nutrients; decomposition rates are slow and a thick layer of leaf litter usually accumulates.

Several layers of vegetation including herbs, shrubs and one or two strata of trees (species composition varies widely) are present.

- Dominant trees are usually oak, birch, hickory, beech, and maple species.
- Plant diversity is high and results from high availability of light, nutrients, and moisture.
- Variety and abundance of food and habitat supports a rich diversity of animal life.
- Logging, clearing, and the introduction of exotic pests and diseases have greatly reduced the amount of these forests.

G.   Taiga

Taiga (coniferous or boreal forest) is characterized by harsh winters and short, wet (and occasionally warm) summers.

- This is the largest terrestrial biome and occurs in North America, Europe, Asia and at high elevations in more temperate latitudes.

- There is considerable precipitation, mostly in the form of snow that insulates the soil and prevents the permafrost; also provides a protective layer for small mammals.
- Conifers like spruce, pine, fir or hemlock and deciduous oak, birch, willow, alder and aspen are common; usually one or very few species are present in dense stands.
  ⇒ Very little undergrowth is present.
- Animals include: seed eaters such as squirrels, birds, and insects; browsers such as insects, elk, moose, deer, beaver, and porcupine; predators such as grizzly bears, wolves, wolverines and lynxes.
- Soil is thin, acidic and forms slowly due to low temperatures and slow decomposition of waxy needles.

Coastal coniferous forests are similar to taiga in that they are dominated by one or a few species.

- These are the temperate rain forests and redwood forests of the Pacific Northwest.
- Their proximity to the ocean makes them much warmer and moister than the taiga.

H. Tundra

Tundra is at the northern-most limits of plant growth and plant forms are limited to low shrubby or matlike vegetation. (See Campbell, Figure 46.20)

There are two types of tundra:

- *Arctic tundra* encircles the North Pole and extends southward to the taiga.
  ⇒ The climate is very cold with little light for long periods. Some areas are characterized by *permafrost* (permanently frozen ground) which contributes to the absence of taller plant forms.
  ⇒ Little precipitation may occur although permafrost, low temperatures, and low evaporation results in continually saturated soils – further restricting plant forms.
  ⇒ Dwarf perennial shrubs, sedges, grasses, mosses and lichens are common.
  ⇒ Plant growth and reproduction occur rapidly during the brief warm summer.
- *Alpine tundra* occurs at high elevations in all latitudes at altitudes above where trees can grow.
  ⇒ Near the equator, alpine tundra is confined to the highest mountain tops where freezing temperatures occur nightly.
  ⇒ The flora and fauna are similar to that of the arctic tundra.

Both arctic and alpine tundras have a low diversity of animal species.

- Dipterans (blackflies, mosquitoes) are abundant in the arctic tundra's summer but scarce in the alpine tundra where beetles, grasshoppers, and butterflies are common.
- Insect development is slower in the tundra due to the environmental conditions.
- A large diversity of migratory birds enter the arctic tundra during the summer to feed on insects; herbivorous birds are rare.
- Large herbivorous mammals (muskox, caribou, reindeer) are common in the arctic tundra, as well as many small herbivores (lemmings).
- Predators include the arctic fox, wolves, polar bears, and snowy owl.

VIII. Aquatic ecosystems, consisting of freshwater and marine biomes, occupy the largest part of the biosphere

Life arose in water and evolved there for almost three billion years before moving into terrestrial habitats.

- Aquatic habitats still occupy the largest part of the biosphere.

Freshwater and marine biomes are distinguished on the basis of physical and chemical differences that influence the communities occupying these habitats.

- Freshwater biomes have a salt concentration less than 1%; marine biomes average 3% salt concentration.

Freshwater biomes are closely linked to the terrestrial biomes where they are located or through which they flow.

- Runoff from terrestrial habitats creates streams and rivers.
- Accumulated runoff creates ponds and lakes.
- Also influenced by the patterns and speed of water flow, and the climate to which they are exposed.

The oceans cover 75% of Earth's surface and contain the *marine biomes*.

- Evaporation from the oceans provides most of the rainfall.
- Ocean temperatures greatly effect world climate and wind patterns.
- Marine algae consume a large amount of the atmospheric carbon dioxide and produce a major portion of the world's oxygen.

A. Ponds and Lakes

These standing bodies of water vary greatly in size with *ponds* being smaller than *lakes*.

Ponds and lakes usually exhibit a significant vertical stratification in light penetration and water temperature.

- There is a decrease in light intensity with increasing depth as light is absorbed by the water and suspended microorganisms. This divides the pond or lake into two layers:
    ⇒ The *photic zone* is the upper layer where light is sufficient for photosynthesis.
    ⇒ The lower *aphotic zone* receives little light and no photosynthesis occurs.
- Temperature stratification also occurs in deeper ponds and lakes during summer in temperate zones.
    ⇒ Heat energy from sunlight warms the upper layers of water as far as it penetrates; the deeper waters remain cold.
- As depth increases, the separation point of the warmer upper water form the lower colder water is noticeable as the *thermocline*.
    ⇒ The thermocline is a narrow vertical zone between the warmer and colder waters where a rapid temperature change occurs.

The distribution of plants and animals within a pond or lake also shows a stratification based on water depth and distance from the shore.

- The *littoral zone* is shallow, well-lighted, warm water close to shore.
    - ⇒ Characterized by the presence of rooted and floating vegetation, and a diverse attached algal community (especially diatoms).
    - ⇒ There is a diverse animal fauna including suspension feeders (clams); herbivorous grazers (snails); and herbivorous and carnivorous insects, crustaceans, fishes, and amphibians.
    - ⇒ Some reptiles, water fowl, and mammals also frequent this zone.
- The *limnetic zone* is the open, well-lighted waters away from shore.
    - ⇒ Occupants include photosynthetic phytoplankton (algae and cyanobacteria), zooplankton (rotifers and small crustaceans) that graze on phytoplankton, and small fish which feed on the zooplankton.
    - ⇒ Occasional visitors to this zone are large fish, turtles, snakes, and piscivorous birds.
- The *profundal zone* is the deep, aphotic zone lying beneath the limnetic zone.
    - ⇒ This is an area of decomposition where *detritus* (dead organic matter that drifts in from above) is broken down.
    - ⇒ Water temperature is usually cold and oxygen is low due to cellular respiration of decomposers.
    - ⇒ Mineral nutrients are usually plentiful due to decomposition of detritus.

Waters of the profundal zone usually do not mix with surface waters due to density differences related to temperature.

- Mixing of these layers usually occurs twice each year in temperate lakes and ponds; thus, oxygen enters the profundal zone and nutrients are cycled into the limnetic zone.

Lakes are often classified as *oligotrophic* or *eutrophic*, depending on the amount of organic matter produced.

- *Oligotrophic lakes* are deep, nutrient-poor lakes in which the phytoplankton is not very productive.
    - ⇒ The water is usually clear and the profundal zone has a high oxygen concentration since little detritus is produced in the limnetic zone to be decomposed.
- *Eutrophic lakes* are usually shallow, nutrient-rich lakes with very productive phytoplankton.
    - ⇒ The waters are usually murky due to large phytoplankton populations and the large amounts of detritus being decomposed may result in oxygen depletion in the profundal zone during the summer.

Oligotrophic lakes may develop into eutrophic lakes over time.

- Runoff from surrounding terrestrial habitats brings in mineral nutrients and sediments.

- Human activities increase the nutrient content of runoff due to lawn and agricultural fertilizers; municipal wastes dumped into lakes dramatically enriches the nitrogen and phosphorus concentrations which increases phytoplankton and plant growth.
- Algal blooms and increased plant growth results in more detritus and can lead to oxygen depletion due to increased decomposition.
    ⇒ This cultural eutrophication usually makes the water unusable.

B.   Streams and Rivers

Streams and rivers are bodies of water that move continuously in one direction.

There is a change in structure of these bodies of water from their headwaters (point of origin) to their mouths (where they empty into a larger body of water).

- At the headwaters, the water is cold and clear, carries little sediment, and has few mineral nutrients.
    ⇒ The channel is narrow with a rocky substrate and the water flows swiftly.
- Near the mouth, water moves slowly and is more turbid due to sediment entering from other streams and erosion; the nutrient content is also higher.
    ⇒ The channel is usually wider with a silty substrate that has resulted from deposition of silt.

The nutrient and oxygen content, turbidity, and rate of flow in rivers and streams are influenced by many factors.

- Rough, shallow bottoms produce rapid turbulent flow known as *riffles*; smooth, deep bottoms result in a slower, smooth flow in areas called *pools*.
- Nutrient content of the water is higher in streams and rivers flowing through densely vegetated regions and where erosion takes place.
    ⇒ Leaves and other vegetation entering the water add organic matter.
    ⇒ Erosion of rocks in the streambed increases inorganic nutrient content.
- Oxygen content of the water is also affected by the flow rate.
    ⇒ Turbulent flow constantly oxygenates the water while slow, murky waters contain relatively little oxygen.
- Turbidity reflects the amount of material suspended in the water.
    ⇒ Streams and rivers flowing through areas of high erosion will have more suspended materials than those surrounded by hard substrates.
    ⇒ Large amounts of suspended organic matter also increases turbidity.

The biological communities found in rivers and streams differ from headwaters to mouth; they also differ from those found in ponds and lakes.

- Due to the current, large plankton communities are not found in rivers and streams.
    ⇒ Photosynthesis which supports the food chains is carried out by attached algae and rooted plants.
- Organic material washed into the system also provides an important food source, especially where dense vegetation along the shore blocks out sunlight or high turbidity in slow waters prevents light penetration.

- In upstream areas where water is cool, clear, and has a high oxygen content, there are many aquatic insects and fish such as trout.
- Near the mouth where water is murky and warmer with a lower oxygen content, fish such as carp and catfish are prominent; a different group of insects is also found in this area.

Various adaptations are found in organisms which permit them to inhabit their specific areas of rivers and streams.

- Small animals typically have flattened bodies and can attach temporarily to rocks.
- Many insects and other invertebrates live on the underside or downstream sides of rocks to escape the current.
- Some insects produce silken nets used to filter food from the water.

Human activities have greatly affected many streams and rivers.

- Channelization (increases flow rate) and damming (slows flow rate) alter the natural characteristics of flowing waters.
- Pollutants may be taken up by or destroy the natural flora and fauna.

C. Wetlands

A *wetland* is an area covered by water that supports aquatic vegetation. (See Campbell, Figure 46.24)

- Includes a broad range of habitats from periodically flooded regions to soil that is permanently saturated during growing season.
- Conditions favor *hydrophytes* which are plants specially adapted to grow in water or soil periodically anaerobic due to being saturated with water.
    ⇒ Cattails, pond lilies, and sedges are examples of hydrophytes.
- Both the hydrology and the vegetation are important determinants in wetland classification.

A wide variety of wetlands has been recognized although they form in one of only three topographic situations:

- *Basin wetlands* develop in shallow basins ranging from upland depressions to lakes and ponds that have filled in.
- *Riverine wetlands* develop along shallow and periodically flooded banks of streams and rivers.
- *Fringe wetlands* are found along coasts of large lakes and seas where rising lake levels or tides cause water to flow back and forth.
    ⇒ Fringe wetlands include both freshwater and marine habitats.
    ⇒ Marine wetlands are closely linked to estuaries.

The types of plants present in a wetland are determined by the flow of water through the habitat.

- The duration, frequency, and depth of water, as well as, the flood season are also important.

Wetlands are among the richest and valuable of biomes.

- A diverse invertebrate community is present which supports a wide variety of birds.
- A variety of herbivorous species consume the algae, detritus, and plants.
- They provide water storage basins that reduce the intensity of flooding.
- They improve water quality by filtering pollutants.

Although many wetlands have been destroyed for agriculture and development, a move to protect the remaining wetlands is underway.

D. Estuaries

An *estuary* is the area where a freshwater stream or river merges with the ocean.

- Often bordered by salt marshes or intertidal mudflats.
- Salinity varies spatially within the estuary from nearly fresh water to ocean water; varies daily in these areas due to rise and fall of tides.
- Estuaries are very productive due to nutrients brought in by rivers.

Because of their productivity, estuaries have a diverse flora and fauna.

- Salt marsh grasses, algae, and phytoplankton are the major producers.
- Many species of annelids, oysters, crabs, and fish are also present.
- Many marine invertebrates and fish breed in estuaries or migrate through them to freshwater habitats upstream.
- A large number of water fowl and other semiaquatic vertebrates use estuaries as feeding areas.

Human activities are having a large impact on estuaries.

- Estuaries receive the pollutants dumped into the streams and rivers that feed them.
- Residential and commercial development not only adds to pollution but eliminates some estuaries due to land filling.

E. Marine Zones: An Introduction

Marine communities are classified based on the depth at which they occur, their distance from shore, and light penetration. (See Campbell, Figure 46.26)

- A photic zone is present and extends to the depth at which light penetration supports photosynthesis; occupied by phytoplankton, zooplankton, and many fish species.
- The aphotic zone is below the level of effective light penetration and represents a majority of the ocean's volume.
- The *intertidal zone* is the shallow zone where terrestrial habitat meets the ocean's water.
- The *neritic zone* extends from the intertidal zone, across the shallow regions, to the edge of the continental shelf.
- The *oceanic zone* extends over deep water from one continental shelf to another; reaches great depth.
- *Pelagic zone* refers to open waters of any depth.
- *Benthic zone* refers to the seafloor.

F.   The Intertidal Zones

Intertidal zones, where land and sea meet, are alternately submerged and exposed by daily tide cycles.

- Organisms in this zone are exposed to variations in temperature and availability of sea water.
- These organisms are also subjected to the mechanical forces of wave action.

Rocky intertidal zones are vertically stratified and inhabited by organisms that possess structural adaptations that allow them to remain attached in this harsh environment. (See Campbell, Figure 46.27)

- The uppermost zone is submerged only by the highest tides and is occupied by relatively few species of algae, grazing mollusks, and suspension-feeding barnacles.
  ⇒ These organisms have various adaptations to prevent dehydration.
- The middle zone is exposed at low tide and submerged at high tide.
  ⇒ Many species of algae, sponges, sea anemones, barnacles, mussels, and other invertebrates are found in this area.
  ⇒ The diversity is greater here due to the longer time spans this area is submerged.
- *Tide pools* are often found in the middle zone.
  ⇒ These are depressions which are covered during high tide and remain as pools during low tide.
  ⇒ Tidepool organisms face dramatic salinity increases as water evaporates at low tide.
- The low intertidal zone is exposed only during the lowest tides and shows the greatest diversity of invertebrates, fishes, and seaweeds.

Sandy intertidal zones and *mudflats* do not show a clear stratification.

- Wave action continually shifts sand or mud particles; few algae or plants are present.
- Predatory crustaceans and many suspension-feeding worms and clams burrow into the sand or mud and feed when the tide submerges the area.
- Predatory or scavenging crabs and shorebirds often feed on burrowing organisms in these areas.

The diversity of intertidal zones is being reduced by human impact.

- Oil pollution destroys many species.
- Polluted water, old fishing lines, and plastic debris is harmful to most species.
- Recreational use of these areas has greatly reduced the number of beach-nesting birds and turtles.

G.   Coral Reefs

Coral reefs are found in the neritic zone of warm tropical waters where sunlight penetrates to the ocean floor.

- Sunlight penetration permits photosynthesis and a constant supply of nutrients is provided by currents and waves.

Coral reefs are dominated by the structure of the coral. This diverse group of cnidarians secretes a hard, calcium carbonate external skeleton that provides a substrate on which other corals, sponges, and algae grow. (See Campbell, Figure 46.28)

- The cnidarian coral animals feed on microscopic organisms and organic debris even though they are dependent on the photosynthetic products of their symbiotic dinoflagellate algae.
- Coral reefs are very diverse and productive.
  ⇒ A large variety of microorganisms, invertebrates, and fish are found among the coral and algae.
- Many herbivorous snails, sea urchins, and fish are present along with predators such as the octopus, sea stars, and many carnivorous fish.

Although many coral reefs are very large, they are delicate and can be severely damaged or destroyed by pollution, human induced damage, or introduced predators.

H. The Oceanic Pelagic Biome

The *oceanic pelagic biome* consists of the open waters far from shore.

- The area is constantly mixed by circulating ocean currents.
- Nutrient content is usually very low due to the sinking of the remains of organisms out of the zone to the lower benthic regions.
  ⇒ In tropical waters, nutrient content of this area is very low due to a permanent thermal stratification which prevents nutrient exchange with the deep waters.
  ⇒ In some areas of the temperate oceans, periodic upwellings carry nutrients from the bottom to the surface during the spring.
- Waters are generally cold, although temperatures vary with latitude and depth.

Plankton is prevalent in this zone.

- Photosynthetic phytoplankton grow and reproduce in the photic region (top 100m) of this biome.
- Zooplankton (which graze on the phytoplankton) includes protozoans, copepods, krill, jellyfish, and larvae of many invertebrates and fishes.
- Most species stay afloat in this zone through the aid of morphological structures like bubble-trapping spines, lipid droplets, gelatinous capsules, and air bladders.

Nekton (free-swimming animals) are also found in the oceanic pelagic biome.

- Large squid, fishes, sea turtles, and marine mammals feed in this area.
- Many fish are adapted to and live in the aphotic region of the pelagic zone.
  ⇒ Some have large eyes allowing them to see in dim light, while others have luminescent organs used to attract mates and prey.
- Many pelagic bird species also feed on fish in this region.
- Many migratory marine animals move through this region either following available food or moving from summer breeding to winter feeding areas.

I.   **Benthos**

Benthos refers to those organisms which inhabit the benthic zone, the ocean bottom below both the neritic and pelagic zones.

- The benthic zone receives nutrients in the form of detritus which settles into the area from the waters above.
- Light and temperature decline rapidly from shallow, near-shore benthic areas to the ocean's depths.
- The benthic zone substrate may be sand or very fine sediment composed of silt and shells of dead microscopic organisms.

Neritic benthic communities are diverse and productive.

- Many bacteria, fungi, seaweeds, filamentous algae, and sponges are found here.
- Sea anemones, worms, clams, various echinoderms and fishes are usually present.
- Many of the species present are burrowing forms.
- The composition of the community varies with distance from shore, water depth, and bottom composition.

Deep benthic communities living in the *abyssal zone* are exposed to very different conditions than those under the neritic and pelagic zones.

- Water temperature is continuously cold (3°C), water pressure is extremely high, there is very little (if any) light, and low nutrient concentrations are typical.
- Oxygen is usually present and a fairly diverse community of invertebrates and fishes are found.
- The deep-sea hydrothermal vent communities are found along midocean ridges in this region.
  - ⇒ These vent communities include chemoautotrophic bacteria as the primary producers in place of photosynthesizing organisms.
  - ⇒ These bacteria obtain energy by oxidizing H_2S which forms by a reaction of the hot vent water with dissolved sulfate.
  - ⇒ The bacteria are consumed by a variety of giant polychaete worms, arthropods, echinoderms, and fishes. (See Campbell, Figure 46.29)

REFERENCES

Brown, James H., and Arthur C. Gibson. *Biogeography*. St. Louis, Missouri: The C.V. Mosby Company, 1983.

Campbell, N. *Biology*. 4th ed. Menlo Park, California: Benjamin/Cummings, 1996.

Ricklefs, R.E. *Ecology*. 3rd ed. New York: Chiron Press, 1986.

Smith, Robert Leo. *Ecology and Field Biology*. 3rd ed. New York: Harper & Row, Publishers, 1980.

CHAPTER 47

POPULATION ECOLOGY

OUTLINE

OBJECTIVES

After reading this chapter and attending lecture, the student should be able to:

1. Define the scope of population ecology.
2. Distinguish between density and dispersion.
3. Explain how ecologists measure density of a species.
4. Describe conditions which may result in clumped dispersion, random dispersion, and uniform dispersion of populations.
5. Explain how age structure, generation time, and sex structure of populations can affect population growth.
6. Describe the characteristics of populations which exhibit Type I, Type II, and Type III survivorship curves.
7. Explain how carrying capacity of the environment affects the intrinsic rate of increase of a population.
8. Explain how density-dependent factors affect population growth.
9. Describe how weather and climate can function as density-independent factors in controlling population growth.
10. Explain how density-dependent and density-independent factors may work together to control a population's growth.
11. List the three major characteristics of a life history and explain how each affects the:
    a. Number of offspring produced by an individual.
    b. Population's growth.
12. Explain how predation can affect life history through natural selection.
13. Distinguish between r-selected populations and K-selected populations.
14. Explain how a "stressful" environment may alter the standard r-selection and K-selection characteristics.

KEY TERMS

population
density
dispersion
geographical range
clumped dispersion
random dispersion
uniform dispersion
biogeography
demography

age structure
fecundity
generation time
life table
cohort
survivorship curve
life history
semelparity
iteroparity

intrinsic rate of
  increase
exponential population
  growth
carrying capacity
logistic population
  growth
density-dependent
  factor

Allee effect
intraspecific
  competition
territoriality
density-independent
  factor
r-selected species
K-selected species

LECTURE NOTES

Population ecology is concerned with measuring changes in population size and composition and identifying the factors that cause these changes.

<u>Population</u> = Individuals of one species simultaneously occupying the same general area, utilizing the same resources, and influenced by similar environmental factors.

No population can continue to grow indefinitely.
- Many populations remain relatively stable over time with only minor increases and decreases.
- Other populations show dramatic increases followed by equally dramatic decreases.

I. **Two important characteristics of any population are density and the spacing of individuals**

Every population has geographical boundaries and a population size.
- A population will exhibit two characteristics within its boundaries: *density* and *dispersion*.

<u>Population density</u> = The number of individuals per unit area or volume.

<u>Population dispersion</u> = The pattern of spacing among individuals within the geographical boundaries of the population.

A. **Measuring Density**

It is usually impractical or impossible to count all individuals in a population, so ecologists use a variety of sampling techniques to estimate densities and total population size.
- May count all individuals in a sample of representative plots. Estimates become more accurate as sample plots increase in size or number.
- May estimate by indirect indicators such as number of nests or burrows, or by droppings or tracks.
- May use the *mark-recapture method*.

In the *mark-recapture method*, animals are trapped within boundaries, marked in some way and after time retrapped.
- The number of individuals in a population (N) is estimated by the formula:

$$N = \frac{(\text{number marked}) \times (\text{total catch the second time})}{\text{number of marked recaptures}}$$

- Assumes marked individuals have same probability of being trapped as unmarked individuals. This assumption is not always valid.

B.   Patterns of Dispersion

A population's geographical range is the geographic limits within which a population lives.

- Local densities may vary substantially because not all areas of a range provide equally suitable habitat.

Individuals exhibit a continuum of three general patterns of spacing in relation to other individuals: *clumped*, *uniform*, and *random*.

1. A *clumped* pattern is when individuals are aggregated in patches.
   - May result from the environment being heterogeneous, with resources concentrated in patches.
   - May be associated with mating or other social behavior in animals.
   - May also be associated with defense against predators.

2. A *uniform* pattern is when the spacing of individuals is even.
   - May result from antagonistic interactions between individuals of the population.
   - For example, competition for some resource or social interactions that set up individual territories for feeding, breeding or nesting.

3. A *random* pattern is when individual spacing varies in an unpredictable way.
   - Occurs in the absence of strong attractions or repulsions among individuals.
   - Not very common in nature.

While the above applies to local dispersion patterns within populations, it is important to remember that populations within a species also show dispersion patterns.

- Such populations often concentrate in clusters within the species' range.
- *Biogeography* is the study of factors that influence the distribution of a species over its range.

II.   Demography is the study of factors that affect birth and death rates in a population

Demography is the study of the vital statistics affecting population size.

- Reflects relative rates of processes that add individuals to a population (birth, immigration) versus processes that eliminate individuals (death, migration).
- Birth and death rates vary among population subgroups depending on age and sex.

A.   Age Structure and Sex Ratio

Many populations have overlapping generations where individuals of more than one generation coexist.

- Exceptions include species in which all of the adults reproduce at the same time, then die (annual plants, many insects).

The coexistence of generations produces an age structure in most populations.

Age structure = Relative numbers of individuals of each age in a population.

- Generations overlap when average life span is greater than the time it takes to mature and reproduce.
- Every age group has a characteristic birth and death rate.
  ⇒ Birth rate greatest for those of intermediate age.
  ⇒ Death rate often greatest for the very young and very old.
- In general, a population with more older, nonreproductive individuals will grow more slowly, than a population with a larger percentage of young, reproductive age individuals.

Generation time is an important demographic feature related to age structure.

Generation time = The average span of time between the birth of individuals and the birth of their offspring.

- Strongly related to body size. (See Campbell, Figure 47.3)
- A shorter generation time usually results in faster population growth, assuming birth rate is greater than death rate and all other factors being equal.

Another important demographic factor affecting population growth is the *sex ratio*.

Sex ratio = The proportion of individuals of each sex found in a population.

- The number of females is usually directly related to the expected number of births.
- The number of males may be less significant since one male may mate with several females.
- In strictly monogamous species, the number of males is more significant in affecting the birth rate than in nonmonogamous species.

B.   Life Tables and Survivorship Curves

Life tables describe how birth rates and death rates vary with age over a time period corresponding to maximum life span. (See Campbell, Table 47.1)

- Constructed by following the fate of a *cohort*, a group of individuals of the same age, from birth until all are dead or by using the age specific birth and death rates in a population during a specified time.

Various information about change in population size can be obtained from life tables. For example, in the great tit population represented in Table 47.1,

- Fewer than 20% of the original females survived to an age of 3 years, and a 3 year-old female has only about 44% chance of surviving to age 4.
- Mortality rates for females are relatively constant during the middle years, but higher in young (1 year) and older (≥5 years) birds.
- The birth rate is highest for 4 year old birds and lower in very young and very old females.

Survivorship curves plot the numbers in a cohort still alive at each age. Organisms may exhibit one of three general types of survivorship curves: (See Campbell, Figure 47.4)

1. Type I curves are flat during early and middle life and drops suddenly as death rates increase among the older individuals.
   - Associated with species such as humans and other large mammals that produce few offspring that are well cared for.

2. Type II curves are intermediate with mortality being more constant over the life span.
   - Seen in *Hydra*, gray squirrels, and some lizards.

3. Type III curves show very high death rates for the young followed by lower death rates after individuals have survived to a certain critical age.
   - Associated with organisms, such as oysters, that produce very large numbers of offspring but provide little or no care.

Many species exhibit curves between the basic types of survivorship curves and some have more complex curves.

- Great tits show a high mortality rate in young birds (Type III) but a fairly constant mortality (Type II) in adults.
- Some invertebrates show a "stair-stepped" curve with brief periods of high mortality during molts, followed by periods of lower mortality when the exoskeleton is hard (i.e. crabs).

III. The traits that affect an organism's schedule of reproduction and death make up its life history

A. Variation in Life Histories

Natural selection, working over evolutionary time, results in traits that affect an organism's *life history*.

Life history = An organism's schedule of reproduction and death.

- Life histories are important in population ecology because these traits affect population growth over time.

There is a diversity in life histories due to the varying pressures of natural selection.

- Some animal species hatch in one type of biome, migrate to another where they mature for several years, then return to the initial biome for a single massive reproductive effort, then die (i.e. salmon).
- Other animal species hatch and mature rapidly in a single habitat, then have small reproductive efforts each year for several years (i.e. lizards).
- Life history characteristics may vary significantly among populations of the same species or even among individuals within a population.

Even though life history traits vary widely, there are some basic patterns:

- Life histories often vary in parallel with environmental factors. For example, clutch sizes increase with latitude.

    ⇒ Tropical birds lay fewer eggs than those in higher latitudes which reflects the number of offspring that can be successfully fed.

    ⇒ Since daylengths are longer at higher latitudes than the tropics during offspring-rearing season, parent birds can gather more food and feed more offspring.

    ⇒ The clutch size-latitude variation is also found in many mammals, lizards, and insects.

- Life history traits often vary in relation to one another. For example, fecundity and mortality tend to vary in close association among birds.

    ⇒ Albatrosses only average one surviving offspring every five years but adult birds have only a 5% chance of dying from one breeding season to the next.

    ⇒ Tree sparrows average six surviving offspring each year but have over a 50% chance of dying between breeding seasons.

- Delayed maturation and high parental investment in offspring tend to be correlated with low fecundity and low mortality.

B. Allocation of Limited Resources

A life history based on heritable traits that result in producing the most reproductively successful descendants will become more prevalent in a population.

A successful life history resolves the conflict between limited resources and competing functions.

- Time, energy, and nutrients used for one function are not available for other functions.

- The integrated life histories seen in natural populations balance the investment in the number of offspring produced against the prospects of future reproductions.

    ⇒ In general, organisms that produce fewer offspring during a reproductive effort survive longer and have more reproductive episodes.

- The life history traits exhibited by an organism are evolutionary outcomes reflected in the development and physiology of that organism.

C. Number of Reproductive Episodes Per Lifetime

Two extremes are found in life history strategies where there is a trade-off between fecundity and survival probability.

- *Semelparity* is a type of life history in which organisms invest most of their energy in growth and development, then expend that energy in a single reproductive effort before dying.

- *Iteroparity* is a type of life history in which organisms produce fewer offspring at a time over many reproductive seasons.

Semelparity is expected when there is a high cost to parents to stay alive between broods or if there is a trade-off between fecundity and survival.

- Seen in annual plants, salmon, and some perennial plants such as bamboo and the century plant.
- In the harsh desert climates, most plants live only a single season and put all their energy into one reproductive effort.
- Rarely found in animals or plants that live more than one or two years, although some (i.e. century plants) live for several seasons before investing all their energy into a single reproduction.
  - ⇒ Century plants live in arid climates with unpredictable rainfall.
  - ⇒ They grow vegetatively and store nutrients for several years.
  - ⇒ The stored nutrients are used to produce seeds when an unusually wet year occurs that favors offspring survival.

Iteroparity is expected when established individuals are likely to survive but mortality is high in immature individuals.

- Iteroparous plants are more common in the tropics where competition and predation hinder seedling establishment, but mature plants survive for many seasons.
- More resources are invested in survival (i.e. root system, resistant buds) to prepare for multiple breeding episodes than is allocated by single season plants.

D.   Clutch Size

Clutch size is the number of offspring produced at each reproductive episode.

- Organisms with a low probability of surviving to the next reproductive season usually invest more energy into producing a large number of offspring.
- Organisms with a high probability of survival invest less energy and produce fewer offspring.
- Clutch size may vary seasonally within a single population in some species.

There is usually a trade-off between the number and quality of offspring produced.

- In general, organisms with large clutch sizes produce small eggs or offspring.
  - ⇒ Each offspring thus starts with a limited amount of energy.
  - ⇒ Large clutch size and small offspring are typical of organisms with a Type III survivorship curve.
  - ⇒ Typical of organisms that colonize disturbed or harsh environments or are subject to high levels of predation.
- Organisms which produce small clutches generally have larger eggs or offspring.
  - ⇒ Offspring have a longer initial amount of energy which improves their chance of survival.
  - ⇒ Small clutches of large offspring are typical of organisms with Type I and II survivorship curves.
  - ⇒ Parental investment is higher but is off-set by increased survival of offspring.

E.   Age at First Reproduction

The timing of the first reproduction greatly influences the female's lifetime reproductive output in organisms that have several reproductive episodes during their lifespan.

- Balances the cost between current reproduction and survival plus future reproduction.
- Reproduction at a younger-than-average age may reduce a female's reproductive potential by reducing the amount of energy available for growth and maintenance.
- Female's that delay reproduction tend to be larger due to energy used for growth and maintenance; older (larger) females produce larger clutches and appear to maximize their reproductive output by delaying.

IV.   A mathematical model for exponential growth describes an idealized population in an unlimited environment

Indefinite increases in population size do not occur.

- A population may increase rapidly from a low level under favorable environmental conditions, but this increase in numbers will eventually approach the level where resources cannot support continued increases.
- The combination of limited resources and other factors will stop the growth of the population.

A combination of observation, experimentation, and mathematical modeling is used to determine answers to many ecological questions.

- Birth rates and death rates can be quantified, by measurement or observation, in many populations and used to predict changes in the population size.
- Laboratory studies on small animals can determine how various factors affect population growth rates and a few natural populations can be manipulated experimentally to answer a range of questions.
- Mathematical models can test hypotheses about the effects of various factors on population growth in organisms that are difficult or impossible to study experimentally.

A population consisting of a few individuals which lives in an environment without limiting factors (no restrictions on available energy, growth, or reproduction) will increase over time in proportion to the birth and death rates:

$$\text{Change in population size during time interval} = \text{Births during time interval} - \text{Deaths during time interval}$$

In a mathematical form, this equation becomes:

$$\frac{\Delta N}{\Delta t} = B - D$$

where N = population size, t = time, B = absolute number of births during the time interval, and D = absolute number of deaths during the time interval, ΔN = change in population size, Δt = the time interval (lifespan or generation time).

Since populations differ in size, the basic model must be altered in order to apply it to *any* population. This alteration involves a conversion from absolute birth and death rates to average numbers of births and deaths per individual during the specified time interval.

- If *b* = the annual per capita birth rate, its value would equal 0.034 in a population of 1000 individuals where 34 births occurred (34/1000 = 0.034) in a year.
- If *d* = per capita death rate, its value would equal 0.016 in a population of 1000 individuals where 16 deaths occurred (16/1000 = 0.016) in a year.

Including average (or per capita) birth and death rates alters the previous equation so that expected numbers of births and deaths can be predicted for a population of any size:

$$\frac{\Delta N}{\Delta t} = bN - dN$$

- The change in size of our 1000 individual population would be:

  $\Delta N/\Delta t = (0.034)(1000) - (0.016)(1000)$

  $\Delta N/\Delta t = 34 - 16$

  $\Delta N/\Delta t = 18$

- If another population of the same species (same *b* and *d* values) contained 1500 individuals, then:

  $\Delta N/\Delta t = (0.034)(1500) - (0.016)(1500)$

  $\Delta N/\Delta t = 51 - 24$

  $\Delta N/\Delta t = 27$

Population ecologists study overall changes in population sizes and use *r* to represent the difference between per capita birth rates and per capita death rates:

$$r = b - d \qquad \text{thus} \qquad \frac{\Delta N}{\Delta t} = rN$$

- The value *r* is thus the per capita population growth rate and can be used to determine if a population is growing or declining.
- *Zero population growth* (ZPG) occurs where per capita birth and death rates are equal, thus $r = 0$. (Note that births and deaths still occur, but are equal in number.)
- A population is increasing in size if $r > 0$ (more births than deaths); it is decreasing in size if $r < 0$ (fewer births than deaths).

Many population ecologists use a slightly different equation based on differential calculus to express population growth in terms of instantaneous growth rates:

$$\frac{dN}{dt} = rN$$

- This expresses the population growth over very short time intervals.

A population living under ideal conditions will increase at the fastest rate possible; nutrients are abundant and only the physiological capacity of the individuals limits reproduction.

- The maximum population growth rate is called the *intrinsic rate of increase* and is symbolized by r_{max}.

Exponential population growth is the population increase under ideal conditions due to intrinsic rate of increase. It is expressed as:

$$\frac{dN}{dt} = r_{max}N$$

- The size of the population increases rapidly due to ideal conditions of unlimited resources.
- Produces a J-shaped growth curve. (See Campbell, Figure 47.11)
- Although the intrinsic rate of increase is constant, more new individuals accumulate when the population is large than when it is small; this is due to the fact that N gets larger.

A population with a higher intrinsic rate of increase will grow faster than one with a lower rate of increase. (See Campbell, Figure 47.11)

- A population's r_{max} value is influenced by its life history features (age of first reproduction, clutch size, offspring survival rate).
- Generation time and r_{max} are usually inversely related over a range of species.
- Exponential growth is characteristic of populations introduced into a new or unfilled environment or whose numbers have been decimated by a catastrophe and are rebounding.

V.   A logistic model of population growth incorporates the concept of carrying capacity

No population can grow exponentially for an indefinite period of time. As a population increases in size, the higher density may influence the ability of individuals to obtain resources necessary for maintenance, growth and reproduction.

- Populations inhabiting environments which contain a finite amount of available resources reach a size where further increases in number reduces the share of resources available to each individual.
- The *carrying capacity* (K) of a habitat is the maximum stable population size that the particular environment can support over a relatively long time period.
- Carrying capacity is an environmental property that varies over space and time with the abundance of limiting resources.

Carrying capacities can be determined by many factors:

- Energy limitations (food resources) is the most common determinant of K.
- Other factors include: availability of specialized nesting sites required by some birds; roosting sites as for some bats; shelters and refuges from potential predators.

Crowding and resource limitation can greatly effect the population growth rate.

- Insufficient resources may reduce per capita birth rates in a population (lower b).
- If enough energy cannot be obtained for maintenance, per capita death rates increase (higher d).
- A decrease in b and/or an increase in d results in a lower overall population growth rate (smaller r).

A. The Logistic Growth Equation

A *logistic population growth* model assumes the rate of population growth (r) slows as the population size (N) approaches the carry capacity (K) of the environment.

- A mathematical model for logistic population growth incorporates the effect of population density on r, allowing it to vary from r_{max} when resources are plentiful to zero when the carrying capacity is reached.

The equation for logistic population growth is:

$$\frac{dN}{dt} = r_{max}N\left(\frac{K-N}{K}\right)$$

- K = carrying capacity; the maximum sustainable population.
- $K - N$ = the number of new individuals the environment can accommodate.
- $(K - N)/K$ = percentage of K available for population growth.
- Multiplying r_{max} by $(K - N)/K$ reduces the value of r as N increases. (See Campbell, Table 47.2)
- The actual growth rate of a population of any size is r_{max} N$(K - N)/K$.

The implications of the logistic growth equation at varying population sizes for a growing population are:

- When N is low, $(K - N)/K$ is large and r is only slightly changed from r_{max}.
- When N is large and resources are limiting, $(K - N)/K$ is small; this reduces r substantially from r_{max}.
- When N=K, $(K - N)/N$ is 0 and $r = 0$; this means the number of births is equal to the number of deaths and zero population growth occurs.

The logistic model of population growth produces a sigmoid (S-shaped) growth curve. (See Campbell, Figure 47.14)

- Intermediate population sizes add new individuals most rapidly since the breeding population is of a substantial size and the habitat contains plentiful amounts of resources and available space.
- As N approaches K, the population growth rate slows due to limitations in available resources.
- The logistic model is density-dependent since the rate at which a population grows changes as the density of the population changes.

B. How Well Does the Logistic Model Fit the Growth of Real Populations?

Laboratory populations of paramecia and *Daphnia*, as well as other animals, show population growth rates which fit the predicted S-shaped curve fairly well. (See Campbell, Figures 47.15a and 47.15b)

- These represent relatively unnatural situations of idealized conditions without predators and other species.
  - ⇒ Even under these conditions, some populations show deviations from smooth, sigmoid curves and do not stabilize at a clear carrying capacity.

- Studies of wild populations which have been introduced into new habitats and populations rebounding after near elimination by disease or hunting, provide general support for logistic population growth. (See Campbell, Figure 47.15c)

Some assumptions of the logistic model do not hold true for all populations. For example, the logistic model assumes that:

1. Even at low levels, each individual added may have the same negative effect on population growth rate; any increase in N reduces (K – N)/N.

    - The *Allee effect* points out that individuals may benefit by population increase which improves the chances for survival or reproduction.

        ⇒ For example, a plant standing alone would suffer from excessive wind, but be protected from the wind in a clump of individuals.

        ⇒ Solitary animals like rhinoceros have a greater chance of locating a mate during breeding season if populations are higher.

    - When a population is at low levels, there is a greater possibility that chance events will eliminate all individuals or inbreeding will lead to reduction in fitness.

2. Populations approach carrying capacity smoothly.

    - Often see a lag time before the negative effects of an increasing population are realized, this causes the population to overshoot carrying capacity.

        ⇒ Eventually, deaths exceed births and population size drops below carrying capacity.

        ⇒ Many populations thus appear to oscillate about a general carrying capacity.

Populations do not necessarily remain at, or even reach, levels where population density is an important factor. In these cases, the idea of carrying capacity does not really apply.

- Seen in short-lived, quickly reproducing insects and other small organisms sensitive to environmental fluctuations

Although the logistic model fits few if any real populations, it incorporates ideas that apply to many.

C. Population Growth Models and Life Histories

The logistic population growth model predicts that there will be different growth rates for populations with high and low densities in relation to the environmental carrying capacity.

- Each individual has fewer resources available at high densities and the population is growing slowly.

- Abundant resources are available to individuals at low densities and the population is growing rapidly.

The concept that different life history adaptations would be favored under high densities and low densities was introduced by Martin Cody in the late 1960s. For example,

- Under high population densities, selection would favor adaptations that enhanced survival and reproduction with few resources.

    ⇒ Competitive ability and maximum efficiency of resource utilization would be favored in a population that was maintained at or near the carrying capacity.

- Under low population densities, selection would favor adaptations that enhanced rapid and high rates of reproduction regardless of efficiency.
    ⇒ Increased fecundity and reaching maturity quickly would be favored.

The development of Cody's concept lead to designations of *K-selected* and *r-selected* traits of life history strategies. (See Campbell, Table 47.3)
- The designations *r* and *K* refer to variables in the logistic growth model equation.
- K-selected populations are those living at a density near the limits of their resources (K).
- r-selected populations are more likely to be found in variable environments where population densities fluctuate or in open or disturbed habitats where individuals have little competition.
- Most populations show a mixture of r-selected and K-selected traits since life history evolves in a natural setting of complex factors.

VI. Both density-dependent and density-independent factors can affect population growth

Populations are regulated by density-dependent factors and density independent factors, either separately or in combination.
- The relative importance of these factors differs among species and their specific circumstances.

A. Density-Dependent Factors

The prime implication of the logistic growth model is that increasing population density reduces resource availability and resource limitations ultimately limits population growth.
- This model can thus be applied to *intraspecific competition*: the reliance of two or more individuals of the same species on the same limited resource.
- Competition becomes more intense as population size increases, and *r* is reduced in proportion to the intensity of competition.

In restricting population growth, a *density-dependent factor* intensifies as the population size increases, affecting each individual more strongly. They also affect a greater percentage of individuals in a population as the number of individuals increases.
- Population growth declines because death rate increases, birth rate decreases or both.
- Resource limitation is one such factor.
    ⇒ A reduction in available food often limits reproductive output as each individual produces fewer eggs or seeds.
- Resources other than nutrients may also limit populations.
    ⇒ *Territoriality* (the defense of a well-bounded physical space) is a behavioral mechanism to reduce intraspecific competition since each individual protects resources only within their own territory.
    ⇒ Competition still occurs when individuals compete for space to establish their territories.

- Health and survivorship also decrease as crowding results in smaller, less robust individuals.
- Many predators concentrate on a particular prey when its population density is high, taking a greater percentage than usual.
  ⇒ Predators may switch to other more dense prey populations if energy expenditure to capture prey increases.
- The accumulation of toxic metabolic wastes may also limit a population.

Intrinsic factors may also play a role in regulating population size.

- Population growth rate decreases may occur even when food and shelter are abundant.
- High densities may cause stress syndromes resulting in hormonal changes that delay sexual maturation or otherwise inhibit reproduction.
- High densities have also been shown to produce a stress syndrome which suppresses the immune system.
- High densities can thus reduce birth rates and increase death rates.

B. Density-Independent Factors

Density-independent factors are unrelated to population size and affect the same percentage of individuals regardless of the size of the population.

- Weather, climate and natural disasters such as freezes, seasonal changes, hurricanes and fires are examples.
  ⇒ The severity and time of occurrence is the determining factor on what proportion of the population is affected.
- In some natural populations, these effects routinely control population size before density-dependent factors become important.

C. The Interaction of Regulating Factors

Many populations remain fairly stable in size, close to the carrying capacity determined by density-dependent factors.

- Although exhibiting stability, they often display short term fluctuations due to density-independent factors.
- Long term population size may remain the same, masking short term effects of some factor such as an extremely cold winter.

Density-dependent and density-independent factors sometimes work together to regulate a population, although the relative importance of each may vary seasonally.

- A severe winter may greatly reduce a population due to cold temperatures (density-independent) and intraspecific competition for limited food (density-dependent).
- This reduction in population size may benefit the surviving adults by reducing competition for food in the following spring.

D.   Population Cycles

Some bird, mammal and insect populations show a regular fluctuation in density. For example,

- Lemmings show a 4-year cycle.
- Snowshoe hares show a ten-year cycle.

Several hypotheses have been proposed to explain population cycles:

- Crowding may regulate cyclical population by affecting the organisms endocrine systems.
    - ⇒ Stress from high density may alter hormone balance and reduce fertility, increase aggressiveness, and induce mass emigrations.
- Population cycles may result from a time lag in the response to density-dependent factors. This lag causes the population to overshoot and undershoot the carrying capacity.
    - ⇒ A high density of snowshoe hares may cause a deterioration of food quality which makes it unfit for consumption.
    - ⇒ Predation may also play a role if predators take enough prey to cause a decrease in prey population density.
- Much longer cycles are also known. The periodical cicadas have population cycles of 13 or 17 years.
    - ⇒ May be an adaptation to reduce predation.
    - ⇒ Cicada populations are controlled in some local regions by a fungus whose spores can survive in soil for the years between cicada outbreaks.

VII.   The human population has been growing exponentially for centuries but will not be able to do so indefinitely

The exponential growth of the human population has caused severe environmental degradation.

- Until 1650, the human population increased slowly.
- The population doubled by 1850 (200 years), doubled again by 1920 (80 years), and doubled yet again by 1975 (45 years). (See Campbell, Figure 47.20)

Human population growth is affected by the same parameters (birth and death rates) that affect other plant and animal populations.

- The advent of agriculture 10,000 years ago increased birth rates and decreased death rates.
- Since the Industrial Revolution, exponential growth resulted mainly from a drop in deaths, especially infant mortality.
    - ⇒ The decrease was due to improved nutrition, better medical care, and improved sanitation.
- The population growth rates in most developing countries are actually increasing due to relatively high birth rates being coupled with decreasing mortality.

Population ecologists can not agree on Earth's carrying capacity for the human population or what factor will eventually limit the population.

- Limited nutrients is the usual factor in restraining populations of other animals.
  - ⇒ Agricultural technology has advanced to the point that food supplies have maintained an equivalent pace to population increases.
- Space limitations may play a role.
  - ⇒ As the human population continues to increase, conflicts over land use will intensify.
- Resources other than nutrients and space may play a role.
  - ⇒ How will the disappearance of nonrenewable fossil fuels and metals affect the human population?
  - ⇒ Will the accumulation of pollutants and other forms of wastes eventually reach toxic levels?

Worldwide population growth is a mosaic with some countries having near zero population growth while others have relatively high growth rates.

- Although variation in growth rates is found, the human population as a whole continues to grow.

Age structure within each country appears to be a major factor in the variation of population growth rates. (See Campbell, Figure 47.22)

- A relatively uniform age distribution, where individuals of reproductive age or younger are not disproportionately represented, lends itself to a stable population size.
- A population which contains a large proportion of reproductive age or younger individuals will face a sudden increase in the rate of population growth in the future.

A unique feature of human reproduction is the ability to be consciously controlled by voluntary contraception or government sponsored family planning.

- Social change, such as women delaying reproduction in favor of employment and advanced education, can also result in decreases in population growth rates.

Human cultural evolution has had an impact on Earth's carrying capacity.

- The development of agriculture and industrial technology has twice increased the Earth's carrying capacity for humans.
- What Earth's carrying capacity is and how we approach it are questions generating concern and debate.

The human population will eventually stop growing due to social changes, individual choice, government intervention, or increased mortality due to environmental limitations.

- Hopefully, if the human population reaches K, it will approach it smoothly and level off.
  - ⇒ This will occur when birth rates and death rates are equal.
- If the population fluctuates around K, periods of increase will be followed by times of mass death due to famine, disease, or conflict.

REFERENCES

Campbell, N. *Biology*. 4th ed. Menlo Park, California: Benjamin/Cummings, 1996.

Hutchinson, G. Evelyn. *An Introduction to Population Ecology*. New Haven, Connecticut: Yale University Press, 1979.

Ricklefs, R.E. *Ecology*. 3rd ed. New York: Chiron Press, 1986.

Smith, Robert Leo. *Ecology and Field Biology*. 3rd ed. New York: Harper & Row, Publishers, 1980.

CHAPTER 48

COMMUNITY ECOLOGY

OUTLINE

I. The interactive and individualistic hypotheses pose alternative explanations of community structure: *science as a process*

II. Community interactions can provide strong selection factors in evolution

III. Interspecific interactions may have positive, negative, or neutral effects on a population's density: *an overview*

IV. Predation and parasitism are (+ −) interactions: *a closer look*
   A. Predation
   B. Parasitism

V. Interspecific competitions are (− −) interactions: *a closer look*
   A. The Competitive Exclusion Principle
   B. Ecological Niches
   C. Evidence for Competition in Nature

VI. Commensalism and mutualism are (+ 0) and (+ +) interactions, respectively: *a closer look*
   A. Commensalism (+ 0)
   B. Mutualism (+ +)

VII. A community's structure is defined by the activities and abundance of its diverse organisms
   A. Feeding Relationships Within Communities
   B. Species Richness, Relative Abundance, and Diversity
   C. Disturbances and Community Stability

VIII. The factors that structure communities include competition, predation, and environmental patchiness
   A. The Role of Competition
   B. The Role of Predation
   C. The Role of Environmental Patchiness
   D. Evaluating Causative Factors

IX. Succession is the sequence of changes in a community after a disturbance
   A. Causes of Succession
   B. Natural and Human Disturbance
   C. Community Equilibrium, Disturbance, and Species Diversity

X. Biogeography complements community ecology in the analysis of species distribution
   A. Limits of Species Ranges
   B. Global Clines in Species Diversity
   C. Island Biogeography

XI. Lessons from community ecology and biogeography can help us preserve biodiversity

OBJECTIVES

After reading this chapter and attending lecture, the student should be able to:
1. Compare and contrast the individualistic hypothesis of H.A. Gleason and the interactive hypothesis of F.E. Clements with respect to communities.
2. Explain the relationship between species richness, relative abundance and diversity.
3. List four properties of a community and explain the importance of each.
4. Explain how interspecific competition may affect community structure.
5. Describe the Competitive Exclusion Principle and explain how competitive exclusion may affect community structure.
6. Distinguish between an organism's fundamental niche and realized niche.
7. Explain how resource partitioning can affect species diversity.
8. Describe the defense mechanisms evolved by plants to reduce predation by herbivores.
9. Explain how cryptic coloration and aposematic coloration aid an animal in avoiding predators.
10. Distinguish between Batesian mimicry and Mullerian mimicry.
11. Describe how predators use mimicry to obtain prey.
12. Explain the role of predators in community structure.
13. Distinguish among parasitism, mutualism and commensalism.
14. Explain why it is difficult to determine what factor is most important in structuring a community.
15. Distinguish between primary succession and secondary succession.
16. Explain how inhibition and facilitation may be involved in succession.
17. Describe how natural and human disturbances can affect community succession.
18. Explain how the intensity of disturbances can affect equilibrium and species diversity.
19. List the factors involved in limiting a species to a particular range.
20. Describe the mechanisms which contribute to the global clines in diversity.
21. Explain the factors which determine what species eventually inhabit islands.

KEY TERMS

community
individualistic
  hypothesis
interactive
  hypothesis
H.A. Gleason
F.E. Clements
coevolution
predator
prey
secondary
  compounds
cryptic coloration
aposematic
  coloration

mimicry
Batesian mimicry
Mullerian mimicry
parasitism
brood parasite
interspecific
  competition
competitive
  exclusion principle
A.J. Lotka
V. Volterra
G.F. Gause
ecological niche
fundamental niche
realized niche

resource partitioning
Robert Paine
symbiosis
commensalism
mutualism
relative abundance
diversity
species richness
trophic structure
stability
community resilience
exotic species
keystone predator
ecological succession

primary succession
secondary succession
climax community
fugitive species
inhibition
facilitation
intermediate
  disturbance
  hypothesis
biogeography
Robert MacArthur
E.O. Wilson
D.S. Simberloff
Thomas Lovejoy

LECTURE NOTES

A *community* consists of all the organisms inhabiting a particular area; an assemblage of populations of different species living close enough together for potential interaction.

I. **The interactive and individualistic hypotheses pose alternative explanations of community structure:** *science as a process*

Among the pioneers of community ecology, there were two divergent views on why certain combinations of species are found together as members of a community: the *individualistic hypothesis* and the *interactive hypothesis*.

The *individualistic hypothesis* was proposed by H.A. Gleason. This hypothesis depicted a community as a chance assemblage of species found in an area because they have similar environmental requirements.

- Emphasizes studying single species.

- Predicts each species will have an independent distribution along an environmental gradient and there are no discrete geographical boundaries between communities.

- In most cases, the composition of communities would change continuously along some environmental gradient due to addition or loss of particular species.

The *interactive hypothesis* proposed by F.E. Clements saw each community as an assemblage of closely linked species having mandatory biotic interactions that cause the community to function as an integrated unit.

- Based on the fact that certain plant species are consistently found together.
- Emphasizes entire assemblages of species as the essential units for understanding the interrelationships and distributions of organisms.
- Predicts species should be clustered, with discrete boundaries between communities. Since the presence or absence of one species is governed by the presence or absence of other species with which it interacts.
- Not generally seen in plants, especially where broad regions are characterized by gradients of environmental variation.

Animals in a community are often linked more rigidly to other organisms.

- Some animals feed primarily on certain food items so their distribution is linked to distribution of their food source.
- Other animals feed on a variety of food items and tend to be distributed in a variety of communities.

Distribution of almost all organisms is affected by both abiotic gradients and interactions with other species.

II.   Community interactions can provide strong selection factors in evolution

Interactions between species are as important to adaptation by natural selection as the physical and chemical features of the environment.

Coevolution = A change in one species acts as a new selective force on another species, and counteradaptation of the second species, in turn, affects selection of individuals in the first species.

- Studied most extensively in predator-prey relationships and in mutualism.
- Often, what appears to be coevolution may actually result from interactions with many species (not just the obvious two).

The association between passionflower vines (*Passiflora*) and the butterfly *Heliconius* is believed to be an example of coevolution.

- The vines produce toxic chemicals to reduce damage to young shoots and leaves by herbivorous insects.
- Butterfly larvae of *Heliconius* can tolerate these chemicals due to digestive enzymes which break down the toxic chemicals (a counteradaptation).
- Females of some *Heliconius* species avoid laying eggs (which are bright yellow) on leaves where other yellow egg clusters have been laid.
  ⇒ This reduces intraspecific competition on individual leaves.
- Some species of passionflowers develop large, yellow nectaries which resemble *Heliconius* eggs; an adaptation that may divert egg-laying butterflies to other plants.
- These nectaries, as well as smaller ones, also attract ants and wasps which prey on butterfly eggs and larvae.

Biologists agree that adaptation of organisms to other species in a community is a fundamental characteristic of life.

III. Interspecific interactions may have positive, negative, or neutral effects on a population's density: *an overview*

While adaptations to abiotic factors largely determine the geographic distributions of many species, all organisms are influenced by biotic interactions with other individuals.

- Intraspecific competition for resources was discussed in Chapter 47.
- *Interspecific interactions* are those that occur between populations of different species living in a community.
  ⇒ May take a variety of forms.

The two species involved in the interaction may have a positive (+), negative (–), or neutral (0) effect on their population densities.

- Both population densities may increase (+ +), one may increase while the other decreases (+ –), one may increase while the other is not affected (+ 0), or both may decrease (– –). (See Campbell, Table 48.1)

IV. Predation and parasitism are (+ –) interactions: *a closer look*

There are several (+ –) interactions, some are obvious, some are not.

- *Predation* involves a *predator* which eats its *prey*.
- *Parasitism* involves predators which live in or on their *hosts* and seldom involves outright host death.
- *Parasitoidism* involves insects such as small wasps who lay their eggs on living hosts – after hatching, the larvae feed within the host's body and eventually cause its death.
- *Herbivory* occurs when animals eat plants. May only cause damage to a portion of the plant (grazing) or may kill an entire plant (seed eating).

A. Predation

Predation = A community interaction where one species, the *predator*, eats another, the *prey*. Includes both animal-animal interactions and animal-plant interactions.

Adaptations of predators to this interaction are obvious and familiar. Predators have:

- Acute senses that are used to locate and identify prey items (e.g. heat-sensing pits of rattlesnakes, and chemical sensors of herbivorous insects).
- Structures such as claws, teeth, fangs, stingers, and poisons which function to catch, subdue, or chew the prey item.
- Speed and agility to pursue prey or camouflage which permits them to ambush prey.

Various defensive adaptations have evolved in prey species as a result of repeated encounters with predators over evolutionary time.

1. **Plant Defenses Against Herbivores**

   Since plants cannot escape herbivores, they have evolved several ways to protect themselves from predation.

   Plants have mechanical and chemical defenses against herbivores.
   - Thorns or microscopic hooks, spines, and crystals may be present to discourage herbivores.
   - May produce *secondary compounds* (formed as by-products of major metabolic pathways) that make vegetation distasteful or harmful.
   - May produce analogues of insect hormones that cause abnormal development in insects that prey on them.

   Plant defenses may act as selective factors leading to counteradaptations in herbivores that can then nullify those defenses and consume the plant.

2. **Animal Defenses Against Predators**

   Animal defenses against predators may be passive (hiding), active (escaping and physical defense), or chemical based.

   Other forms of defense involving adaptive coloration have evolved repeatedly among animals.
   - *Cryptic coloration* (coloration making prey difficult to spot against its surroundings) is common and only requires the animal to remain still to avoid detection.
   - Deceptive marking such as large, fake eyes can startle predators, allowing prey to escape, or cause predators to strike a nonvital area.
   - Some mechanical and chemical defenses actively discourage predators (e.g. porcupine quills and skunk spray).
   - Some insects acquire chemical defense passively by accumulating toxins from plants they eat; these make them distasteful to predators.
   - *Aposematic coloration* (bright coloration that acts as a warning of effective physical or chemical defense) appears to be adaptive since predators quickly learn to avoid prey with this coloration.

3. **Mimicry**

   Mimicry is a phenomenon in which a *mimic* bears a superficial resemblance to another species, the *model*.
   - Occurs in both predatory and prey species.
   - Defensive mimicry in prey usually involves aposematic models.

   Batesian mimicry = A palatable species mimics an unpalatable or harmful model.
   - Mimic must be much less abundant than the model to be effective since predators must learn the coloration indicates a bad or harmful food item.

<u>Mullerian mimicry</u> = Two or more unpalatable, aposematically colored species resemble each other.

- Each gains additional advantage since predators learn more quickly to avoid prey with this coloration.

Predators also use mimicry to lure prey.

- The tongue of snapping turtles resembles a wriggling worm which attracts small fish into capture range.

B.   Parasitism

Parasitism is a (+ −) interaction in which one organism, the *parasite*, derives its nourishment from another organism, the *host*, which is harmed in some way.

- *Endoparasites* live within the host tissues or body cavities.
- *Ectoparasites* attach to or briefly feed on the external surface of the host.

Natural selection favors parasites that are best able to locate and feed on a host.

- Infection by a parasite (locating a host) may be passive as when an endoparasite's egg is swallowed by a host.
- Active location of a host may involve thermal or chemical cues that help the parasite identify the host.

Defensive capabilities in potential hosts have also been selected for by natural selection.

- Secondary plant products toxic to herbivores may also be toxic to parasitic fungi and bacteria.
- The vertebrate immune system provides defense against endoparasites.
- Many parasites are adapted to specific hosts, often a single species.
- Coevolution generally results in stable relationships in which the host is not killed (host death would also kill the parasite).

Some forms of parasitism do not involve the consumption of the host, but are the exploitation of the host's behavior by the parasite. For example, in *brood parasitism*:

- Some species of birds (cowbirds, European cuckoos) lay their eggs in the nests of other species.
- Often a newly hatched brood parasite will move other eggs out of the nest.
- Host parents invest their energy in feeding the brood parasite instead of their own offspring – a (+ −) interaction.
- An evolutionary adaptation in many host species is the ability to detect parasite eggs which are removed or the nest is abandoned.
    ⇒ A counter-adaptation to this host defense is egg mimicry in some brood parasites.

V. Interspecific competitions are (– –) interactions: *a closer look*

Interspecific competition occurs when two or more species in a community rely on similar limiting resources.

- May take the form of *interference competition* (actual fighting) or *exploitative competition* (consumption or use).
- As the population density of one species increases, it may limit the density of the competing species as well as its own.

A. The Competitive Exclusion Principle

The *competitive exclusion principle* predicts that two species competing for the same limiting resources cannot coexist in the same community. One will use resources more efficiently, thus reproducing more rapidly and eliminating the inferior competitor.

- Derived independently by A.J. Lotka and V. Volterra who modified the logistic model of population growth to incorporate interspecific competition.
- Experiments by G.F. Gause confirmed competitive exclusion among species of *Paramecium*.
- Laboratory experiments have confirmed the principle for other organisms.

B. Ecological Niches

An *ecological niche* is the sum total of an organism's use of biotic and abiotic resources in its environment; how it "fits into" an ecosystem.

- A *fundamental niche* is the resources a population is theoretically capable of using under ideal circumstances.
- Biological constraints (competition, predation, resource limitations) restrict organisms to their *realized niche*: the resources a population actually uses.
- Two species cannot coexist in a community if their niches are identical.
- Ecologically similar species can coexist in a community if there are one or more significant differences in their niches.

C. Evidence for Competition in Nature

If competition is as important as indicated by the competitive exclusion principle, it should be rare in natural communities, since only two outcomes are possible:

- The weaker competitor will become extinct.
- One of the species will evolve to the point of using a different set of resources.

Several lines of evidence, including *resource partitioning* and *character displacement*, indicate that competition in the past has had a major role in shaping current ecological relationships.

Resource partitioning is well documented among animals.

- Sympatric species consume slightly different foods or use other resources in different ways. For example,
    ⇒ In the Dominican Republic, several species of *Anolis*, small arboreal lizards, are sympatric.
    ⇒ They feed on small arthropods that land within their territories.
    ⇒ Each species has a characteristic perching site which reduces competition.
    ⇒ Natural selection appears to have favored perch site selection since each species has morphological characteristics (body size, leg length) which are adaptive to their microhabitats.

Character displacement is the tendency for characteristics to be more divergent in sympatric populations of two species than in allopatric populations of the same species.

- Often occurs in areas where populations of closely related species which are normally allopatric become sympatric.
- Allopatric populations of such species are similar in structure and use similar resources.
- Sympatric populations show morphological differences and use different resources.
- Two closely related species of Galapagos finches (*Geospiza fuliginosa* and *G. fortis*) have beaks of similar size when the populations are allopatric; however, on the island where they are sympatric, a significant difference in beak depth has occurred.
    ⇒ Indicates they feed on seeds of different sizes when sympatric.

Controlled field experiments also provide evidence for the influence of one species on the density and distribution of another species. J. H. Connell manipulated the densities of two species of barnacles that compete for attachment space in the rocky intertidal zone. (See Campbell, Figure 48.12)

- *Balanus balanoides* is usually concentrated in the lower portions of the rocky intertidal zone.
- *Chthamalus stellatus* is usually found in the upper portions of the zone.
- Larvae of both settle at random and begin to develop.
- *Balanus* fails to survive in the upper region of the zone due to a lower tolerance to desiccation.
- *Chthamalus* can survive in both the upper and lower regions of the zone.
- In the lower part of the zone, *Balanus* out competes *Chthamalus* for attachment sites, thus limiting it to primarily the upper part of the rocks.

VI. Commensalism and mutualism are (+ 0) and (+ +) interactions, respectively: *a closer look*

Symbiosis is a form of interspecific interaction in which a *host* species and a *symbiont* maintain a close association.

- May be parasitism, commensalism, or mutualism.
- All are important determinants of community structure.

A.   **Commensalism**

Commensalism is a (+ 0) interaction in which the symbiont benefits and the host is unaffected.

- Few examples since it is unusual for one species not to be affected in some way by a close association with another species.

A good example of commensalism is that of the cattle egret and grazing cattle. (See Campbell, Figure 48.13)

- Cattle egrets feed on insects and other small animals.
- The birds concentrate their feeding activity near grazing cattle whose movements flush prey items from the vegetation.
- The egrets benefit by an increase in feeding rates by following the cattle.
- The cattle are not affected by the activity of the egrets.

Since only one species in a commensalistic association is benefited, any evolutionary change in the relationship will likely occur in the beneficiary.

B.   **Mutualism**

Mutualism is a (+ +) interaction requiring the evolution of adaptations in both species.

- A change in either species is likely to affect the survival and reproduction of the other.
- Examples include:
  ⇒ Nitrogen fixation by bacteria in the root nodules of legumes.
  ⇒ Cellulose digestion by microorganisms in the alimentary canals of termites and ruminant mammals.
  ⇒ Photosynthesis by algae in the tissues of corals.

Mutualistic relationships may have evolved from predator-prey or host-parasite relationships.

- Some angiosperms have adaptations that attract animals for pollination or seed dispersal.
  ⇒ May represent counter-adaptations to herbivores feeding on seeds or pollen.
  ⇒ Pollen is spared when the pollinator can feed on nectar and seeds are dispersed when animals eat fruits.
- Plants that received a benefit from sacrificing tissues other than pollen or seeds would increase their reproductive success.

VII.   A community's structure is defined by the activities and abundance of its diverse organisms

The relationships of organisms within a community are tied together in a complex web of community structure and processes.

Three features of community structure are:
- The feeding relationships among organisms.
- Patterns of abundance and diversity among species.
- The ways a community is affected by disturbances.

A.   Feeding Relationships Within Communities

The various feeding relationships found within a community is the *trophic structure* of that community – put simply, it's what eats what.
- Determined by predator-prey, host-parasite, and plant-herbivore interactions.
- Some interactions represent competition for food.

Community trophic structure may be studied in two ways: *functional groupings* or *food webs*.
- The *functional groupings* approach involves placing species in groups with similar trophic positions.
  - ⇒ All plants are grouped together as producers; herbivores, regardless of size, are primary consumers; organisms that feed on primary consumers are secondary consumers, etc.
  - ⇒ A problem arises in this method since lumping species into the functional categories may mask some aspects of the community structure.
- *Food web analysis* includes species-level information about the community and emphasizes species-species connections.
  - ⇒ Can be very complex.
  - ⇒ Presents a detailed knowledge of feeding relationships within the community structure.

B.   Species Richness, Relative Abundance, and Diversity

Communities , even those that appear similar, vary with regards to their *species diversity*.

Species diversity = The number and relative abundance of species in a biological community.
- Based on both species richness and relative abundance.
  - ⇒ *Species richness* is the number of species in a community.
  - ⇒ *Relative abundance* is a measure of the proportion of a species in the community as a whole.
- A community with a certain species richness with equal relative abundance of each species would have a greater diversity than a second community with the same species richness with a few common species and many rare ones.

C.   **Disturbances and Community Stability**

Communities vary in their responses to both natural and human-induced disturbances.

- The *stability* of a community is its tendency to reach and maintain an equilibrium (or relatively constant condition) in response to disturbance.
    ⇒ Depends on the type of community and nature of the disturbance.
- *Community resilience* is the ability of a community to persist in the face of a disturbance.
    ⇒ Related to stability but does not imply stability of individual populations.
    ⇒ Populations may show large fluctuations in response to a disturbance, but the community persists.

VIII.   The factors that structure communities include competition, predation, and environmental patchiness

Population interactions such as competition, predation, and environmental patchiness influence the characteristics of a community.

A.   **The Role of Competition**

Competition is recognized as an important factor in determining community structure but its importance as the major factor is open to debate.

- It is generally recognized that competition becomes more important as populations sizes approach carrying capacities and resources are limiting.

Many studies have documented interspecific competition, but the competition did not always result in competitive exclusion, competing species sometimes coexisted at reduced densities.

- The competition for space between barnacles discussed earlier resulted in exclusion of *Chthamalus* from the lower intertidal zone but it remained in the upper zone.
- Exotic species (species introduced into new communities by humans) often out compete native species and alter the community structure.

B.   **The Role of Predation**

Predation plays a complex role in helping shape community structure and may actually help maintain diversity.

- Predators do not always reduce diversity.
- A *keystone predator* may exert a regulating effect on other species and maintain a higher community species diversity by reducing the densities of strong competitors.
    ⇒ This would prevent competitive exclusion of poorer competitors and maintain higher community diversity.
- Defensive abilities of prey prevent them from becoming extinct.
- Prey switching also protects prey species as predators change to different prey as the population of one is depleted due to predation.

C.   **The Role of Environmental Patchiness**

The more heterogeneous the habitat, the more diverse the community.

- The greater the heterogeneity, the more ecological niches available to organisms.
- Heterogeneity may be *spatial* or *temporal*.

Spatial heterogeneity is affected by the vegetation structure in the community.

- More structurally complex vegetation provides a larger number of microhabitats.
- Vegetation greatly influences the types of animals found in a community. Animal populations can more easily partition a structurally complex system than one of less complexity.

Environmental patchiness is another type of spatial heterogeneity.

- All environments are patchy over time and space. For example,
    ⇒ Chemical composition of rocks influences the mineral content of the soil derived from their erosion.
    ⇒ Soil moisture varies with topography.
- Environmental patchiness can increase community diversity by facilitating resource partitioning among potential competitors adapted to different local conditions.
- The importance of patchiness varies with the size and lifespan of the organisms.

Community diversity is also affected by the *temporal* use of habitats.

- Nocturnal animals utilize a habitat at night and are replaced by diurnal animals when it becomes day time.
- A community will contain a variety of plants which bloom during different seasons.

D.   **Evaluating Causative Factors**

Evaluating what factors are important in developing the characteristics of a community is difficult since interspecific interactions and abiotic factors that cause patchiness all have significant impact.

- Communities are structured by multiple interactions of organisms with the biotic and abiotic factors in their environment.
- Which types of interactions are most important varies from one type of community to another, and even among different components of the same community.

IX.   **Succession is the sequence of changes in a community after a disturbance**

Ecological succession is the transition in species composition over ecological time.

- Called *primary succession* if it begins in areas essentially barren of life due to lack of formed soil (e.g. volcanic formations) or on rubble (e.g. retreat of glaciers).
- Called *secondary succession* if an existing community has been cleared by some disturbance that leaves the soil intact (e.g. abandoned agricultural fields).
- The traditional view was that in some cases, the community passes through a series of predictable transitional stages to reach a relatively stable state, the *climax community*.
    ⇒ Views on the validity of the climax community are changing.

A. **Causes of Succession**

A variety of interrelated factors determines the course of succession.

- Individual species compete for available resources and replace each other.
- Different species compete better at different successional stages since resource availability changes over the course of succession.
- Early stages are typically characterized by *r-selected* species that are good colonizers.
  ⇒ Have a high fecundity and excellent dispersal mechanisms.
  ⇒ Others may be fugitive species that do not compete well in established communities.
- Abiotic conditions of the area, which can differ on a fairly local level, also affect succession.

The species composition during early successional stages is affected by the abiotic conditions in a barren area.

- *K-selected* species may colonize an area but fail to become abundant because of environmental conditions which are outside of their tolerance limits.
- Variation in growth rates and maturation times of colonizers impact the community. For example,
  ⇒ Herbaceous species are more prevalent in early successional stages than trees because they grow faster and mature quicker.

Changes in community structure during succession may be induced by the organisms themselves.

- Direct biotic interactions include *inhibition* of some species by others through exploitative or interference competitions (or both).
- Organisms also affect the abiotic environment by modifying local conditions.
  ⇒ In *facilitation*, the group of organisms representing one successional stage "paves the way" for species typical of the next stage (e.g. alder leaves lower soil pH as they decompose, and lower soil pH facilitates the movement of spruce and hemlock into the area).
  ⇒ Sometimes facilitation for other species makes the environment unsuitable for the very species responsible for the changes.
- Inhibition and facilitation may be involved throughout the stages of succession.

At *climax* stage, environmental conditions are such that one set of species can continue to maintain themselves.

- Communities may never truly reach a climax, but just appear to because further change is slow.
- Many communities are also routinely disturbed by outside factors during the course of succession.

B.   **Natural and Human Disturbance**

Disturbances are events that disrupt communities.

- They change resource availability and create opportunities for new species to become established.
- The magnitude of the disturbance's impact depends on the size, frequency, and severity of the disturbance.
- Natural disturbances are fire, drought, wind, and moving water.
- Animals, including humans, can also cause disturbances.

Human disturbances have the greatest impact on communities.

- Logging and clearing for farmland have reduced and disconnected forests.
- Agricultural development has disrupted grasslands.
- After a community disturbance, the early stages of succession, characterized by weedy and shrubby vegetation, may persist for many years.
- Centuries of overgrazing and agricultural disturbance have contributed to current famine in parts of Africa by turning grasslands into barren areas.

It is possible for disturbances to have *positive* impact on a community.

- Small-scale natural disturbances may help maintain species diversity in a community.
    ⇒ When patches are formed during different successional stages, habitat heterogeneity is increased.
- Frequent small-scale disturbances can prevent large-scale disturbances of greater negative impact.
    ⇒ Small fires prevent the accumulation of more flammable materials which would feed a more damaging large fire.

C.   **Community Equilibrium, Disturbance, and Species Diversity**

The traditional concept of ecological succession predicts that species diversity reaches an equilibrium in the climax community.

- K-selected species (better competitors) replace r-selected species as population density increases and vegetation modifies the site.
    ⇒ The rate of species replacement would slow once the longer-lived K-selected species become established.
- Although turnover may occur, the addition of new colonists would be balanced by localized extinctions.
- Emphasizes the importance of interspecific interactions in shaping community structure.
- Interactions (predation, competition, symbiosis) become more extensive and varied, increasing diversity.
- The climax is reached when the web of interactions is so intricate that the community is saturated.
    ⇒ No new species can enter unless a localized extinction occurs.

An opposing view sees communities as being in continual flux.

- The identity and numbers of species change in all successional stages, even the climax.
- Emphasizes the importance of less predictable factors (dispersal, disturbance) in shaping community composition and structure.
- The course of succession may vary with the identity of colonizing species.
- Disturbances may prevent the community from reaching equilibrium.
- A mature community is an unpredictable mosaic of patches at different successional stages.
- Local environmental heterogeneity contributes to the mosaic structure as different species inhabit different patches.

Intermediate disturbance hypothesis states that species diversity is greatest where disturbances are moderate in frequency and severity because organisms typical of several successional stages will be present.

- Supported by studies of clearings formed by trees falling in tropical rain forests.
  - ⇒ In these disturbed areas, immigrations and extinctions occur rapidly and species from different successional stages coexist.

X.   Biogeography complements community ecology in the analysis of species distribution

<u>Biogeography</u> = The study of the past and present distribution of individual species, as well as entire communities.

- Provides a different approach to understanding community properties by analyzing global *and* local phenomena.
- Traditionally concerned with the identities of species comprising particular communities.
- Currently also applies the principles of community ecology to the analysis of geographical distribution.

Terrestrial life can be divided into biogeographical realms having boundaries associated with continental drift patterns. (See Campbell, Figure 48.19)

- Thus species distribution reflects past history as well as present interactions of organisms with their environment.

A.   Limits of Species Ranges

Limitation of a species to a particular range may be due to:

1. Failure of the species to disperse beyond its current range.
2. Pioneer individuals that spread beyond the observed range failing to survive.
3. The species having retracted from a once larger range to its current boundaries.

The first two limitations are distinguished by transplant experiments in which specimens of a species are moved to similar environments outside of their range.

- If the transplants survive, they probably never dispersed there; if they die, it indicates an intolerance to abiotic conditions or an inability to compete limits their range.

- Most species fail to survive outside their range, but there are notable exceptions.

The third limitation is well supported by data from paleontology and historical biogeography.

- The fossil record includes evidence that close relatives of some living animals with limited ranges once had much wider ranges.

  ⇒ For example, fossils of close relatives of elephants and camels have been found in North America.

B. Global Clines in Species Diversity

The presence of clines (gradual variation) in species diversity with major geographical gradients has long been recognized.

- Generally, diversity increases in moving from far northern and southern latitudes toward the equator.

- Diversity of terrestrial organisms at a given latitude decreases with altitude.

- In the ocean benthic community, diversity increases from shallow coastal waters toward deep sea.

Many hypotheses have been proposed as to why tropical communities have such a high diversity and why latitudinal gradients exist.

1. Tropical communities are very old and rarely experience major natural disturbance. This fosters rapid and frequent speciation, and their age has allowed complex population interactions to evolve and develop to a greater extent than in temperate zones.

2. Tropical regions usually experience intermediate levels of disturbance and have greater environmental patchiness. This permits a greater diversity of plant species which forms a resource base for diverse animal communities.

3. Stability and predictability in tropical climates may permit organisms to specialize on a narrow range of resources. Small niches would reduce competition and permit a finer level of resource partitioning, both of which increases species diversity.

4. Increased solar radiation in the tropics increases the photosynthetic activity of plants. This provides a larger resource base for animals.

5. The structural complexity of tropical forests creates a great variety of microhabitats that other plants and animals can partition.

6. Diversity is self-propagating because complex predator-prey and symbiotic interactions in a diverse community prevent any populations from becoming dominant.

While most of these hypotheses will apply to some terrestrial organisms and communities, complex combinations of these factors probably operate under most circumstances.

- None of these hypotheses apply to the increased diversity of marine benthic communities with increased depth; long term stability in these habitats may explain the many relationships among these organisms.

C. Island Biogeography

Islands provide opportunities to study factors affecting species diversity of communities due to their isolation and limited size.

- An island is any area surrounded by an environment not suitable for the "island" species.
- Islands can be oceanic islands or habitat islands on land (lakes or mountain peaks separated by lowlands).

Robert MacArthur and E.O. Wilson developed a general theory of island biogeography that predicts species diversity on an island will reach an equilibrium directly proportional to island size and inversely proportional to the island's distance from a mainland.

- Number of species to inhabit islands is determined by immigration and extinction rates, themselves affected by island size and distance from the mainland.
- The smaller the island and the more distant from mainland, the lower the immigration rate due to difficulty in finding the island and the distance colonizers must travel.
- Smaller islands also have higher extinction rates since they have fewer resources and less diverse habitat to be partitioned; these factors increase the probability of competitive exclusion.
- Immigration and extinction rates also are affected by the number of species already present.
  ⇒ As species numbers increase, immigration rate of new species decreases since new arrivals are less likely to represent a new species.
  ⇒ Extinction rate increases due to increased chance of competitive exclusion.
- Eventually, an equilibrium will be reached where immigration rates match extinction rates.
  ⇒ The number of species present at the equilibrium point being correlated to island size and distance from the mainland. (See Campbell, Figure 48.21)
- The species composition may change since equilibrium is dynamic with immigrations and extinctions continuing.
- Applies over a short period while colonization is the most important process determining species composition.
- Over longer periods, the species composition and community structure will be affected by speciation and evolutionary change in the island species.
- Theory supported from experimental research by E.O. Wilson and D. Simberloff who studied small mangrove islands off the tip of South Florida.

XI. **Lessons from community ecology and biogeography can help us preserve biodiversity**

The global support for conservation efforts has developed out of:

- Esthetic appreciation for other life forms.

- Recognition that yet undiscovered species may provide useful products.

- The understanding that natural environments are necessary to maintain proper functioning of the biosphere.

- Conservation efforts are complicated by religious and cultural opposition to human population control and the economic problems facing humans in every region.

Conservation efforts currently focus on creating natural preserves which are protected from development and pollution and are designed to preserve not only endangered species but also entire ecosystems. Information from community ecology and biogeography can be applied to the preserves in order to maximize the conservation of biodiversity.

1. Preserves can be viewed as islands surrounded by disturbed habitat.

    - Studies of island biogeography provide advanced knowledge about what species a preserve with a particular set of characteristics can successfully protect.

    - For example, we know the number of species that can be protected in a preserve is directly related to the size of the preserve. Species in large preserves are less likely to encounter competitive exclusion. (See Campbell, Figure 48.22)

2. Detailed knowledge of species requirements is essential to preserve design.

    - Minimum required sizes of preserves are known for the successful preservation of some species.

    - Thomas Lovejoy used the principles of island biogeography to study "islands" of tropical rain forest left in areas of the Amazon region which have been converted to pastureland.

        ⇒ These islands range in size from 1 to 1000 hectares.

        ⇒ Teams monitor diversity at each site.

        ⇒ Diversity of small islands is decreasing most rapidly; many extinctions can be attributed to wind penetration which dessicates areas 100 m or more from the edge.

        ⇒ Colonies of army ants require 10 hectares to maintain a work force.

        ⇒ In <40 hectare plots, army ants disappear first, and then five species of birds that feed on insects driven into the open by ant swarms, soon disappear.

3. Would one large preserve be more beneficial to conservation of species than several small preserves representing the same area?

    - One large preserve would be preferable if the smaller preserves all contained the same species.

    - Several small preserves would collectively retain a larger number of species, if the community being preserved included environmental heterogeneity resulting in different species occupying different areas.

- Several small preserves would also be less susceptible to disease epidemics which might not be able to cross unsuitable habitat.
- Spatial relationships between small preserves must also be considered. Closely spaced preserves or those connected by land corridors might encourage the spread of protected species among the preserves or permit recolonization by a protected species in case of local extinction in one preserve.

REFERENCES

Campbell, N. *Biology*. 4th ed. Menlo Park, California: Benjamin/Cummings, 1996.

MacArthur, Robert H., and Edward O. Wilson. *The Theory of Island Biogeography*. Princeton, New Jersey: Princeton University Press, 1969.

Ricklefs, R.E. *Ecology*. 3rd ed. New York: Chiron Press, 1986.

Smith, Robert Leo. *Ecology and Field Biology*. 3rd ed. New York: Harper & Row, Publishers, 1980.

OUTLINE

I. Trophic structure determines an ecosystem's routes of energy flow and chemical cycling
    A. Producers
    B. Consumers
    C. Decomposers (Detritivores)

II. An ecosystem's energy budget depends on primary productivity
    A. The Global Energy Budget
    B. Primary Productivity

III. As energy flows through an ecosystem, much is lost at each trophic level
    A. Secondary Productivity
    B. Ecological Efficiency and Ecological Pyramids

IV. Matter cycles within and between ecosystems

V. A combination of biological and geological processes drives chemical cycles
    A. The Water Cycle
    B. The Carbon Cycle
    C. The Nitrogen Cycle
    D. The Phosphorus Cycle
    E. Variations in Nutrient-Cycling Time

VI. Field experiments reveal how vegetation regulates chemical cycling: *science as a process*

VII. The human population is disrupting chemical cycles throughout the biosphere
    A. Agricultural Effects on Nutrient Cycling
    B. Accelerated Eutrophication of Lakes
    C. Poisons in Food Chains
    D. Intrusions in the Atmosphere
    E. Carbon Dioxide Emissions and the Greenhouse Effect
    F. Depletion of Atmospheric Ozone

VIII. Human activities are altering species distribution and reducing biodiversity
    A. The Introduction of Exotic Species
    B. Habitat Destruction and the Biodiversity Crisis

IX. The Sustainable Biosphere Initiative is reorienting ecological research

OBJECTIVES

After reading this chapter and attending lecture, the student should be able to:

1. Explain the importance of autotrophic organisms with respect to energy flow and nutrient cycling in ecosystems.
2. List and describe the importance of the four consumer levels found in an ecosystem.
3. Explain how gross primary productivity is allocated by the plants in an ecosystem.
4. List the factors that can limit productivity of an ecosystem.
5. Explain why productivity declines at each trophic level.
6. Distinguish between energy pyramids and biomass pyramids.
7. Describe the hydrologic (water) cycle.
8. Describe the carbon cycle and explain why it is said to result from the reciprocal processes of photosynthesis and cellular respiration.
9. Describe the nitrogen cycle and explain the importance of nitrogen fixation to all living organisms.
10. Explain how phosphorus is recycled locally in most ecosystems.
11. Explain why the soil in tropical forests contains lower levels of nutrients than soil in temperate forests.
12. Describe how agricultural practices can interfere with nitrogen cycling.
13. Describe how deforestation can affect nutrient cycling within an ecosystem.
14. Describe how the carbon cycle differs in terrestrial and aquatic systems.
15. Explain how "cultural eutrophication" can alter freshwater ecosystems.
16. Explain why toxic compounds usually have the greatest effect on top-level carnivores.
17. Describe how increased atmospheric concentrations of carbon dioxide could affect the Earth.
18. Describe the Gaia hypothesis and explain how human interference might alter the biosphere.

KEY TERMS

ecosystem	food chain	standing crop	nitrification
trophic structure	food web	biomass	ammonification
trophic level	primary productivity	limiting nutrient	denitrification
primary producers	gross primary	secondary	eutrophic
primary consumers	productivity	productivity	biological
secondary consumers	net primary	ecological efficiency	magnification
tertiary consumers	production	energy pyramid	greenhouse effect
decomposers	biomass	biogeochemical cycle	biodiversity

LECTURE NOTES

<u>Ecosystem</u> = All organisms living in a given area along with the abiotic factors with which they interact.

- Boundaries are usually not discrete.
- The most inclusive level of biological organization.
- Involves two processes which cannot be described at lower levels: energy flow and chemical cycling.

I.   Trophic structure determines an ecosystem's routes of energy flow and chemical cycling

Each ecosystem has a *trophic structure* of feeding relationships that determine the paths of energy flow and chemical cycling.

- Ecologists divide the species in a community or ecosystem into different *trophic levels* based on their main source of nutrition.

The five trophic levels typically recognized are:

<u>Primary producers</u> = Autotrophs (usually photosynthetic) that support all other trophic levels either directly or indirectly by synthesizing sugars and other organic molecules using light or chemical energy.

<u>Primary consumers</u> = Herbivores that consume primary producers (plants and algae).

<u>Secondary consumers</u> = Carnivores that eat herbivores.

<u>Tertiary consumers</u> = Carnivores that eat other carnivores.

<u>Decomposers (detritivores)</u> = Consumers that derive energy from detritus (organic wastes) and dead organisms from other trophic levels.

<u>Food chain</u> = The pathway along which food is transferred from trophic level to trophic level. (See Campbell, Figure 49.1)

- Rarely are unbranched since several different primary consumers may feed on the same plant species and a primary consumer may eat several species of plants.
- Feeding relationships are usually woven into elaborate *food webs* within an ecosystem. (See Campbell, Figure 49.2)

A.   Producers

The main producers will vary depending on the ecosystem.

- Plants are the main producers in most terrestrial ecosystems.
   ⇒ Debris falling from terrestrial plants that reaches streams (directly or through run off) is a major source of organic material.
- Phytoplankton are the most important autotrophs in the limnetic zone of lakes and in the open ocean.
- Multicellular algae and aquatic plants are most important in the shallow, near-shore areas of freshwater and marine ecosystems.

- The aphotic zone of the deep sea receives energy and nutrients (dead plankton, detritus) from the overlying photic zone.
- Organisms in communities surrounding the hot water vents on the deep-sea floor depend more on chemical energy than solar energy.
  ⇒ The main producers are chemoautotrophic bacteria that derive energy from the oxidation of hydrogen sulfide.

B. Consumers

Consumers also vary with the type of ecosystem.
- Primary consumers in terrestrial ecosystems are mostly insects, snails, grazing mammals, and seed-eating and fruit-eating birds and mammals.
- In aquatic systems, the primary consumers are the zooplankton (heterotrophic protists, small invertebrates, numerous larval stages) and some fish.
- Secondary consumers in terrestrial habitats are spiders, frogs, insect-eating birds, carnivorous mammals, and animal parasites.
- Secondary consumers in aquatic habitats are fish as well as benthic forms such as sea stars and other carnivorous invertebrates.

C. Decomposers (Detritivores)

Decomposers are a critical component in any ecosystem.
- Are key components of the material recycling.
- Decomposers form a major link between primary producers and higher-level consumers.
- The most important decomposers are bacteria and fungi which digest materials externally and then absorb the products.
- Earthworms and a variety of scavengers are also decomposers.

II. An ecosystem's energy budget depends on primary productivity

Energy for growth, maintenance, and reproduction is required by all organisms; some species also require energy for locomotion.
- Light energy is used by primary producers to synthesize organic molecules (photosynthesis) which are later broken down to produce ATP (cellular respiration).
- Consumers obtain energy in the form of organic molecules that were produced at the previous trophic level. Thus, energy flows to higher trophic levels through food webs.
- Since only primary producers can directly utilize solar energy, an ecosystem's entire energy budget is determined by the photosynthetic activity of the system.

A.   The Global Energy Budget

The amount of solar radiation striking the Earth's surface shows dramatic regional variation which limits the photosynthetic output of ecosystems in different places.

- Earth receives an estimated 10^{22} joules (J) of solar radiation each day.
- Most of the solar radiation is reflected, absorbed, or scattered by the atmosphere, clouds, and dust particles in the air; this amount varies over different regions.
- The intensity of solar radiation also varies with latitude resulting in the tropics receiving the most input.
- Only a fraction of the solar radiation which reaches the biosphere strikes plants and algae (much hits bare ground or is absorbed or reflected by water) and these primary producers can only use some wavelengths for photosynthesis.
- Only about 1% – 2% of the visible light reaching primary producers is converted to chemical energy by photosynthesis.
  ⇒ The photosynthetic efficiency also varies with the type of plant, light levels, and other factors.
- Even with all the variations mentioned above, primary production of Earth collectively creates about 170 billion tons of organic material each year.

B.   Primary Productivity

<u>Primary productivity</u> = The amount of light energy converted to chemical energy by autotrophs of an ecosystem.

- The total is known as *gross primary productivity* (GPP) which may be determined by measuring the total oxygen produced by photosynthesis.

<u>Net primary productivity (NPP)</u> = GPP - Rs (energy used by producers for respiration)

- NPP accounts for the organic mass of plants (growth) and represents storage of chemical energy available to consumers.
- The NPP:GPP ratio is generally smaller for large producers with elaborate nonphotosynthetic structures (such as trees) which support large metabolically active stem and root systems.
- Can be expressed as biomass (expressed as dry weight since water contains no usable energy) added to an ecosystem per unit area per unit time ($g/m^2/yr$) or as energy per unit time ($J/m^2/yr$).

Primary productivity should not be confused with *standing crop biomass*.

- Primary productivity is the *rate* at which *new* biomass is synthesized by vegetation.
- Standing crop biomass is the total biomass of plants present at a given time which may have accumulated over several growing seasons.

Primary productivity varies among ecosystems, and an ecosystem's size affects its contribution to the Earth's total productivity. (See Campbell, Figure 49.3)

- Tropical rain forests are very productive and contribute a large proportion to the planet's overall productivity since they cover a large portion of the Earth's surface.
- Estuaries and coral reefs are also very productive but make only a small contribution to planetary productivity since they do not cover an extensive area.
- The open ocean has a relatively low productivity but makes the largest contribution to overall productivity of any ecosystem due to its very large size.
- Deserts and tundra also have low productivity.

Factors important in limiting productivity depend on the type of ecosystem and temporal changes such as seasons.

1. Generally, precipitation, temperature, and light intensity are factors limiting productivity in terrestrial ecosystems.
   - Productivity increases as latitudes approach the equator because availability of water, heat, and light increases in the tropics.

2. Productivity in terrestrial ecosystems may also be limited by availability of inorganic nutrients.
   - Plants require a variety of nutrients, some in large quantities and some in small quantities.
   - Primary production sometimes removes nutrients from the system faster than they can be replenished.
   - If a nutrient is removed in such quantities that sufficient amounts are no longer available, it becomes the *limiting nutrient*.
   - The limiting nutrient is one present in insufficient quantities to support further primary production.
   - Nitrogen and phosphorus are usually limiting nutrients since they are needed in large quantities but are often present in small or moderate amounts in natural environments.
   - Carbon dioxide availability sometimes limits productivity.

3. An aquatic ecosystem's productivity is usually determined by light intensity, water temperature, and availability of inorganic nutrients.
   - Productivity is greatest in shallow waters near continents and along coral reefs due to abundant nutrients and sunlight.
   - Light intensity and temperature affect primary productivity of phytoplankton in the open oceans; productivity is highest near the surface and decreases with depth.
   - Inorganic nutrients are limiting at the surface of open ocean waters with nitrogen and phosphorus in especially short supply.
     ⇒ This is a primary reason for the relatively low productivity of open oceans.
   - Marine phytoplankton is most productive where upwellings bring nutrient rich waters to the surface.
     ⇒ These areas (usually in polar seas) are more productive than tropical seas.

⇒ Thermal vent communities are also very productive though they are not very wide-spread and contribute little to marine productivity.

- Freshwater ecosystem productivity also varies from the surface to the depths in relation to light intensity.

⇒ Availability of inorganic nutrients is sometimes limiting, but biannual turnovers bring nutrients to the surface waters.

III. As energy flows through an ecosystem, much is lost at each trophic level

The transfer of energy from one trophic level to another is much less efficient than many realize. The biomass of producers is much greater than that of herbivores and the relationship holds through subsequent consumer levels.

A. Secondary Productivity

Secondary productivity is the rate at which consumers convert the chemical energy in the food they eat into their own biomass.

As an example, consider herbivores which consume only a small fraction of available plant material and they can not digest all of the organic compounds in what they do ingest. (See Campbell, Figure 49.4)

- About 1/6 of the calories is used for growth which adds biomass to the trophic level.
- The remaining organic material consumed is used for cellular respiration or is passed out of the body as feces.
- The energy in the feces stays in the system and is consumed by decomposers.
- The energy used in cellular respiration is lost from the system.
- Carnivores are more efficient at converting food into biomass but more is used for cellular respiration, further decreasing energy available to the next trophic level.
- Consequently, *energy flows* through an ecosystem, it does not cycle within the ecosystem.

B. Ecological Efficiency and Ecological Pyramids

Ecological efficiency is the ratio of net productivity at one trophic level compared to net productivity at the level below – the percentage of energy transferred from one trophic level to the next.

- Efficiencies can vary greatly depending on the organisms involved, but usually ranges from 5% – 20%.
- This means that 85% – 95% of the energy available at one trophic level never transfers to the next.

Loss of energy in a food chain can be represented diagrammatically:

1. A *pyramid of productivity* has trophic levels stacked in blocks proportional in size to the productivity of each level.
    - Usually bottom heavy since ecological efficiencies are low. (See Campbell, Figure 49.5)

2. A *biomass pyramid* has each tier symbolizing the total dry weight of all organisms (standing crop biomass) in an ecosystem's levels. (See Campbell, Figure 49.6)
    - Most narrow sharply from producers at the base to top-level carnivores at the top because of low ecological efficiencies.
    - Some aquatic systems are inverted since producers can have high turnover rates. They grow rapidly but are consumed rapidly, leaving little standing crop biomass.

3. A *pyramid of numbers* is comprised of blocks whose size is proportional to the numbers of individuals present at each level. (See Campbell, Figure 49.7)
    - Biomass of top-level carnivores is usually small compared to the total biomass of producers and lower-level consumers.
    - Only about 1/1000 of the chemical energy fixed by photosynthesis flows through a food web to a tertiary consumer.
    - Only 3 – 5 trophic levels can be supported since biomass at the apex is insufficient to support another level.
    - Predators (top-level consumers) are highly susceptible to extinction when their ecosystem is disturbed due to their small population and wide spacing within the habitat.

IV. Matter cycles within and between ecosystems

Despite an inexhaustible influx of energy in the form of sunlight, continuation of life depends on recycling of essential chemical elements.
- These elements are continually cycled between the environment and living organisms as nutrients are absorbed and wastes released.
- Decomposition of wastes and the remains of dead organisms replenishes the pool of inorganic nutrients available to autotrophs.

Biogeochemical cycles = Nutrient circuits involving both biotic and abiotic components of ecosystems.
- Elements such as carbon, oxygen, sulfur, and nitrogen have gaseous forms, thus, their cycles are global in character and the atmosphere serves as a reservoir.
- Elements less mobile in the environment like phosphorus, potassium, calcium and trace elements generally cycle on a more localized scale over the short term. The soil serves as the main reservoir for these elements.

A general scheme of nutrient cycling includes the four main *reservoirs* of elements and the *processes that transfer elements* between reservoirs. (See Campbell, Figure 49.8)

Reservoirs are defined by two characteristics: whether they contain organic or inorganic materials; and whether or not the materials are directly available for use by organisms.

- The *available organic reservoir* contains the living organisms and detritus.
    ⇒ The nutrients are readily available when organisms feed on one another.
- The *unavailable organic reservoir* is comprised of coal, oil, and peat which formed from organisms that died and were buried millions of years ago.
    ⇒ These nutrients are unavailable since they cannot be directly assimilated.
- The *available inorganic reservoir* includes all matter (elements and compounds) present in the soil or air and those dissolved in water.
    ⇒ Organisms can directly assimilate these nutrients from the soil, air, or water.
- The *unavailable inorganic reservoir* contains nutrients tied up in limestone and minerals of other rocks.
    ⇒ These nutrients cannot be assimilated until released by weathering or erosion.

Various *processes* are involved in the transfer of nutrients between the four reservoirs which form the basis for biogeochemical cycling. The general schemes were determined by adding small amounts of radioactive tracers to systems in order to follow the movement of elements.

- Weathering and erosion are the primary processes which move nutrients from the unavailable inorganic reservoir to the available inorganic reservoir.
- Erosion is also important, along with the burning of fossil fuels, in moving nutrients from the unavailable organic reservoir to the available inorganic reservoir.
- Nutrients are transferred from the available organic reservoir to the unavailable organic reservoir only by the covering of detritus by sediments and its eventual fossilization to oil, coal, or peat.
- Sedimentary rock formation is the process which moves nutrients from the available inorganic reservoir to the unavailable inorganic reservoir.
- Nutrients enter the available organic reservoir from the available inorganic reservoir through photosynthesis and assimilation by living organisms.
- Nutrients are transferred from the available organic reservoir to the available inorganic reservoir by respiration, decomposition, excretion, and leaching.

V.   A combination of biological and geological processes drives chemical cycles

The cycling of materials through an ecosystem depends on both biological and geological processes.

A.   The Water Cycle

The essential nature of water to living organisms has many facets:
- It is essential to maintaining homeostasis in every organism.
- It contributes to the fitness of the environment.
- It's movement within and between ecosystems transfers other materials in several biogeochemical cycles.

Most of the water cycle occurs between the oceans and the atmosphere.

- Solar energy results in evaporation from the oceans.
- Water vapor rises, cools, and falls as precipitation.
- Over the oceans, evaporation exceeds precipitation; the excess water vapor is moved onto land by winds.
- Precipitation exceeds evaporation and transpiration over land; runoff and ground water balance the net flow of water vapor to land.

The water cycle differs from other cycles in that it occurs primarily due to *physical* processes, not chemical processes.

B.   The Carbon Cycle

During the carbon cycle, autotrophs acquire carbon dioxide (CO_2) from the atmosphere by diffusion through leaf stomata, incorporating it into their biomass. Some of this becomes a carbon source for consumers, and respiration returns CO_2 to the atmosphere.

- Photosynthesis and cellular respiration form a link between the atmosphere and terrestrial environments. (See Campbell, Figure 49.10)
- Carbon cycles fast. Plants have a high demand for CO_2, yet CO_2 is present in the atmosphere at a low concentration (0.03%).
- Carbon loss by photosynthesis is balanced by carbon release during respiration.

Some carbon is diverted from cycling for longer periods of time, as when it accumulates in wood or other durable organic material.

- Decomposition eventually recycles this carbon to the atmosphere.
- Can be diverted for millions of years, such as in the formation of coal and petroleum.

The amount of atmospheric CO_2 decreases in the Northern Hemisphere in summer due to increased photosynthetic activity.

- Amounts increase in the winter when respiration exceeds photosynthesis.

Atmospheric CO_2 is increased by combustion of fossil fuels by humans, disturbing the balance.

- The ocean may act as a buffer to absorb excess CO_2.

In aquatic environments photosynthesis and respiration are also important but carbon cycling is more complex due to interaction of CO_2 with water and limestone.

- Dissolved CO_2 reacts with water to form carbonic acid, which reacts with limestone to form bicarbonates and carbonate ions.
- As CO_2 is used in photosynthesis, bicarbonates convert back to CO_2; thus bicarbonates serve as a CO_2 reservoir and some aquatic autotrophs can use dissolved bicarbonates directly as a carbon source.
- The ocean contains about 50 times the amount of carbon (in various inorganic forms) as is available in the atmosphere.

C.   The Nitrogen Cycle

Nitrogen is a key chemical in ecosystems as it is found in all amino acids which comprise the proteins of organisms.

- Although the Earth's atmosphere is almost 80% N_2, it is unavailable to plants since they cannot assimilate this form.
- Nitrogen is available to plants in only two forms: ammonium (NH_4^+) and nitrate (NO_3)
- Nitrogen enters ecosystems by either atmospheric deposition or nitrogen fixation.

Atmospheric deposition accounts for only 5 – 10% of the usable nitrogen that enters an ecosystem.

- NH_4^+ and NO_3 are added to the soil by being desolved in rain or by settling as part of fine dust or other paritculates.

Nitrogen fixation is the reduction of atmospheric nitrogen (N_2) to ammonia (NH_3) which can be used to synthesize nitrogenous organic compounds such as amino acids.

- Only certain prokaryotes can fix nitrogen.
    ⇒ In terrestrial ecosystems some nonsymbiotic soil bacteria and some symbiotic (*Rhizobium*) bacteria fix nitrogen.
    ⇒ Cyanobacteria fix nitrogen in aquatic ecosystems.
- Nitrogen fixing prokaryotes are fulfilling their own metabolic needs, but other organisms benefit since excess ammonia is released into the soil or water.
- Industrial fixation in the form of fertilizer makes significant contributions to the nitrogen pool in agricultural regions.
- The slightly acidic nature of soil results in NH_3 being protonated to ammonium (NH_4^+).
    ⇒ NH_3 is a gas and can evaporate quickly to the atmosphere.
    ⇒ NH_4^+ can be used directly by plants.

The nitrogen cycle involves three processes in addition to nitrogen fixation: nitrification, denitrification, and ammonification. (See Campbell, Figure 49.11)

Nitrification is a metabolic process by which certain aerobic soil bacteria use ammonium (NH_4^+) as an energy source by first oxidizing it to nitrite (NO_2^-) and then to nitrate (NO_3^-).

- While plants can use NH_4^+ directly, the nitrifying bacteria use most of the available NH_4^+ as an energy source.
- Plants assimilate the NO_3^- released from these bacteria and convert it to organic forms like amino acids and proteins.
- Animals can only assimilate organic nitrogen which they obtain by eating plants and other animals.

Denitrification is a process that returns nitrogen to the atmosphere by converting NO_3^- to N_2.

- Some soil bacteria obtain the oxygen necessary for their metabolism from NO_3^- rather than O_2.

Ammonification is the decomposition of organic nitrogen back into ammonia.
- Carried out by many decomposer anaerobic and aerobic bacteria and fungi.
- Process is especially important since it recycles large amounts of nitrogen to the soil.

Some important aspects of the Nitrogen Cycle to note:
- Prokaryotes serve as vital links at several points in the cycle.
- Most of the nitrogen cycling involves nitrogenous compounds in the soil and water.
- While atmospheric nitrogen is plentiful, nitrogen fixation contributes only a small fraction of the nitrogen assimilated by plants; however, many species of plants depend on symbiotic, nitrogen-fixing bacteria in their root nodules as a source of nitrogen in a form that can be assimilated.
- Denitrification returns only a small amount of N_2 to the atmosphere.
- Most assimilated nitrogen comes from nitrate which is efficiently recycled from organic forms by ammonification and nitrification.
- The majority of nitrogen in most ecosystems is recycled locally by decomposition and reassimilation, although nitrogen exchange between the soil and atmosphere are of long-term importance.

D.   The Phosphorus Cycle

Phosphorus is a major component of nucleic acids, phospholipids, ATP, and a mineral in bones and teeth.

The phosphorus cycle is relatively simple since it does not have a gaseous form and occurs in only one important inorganic form, phosphate.

Phosphorus cycles locally as follows: (See Campbell, Figure 49.12)
- Weathering of rock adds phosphate to the soil.
- Producers absorb the soil phosphate and incorporate it into molecules.
- Phosphorus is transferred to consumers in organic form.
- Phosphorus is added back to the soil by excretion and decomposition of detritus.

Phosphorus cycling is localized since humus and soil particles bind phosphate.
- Some leaching does occur and phosphate is lost to the oceans through the water table.
- Weathering of rocks keeps pace so terrestrial systems are not depleted.
- Phosphate that reaches the oceans accumulates in sediments and becomes incorporated into rocks which may eventually be exposed to weathering.

Phosphorus may limit algal productivity in aquatic habitats.
- Production in these habitats is stimulated by the introduction of phosphorus in the form of sewage or runoff from fertilized agricultural areas.

E.   Variations in Nutrient-Cycling Time

The rate of decomposition has a great impact on the time table for nutrient cycling.

- The rate of decomposition (and thus nutrient cycling) is affected by water availability, oxygen, and temperature.
- Decomposition of organic material in the tropical forests usually occurs in a few months to a few years.
- It takes an average of four to six years for decomposition to occur in temperate forests.
- Decomposition in the tundra may take 50 years.
- In aquatic ecosystems that become anaerobic, decomposition may occur even more slowly than in the tundra.

Soil chemistry and the frequency of fires also influence nutrient cycling times.

Some key nutrients are present in the soil of tropical rain forests in levels much lower than those found in temperate forests. Several conditions influence this paradox:

- There is rapid decomposition in tropical areas due to warm temperature and abundant water.
- The large biomass of tropical rain forests creates a high demand for nutrients which are absorbed as soon as they become available for decomposers.
- Relatively little organic material accumulates as litter due to the rapid decomposition.
  - ⇒ About 10% of the nutrients are in the soil; 75% are present in the woody parts of trees.
- The rapid cycling time results in the low nutrient content of the soil.

The soil in temperate forests may contain 50% of all organic material in the ecosystem.

- The rate of decomposition is slow.
- The nutrients present in detritus and soil may stay in the soil for long periods before being assimilated.

The sediments of aquatic systems form a nutrient sink and there must be an interchange between the bottom layers of water and the surface for the ecosystem to be productive.

- The rate of decomposition in the sediments is very slow.
- Algae and aquatic plants usually assimilate their nutrients directly from the water.

VI. **Field experiments reveal how vegetation regulates chemical cycling:** *science as a process*

Long-term ecological research (*LTER*) is being used to examine the dynamics of many natural ecosystems.

Scientists have been studying nutrient cycling in a forest ecosystem under natural conditions and after vegetation is removed. The study began in 1963 at the Hubbard Brook Experiment Forest in New Hampshire.

- The team first determined mineral budgets of six valleys by measuring inflow and outflow of several key nutrients.
- Rainwater was collected to measure amounts of water and dissolved minerals added to the ecosystem.
- Water and mineral loss were monitored by using concrete weirs across the creek at the bottom of each valley.
  ⇒ Found that 60% of the water added by rainfall exits through streams and 40% is lost by plant transpiration and evaporation from soil.
- Also found that mineral inflow and outflow were nearly balanced and were small compared to minerals being recycled within the forest ecosystem.
  ⇒ Only about 0.3% more Ca^{++} exited a valley through its creek than was added by rainwater.
  ⇒ Net mineral losses were probably replaced by chemical decomposition of bedrock.
- During most years, some net gains of a few mineral nutrients occurred.

After logging an experimental area and preventing reforestation, comparisons were made over a three year period.

- Water runoff increased by 30 – 40% (no trees were left to absorb and transpire water).
- Net losses of minerals were very large:
  ⇒ Nitrate loss increased sixty fold (water nitrate levels made the water unsafe for drinking).
  ⇒ Calcium loss increased 400%
  ⇒ Potassium loss increased 1500%.

The study demonstrated the importance of plants in retaining nutrients within an ecosystem and the effects of human intrusion into a system.

VII. **The human population is disrupting chemical cycles throughout the biosphere**

The ever increasing human population has intruded into the dynamics of most ecosystems through human activities or technology.

- Some natural systems are totally destroyed while others have had major components (trophic structure, energy flow, chemical cycling) disrupted.
- Most effects are local or regional, while others are global in scale (i.e. acid rain).

Human activity often removes nutrients from one part of the biosphere and adds them to another.

- May deplete one area of key nutrients while creating an excess in another area.
- These occurrences disrupt the natural equilibrium in both areas.

Farming exhausts the natural store of nutrients as crop biomass is removed from an area, this greatly reduces the amount of nutrients recycled. Supplements must then be added in the form of fertilizer.

- Nutrients in crops soon appear in human and livestock wastes, and then turn up in lakes and streams through sewage discharge and field run-off.
- Once in aquatic systems, these nutrients may stimulate excessive algal growth which degrades the system.
- Consequently, disruptions can flow from one system to another.

A.   **Agricultural Effects on Nutrient Cycling**

As the human population continues to grow, greater demands for production of food will result in more natural habitats being converted to agricultural use. This will result in more:

- Intrusions into the cycling of nutrients.
- Over-harvesting of natural populations of food-species.
- Introductions of toxic compounds into ecosystems in the form of pesticides.

The time period during which no additional nutrients need to be added to new agricultural ecosystems vary.

- Nutrient reserves in the soil will support crops for sometime after the natural vegetation has been removed.
- These nutrients are not recycled locally since they are removed from the system as crop biomass.
- New farmlands in the tropics can usually be farmed for one or two years.
    - ⇒ Remember, in the tropical rain forests only about 10% of the nutrients are in the soil.
- In temperate areas, crops may be grown for many years due to the nutrients present in the soil.
- When nutrients are added, it is normally in the form of industrially synthesized fertilizers.

The nitrogen cycle of an area is greatly impacted by agriculture.

- Breaking up and mixing the soil increase the rate of decomposition of organic matter.
- This releases usable nitrogen which is taken up by the crop and exported from the system at harvest.
    - ⇒ Nitrates remaining in the soil are quickly leached out of the system.
- Fertilizers are applied to replace the lost nitrogen.
    - ⇒ Results in more N_2 being released into the atmosphere by upsetting the balance between denitrification and nitrogen fixation.

- Leaching to the water table and surface runoff carries much of the nitrate in fertilizers into aquatic systems.

    ⇒ Excess algal growth typically results from an over-abundance of nitrates entering an aquatic system.

B. Accelerated Eutrophication of Lakes

Lakes undergo a naturally occurring change in chemical composition and character.

- Oligotrophic lakes have low primary productivity because mineral nutrient levels will not support large phytoplankton populations.

- The lake will gradually mature (become more *eutrophic*) due to runoff from surrounding land which contains nutrients.

- These additional nutrients are assimilated by phytoplankton and productivity increases; once in the lake the nutrients are continually recycled through the lake's food webs.

- Lakes reach an equilibrium where nutrient input is balanced by losses due to outflow and sedimentation.

Sewage, factory wastes, livestock runoff, and fertilizer leaching increases inorganic nutrient levels in waters and results in *cultural eutrophication*.

- This enrichment often results in explosive growth of photosynthetic organisms.

    ⇒ Large algal blooms occur; shallow areas become choked with weeds.

- As these producers die, metabolism of decomposers consumes all the oxygen in the water and many species die.

C. Poisons in Food Chains

A variety of toxic chemicals, including unnatural synthetics, have been and are dumped into ecosystems.

- Many cannot be degraded by microbes and persist for years or decades.

- Some are harmless when released but are converted to toxic poisons by reactions with other substances or by the metabolism of microbes.

Organisms acquire toxic substances along with nutrients or water, some are metabolized and excreted wile others accumulate in their tissues.

Biological magnification = Toxins become more concentrated with each successive trophic level of a food web. Results from biomass at each trophic level being produced from a much larger biomass ingested from the level below.

- Top level carnivores are usually most severely affected by toxic compounds released into the environment.

The pesticide DDT is a well known example of biological magnification. (See Campbell, Figure 49.16)

- Used to control mosquitoes and agricultural pests.

- DDT persists in the environment and is transported by water to areas away from the point of application.

- It is lipid-soluble and collects in fatty tissues of animals.
- The concentration is magnified at each trophic level and reached such high concentrations (10×10^6 increase) in top-level carnivorous birds that calcium deposition in eggshells was disrupted.
- Reproductive rates declined dramatically since the weight of nesting birds broke the weakened shells.
- DDT use was banned in the United States in 1971 and the affected bird populations have recovered.
- The use of DDT still continues in other parts of the world.

D. Intrusions in the Atmosphere

Human activities have resulted in the release of many gaseous waste products into the atmosphere.

- One problem directly related to nutrient cycling is the rising levels of carbon dioxide.

E. Carbon Dioxide Emissions and the Greenhouse Effect

Carbon dioxide emissions have caused atmospheric CO_2 concentrations to increase 13% since 1958. This increase is due to combustion of fossil fuels and burning of wood removed by deforestation.

Some effects of increased carbon dioxide levels might appear to be beneficial while others are definitely detrimental.

- Increased productivity by vegetation would occur with increased CO_2.
  - $\Rightarrow$ C_3 plants are more limited than C_4 plants by CO_2, so spread of C_3 species into habitats previously favoring C_4 species may have important natural and agricultural implications.
- Temperature increases with increased CO_2 concentration since CO_2 and water vapor absorb infrared radiation and slows its escape from Earth.
  - $\Rightarrow$ Called the *greenhouse effect*.
  - $\Rightarrow$ Some predict warming near the poles resulting in melting of polar ice and flooding of current coastal areas.
  - $\Rightarrow$ A warming trend will probably alter geographical distribution of precipitation which could have major agricultural implications.
- Some predictions about how the climate in each region would be are being made by scientists studying the past.
  - $\Rightarrow$ Paleoecologists are studying the records of pollen cores to determine how past temperature changes have affected vegetation.

F.   Depletion of Atmospheric Ozone

Depletion of atmospheric *ozone* weakens a protective layer in the stratosphere which absorbs ultraviolet radiation.

- Much of the ultraviolet radiation is absorbed by an ozone layer 17 to 25 km above the Earth's surface.
- Destruction of the ozone layer is largely due to accumulation of chlorofluorocarbons used as aerosol propellants and in refrigeration.
- Breakdown products of chlorofluorocarbons include chlorine which rises to the stratosphere where it reacts with ozone (O_3) and reduces it to atmospheric oxygen (O_2).
  - $\Rightarrow$ The chlorine is released in other reactions and reacts with additional ozone molecules.
- Ozone depletion is best documented over Antarctica but levels in the middle latitudes have decreased 2% – 10% in the last 20 years.

Ozone depletion could have serious consequences.

- Increases are expected in lethal and nonlethal forms of skin cancer and cataracts among humans.
- Unpredictable effects on crops and natural communities (especially phytoplankton) are expected.

VIII.   Human activities are altering species distribution and reducing biodiversity

The growth of human populations, human activities, and our technological capabilities have disrupted the trophic structure, energy flow, and chemical cycling of ecosystems in most areas of the world.

- Some effects are local while others affect the biosphere's distribution and diversity of organisms.

A.   The Introduction of Exotic Species

The *introduction of exotic species* can cause a variety of problems. Although most transplanted species fail to survive, there are notable exceptions.

- Starlings and English sparrows were imported into the United States from Europe in the 1800s. Both species are now very abundant and have displaced native bird species in many areas.
- While some exotic species have little or no impact on the new communities, others, such as fire ants, have a tremendous impact.
  - $\Rightarrow$ Fire ants were accidentally introduced into the United States from Brazil in the 1940s. (See Campbell, Figure 49.18)
  - $\Rightarrow$ Have spread across the Southern U.S. and reduced the number of native species dramatically in some areas.

The release of genetically engineered organisms into natural and agricultural ecosystems poses some potential ecological problems.

- Genetically engineered plants might escape the plots and become weeds or pests in other areas.

- Genetically engineered plants and animals might expand into new areas and out compete native species.

- Engineered plants might interbreed with wild plants and pass on their "special" traits (such as herbicide resistance) to offspring that could not be controlled by standard means.

- Genetically engineered insect resistance might result in a coevolutionary change in insect pests.

B.   Habitat Destruction and the Biodiversity Crisis

The destruction of natural systems due to human encroachment has resulted in only a small proportion of natural, undisturbed habitat remaining in existence.

- Only 15% of the original primary forest and just 1% of original tallgrass prairie remain in the United States.

- Tropical rain forests are being cut at a rate of 500,000 Km^2 per year and will be eliminated in a few decades.

- Human activities which disrupt entire systems include: development, logging, war, oil spills, etc.

One result of the destruction of natural habitat will be the loss of biodiversity.

- 73% of the IUCNs designations of extinct, endangered, vulnerable, and rare species are related to destruction of natural habitats.

- Protection of endangered species is made difficult by a variety of problems.

  ⇒ Minimum viable population sizes and minimal habitat size requirements are not known for many species.

  ⇒ Protection for migratory species requires international cooperation and habitat preservation in two areas.

It is difficult to accurately estimate the magnitude of the biodiversity crisis.

- Only a fraction of the Earth's species has been identified and described by taxonomists.

- Extinction is an obscure, local process that can occur unnoticed because the exact distributions and habitats of many species are unknown.

The absence of clear documentation of the rate of extinction has led some to argue that there is no reason to worry at this time. There is, however, evidence that the extinction of some well-known organisms is occurring at a rapid rate.

- 11% of the 9040 known bird species in the world are endangered.

  ⇒ Population densities of migratory songbirds in the Mid-Atlantic United States has decreased by 50% in the last 40 years.

- At least 20% of the known species of freshwater fishes in the world have become extinct in historical times or are seriously threatened.

⇒ Only 122 of 266 known fish species in lowland Malaysia were observed during a recent search.

- 680 of the 20,000 species of plants known in the United States are in danger of becoming extinct by the year 2000.

The above list represents only a few of the many documented examples of extinction and some individuals argue that extinction is a natural phenomenon that has occurred since life first evolved.

- What must be realized is that the *rate* of extinction is rapidly increasing.
- Species are being lost on a worldwide scale at a rate of about 50 times greater than at any time in the past 100,000 years.

⇒ Ecological models estimate that human activity in tropical rain forests has increased extinction between 1000 and 10,000 times over the normal rate of one in every one million species per year.

The question many ask is "Why should we care about the loss of biodiversity?" Answers to this question range from general to specific:

- *Biophilia*, the human sense of connection to nature and other forms of life, is centered around aesthetics and ethics.
- Biodiversity is a crucial natural resource and threatened species could provide crops, fibers, and medicines.

⇒ 25% of the prescriptions dispensed from pharmacies in the United States contain substances derived from plants.

⇒ Alkaloids isolated from the rose periwinkle of Madagascar in 1970 were found to inhibit cancer cell growth and result in remission for victims of childhood leukemia and Hodgkin's disease.

⇒ One of the five other periwinkle species on Madagascar is approaching extinction.

IX. The Sustainable Biosphere Initiative is reorienting ecological research

A great deal of basic knowledge is needed before sensible decisions can be made on how to lessen the negative impacts of humans on ecosystems.

- While each individual can play a role, an understanding of the complex interconnections of the biosphere is needed.
- The Sustainable Biosphere Initiative is a new research agenda designed to obtain this knowledge.

The goal of the Sustainable Biosphere Initiative is to acquire the basic ecological information needed for intelligent development, management, and conservation of the Earth's resources. This will include studies of:

- Global change including interactions between climate and ecological processes.
- Biological diversity and its role in maintaining ecological processes.
- The ways in which productivity of natural and artificial ecosystems can be sustained.

The nature of ecological research will have to be reoriented but the importance can not be overstated due to the current state of the biosphere.

REFERENCES

Campbell, N. *Biology*. 4th ed. Menlo Park, California: Benjamin/Cummings, 1996.

Cohen, Joel E. *Food Webs and Niche Space*. Princeton, New Jersey: Princeton University Press, 1978.

Odum, H.T. *Systems Ecology: And Introduction*. New York: Wiley, 1983.

Ricklefs, R.E. *Ecology*. 3rd ed. New York: Chiron Press, 1986.

Smith, Robert Leo. *Ecology and Field Biology*. 3rd ed. New York: Harper & Row, Publishers, 1980.

OUTLINE

I.   Behavior is what an animal does and how it does it

II. Behavioral ecology emphasizes evolutionary hypotheses: *science as a process*

III. A behavior has both an ultimate and a proximate cause

IV. Certain stimuli trigger innate behaviors called fixed action patterns
    A.   Fixed-Action Patterns
    B.   The Nature of Sign Stimuli

IV. Learning is experience-based modification of behavior
    A.   Nature Versus Nurture
    B.   Learning Versus Maturation
    C.   Habituation
    D.   Imprinting
    E.   Classical Conditioning
    F.   Operant Conditioning
    G.   Observational Learning
    H.   Play
    I.   Insight
    J.   Animal Cognition

V.   Rhythmic behaviors synchronize an animal's activities with daily and seasonal changes in the environment

VI. Environmental cues guide animal movement
    A.   Kinesis and Taxis
    B.   Migration Behavior

VII. Behavioral ecologists are using cost/benefit analysis to study foraging behavior

VIII. Sociobiology places social behavior in an evolutionary context

IX. Competitive social behaviors often represent contests for resources
    A. Agonistic Behavior
    B. Dominance Hierarchies
    C. Territoriality

X. Mating behavior relates directly to an animal's fitness
    A. Courtship
    B. Mating Systems

XI. Social interactions depend on diverse modes of communication

XII. The concept of inclusive fitness can account for most altruistic behavior

XIII. Human sociobiology connects biology to the humanities and social sciences

OBJECTIVES

After reading this chapter and attending lecture, the student should be able to:

1. Explain the difference between innate and learned behaviors.
2. Describe the evolutionary basis for behavioral ecology.
3. Explain the difference between ultimate and proximate causations of behavior.
4. Describe a fixed-action pattern and a sign stimulus.
5. Explain the nature versus nurture controversy.
6. Explain the effect of maturation on behavioral improvement.
7. Define habituation.
8. Discuss imprinting, imprinting stimulus and critical period.
9. Define associative learning.
10. Distinguish among classical conditioning, operant conditioning and observational learning.
11. Describe two hypotheses for the evolution of play behavior.
12. Define insight learning.
13. Discuss the ultimate bases of learning.
14. Explain how exogenous and endogenous cues may influence behavioral rhythms.
15. Describe and define kinesis, taxis and migration.
16. Explain the differences among piloting, orientation and navigation.
17. Compare generalist and specialist foraging strategies.
18. Explain how a search image is adaptive.
19. Describe optimal foraging strategies in terms of energetics and prey densities.
20. Describe agonistic behavior.
21. Explain what is meant by a ritual behavior and describe the evolutionary advantage of ritual behavior.
22. Describe a dominance hierarchy and explain the advantages to individuals in the hierarchy.
23. Explain how dominance hierarchies and territories may stabilize population densities.
24. Describe the advantages of courtship.

25. Explain how ritualized courtships may have evolved.
26. Define parental investment.
27. Discuss the ultimate bases for mate selection.
28. Compare and contrast the three main mating systems.
29. Describe the differences between polygyny and polyandry.
30. Discuss how the needs of the young influence the development of mating systems.
31. Describe how the certainty of paternity influences the development of mating systems.
32. Describe the various modes of communication.
33. Relate an animal's mode of communication with its lifestyle.
34. Discuss why altruistic behavior might evolve.
35. Define inclusive fitness and kin selection.
36. Define reciprocal altruism.
37. Define cognition.
38. Describe the premise of sociobiology.

KEY TERMS

behavior
optimal behavior
behavioral ecology
ultimate causation
proximate causation
ethology
Konrad Lorenz
Niko Tinbergen
Karl von Frisch
fixed-action pattern
sign stimulus
releaser
supernormal stimulus
innate
learning
imprinting
imprinting stimulus

critical period
associative learning
classical conditioning
Ivan Pavlov
operant conditioning
B.F. Skinner
operational learning
maturation
habituation
play
insight learning
Nature versus
  Nurture
social behavior
agonistic behavior
ritual
dominance hierarchy

territory
inclusive fitness
kin selection
altruism
reciprocal altruism
mating behavior
courtship
parental investment
mating systems
promiscuous
monogamous
polygamous
polygyny
polyandry
exogenous
endogenous

biological clock
communication
pheromones
kinesis
taxis
migration
piloting
orientation
navigation
optimal foraging
generalists
specialists
search image
cognitive ethology
behaviorism
sociobiology

LECTURE NOTES

I.   Behavior is what an animal does and how it does it

Behavior is influenced by innate and learned factors.

- Baby sounds (human and bird) are coded by innate neural templates, but young animals must learn and practice the sounds which make up adult repertoires.
- Motor and memory elements are essential to behavior.
- Animal behavior is one of the oldest studied areas in biology. Pre-agricultural societies needed to understand animal behavior in order to survive.

Innate = Inborn or present at birth.

Behavior = What an animal does and *how* it does it.

II.   Behavioral ecology emphasizes evolutionary hypotheses: science as a process

Fitness is a central concept in animal behavior. Since natural selection works on genetic variation caused by mutation and recombination, organisms should have features that maximize fitness over time. Animals are expected to engage in *optimal behaviors*.

Optimal behavior = A behavior that maximizes individual fitness.

- Optimal behavior is a valid concept if behavior is genetically influenced and subject to natural selection.
- Genes influence behavior. (See Campbell, Figure 50.1)

Experiment: Two species of lovebirds were interbred. Female fischer's lovebirds cut long strips of nesting material which are carried individually to the nest. Female peach-faced lovebirds cut short strips and carries several at a time by tucking them into her back feathers.

- *Results:* Hybrid females cut intermediate length strips and tried, but failed, to transport them by tucking into back feathers. They learned to carry strips in their beaks, but never gave up all tucking behavior.
- *Conclusions:* Phenotypic differences in the behavior of the two species are based on different genotypes. Innate behavior can be modified by experience.

Learned behaviors are typically based upon gene created neural systems that are receptive to learning.

Behavioral ecologists assume optimal behavior increases fitness.

- Because behavioral ecology is based on the assumption that behavior increases fitness, it leads logically into hypothesis testing and the scientific method for problems in behavior research.

Behavioral ecology = A field of study that assumes animals increase fitness through optimal behavior.

III. A behavior has both an ultimate and a proximate cause

Because behavior is assumed to increase fitness, questions about *ultimate causation*, or the reasons <u>why</u> behaviors exist are evolutionary questions.

<u>Ultimate causation</u> = The evolutionary reason for the existence of a behavior.

The immediate mechanism or <u>how</u> a behavior is expressed is the *proximate cause*. Proximate causes may be internal processes or environmental stimuli.

<u>Proximate causation</u> = The immediate cause and/or mechanism underlying a behavior.
- Proximate causation limits the range of behaviors upon which natural selection can act.
- Proximate mechanisms produce behaviors that ultimately evolved because they increased fitness.

<u>Example</u>: Behavior - Bluegill sunfish breed in spring and early summer.
   Proximate cause: Breeding is triggered by the effect of increased day length on a fish's pineal gland.
   Ultimate cause: Breeding is most successful when water temperatures and food supplies are optimal.

<u>Example</u>: Behavior - Human "sweet tooth".
   Proximate cause: Sweet taste buds are a proximate mechanism that increases the chances of eating high-energy foods.
   Ultimate cause: Sweet, high-energy, foods were rare prior to mechanized agriculture. Increased fitness associated with consuming these foods is the ultimate reason for the natural selection of a "sweet tooth".

IV. Certain stimuli trigger innate behaviors called fixed action patterns

Ethology pre-dates behavioral ecology. Relying on descriptive studies, ethologists discovered many behaviors were innate.

<u>Ethology</u> = Descriptive science based on studies of animals in the natural environment.

Innate behaviors may seem purposeful, but animals with innate behaviors are unaware of the significance of their actions.
- Some cuckoos are brood parasites. Females lay eggs in the nests of other species. Within hours of hatching, the otherwise helpless cuckoo pushes its host's eggs and/or chicks out of the nest.

Konrad Lorenz, Niko Tinbergen and Karl von Frisch shared the 1973 Nobel Prize for physiology or medicine for their work in ethology.

A. **Fixed-Action Patterns**

The most fundamental concept in classical ethology is the fixed-action pattern (FAP).

Fixed-action pattern = A highly stereotyped, innate, behavior.
- Once a fixed-action pattern is triggered, the behavior continues until completion even in the presence of other stimuli or if the behavior is inappropriate.
- A fixed-action pattern is triggered by an external sign stimulus or releaser. (See Campbell, Figure 50.5)
- Fixed-action patterns are adaptive responses to <u>natural</u> stimuli. Strange responses can be initiated by presenting unnatural situations to animals with fixed-action patterns.

Sign stimulus = An external sensory stimulus which triggers a fixed-action pattern.

Example: Niko Tinbergen noticed male three-spined stickleback fish responded aggressively to red trucks passing by their tank.

Fixed-Action Pattern: Male sticklebacks attack other males that enter their territories.

Sign stimulus: The red belly of the invading male. Sticklebacks attacked nonfish-like models with red on the ventral surface.

Example: Parent/young feeding behavior in birds.

Fixed-Action Pattern: The begging behavior of newly hatched chicks (raised heads, open mouths, loud cheeps).

Sign stimulus: Parent landing at the nest.

Example: Greylag goose egg retrieval behavior: (See Campbell, Figure 50.6)

Fixed-Action Pattern: Rolls the egg back to the nest using side-to-side head motions.

Sign stimulus: The appearance of an object near the nest.
- If the goose loses the egg during the retrieval process, it stops the head motion, but continues the "pulling" motion of retrieval. It must sit down before it notices the egg at which time another retrieval FAP is initiated.
- If an inappropriate object (toy dog, doorknob) is placed near the nest, the goose retrieves it but may not keep or incubate it.

Example: Protective behavior in hen turkeys:

Fixed-Action Pattern: Mothering behavior.

Sign stimulus: Cheeping sound of chicks.
- A deaf turkey cannot hear the sign stimulus releasing mothering behavior and kills her chicks.

Example: The human infant grasping response is a fixed-action pattern released by a tactile stimulus. Human babies smile when they hear certain sounds, or see a figure consisting of two dark spots on a circle (rudimentary representation of a face).

Example: Female digger wasps place a paralyzed cricket in their nests which serves as food for the young wasp after it hatches.

Fixed-Action Pattern: She places the cricket 2.5 cm from the nest.

Sign stimulus: Nest site.

Fixed-Action Pattern: She enters the nest and inspects it.

Sign stimulus: Presence of cricket 2.5 cm from nest.

Fixed-Action Pattern: She exits the nest and retrieves the cricket.

Sign stimulus: Presence of cricket 2.5 cm from nest.

- If the cricket is moved during the nest inspection stage, the wasp will retrieve the cricket and repeat the FAP from the first step. She cannot get past the inspect the nest step if the cricket is not where she left it when she tries to retrieve it.

B.   **The Nature of Sign Stimuli**

Sign stimuli are usually simple characteristics (e.g. ultrasonic bat sounds trigger avoidance behavior in moths).

Sign stimuli may be specific choices from an array of possibilities.

<u>Example</u>: Herring gull chick feeding behavior. The adult lowers its head and moves its beak. The chick pecks the red spot on the beak, causing the adult to regurgitate.

Fixed-action pattern: Pecking the red spot on the beak.

Sign stimulus: Red spot swung horizontally at the end of a long, vertical object.

Natural selection favors cues associated with the relevant behavior or object. Some randomness is probable in fixing upon one of many possible sign stimuli for a FAP.

An animal's sensitivity to general stimuli and its sign stimuli are correlated.

- For example, frogs' retinal cells are sensitive to movement. Movement is the sign stimulus that releases the tongue-shooting FAP in frogs.
- A frog starves if surrounded by motionless flies.

A supernormal stimulus may elicit stronger responses than natural stimuli.

- For example, when given a choice between an egg and a volleyball, a greylag goose ignores the egg and tries to retrieve the volleyball.

<u>Supernormal stimulus</u> = Artificial stimulus that elicits a stronger response than occurs naturally.

V. Learning is experience-based modification of behavior

Learning = The modification of behavior by experience.

A. Nature Versus Nurture

While European ethologists were discovering innate behaviors in nature, American psychologists were finding learning abilities in lab animals. Thus began the debate over nature versus nurture.

Nature versus nurture = The debate over whether instinct or learning is of primary importance in animal behavior.

- Modern biologists feel that behavior is a result of genetic and environmental factors.

Some behavioral aspects seem to be due primarily to either instinct or learning.

- Language differences among people are learned behavioral variations on an innate ability to vocalize.
- Songbirds have regional "dialects".
- Most *Drosophila* (fruit flies) have an activity cycle of 24 hours. There are individuals with 19 hour cycles. The difference is due to genetic variation.

B. Learning Versus Maturation

Individuals may improve behaviors over time. This is often attributed to learning but, in some cases, may be due to developmental changes in neuro-muscular systems as animals mature.

Maturation = Development of neuro-muscular systems that allows behavioral improvement.

C. Habituation

Animals stop responding to stimuli that do not provide appropriate feedback.

- Gray squirrels respond to the alarm calls of other squirrels. They stop responding if the calls are not followed by an attack ("cry-wolf" effect).

Habituation = Learning to ignore irrelevant stimuli or stimuli that do not provide proper feedback.

D. Imprinting

Imprinting is a form of learning closely associated with innate behavior.

Konrad Lorenz conducted an experiment with greylag geese. (See Campbell, Figure 50.8)

Experiment: A clutch of goose eggs was divided between the mother and an incubator.

Results: Goslings reared by the mother behaved normally and mated with other geese. The incubator goslings spent their first hours of life with Lorenz and preferred humans for the rest of their lives. They even tried to mate with humans.

Conclusions: Greylags have no innate sense of "mother" or "gooseness". They identify with and respond to the first object with certain characteristics they encounter. The ability or tendency to respond is innate.

- The object to which the response is directed is the imprinting stimulus. For Lorenz's geese, the imprinting stimulus was movement of an object away from the young.

Imprinting stimulus = An object in the environment to which the response is directed.

- For example, salmon return to the stream they were hatched in to spawn. The imprinting stimulus is the unique chemical composition (odor) of the hatching stream.
- Imprinting is distinguished from other types of learning by its irreversibility and the presence of a critical period during which imprinting can occur.

1. **Critical Period**

   Critical period = A limited time during which imprinting can occur.

   - Imprinting is usually thought to involve very young animals and short critical periods.
   - Imprinting may occur at different ages with critical periods of varying durations.
   - Adults require a critical period to "know" their young. Prior to imprinting on their young, adult herring gulls defend strange chicks. After the imprinting period, they kill and eat strange chicks.
   - Sexual imprinting (or species identity) occurs later than parental imprinting and has a longer critical period. Male finches cross-fostered by two different finch species imprinted on the wrong species when they developed a sexual identity.
   - While irreversibility and critical period characterize imprinting, they are not always fixed. The cross-fostered finches eventually mated with females of their own species.

2. **Song Development in Birds: A Study of Imprinting**

   Song development is highly responsive to selective pressures.

   In male white-crowned sparrows:
   - Birds raised in soundproof chambers developed abnormal songs.
   - Birds exposed to normal song recordings at 10-50 days old developed normal songs.
   - Birds deafened after they were exposed to the recordings, but before they began to sing, developed songs more abnormal than birds reared in isolation.
   - White-crowned sparrows learn to sing by hearing adults and matching those songs while listening to themselves.
   - 10 – 50 days was the critical period for hearing adult song. Birds who heard a recording at 50+ days never sang properly.
   - Birds exposed to other species' recordings did not learn those songs. There was an innate, genetic predisposition, for their own song.

- The innate predisposition was overcome by social interaction. Sparrows 50+ days old, when in social contact with another species, learned that species' song.
- The critical period may be flexible. It was longer for a live stimulus than for a recording.

Song development may be more or less fixed in other species:

- Song sparrows reared in isolation develop normal songs although males have larger repertoires if they hear other birds.
- Mockingbirds have repertoires of 150+ songs. Genetic programming places few or no restrictions on their song development.

E.   Classical Conditioning

Associative learning = The learning process where animals associate one stimulus with another.

Classical conditioning is a type of associative learning where a specific, or conditioned, response is elicited by a specific stimulus.

- A Russian physiologist, Ivan Pavlov induced dogs to salivate when they heard a bell. The dogs learned to associate the bell with powdered meat which was the normal stimulus for the salivary response.

Classical conditioning = A process in which an animal learns to respond to an external stimulus which does not normally elicit that response.

F.   Operant Conditioning

Operant conditioning, or trial-and-error learning, is another type of associative learning.

- B.F. Skinner put lab animals in a box with a variety of levers. Test animals learned to choose only those levers which yielded food.

Operant conditioning = A process where an animal learns to associate one of its behaviors with a reward or punishment and then tends to repeat or avoid that behavior.

- English tits learned to open milk bottles left on doorsteps and drink the cream.
- One or more of the birds discovered its probing behavior was rewarded when directed at the bottles.

Operant conditioning is the basis for most animal training and is common in nature.

- Genes influence the outcome of operant conditioning.

G.   Observational Learning

Observational learning = The ability of animals to learn by observing the actions of others.

- Observational learning allows new behaviors to become established and passed on to succeeding generations.
- The speed with which the tit milk drinking behavior spread throughout England indicates observational learning on the part of the birds.
- Song development in many birds involves observational learning as individuals listen to and eventually adopt the songs of older birds.

H. Play

Play has no apparent goal but uses movements closely associated with goal-directed behaviors.

- Young predators playfully stalk and attack each other using motions similar to those used to capture and kill prey.
- Play occurs in the absence of distracting external stimuli.

Play is potentially dangerous or costly.

- Young vervet monkeys are at higher risk of being caught and eaten by baboons when they are at play.
- In a study of young goats, 1/3 sustained play injuries that resulted in limps.

What is the selective advantage of play?

- Practice hypothesis - play is a type of learning that allows the perfection of survival behaviors. However, play movements rarely improve after the first few practices.
- Exercise hypothesis - play keeps the cardiovascular and muscular systems in condition and is common in young animals because they do not exert themselves in other ways while they are under parental care.

I. Insight

Insight learning is also called reasoning.

Insight learning = The ability of animals to perform appropriate behaviors on the first attempt in situations with which they have no prior experience.

- Insight learning is best developed in primates.
- A chimpanzee placed in an area where a banana is hung too high to reach, but where boxes are scattered about, will "size" up the situation, stack the boxes, climb up and retrieve the banana.

J. Animal Cognition

Cognition = Knowing, including awareness and judgment.

- It is difficult to determine if animals are aware of themselves and their surroundings.

Behaviorism = A mechanistic approach which describes behavior in terms of stimulus and response.

- Behaviorism, does not test for cognitive function.

Cognitive ethologists think cognitive ability arises through natural selection and forms a phylogenetic continuum stretching into evolutionary history.

Cognitive ethology = A view that sees conscious thinking as an inherent part of animal behavior.

VI. **Rhythmic behaviors synchronize an animal's activities with daily changes in the environment**

Animals repeat behaviors at regular intervals (daily, seasonally).

Animals carry out behaviors when their particular ecological niches can be exploited most profitably from a survival and fitness standpoint. This can cause behavioral rhythms.

The proximate bases for rhythms are not obvious.
- Circadian (daily) rhythms are regulated by environmental cues.
- Rhythmic behavior is based on exogenous (external) and endogenous (internal) factors.
- Circadian rhythms are strongly influenced by endogenous factors, or the biological clock. Internal rhythms do not match environmental time so exogenous cues are used to keep the cycles in sync.
- Light is the most common exogenous cue.

Experiment: Flying squirrels are normally active at night. Flying squirrels were placed in constant darkness and their activity monitored. (See Campbell, Figure 50.13)

Results: Rhythmic activity continued. Since normal cycles deviate from the 24 hour clock, the test squirrels eventually were out of phase with the environment.

Conclusions: Biological clocks are intrinsic, but the timing of rhythmic behavior must be adjusted to environmental cues to remain in sync.

Even when endogenous factors are present, the internal, proximate, timing mechanism is unknown. A biochemical, possibly molecular, clock has been hypothesized.

It is unclear if endogenous clocks are important in longer cycle rhythmic behaviors. Breeding and hibernation are partially based on physiological and hormonal changes linked to exogenous factors such as day length.

VII. **Environmental cues guide animal movement**

Studies have concentrated on the proximate mechanisms animals use to detect and respond to external cues that guide movements.

The ultimate bases for oriented movements are related to animal adaptations to different environments and the development of behaviors that bring them to those environments at the proper time of the year.

A. **Kinesis and Taxis**

Animals use a variety of cues to guide them in their movements.

Kinesis involves a change in activity rate in response to a stimulus.
- Sowbugs are more active in dry areas and less active in humid regions. This behavior tends to keep them in moist areas.

Kinesis = A randomly directed change in activity rate in response to an environmental stimulus.

Taxis is a directed movement toward or away from a stimulus.

- Housefly larvae are negatively phototactic after feeding. Presumably this makes them less visible to predators.
- Trout are positively rheotactic. Swimming against the current keeps them from being swept downstream.

Taxis = A semi-automatic, oriented, movement toward or away from a stimulus.

B.   Migration Behavior

Migration is the most commonly known type of oriented animal movement.

- Migrants generally make an annual round trip between two regions (e.g. birds, whales, some butterflies, some pelagic fish). (See Campbell, Figures 50.14 and 50.15)

Migration = The seasonal movement of animals over relatively long distances.

Migrating animals use three mechanisms: *piloting, orientation* and *navigation*.

Piloting = Movement of animals from one landmark to another.

- Is used over short distances.
- It is not useful at night or over the ocean.

Orientation = Movement of animals along a compass line.

- Animals that use orientation can detect compass directions and travel in a straight-line to a destination.

Navigation = The ability of animals who can orient along compass lines to determine their location in relation to their destination.

- Migrant starlings captured in the Netherlands were released in Switzerland. Juvenile birds oriented in a straight-line to Spain. Adults navigated a new route to their wintering grounds in northern Europe.
- Many birds use celestial points for orientation and navigation. These animals need an internal clock to compensate for the movement of the sun and stars. The indigo bunting avoids the need for an internal clock by fixing on the north star.
- Some birds, bees and bacteria orient to the Earth's magnetic field. The mechanisms are poorly known, but magnetite, an iron-containing ore, has been found in animals that orient to the magnetic field.

VIII. Behavioral ecologists are using cost/benefit analysis to study foraging behavior

Generalists feed on many items. They are not efficient collectors of any single food, but take advantage of multiple options when foods are scarce.

- Generalists concentrate on abundant prey.
- Generalists develop a search image for a favored item. If the item becomes scarce, a new search image is developed. Search images let generalists combine efficient short-term specialization with generalist flexibility.

Search image = The ability of a generalist feeder to learn the key visual characters of a prey item.

Specialists feed on specific items and usually have highly specific morphological and behavioral adaptations. They are extremely efficient foragers.

Food habits are fundamental to an animal's niche and may be shaped by interspecific competition and evolutionary factors.

Natural selection should favor foraging strategies that maximize gains and minimize costs in terms of calories gained and expended. Other criteria such as nutrient gain may be equally important.

- Foraging costs include: energy needed to locate, catch and eat food; the risk of being caught by a predator while feeding; time taken from other activities such as courtship and breeding.
- Behavioral ecologists analyze tradeoffs to predict optimal foraging strategies.
- Tradeoffs include density and size of prey versus foraging distances or prey catchability.
- For example, small mouth bass eat minnows and crayfish. Since no preference is shown, each may be optimal under different conditions. Minnows have more useable energy per unit weight than crayfish, but are harder to catch. Crayfish are easier to catch, but more difficult to subdue.

Animals modify behavior to keep the ratio of energy gain to loss high.

- This ability is probably innate, although learning may be involved.

Experiment: Bluegill sunfish eat small crustaceans. (See Campbell, Figure 50.16) Optimal foraging theory predicts the proportion of small to large prey will vary with the density of the prey population. At low densities, sunfish should not be selective but at high densities, they should concentrate on larger prey.

Results: Sunfish were more selective at higher prey densities, but not to the extent predicted. Young fish were less efficient than adults. Younger fish may be less able to judge size and distance due to incompletely developed neural systems.

Conclusion: Maturation and learning may result in increased foraging efficiency in adults.

IX. Sociobiology places social behavior in an evolutionary context

Most sexually reproducing species must be social for part of their life cycle in order to reproduce; some species spend most of their lives in close association with conspecifics.

Social behavior = Any interaction between two or more animals, usually of the same species.
- Includes aggression, courtship, cooperation, and even deception.
- Has both costs and benefits to members of species that interact extensively.

Sociobiology = Study of social behavior that has evolutionary theory as its conceptual framework.
- In 1975, E.O. Wilson published *Sociobiology: The New Synthesis* which helped form sociobiology into a coherent method of analysis and interpretation.
- From the sociobiologist's point of view, the genetic basis of behavior and fitness can be used to analyze the evolution and maintenance of social behavior.

Because members of a population share a common niche, there is potential for conflict, especially among members of species that maintain densities near carrying capacity.
- Sometimes social behavior involves cooperative effort, as when a group accomplishes something more efficiently than a single individual.
- Even when cooperation seems to be mutually beneficial, each participant usually acts to maximize its fitness, even at a cost to the other participant.

X. Competitive social behaviors often represent contests for resources

A. Agonistic Behavior

Agonistic behavior = A contest of threat displays which continues until a participant submits and yields access to a resource (e.g. mate, food, territory).
- Ritual behaviors are prevalent so it is rare that participants are seriously injured. (See Campbell, Figure 50.18)

Ritual = Symbolic behavior that minimizes the possibility of serious injury to the antagonists.
- Natural selection favors ending a contest as soon as a winner is established because further conflict could injure the victor as well as the vanquished.
- Canines show agonistic behavior by trying to look larger. They bare teeth; erect ears, tail and fur; stand upright; and make eye contact. The loser submits by sleeking its fur, tucking its tail and looking away.

B.   **Dominance Hierarchies**

The top ranked member of a social group controls the behavior of the members of the group. The second ranked animal controls everyone except the top individual and so on down the line to the lowest ranked animal.

Dominance hierarchy = A linear social organization within a group.

- Top ranked animals are assured access to resources. Low ranked animals do not waste energy or risk harm in combat.
- Wolf packs typically have a female dominance hierarchy. The top female relies on food availability to control mating in the pack.

C.   **Territoriality**

Territories are defended areas typically used for feeding, mating, rearing young etc.

Territory = An area defended from conspecifics.

- Territory size varies with the species, territory function, and the amount of resources available. (See Campbell, Figure 50.19)
- Some species defend territories during the breeding season and form social groups at other times of the year.
- Territories are not home ranges. Home ranges are areas which animals inhabit but do not defend. Territories and home ranges may overlap.

Territories are usually successfully defended by their owners.

- Owners usually win because a territory is more valuable to the owner since he is familiar with it.
- Ownership is continually proclaimed - a primary function of bird song, red squirrel chattering and the bellowing of sea lions. Others use scent marks or patrols to announce their presence. (See Campbell, Figure 50.20)
- Defense is usually directed at conspecifics who are most likely to compete directly for the same resources.

Dominance hierarchies and territoriality tend to stabilize population densities by assuring enough individuals reproduce to result in relatively stable populations from year to year.

XI.   **Mating behavior relates directly to an animal's fitness**

The correlation between mating behavior and reproductive fitness is vital to behavioral ecology.

A.   **Courtship**

Species often have a complex courtship ritual unique to that species.

- Courtships are a series of fixed-action patterns alternately triggered by the participants.
- Ritual courtships probably evolved from behaviors that once had a direct meaning.

For example, male balloon flies spin oval silk balloons which they carry while flying in a swarm. The swarm is approached by females seeking mates. A female accepts a male's balloon when they fly off to copulate.

- In a related species, the male brings a dead insect for the female to eat while they mate.
- In another species, the insect is presented inside a silk balloon, possibly because silk helps subdue the insect or makes it look larger.
- Balloon flies eat nectar. The ritual has evolved into bringing something that was once associated with food.
- It is as if, over evolutionary time, a suitor wooed a lady with diamonds, then with a box containing diamonds and finally with an empty box.

Courtship assures each partner that the potential mate is not a threat, is the proper species, the proper sex and in the correct physiological condition.

- Courtship may allow one or both sexes to choose a mate from a number of candidates.

Females are usually more discriminating than males because they normally have a greater parental investment.

- Eggs are usually larger and more costly to produce than sperm.
- Gametes of placental mammals are closer in size, but females invest considerable time and energy carrying young before birth.

Parental investment = The time and resources an individual expends to produce an offspring.

Competition among individuals of the same sex (usually males) may determine which individuals of that sex will mate.

- Most males mate with as many females as possible. They compete with other males for mates and may try to impress females.
- Males perform more intense courtship displays than females.
- Secondary sex characteristics may be highly developed in males (e.g. deer antlers, bird colors).

There are two ultimate bases for mate selection.

1. If the other sex gives parental care, it is best to choose the most competent mate.
   - For example, male common terns bring fish to potential mates as part of the courtship ritual. This behavior may be a proximate indicator of his ability to feed the chicks.
   - Some females prefer males with the most extreme and energetic courtship displays or secondary sex characteristics. These characteristics may be proximate indicators of the male's health.

2. Genetic quality is important when males provide no parental care and sperm are their only contribution to offspring.
   - Lek species have a communal area where males display. Females visit the lek and choose a mate. The proximate basis for her choice is a preference for males

that court the most vigorously and have the most extreme secondary sex characteristics.

It may be difficult to determine if differential mating success among males is due to male-male competition, female choice or both.

- Three-spined stickleback courtship is based on stereotyped releasers and FAPs. (See Campbell, Figure 50.21) Despite this, the female can back out of the courtship anytime.
- Female choice is probably ultimately based on the quality of the male's parental care, because only male sticklebacks give parental care.

B.   Mating Systems

Promiscuous = A mating system with no strong pair-bonds or lasting relationships.

Monogamous = A mating system where one male mates with one female.

Polygamous = A mating system where an individual of one sex mates with several of the other.

- *Polygyny* is a mating sub-system where one male mates with multiple females.
- *Polyandry* is a mating sub-system where one female mates with multiple males.

The needs of the young are an important ultimate factor in the evolution of mating systems.

- Most birds are monogamous. Young birds often require significant parental care. A male may ultimately increase his reproductive fitness by helping a single mate rear a brood than by seeking additional mates.
- Polygyny is common in birds where the young are able to care for themselves soon after hatching. Males can maximize their fitness by seeking additional mates.

Another factor influencing mating systems and parental care is the certainty of paternity. (See Campbell, Methods Box, page 1197)

- Young born or eggs laid by a female definitely contain the female's genes, but even in monogamous species, the young could have been fathered by a male other than the female's normal mate.
- The certainty of paternity is relatively low in species with internal fertilization because mating and birth (or egg laying) are separated over time. Exclusive male parental care is rare in birds or mammals.
- Certainty of paternity is higher when egg laying and mating occur together, as in external fertilization. Parental care, when present, in fishes and amphibians is as likely to be by males as by females.
- When parental care is given by males, the mating system may be polygynous with multiple females laying eggs in a nest tended by a male.

XII. Social interactions depend on diverse modes of communication

<u>Communication</u> = The intentional transmission of information between individuals.

Behavioral ecologists assume communication has occurred when an act by a "sender" produces a change in the behavior of another individual, the "receiver".

Ethologists assumed communication evolved to maximize the quantity and accuracy of information. Behavioral ecologists argue that communication evolved to maximize the fitness of communicators.

Animals lie. Mimicry often is adaptive to the sender and maladaptive for the receiver.

- Male and female *Photinus* fireflies communicate by a characteristic pattern of flashes. Females of the predatory firefly genus *Photurus* mimic the female *Photinus* flash pattern, attracting male *Photinus* fireflies which they kill and eat.
- In some mammals, a new dominant male kills young born too soon to be his offspring. Without dependent young, females ovulate sooner, allowing the new male to father their young. Hanuman langur females in the early stages of pregnancy solicit copulations from new dominant males. When they give birth shortly before young fathered by these males would appear, they may deceive the male into treating their young as his own.

An evolutionary consideration is the mode used to transmit information. Animals use visual, auditory, chemical, tactile and electrical signals.

The mode used to transmit information is related to an animal's lifestyle.

- Most mammals are nocturnal and use olfactory and auditory signals.
- Animals that communicate by odors emit chemical signals called pheromones. Pheromones are important releasers for specific courtship behaviors and are the cues ant scouts release that guide other ants to food.
- Birds are mostly diurnal and use visual and auditory signals. Diurnal humans also use visual and auditory signals. If we could detect the chemical signals of mammals, then mammal sniffing might be as popular as bird watching.

A complex communication system is found in honeybees. (See Campbell, Figure 50.22)

- To maximize foraging efficiency, workers communicate the location of food sources which change as flowers bloom and new patches are found.
- Karl von Frisch studied honeybee communication. He found individual bees communicated to other bees when they returned to the hive.
- Returning bees "dance" to indicate the location of food.
- If the source is < 50m, the bee does the "round dance", moving rapidly sideways in tight circles and regurgitating nectar. Workers leave the hive and forage nearby.
- If the food is farther away, the bee does a "waggle dance", a half-circle swing in one direction, followed by a straight run and then a half-circle swing in the other direction. This dance indicates location in two ways:
  1. The angle of the run in relation to the vertical surface of the hive is the same as the horizontal angle of the food in relation to the sun.
  2. Distance to the food is indicated by variations in the speed at which a bee wags its abdomen during the straight run.

XIII. The concept of inclusive fitness can account for most altruistic behavior

Behavior that maximizes individual reproductive success will be favored by selection, regardless of how much damage such behavior does to another individual, local population, or species.

Animals occasionally exhibit apparently unselfish or altruistic behavior.

<u>Altruistic behavior</u> = A behavior that reduces an individual's personal welfare but benefits others.
- When parents sacrifice their well-being to produce and aid offspring, they increase their fitness because it maximizes their genetic representation in the population.
- Like parents and offspring, siblings share 1/2 their genes, so selection might favor helping one's parents produce more siblings, or even helping siblings directly.
- Selection might result in animals increasing their genetic representation in the next generation by "altruistically" helping close relatives.
- Inclusive fitness includes assistance to relatives that maximizes individual fitness.
- *Coefficient of relatedness* is the proportion of genes that are identical in two individuals because of common ancestry. The higher the coefficient of relatedness, the more likely an individual is to aid a relative.

<u>Inclusive fitness</u> = The reproductive fitness of an individual as measured by its offspring and assistance to the reproductive efforts of close relatives.

<u>Kin selection</u> = The mechanism of increasing inclusive fitness.
- The contribution of kin selection to inclusive fitness varies among species. It may be rare or nonexistent in species that are not social or disperse widely.
- In predicting if an individual will aid relatives, behavioral ecologists have derived a formula that combines coefficients of relatedness, costs to the altruist and benefits to the recipient.

If kin selection explains altruism, then examples of unselfish behavior should involve close relatives.
- Belding's groundsquirrels give alarm calls when danger appears. These calls alert other squirrels but increase the risk to the alarm givers. Females remain near their birth sites and are usually related to other members of the group. Only females give alarm calls. (See Campbell, Figure 50.23)
- Worker bees are sterile. They labor on behalf of a single fertile queen. Workers sting intruders, a behavior that defends the hive but results in the death of the worker. The queen is the mother of all the bees in the hive.
- Nesting red-cockaded woodpeckers are aided by 2 – 4 nonbreeders that assist in all aspects including incubation and feeding young. Nest helpers are older offspring of the breeding pair or siblings of one parent that have been unable to establish a breeding territory. Helpers may eventually inherit the territory.

Altruistic behavior toward non-relatives sometimes occurs. This behavior is adaptive if there is a reasonable chance of the aid being returned in the future.
- Reciprocal altruism only occurs in stable social groups where individuals have many opportunities to exchange aid.

XIV. Human sociobiology connects biology to the humanities and social sciences

The book, *Sociobiology* by E.O. Wilson (1975) presented the thesis that social behavior has an evolutionary basis.

- Behavioral characteristics are expressions of genes favored by natural selection.
- Rekindled the nature-versus-nurture controversy.

An example of the debate involves cultural taboos on incest.

- Incest avoidance is adaptive because inbreeding may increase the frequency of genetic disorders.
- Many species avoid incest.
- Most human cultures have taboos forbidding incest.

Is there an innate aversion to incest or is this an acquired behavior?

- The argument in favor of the "nurture or learned behavior" position is: cultural taboos are unnecessary if the behavior is innate, therefore incest avoidance is a learned behavior and the social stigma attached to incest is based on experience.
- The argument in favor of the "nature or genetic behavior" position is: the occurrence of incest taboos in many cultures is evidence for an innate component and taboos are simply proximate mechanisms that reinforce a behavior that ultimately evolved because of its effect on fitness.

REFERENCES

Alcock, J. *Animal Behavior: An Evolutionary Approach*. 4th ed. Sunderland, Mass.: Sinauer, 1989.

Campbell, N. *Biology*. 4th ed. Menlo Park, California: Benjamin/Cummings, 1996.

Krebs, J.R. and N.B. Davies (editors). *Behavioral Ecology: An Evolutionary Approach*. 1st ed. Sunderland, Mass.: Sinauer, 1978.

Krebs, J.R. and N.B. Davies (editors). *Behavioral Ecology: An Evolutionary Approach*. 2nd ed. Sunderland, Mass.: Sinauer, 1984.

Wilson, E.O. *Sociobiology: The New Synthesis*. Cambridge, Mass.: Harvard University Press, 1975.